T0191665

# Advances in Intelligent and Soft Computing

**100**

Editor-in-Chief: J. Kacprzyk

# Advances in Intelligent and Soft Computing

**Editor-in-Chief**

Prof. Janusz Kacprzyk
Systems Research Institute
Polish Academy of Sciences
ul. Newelska 6
01-447 Warsaw
Poland
E-mail: kacprzyk@ibspan.waw.pl

Further volumes of this series can be found on our homepage: springer.com

Vol. 88. Y. Demazeau, M. Pĕchouček,
J.M. Corchado, and J.B. Pérez (Eds.)
*Advances on Practical Applications of Agents
and Multiagent Systems, 2011*
ISBN 978-3-642-19874-8

Vol. 89. J.B. Pérez, J.M. Corchado,
M.N. Moreno, V. Julián, P. Mathieu,
J. Canada-Bago, A. Ortega, and
A.F. Caballero (Eds.)
*Highlights in Practical Applications of Agents
and Multiagent Systems, 2011*
ISBN 978-3-642-19916-5

Vol. 90. J.M. Corchado, J.B. Pérez,
K. Hallenborg, P. Golinska, and
R. Corchuelo (Eds.)
*Trends in Practical Applications of Agents
and Multiagent Systems, 2011*
ISBN 978-3-642-19930-1

Vol. 91. A. Abraham, J.M. Corchado,
S.R. González, J.F. de Paz Santana (Eds.)
*International Symposium on Distributed
Computing and Artificial Intelligence, 2011*
ISBN 978-3-642-19933-2

Vol. 92. P. Novais, D. Preuveneers, and
J.M. Corchado (Eds.)
*Ambient Intelligence - Software and
Applications, 2011*
ISBN 978-3-642-19936-3

Vol. 93. M.P. Rocha, J.M. Corchado,
F. Fernández-Riverola, and
A. Valencia (Eds.)
*5th International Conference on Practical
Applications of Computational Biology &
Bioinformatics 6-8th, 2011*
ISBN 978-3-642-19913-4

Vol. 94. J.M. Molina, J.R. Casar Corredera,
M.F. Cátedra Pérez, J. Ortega-García, and
A.M. Bernardos Barbolla (Eds.)
*User-Centric Technologies and
Applications, 2011*
ISBN 978-3-642-19907-3

Vol. 95. Robert Burduk, Marek Kurzyński,
Michał Woźniak, and Andrzej Żołnierek (Eds.)
*Computer Recognition Systems 4, 2011*
ISBN 978-3-642-20319-0

Vol. 96. A. Gaspar-Cunha, R. Takahashi,
G. Schaefer, and L. Costa (Eds.)
*Soft Computing in Industrial Applications, 2011*
ISBN 978-3-642-20504-0

Vol. 97. W. Zamojski, J. Kacprzyk,
J. Mazurkiewicz, J. Sugier,
and T. Walkowiak (Eds.)
*Dependable Computer Systems, 2011*
ISBN 978-3-642-21392-2

Vol. 98. Z.S. Hippe, J.L. Kulikowski, and
T. Mroczek (Eds.)
*Human – Computer Systems Interaction:
Backgrounds and Applications 2, 2011*
ISBN 978-3-642-23186-5

Vol. 99. Z.S. Hippe, J.L. Kulikowski, and
Tteresa Mroczek (Eds.)
*Human – Computer Systems Interaction:
Backgrounds and Applications 2, 2011*
ISBN 978-3-642-23171-1

Vol. 100. Shoumei Li, Xia Wang,
Yoshiaki Okazaki, Jun Kawabe,
Toshiaki Murofushi, and Li Guan (Eds.)
*Nonlinear Mathematics for Uncertainty and
its Applications, 2011*
ISBN 978-3-642-22832-2

Shoumei Li, Xia Wang, Yoshiaki Okazaki,
Jun Kawabe, Toshiaki Murofushi,
and Li Guan (Eds.)

# Nonlinear Mathematics for Uncertainty and its Applications

Springer

## Editors

Shoumei Li
Beijing University of Technology
College of Applied Sciences
100 Pingleyuan
Chaoyang District
Beijing, 100124
P.R. China
E-mail: lisma@bjut.edu.cn

Xia Wang
Beijing University of Technology
College of Applied Sciences
100 Pingleyuan
Chaoyang District
Beijing, 100124
P.R. China
E-mail: wangxia@bjut.edu.cn

Yoshiaki Okazaki
Kyushu Institute of Technology
Department of Systems Design and Informatics
680-4, Kawazu
Iizuka 820-8502
Japan
E-mail: okazaki@ces.kyutech.ac.jp

Jun Kawabe
Shinshu University
Department of Mathematics
4-17-1 Wakasato
Nagano 380-8553
Japan
E-mail: jkawabe@shinshu-u.ac.jp

Toshiaki Murofushi
Tokyo Institute of Technology
Department of Computational Intelligence and
Systems Science
4259-G3-47, Nagatsuta, Midori-ku
Yokohama 226-8502
Japan
E-mail: murofusi@dis.titech.ac.jp

Li Guan
Beijing University of Technology
College of Applied Sciences
100 Pingleyuan
Chaoyang District
Beijing, 100124
P.R. China
E-mail: guanli@bjut.edu.cn

ISBN 978-3-662-52038-3          ISBN 978-3-642-22833-9 (eBook)

DOI 10.1007/978-3-642-22833-9

Advances in Intelligent and Soft Computing          ISSN 1867-5662

*Typeset & Cover Design:* Scientific Publishing Services Pvt. Ltd., Chennai, India

Printed on acid-free paper
5 4 3 2 1 0
springer.com

# Preface

This volume is a collection of papers presented at the international conference on Nonlinear Mathematics for Uncertainty and Its Applications (NLMUA2011), held at Beijing University of Technology during the week of September 7–9, 2011.

Over the last fifty years there have been many attempts in extending the theory of classical probability and statistical models to the generalized one which can cope with problems of inference and decision making when the model-related information is scarce, vague, ambiguous, or incomplete. Such attempts include the study of nonadditive measures and their integrals, imprecise probabilities and random sets, and their applications in information sciences, economics, finance, insurance, engineering, and social sciences.

Possibility measures, belief functions, Choquet capacities, and fuzzy measures are all nonadditive measures, and their related integrals are nonlinear. Imprecise probability allows us to measure chance and uncertainty with a family of classical probability measures. Their lower and upper expectations or previsions are nonlinear again. Theory of random sets and related subjects extend the horizon of classical probability and statistics to set-valued and fuzzy set-valued cases.

The conference brought together more than one hundred participants from fourteen different countries. It gathered researchers and practitioners involved with all aspects of nonlinear mathematics for uncertainty and its applications. Researchers in probability theory and statistics were also invited since the scope of the conference is the heart of current interests in new mathematical perspectives for quantifying appropriately risk measures in financial econometrics as well as formulating realistic models for prediction. They exchanged their research results and discussed open problems and novel applications. They also fostered the collaboration in future research and applied projects in the area of nonlinear mathematics and uncertainty management.

During the meeting the following five principal speakers delivered plenary lectures.

Michio Sugeno, Japan
Zengjing Chen, Shaodong University, China
Thierry Denoeux, Universite de Technologie de Compiegne, France
Jianming Xia, Academy of Mathematics and Systems Science, Chinese
Academy of Sciences, China
Gert de Cooman, Ghent University, Belgium

In addition to plenary lectures, there were also eighty-six talks organized in parallel sessions.

The proceedings presented here contain eighty-five papers; some of them are expository articles, while others are original papers. The papers included in this volume were carefully evaluated and recommended for publication by reviewers. Some of the topics include nonadditive measures and nonlinear integrals, Choquet, Sugeno and other types of integrals, possibility theory, Dempster-Shafer theory, random sets, fuzzy random sets and related statistics, set-valued and fuzzy stochastic processes, imprecise probability theory and related statistical models, fuzzy mathematics, nonlinear functional analysis, information theory, mathematical finance and risk managements, decision making under various types of uncertainty, information fusion and knowledge integration in uncertain environments, soft computing and intelligent data analysis, applications in economics, finance, insurance, biology, engineering and others.

The publication of this volume was supported by grants PHR (No. 2010061 02), NSFC (No. 10971007), China.

We would like to express our appreciation to all the members of Program Chairs, Publication Chairs, International Program Committee, and the external referees for their great help and support.

On behalf of the organizing committee we would like to take this opportunity to thank all our colleagues, secretarial staff members, and graduate students from College of Applied Sciences, Beijing University of Technology. Our acknowledgement of appreciation goes to their assistance in organizing the meetings, chairing the sessions, and helping with registration and many other necessary undertakings.

Beijing, China                                                    Shoumei Li
September 2011                                                      Xia Wang
                                                             Yoshiaki Okazaki
                                                                 Jun Kawabe
                                                           Toshiaki Murofushi
                                                                    Li Guan

# Members of Committees

## Honorary Chairs

Michio Sugeno      Japan, at European Centre for
Soft Computing, Spain
Jia-An Yan      Chinese Academy of Sciences, China
Hung T. Nguyen      New Mexico State University, USA

## General Chairs

Shoumei Li      Beijing University of Technology, China
Yoshiaki Okazaki      Kyushu Institute of Technology, Japan
Jun Kawabe      Shinshu University, Japan
Toshiaki Murofushi      Tokyo Institute of Technology, Japan
Caozong Cheng      Beijing University of Technology, China

## Program Chairs

Li Guan      Beijing University of Technology, China
Katsushige Fujimoto      Fukushima University, Japan

## Publication Chairs

Xia Wang      Beijing University of Technology, China
Aoi Honda      Kyushu Institute of Technology, Japan

## International Program Committee

| | |
|---|---|
| Zengjing Chen | Shandong University, China |
| Thierry Denoeux | Universite de Technologie de Compiegne, France |
| Sompong Dhompongsa | Chiang Mai University, Thailand |
| Didier Dubois | Universite Paul Sabatier, France |
| Zhoutian Fan | Beijing University of Technology, China |
| Weiyin Fei | Anhui Polytechnic University, China |
| Maria Gil | University of Oviedo, Spain |
| Gert de Gooman | Ghent University, Belgium |
| Michel Grabisch | Université Paris I, Panthéon-Sorbonne, France |
| Peijun Guo | Yokohama National University, Japan |
| Liangjian Hu | Donghua University, China |
| Hiroshi Inoue | Tokyo University of Science, Japan |
| Janusz Kacprzyk | Polish Acadamy of Sciences, Poland |
| Yun Kyong Kim | Dongshin University, Korea |
| Michal Kisielewicz | University of Zielona Gora, Poland |
| Rudolf Kruse | Otto-von-Guericke University of Magdeburg, Germany |
| Coeraad Labuschagne | University of the Witwatersrand, South Africa |
| Jonathan Lawry | Bristol University, UK |
| Hongxing Li | Dalian University of Technology, China |
| Jun Li | Communication University of China, China |
| Zhenquan Li | Charles Sturt University, Aulbury, Australia |
| Baoding Liu | Tsinghua University, China |
| Jie Lu | University of Technology, Australia |
| Motoya Machida | Tennessee Technological University, USA |
| Radko Mesiar | Slovak University of Technology, Slovakia |
| Mariusz Michta | University of Zielona Gora, Poland |
| Enrique Miranda | University of Oviedo, Spain |
| Yasuo Narukawa | Tokyo Institute of Technology, Japan |
| Endre Pap | University of Novi Sad, Serbia and Montenegro |
| Dan Ralescu | University of Cincinnati, USA |
| Da Ruan | Belgian Nuclear Research Centre, Belgium |
| Pedro Teran | University of Oviedo, Spain |
| Matthias Troffaes | Durham University, UK |
| Makoto Tsukada | Toho University, Japan |
| Dabuxilatu | Wang Guangzhou University, China |
| Berlin Wu | National Chengchi University, Taiwan |
| Jianming Xia | Chinese Academy of Sciences, China |
| Bing Xu | Zhejiang Gongshang University, China |
| Kenjiro Yanagi | Yamaguchi University, Japan |
| Masami Yasuda | Chiba University, Japan |

| | |
|---|---|
| Guangquan Zhang | University of Technology, Australia |
| Guoli Zhang | North China Electric Power University, China |
| Jinping Zhang | North China Electric Power University, China |
| Qiang Zhang | Beijing Institute of Technology, China |

## Local Organizers (Beijing University of Technology)

Zhongzhan Zhang
Liugen Xue
Weihu Cheng
Xuejing Li
Tianfa Xie
Hongxia Wang
Junfei Zhang
Xu Zhang

# Contents

**Ordinal Preference Models Based on S-Integrals and Their Verification** .......................................... 1
*Michio Sugeno*

**Strong Laws of Large Numbers for Bernoulli Experiments under Ambiguity** ................................. 19
*Zengjing Chen, Panyu Wu*

**Comparative Risk Aversion for $g$-Expected Utility Maximizers** ....................................... 31
*Guanqyan Jia, Jianming Xia*

**Riesz Type Integral Representations for Comonotonically Additive Functionals** ............................... 35
*Jun Kawabe*

**Pseudo-concave Integrals** ...................................... 43
*Radko Mesiar, Jun Li, Endre Pap*

**On Spaces of Bochner and Pettis Integrable Functions and Their Set-Valued Counterparts** ........................... 51
*Coenraad C.A. Labuschagne, Valeria Marraffa*

**Upper Derivatives of Set Functions Represented as the Choquet Indefinite Integral** ............................. 61
*Shinsuke Nakamura, Toshiyuki Takasawa, Toshiaki Murofushi*

**On Regularity for Non-additive Measure** ...................... 69
*Toshikazu Watanabe, Tamaki Tanaka*

**Autocontinuity from below of Set Functions and Convergence in Measure** ................................. 77
*Jun Li, Masami Yasuda, Ling Zhou*

**On Lusin's Theorem for Non-additive Measure** ............... 85
*Tamaki Tanaka, Toshikazu Watanabe*

**Multiple-Output Choquet Integral Models and Their
Applications in Classification Methods** ........................ 93
*Eiichiro Takahagi*

**On Convergence Theorems of Set-Valued Choquet Integrals** .. 101
*Hongxia Wang, Shoumei Li*

**On Nonlinear Correlation of Random Elements** ............... 109
*Hung T. Nguyen*

**On Fuzzy Stochastic Integral Equations–A Martingale
Problem Approach** ............................................ 117
*Mariusz Michta*

**Set-Valued Stochastic Integrals with Respect to Poisson
Processes in a Banach Space** ................................. 125
*Jinping Zhang, Itaru Mitoma*

**The Convergence Theorems of Set-Valued Pramart in a
Banach Space** ............................................... 135
*Li Guan and Shoumei Li*

**Fuzzy Stochastic Integral Equations Driven by Martingales** .. 143
*Marek T. Malinowski, Mariusz Michta*

**Moderate Deviations of Random Sets and Random Upper
Semicontinuous Functions** ................................... 151
*Xia Wang*

**Solution of Random Fuzzy Differential Equation** ............. 161
*Jungang Li, Jinting Wang*

**Completely Monotone Outer Approximations of Lower
Probabilities on Finite Possibility Spaces** .................... 169
*Erik Quaeghebeur*

**Belief Networks and Local Computations** .................... 179
*Radim Jiroušek*

**Some New Entropies on the Interval-Valued Fuzzy Set** ........ 189
*Wenyi Zeng, Hongxing Li, Shuang Feng*

**Bottleneck Combinatorial Optimization Problems with
Fuzzy Scenarios** ............................................. 197
*Adam Kasperski, Paweł Zieliński*

About the Probability-Field- Intersections of
Weichselberger and a Simple Conclusion from Least
Favorable Pairs ................................................ 205
*Martin Gümbel*

Using Imprecise and Uncertain Information to Enhance
the Diagnosis of a Railway Device ........................... 213
*Zohra L. Cherfi, Latifa Oukhellou, Etienne Côme, Thierry Denœux,
Patrice Aknin*

Pricing CDS with Jump-Diffusion Risk in the
Intensity-Based Model .......................................... 221
*Ruili Hao, Zhongxing Ye*

Pricing Convertible Bonds Using the CGMY Model ......... 231
*Coenraad C.A. Labuschagne, Theresa M. Offwood*

The Dividend Problems for Compound Binomial Model
with Stochastic Return on Investments ...................... 239
*Jiyang Tan, Xiangqun Yang, Youcai Zhang, Shaoyue Liu*

Pricing Formulas of Compound Options under the
Fractional Brownian Motion ................................... 247
*Chao Zhang, Jizhou Zhang, Dongya Tao*

Influence of Risk Incentive by Limited Dividend Provision .. 255
*Hiroshi Inoue, Masatoshi Miyake, Li Guan*

An Application of the Forward Integral to an Insider's
Optimal Portfolio with the Dividend ......................... 263
*Yong Liang, Weiyin Fei, Hongjian Liu, Dengfeng Xia*

The Nonlinear Terminal-Boundary Problems for Barrier
Options ........................................................ 271
*Yulian Fan*

European Option Pricing with Ambiguous Return Rate
and Volatility ................................................. 279
*Junfei Zhang, Shoumei Li*

Knightian Uncertainty Based Option Pricing with Jump
Volatility ..................................................... 287
*Min Pan, Liyan Han*

Conditional Ruin Probability with a Markov Regime
Switching Model ............................................... 295
*Xuanhui Liu, Li-Ai Cui, Fangguo Ren*

**A Property of Two-Parameter Generalized Transition
Function** .................................................................... 301
*Yuquan Xie*

**The Extinction of a Branching Process in a Varying or
Random Environment** ................................................... 309
*Yangli Hu, Wei Hu, Yue Yin*

**Metric Adjusted Skew Information and Metric Adjusted
Correlation Measure** .................................................... 317
*Kenjiro Yanagi, Shigeru Furuichi*

**Integral-Based Modifications of OWA-Operators** .............. 325
*Erich Peter Klement, Radko Mesiar*

**Fuzzy Similarity Measure Model for Trees with Duplicated
Attributes** ................................................................. 333
*Dianshuang Wu, Guangquan Zhang*

**Linking Developmental Propensity Score to Fuzzy Sets:
A New Perspective, Applications and Generalizations** ........ 341
*Xuecheng Liu, Richard E. Tremblay, Sylvana Cote, Rene Carbonneau*

**Chaos in a Fractional-Order Dynamical Model of Love and
Its Control** ................................................................ 349
*Rencai Gu, Yong Xu*

**The Semantics of *wlp* and *slp* of Fuzzy Imperative
Programming Languages** ............................................... 357
*Hengyang Wu, Yixiang Chen*

**Public-Key Cryptosystem Based on $n$-th Residuosity of
$n = p^2 q$** ................................................................ 365
*Tomoko Adachi*

**Shrinking Projection Method for a Family of
Quasinonexpansive Mappings with a Sequence of
Subsets of an Index Set** ................................................ 371
*Yasunori Kimura*

**Note on Generalized Convex Spaces** ............................. 379
*Xiaodong Fan, Yue Cheng*

**Halpern's Iteration for a Sequence of Quasinonexpansive
Type Mappings** ........................................................... 387
*Koji Aoyama*

Contents

Convergence of Iterative Methods for an Infinite Family of
Pseudo-contractions ........................................... 395
Yuanheng Wang, Jiashuai Dong

Existence of Fixed Points of Nonspreading Mappings with
Bregman Distances............................................. 403
Fumiaki Kohsaka

The $(h, \varphi)$–Generalized Second-Order Directional
Derivative in Banach Space ................................... 411
Wenjuan Chen, Caozong Cheng, Caiyun Jin

Estimation of Bessel Operator Inversion by Shearlet .......... 419
Lin Hu, Youming Liu

Globally Convergent Inexact Smoothing Newton Method
for SOCCP .................................................... 427
Jie Zhang, Shao-Ping Rui

Existence of Positive Solution for the Cauchy Problem for
an Ordinary Differential Equation ............................ 435
Toshiharu Kawasaki, Masashi Toyoda

Stability of a Two Epidemics Model ........................... 443
T. Dumrongpokaphan, W. Jaihonglam, R. Ouncharoen

The Impulsive Synchronization for m-Dimensional
Reaction-Diffusion System .................................... 453
Wanli Yang, Suwen Zheng

A New Numerical Method for Solving Convection-Diffusion
Equations .................................................... 463
Hengfei Ding, Yuxin Zhang

Pullback Attractor for Non-autonomous P-Laplacian
Equation in Unbounded Domain ................................ 471
Guangxia Chen

Monte-Carlo Simulation of Error Sort Model .................. 479
Ryoji Fukuda, Kaoru Oka

A Fuzzy Estimation of Fuzzy Parameters with Rational
Number Cuts .................................................. 487
Dabuxilatu Wang, Jingjing Wang

Regularized REML for Estimation in Heteroscedastic
Regression Models ............................................ 495
Dengke Xu, Zhongzhan Zhang

Testing of Relative Difference under Inverse Sampling ........ 503
*Shaoping Jiang, Yanfang Zhao*

Adaptive Elastic-Net for General Single-Index Regression
Models .................................................................. 509
*Xuejing Li, Gaorong Li, Suigen Yang*

Some Remark about Consistency Problem of Parameter
Estimation ............................................................. 517
*Mingzhong Jin, Minqing Gong, Hongmei Liu*

Variable Selection for Semiparametric Isotonic Regression
Models .................................................................. 525
*Jiang Du, Zhongzhan Zhang, Tianfa Xie*

Study of Prognostic Factor Based on Factor Analysis and
Clustering Method .................................................... 533
*Zheng Liu, Liying Fang, Mingwei Yu, Pu Wang*

Applying Factor Analysis to Water Quality Assessment:
A Study Case of Wenyu River ...................................... 541
*Chen Gao, Jianzhuo Yan, Suhua Yang, Guohua Tan*

On the Profit Manipulations of Chinese Listed Companies .. 549
*Shuangjie Li, Xingxing Chen*

Fuzzy Portfolio Optimization Model with Fuzzy Numbers .... 557
*Chunquan Li, Jianhua Jin*

Soft Computing Techniques and Its Application in the
Intelligent Capital Compensation ............................... 567
*Kuo Chang Hsiao, Berlin Wu, Kai Yao*

Geometric Weighting Method and Its Application on Load
Dispatch in Electric Power Market .............................. 579
*Guoli Zhang, Chen Qiao*

Application of Cross-Correlation in Velocity Measurement
of Rainwater in Pipe ................................................. 587
*Zhanpeng Li, Bin Fang*

A Decision-Making Model of Adopting New Technology by
Enterprises in Industrial Clusters under the Condition of
Interaction ............................................................. 595
*Xianli Meng, Xin-an Fan, Baomin Hu, Xuan Li*

Study of Newsboy Problem with Fuzzy Probability
Distribution ............................................................ 603
*Xinshun Ma, Ting Yuan*

**Blogger's Interest Mining Based on Chinese Text Classification** ...................................................... 611
*Suhua Yang, Jianzhuo Yan, Chen Gao, Guohua Tan*

**Characterization of Generalized Necessity Functions in Łukasiewicz Logic** ............................................... 619
*Tommaso Flaminio, Tomáš Kroupa*

**On Generating Functions of Two-Dimensional T-Norms** ...... 627
*Masaya Nohmi, Aoi Honda, Yoshiaki Okazaki*

**Approximating a Fuzzy Vector Given Its Finite Set of Alpha-Cuts** .......................................................... 635
*Xuecheng Liu, Qinghe Sun*

**A New Fuzzy Linear Programming Model and Its Applications** ............................................................ 643
*Hsuan-Ku Liu, Berlin Wu*

**Interval Relaxation Method for Linear Complementarity Problem** ............................................................ 651
*Juan Jiang*

**The Representation Theorem on the Category of FTML($\mathcal{L}$)** .. 659
*Jie Zhang, Xiaoliang Kou, Bo Liu*

**L Fuzzy Subalgebras and L-Fuzzy Filters of $R_0$-Algebras** ...... 667
*Chunhui Liu, Luoshan Xu*

**Completely Compact Elements and Atoms of Rough Sets** .... 675
*Gaolin Li, Luoshan Xu*

**Weak Approximable Concepts and Completely Algebraic Lattices** ............................................................. 683
*Hongping Liu, Qingguo Li, Lankun Guo*

**Ideal-Convergence in Quantales** ................................. 691
*Shaohui Liang*

**On the Factorization Theorem of a Monotone Morphism in a Topos** ............................................................. 699
*Tao Lu, Hong Lu*

**Author Index** ...................................................... 707

# Ordinal Preference Models Based on S-Integrals and Their Verification

Michio Sugeno

**Abstract.** This paper discusses the ordinal preference models based on S-integrals. In parallel with the cardinal preference models based on Choquet integrals, there are various S-integrals: S-integral, SS-integral, CPTS-integral and BCS-integral.

First, we verify the ordinal models. To this aim, two psychological experiments have been conducted, where all the preferences of subjects could be modeled by S-, SS-, CPTS- and BCS-integrals. A counter example to BCS-integral models is also shown.

Next, we consider Savage's Omelet problem in multi-criteria decision making. There are many admissible preference orders of acts depending on the consequents of acts. We find that there exist some preferences which cannot be modeled even by the BCS-integral.

Finally, to breakthrough the difficulty of BCS-integral models, we propose hierarchical preference models by which we can model the above counter examples.

**Keywords:** S-integrals, Ordinal preference, Savage's omelet problem, Hierarchical preference.

## 1 Introduction

There are two streams in theoretical models in preference theory. One is to use cardinal models with addition and multiplication on real numbers, where there are two cases: preference under risk when objective probabilities are available and preference under uncertainty when only subjective probabilities

Michio Sugeno (Japan)
European Centre for Soft Computing, Asturias, Spain
e-mail: michio.sugeno@gmail.com

S. Li (Eds.): Nonlinear Maths for Uncertainty and its Appli., AISC 100, pp. 1–18.
springerlink.com

are available. The other is to use ordinal models with max and min operations on ordinal numbers.

Concerning the cardinal preference theory, we can go back to St. Petersburg's paradox in which Bernoulli presented the seminal idea of utility functions in 1738 [2]. Many years later, von Neumann and Morgenstern initiated the expected utility theory with objective probabilities in 1944 [12]. Soon later, Allais showed a counter example to their theory in 1953 [1]. In case that objective probabilities are not available, Savage suggested to use subjective probabilities in 1954 [13]. Then Ellsberg showed the well-known Ellsberg paradox in 1961 [4] that is a counter example to the Savage's theory.

These difficulties can be solved by Schmeidler's concept of nonlinear expectation in 1989 [14] based on Choquet integrals with distorted probabilities (a special case of fuzzy measures) [3] that are obtained by the monotonic transformation of probabilities, where the Choquet integral with respect to fuzzy measures [15] was first suggested by Höhle in 1982 [8] and also studied by Murofushi and Sugeno in 1989 [10]. Then, based on Choquet integrals, Tversky and Kahneman presented Cumulative Prospect Theory in 1992 [16] in which they used the so-called Cumulative Prospect Theory (CPT) Choquet integrals with two distorted probabilities on a bipolar scale. Their theory stating that people behave in a case of losses differently from a case of gains was experimentally verified. However, a counter example to Cumulative Prospect Theory was shown by Grabisch and Labresuche, and to solve this difficulty, the concept of Bi-capacity (BC) Choquet integrals was suggested as an extension of CPT-Choquet integrals in 2002 [5]. The concept of Bipolar capacities was also presented by Greco et al. in 2002 [7]. Again, a counter example to BC-Choquet integral models was shown in 2007 [9], [6]. That is, there exists a reasonable preference that cannot be modeled even by a BC-Choquet integral model.

## 2   Ordinal Preference Theory Based on S-Integrals

While the cardinal theory has a long history, the ordinal theory started very recently. It is considered that the ordinal theory based on only comparisons matches human subjective preferences better than the cardinal theory since it is hardly assumed that ordinary people make numerical calculations in their brains. In parallel with various Choquet integrals, we can define S-integral, Symmetric S-integral, CPT S-integral, and Bi-capacity S-integral [8], [6].

Define a finite set: $X = \{x_1, \cdots, x_n\}$, an ordinal scale: $L^+ = \{0, \cdots, \ell\}$ Let $f$ be a function $f : X \to L$, where $f(x)$ can be expressed as a row vector $f = (f_1, \cdots, f_n)$ with $f_i = f(x_i)$. Without loss of generality, we assume $f_1 \leq \cdots \leq f_n$; if not, we can rearrange it in an increasing order. Let $\mu$ be a fuzzy measure on $X$, $\mu: 2^X \longrightarrow L^+$ such that

(i)  $\mu(\phi) = 0$,
(ii) $\mu(A) \leq \mu(B), for\, A \subset B$.

The S-integral of $f$ with respect to a fuzzy measure $\mu$ is defined as

$$S_\mu(f) = S_\mu(f_1, \cdots, f_n) = \bigvee_{i=1}^{n} [f_i \wedge \mu(A_i)], \qquad (1)$$

where $A_i = \{x_i, \cdots, x_n\}$.

Define $L^- = \{-\ell, \cdots, -1\}$ and a bipolar scale $L = L^- \cup L^+$. We extend max $\vee$ and min $\wedge$ operations to $L$ such that for $a, b \in L$,

$$a \curlyvee b = \text{sign}(a + b)(|a| \vee |b|),$$
$$a \curlywedge b = \text{sign}(a \cdot b)(|a| \wedge |b|). \qquad (2)$$

where neither $\curlyvee$ nor $\curlywedge$ is associative, and $\curlywedge$ is not distributive over $\curlyvee$. However, these do not cause any difficulties in the following applications.

The symmetric S-integral of $f$ with respect to a fuzzy measure $\mu$ is defined as

$$SS_\mu(f) = S_\mu(f^+) \curlyvee -S_\mu(f^-), \qquad (3)$$

where $f^+ = \max\{f, 0\}$ and $f^- = -\min\{f, 0\}$.

The Cumulative Prospect Theory S-integral of $f$ with respect to a pair of fuzzy measures $\mu^+$ and $\mu^-$ is defined as

$$CPTS_{\mu^+, \mu^-}(f) = S_{\mu^+}(f^+) \curlyvee -S_{\mu^-}(f^-), \qquad (4)$$

where $\mu^+$ is a fuzzy measure for the positive part of $f$ and $\mu^-$ for the negative part of $f$.

Now define a bi-capacity $v$ as a set function with two arguments $v : 2^X \times 2^X \to L$ such that

(i) $v(\phi, \phi) = 0$,

(ii) $v(A, B) \le v(A', B), for A \subset A'$,

(iii) $v(A, B) \ge v(A, B'), for B \supset B'$.

Then, the Bi-Capacity S-integral of $f$ with respect to a bi-capacity $v$ is defined as

$$BCS_v(f) = \bigcurlyvee_{i=1}^{n} [|f_i| \curlywedge v(A_i \cap X^+, A_i \cap X^-)], \qquad (5)$$

where $|f_1| \le \cdots \le |f_n|$, $X^+ = \{x | f \ge 0\}$ and $X^- = \{x | f < 0\}$.

Obviously, SS-integral is an extension of S-integral and CPTS-integral is a further extension, but BCS-integral is not exactly an extension of CPTS-integral; there exist a CPTS-integral that cannot be expressed by a BCS-integral.

# 3  Experimental Verifications of S-Integral Models

We have conducted two psychological experiments to verify S-integral models for human subjective preferences.

## 3.1  Preferences of Apartments

We have prepared five samples of apartments for students and married adults with four attributes: (1) time to the nearest station, (2) life environment such as shops, restaurants, parks, etc., (3) plan of apartment, and (4) direction of the main room. The samples are shown in Tables 1 and 2.

**Table 1** Samples of apartments for students

| Apartment | Time | Life | Plan | Direction |
|-----------|---------|-------------|------|-----------|
| A | 20 min. | good | 1K | SE |
| B | 3 min. | not so good | 2DK | E |
| C | 10 min. | not good | 1R | S |
| D | 18 min. | very good | 1DK | SW |
| E | 7 min. | medium | 2K | W |

**Table 2** Samples of apartments for married adults

| Apartment | Time | Life | Plan | Direction |
|-----------|---------|-------------|------|-----------|
| A | 16 min. | not so good | 2LDK | SW |
| B | 3 min. | medium | 1LDK | W |
| C | 20 min. | very good | 2DK | S |
| D | 12 min. | good | 3DK | NW |
| E | 7 min. | not good | 2LK | E |

We made a questionnaire about preference orders on the attributes and the apartments, where subjects were 24 students aged 21-22 and 21 married adults aged around 30-40. We show an example of the obtained answers; here, life environment is a priori ordered as is seen in Tables 1 and 2, where 1K: one bedroom and one kitchen; 2DK: two bedroom and one dining kitchen; 2LDK: two bedrooms, one living room and one dining kitchen; 1R: one bedroom; 2LK: two bedrooms, one living room and one kitchen; E: East; S: South; SE: South East.

Time:      $B > E > C > D > A$: $5 > 4 > 3 > 2 > 1$,
Plan:      $B > D = E > A > C$: $5 > 4 = 4 > 2 > 1$,
Direction: $A > B > C = D > E$: $5 > 4 > 3 = 3 > 1$,
Apartment: $B = D > A > E > C$: $5 = 5 > 3 > 2 > 1$.

As is shown above, preference orders are transformed to those in the scale $L^+ = \{0, \cdots, 5\}$.

We show some modeling examples of preferences. For modeling, we consider an apartment, e.g., $A$ as a map from $X$ (a set of attributes) to $L^+$ (a unipolar scale) and also express it by a vector $(1, 4, 3, 1)$ as is seen in Table 3, where the preference of apartments by the student No.13 is $B > E > A = D > C$.

**Table 3** Preference by student No.13: $B > E > A = D > C$

| Apart. | $x_1$: Time | $x_2$: Life | $x_3$: Plan | $x_4$: Direc. |
|--------|-------------|-------------|-------------|---------------|
| A | 1 | 4 | 3 | 1 |
| B | 5 | 2 | 5 | 2 |
| C | 4 | 1 | 1 | 5 |
| D | 1 | 5 | 5 | 4 |
| E | 4 | 3 | 3 | 4 |

For the sake of simplicity, we express, for instance, $\mu(\{x_1, x_3\})$ as $\mu(13)$, By setting $\mu(13) = 5$, $\mu(14) = 2$ and $\mu(234) = 2$, where other values are free as far as they satisfy the monotonicity of a fuzzy measure, we easily obtain $S_\mu(A) = 2, S_\mu(B) = 5, S_\mu(C) = 1, S_\mu(D) = 2, S_\mu(E) = 3$. From these follows $B > E > A = D > C$. That is, this preference is modeled by a S-integral.

Another example is shown in Table 4 (a). This example is found to be modeled neither by a S-integral nor by a SS-integral. We apply a CPTS-integral. To this aim we choose a neutral value 3 and subtract it from the orders of the attributes expressed in $L^+$ as is seen in Table 4 (b), where a neutral value is chosen from $\{0, \cdots, 4\}$.

**Table 4** Preference by student No. 2: $E > D > B > C > A$

(a)                                    (b) :subtract 3 from(a)

| | $x_1$ | $x_2$ | $x_3$ | $x_4$ | | $x_1$ | $x_2$ | $x_3$ | $x_4$ |
|---|---|---|---|---|---|---|---|---|---|
| A | 1 | 4 | 2 | 3 | A | -2 | 1 | -1 | 0 |
| B | 3 | 2 | 3 | 1 | B | 0 | -1 | 0 | -2 |
| C | 5 | 1 | 1 | 5 | C | 2 | -2 | -2 | 2 |
| D | 1 | 5 | 4 | 5 | D | -2 | 2 | 1 | 2 |
| E | 5 | 3 | 5 | 2 | E | 2 | 0 | 2 | -1 |

Then we normalize these in $L = \{-5, \cdots, 5\}$. As a result, we obtain the normalized orders of the attributes expressed on a bipolar scale $L$ as is shown in Table 5.

**Table 5** Normalized preference table of student No. 2 with n = 3

| | $x_1$ | $x_2$ | $x_3$ | $x_4$ |
|---|---|---|---|---|
| A | -5 | 3 | -3 | 0 |
| B | 0 | -3 | 0 | -5 |
| C | 5 | -5 | -5 | 5 |
| D | -5 | 5 | 3 | 5 |
| E | 5 | 0 | 5 | -3 |

Setting $\mu^+(2) = 2, \mu^+(13) = 4, \mu^+(14) = 1, \mu^+(234) = 3$, and $\mu^-(1) = 2, \mu^-(4) = 1, \mu^-(13) = 3, \mu^-(23) = 2, \mu^-(24) = 1$, we obtain that

$$CPTS_{\mu^+,\mu^-}(A) = [3 \wedge \mu^+(2)] \curlyvee -[(3 \wedge \mu^-(13)) \vee (5 \wedge \mu^-(1))] = -3,$$
$$CPTS_{\mu^+,\mu^-}(B) = -[(3 \wedge \mu^-(24)) \vee (5 \wedge \mu^-(4))] = -1,$$
$$CPTS_{\mu^+,\mu^-}(C) = [(5 \wedge \mu^+(14))] \curlyvee -[(5 \wedge \mu^-(23))] = -2,$$
$$CPTS_{\mu^+,\mu^-}(D) = [(3 \wedge \mu^+(234)) \vee (5 \wedge \mu^+(24))] \curlyvee -[(5 \wedge \mu^-(1))] = 3,$$
$$CPTS_{\mu^+,\mu^-}(E) = [5 \wedge \mu^+(13)] \curlyvee -[3 \wedge \mu^-(4)] = 4.$$

From these follows $E > D > B > C > A$. Therefore this preference can be modeled by a CPTS-integral.

The results of modeling are summarized in Table 6. As is seen, all the preferences in the experiments are modeled by S-, SS-, or CPTS-integrals.

**Table 6** Modeling results of preferences for apartments

| | S-integral | SS-integral | CPTS-integral | Total |
|---|---|---|---|---|
| Students | 10 (42%) | 8 (33%) | 6 (25%) | 24 |
| Adults | 7 (33%) | 8 (38%) | 6 (29%) | 21 |
| Total | 17 (38%) | 16 (36%) | 12 (27%) | 45 |

## 3.2 Preferences of Part-Time Jobs

We consider the preferences of part-time jobs. Table 7 shows five samples of part-time jobs with four attributes: (1) hourly wage, (2) number of times per week, (3) access time to work place, and (4) hardness of job. We conducted similar experiments where the subjects were 10 male students and 10 female students. The preference by the student No. 6 modified with a neutral value 3 is shown in Table 8. It is found that this preference cannot be modeled for any neutral value by a CPTS-integral. Therefore, we apply a BCS-integral to the preference table.

**Table 7** Part-time jobs for students

| Job | Hourly Wage | Times/Week | Access Time | Hardness of Job |
|-----|-------------|------------|-------------|-----------------|
| A | 1100 yen | 2 | 20 min | rather busy |
| B | 900 yen | 4 | 15 min | average |
| C | 1100 yen | 2 | 25 min | busy |
| D | 700 yen | 6 | 10 min | rather not busy |
| E | 700 yen | 6 | 5 min | not busy |

**Table 8** Preference expressed in $L$ with n = 3 by student No. 6: $A > C > B > E > D$

|  | $x_1$ | $x_2$ | $x_3$ | $x_4$ |
|-----|-------|-------|-------|-------|
| A | 5 | 5 | -3 | -3 |
| B | 0 | 0 | 0 | 0 |
| C | 5 | 5 | -5 | -5 |
| D | -5 | -5 | 3 | 3 |
| E | -5 | -5 | 5 | 5 |

Setting a bi-capacity as $v(12, \phi) = 5, v(12, 34) = 3, v(34, 12) = -3$, and $v(\phi, 12) = -5$, we obtain that

$$BCS_v(A) = [3 \curlywedge v(12, 34)] \curlyvee [5 \curlywedge v(12, \phi)] = 5,$$
$$BCS_v(B) = 0,$$
$$BCS_v(C) = (5 \curlywedge v(12, 34)) = 3,$$
$$BCS_v(D) = [3 \curlywedge v(34, 12)] \curlyvee [5 \curlywedge v(\phi, 12)] = -5,$$
$$BCS_v(E) = (5 \curlywedge v(34, 12)) = -3.$$

From these, the preference order by the student No. 6: $A > C > B > E > D$ is derived.

The results of modeling are summarized in Table 9 that shows all the preferences are modeled by S-, SS-, CPTS-, or BCS-integrals. The results suggest that the female students make more sophisticated preferences than the male students, since compared to the male students; there are no preferences modeled by S-integrals and are more preferences modeled by BCS-integrals.

Hinted by the above examples, we present a counter example to BCS-integral models as is shown in Table 10 (a). We apply the following preference rules to it: (1) first evaluate $x_1$ and $x_2$ roughly and then (2) evaluate $x_3$ and $x_4$ in detail. From the rule 1, we obtain $\{A, B\} > C > \{D, E\}$ and then, from the rule (2) follows $A > B > C > D > E$. This reasonable preference

**Table 9** Modeling results of preferences for part-time jobs

|  | S | SS | CPTS | BCS | Total |
|---|---|---|---|---|---|
| Male | 7 | 0 | 1 | 2 | 10 |
| Female | 0 | 3 | 1 | 6 | 10 |
| Total | 7 | 3 | 2 | 8 (40%) | 20 |

cannot be modeled even by BCS-integral models with any neutral values. For instance, we transform (a) in Table 10 to (b) with $n = 3$.

**Table 10** A counter example to BCS-integral models

<table>
<tr><td></td><td colspan="4">(a)</td><td colspan="4">(b) with $n = 3$</td></tr>
<tr><td></td><td>$x_1$</td><td>$x_2$</td><td>$x_3$</td><td>$x_4$</td><td>$x_1$</td><td>$x_2$</td><td>$x_3$</td><td>$x_4$.</td></tr>
<tr><td>A</td><td>4</td><td>5</td><td>2</td><td>1</td><td>4</td><td>5</td><td>-4</td><td>-5</td></tr>
<tr><td>B</td><td>5</td><td>5</td><td>1</td><td>1</td><td>5</td><td>5</td><td>-5</td><td>-5</td></tr>
<tr><td>C</td><td>3</td><td>3</td><td>3</td><td>3</td><td>0</td><td>0</td><td>0</td><td>0</td></tr>
<tr><td>D</td><td>1</td><td>1</td><td>5</td><td>5</td><td>-5</td><td>-5</td><td>5</td><td>5</td></tr>
<tr><td>E</td><td>2</td><td>1</td><td>4</td><td>5</td><td>-4</td><td>-5</td><td>4</td><td>5</td></tr>
</table>

Applying BCS-integral, we obtain that

$$BCS_v(A) = [4 \curlywedge v(12, 34)] \curlyvee [(5 \curlywedge v(2, 4))],$$
$$BCS_v(B) = 5 \curlywedge v(12, 34),$$
$$BCS_v(C) = 0,$$
$$BCS_v(D) = 5 \curlywedge v(34, 12),$$
$$BCS_v(E) = [4 \curlywedge v(34, 12)] \curlyvee [(5 \curlywedge v(4, 2))].$$

It is easily proved that there do not exist $v(12, 34)$, $v(2, 4)$, $v(34, 12)$ and $v(4, 2)$ satisfying that $BCS_v(A) > BCS_v(B) > 0 > BCS_v(D) > BCS_v(E)$.

## 4  Savage's Omelet Problem

In his book, "The foundation of statistics" in 1954 [13], Savage gave an interesting example for multi-criteria decision making known as Savage's omelet problem that consists of three acts and two states as we shall see. We are making an omelet with six eggs and so far five fresh eggs have been broken into a bowl. For a sixth egg that may be not fresh, we can take three acts.

$a_1$ : break it into a bowl;
$a_2$ : break it into a saucer to examine its quality;
$a_3$ : throw it away.

If we take $a_1$, we may lose the whole omelet when a 6th egg is rotten; if we take $a_2$, we have a saucer to wash; and if we take $a_3$, we may lose a fresh egg. Then we have the following decision (or preference in our terminology) table, where there are two states: fresh and rotten. Many scientists have ever used this problem to explain decision making problems, but nobody has ever tried to fully solve the problem. The reason for this is, perhaps, that the problem looks very simple. It is, however, not the case as we shall see. In the sequel, we will fully analyze this problem and give a complete solution to it.

We denote 6E: 6-egg-omelet; 6EW: 6-egg-omelet and a saucer to wash; 5EW: 5-egg-omelet and a saucer to wash; 5EL: 5-egg-omelet and one egg lost; 5E: 5-egg-omelet; NE: no omelet. Then the preference table of the Savage's omelet problem is shown in Table 11.

**Table 11**   Preference table of Savage's omelet problem

|  | fresh | rotten |
|---|---|---|
| $a_1$: break into bowl | 6E: 6-egg-omlet | NE: no omlet |
| $a_2$: break into saucer | 6EW: 6-egg-omlet & saucer to wash | 5EW: 5-egg-omlet with saucer to wash |
| $a_3$: throw away | 5EL: 5-egg-omlet & saucer to wash | 5E: 5-egg-omlet |

First we may assume in general that (i) $6E > any > $ NE, (ii) $6EW > 5EW$, and (iii) $5E > 5EW$ or $5EL$. These assumptions seem quite reasonable from our common sense though there may be a person who states that all the consequents are the same.

Now we find 11 admissible orders for six consequents under the above assumption, for instance, #1: $6E > 6EW > 5E > 5EW > 5EL > NE$, #8: $6E > 5E > 6EW > 5EW = 5EL > NE$, and so on. We express these orders in $L^+ = \{0, \cdots, 5\}$. Then we can list up 13 possible orders of three acts, for instance, ① $a_1 > a_2 > a_3$, ⑤ $a_2 > a_1 > a_3$, ⑩ $a_3 > a_1 = a_2$, ⑬ $a_1 = a_2 = a_3$ and so on. We note that not all orders of acts are admissible with respect to those of consequents. For example, consider the following preference order of consequents:

$a_1$: 5(6E)/fresh, 0(NE)/rotten;
$a_2$: 3(6EW)/fresh, 1(5EW)/rotten;
$a_3$: 3(5EL)/fresh, 4(5E)/rotten.

In this case, the order $a_2 > a_3$ is not admissible since we have to assume monotonicity: $a_2 \leq a_3$; thus there are only 8 admissible orders about this preference table. Then we find that there are altogether 130 admissible orders of acts concerned with 11 preference tables corresponding to the orders of consequents. Taking into account that we may have to consider a neutral value chosen in $\{0, \cdots, 4\}$ for modeling, there are 650 ($130 \times 5$) cases altogether.

We consider the preference table #6 shown in Table 12 with five neutral values, where 0 implies the original table.

**Table 12** Preference table #6 with 13 admissible orders of acts

| | n=0 | | n=1 | | n=2 | | n=3 | | n=4 | |
|---|---|---|---|---|---|---|---|---|---|---|
| $a_1$ | 5 | 0 | 5 | -2 | 5 | -4 | 4 | -5 | 2 | -5 |
| $a_2$ | 4 | 1 | 4 | 0 | 4 | -3 | 3 | -4 | 0 | -4 |
| $a_3$ | 2 | 4 | 2 | 4 | 0 | 4 | -3 | 3 | -3 | 0 |

The six orders ① $a_1 > a_2 > a_3$, ② $a_1 > a_2 = a_3$, ④ $a_1 = a_2 > a_3$, ⑩ $a_3 > a_1 = a_2$, ⑪ $a_3 > a_2 > a_1$, ⑬ $a_1 = a_2 = a_3$ are modeled by S-integrals with $n = 0$; the two orders ⑧ $a_2 = a_3 > a_1$ and ⑫ $a_3 = a_1 > a_2$ are modeled by SS-integral with $n = 1$ and $n = 2$, respectively; the two orders ⑥ $a_2 > a_1 = a_3$, ⑦ $a_2 > a_3 > a_1$ are modeled by CPTS-integrals with $n = 2$; the two orders ③ $a_1 > a_3 > a_2$, ⑨ $a_3 > a_1 > a_2$ are also modeled by CPTS-integrals with $n = 3$, and finally, the order ⑤ $a_2 > a_1 > a_3$ cannot be modeled even by a BCS-integral. For instance, we can calculate the BCS-integral of the table with $n = 4$ as follows:

$$BCS(a_1) = [2 \curlywedge v(1,2)] \curlyvee [(5 \curlywedge v(\phi,2))],$$
$$BCS(a_2) = 4 \curlywedge v(\phi,2),$$
$$BCS(a_3) = 3 \curlywedge v(1,2).$$

It is easily proved that there do not exist $v(1,2)$ and $v(\phi,2)$ satisfying that $BCS(a_2) > BCS(a_1) > BCS(a_3)$: ⑤ $a_2 > a_1 > a_3$; it is also the case for the other neutral values.

We examined all the cases with S-, SS-, CPTS-, and BCS-integrals and five neutral values including 0. The results are summarized in Table 13 for the 130 preference orders of acts corresponding to the 11 preference tables. As is seen, there are 16 preference orders that cannot be modeled even by BCS-integrals.

**Table 13** Modeling results of Savage's omelet problem

| S | SS | CPTS | BCS | No model | Total |
|---|---|---|---|---|---|
| 80 | 13 | 19 | 2 | 16 | 130 |
| 62% | 10% | 14% | 2% | 12% | 100% |

## 5  Hierarchical Preference Models

In order to breakthrough the difficulties of BCS-integrals, we propose hierarchical preference models. We begin with the definition of a preference table.

Define a preference table as a triplet $(A, X, L^+)$, where $A = \{a_1, \cdots, a_p\}$: a set of objects, $X = \{x_1, \cdots, x_q\}$: a set of attributes, and $L^+ = \{0, \cdots, \ell\}$: a unipolar scale. Here $a_i$ is also regarded as a map $a_i : X \to L^+$ assigning the orders of consequents to attributes, alternatively, $a_i$ can be expressed for a finite set $X$ by a row vector $(a_{i1}, \cdots, a_{iq}) \in (L^+)^{1 \times q}$, where $a_{ij} = a_i(x_j)$. As has been stated before, the S-integral of $a_i$ with respect to a fuzzy measure $\mu$ is written as $S_\mu(a_i) = S_\mu(a_{i1}, \cdots, a_{iq})$. Define a column vector of acts $[a] = (a_1, \cdots, a_p)^t$ where $t$ means transpose. Here, $[a]$ can be expressed as a matrix $\{a_{ij}\} \in L^{p \times q}$. First, we extend $[a]$ to $[a]_n$ on a bipolar scale $L = \{-\ell, \cdots, 0, \cdots, \ell\}$ by subtracting a neutral value $n \in \{0, \cdots, \ell - 1\}$ from each element of a matrix $[a]$ and then normalize it in $L$ as we have seen. This process is represented by a map

$$n : (L^+)^{p \times q} \to L^{p \times q} : [a] \longmapsto [a]_n, \tag{6}$$

We call $[a]_n$ a preference matrix with a neutral value $n$, where $[a]_0$ implies an original preference matrix.

Now define a general S-integral $GS$ as an element of $\{S, SS, CPTS, BCS\}$ and a general fuzzy measure $g\mu$ on $X$ as an element of $\{$ordinary fuzzy measure $\mu$, a pair of CPT-type fuzzy measures $(\mu^+, \mu^-)$, a bi-capacity $v\}$. Consider a general S-integral as a map

$$GS_{g\mu} : L^{1 \times q} \to L :$$
$$(a_i)_n = (a_{i1}, \cdots, a_{iq})_n \longmapsto GS_{g\mu}((a_i)_n) = GS_{g\mu}((a_{i1}, \cdots, a_{iq})_n), \tag{7}$$

where $(a_i)_n$ is the i-th row vector of a matrix $[a]_n$.

Preference model at level 1
Extending the domain of a map $GS_{g\mu}$ from $L^{1 \times q}$ to $L^{p \times q}$, we define a general S-integral of $[a]_n$ with respect to a general fuzzy measure $g\mu$ as a map

$$GS_{g\mu} : L^{p \times q} \to L^{p \times 1} :$$
$$[a]_n = ((a_1)_n, \cdots, (a_p)_n)^t \longmapsto GS_{g\mu}([a]_n) = (GS_{g\mu}(a_1), \cdots, GS_{g\mu}(a_p))^t. \tag{8}$$

We call it preference model of the original preference matrix $[a] \in (L^+)^{p \times q}$, based on $[a]_n$ with a neutral value $n$.

Remark:
We can consider $GS_{g\mu}([a]_n)$ as $(GS_{g\mu} \circ n)([a])$, i.e., the composition of two maps $n : (L^+)^{p \times q} \to L^{p \times q}$ and $GS_{g\mu} : L^{p \times q} \to L^{p \times 1}$. That is,

$$GS_{g\mu} \circ n : (L^+)^{p \times q} \to L^{p \times 1} : [a] \longmapsto GS_{g\mu}([a]_n). \tag{9}$$

Let $N$ be a set of neutral values: $N = \{n_1, \cdots, n_r\}$ and define $[a]_N = ([a]_{n1}, \cdots, [a]_{nr}) \in (L^{p \times q})^r$ : a vector of matrices. Let GM be a set of general fuzzy measure corresponding to $N$ at level 1: $GM = \{g\mu_1, \cdots, g\mu_r\}$. Here again, we consider $N$ as a map

$$N : (L^+)^{p \times q} \to (L^{p \times q})^r : [a] \longmapsto [a]_N = ([a]_{n_1}, \cdots, [a]_{n_r}), \tag{10}$$

Then define a general S-integral with respect to a set of general fuzzy measures $GM$ as a map

$$GS_{GM} : (L^{p \times q})^r \to L^{p \times r} :$$
$$[a]_N \longmapsto GS_{GM}([a]_N) = (GS^1_{g\mu_1}([a]_{n_1}), \cdots, GS^r_{g\mu_r}([a]_{n_r})), \tag{11}$$

where $GS^k$, $1 \leq k \leq r$, implies a certain general S-integral and $GS_{GM}([a]_N)$ is expressed as the composition of two maps

$$GS_{GM} \circ N : (L^+)^{p \times q} \to L^{p \times r} : [a] \longmapsto GS_{GM}([a]_N), \tag{12}$$

We call $GS_{GM}([a]_N)$ preference matrix at level 2. Now we define a set $GMN = \{(g\mu_k, n_k) | g\mu_k \in GM, n_k \in N, 1 \leq k \leq r\}$. Noting the expression at Eq. (11) and $[a]_N = ((a_1)_N, \cdots, (a_p)_N)^t$, we can consider $GS_{GM}([a]_N)$ as a matrix shown at Fig. 1, where $(a_i)_N$, the i-th element of $[a]_N$, is regarded as a map (note: at level 1, $(a_i)_n$ is regarded as a map from $X$ to $L$)

$$(a_i)_N : GMN \longrightarrow L : (g\mu_k, n_k) \longmapsto GS_{g\mu_k}((a_i)_{n_k}), \tag{13}$$

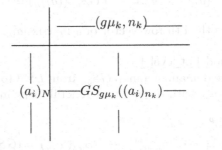

**Fig. 1** Preference matrix $GS_{GM}([a]_N)$ at level 2

Preference model at level 2

Let $g\sigma$ be a general fuzzy measure on $GMN$ and $GS_{g\sigma}$ be a general S-integral with respect to $g\sigma$. We define a preference model of the preference matrix $[a] \in (L^+)^{p \times q}$ based on $[a]_N$ with a set of neutral values $N$ at level 2 as

$$GS_{g\sigma} : L^{p\times r} \to L^{p\times 1} : GS_{GM}([a]_N) \longmapsto GS_{g\sigma}(GS_{GM}([a]_N)), \qquad (14)$$

where $GS_{g\sigma}$ is a general S-integral of $GS_{GM}([a]_N)$ with respect to a general fuzzy measure $g\sigma$, and $GS_{g\sigma}(GS_{GM}([a]_N))$ is expressed as the composition of three maps

$$GS_{g\sigma} \circ GS_{GM} \circ N : (L^+)^{p\times q} \to L^{p\times 1} : [a] \longmapsto GS_{g\sigma}(GS_{GM}([a]_N)), \quad (15)$$

We illustrate the hierarchi cal modeling process at Fig. 2.

Now we are ready to apply a hierarchical preference model to the Savage's omelet problem. Consider the preference table #6. We have

$$[a]_0 = \begin{pmatrix} 5 & 0 \\ 4 & 1 \\ 2 & 4 \end{pmatrix}, \quad [a]_2 = \begin{pmatrix} 5 & -4 \\ 4 & -3 \\ 0 & 4 \end{pmatrix}.$$

Let $\mu_1(1) = 3$, $\mu_1(2) = 2$ and $\mu_2^+(1) = 4$, $\mu_2^+(2) = 0$, $\mu_2^-(2) = 4$. Then we obtain

$$S_{\mu_1}([a]_0) = S_{\mu_1} \begin{pmatrix} 5 & 0 \\ 4 & 1 \\ 2 & 4 \end{pmatrix} = \begin{pmatrix} 3 \\ 3 \\ 2 \end{pmatrix},$$

$$CPTS_{\mu_2^+,\mu_2^-}([a]_2) = CPTS_{\mu_2} \begin{pmatrix} 5 & -4 \\ 4 & -3 \\ 0 & 4 \end{pmatrix} = \begin{pmatrix} 0 \\ 4 \\ 0 \end{pmatrix},$$

where $S_{\mu_1}([a]_0)$ gives $a_1 = a_2 > a_3$ and $CPTS_{\mu_2^+,\mu_2^-}([a]_2)$ gives $a_2 > a_1 = a_3$. Now setting that $\sigma(1) = 3$ and $\sigma(2) = 4$, we obtain that

$$S_\sigma(S_{\mu_1}([a]_0), CPTS_{\mu_2^+,\mu_2^-}([a]_2)) = S_\sigma \begin{pmatrix} 3 & 0 \\ 3 & 4 \\ 2 & 0 \end{pmatrix} = \begin{pmatrix} 3 \\ 4 \\ 2 \end{pmatrix}.$$

Remark:
As we see in the above example, $S_\sigma$ is chosen for $GS_{g\sigma}$, and $GS^1 = S$ with $g\mu_1 = \mu_1, GS^2 = CPTS$ with $g\mu_2 = (\mu_2^+, \mu_2^-)$.

That is, we can model the preference order ⑤ $a_2 > a_1 > a_3$ for the preference table #6 by a hierarchical preference model. Similarly we can show that all the preference orders of the Savage's omelet problem to which BCS-integral models are not applicable are modeled by hierarchical preference models. The above hierarchical preference model is illustrated at Fig. 3.

**Level 1**

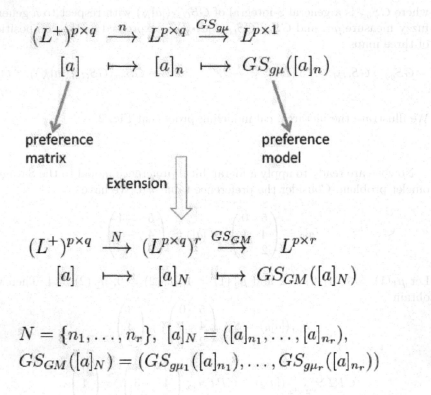

$$(L^+)^{p\times q} \xrightarrow{\ n\ } L^{p\times q} \xrightarrow{GS_{g\mu}} L^{p\times 1}$$
$$[a] \longmapsto [a]_n \longmapsto GS_{g\mu}([a]_n)$$

preference matrix

preference model

Extension

$$(L^+)^{p\times q} \xrightarrow{\ N\ } (L^{p\times q})^r \xrightarrow{GS_{GM}} L^{p\times r}$$
$$[a] \longmapsto [a]_N \longmapsto GS_{GM}([a]_N)$$

$$N = \{n_1, \ldots, n_r\}, \ [a]_N = ([a]_{n_1}, \ldots, [a]_{n_r}),$$
$$GS_{GM}([a]_N) = (GS_{g\mu_1}([a]_{n_1}), \ldots, GS_{g\mu_r}([a]_{n_r}))$$

**Level 2**

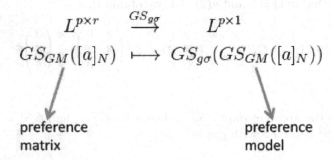

$$L^{p\times r} \xrightarrow{GS_{g\sigma}} L^{p\times 1}$$
$$GS_{GM}([a]_N) \longmapsto GS_{g\sigma}(GS_{GM}([a]_N))$$

preference matrix

preference model

**Fig. 2** Hierarchical modeling of preferences

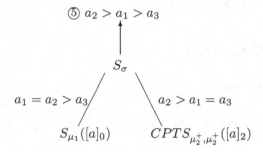

**Fig. 3** Hierarchical model of preference order ⑤ for preference table #6

Next we consider the counter example shown in Table 10 concerning preferences on part-time jobs: $a_1 > a_2 > a_3 > a_4 > a_5$. We have its preference matrices $[a]_0$ and $[a]_3$ such that

$$[a]_0 = \begin{pmatrix} 4\,5\,2\,1 \\ 5\,5\,1\,1 \\ 3\,3\,3\,3 \\ 1\,1\,5\,5 \\ 2\,1\,4\,5 \end{pmatrix}, \quad [a]_3 = \begin{pmatrix} 4 & 5 & 5 & -5 \\ 5 & 5 & -5 & -5 \\ 0 & 0 & 0 & 0 \\ -5 & -5 & 5 & 5 \\ -4 & -5 & 4 & 5 \end{pmatrix}$$

It is found that this counter example can be modeled by a hierarchical preference model with 4 levels as is illustrated at Fig. 4; we omit the detailed modeling process. Note that this hierarchical model evaluates at level 2 the outputs from the level 1, but at level 3 the outputs from both the levels 1 and 2, and so on.

Finally, we consider a counter example to Bi-capacity Choquet integral models presented by Labreuche and Grabisch [9], [6]. First we need the definitions of various Choquet integrals.

Let $f : X \longrightarrow R^+$ be a non-negative real valued function where $X$ is a finite set $\{x_1, \cdots, x_n\}$, and $\mu$ be a fuzzy measure on $X$ with a range $R^+$. Then, the Choquet integral of $f$ with respect of $\mu$ is defined as

$$C_\mu(f) = \sum_{i=0}^{n} (f_i - f_{i-1}), \tag{16}$$

where $f_i \leq \cdots \leq f_n, f_i = f(x_i), f_0 = 0$ and $A_i = \{x_i, ., x_n\}$.

Let $f : X \longrightarrow R, f^+ = \max(f,0)$ and $f^- = \min(f,0)$, then the Symmetric Choquet integral (or Šipoš integral) is defined as

$$SC_\mu(f) = C_\mu(f^+) - C_\mu(f^-), \tag{17}$$

Similarly the Cumulative Prospect Theory Choquet integral is defined with respect to two fuzzy measures $\mu^+$ and $\mu^-$ as

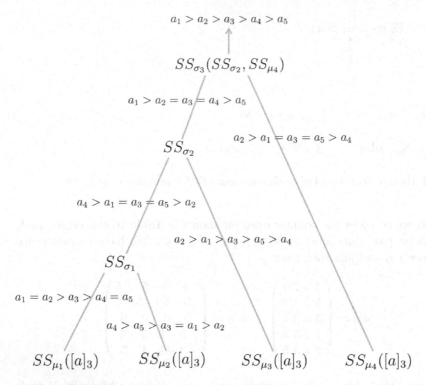

$$a_1 > a_2 > a_3 > a_4 > a_5$$

$$SS_{\sigma_3}(SS_{\sigma_2}, SS_{\mu_4})$$

$$a_1 > a_2 = a_3 = a_4 > a_5$$

$$SS_{\sigma_2}$$

$$a_2 > a_1 = a_3 = a_5 > a_4$$

$$a_4 > a_1 = a_3 = a_5 > a_2$$

$$a_2 > a_1 > a_3 > a_5 > a_4$$

$$SS_{\sigma_1}$$

$$a_1 = a_2 > a_3 > a_4 = a_5$$

$$a_4 > a_5 > a_3 = a_1 > a_2$$

$$SS_{\mu_1}([a]_3) \qquad SS_{\mu_2}([a]_3) \qquad SS_{\mu_3}([a]_3) \qquad SS_{\mu_4}([a]_3)$$

**Fig. 4** Hierarchical model for a counter example to BCS-integrals with four levels

$$CPTC_\mu(f) = C_{\mu^+}(f^+) - C_{\mu^-}(f^-), \tag{18}$$

Let $v$ be a bi-capacity such that $v : 2^X \times 2^X \longrightarrow [-1, 1]$, then the Bi-capacity Choquet integral is defined as

$$BCC_\mu(f) = \sum_{i=0}^{n}(|f_i| - |f_{i-1}|)v(A_i \cap X^+, A_i \cap X^-), \tag{19}$$

where $|f_1| \leq \cdots \leq |f_n|, |f_0| = 0, X^+ = \{x|f \geq 0\}$ and $X^- = \{x|f < 0\}$.

Table 14 shows the preference table with consequents in $R$ of the counter example. In this table, $a_1 \sim a_4$ are four high school students and $x_1, x_2$ and $x_3$ are the test scores of mathematics, statistics, and language, respectively. A teacher applies the rules; if mathematics is well satisfied, consider language, and if mathematics is ill-satisfied, consider statistics since it is usually known that a student good at mathematics is also good at statistics. Then the preference order of the students is given as $a_1 > a_2 > a_3 > a_4$. This order looks reasonable, but it cannot be modeled even by a BCC-integral.

**Table 14** A counter example to BCC-integrals: $[a]_*$

|       | $x_1$ | $x_2$ | $x_3$ |
|-------|-------|-------|-------|
| $a_1$ | 10    | 10    | -20   |
| $a_2$ | 10    | 10    | -30   |
| $a_3$ | -10   | 20    | -30   |
| $a_4$ | -10   | 10    | -20   |

We can calculate BCC integrals as

$$BCC_v(a_1) = 10v(12,3) + 10v(\phi,3),$$
$$BCC_v(a_2) = 10v(12,3) + 10v(2,3) + 10v(\phi,3),$$
$$BCC_v(a_3) = 10v(2,13) + 10v(2,3) + 10v(\phi,3),$$
$$BCC_v(a_4) = 10v(2,13) + 10v(\phi,3).$$

Then, $BCC_v(a_1) > BCC_v(a_2)$ implies that $v(2,3) < 0$, and $BCC_v(a_3) > BCC_v(a_4)$ implies that $v(2,3) > 0$. This is a contradiction and so the preference cannot be modeled even by a BCC-integral.

Now we apply a hierarchical preference model to this counter example. We define a preference matrix $[a]_*$ for the preference table shown in Table 14, where $*$ implies a certain neutral value with which the table is derived from the original test scores.

Setting $\mu_1(2) = 0.5, \mu_1(3) = 0$ and $\mu_1(12) = \mu_1(13) = 1$, we have $SC_{\mu_1}([a]*) = (10, 15, 0, -5)^t$ that implies $a_2 > a_1 > a_3 > a_4$. Also setting $\mu_2(2) = 0, \mu_2(3) = 0.2$ and $\mu_2(12) = \mu_2(13) = 1$, we have $SC_{\mu_2}([a]*) = (6, 4, -14, -12)^t$ that implies $a_1 > a_2 > a_4 > a_3$. Next setting $\sigma(1) = \sigma(2) = 0$ at level 2, we obtain

$$SC_\sigma(SC_{\mu_1}([a]*), SC_{\mu_2}([a]*)) = \begin{pmatrix} 10 & 6 \\ 15 & 4 \\ 0 & -14 \\ -5 & -12 \end{pmatrix} = \begin{pmatrix} 6 \\ 4 \\ 0 \\ -5 \end{pmatrix}.$$

This implies $a_1 > a_2 > a_3 > a_4$. Therefore we can model the counter example by a hierarchical model. An idea similar to a hierarchical model is discussed in [11].

## Conclusions

In his paper, we discussed the ordinal preference models based on various S-integrals. First, we verified these models by conducting psychological experiments. Then, we analyzed the Savage's omelet problem under the framework of S-integral models and gave a complete solution to it. In order to breakthrough the difficulty of Bi-capacity S-integral models, we proposed hierarchical preference models based on S-integrals. It has been shown that all the counter examples to S-integral models can be modeled by hierarchical preference models. It has been also shown that a counter example to Bi-capacity Choquet integral models can be modeled by the same scheme.

# References

1. Allais, M.: Le comportement de l'homme rationnel devant le risque: Critique des postulats et axiomes de l'école américaine. Econometrica 21, 503–546 (1953)
2. Bernoulli, D.: Presentation at St Petersburg Academy (1738)
3. Edwards, W.: Probability-preferences in gambling. American Journal of Psychology 66, 349–364 (1953)
4. Ellsberg, D.: Risk, ambiguity, and the Savage axioms. Quarterly Journal of Economics 75, 643–669 (1961)
5. Grabish, M., Labreuche, C.: Bi-capacities for decision making on bipolar scales. In: Proc. of EUROFUSE Workshop on Information Systems, Varenna, pp. 185–190 (September 2002)
6. Grabisch, M., Labreuche, C.: A decade of application of the Choquet and Sugeno integrals in multi-criteria decision aid. 4OR - Quarterly Journal of Operations Research 6, 1–44 (2008)
7. Greco, S., Motarazzo, B., Slowinski, R.: Bipolar Sugeno and Choquet integrals. In: EUROFUSE Workshop on Information Systems, Varenna, pp. 191–196 (2002)
8. Höhle, U.: Integration with respect to fuzzy measures. In: Proc. of IFAC Symposium on Theory and Applications of Digital Control, New Delhi, pp. 35–37 (January 1982)
9. Labreuche, C., Grabisch, M.: The representation of conditional relative importance between criteria. Annals of Operations Research 154, 93–112 (2007)
10. Murofushi, T., Sugeno, M.: An interpretation of fuzzy measures and the Choquet integral as an integral with respect to fuzzy measures. Fuzzy Sets and Systems 42, 201–227 (1989)
11. Narukawa, Y., Torra, V.: Meta-knowledge Choquet integral and bi-capacity models: links and relations. International Journal of Intelligent Systems 20, 1017–1036 (2005)
12. von Neumann, J., Morgenstern, O.: Theory of Games and Economical Behavior. Princeton University Press, Princeton (1947)
13. Savage, L.: The Foundations of Statistics. Wiley, NY (1954)
14. Schmeidler, D.: Subjective probability and expected utility without additivity. Econometrica 57, 517–587 (1989)
15. Sugeno, M.: Theory of fuzzy integrals and its applications, Dr. Eng. Thesis. Tokyo Institute of Technology (1974)
16. Tversky, A., Kahneman, D.: Advances in prospect theory: cumulative representation of uncertainty. Journal of Risk and Uncertainty 5, 297–323 (1992)

# Strong Laws of Large Numbers for Bernoulli Experiments under Ambiguity

Zengjing Chen and Panyu Wu

**Abstract.** In this paper, we investigate the strong laws of large numbers on upper probability space which contains the case of Bernoulli experiments under ambiguity. Our results are natural extensions of the classical Kolmogorov's strong law of large numbers to the case where probability measures become to imprecise. Finally, an important feature of these strong laws of large numbers is to provide a frequentist perspective on capacities.

**Keywords:** Bernoulli experiments, Capacity, Strong law of large number, Upper-expectation.

## 1 Introduction

One of the key concepts in probability theory is the notion of independence. Using independence, we can decompose a complex problem into simpler components and build a global model from smaller sub-models. The concept of independence is essential for imprecise probabilities too, but there is disagreement about how to define it (one can see [2] to get the different definitions of independence under imprecise probability).

Recently, Peng introduced a definition of independence in upper probability space which is very popular (see eg. [6]), many results such as the law of large number (for short LLN) [6], central limit theorem [6], theory of large deviations [5] have been established and have been used to many fields for

Zengjing Chen
Department of Mathematics, Shandong University, Jinan, China
Department of Financial Engineering, Ajou University, Suwon, Korea
e-mail: zjchen@sdu.edu.cn

Panyu Wu
Department of Mathematics, Shandong University, Jinan, China
e-mail: wupanyu@mail.sdu.edu.cn

S. Li (Eds.): Nonlinear Maths for Uncertainty and its Appli., AISC 100, pp. 19–30.
springerlink.com                                    © Springer-Verlag Berlin Heidelberg 2011

example, finance, control theory, statistics, et al.. Using Peng's definition of IID, Chen [1] obtains the strong laws of large number (for short SLLN) on upper probability space. In this paper, we give a new definition of independence in upper probability space which contains the Bernoulli experiments under ambiguity, and the Peng's independence is our special case. Using this new independence, we investigate the SLLN which need not the condition some like identically distributed on upper probability space. The conclusion of Chen [1] will become a direct corollary of our SLLN. Note that we do not use Peng's LLN as in proof of [1], but fortunately we overcome this gap by exponential function. Finally, we give the invariance principle of SLLN under upper probability space.

## 2  Notation and Lemmas

Let $(\Omega, \mathcal{F})$ be a measurable space, and $\mathcal{M}$ the set of all probability measures on $\Omega$. Every non-empty subset $\mathcal{P} \subseteq \mathcal{M}$ defines an upper probability

$$\mathbb{V}(A) = \sup_{P \in \mathcal{P}} P(A), \quad A \in \mathcal{F},$$

and a lower probability

$$v(A) = \inf_{P \in \mathcal{P}} P(A), \quad A \in \mathcal{F}.$$

Obviously $\mathbb{V}$ and $v$ are conjugate to each other, that is

$$\mathbb{V}(A) + v(A^c) = 1,$$

where $A^c$ is the complement set of $A$.

**Definition 1.** $V(\cdot)$ is a set function from $\mathcal{F}$ to $[0,1]$, $V(\cdot)$ is called a capacity if it satisfying the following (1)(2) and is called a lower/upper continuous capacity if it further satisfying the following(3)/(4):

(1) $V(\phi) = 0, V(\Omega) = 1$.
(2) $V(A) \leq V(B)$, whenever $A \subset B$ and $A, B \in \mathcal{F}$.
(3) $V(A_n) \uparrow V(A)$, if $A_n \uparrow A$, where $A_n, A \in \mathcal{F}$.
(4) $V(A_n) \downarrow V(A)$, if $A_n \downarrow A$ and $A_n$, where $A_n, A \in \mathcal{F}$.

*Remark 1.* It is easy to check that $\mathbb{V}(\cdot)$ is always a lower continuous capacity and $v(\cdot)$ is always a upper continuous capacity. And $\mathbb{V}(\cdot)$ being upper continuous is equivalent to $v(\cdot)$ being lower continuous.

**Definition 2.**  [3] **quasi-surely**
A set $D$ is polar set if $\mathbb{V}(D) = 0$ and a property holds "quasi-surely" (q.s. for short) if it holds outside a polar set.

Now we define the upper expectation $\mathbb{E}[\cdot]$ and the lower expectation $\mathcal{E}(\cdot)$ on $(\Omega, \mathcal{F})$, for each $X \in L^0(\Omega)$ (the space of all $\mathcal{F}$ measurable real random variables on $\Omega$),

$$\mathbb{E}[X] = \sup_{P \in \mathcal{P}} E_P[X], \quad \mathcal{E}[X] = \inf_{P \in \mathcal{P}} E_P[X].$$

It is easy to check that $\mathcal{E}[X] = -\mathbb{E}[-X]$ and $\mathbb{E}[\cdot]$ is a sub-linear expectation (more details can see [6]) on $(\Omega, \mathcal{F})$, that is $\mathbb{E}[\cdot]$ satisfies the following properties $(a)$-$(d)$: for all $X, Y \in L^0(\Omega)$,

(a) Monotonicity: $X \geq Y$ implies $\mathbb{E}[X] \geq \mathbb{E}[Y]$.
(b) Constant preserving: $\mathbb{E}[c] = c, \forall c \in \mathbb{R}$.
(c) Positive homogeneity: $\mathbb{E}[\lambda X] = \lambda \mathbb{E}[X], \forall \lambda \geq 0$.
(d) Sub-additivity: $\mathbb{E}[X + Y] \leq \mathbb{E}[X] + \mathbb{E}[Y]$.

Now we give the new definition of independence on upper probability space $(\Omega, \mathcal{F}, \mathbb{E})$.

**Definition 3. Independence:** Suppose that $X_1, X_2, \cdots, X_n$ is a sequence of real measurable random variables on $(\Omega, \mathcal{F})$. $X_n$ is said to be independent of $(X_1, \cdots, X_{n-1})$ under $\mathbb{E}$ (or $\mathbb{V}$), denoted by $X_n \perp (X_1, \cdots, X_{n-1})$ if for each nonnegative continuous function $\varphi_i(\cdot)$ on $R$ with $\varphi(X_i) \in L^0(\Omega)$ for each $i = 1, \cdots, n$, we have

$$\mathbb{E}[\prod_{i=1}^n \varphi_i(X_i)] = \mathbb{E}[\prod_{i=1}^{n-1} \varphi_i(X_i)] \mathbb{E}[\varphi_n(X_n)].$$

**Independence random variables sequence:** $(X_i)_{i=1}^\infty$ is said to be a sequence of independent random variables, if $X_{i+1}$ is independent of $(X_1, \cdots, X_i)$ for each $i \in \mathbb{N}$.

*Example 1.* If $X \perp Y$, from the definition of 3 and Jensen's inequality, we have

$$\mathbb{E}[e^{X+Y}] = \mathbb{E}[e^X] \cdot \mathbb{E}[e^Y] \geq e^{\mathbb{E}[X]} \cdot e^{\mathbb{E}[Y]}.$$

*Example 2.* If $X \perp Y$, then $X + a \perp Y + b$, and $X I_{X \leq a} \perp Y I_{Y \leq b}$, for any $a, b \in \mathbb{R}$.

## 3 The Main Results

In this section, we give the strong laws of large numbers in upper probability space and list some applications. The proof will be in Appendix.

**Theorem 1. SLLN:** *Let* $(X_i)_{i=1}^\infty$ *be a sequence of independence random variables for upper expectation* $\mathbb{E}$. *Suppose* $\sup_{i \geq 1} \mathbb{E}[|X_i|^{1+\alpha}] < \infty$ *for some* $\alpha > 0$, *and* $\mathbb{E}[X_i] \equiv \overline{\mu}$, $\mathcal{E}[X_i] \equiv \underline{\mu}$. *Set* $S_n := \sum_{i=1}^n X_i$. *Then*

*(I)*

$$\mathbb{V}\left(\{\liminf_{n\to\infty} S_n/n < \underline{\mu}\}\bigcup\{\limsup_{n\to\infty} S_n/n > \overline{\mu}\}\right) = 0, \tag{1}$$

*and*

$$v\left(\underline{\mu} \le \liminf_{n\to\infty} S_n/n \le \limsup_{n\to\infty} S_n/n \le \overline{\mu}\right) = 1. \tag{2}$$

*If further assume $v(\cdot)$ is lower continuous, and for any subsequence of $\mathbb{N}$ denoted by $(n_i)_{i=1}^\infty$, $(S_{n_i} - S_{n_{i-1}})_{i=1}^\infty$ are pairwise independence under $v(\cdot)$, and also there exist $n_0 \in \mathbb{N}$, $c_0 > 0$ such that*

$$\frac{1}{n}|\sum_{i=1}^{n} X_i| \le c_0 \ln(n+1), \quad \forall n > n_0. \tag{3}$$

*Then we have*

*(II)*

$$\mathbb{V}\left(\limsup_{n\to\infty} S_n/n = \overline{\mu}\right) = 1, \quad \mathbb{V}\left(\liminf_{n\to\infty} S_n/n = \underline{\mu}\right) = 1.$$

*Remark 2.* If $(X_i)_{i=1}^\infty$ is bounded, then it is evident that (3) holds true.

We can easily obtain the following two corollaries from the above theorem.

**Corollary 1.** *Under the whole assumptions of theorem 1, $v(\lim_{n\to\infty} S_n/n = a) = 1$ holds if and only if $a = \overline{\mu} = \underline{\mu}$.*

**Corollary 2.** *For any continuous function $\varphi(\cdot)$ on $\mathbb{R}$, under the assumptions of theorem 1 (I), we have*

$$\mathbb{V}\left(\{\liminf_{n\to\infty} \varphi(S_n/n) < \inf_{u\in[\underline{\mu},\overline{\mu}]} \varphi(u)\}\bigcup\{\limsup_{n\to\infty} \varphi(S_n/n) > \sup_{u\in[\underline{\mu},\overline{\mu}]} \varphi(u)\}\right) = 0,$$

*and* $v\left(\inf_{u\in[\underline{\mu},\overline{\mu}]} \varphi(u) \le \liminf_{n\to\infty} \varphi(S_n/n) \le \limsup_{n\to\infty} \varphi(S_n/n) \le \sup_{u\in[\underline{\mu},\overline{\mu}]} \varphi(u)\right) = 1.$

Now we give a example of SLLN on upper probability space which here we called the Bernoulli experiments under ambiguity. In classical probability theory, repeated independent experiments are called Bernoulli experiments if there are only two possible outcomes for each experiment and their probabilities remain the same throughout the experiments [4].

*Example 3.* We consider a countable infinity Ellsberg urns, ordered and indexed by the set $\mathbb{N} = \{1, 2, \cdots\}$. You are told that each urn contains 100 balls (actually the total balls can be different, here we ask all 100 balls just to satisfying $\mathcal{P}$ is the same) that are either red or blue, thus $\Omega_i = \{R, B\}$. You may also be given additional information, symmetric across urns, but it does not pin down either the precise composition of each urn or the relationship between urns. However you are told that for any $i$, the proportion of red balls $p_i$ in the i-th urn is in the same interval, without loss of generality, we suppose $p_i \in [1/4, 1/2]$. One ball will be drawn from each urn, which defines "experiment". Let define the random variable on $\Omega_i$ by

$$Y_i(\omega_i) = \begin{cases} 1 & \omega_i = R \text{ i.e. if ball drawn from the i-th urn is red;} \\ 0 & \omega_i = B \text{ i.e. if ball drawn from the i-th urn is blue.} \end{cases}$$

The full state space is

$$\Omega = \prod_{i=1}^{\infty} \Omega_i = \Omega_1 \times \Omega_2 \times \cdots.$$

Denote $\mathcal{F}$ the product $\sigma$-algebra on $\Omega$. The set of probability $\mathcal{P}$ on measure space $(\Omega, \mathcal{F})$ is generated by

$$\mathcal{P} = \prod_{i=1}^{\infty} \mathcal{P}_i = \mathcal{P}_1 \times \mathcal{P}_2 \times \cdots,$$

where $\mathcal{P}_i$ is generated by $p_i \in [1/4, 1/2]$. A random variables sequence $(X_i)_{i=1}^{\infty}$ on $(\Omega, \mathcal{F})$ is defined by

$$X_i(\omega) = Y_i(\omega_i), \quad \forall i \in \mathbb{N}.$$

It is easy to check that $(X_i)_{i=1}^{\infty}$ is independence under definition 3, and since $X_i$ is bounded, from remark 2 we know that $(X_i)_{i=1}^{\infty}$ satisfying condition (3). It is also obvious that the lower probability of $\mathcal{P}$ is lower continuous, and for any subsequence of $\mathbb{N}$ denoted by $(n_i)_{i=1}^{\infty}$, $(S_{n_i} - S_{n_{i-1}})_{i=1}^{\infty}$ are pairwise independence under $v(\cdot)$. Then we have

$$\mathbb{V}\left(\{\liminf_{n \to \infty} S_n/n < 1/4\} \bigcup \{\limsup_{n \to \infty} S_n/n > 1/2\}\right) = 0,$$

and $v\left(1/4 \leq \liminf_{n \to \infty} S_n/n \leq \limsup_{n \to \infty} S_n/n \leq 1/2\right) = 1,$

also $\mathbb{V}\left(\limsup_{n \to \infty} S_n/n = 1/2\right) = 1, \quad \mathbb{V}\left(\liminf_{n \to \infty} S_n/n = 1/4\right) = 1.$

## Appendix: Proof of Theorem 1

Before we prove Theorem 1, let us first introduce the following Lemmas which will be important in the proof of Theorem 1. First is the Borel-Cantelli Lemma in upper probability space. The second lemma is motivated by Lemma 4 in [1], and both proofs are actually the same for it just need the independence of exponential functions. So we omit the proof.

**Lemma 1.** *[1] **Borel-Cantelli Lemma** Let $\{A_n, n \geq 1\}$ be a sequence of events in $\mathcal{F}$ and $(\mathbb{V}, v)$ be a pair of upper and lower probability generated by $\mathcal{P}$.*

*(1) If $\sum\limits_{n=1}^{\infty} \mathbb{V}(A_n) < \infty$, then $\mathbb{V}\left( \bigcap\limits_{n=1}^{\infty} \bigcup\limits_{i=n}^{\infty} A_i \right) = 0$.*

*(2) If further $v(\cdot)$ is lower continuous and $\{A_n, n \geq 1\}$ are pairwise independent with respect to $v(\cdot)$, that is for any $n \in \mathbb{N}$, $v\left( \bigcap\limits_{i=n}^{\infty} A_i^c \right) =$ $\prod_{i=n}^{\infty} v(A_i^c)$. Also if $\sum\limits_{n=1}^{\infty} \mathbb{V}(A_n) = \infty$, then*

$$\mathbb{V}\left( \bigcap\limits_{n=1}^{\infty} \bigcup\limits_{i=n}^{\infty} A_i \right) = 1.$$

**Lemma 2.** *Given upper expectation $\mathbb{E}[\cdot]$, let $(X_i)_{i=1}^{\infty}$ be a sequence of independent random variables such that $\sup\limits_{i \geq 1} \mathbb{E}[|X_i|^{1+\alpha}] < \infty$ for some constant $\alpha > 0$. Suppose that there exists a constant $c > 0$ such that*

$$|X_n - \mathbb{E}[X_n]| \leq c\frac{n}{\ln(1+n)}, \quad n = 1, 2, \cdots.$$

*Then for any $m > 1$ we have*

$$\sup\limits_{n \geq 1} \mathbb{E}\left[ \exp\left( \frac{m\ln(1+n)}{n} \sum\limits_{i=1}^{n} [X_i - \mathbb{E}[X_i]] \right) \right] < \infty.$$

Now we begin the proof of Theorem 1.

*Proof.* (I). Since $\mathbb{V}$ and $v$ is conjugate, (2) is equivalent to (1). And using the monotonicity and sub-additivity of $\mathbb{V}$, it is easy to check that argument (1) is equivalent to the conjunction of

$$\mathbb{V}\left( \limsup\limits_{n \to \infty} S_n/n > \overline{\mu} \right) = 0, \tag{4}$$

$$\mathbb{V}\left( \liminf\limits_{n \to \infty} S_n/n < \underline{\mu} \right) = 0. \tag{5}$$

Now we prove (4) by two steps.

**Step 1.** Assume that there exists a constant $c > 0$ such that $|X_n - \overline{\mu}| \leq \frac{cn}{\ln(1+n)}$ for $n \geq 1$. Thus, $(X_i)_{i=1}^{\infty}$ satisfies the assumptions of Lemma 2.

To prove (4), we shall show that for any $\epsilon > 0$,

$$\mathbb{V}\left(\bigcap_{n=1}^{\infty}\bigcup_{k=n}^{\infty}\{S_k/k \geq \overline{\mu} + \epsilon\}\right) = 0. \tag{6}$$

In fact, by Lemma 2, for $\epsilon > 0$, let us choose $m > 1/\epsilon$ we have

$$\sup_{n \geq 1}\mathbb{E}\left[\exp\left(\frac{m\ln(1+n)}{n}\sum_{k=1}^{n}(X_k - \mathbb{E}[X_k])\right)\right] < \infty.$$

By Chebyshev's inequality,

$$\mathbb{V}(S_n/n \geq \overline{\mu} + \epsilon) = \mathbb{V}(\tfrac{S_n - n\overline{\mu}}{n} \geq \epsilon)$$
$$= \mathbb{V}\left(\frac{m\ln(1+n)}{n}\sum_{k=1}^{n}(X_k - \overline{\mu}) \geq \epsilon m\ln(1+n)\right)$$
$$\leq e^{-\epsilon m\ln(1+n)}\mathbb{E}\left[\exp\left(\frac{m\ln(1+n)}{n}\sum_{k=1}^{n}(X_k - \overline{\mu})\right)\right]$$
$$\leq \frac{1}{(1+n)^{\epsilon m}}\sup_{n \geq 1}\mathbb{E}\left[\exp\left(\frac{m\ln(1+n)}{n}\sum_{k=1}^{n}(X_k - \overline{\mu})\right)\right].$$

Since $\epsilon m > 1$, $\sup_{n \geq 1}\mathbb{E}\left[\exp\left(\frac{m\ln(1+n)}{n}\sum_{k=1}^{n}(X_k - \overline{\mu})\right)\right] < \infty$, following from the convergence of $\sum_{n=1}^{\infty}\frac{1}{(1+n)^{\epsilon m}}$, we have

$$\sum_{n=1}^{\infty}\mathbb{V}(S_n/n \geq \overline{\mu} + \epsilon) < \infty.$$

Using the first Borel-Cantelli Lemma, we have

$$\mathbb{V}\left(\limsup_{n \to \infty} S_n/n \geq \overline{\mu} + \epsilon\right) = 0 \quad \forall \epsilon > 0.$$

By the lower continuous of $\mathbb{V}(\cdot)$, we have

$$\mathbb{V}\left(\limsup_{n \to \infty} S_n/n > \overline{\mu}\right) = 0, \quad \text{and} \quad v\left(\limsup_{n \to \infty} S_n/n \leq \overline{\mu}\right) = 1.$$

**Step 2.** For any fixed $c > 0$, set

$$\overline{X}_n := (X_n - \overline{\mu})I_{\{|X_n - \overline{\mu}| \leq \frac{cn}{\ln(1+n)}\}} - \mathbb{E}\left[(X_n - \overline{\mu})I_{\{|X_n - \overline{\mu}| \leq \frac{cn}{\ln(1+n)}\}}\right] + \overline{\mu}.$$

Immediately, $\mathbb{E}[\overline{X}_n] \equiv \overline{\mu}$, and for each $n \geq 1$,

$$|\overline{X}_n - \overline{\mu}| \leq \frac{2c\, n}{\ln(1+n)}.$$

Meanwhile, for each $n \geq 1$, it easy to check that

$$|\overline{X}_n - \overline{\mu}| \leq |X_n - \overline{\mu}| + \mathbb{E}[|X_n - \overline{\mu}|].$$

Then

$$\mathbb{E}[|\overline{X}_n - \overline{\mu}|^{1+\alpha}] \leq 2^{1+\alpha}\left(\mathbb{E}[|X_n - \overline{\mu}|^{1+\alpha}] + (\mathbb{E}[|X_n - \overline{\mu}|])^{1+\alpha}\right) < \infty.$$

So $(\overline{X}_i)_{i=1}^{\infty}$ satisfies the assumptions in Lemma 2.

Setting $\overline{S}_n := \sum_{i=1}^{n} \overline{X}_i$, immediately,

$$\frac{1}{n}S_n \leq \frac{1}{n}\overline{S}_n +$$

$$\frac{1}{n}\sum_{i=1}^{n}|X_i - \overline{\mu}|I_{\{|X_i-\overline{\mu}|>\frac{c\,i}{\ln(1+i)}\}} + \frac{1}{n}\sum_{i=1}^{n}\mathbb{E}\left[|X_i - \overline{\mu}|I_{\{|X_i-\overline{\mu}|>\frac{c\,i}{\ln(1+i)}\}}\right] \quad (7)$$

Applying the Hölder and Cheyshev's inequality, we have

$$\sum_{i=1}^{\infty}\frac{\mathbb{E}\left[|X_i-\overline{\mu}|I_{\{|X_i-\overline{\mu}|>\frac{c\,i}{\ln(1+i)}\}}\right]}{i}$$

$$\leq \sum_{i=1}^{\infty}\frac{1}{i}(\mathbb{E}[|X_i - \overline{\mu}|^{1+\alpha}])^{1/(1+\alpha)}(\mathbb{E}[I_{\{|X_i-\overline{\mu}|>\frac{c\,i}{\ln(1+i)}\}}])^{\alpha/(1+\alpha)}$$

$$\leq \sum_{i=1}^{\infty}\frac{[\ln(1+i)]^{\alpha}}{c^{\alpha}\,i^{1+\alpha}}\mathbb{E}[|X_i - \overline{\mu}|^{1+\alpha}]$$

$$\leq \sup_{i\geq 1}\mathbb{E}[|X_i - \overline{\mu}|^{1+\alpha}]\left(\frac{1}{c}\right)^{\alpha}\sum_{i=1}^{\infty}\frac{[\ln(1+i)]^{\alpha}}{i^{1+\alpha}} < \infty.$$

By Kronecker Lemma,

$$\frac{1}{n}\sum_{i=1}^{n}\mathbb{E}\left[|X_i - \overline{\mu}|I_{\{|X_i-\overline{\mu}|>\frac{c\,i}{\ln(1+i)}\}}\right] \to 0. \quad (8)$$

Next we want to prove

$$\frac{1}{n}\sum_{i=1}^{n}\left(|X_i - \overline{\mu}|I_{\{|X_i-\overline{\mu}|>\frac{c\,i}{\ln(1+i)}\}}\right) \to 0, \quad q.s.. \quad (9)$$

Similarly by the Kronecker Lemma, we just need to show that

$$\sum_{i=1}^{\infty}\frac{|X_i - \overline{\mu}|I_{\{|X_i-\overline{\mu}|>\frac{c\,i}{\ln(1+i)}\}}}{i} < \infty, \quad q.s..$$

Set $A_i := \{|X_i - \overline{\mu}| > \frac{c\,i}{\ln(1+i)}\}$ for $i \geq 1$. It suffices to prove that

$$v\left(\bigcup_{n=1}^{\infty} \bigcap_{i=n}^{\infty} A_i^c\right) = 1 \quad \text{that is} \quad \mathbb{V}\left(\bigcap_{n=1}^{\infty} \bigcup_{i=n}^{\infty} A_i\right) = 0.$$

In fact, by Chebyshev's inequality,

$$\mathbb{V}\left(|X_i - \overline{\mu}| > \frac{c\,i}{\ln(1+i)}\right) \leq \left(\frac{\ln(1+i)}{c\,i}\right)^{1+\alpha} \mathbb{E}[|X_i - \overline{\mu}|^{1+\alpha}]$$

Hence,

$$\sum_{i=1}^{\infty} \mathbb{V}\left(|X_i - \overline{\mu}| > \frac{c\,i}{\ln(1+i)}\right) < \infty$$

and by the first Borel-Cantelli Lemma, we have $\mathbb{V}\left(\bigcap_{n=1}^{\infty} \bigcup_{i=n}^{\infty} A_i\right) = 0$. So we get (9).

Taking $\limsup_{n\to\infty}$ on both side of (7), then by (8) and (9), we have

$$\limsup_{n\to\infty} S_n/n \leq \limsup_{n\to\infty} \overline{S}_n/n, \quad q.s..$$

Since $(\overline{X}_n)_{n=1}^{\infty}$ satisfies the assumption of Step 1, by Step 1 and above, we get

$$\mathbb{V}(\limsup_{n\to\infty} S_n/n > \overline{\mu}) = 0, \quad \text{and also} \quad v(\limsup_{n\to\infty} S_n/n \leq \overline{\mu}) = 1.$$

Similarly, considering the sequence $(-X_i)_{i=1}^{\infty}$, from $\mathbb{E}[-X_i] \equiv -\underline{\mu}$, we obtain

$$\mathbb{V}\left(\limsup_{n\to\infty}(-S_n)/n > -\underline{\mu}\right) = 0.$$

Hence,

$$\mathbb{V}\left(\liminf_{n\to\infty} S_n/n < \underline{\mu}\right) = 0, \quad \text{also} \quad v\left(\liminf_{n\to\infty} S_n/n \geq \underline{\mu}\right) = 1.$$

Therefore, the proof of (I) is complete.

(II). If $\overline{\mu} = \underline{\mu}$, it is trivial. Suppose $\overline{\mu} > \underline{\mu}$, we only need to prove that there exists an increasing subsequence $(n_k)$ of $\mathbb{N}$ such that for any $0 < \epsilon < \overline{\mu} - \underline{\mu}$,

$$\mathbb{V}\left(\bigcap_{m=1}^{\infty} \bigcup_{k=m}^{\infty} \{S_{n_k}/n_k \geq \overline{\mu} - \epsilon\}\right) = 1. \tag{10}$$

Because $v(\cdot)$ is lower continuous, we have $\mathbb{V}(\cdot)$ is upper continuous, then

$$\mathbb{V}\left(\limsup_{k \to \infty} S_{n_k}/n_k \geq \overline{\mu}\right) = 1.$$

This with (I) suffices to yield the desired result (II). Because the previous finite $X_i$ have no impact on the limit of $(X_i)_{i=1}^{\infty}$, so we may assume that there exist $c_1 > 0$ such that

$$\frac{1}{n}|\sum_{i=1}^{n}(X_i - \overline{\mu})| \leq c_1 \ln(n+1), \quad \forall n \in \mathbb{N}.$$

Choosing $n_k = k^k$ for $k \geq 1$ and setting $\overline{Z}_n := \sum_{i=1}^{n}(X_i - \overline{\mu})$, we have

$$\frac{1}{n_k - n_{k-1}}|Z_{n_k} - Z_{n_{k-1}}|$$

$$\leq \frac{n_k}{n_k - n_{k-1}} \cdot \frac{1}{n_k}|Z_k| + \frac{n_{k-1}}{n_k - n_{k-1}} \cdot \frac{1}{n_{k-1}}|Z_{k-1}|$$

$$\leq \frac{n_k}{n_k - n_{k-1}} \cdot c_1 \ln(n_k + 1) + \frac{n_{k-1}}{n_k - n_{k-1}} \cdot c_1 \ln(n_{k-1} + 1)$$

$$\leq \frac{2c_1 n_k \ln(n_k + 1)}{n_k - n_{k-1}}.$$

Notice the fact that for any $\lambda > 0$, $I_{Y \geq -\epsilon} \geq \frac{e^{\lambda Y} - e^{-\lambda \epsilon}}{e^{\lambda M}}$, where $Y$ is a random variable bounded by $M$, we have

$$\mathbb{V}\left(\frac{S_{n_k} - S_{n_{k-1}}}{n_k - n_{k-1}} \geq \overline{\mu} - \epsilon\right)$$

$$= \mathbb{V}\left(\frac{S_{n_k} - S_{n_{k-1}} - (n_k - n_{k-1})\overline{\mu}}{n_k - n_{k-1}} \geq -\epsilon\right)$$

$$= \mathbb{V}\left(\frac{Z_{n_k} - Z_{n_{k-1}}}{n_k - n_{k-1}} \geq -\epsilon\right)$$

$$= \mathbb{E}[I_{\frac{Z_{n_k} - Z_{n_{k-1}}}{n_k - n_{k-1}} \geq -\epsilon}]$$

$$\geq \mathbb{E}\left[\left(\exp(\lambda \frac{Z_{n_k} - Z_{n_{k-1}}}{n_k - n_{k-1}}) - \exp(-\lambda \epsilon)\right)\exp(-\lambda \frac{2c_1 n_k \ln(n_k+1)}{n_k - n_{k-1}})\right]$$

$$\geq (n_k + 1)^{\frac{-2\lambda c_1 n_k}{n_k - n_{k-1}}}\left(\prod_{i=n_{k-1}+1}^{n_k} \mathbb{E}[\exp \lambda(X_i - \overline{\mu})] - \exp(-\lambda \epsilon)\right)$$

$$\geq (n_k + 1)^{\frac{-2\lambda c_1 n_k}{n_k - n_{k-1}}}(1 - e^{-\lambda \epsilon}),$$

where the last inequality is coming from the Jensen's inequality. We choose $\lambda$ such that $2\lambda c_1 < 1$, because of $\frac{n_k}{n_k - n_{k-1}} \to 1$, we get

$$\sum_{k=1}^{\infty} \mathbb{V}\left(\frac{S_{n_k} - S_{n_{k-1}}}{n_k - n_{k-1}} \geq \overline{\mu} - \epsilon\right) \geq \sum_{k=1}^{\infty}(n_k + 1)^{\frac{-2\lambda c_1 n_k}{n_k - n_{k-1}}}(1 - e^{-\lambda \epsilon}) = \infty.$$

By assuming $(S_{n_k} - S_{n_{k-1}})_{k \geq 1}$ are pairwise independence under $v(\cdot)$, using the second Borel-Cantelli Lemma, we have

$$\mathbb{V}\left(\bigcap_{i=1}^{\infty} \bigcup_{k=i}^{\infty} \left\{ \frac{S_{n_k} - S_{n_{k-1}}}{n_k - n_{k-1}} \geq \overline{\mu} - \epsilon \right\}\right) = 1.$$

Thus

$$\mathbb{V}\left(\limsup_{k \to \infty} \frac{S_{n_k} - S_{n_{k-1}}}{n_k - n_{k-1}} \geq \overline{\mu} - \epsilon\right) = 1.$$

But

$$\frac{S_{n_k}}{n_k} \geq \frac{S_{n_k} - S_{n_{k-1}}}{n_k - n_{k-1}} \cdot \frac{n_k - n_{k-1}}{n_k} - \frac{|S_{n_{k-1}}|}{n_{k-1}} \cdot \frac{n_{k-1}}{n_k},$$

and

$$\frac{n_{k-1}}{n_k} \to 0, \text{ as } k \to \infty,$$

from

$$\limsup_{n \to \infty} S_n/n \leq \overline{\mu}, \quad \limsup_{n \to \infty}(-S_n)/n \leq -\underline{\mu} \quad q.s.,$$

we have

$$\limsup_{n \to \infty} |S_n|/n \leq \max\{|\overline{\mu}|, |\underline{\mu}|\}, \quad q.s..$$

Hence,

$$\limsup_{k \to \infty} \frac{S_{n_k}}{n_k} \geq \limsup_{k \to \infty} \frac{S_{n_k} - S_{n_{k-1}}}{n_k - n_{k-1}} \lim_{k \to \infty} \frac{n_k - n_{k-1}}{n_k} - \limsup_{k \to \infty} \frac{|S_{n_{k-1}}|}{n_{k-1}} \lim_{k \to \infty} \frac{n_{k-1}}{n_k}.$$

We conclude that

$$\mathbb{V}\left(\limsup_{k \to \infty} \frac{S_{n_k}}{n_k} \geq \overline{\mu} - \epsilon\right) = 1.$$

Since $\epsilon$ is arbitrary and $\mathbb{V}(\cdot)$ is upper continuous, we have

$$\mathbb{V}\left(\limsup_{k \to \infty} S_{n_k}/n_k \geq \overline{\mu}\right) = 1.$$

By (I), we know $\mathbb{V}\left(\limsup_{n \to \infty} S_n/n > \overline{\mu}\right) = 0$, thus

$$\mathbb{V}\left(\limsup_{n \to \infty} S_n/n = \overline{\mu}\right) = \mathbb{V}\left(\limsup_{n \to \infty} S_n/n = \overline{\mu}\right) + \mathbb{V}\left(\limsup_{n \to \infty} S_n/n > \overline{\mu}\right)$$

$$\geq \mathbb{V}\left(\limsup_{n \to \infty} S_n/n \geq \overline{\mu}\right) = 1.$$

Considering the sequence of $(-X_n)_{n=1}^{\infty}$, from $\mathbb{E}[-X_i] \equiv -\underline{\mu}$, we have

$$\mathbb{V}\left(\limsup_{n\to\infty}(-S_n)/n = -\underline{\mu}\right) = 1.$$

Therefore,

$$\mathbb{V}\left(\liminf_{n\to\infty} S_n/n = \underline{\mu}\right) = 1.$$

The proof of (II) is complete.                                                    □

**Acknowledgements.** This work is supported partly by the National Basic Research (973)Program of China (No. 2007CB814901) and WCU(World Class University) program of the Korea Science and Engineering Foundation (R31-20007).

# References

1. Chen, Z.: Strong laws of large numbers for capacities. arXiv: 1006.0749v1 (math.PR) (June 3, 2010)
2. Couso, I., Moral, S., Walley, P.: Examples of Independence for Imprecise Probabilities. In: 1st International Symposium on Imprecise Probabilities and Their Applications, Ghent, Belgium (1999)
3. Denis, L., Hu, M., Peng, S.: Function spaces and capacity related to a sublinear expectation: application to G-Brownian Motion. Paths. Potential Analy. 34, 139–161 (2011)
4. Feller, W.: An Introduction to Probability Theory and its Applications, vol. 1. John Wiley and Sons, Chichester (1968)
5. Gao, F., Jiang, H.: Large deviations for stochastic differential equations driven by G-Brownian motion. Stoch. Proc. Appl. 120, 2212–2240 (2010)
6. Peng, S.: Survey on normal distributions, central limit theorem, Brownian motion and the related stochastic calculus under sublinear expectations. Sci. China Ser. A 52, 1391–1411 (2009)

# Comparative Risk Aversion for g-Expected Utility Maximizers

Guangyan Jia and Jianming Xia

**Abstract.** An index is introduced to measure the risk aversion of a g-expected utility maximizer.

In the following we just introduce the main idea and the main results of our on-going manuscript [3], in which the detailed proofs are provided.

The comparative risk aversion for expected utility maximizers was carried out by Arrow [1] and Pratt [6]. They derived the famous Arrow-Pratt index $-u''/u'$ of risk aversion for a utility function $u$.

We now consider a kind of time-consistent nonlinear expectation, called g-expectation, instead of the classical linear expectation. The notion of g-expectation was introduced and developed by Peng [4, 5]. It is defined as a solution of a backward stochastic differential equation (BSDE) with a generator $g$. More precisely, consider a BSDE:

$$dy_t = g(t, z_t)dt - z_t dB_t, \quad y_T = \xi.$$

Here $B_t$ is a standard Brownian motion defined on a completed and filtered probability space $(\Omega, \mathcal{F}, (\mathcal{F}_t), \mathbb{P})$, $\mathcal{F}_t$ is the augmented natural filtration generated by $B_t$, $\mathcal{F} = \mathcal{F}_T$, $g : [0, T] \times \mathbb{R} \to \mathbb{R}$ satisfies the usual standard conditions such as Lipschitz continuity and linear growth w.r.t. $z$ and $g(t, 0) = 0$, and

Guangyan Jia
School of Mathematics, Shandong University, Ji'nan 250100, Shandong, China
e-mail: jiagy@sdu.edu.cn

Jianming Xia
Laboratory of Random Complex Structures and Data Science,
Academy of Mathematics and Systems Science, Chinese Academy of Sciences,
Beijing 100190, China
e-mail: xia@amss.ac.cn

S. Li (Eds.): Nonlinear Maths for Uncertainty and its Appli., AISC 100, pp. 31–34.
springerlink.com                    © Springer-Verlag Berlin Heidelberg 2011

$\xi \in L^2(\Omega, \mathcal{F}_T, \mathbb{P})$. The $g$-expectation of $\xi$ is given by $\mathbb{E}^g[\xi] = y_0$ and, more generally, the conditional $g$-expectation of $\xi$ at time $t$ is given by $\mathbb{E}^g_t[\xi] = y_t$.

A risk is a random variable $X \in \mathcal{F}$. Assume a decision maker's preference can be represented by $g$-expected utility, that is, risk $X$ is preferred to risk $Y$ if and only if $\mathbb{E}^g[u(X)] \geq \mathbb{E}^g[u(Y)]$, where utility function $u : \mathbb{R} \to \mathbb{R}$ is twice continuously differentiable and satisfies $u'(x) > 0$ for all $x \in \mathbb{R}$. Hereafter, such a decision maker is simply denoted by $(g, u)$.

**Definition 1 ($g$-risk averse).** A decision maker $(g, u)$ is called $g$-risk averse if $\mathbb{E}^g[u(X)] \leq u(\mathbb{E}^g[X])$ for all $X \in L^\infty$.

For any utility function $u$, set

$$P^g_u(t, x, z) = -\frac{1}{2}\frac{u''(x)}{u'(x)}|z|^2 - \frac{g(t, u'(x)z)}{u'(x)} + g(t, z).$$

It has already been essentially reported in [2] that $(g, u)$ is $g$-risk averse if and only if $P^g_u(t, x, z) \geq 0$ for all $(t, x, z)$. In this case, function $u$ is called $g$-concave.

**Definition 2.** Given two utility functions $u$ and $v$, we say $u$ is more $g$-risk averse than $v$ if for any constant $x \in \mathbb{R}$ and risk $X \in L^\infty$, $\mathbb{E}^g[v(X)] \leq v(x)$ implies $\mathbb{E}^g[u(X)] \leq u(x)$.

One of our main results is

**Theorem 1.** *Given two utility functions $u$ and $v$, then the following conditions are equivalent:*

*(i)   $u$ is more $g$-risk averse than $v$.*
*(ii)   There exists an increasing and $g$-concave function $\psi$ such that $u(x) = \psi(v(x))$ for all $x$; that is, $u$ is a $g$-concave transformation of $v$.*
*(iii)   $A^g_u(t, x, z) \geq A^g_v(t, x, z)$ for all $t, x, z$, where*

$$A^g_w(t, x, z) = -\frac{1}{2}\frac{w''(x)}{w'(x)}|z|^2 - \frac{g(t, w'(x)z)}{w'(x)}, \quad w \in \{u, v\}.$$

From the preceding theorem, the index $A^g_u$ measures the risk aversion of a decision maker $(g, u)$.

In order to compare risk aversion across individuals with heterogenous expectations, we need extend the concept $g$-concavity to adopt the heterogeneity of expectations.

**Definition 3.** A function $u$ is called $(g, h)$-concave if for all $t, x, z$,

$$-\frac{1}{2}u''(x)|z|^2 - g(t, u'(x)z) + u'(x)h(t, z) \geq 0.$$

The next lemma can be easily proved by mimicking the arguments of [2].

**Lemma 1.** *The following properties are equivalent:*

*(i)    $u$ is $(g,h)$-concave.*
*(ii)    For any risk $X \in L^\infty$, $\mathbb{E}^g[u(X)] \leq u(\mathbb{E}^h[X])$.*

**Definition 4.** We say a decision maker $(g,u)$ is more risk averse than $(h,v)$ if for any constant $x \in \mathbb{R}$ and risk $X \in L^\infty$, $\mathbb{E}^h[v(X)] \leq v(x)$ implies $\mathbb{E}^g[u(X)] \leq u(x)$.

Another main result of ours is

**Theorem 2.** *The following properties are equivalent:*

*(i)    $(g,u)$ is more risk averse than $(h,v)$.*
*(ii)    $u(x) = \psi(v(x))$ for some increasing and $(g,h)$-concave function $\psi$.*
*(iii)    $A_u^g(t,x,z) \geq A_v^h(t,x,z)$ for all $t,x,z$.*

**Definition 5.** Decision makers $(g,u)$ and $(h,v)$ are called equivalent if for any risk $X$ and $Y$, $\mathbb{E}^g[u(X)] \leq \mathbb{E}^g[u(Y)]$ if and only if $\mathbb{E}^h[v(X)] \leq \mathbb{E}^h[v(Y)]$.

Obviously, the equivalence of the decision makers implies that they are as risk averse as each other. The following theorem shows that the conversion is also true.

**Theorem 3.** *The following properties are equivalent:*

*(i)    $(g,u)$ and $(h,v)$ are equivalent.*
*(ii)    $A_u^g = A_v^h$.*
*(iii)    There exist constants $\alpha > 0$ and $\beta \in \mathbb{R}$ such that $u(x) = \alpha v(x) + \beta$ (that is, $u$ is a positively affine transformation of $v$) and $g(t, \alpha z) = \alpha h(t, z)$.*
*(iv)    There exist constants $\alpha > 0$ and $\beta \in \mathbb{R}$ such that $\mathbb{E}_t^h[u(X)] = \alpha \mathbb{E}_t^g[v(X)] + \beta$ for any risk $X$.*
*(v)    There exist constants $\alpha > 0$ and $\beta \in \mathbb{R}$ such that $\mathbb{E}^h[u(X)] = \alpha \mathbb{E}^g[v(X)] + \beta$ for any risk $X$.*

**Acknowledgements.** This work is supported partially by the National Basic Research Program of China (973 Program) under grants 2007CB814901 and 2007CB814902. We would like to thank Prof Shige Peng for helpful and fruitful suggestions. All remaining errors are our own.

# References

1. Arrow, K.: Essays in the Theory of Risk Bearing. North Holland, London (1970)
2. Jia, G., Peng, S.: Jensen's inequality for g-convex function under g-expectation. Probab. Theory Relat. Fields 147, 217–239 (2010)
3. Jia, G., Xia, J.: Comparative risk aversion for g-expected utility maximizers. On-going manuscript (2010)

4. Peng, S.: BSDE and related $g$-expectation. In: El Karoui, N., Mazliak, L. (eds.) Backward Stochastic Differential Equations. Pitman Research Notes in Mathematics Series, vol. 364, pp. 141–159 (1997a)
5. Peng, S.: BSDE and stochastic optimizations. In: Yan, J.A., Peng, S., Fang, S., Wu, L. (eds.) Topics in Stochastic Analysis. Science Press, Beijing (1997b) (in Chinese)
6. Pratt, J.: Risk aversion in the small and in the large. Econometrica 32, 122–136 (1964)

# Riesz Type Integral Representations for Comonotonically Additive Functionals

Jun Kawabe

**Abstract.** The Daniell-Stone type representation theorem of Greco leads us to another proof and an improvement of the Riesz type representation theorem of Sugeno, Narukawa, and Murofushi for comonotonically additive, monotone functionals.

**Keywords:** nonadditive measure, Choquet integral, Comonotonic additivity, Greco theorem, Riesz type integral representation theorem.

## 1 Introduction

Let $X$ be a locally compact Hausdorff space. Let $C_{00}^+(X)$ denote the space of all nonnegative, continuous functions on $X$ with compact support and let $C_0^+(X)$ denote the space of all nonnegative, continuous functions on $X$ vanishing at infinity. In [8], Sugeno et al. succeeded in proving an analogue of the Riesz type integral representation theorem in nonadditive measure theory. More precisely, they gave a direct proof of the assertion that every comonotonically additive, monotone functional on $C_{00}^+(X)$ can be represented as the Choquet integral with respect to a nonadditive measure on $X$ with some regularity properties. Their theorem gives a functional analytic characterization of the Choquet integrals and is inevitable in order to develop nonadditive measure theory based on the topology of the underlying spaces on which measures are defined.

In this paper, we first give another proof and an improvement of the above theorem with the help of the Greco theorem [4], which is the most general Daniell-Stone type integral representation theorem for comonotonically additive, monotone functionals on function spaces. By using the same approach,

Jun Kawabe
Shinshu University, 4-17-1 Wakasato, Ngano 3808553, Japan
e-mail: jkawabe@shinshu-u.ac.jp

S. Li (Eds.): Nonlinear Maths for Uncertainty and its Appli., AISC 100, pp. 35–42.
springerlink.com      © Springer-Verlag Berlin Heidelberg 2011

we also give a Riesz type integral representation theorem for a bounded functional on $C_0^+(X)$.

## 2   Notation and Preliminaries

Let $X$ be a non-empty set and let $2^X$ denote the family of all subsets of $X$. For each $A \subset X$, let $\chi_A$ denote the characteristic function of $A$. Let $\mathbb{R}$ and $\mathbb{R}^+$ denote the set of all real numbers and the set of all nonnegative real numbers, respectively. Also let $\overline{\mathbb{R}}$ and $\overline{\mathbb{R}}^+$ denote the set of all extended real numbers and the set of all nonnegative extended real numbers, respectively. Let $\mathbb{N}$ denote the set of all natural numbers. For any functions $f, g : X \to \overline{\mathbb{R}}$, let $f \vee g := \max(f, g)$ and $f \wedge g := \min(f, g)$. For any bounded $f$, let $\|f\|_\infty := \sup_{x \in X} |f(x)|$.

**Definition 1.** A set function $\mu : 2^X \to \overline{\mathbb{R}}^+$ is called a *nonadditive measure* on $X$ if $\mu(\emptyset) = 0$ and $\mu(A) \leq \mu(B)$ whenever $A \subset B$.

Let $\mu$ be a nonadditive measure on $X$ and let $f : X \to \overline{\mathbb{R}}^+$ be a function. Since the function $t \in \mathbb{R}^+ \mapsto \mu(\{f > t\})$ is non-increasing, it is Lebesgue integrable on $\mathbb{R}^+$. Therefore, the following formalization is well-defined; see [2] and [7].

**Definition 2.** Let $\mu$ be a nonadditive measure on $X$. The *Choquet integral* of a nonnegative function $f : X \to \overline{\mathbb{R}}^+$ with respect to $\mu$ is defined by

$$(C) \int_X f d\mu := \int_0^\infty \mu(\{f > t\}) dt,$$

where the right hand side of the above equation is the usual Lebesgue integral.

*Remark 1.* For any nonadditive measure $\mu$ on $X$ and any function $f : X \to \overline{\mathbb{R}}^+$, the two Lebesgue integrals $\int_0^\infty \mu(\{f > t\}) dt$ and $\int_0^\infty \mu(\{f \geq t\}) dt$ are equal, since $\mu(\{f \geq t\}) \geq \mu(\{f > t\}) \geq \mu(f \geq t + \varepsilon\})$ for every $\varepsilon > 0$ and $0 \leq t < \infty$. This fact will be used implicitly in this paper.

See [3], [6], and [9] for more information on nonadditive measures and Choquet integrals.

For the reader's convenience, we introduce the Greco theorem [4, Proposition 2.2], which is the most general Choquet integral representation theorem for comonotonically additive, monotone, extended real-valued functionals. Recall that two functions $f, g : X \to \overline{\mathbb{R}}$ are *comonotonic* and is written by $f \sim g$ if, for every $x, x' \in X$, $f(x) < f(x')$ implies $g(x) \leq g(x')$.

**Theorem 1 (The Greco theorem).** *Let $\mathcal{F}$ be a non-empty family of functions $f : X \to \overline{\mathbb{R}}$. Assume that $\mathcal{F}$ satisfies*

(i) $0 \in \mathcal{F}$,
(ii) $f \geq 0$ for every $f \in \mathcal{F}$ (nonnegativity), and
(iii) if $f \in \mathcal{F}$ and $c \in \mathbb{R}^+$, then $cf$, $f \wedge c$, $f - f \wedge c = (f - c)^+ \in \mathcal{F}$ (the Stone condition).

Assume that a functional $I : \mathcal{F} \to \overline{\mathbb{R}}$ satisfies

(iv) $I(0) = 0$,
(v) if $f, g \in \mathcal{F}$ and $f \leq g$, then $I(f) \leq I(g)$ (monotonicity),
(vi) if $f, g \in \mathcal{F}, f + g \in \mathcal{F}$, and $f \sim g$, then $I(f + g) = I(f) + I(g)$ (comonotonic additivity),
(vii) $\lim_{a \to +0} I(f - f \wedge a) = I(f)$ for every $f \in \mathcal{F}$, and
(viii) $\lim_{b \to \infty} I(f \wedge b) = I(f)$ for every $f \in \mathcal{F}$.

For each $A \subset X$, define the set functions $\alpha, \beta : 2^X \to \overline{\mathbb{R}}^+$ by

$$\alpha(A) := \sup\{I(f) : f \in \mathcal{F}, \ f \leq \chi_A\},$$
$$\beta(A) := \inf\{I(f) : f \in \mathcal{F}, \ \chi_A \leq f\},$$

where let $\inf \emptyset := \infty$.

(1) The set functions $\alpha$ and $\beta$ are nonadditive measures on $X$ with $\alpha \leq \beta$.
(2) For any nonadditive measure $\lambda$ on $X$, the following two conditions are equivalent:

  (a) $\alpha < \lambda < \beta$.
  (b) $I(f) = (C)\int_X f d\lambda$ for every $f \in \mathcal{F}$.

Remark 2. The functional $I$ given in Theorem 1 is nonnegative, that is, $I(f) \geq 0$ for every $f \in \mathcal{F}$, and positively homogeneous, that is, $I(cf) = cI(f)$ for every $f \in \mathcal{F}$ and $c \in \overline{\mathbb{R}}^+$. See, for instance, [3, page 159] and [5, Proposition 4.2].

## 3 Riesz Type Integral Representation Theorems

In this section, we first give another proof and an improvement of the Sugeno-Narukawa-Murofushi theorem [8, Theorem 3.7]. This can be done by the effective use of the Greco theorem and the following technical lemma.

**Lemma 1.** Let $\mathcal{F}$ and $I$ satisfy the same hypotheses as Theorem 1.

(1) Assume that, for any $f \in \mathcal{F}$, there is a $g \in \mathcal{F}$ such that $\chi_{\{f>0\}} \leq g$ and $I(g) < \infty$ (in particular, $1 \in \mathcal{F}$ and $I(1) < \infty$). Then, condition (vii) of Theorem 1 holds.
(2) Assume that every $f \in \mathcal{F}$ is bounded. Then, condition (viii) of Theorem 1 holds.

(3) *Assume that every $f \in \mathcal{F}$ is bounded. Also assume that $I$ is bounded, that is, there is a constant $M > 0$ such that $I(f) \leq M\|f\|_\infty$ for every $f \in \mathcal{F}$. Then, conditions* (vii) *and* (viii) *of Theorem 1 hold.*

*Proof.* (1) Let $f \in \mathcal{F}$, and chose $g \in \mathcal{F}$ such that $\chi_{\{f>0\}} \leq g$ and $I(g) < \infty$. Since $f \wedge a \leq ag$ for every $a > 0$, the monotonicity and the positive homogeneity of $I$ imply $I(f \wedge a) \leq I(ag) = aI(g)$. Since $f - f \wedge a \sim f \wedge a$, by the comonotonic additivity of $I$, we have $I(f) \geq I(f - f \wedge a) = I(f) - I(f \wedge a) \geq I(f) - aI(g)$. Thus, $\lim_{a \to +0} I(f - f \wedge a) = I(f)$.

(2) Let $f \in \mathcal{F}$. Then $f \wedge b = f$ for a sufficiently large $b > 0$, so that $\lim_{b \to \infty} I(f \wedge b) = I(f)$.

(3) Fix $f \in \mathcal{F}$, and let $g_n := f - f \wedge (1/n)$ for each $n \in \mathbb{N}$. Since $g_n \sim f \wedge (1/n)$, the comonotonic additivity of $I$ implies $I(f) = I(g_n) + I(f \wedge (1/n))$. Since $\|f \wedge (1/n)\|_\infty \leq 1/n$ for all $n \in \mathbb{N}$, by the boundedness of $I$, $I(g_n) \to I(f)$ as $n \to \infty$.

Take $\varepsilon > 0$ arbitrarily, and chose $\delta > 0$ such that $\delta < 1/n_0$ and $I(f) - I(g_{n_0}) < \varepsilon$. Let $0 < a < \delta$. Then, $f - f \wedge a \geq g_{n_0}$, so that $I(f - f \wedge a) \geq I(g_{n_0}) > I(f) - \varepsilon$. Thus, we have $\lim_{a \to +0} I(f - f \wedge a) = I(f)$. □

From this point forwards, $X$ is a locally compact Hausdorff space. For any real-valued function $f$ on $X$, let $S(f)$ denote the support of $f$, which is defined by the closure of $\{f \neq 0\}$.

The following regularity properties give a tool to approximate general sets by more tractable sets such as open and compact sets. They are still important in nonadditive measure theory.

**Definition 3.** Let $\mu$ be a nonadditive measure on $X$.

(1) $\mu$ is said to be *outer regular* if, for every subset $A$ of $X$, $\mu(A) = \inf\{\mu(G) : A \subset G, G \text{ is open}\}$.
(2) $\mu$ is said to be *quasi outer regular* if, for every compact subset $K$ of $X$, $\mu(K) = \inf\{\mu(G) : K \subset G, G \text{ is open}\}$.
(3) $\mu$ is said to be *inner Radon* if, for every subset $A$ of $X$, $\mu(A) = \sup\{\mu(K) : K \subset A, K \text{ is compact}\}$.
(4) $\mu$ is said to be *quasi inner Radon* if, for every open subset $G$ of $X$, $\mu(G) = \sup\{\mu(K) : K \subset G, K \text{ is compact}\}$.

The following theorem is an improvement of [8, Theorem 3.7] and it has essentially been derived from the Greco theorem.

**Theorem 2.** *Let a functional $I : C_{00}^+(X) \to \mathbb{R}$ satisfy the following conditions:*

(i) *if $f, g \in C_{00}^+(X)$ and $f \leq g$, then $I(f) \leq I(g)$ (monotonicity), and*
(ii) *if $f, g \in C_{00}^+(X)$ and $f \sim g$, then $I(f + g) = I(f) + I(g)$ (comonotonic additivity).*

*For each $A \subset X$, define the set functions $\alpha, \beta, \gamma : 2^X \to \overline{\mathbb{R}}^+$ by*

$$\alpha(A) := \sup\{I(f) : f \in C_{00}^+(X), f \le \chi_A\},$$
$$\beta(A) := \inf\{I(f) : f \in C_{00}^+(X), \chi_A \le f\},$$
$$\gamma(A) := \sup\{I(f) : f \in C_{00}^+(X), 0 \le f \le 1, S(f) \subset A\},$$

*where let* $\inf \emptyset := \infty$, *and their regularizations* $\alpha^*, \beta^*, \gamma^* : 2^X \to \overline{\mathbb{R}}^+$ *by*

$$\alpha^*(A) := \inf\{\alpha(G) : A \subset G, G \text{ is open}\},$$
$$\beta^{**}(A) := \sup\{\beta(K) : K \subset A, K \text{ is compact}\},$$
$$\gamma^*(A) := \inf\{\gamma(G) : A \subset G, G \text{ is open}\}.$$

(1) *The set functions $\alpha, \beta, \gamma, \alpha^*, \beta^{**}$, and $\gamma^*$ are nonadditive measures on $X$.*
(2) *For any nonadditive measure $\lambda$ on $X$, the following two conditions are equivalent:*

   (a) $\alpha \le \lambda \le \beta$.

   (b) $I(f) = (C)\int_X f d\lambda$ *for every $f \in C_{00}^+(X)$.*

(3) $\gamma^*(K) = \beta(K) < \infty$ *for every compact subset $K$ of $X$.*
(4) $\gamma^*$ *is quasi inner Radon and outer regular.*
(5) $\beta^{**}$ *is inner Radon and quasi outer regular.*
(6) $\beta^{**}(G) = \gamma(G)$ *for every open subset $G$ of $X$.*
(7) *The defined nonadditive measures are comparable, that is, $\alpha = \gamma < \beta^{**} < \alpha^* = \gamma^* \le \beta$, so that any of them is a representing measure of $I$.*

*Proof.* (1) This assertion is obvious.

(2) Let $\mathcal{F} := C_{00}^+(X)$ in Theorem 1 and Lemma 1. Conditions (i)–(iii), (v), and (vi) of Theorem 1 are easily verified. Consequently, in order to prove this assertion, we have only to check the rest of the conditions.

(iv): Since $0 \in C_{00}^+(X)$ and $0 \sim 0$ and since $I$ is comonotonic, $I(0) = I(0 + 0) = I(0) + I(0)$, so $I(0) = 0$.

(vii): Let $f \in C_{00}^+(X)$. Since $S(f)$ is compact, by [1, Theorem 54.3], there is $g \in C_{00}^+(X)$ such that $\chi_{S(f)} \le g$. Thus, we have $\chi_{\{f>0\}} \le g$, so that (vii) follows from Lemma 1.

(viii): This condition follows from Lemma 1 since every $f \in C_{00}^+(X)$ is bounded.

(3) Let $K$ be a compact subset of $X$. By [1, Theorem 54.3], there is $f_0 \in C_{00}^+(X)$ with $\chi_K \le f_0$, so we have $\beta(K) \le I(f_0) < \infty$.

Next we prove $\gamma^*(K) \le \beta(K)$. Take $f \in C_{00}^+(X)$ with $\chi_K \le f$ arbitrarily. Let $0 < r < 1$ and let $G_r := \{f > r\}$. Then $K \subset G_r$. For any $g \in C_{00}^+(X)$ such that $0 \le g \le 1$ and $S(g) \subset G_r$, we have $rg \le f$, so that $\gamma^*(K) \le \gamma(G_r) \le I(f/r) = I(f)/r$. Letting $r \uparrow 1$, we have $\gamma^*(K) \le I(f)$, so $\gamma^*(K) \le \beta(K)$ follows.

Finally we prove the reverse inequality. Take $\varepsilon > 0$ arbitrarily, and chose an open set $G$ such that $G \supset K$ and $\gamma(G) \leq \gamma^*(K) + \varepsilon$. By [1, Theorem 54.3], there is $f_0 \in C_{00}^+(X)$ with $0 \leq f_0 \leq 1$, $S(f_0) \subset G$ and $\chi_K \leq f_0 \leq \chi_G$. Therefore, we have $\gamma^*(K) + \varepsilon \geq I(f_0) \geq \beta(K)$. Letting $\varepsilon \downarrow 0$, the reverse inequality $\gamma^*(K) \geq \beta(K)$ follows.

(4) The outer regularity of $\gamma^*$ is obvious. So, we prove that $\gamma^*$ is quasi inner Radon. Let $G$ be an open subset of $X$. Then $\gamma^*(G) \geq \sup\{\gamma^*(K) : K \subset G, K \text{ is compact}\}$ and, hence, we prove the reverse inequality.

Take $r < \gamma^*(G) = \gamma(G)$ arbitrarily, and chose $f_0 \in C_{00}^+(X)$ such that $0 \leq f_0 \leq 1$, $S(f_0) \subset G$ and $r < I(f_0)$. Let $K_0 := S(f_0)$. For any open $H \supset K_0$, we have $\gamma(H) \geq I(f_0) > r$, and this implies $\gamma^*(K_0) \geq r$. Since $K_0 \subset G$ is compact, $r \leq \sup\{\gamma^*(K) : K \subset G, K \text{ is compact}\}$. Letting $r \uparrow \gamma^*(G)$, the reverse inequality follows.

(5) It is obvious that $\beta^{**}$ is inner Radon. So, we prove that $\beta^{**}$ is quasi outer regular. Let $K$ be a compact subset of $X$. Then $\beta^{**}(K) \leq \inf\{\beta^{**}(G) : K \subset G, G \text{ is open}\}$ and, hence, we prove the reverse inequality.

Take $0 < r < 1$ arbitrarily, and chose $f_0 \in C_{00}^+(X)$ such that $\chi_K \leq f_0$ and $I(f_0) \leq \beta(K)+r$. Let $G_r := \{f_0 > 1-r\}$. Then $K \subset G_r$ and $\chi_{G_r} \leq f_0/(1-r)$. For any compact $L \subset G_r$, we have $\chi_L \leq f_0/(1-r)$, and this implies $\beta(L) \leq I(f_0/(1-r)) = I(f_0)/(1-r)$. Thus, $\beta^{**}(G_r) \leq I(f_0)/(1-r)$ and, hence, $\inf\{\beta^{**}(G) : K \subset G, G \text{ is open}\} \leq (\beta(K)+r)/(1-r) = (\beta^{**}(K)+r)/(1-r)$. Letting $r \downarrow 0$, the reverse inequality follows.

(6) This assertion follows from (3) and (4) of this theorem.

(7) It is readily seen that the inequalities $\gamma \leq \alpha \leq \beta$, $\alpha \leq \alpha^*$, $\beta^{**} \leq \beta$, and $\gamma \leq \gamma^* \leq \alpha^*$ hold. Therefore, we have only to prove the inequalities $\beta^{**} \leq \gamma^*$, $\alpha \leq \gamma$, $\alpha \leq \beta^{**}$, and $\alpha^* \leq \beta$.

Let $A \subset X$. For any compact $K$ and open $G$ with $K \subset A \subset G$, by [1, Theorem 54.3], there is $f_0 \in C_{00}^+(X)$ such that $0 \leq f_0 \leq 1$, $S(f_0) \subset G$, and $\chi_K \leq f_0 \leq \chi_G$. Thus, $\beta(K) \leq \gamma(G)$, and this implies $\beta^{**} \leq \gamma^*$.

By (2) and (6) of this theorem, for every $f \in C_{00}^+(X)$, $I(f) = \int_X f d\beta = \int_0^\infty \beta(\{f > t\})dt = \int_0^\infty \beta(\{f \geq t\})dt = \int_0^\infty \beta^{**}(\{f \geq t\})dt = \int_0^\infty \beta^{**}(\{f > t\})dt = \int_0^\infty \gamma(\{f > t\})dt = \int_X f d\gamma$. Thus, again by (2), we have $\alpha \leq \gamma$.

We can prove $\alpha \leq \beta^{**}$ and $\alpha^* \leq \beta$ in a similar fashion.  $\square$

*Remark 3.* Define the functional $I : C_{00}^+(\mathbb{R}) \to \mathbb{R}$ by $I(f) := \int_{-\infty}^\infty f(t)dt$ for every $f \in C_{00}^+(\mathbb{R})$. Then $I$ satisfies (i) and (ii) of Theorem 2, but it is not bounded. So, Theorem 2 does not follow from (3) of Lemma 1.

From Theorem 2 and Lemma 1, we can derive a representation theorem for *bounded*, comonotonically additive, monotone functionals on $C_0^+(X)$.

**Theorem 3.** *Let a functional* $I : C_0^+(X) \to \mathbb{R}$ *satisfy*

(i) *if* $f, g \in C_0^+(X)$ *and* $f \leq g$, *then* $I(f) \leq I(g)$ *(monotonicity),*
(ii) *if* $f, g \in C_0^+(X)$ *and* $f \sim g$, *then* $I(f + g) = I(f) + I(g)$ *(comonotonic additivity), and*

(iii) *there is a constant $M > 0$ such that $I(f) \leq M\|f\|_\infty$ for every $f \in C_0^+(X)$ (boundedness).*

*For each $A \subset X$, define the set functions $\alpha, \beta, \gamma : 2^X \to \overline{\mathbb{R}}^+$ by*

$$\alpha(A) := \sup\{I(f) : f \in C_0^+(X), f \leq \chi_A\},$$
$$\beta(A) := \inf\{I(f) : f \in C_0^+(X), \chi_A \leq f\},$$
$$\gamma(A) := \sup\{I(f) : f \in C_0^+(X), 0 \leq f \leq 1, S(f) \subset A\},$$

*where let $\inf \emptyset := \infty$, and their regularizations $\alpha^*, \beta^{**}, \gamma^* : 2^X \to \overline{\mathbb{R}}^+$ by*

$$\alpha^*(A) := \inf\{\alpha(G) : A \subset G, G \text{ is open}\},$$
$$\beta^{**}(A) := \sup\{\beta(K) : K \subset A, K \text{ is compact}\},$$
$$\gamma^*(A) := \inf\{\gamma(G) : A \subset G, G \text{ is open}\}.$$

(1) *The set functions $\alpha$, $\beta$, $\gamma$, $\alpha^*$, $\beta^{**}$, and $\gamma^*$ are nonadditive measures on $X$ and they satisfy properties (3)–(7) in Theorem 2.*
(2) *For any nonadditive measure $\lambda$ on $X$, the following two conditions are equivalent:*

   (a) $\alpha \leq \lambda \leq \beta$.
   (b) $I(f) = (C)\int_X f \, d\lambda$ *for every* $f \in C_0^+(X)$.

(3) $\alpha(X) = \gamma(X) = \alpha^*(X) = \gamma^*(X) < \infty.$
(4) *Let $\lambda$ be a nonadditive measure on $X$ with $\lambda(X) < \infty$. Define the functional $I : C_0^+(X) \to \mathbb{R}$ by $I(f) := (C)\int_X f \, d\lambda$ for every $f \in C_0^+(X)$. Then $I$ satisfies conditions (i)–(iii).*

*Proof.* (1) We first prove that, $\alpha$, $\beta$, and $\gamma$ are equal to those defined in Theorem 2, respectively, that is, for each $A \subset X$,

$$\alpha(A) = \sup\{I(f) : f \in C_{00}^+(X), f \leq \chi_A\}, \tag{1}$$
$$\beta(A) = \inf\{I(f) : f \in C_{00}^+(X), \chi_A \leq f\}, \tag{2}$$
$$\gamma(A) = \sup\{I(f) : f \in C_{00}^+(X), 0 \leq f \leq 1, S(f) \subset A\}. \tag{3}$$

To prove (1), take $f \in C_0^+(X)$ with $f \leq \chi_A$ arbitrarily, and let $g_n := f - f \wedge (1/n)$ for each $n \in \mathbb{N}$. Then, it is easy to see that $0 \leq g_n \leq f$, $S(g_n) \subset \{f \geq 1/n\} \subset S(f)$. Since $\{f \geq 1/n\}$ is compact, $S(g_n)$ is also compact. Thus $g_n \in C_{00}^+(X)$. Since $g_n \sim f \wedge (1/n)$, by the comonotonic additivity of $I$, we have $I(f) = I(g_n) + I(f \wedge (1/n))$, which implies $I(g_n) \to I(f)$ as $n \to \infty$, since the boundedness of $I$ and since $\|f \wedge (1/n)\|_\infty \leq 1/n$ for all $n \in \mathbb{N}$. Thus, a routine argument leads us to (1).

We can prove (3) in a similar way.

To prove (2), take $f \in C_0^+(X)$ with $\chi_A \leq f$ arbitrarily, and let $h_n := (1 + 1/n)f - \{(1 + 1/n)f\} \wedge (1/n)$ for each $n \in \mathbb{N}$. Then, it is easy to verify that $\chi_A \leq h_n$ and $S(h_n) \subset \{f \geq 1/(n+1)\}$. Since $\{f \geq 1/(n+1)\}$ is compact, $S(h_n)$ is also compact. Thus $h_n \in C_{00}^+(X)$. Since $h_n \sim \{(1+1/n)f\} \wedge (1/n)$, by the comonotonic additivity and the positive homogeneity of $I$, we have $(1+1/n)I(f) = I((1+1/n)f) = I(h_n) + I(\{(1+1/n)f\} \wedge (1/n))$, which implies $I(h_n) \to I(f)$ as $n \to \infty$, since the boundedness of $I$ and since $\|\{(1+1/n)f\} \wedge (1/n)\|_\infty \leq 1/n$ for all $n \in \mathbb{N}$. Thus, a routine argument leads us to (2). Consequently, by Theorem 2, the nonadditive measures $\alpha$, $\beta$, $\gamma$, and their regularizations satisfy (3)–(7) in Theorem 2.

(2) This assertion follows from Theorem 1 and (3) of Lemma 1.

(3) Let $M$ be the positive constant given in (iii) of this theorem. For any $f \in C_0^+(X)$ with $f \leq 1$, we have $I(f) \leq M$, so $\alpha(X) \leq M < \infty$. Since $\alpha = \gamma$, the rest of the assertion follows.

(4) This assertion follows from the fundamental properties of Choquet integrals [3, Proposition 5.1].                                                          □

## 4  Conclusions

In this paper, we gave another proof and an improvement of the Riesz type integral representation theorem of Sugeno, Narukawa, and Murofushi by the help of the Daniell-Stone type integral representation theorem of Greco. By using the same approach, we also gave a Riesz type integral representation theorem for a bounded functional on $C_0^+(X)$. Our approach will lead us to various Riesz type integral representation theorems on a wide variety of function spaces and sequence spaces.

## References

1. Berberian, S.K.: Measure and Integration. Macmillan, New York (1965)
2. Choquet, G.: Theory of capacities. Ann. Inst. Fourier Grenoble 5, 131–295 (1953-54)
3. Denneberg, D.: Non-Additive Measure and Integral, 2nd edn. Kluwer Academic Publishers, Dordrecht (1997)
4. Greco, G.H.: Sulla rappresentazione di funzionali mediante integrali. Rend. Sem. Mat. Univ. Padova 66, 21–42 (1982)
5. Narukawa, Y., Murofushi, T., Sugeno, M.: Regular fuzzy measure and representation of comonotonically additive functional. Fuzzy Sets and Systems 112, 177–186 (2000)
6. Pap, E.: Null-Additive Set Functions. Kluwer Academic Publishers, Dordrecht (1995)
7. Schmeidler, D.: Integral representation without additivity. Proc. Amer. Math. Soc. 97, 255–261 (1986)
8. Sugeno, M., Narukawa, Y., Murofushi, T.: Choquet integral and fuzzy measures on locally compact space. Fuzzy Sets and Systems 99, 205–211 (1998)
9. Wang, Z., Klir, G.J.: Generalized Measure Theory. Springer, Heidelberg (2009)

# Pseudo-concave Integrals

Radko Mesiar, Jun Li, and Endre Pap

**Abstract.** The notion of Lehrer-concave integral is generalized taking instead of the usual arithmetic operations of addition and multiplication of reals more general real operations called pseudo-addition and pseudo-multiplication.

**Keywords:** Pseudo-addition, Pseudo-multiplication, Lehrer integral, Choquet integral.

## 1 Introduction

The integration theory makes a fundament for the classical measure theory, see [19]. The Riemann and the Lebesgue integrals are related to the additive measure (more precisely countably additive measure), which makes the base also for the probability theory. The first integral based on non-additive

Radko Mesiar
Department of Mathematics and Descriptive Geometry,
Faculty of Civil Engineering, Slovak University of Technology, Radlinského 11,
813 68 Bratislava, Slovakia
UTIA CAS, P.O. Box 18, 18208 Prague, Czech Republic
e-mail: mesiar@math.sk

Jun Li
School of Science, Communication University of China, Beijing, 100024,
People's Republic of China
e-mail: lijun@cuc.edu.cn

Endre Pap
Department of Mathematics and Informatics, University of Novi Sad, Trg Dositeja
Obradovića 4, 21000 Novi Sad, Serbia
Óbuda University, Becsi út 96/B, H-1034 Budapest, Hungary
e-mail: pape@eunet.rs

S. Li (Eds.): Nonlinear Maths for Uncertainty and its Appli., AISC 100, pp. 43–49.
springerlink.com                                    © Springer-Verlag Berlin Heidelberg 2011

measures (first of all on monotone measure or capacity) was the Choquet integral [2], see [3, 18], which also covers the classical Lebesgue integral. On the other side, Sugeno [22] has introduced an integral based also on non-additive measures, but with respect to join (maximum) and meet (minimum) operations instead of the usual addition and product. In these approaches it was important to consider the horizontal representation by means of the level sets of a fuzzy subset of a universe $X$, where the level set for level $t$ consists of all points of $X$ with degree of membership greater than or equal to $t$. The Choquet integral is based also on the horizontal approach. Further generalization of integrals related to the horizontal approach is given by means of the universal integral [8]. The horizontal approach based on level sets has as a consequence that chains of subsets of $X$ play the important role. Lehrer [10, 11] has introduced concave integral which is not based on horizontal approach, but considering all possible set systems, see also [24]. His integral coincides with the Choquet integral only when the monotone measure (capacity) is convex (supermodular).

In this paper we investigate the generalization of the Lehrer integral taking more general real operations than classical plus and product. For this purpose we shall use pseudo-addition $\oplus$ and pseudo-multiplication $\odot$, see [1, 18, 23]. These operations form a semiring structure, which was a base for treating many nonlinear and optimizations problems, see [9, 18, 20], and they are giving common framework for treating many different types of integrals, e.g., Choquet, Sugeno and Imaoka ( [4]) integrals.

## 2   Definitions of Choquet, Sugeno and Lehrer integrals

Let $\Omega$ be a non-empty set, $\mathcal{A}$ a $\sigma$-algebra of subsets of $\Omega$ and $v : \mathcal{A} \to [0, \infty]$ a monotone set function with $v(\varnothing) = 0$. Two well-known nonadditive integrals are Choquet and Sugeno integrals, see [18]. The Choquet integral [2] of a measurable nonnegative function $f$ is given by

$$\mathcal{C}_v(f) = \int_0^\infty v(f \geqslant t) \, dt$$

$$= \sup \left\{ \sum_{i \in I} a_i v(A_i) \mid \sum_{i=1}^N a_i \mathbf{1}_{A_i} \leqslant f, (A_i)_{i=1}^N \subset \mathcal{A} \text{ decreasing}, a_i \geqslant 0, N \in \mathbb{N} \right\},$$

where $A_{i+1} \subseteq A_i$ for every $i = 1, 2, \ldots, N - 1$, and $\mathbf{1}_{A_i}$ is the characteristic function of the set $A_i$.

The Choquet integral has the property of the reconstruction of the measure, i.e., for every $A \in \mathcal{A}$ we have

$$\int_0^\infty \mathbf{1}_A \, dv = v(A).$$

The Choquet integral for finite case (and thus also for simple functions) is given on $\Omega = \{1, \ldots, n\}$ and for $(a_1, a_2, \ldots, a_n) \in [0, \infty]^n$ by

$$\mathcal{C}_v(a_1, a_2, \ldots, a_n) = \sum_{i=1}^{n} (a_{(i)} - a_{(i-1)}) v(A_{(i)})$$

with a permutation $\sigma$ on $\{1, \ldots, n\}$ such that $a_{(1)} \leqslant a_{(2)} \leqslant \cdots \leqslant a_{(n)}$, with the convention $x_{(0)} = 0$, $A_{(i)} = \{(i), \ldots, (n)\}$, and $0 \cdot \infty = 0$. The Choquet integral can be given by an equivalent formula

$$\mathcal{C}_v(a_1, a_2, \ldots, a_n) = \sum_{i=1}^{n} a_{(i)} \big( v(A_{(i)}) - v(A_{(i+1)}) \big),$$

with $A_{(n+1)} = \varnothing$. The Sugeno integral [22] is given on $\Omega = \{1, \ldots, n\}$ and for $(a_1, a_2, \ldots, a_n) \in [0, \infty]^n$ by

$$\mathcal{S}_v(a_1, a_2, \ldots, a_n) = \bigvee_{i=1}^{n} \big( a_{(i)} \wedge v(A_{(i)}) \big),$$

and in the case of a general space $(\Omega, \mathcal{A})$ it is given by

$$\mathcal{S}_v(f) = \sup\{ t \wedge v(\{f \geqslant t\}) \mid t \in [0, \infty] \}.$$

Another integral can also be introduced, namely, the Shilkret integral [21] given by

$$\mathcal{K}_v(f) = \sup\{ t \cdot v(\{f \geqslant t\}) \mid t \in [0, \infty] \}.$$

Observe that again we have the reconstruction property

$$\mathcal{S}_v(\mathbf{1}_A) = \mathcal{K}_v(\mathbf{1}_A) = v(A).$$

A set function $v$ on $2^N$ is supermodular, i.e., 2-monotone or convex, if it satisfies the following inequality for all $A, B \in 2^N$ :

$$v(A \cup B) + v(A \cap B) \geqslant v(A) + v(B).$$

The functional $f \mapsto \mathcal{C}_v(f)$ is concave if and only if $v$ is supermodular, see [12].

Lehrer concave integral is given in the following definition, [11].

**Definition 1.** Concave integral of a measurable function $f : \Omega \to [0, \infty[$ is given by

$$(L) \int f \, dv = \sup \left\{ \sum_{i \in I} a_i v(A_i) \mid \sum_{i \in I} a_i \mathbf{1}_{A_i} \leqslant f, I \text{ is finite }, a_i \geqslant 0 \right\},$$

where $A_i, i \in I$, are measurable.

Note that the equality $(L) \int 1_A \, dv = v(A)$ is violated, in general, and it holds only for supermodular $v$ (for all $A \in \mathcal{A}$).

## 3   Pseudo-operations

We have seen that the Lehrer-concave integral as also the Choquet integral are strongly related to the usual operations of addition $+$ and multiplication $\cdot$ on the interval $[0, \infty]$. Similarly, the Sugeno integral depends on the operations $\vee$ and $\wedge$ on the interval $[0, 1]$.

There were considered generalizations of the previously mentioned operations $[1, 7, 23]$. In order to find a common framework for both Choquet and Sugeno integrals, we have to deal with a general pseudo-addition $\oplus$ and a general pseudo-multiplication $\odot$ which must be fitting to each other. In general, $\oplus$ is supposed to be a continuous generalized triangular conorm (see $[1, 6, 18]$).

**Definition 2.** A binary operation $\oplus : [0, \infty]^2 \to [0, \infty]$ is called a pseudo-addition on $[0, \infty]$ if the following properties are satisfied:

**(PA 1)** $a \oplus b = b \oplus a$ (commutativity)

**(PA 2)** $a \leq a', \, b \leq b' \Rightarrow a \oplus b \leq a' \oplus b'$ (monotonicity)

**(PA 3)** $(a \oplus b) \oplus c = a \oplus (b \oplus c)$ (associativity)

**(PA 4)** $a \oplus 0 = 0 \oplus a = a$ (neutral element)

**(PA 5)** $a_n \to a, b_n \to b \Rightarrow a_n \oplus b_n \to a \oplus b$ (continuity).

*Remark 1.* The structure of the operation $\oplus$ is described in details as an $I$-semigroup, for more details see $[6]$.

By **(PA 5)**, the continuity of the operation $\oplus$, the set of $\oplus$-idempotent elements $C_\oplus = \{a \in [0, \infty] \mid a \oplus a = a\}$ is closed and non-empty since $0, \infty \in C_\oplus$. Two extreme cases are possible: $C_\oplus = \{0, \infty\}$ and $C_\oplus = [0, \infty]$. In the first case the pseudo-addition $\oplus$ is isomorphic with the usual addition on $[0, \infty]$ or isomorphic with the truncated addition on $[0, 1]$. In the second case, $\oplus = \vee$. All other cases are covered by ordinal sums of the first case, $[6, 23]$.

For the integration procedure we need another binary operation $\odot$, which is called pseudo-multiplication. The expected properties of the integral determine the next minimal properties of $\odot$ which we have to require, see $[6, 23]$.

**Definition 3.** Let $\oplus$ be a given pseudo-addition on $[0, \infty]$. A binary operation $\odot : [0, \infty] \times [0, \infty] \to [0, \infty]$ is called a $\oplus$-*fitting pseudo-multiplication* if the following properties are satisfied:

**(PM 1)** $a \odot 0 = 0 \odot b = 0$ (zero element)

**(PM 2)** $a \leq a', \, b \leq b' \Rightarrow a \odot b \leq a' \odot b'$ (monotonicity)

**(PM 3)** $(a \oplus b) \odot c = a \odot c \oplus b \odot c$ (left distributivity)

**(PM 4)** $(\sup_n a_n) \odot (\sup_m b_m) = \sup_{n,m} a_n \odot b_m$ (left continuity).

## 4 Pseudo-concave Integral

For any $a \in \overline{\mathbb{R}}^+$ and any $A \in \mathcal{A}$, the function $b(a, A)$ defined by:

$$b(a, A)(x) = \begin{cases} a \text{ if } x \in A \\ 0 \text{ if } x \notin A. \end{cases}$$

is called *basic (simple) function.*

We can introduce a generalization of Lehrer's integral with respect to pseudo-addition $\oplus$ and pseudo-multiplication $\odot$.

**Definition 4.** Let $\oplus : [0, \infty]^2 \to [0, \infty]$ be a pseudo-addition and $\odot : [0, \infty]^2 \to [0, \infty]$ a $\oplus$-fitting pseudo-multiplication. Pseudo-concave integral of a measurable function $f : \Omega \to [0, \infty]$ is given by

$$(L) \int^{\oplus, \odot} f \, dv = \sup \left\{ \bigoplus_{i \in I} a_i \odot v(A_i) \mid \bigoplus_{i \in I} b(a_i, A_i) \leqslant f, I \text{ is finite }, a_i \geqslant 0 \right\},$$

where $A_i, i \in I$, are measurable.

*Example 1.* (i) Of course that for $\oplus = +$ and $\odot = \cdot$ we obtain the Lehrer's integral from Definition 1.

(ii) In a special case, when $\oplus = \vee$, the corresponding pseudo-multiplication have to be non-decreasing, and then we have

$$(L) \int^{\vee, \odot} f \, dv = \bigvee_{a \in [0, \infty]} a \odot v(f \geqslant a)).$$

This case cover Sugeno $\mathcal{S}_v$ and Shilkret $\mathcal{K}_v$ integrals, taking for the pseudo-multiplication $\odot$ minimum $\wedge$ and product $\cdot$, respectively.

(iii) In another special case, when $\oplus$ is strict (strictly monotone), then there exists an increasing bijection $g : [0, \infty] \to [0, \infty]$ such that

$$a \oplus b = g^{-1}(g(a) + g(b)),$$

see [17, 18, 23]. The only left distributive pseudo-multiplication $\odot$ is given by

$$a \odot b = g^{-1}(g(a)h(b)),$$

where $h : [0, \infty] \to [0, \infty]$ is a left-continuous non-decreasing function, $h(0) = 0$, see [15, 16, 17, 18, 20]. Then we have

$$(L) \int^{\oplus,\odot} f \, dv = \sup \left\{ g^{-1} \left( \sum_{i \in I} a_i h(v(A_i)) \right) \mid g^{-1} \left( \sum_{i \in I} b(g(a_i), A_i) \right) \leqslant f, \right.$$

$$\left. I \text{ is finite} , a_i \geqslant 0 \right\}$$

$$= g^{-1} \left( (L) \int g \circ f \, dh \circ v \right).$$

For $h = g$ we obtain convex $g$-integral, for $g$-like integrals see [13].

Depending on additional properties of pseudo-multiplication $\odot$ we can obtain useful properties of pseudo-concave integral. For example, if $\odot$ is associative we obtain the positive homogeneity of pseudo-concave integral. Generally, we have for two measurable functions $f_1$ and $f_2$

$$(L) \int^{\oplus,\odot} (f_1 \oplus f_2) \, dv \geqslant (L) \int^{\oplus,\odot} f_1 \, dv \oplus (L) \int^{\oplus,\odot} f_2 \, dv.$$

## 5    Concluding Remarks

Our proposal of pseudo-concave integrals can be seen as a starting point for a deeper investigation of properties of this interesting functional, promising fruitful applications in the area of multicriteria decision making and related areas. Note that a related concept generalizing Lebesgue integral was recently proposed and discussed in [26], and it would be interesting to see the connections with the pseudo-concave integral. Another line for deeper study of pseudo-concave integrals can be done for integral inequalities and relations with some classical integrals, in the spirit of our recent work [14].

**Acknowledgements.** The first author was supported by the grants APVV-0073-10 and P402/11/0378. The second author was supported by NSFC Grant No.70771010. The third author was supported by the national grants MNTRS (Serbia, Project 174009), and "Mathematical models of intelligent systems and their applications" by Provincial Secretariat for Science and Technological Development of Vojvodina.

## References

1. Benvenuti, P., Mesiar, R., Vivona, D.: Monotone set functions-based integrals. In: Pap, E. (ed.) Handbook of Measure Theory, vol. II, pp. 1329–1379. Elsevier Science, Amsterdam (2002)
2. Choquet, G.: Theory of capacities. Ann. Inst. Fourier (Grenoble) 5, 131–295 (1954)
3. Denneberg, D.: Non–Additive Measure and Integral. Kluwer Academic Publishers, Dordrecht (1994)

4. Imaoka, H.: On a subjective evaluation model by a generalized fuzzy integral. Int. J. Uncertainty Fuzziness Knowl.-Based Syst. 5, 517–529 (1997)
5. Klement, E.P., Mesiar, R., Pap, E.: On the relationship of associative compensatory operators to triangular norms and conorms. Internat. J. Uncertainty, Fuzziness and Knowledge-Based Systems 4, 25–36 (1996)
6. Klement, E.P., Mesiar, R., Pap, E.: Triangular Norms. Kluwer Academic Publishers, Dordrecht (2000a)
7. Klement, E.P., Mesiar, R., Pap, E.: Integration with respect to decomposable measures, based on a conditionally distributive semiring on the unit interval. Internat. J. Uncertain. Fuzziness Knowledge-Based Systems 8, 701–717 (2000b)
8. Klement, E.P., Mesiar, R., Pap, E.: A universal integral as common frame for Choquet and Sugeno integral. IEEE Transactions on Fuzzy Systems 18, 178–187 (2010)
9. Kolokoltsov, V.N., Maslov, V.P.: Idempotent Analysis and Its Applications. Kluwer, Dordrecht (1997)
10. Lehrer, E.: A new integral for capacities. Econ. Theory 39, 157–176 (2009)
11. Lehrer, E., Teper, R.: The concave integral over large spaces. Fuzzy Sets and Systems 159, 2130–2144 (2008)
12. Lovász, L.: Submodular functions and convexity. In: Bachem, A., Grötschel, M., Korte, B. (eds.) Mathematical Programming. The state of the art, pp. 235–257. Springer, Heidelberg (1983)
13. Mesiar, R.: Choquet-like integral. J. Math. Anal. Appl. 194, 477–488 (1995)
14. Mesiar, R., Li, J., Pap, E.: The Choquet integral as Lebesgue integral. Kybernetika 46, 931–934 (2010)
15. Mesiar, R., Rybárik, J.: PAN–operations. Fuzzy Sets and Systems 74, 365–369 (1995)
16. Murofushi, T., Sugeno, M.: Fuzzy t-conorm integral with respect to fuzzy measures: Generalization of Sugeno integral and Choquet integral. Fuzzy Sets and Systems 42, 57–71 (1991)
17. Pap, E.: An integral generated by a decomposable measure. Univ. u Novom Sadu Zb. Rad. Prirod. Mat. Fak. Ser. Mat. 20, 135–144 (1990)
18. Pap, E.: Null–Additive Set Functions. Kluwer, Dordrecht (1995)
19. Pap, E. (ed.): Handbook of Measure Theory. Elsevier, Amsterdam (2002)
20. Pap, E.: Pseudo-additive measures and their applications. In: Pap, E. (ed.) Handbook of Measure Theory, vol. II, pp. 1403–1465. Elsevier Science, Amsterdam (2002)
21. Shilkret, N.: Maxitive measure and integration. Indag. Math. 33, 109–116 (1971)
22. Sugeno, M.: Theory of fuzzy integrals and its applications. Ph. D. Thesis. Tokyo Institute of Technology (1974)
23. Sugeno, M., Murofushi, T.: Pseudo-additive measures and integrals. J. Math. Anal. Appl. 122, 197–222 (1987)
24. Teper, R.: On the continuity of the concave integral. Fuzzy Sets and Systems 160, 1318–1326 (2009)
25. Wang, Z., Klir, G.J.: Generalized Measure Theory. Springer, Boston (2009)
26. Zhang, Q., Mesiar, R., Li, J., Struk, P.: Generalized Lebesgue integral. Int. J. Approximate Reasoning 52, 427–443 (2011)

# On Spaces of Bochner and Pettis Integrable Functions and Their Set-Valued Counterparts

Coenraad C.A. Labuschagne and Valeria Marraffa

**Abstract.** The aim of this paper is to give a brief summary of the Pettis and Bochner integrals, how they are related, how they are generalized to the set-valued setting and the canonical Banach spaces of bounded maps between Banach spaces that they generate. The main tool that we use to relate the Banach space-valued case to the set-valued case, is the Rådström embedding theorem.

**Keywords:** Absolutely summing operator, Banach lattice, Banach space, Bochner integral, Pettis integral.

## 1 Introduction

The aim of this paper is to give a brief summary of the Pettis and Bochner integrals, how they are related, how they are generalized to the set-valued setting and the canonical Banach spaces of bounded maps between Banach spaces that they generate. The main tool that we use to relate the Banach space-valued case to the set-valued case, is the Rådström embedding theorem.

In Section 2 we consider the spaces of Pettis and Bochner integrable functions and the relationship between them via the absolutely summing maps. The purpose of Section 3 is the space of set-valued Pettis intgrable functions.

Coenraad C.A. Labuschagne
School of Computational and Applied Mathematics,
University of the Witwatersrand, Private Bag 3, P O WITS 2050, South Africa
e-mail: Coenraad.Labuschagne@wits.ac.za

Valeria Marraffa
Dipartimento di Matematica e Informatica, Università Degli Studi Di Palermo,
Via Archirafi, 34, 90123 Palermo, Italy
e-mail: marraffa@math.unipa.it

S. Li (Eds.): Nonlinear Maths for Uncertainty and its Appli., AISC 100, pp. 51–59.
springerlink.com      © Springer-Verlag Berlin Heidelberg 2011

We show that there are two canonical metrics on this space, which coincides with the canonical norms in the Banach space case. In Section 4 and Section 5 we consider the integrably bounded functions and the canonical set-valued maps that they generate. In Section 6 we briefly consider the connection between the space of order-Pettis integrable functions and the space of Bochner integrable functions and discuss their relationship using cone absolutely summing maps.

## 2   Connecting Pettis to Bochner Integrable Functions

Let $(\Omega, \Sigma, \mu)$ be a finite measure space and let $X$ and $Y$ be Banach spaces, with duals $X^*$ and $Y^*$.

We first recall the definition of Pettis integral.

**Definition 1.** Let $X$ be a Banach space. Then $f : \Omega \to X$ is called *Pettis integrable* if $\int_\Omega |x^* f| \, d\mu < \infty$ for all $x^* \in X^*$, and for all $A \in \Sigma$ there exists a vector $f_A \in X$ such that $x^*(f_A) = \int_A x^* f \, d\mu$ for all $x^* \in X^*$.

The set function $f_A : \Sigma \to X$ is called the indefinite Pettis integral of $f$. From the integral point of view, the functions with the same indefinite Pettis integrals are non-distinguishable, they are weakly equivalent. Thus denote by $P(\mu, X)$ the space of all weakly equivalent Pettis integrable functions $f : \Omega \to X$. If $B(X^*) = \{x^* \in X^* : \|x^*\| \le 1\}$, it is well known that both

$$\|f\|_{\text{Pettis}} = \sup_{x^* \in B(X^*)} \int_\Omega |x^* f| \, d\mu \quad \text{and} \quad \|f\|_{\text{P}} = \sup_{A \in \Sigma} \|f_A\|_X$$

define norms on $P(\mu, X)$ that are equivalent. Moreover, $(P(\mu, X), \|\cdot\|_{\text{Pettis}})$ is not complete. It is well known that $(P(\mu, X), \|\cdot\|_{\text{Pettis}})$ can be embedded in $L^1(\mu) \otimes^\vee X$, which denotes the norm completion of $L^1(\mu) \otimes X$ with respect to the injective norm (see [5]).

For $1 \le p < \infty$, let $L^p(\mu, X)$ denote the space of (classes of a.e. equal) Bochner $p$-integrable functions $f : \Omega \to X$ and denote the Bochner norm on $L^p(\mu, X)$ by $\Delta_p$, i.e. $\Delta_p(f) = \left( \int_\Omega \|f\|_X^p \, d\mu \right)^{1/p}$.

For $1 \le p < \infty$, the space $L^p(\mu, X)$ is isometrically isomorphic to the norm completion $L^p(\mu) \widetilde{\otimes}_{\Delta_p} X$ of $L^p(\mu) \otimes_{\Delta_p} X$, where $\Delta_p$ denotes the induced Bochner norm (see [5]).

We use the following few notions to describe a connection between Pettis integrable functions and Bochner integrable functions.

Let $\mathcal{L}(X, Y) := \{T : X \to Y : T \text{ is linear and bounded}\}$ be the Banach space with norm defined by $\|T\| := \sup\{\|Tx\| : \|x\| \le 1\}$ for all $T \in \mathcal{L}(X, Y)$.

**Definition 2.** Let $T \in \mathcal{L}(X, Y)$. Then $T$ is called *absolutely summing* if for every summable sequence $(x_n)$ in $X$, the sequence $(Tx_n)$ is absolutely summable in $Y$ (see [4]).

The space $\mathcal{L}^{as}(X,Y) = \{T : X \to Y : T \text{ is absolutely summing}\}$ is a Banach space with respect to the norm defined by

$$\|T\|_{as} = \sup\left\{\sum_{i=1}^{n} \|Tx_i\| : x_1, \ldots, x_n \in X, \left\|\sum_{i=1}^{n} x_i\right\| = 1, \, n \in \mathbb{N}\right\}$$

for all $T \in \mathcal{L}^{as}(X,Y)$.

If composed by absolutely summing maps, strongly measurable Pettis integrable functions may be transformed to Bochner integrable functions as stated in the next result.

**Theorem 1.** ( [3]) *If $T \in \mathcal{L}(X,Y)$, then $T \in \mathcal{L}^{as}(X,Y)$ if and only if $T \circ f \in L^1(\mu,Y)$ for all strongly measurable $f \in \mathcal{P}(\mu,X)$.*

## 3 Set-Valued Pettis Integrable Functions

Let $\mathcal{P}_0(X) := \{A \subseteq X : A \text{ is nonempty}\}$. We restrict ourselves to the subsets

$$\mathrm{cbf}(X) := \{A \in \mathcal{P}_0(X) : A \text{ is convex, bounded and closed}\} \text{ and}$$

$$\mathrm{ck}(X) := \{A \in \mathcal{P}_0(X) : A \text{ is convex and compact}\},$$

of $\mathcal{P}_0(X)$. If $A \in \mathcal{P}_0(X)$ and $x \in X$, the *distance* between $x$ and $A$ is defined by $d(x,A) = \inf\{\|x - y\|_X : y \in A\}$. Define $d_H$ for all $A, B \in \mathrm{cbf}(X)$ by $d_H(A,B) = \sup_{a \in A} d(a,B) \vee \sup_{b \in B} d(b,A)$, and, in particular, $\|A\|_H = d_H(A,\{0\}) = \sup_{a \in A} \|a\|$. Then $d_H$ is a metric on $\mathrm{cbf}(X)$, which is called the *Hausdorff metric*, and $(\mathrm{cbf}(X), d_H)$ is a complete metric space (see [9]). Moreover, $\mathrm{ck}(X)$ is closed in $(\mathrm{cbf}(X), d_H)$.

From here onwards, we assume that $X$ is separable. For every $C \in \mathrm{cbf}(X)$, the *support function of $C$* is denoted by $s(\cdot, C)$ and is defined by $s(x^*, C) = \sup\{\langle x^*, c\rangle : c \in C\}$ for each $x^* \in X^*$. Clearly, the map $x^* \longmapsto s(x^*, C)$ is sublinear on $X^*$ and $-s(-x^*, C) = \inf\{\langle x^*, c\rangle : c \in C\}$, for each $x^* \in X^*$.

The Rådström embedding $R(\mathrm{cbf}(X))$ of $\mathrm{cbf}(X)$ is given by $j : \mathrm{cbf}(X) \to R(\mathrm{cbf}(X))$, where $j(C) = s(\cdot, C)$ for all $C \in \mathrm{cbf(X)}$, and $R(\mathrm{cbf(X)})$ is the closure of the span of $\{s(\cdot, C) : C \in \mathrm{cbf(X)}\}$ in $C(B(X^*))$. Here $C(B(X^*))$ is the Banach space of continuous functions on $B(X^*)$ (the latter is endowed with the weak*-topology), and norm given by $\|f\|_\infty = \sup\{|f(\omega)| : \omega \in B(X^*)\}$ for all $f \in C(B(X^*))$.

A multifunction $\Gamma : \Omega \to \mathrm{cbf}(X)$ is said to be *scalarly measurable* if for every $x^* \in X^*$, the function $s(x^*, \Gamma(\cdot))$ is measurable. We say that $\Gamma$ is *scalarly integrable* if, for every $x^* \in X^*$, the function $s(x^*, \Gamma(\cdot))$ is *integrable* in the Lebesgue sense.

**Definition 3.** Let $X$ be a Banach space. Then $F : \Omega \to \mathrm{ck}(X)$ is called *set-valued Pettis integrable* if $\int_\Omega |s(x^*, F(\omega))| \, d\mu < \infty$ for all $x^* \in X^*$, and for all $A \in \Sigma$ there exists $F_A \in \mathrm{ck}(X)$ such that $s(x^*, F_A) = \int_A s(x^*, F(\omega)) \, d\mu$ for all $x^* \in X^*$.

Thus $F_A$ is called the *Pettis integral* of $F$ on $A$. Let $\mathcal{P}[\Sigma, \mathrm{ck}(X)]$ be the set of all set-valued Pettis integrable $F : \Omega \to \mathrm{ck}(X)$.

There is a canonical metric $d_P$ on $\mathcal{P}[\Sigma, \mathrm{ck}(X)]$ that corresponds to the $\| \cdot \|_{\mathrm{P}}$. Define $d_P : \mathcal{P}[\Sigma, \mathrm{ck}(X)] \times \mathcal{P}[\Sigma, \mathrm{ck}(X)] \to \mathbb{R}_+$ by

$$d_P(F, G) = \sup\{d_H(F_A, G_A) : A \in \Sigma\}.$$

Then $d_P$ is a metric on $\mathcal{P}[\Sigma, \mathrm{ck}(X)]$ which has the property that $d_P(F, \{0\}) = \|F\|_{\mathrm{P}}$.

In order to define the metric $d_{\mathrm{Pettis}}$ on $\mathcal{P}[\Sigma, \mathrm{ck}(X)] \times \mathcal{P}[\Sigma, \mathrm{ck}(X)]$ that we give below, we mention the following result.

**Theorem 2.** ( [1], [6]) *Let* $F : \Omega \to \mathrm{ck}(X)$. *Then* $F$ *is Pettis integrable if and only if* $j(F)$ *is Pettis integrable in* $R(\widetilde{\mathrm{ck}(X)})$.

Thus, if $F \in \mathcal{P}[\Sigma, \mathrm{ck}(X)]$, then $j \circ F : \Omega \to R(\widetilde{\mathrm{ck}(X)})$ is Pettis integrable; hence,

$$\|j \circ F\|_{\mathrm{Pettis}} = \sup\left\{ \int_\Omega |x^*(j(F(\omega)))| \, d\mu : x^* \in (j(\mathrm{ck}(X)))^*, \|x^*\| \le 1 \right\}.$$

Identifying the Pettis integrable multifunctions which have the same indefinite Pettis integral, then it is easy to show that $d_{\mathrm{Pettis}}$, defined by

$$
\begin{aligned}
&d_{\mathrm{Pettis}}(F, G) \\
&= \sup\{ \int_\Omega |x^*(j(F(\omega))) - x^*(j(G(\omega)))| \, d\mu : x^* \in (j(\mathrm{ck}(X)))^*, \|x^*\| \le 1 \}
\end{aligned}
$$

for all $F, G \in \mathcal{P}[\Sigma, \mathrm{ck}(X)]$, is a metric on $\mathcal{P}[\Sigma, \mathrm{ck}(X)] \times \mathcal{P}[\Sigma, \mathrm{ck}(X)]$ and $d_{\mathrm{Pettis}}(F, \{0\}) = \|j(F)\|_{\mathrm{Pettis}}$. Moreover, $d_P$ is equivalent to $d_{\mathrm{Pettis}}$, because $\|F\|_{\mathrm{P}} = \sup\{d_P(F_A, \{0\}) : A \in \Sigma\} = \sup\{\|j(F_A)\|_\infty : A \in \Sigma\} \cong \|j(F)\|_{\mathrm{Pettis}} = d_{\mathrm{Pettis}}(F, \{0\})$, where $\| \cdot \|_\infty$ is the norm on $R(\widetilde{\mathrm{ck}(X)})$.

## 4   Integrably Bounded Functions

Consider $X$ to be a separable Banach space. Let

$$\mathbf{M}[\Sigma, \mathrm{cbf}(X)] := \{F : \Omega \to \mathrm{cbf}(X) : F \text{ is } \Sigma\text{-measurable}\}.$$

In [7], Hiai and Umegaki introduced an analogue of $L^1(\mu, X)$ for set-valued functions.

If $F \in \mathrm{M}[\Sigma, \mathrm{cbf}(X)]$, then $F$ is called *integrably bounded* provided that there exists $\rho \in L^1(\mu)$ such that $\|x\|_X \le \rho(\omega)$ for all $x \in F(\omega)$ and for all $\omega \in \Omega$. In this case, $d_H(F(\omega), \{0\}) = \sup\{\|x\|_X : x \in F(\omega)\} \le \rho(\omega)$ for all $\omega \in \Omega$.

Let $\mathcal{L}^1[\Sigma, \mathrm{cbf}(X)]$ denote the set of all equivalence classes of a.e. equal $F \in \mathrm{M}[\Sigma, \mathrm{cbf}(X)]$ which are integrably bounded. If $\Delta \colon \mathcal{L}^1[\Sigma, \mathrm{cbf}(X)] \times \mathcal{L}^1[\Omega, \mathrm{cbf}(X)] \to \mathbb{R}_+$ is defined by $\Delta(F_1, F_2) = \int_\Omega d_H\big(F_1(\omega), F_2(\omega)\big) d\mu$, then $(\mathcal{L}^1[\Sigma, \mathrm{cbf}(X)], \Delta)$ is a complete metric space (see [7, 9]).

An $\mathbb{R}_+$-linear map $T : \mathrm{ck}(X) \to \mathrm{ck}(Y)$ is called a set-valued *absolutely summing* map if there exists $c \in \mathbb{R}_+$ such that for every finite sequence $X_1, \cdots, X_n \in \mathrm{ck}(X)$,

$$\sum_{i=1}^n d_H(T(X_i), \{0\}) \le c d_H(\bigoplus_{i=1}^n X_i, \{0\}). \tag{1}$$

The following results characterize set-valued absolutely summing maps. For the definition of $\hat{T}$, we recall the following from [14].

If $T : \mathrm{ck}(X) \to \mathrm{ck}(Y)$ is $\mathbb{R}_+$-linear then $\hat{T} : R(\mathrm{ck}(X)) \to R(\mathrm{ck}(Y))$, defined by

$$\hat{T}(Z) = \sum_{i=1}^n \lambda_i s(\cdot, T(C_i)),$$

for all $Z \in \mathrm{ck}(X)$ and $Z = \sum_{i=1}^n \lambda_i s(\cdot, C_i)$. Then, $\hat{T}$ is well defined and linear (see Theorem 3.4 in [14]).

**Theorem 3.** *A function* $T : \mathrm{ck}(X) \to \mathrm{ck}(Y)$ *is a set-valued absolutely summing if and only if* $\hat{T} : R(\widetilde{\mathrm{ck}(X)}) \to R(\widetilde{\mathrm{ck}(Y)})$ *is an absolutely summing map.*

*Proof.* Suppose $T$ is absolutely summing and $c$ satisfies the inequality (1). Let $Z_1, \cdots, Z_n \in \mathrm{ck}(X)$. Suppose that each $Z_k = \sum_{i=1}^{n_k} \lambda_i^k s(\cdot, C_i^k)$. Then

$$\sum_{i=1}^n \left\| \hat{T}(Z_i) \right\|_\infty = \sum_{i=1}^n \left\| \sum_{j=1}^{n_i} \lambda_j^i s(\cdot, T(C_j^i)) \right\|_\infty = \sum_{i=1}^n d_H\Big( \bigoplus_{j=1}^{n_i} \lambda_j^i T(C_j^i), \{0\} \Big)$$

$$\le c d_H(\bigoplus_{i=1}^n \bigoplus_{j=1}^{n_i} \lambda_j^i C_j^i, \{0\}) = \left\| \sum_{i=1}^n \sum_{j=1}^{n_i} \lambda_j^i s(\cdot, C_j^i) \right\|_\infty = c \left\| \sum_{i=1}^n Z_i \right\|_\infty.$$

Since $\hat{T}$ is continuous we can extend it to $\hat{T} : R(\widetilde{\mathrm{ck}(X)}) \to R(\widetilde{\mathrm{ck}(Y)})$. The converse is trivial.

**Theorem 4.** *A set-valued map* $S : \mathrm{ck}(X) \to \mathrm{ck}(Y)$ *is absolutely summing if and only if* $S \circ F \in \mathcal{L}^1[\Sigma, \mathrm{ck}(X)]$ *for all* $F \in \mathcal{P}[\Sigma, \mathrm{ck}(X)]$.

*Proof.* The proof follows easily by Theorems 1, 2 and 3 and using the fact that the space $X$ is separable.

## 5   Set-Valued Cone Absolutely Summing

Using Hörmander's embedding theorem, Hiai and Umegaki obtained in [7] a result similar to the the following:

**Lemma 1.** *Let $X$ be a separable Banach space and $(\Omega, \Sigma, \mu)$ a finite measure space. Then there exists a compact Hausdorff space $\Gamma$ and an embedding $j : \mathrm{cbf}(X) \hookrightarrow C(\Gamma)$, such that if $F \in \mathbf{M}[\Sigma, \mathrm{cbf}(X)]$, then $F \in \mathcal{L}^1[\Sigma, \mathrm{cbf}(X)]$ if and only if $j \circ F \in L^1(\mu, C(\Gamma))$ and $\Delta(F(\cdot), \{0\}) = \|j \circ F\|_{L^1(\mu, C(\Gamma))}$.*

For terminology and notation regarding Banach lattices, the reader may consult [13, 15].

Chaney and Schaefer extended the Bochner norm to the $\| \cdot \|_l$ norm on the tensor product of a Banach lattice $E$ and a Banach space $Y$. The $\| \cdot \|_l$-norm is given by

$$
\|u\|_l = \inf \left\{ \left\| \sum_{i=1}^n \|y_i\| \, |x_i| \right\| : u = \sum_{i=1}^n x_i \otimes y_i \right\}
$$

for all $u = \sum_{i=1}^n x_i \otimes y_i \in E \otimes Y$, and coincides with the Bochner norm on $L^p(\mu) \otimes Y$ for all finite measure spaces $(\Omega, \Sigma, \mu)$ and $1 \le p < \infty$ (see [2, 10, 15]).

We consider a generalization of $l$-tensor products to operators defined on Banach lattices which take their values in Banach spaces.

Let $E$ be a Banach lattice and let $Y$ be a Banach space. A linear map $T : E \to Y$ is called *cone absolutely summing* if for every positive summable sequence $(x_n)$ in $E$, the sequence $(Tx_n)$ is absolutely summable in $Y$ (see [15, Chapter IV, Section 3]). The space $\mathcal{L}^{\mathrm{cas}}(E, Y) = \{T : E \to Y : T \text{ is cone absolutely summing}\}$ is a Banach space with respect to the norm defined by

$$
\|T\|_{\mathrm{cas}} = \sup \left\{ \sum_{i=1}^n \|Tx_i\| : x_1, \dots, x_n \in E_+, \left\| \sum_{i=1}^n x_i \right\| = 1, \, n \in \mathbb{N} \right\}
$$

for all $T \in \mathcal{L}^{\mathrm{cas}}(E, Y)$.

Cone absolutely summing maps extend the Chaney-Schaefer $l$-tensor product in the following sense: Each one of the canonical maps $E^* \otimes_l Y \to \mathcal{L}^{\mathrm{cas}}(E, Y)$, given by $\sum_{i=1}^n x_i^* \otimes y_i =: u \mapsto T_u$, where $T_u x = \sum_{i=1}^n x_i^*(x) y_i$ for all $x \in E$, and $E \otimes_l Y \to \mathcal{L}^{\mathrm{cas}}(E^*, Y)$, given by $\sum_{i=1}^n x_i \otimes y_i =: u \mapsto T_u$, where $T_u x^* = \sum_{i=1}^n x^*(x_i) y_i$ for all $x^* \in E^*$, is an isometry (see [15, Chapter IV, Section 7] and [2, 10]).

Returning to the case at hand, if $F \in \mathcal{L}^1[\Sigma, \mathrm{cbf}(X)]$, then it follows from Lemma 1 that $j \circ F \in L^1(\mu, C(\Gamma))$. By the preceding remarks, the latter defines a cone absolutely summing map $T_{j \circ F} : L^\infty(\mu) \to C(\Gamma)$.

**Definition 4.** Let $E$ be a Banach lattice and $T : E \to \mathrm{cbf}(X)$ a $\mathbb{R}_+$-linear map. Then $T$ is called a *cone absolutely summing* $\mathrm{cbf}(X)$-*valued* map provided that there exists $l \in \mathbb{R}$ such that $\sum_{i=1}^{n} d_H(Tx_i, \{0\}) \leq l \left\| \sum_{i=1}^{n} |x_i| \right\|$ for all $x_1, \cdots, x_n \in E$.

If $V : E \to \mathrm{cbf}(X)$ is an $\mathbb{R}_+$-linear map, let $\|V\|_\circ := \sup\{d_H(Vx, \{0\}) : \|x\| \leq 1\}$, which may be equal to $\infty$. Let

$$\mathbb{R}_+\mathcal{L}[E, \mathrm{cbf}(X)] = \{V : E \to \mathrm{cbf}(X) : V \text{ is } \mathbb{R}_+\text{-linear and } \|V\|_\circ < \infty\}.$$

Since $V : E \to \mathrm{cbf}(X)$ is $\mathbb{R}_+$-linear if and only if $j \circ V : E \to R(\widetilde{\mathrm{cbf}(X)})$ is linear and $d(Vx, \{0\}) = \|(j \circ V)x\|_{R(\widetilde{\mathrm{cbf}(X)})}$ for all $x \in E$, we see that $V \in \mathbb{R}_+\mathcal{L}[E, \mathrm{cbf}(X)]$ if and only if $j \circ T \in \mathcal{L}(E, R(\widetilde{\mathrm{cbf}(X)}))$.

**Theorem 5.** *Let $E$ be a Banach lattice and $l \in \mathbb{R}_+$. Then the following statements are equivalent for $T \in \mathbb{R}_+\mathcal{L}[E, \mathrm{cbf}(X)]$:*

1. *If $x_1, \cdots, x_n \in E$, then $\sum_{i=1}^{n} d_H(Tx_i, \{0\}) \leq l \left\| \sum_{i=1}^{n} |x_i| \right\|$.*
2. *If $x_1, \cdots, x_n \in E$, then $\sum_{i=1}^{n} \|(j \circ T)x_i\|_{R(\widetilde{\mathrm{cbf}(X)})} \leq l \left\| \sum_{i=1}^{n} |x_i| \right\|$.*
3. *There exist an $L$-normed space $L$, a positive linear map $U_1 : E \to L$ and an $\mathbb{R}_+$-linear map $V_1 : L \to \mathrm{cbf}(X)$ such that $T = V_1 \circ U_1$, $\|U_1\| \leq 1$ and $\|V_1\|_\circ \leq l$.*
4. *There exist an AL-space $L$, a positive linear map $U : E \to L$ and a linear map $V : L \to R(\widetilde{\mathrm{cbf}(X)})$ such that $j \circ T = V \circ U$, $\|U\| \leq 1$ and $\|V\| \leq l$.*
5. *There exists $x^* \in E_+^*$ such that $\|x^*\| \leq l$ and $d(Tx, \{0\}) \leq x^*(|x|)$ for all $x \in E$.*
6. *There exists $x^* \in E_+^*$ such that $\|x^*\| \leq l$ and $\|(j \circ T)x\|_{R(\widetilde{\mathrm{cbf}(X)})} \leq x^*(|x|)$ for all $x \in E$.*

Let $\mathbb{R}_+\mathcal{L}^{\mathrm{cas}}[E, \mathrm{cbf}(X)] = \{T : E \to \mathrm{cbf}(X), T \text{ is a cone absolutely summing}\}$ and

$$\|T\|_\circ^{\mathrm{cas}} = \sup\left\{ \sum_{i=1}^{n} d(Tx_i, \{0\}) : x_1, \ldots, x_n \in E_+, \left\| \sum_{i=1}^{n} x_i \right\| = 1, n \in \mathbb{N} \right\}$$

for all $T \in \mathbb{R}_+\mathcal{L}^{\mathrm{cas}}[E, \mathrm{cbf}(X)]$. It is known that $R(\widetilde{\mathrm{ck}(X)})$ is a Banach lattice (see [14]). The next theorem follows in a similar manner to Theorem 3.

**Corollary 1.** *Suppose $E$ is a Banach lattice. If $V : E \to \mathrm{ck}(X)$ is $\mathbb{R}_+$-linear, then $T \in \mathbb{R}_+\mathcal{L}^{\mathrm{cas}}(E, \mathrm{ck}(X))$ if and only if $j \circ T \in \mathcal{L}^{\mathrm{cas}}(E, R(\widetilde{\mathrm{ck}(X)}))$; in which case, $\|T\|_\circ^{\mathrm{cas}} = \|j \circ T\|_{\mathrm{cas}}$.*

## 6  Mapping Order-Pettis Integrable to Integrably Bounded Functions

Let $(\Omega, \Sigma, \mu)$ be a finite measure space, $X$ a Banach space and $E$ a Banach lattice. A function $g : \Omega \to E$ is called *order-Pettis integrable* provided that $g$ is measurable and $|g|$ is Pettis integrable (see [10]). Denote by $\mathcal{P}_{\text{order}}(\mu, E)$ the set of all such order-Pettis integrable functions.

The following description of cone absolutely summing maps was noted in [10, Chapter 3, Theorem 1.1].

**Theorem 6.** ( [10]) *Let $(\Omega, \Sigma, \mu)$ be a finite measure space, $E$ and $F$ Banach lattices and $T \in \mathcal{L}(E, F)$. Then $T$ is cone absolutely summing if and only if $T \circ g \in L^1(\mu, F)$ for all $g \in \mathcal{P}_{\text{order}}(\mu, E)$.*

If $X$ a separable Banach space, let $\mathbb{R}_+ \mathcal{L}_{\mathbf{M}}[E, \text{cbf}(X)]$ denote the set of all $T \in \mathbb{R}_+ \mathcal{L}[E, \text{cbf}(X)]$ which satisfies there exists $(t_i) \subseteq E^* \otimes X$ such that $t_i(z) \in T(z)$ a.e. for all $z \in E$ and $i \in \mathbb{N}$, and $T(z) = \overline{\{t_i(z) : i \in \mathbb{N}\}}$ for all $z \in E$, where the closure is the norm closure in $X$. As in [12] we obtain obtain a set-valued analogue of Theorem 6.

**Theorem 7.** *Let $(\Omega, \Sigma, \mu)$ be a finite measure space, $X$ a separable Banach space, $E$ Banach lattice and $T \in \mathbb{R}_+ \mathcal{L}_{\mathbf{M}}[E, \text{cbf}(X)]$. Then $T$ is a cone absolutely summing $\text{cbf}(X)$-valued map if and only if $T \circ g \in \mathcal{L}^1[\Sigma, \text{cbf}(X)]$ for all $g \in \mathcal{P}_{\text{order}}(\mu, E)$.*

**Acknowledgements.** The authors gratefully acknowledge supported by the National Research Foundation, the Università degli Studi di Palermo and the M.I.U.R. of Italy.

## References

1. Cascales, B., Rodriguez, J.: Birkhoff integral for multi-valued functions. J. Math. Anal. Appl. 297, 540–560 (2004)
2. Chaney, J.: Banach lattices of compact maps. Math. Z. 129, 1–19 (1972)
3. Diestel, J.: An elementary characterization of absolutely summing operators. Math. Ann. 196, 101–105 (1972)
4. Diestel, J., Jarchow, H., Tonge, A.: Absolutely Summing Operators. Cambridge University Press, Providence (1995)
5. Diestel, J., Uhl, J.J.: Vector Measures. A.M.S. Surveys, Providence, Rhode Island, vol. 15 (1977)
6. Di Piazza, L., Musiał, K.: A decomposition theorem for compact-valued Henstock integral. Monatsh. Math. 148, 119–126 (2006)
7. Hiai, F., Umegaki, H.: Integrals, conditional expectations, and martingales of multivalued functiones. Journal of Multivariate Analysis 7, 149–182 (1977)
8. Hu, S., Papageorgiou, N.S.: Handbook of Multivalued Analysis I. Kluwer, Amsterdam (1997)

9. Li, S., Ogura, Y., Kreinovich, V.: Limit Theorems and Applications of Set-Valued and Fuzzy Set-Valued Random Variables. Kluwer Academic Press, Dordrecht (2002)

10. Jeurnink, G.A.M.: Integration of functions with values in a Banach lattice, Thesis. University of Nijmegen, The Netherlands (1982)

11. Labuschagne, C.C.A., Pinchuck, A.L., Van Alten, C.J.: A vector lattice version of Rådström's embedding theorem. Quaestiones Mathematicae 30, 285–308 (2007)

12. Labuschagne, C.C.A., Marraffa, V.: On set-valued cone absolutely summing map. Central European Journal of Mathematics 8, 148–157 (2010)

13. Meyer-Nieberg, P.: Banach Lattices. Springer, Heidelberg (1991)

14. Rådström, H.: An embedding theorem for spaces of convex sets. Proc. Amer. Math. Soc. 3, 165–169 (1952)

15. Schaefer, H.H.: Banach Lattices and Positive Operators. Springer, Heidelberg (1974)

# Upper Derivatives of Set Functions Represented as the Choquet Indefinite Integral

Shinsuke Nakamura, Toshiyuki Takasawa, and Toshiaki Murofushi

**Abstract.** This paper shows that, for a set function $\nu$ represented as the Choquet indefinite integral of a function $f$ with respect to a set function $\mu$, the upper derivative of $\nu$ at a measurable set $A$ with respect to a measure $m$ is, under a certain condition, equal to the difference calculated by subtracting the product of the negative part $f^-$ and the lower derivative of $\mu$ at the whole set with respect to $m$ from the product of the positive part $f^+$ and the upper derivative of $\mu$ at $A$ with respect to $m$.

**Keywords:** Upper and lower derivatives, Set function, Choquet integral.

## 1 Introduction

Morris [4] defines a derivative of a set function by locally approximating it by a linear set function, i.e., a measure. His derivative is a generalization of the Radon-Nikodým derivative. This paper defines upper and lower derivatives of set functions based on his definition, and shows two theorems on the upper derivative of a set function represented as a Choquet indefinite integral.

## 2 Preliminaries

Throughout the paper, we assume that $X$ is the whole set, $2^X$ is the power set of $X$, $(X, \mathscr{X})$ is a measurable space, where $\mathscr{X} \subset 2^X$ is a $\sigma$-algebraC and $M(X, \mathscr{X})$ is the family of measurable functions $f : X \to \mathbb{R}$. Furthermore, $1_A$

Shinsuke Nakamura
Dept. Computational Intelligence and Systems Science, Tokyo Institute of Technology, 4259-G3-47 Nagatsuta, Midori-ku, Yokohama 226-8502, Japan
e-mail: snakamura@fz.dis.titech.ac.jp

S. Li (Eds.): Nonlinear Maths for Uncertainty and its Appli., AISC 100, pp. 61–68.
springerlink.com        © Springer-Verlag Berlin Heidelberg 2011

represents the indicator function of a set $A$, the integral sign $\int$ means the Lebesgue integral, and (C) $\int$ represents the Choquet integral defined below.

**Definition 1.** [2] A *set function* is a function $\mu : \mathscr{X} \to \mathbb{R}$ satisfying $\mu(\emptyset) = 0$.

**Definition 2.** [5] Let $\mu$ be a set function on $(X, \mathscr{X})$. The *total variation* $V(\mu)$ of $\mu$ is defined by

$$V(\mu) = \sup \sum_{i=1}^{n} |\mu(A_i) - \mu(A_{i-1})|,$$

where the supremum in the right-hand side is taken over all finite sequences $\{A_i\}_{i=1}^{n} \subset \mathscr{X}$ of any length satisfying $\emptyset = A_0 \subset A_1 \subset \cdots \subset A_n = X$. A set function $\mu$ is said to be of *bounded variation* if $V(\mu) < \infty$. We denote by $BV(X, \mathscr{X})$ the family of set functions of bounded variation on $(X, \mathscr{X})$ .

*Remark 1.* [5] For every set function $\mu$ on $(X, \mathscr{X})$, it holds that $|\mu(A)| \leq V(\mu)$ for all $A \in \mathscr{X}$. Thus, every set function of bounded variation is bounded.

**Definition 3.** [2] [5] Let $\mu$ be a set function on $(X, \mathscr{X})$. The *Choquet integral* over $A \in \mathscr{X}$ of $f \in M(X, \mathscr{X})$ with respect to $\mu$ is defined by

$$(\text{C}) \int_A f \, d\mu = \int_{-\infty}^{\infty} \mu_{f,A}(r) dr$$

if the integral in the right-hand side exists, where

$$\mu_{f,A}(r) = \begin{cases} \mu(\{f(x) \geq r\} \cap A), & r \geq 0, \\ \mu(\{f(x) \geq r\} \cup A^c) - \mu(X), & r < 0. \end{cases}$$

A measurable function $f$ is called $\mu$-*Choquet integrable* if the Choquet integral of $f$ exists and its value is finite.

*Remark 2.* [5] If a set function $\mu$ is of bounded variation, then $\mu_{f,A}$ is a function of bounded variation, hence it is measurable, and by Remark 1 it is bounded. Furthermore, if $f$ is bounded, then, since $\mu_{f,A}$ vanishes outside the interval $[\inf f, \sup f]$, the function $\mu_{f,A}$ is integrable.

**Definition 4.** [2] [5] Let $f$ and $g$ be real-valued functions. We write $f \sim g$ if

$$f(x) < f(x') \Rightarrow g(x) \leq g(x'), \quad \forall x, x' \in X.$$

When $f \sim g$, functions $f$ and $g$ are said to be *comonotonic*.

# 3 Derivatives of Set Functions

## 3.1 Derivatives, Upper and Lower Derivatives of Set Functions

**Definition 5.** [4] Let $(X, \mathscr{X}, m)$ be a measure space, $\mu$ be a set function on $(X, \mathscr{X})$, and $A_0 \in \mathscr{X}$. If there exists $f \in M(X, \mathscr{X})$ such that

$$\mu(A) = \mu(A_0) + \int_{A \setminus A_0} f \, dm - \int_{A_0 \setminus A} f \, dm + o(m(A \triangle A_0))$$

for every $A \in \mathscr{X}$, then $\mu$ is said to be *differentiable* at $A_0$ with respect to $m$, and the function $f$ is called a *derivative* at $A_0$ with respect to $m$, where $g(x) = o(x)$ means that for each $\varepsilon > 0$ there exists a $\delta > 0$ such that

$$0 < |x| < \delta \Rightarrow \left| \frac{g(x)}{x} \right| < \varepsilon.$$

**Definition 6.** Let $(X, \mathscr{X}, m)$ be a measure space, $\mu$ be a set function on $(X, \mathscr{X})$, $f \in M(X, \mathscr{X})$, and $A_0 \in \mathscr{X}$. Consider the following two conditions:

(UD):   For every $A \in \mathscr{X}$ such that $A \supset A_0$,

$$\mu(A) = \mu(A_0) + \int_{A \setminus A_0} f \, dm + o(m(A \setminus A_0)).$$

(LD):   For every $A \subset \mathscr{X}$ such that $A \subset A_0$,

$$\mu(A) = \mu(A_0) - \int_{A_0 \setminus A} f \, dm + o(m(A_0 \setminus A)).$$

If there exists $f \in M(X, \mathscr{X})$ satisfying condition (UD) [resp. (LD)], then $\mu$ is said to be *upper* [resp. *lower*] *differentiable* at $A_0$ with respect to $m$, and $f$ is called an *upper* [resp. *lower*] *derivative* at $A_0$ with respect to $m$. We denote by $\overline{D}_{A_0}(X, \mathscr{X}, m)$ [resp. $\underline{D}_{A_0}(X, \mathscr{X}, m)$] the family of set functions on $(X, \mathscr{X})$ which are upper [resp. lower] differentiable at $A_0$ with respect to $m$.

*Remark 3.* By definition, the upper [resp. lower] derivative $f$ at $A_0$ can be arbitrary on $A_0$ [resp. $A_0^c$] under the condition that $f$ is measurable.

*Remark 4.* By definition, every differentiable set function at $A_0$ are upper and lower differentiable at $A_0$, and the derivative is an upper and lower derivative. On the other hand, an upper and lower differentiable set function at $A_0$ is not necessarily differentiable at $A_0$ (Example 2 below).

*Example 1.* [4] Let $u : \mathbb{R}^n \to \mathbb{R}$ be a totally differentiable function of $n$ variables, $(X, \mathscr{X}, m)$ be a measure space, $v_i : X \to \mathbb{R}$ $(i = 1, 2, \ldots, n)$ be $m$-integrable functions, and $\mu$ be the set function on $(X, \mathscr{X})$ defined by

$$\mu(A) = u\left(\int_A v_1\, dm, \int_A v_2\, dm, \ldots, \int_A v_n\, dm\right), \quad A \in \mathscr{X}.$$

Then $\mu$ is differentiable at every $A_0 \in \mathscr{X}$ with respect to $m$ and

$$\sum_{i=1}^{n} u_i\left(\int_{A_0} v_1 dm, \int_{A_0} v_2 dm, \ldots, \int_{A_0} v_n dm\right) v_i$$

is the derivative at $A_0$ with respect to $m$, where $u_i$ is the partial derivative of $u$ with respect to the $i$-th variable. Therefore $\mu$ is upper and lower differentiable at every $A_0 \in \mathscr{X}$ with respect to $m$, and the above derivative is an upper and lower derivative at $A_0$ with respect to $m$.

**Definition 7.** [3] Let $(X, \mathscr{X}, m)$ be a measure space. $A \in \mathscr{X}$ is called an *atom* if $m(A) > 0$ and

$$B \subset A,\ B \in \mathscr{X} \Rightarrow m(B) = 0 \text{ or } m(A).$$

A measure space without atoms is said to be *non-atomic*.

*Remark 5.* The values of a derivative $f$ on an atom can be chosen to be arbitrary under the condition that $f$ is measurable. The same holds for upper and lower derivatives.

According to Remark 5 above, in the rest of the paper we deal only with non-atomic measure spaces.

**Proposition 1.** *Let $(X, \mathscr{X}, m)$ be a non-atomic measure space. If $\mu \in \overline{D}_{A_0}(X, \mathscr{X}, m)$ [resp. $\underline{D}_{A_0}(X, \mathscr{X}, m)$], then the upper [resp. lower] derivative is $m$-integrable on every finite-measure subset of $A_0^c$ [resp. $A_0$].*

**Proposition 2.** *Let $(X, \mathscr{X}, m)$ be a non-atomic measure space. If $\mu \in \overline{D}_{A_0}(X, \mathscr{X}, m)$ [resp. $\underline{D}_{A_0}(X, \mathscr{X}, m)$], then the upper [resp. lower] derivative of $\mu$ at $A_0$ with respect to $m$ is $m$-a.e. uniquely determined on $A_0^c$ [resp. $A_0$].*

According to Remark 3 and Proposition 2, we put the following definition.

**Definition 8.** For $\mu \in \overline{D}_{A_0}(X, \mathscr{X}, m)$, we denote by $\frac{d\mu}{dm_+}(A_0)$ the upper derivative at $A_0$ with respect to $m$ such that $\left(\frac{d\mu}{dm_+}(A_0)\right)(x) = 0$ for all $x \in A_0$. Similarly, for $\mu \in \underline{D}_{A_0}(X, \mathscr{X}, m)$, we denote by $\frac{d\mu}{dm_-}(A_0)$ the lower derivative at $A_0$ with respect to $m$ such that $\left(\frac{d\mu}{dm_-}(A_0)\right)(x) = 0$ for all $x \in A_0^c$.

**Proposition 3.** *Let $(X, \mathscr{X}, m)$ be a non-atomic measure space, and $\mu$ be a differentiable set function at $A_0 \in \mathscr{X}$ with respect to $m$. Then $\mu \in$*

$\overline{D}_{A_0}(X, \mathscr{X}, m) \cap \underline{D}_{A_0}(X, \mathscr{X}, m)$, and the derivative $f$ of $\mu$ at $A_0$ with respect to $m$ is represented by

$$f = \frac{d\mu}{dm_+}(A_0) + \frac{d\mu}{dm_-}(A_0).$$

*Example 2.* Let $(X, \mathscr{X}, m)$ be a non-atomic measure space, $A_0 \in \mathscr{X}$ be of positive measure, and $\mu$ be the set function on $(X, \mathscr{X})$ defined by

$$\mu(A) = m(A_0) + \sqrt[3]{m(A \setminus A_0)^3 - m(A_0 \setminus A)^3}, \quad A \in \mathscr{X}.$$

Since

$$\mu(A) = m(A_0) + \int_{A \setminus A_0} 1 \, dm, \quad \forall A \in \mathscr{X}; A \supset A_0,$$

$$\mu(A) = m(A_0) - \int_{A_0 \setminus A} 1 \, dm, \quad \forall A \in \mathscr{X}; A \subset A_0,$$

it follows that $\mu \in \overline{D}_{A_0}(X, \mathscr{X}, m) \cap \underline{D}_{A_0}(X, \mathscr{X}, m)$, and that

$$\frac{d\mu}{dm_+}(A_0) = 1_{A_0^c}, \qquad \frac{d\mu}{dm_-}(A_0) = 1_{A_0}.$$

Assume that $\mu$ is differentiable at $A_0$ with respect to $m$. Then, by Proposition 3, the derivative have to be the constant function 1. However, it follows that for every $A \in \mathscr{X}$

$$\mu(A) - \mu(A_0) - \int_{A \setminus A_0} 1 \, dm + \int_{A_0 \setminus A} 1 \, dm$$
$$= \sqrt[3]{m(A \setminus A_0)^3 - m(A_0 \setminus A)^3} - m(A \setminus A_0) + m(A_0 \setminus A)$$
$$\neq o(m(A \triangle A_0)).$$

Thus, $\mu$ is not differentiable at $A_0$ with respect to $m$.

*Example 3.* Let $(X, \mathscr{X}, m)$ be a non-atomic measure space, and $\mu$ be a finite measure on $(X, \mathscr{X})$ such that $\mu \ll m$. Then, $\mu \in \overline{D}_{A_0}(X, \mathscr{X}, m) \cap \underline{D}_{A_0}(X, \mathscr{X}, m)$, and the Radon-Nikodým derivative $\frac{d\mu}{dm}$ is an upper and lower derivative. Especially, we have

$$\frac{d\mu}{dm_+}(A_0) = \frac{d\mu}{dm} 1_{A_0^c}, \qquad \frac{d\mu}{dm_-}(A_0) = \frac{d\mu}{dm} 1_{A_0}.$$

## 3.2 Two Main Theorems on Upper Derivatives

We state our two main theorems.

**Theorem 1.** Let $(X, \mathscr{X}, m)$ be a non-atomic finite measure space, $\mu \in BV(X, \mathscr{X}) \cap \overline{D}_{A_0}(X, \mathscr{X}, m)$, $f$ be a $\mu$-Choquet integrable non-negative function essentially bounded on $A_0^c$ such that $1_{A_0} \sim f$, and $\nu$ be the set function on $(X, \mathscr{X})$ defined by

$$\nu(A) = (C) \int_A f \, d\mu, \quad A \in \mathscr{X}.$$

Then $\nu \in \overline{D}_{A_0}(X, \mathscr{X}, m)$ and

$$\frac{d\nu}{dm_+}(A_0) = f \frac{d\mu}{dm_+}(A_0), \quad m\text{-a.e.}$$

**Theorem 2.** Let $(X, \mathscr{X}, m)$ be a non-atomic finite measure space, $\mu \in BV(X, \mathscr{X})$, $f$ be a $\mu$-Choquet integrable function essentially bounded on $A_0^c$, and $\nu$ be the set function on $(X, \mathscr{X})$ defined by

$$\nu(A) = (C) \int_A f d\mu, \quad A \in \mathscr{X}.$$

If $f \geq 0$ on $A_0$, $1_{A_0} \sim f$, and $\mu \in \overline{D}_{A_0}(X, \mathscr{X}, m) \cap \underline{D}_X(X, \mathscr{X}, m)$, then $\nu \in \overline{D}_{A_0}(X, \mathscr{X}, m)$ and

$$\frac{d\nu}{dm_+}(A_0) = f^+ \frac{d\mu}{dm_+}(A_0) - f^- \frac{d\mu}{dm_-}(X) 1_{A_0^c}, \quad m\text{-a.e.}$$

If $f \leq 0$ on $A_0$, $1_{A_0^c} \sim f$, and $\mu \in \overline{D}_\emptyset(X, \mathscr{X}, m) \cap \underline{D}_{A_0^c}(X, \mathscr{X}, m)$, then $\nu \in \overline{D}_{A_0}(X, \mathscr{X}, m)$ and

$$\frac{d\nu}{dm_+}(A_0) = f^+ \frac{d\mu}{dm_+}(\emptyset) 1_{A_0^c} - f^- \frac{d\mu}{dm_-}(A_0^c), \quad m\text{-a.e.}$$

*Example 4.* Let $(X, \mathscr{X}, m)$ be a non-atomic measure space, $\mu$ be a measure on $(X, \mathscr{X})$ such that $\mu \ll m$, $f$ be a $\mu$-integrable function, and $\nu$ be the set function defined by

$$\nu(A) = \int_A f \, d\mu, \quad A \in \mathscr{X}.$$

Then, the Radon-Nikodým derivatives $\frac{d\nu}{dm}$ and $\frac{d\mu}{dm}$ satisfy the following equation [3]:

$$\frac{d\nu}{dm} = f \frac{d\mu}{dm}, \quad m\text{-a.e.} \tag{1}$$

Let $A_0 \in \mathscr{X}$, $f$ be essentially bounded on $A_0^c$, and $1_{A_0} \sim f$. First, we consider the case where $f$ is a non-negative function. Since $\mu \in BV(X, \mathscr{X}) \cap \overline{D}_{A_0}(X, \mathscr{X}, m)$, it follows from Theorem 1 that

$$\frac{d\nu}{dm_+}(A_0) = f \frac{d\mu}{dm_+}(A_0), \quad m\text{-a.e.} \tag{2}$$

Since by Example 3 we have

$$\frac{d\nu}{dm_+}(A_0) = \frac{d\nu}{dm}1_{A_0^c}, \quad \frac{d\mu}{dm_+}(A_0) = \frac{d\mu}{dm}1_{A_0^c},$$

by substituting the above two formulae into Eq. (2), we obtain

$$\frac{d\nu}{dm}1_{A_0^c} = f\frac{d\mu}{dm}1_{A_0^c}, \quad m\text{-a.e.},$$

which is just Eq. (1) multiplied by $1_{A_0^c}$.

Next, we consider the case where $f$ is not necessarily non-negative. Assume that $f \geq 0$ on $A_0$. Then, since $\mu \in \overline{D}_{A_0}(X, \mathscr{X}, m) \cap \underline{D}_X(X, \mathscr{X}, m)$, it follows from Theorem 2 that

$$\frac{d\nu}{dm_+}(A_0) = f^+\frac{d\mu}{dm_+}(A_0) - f^-\frac{d\mu}{dm_-}(X)1_{A_0^c}. \tag{3}$$

Since by Example 3 we have

$$\frac{d\nu}{dm_+}(A_0) = \frac{d\nu}{dm}1_{A_0^c}, \quad \frac{d\mu}{dm_+}(A_0) = \frac{d\mu}{dm}1_{A_0^c}, \quad \frac{d\mu}{dm_-}(X) = \frac{d\mu}{dm}1_X = \frac{d\mu}{dm},$$

by substituting the above three formulae into Eq. (3), we obtain

$$\frac{d\nu}{dm}1_{A_0^c} = f^+\frac{d\mu}{dm}1_{A_0^c} - f^-\frac{d\mu}{dm}1_{A_0^c} = (f^+ - f^-)\frac{d\mu}{dm}1_{A_0^c} = f\frac{d\mu}{dm}1_{A_0^c}, \quad m\text{-a.e.}$$

which is just Eq. (1) multiplied by $1_{A_0^c}$.

The same argument is valid when $1_{A_0^c} \sim f$ and $f \leq 0$ on $A_0$. Therefore, Theorems 1 and 2 can be regarded as generalizations of the existing result (1) in measure theory.

*Example 5.* Let $(X, \mathscr{X}, m)$ be a non-atomic space, $n \in \mathbb{N}$, $v_i : X \to \mathbb{R}^+$ be an $m$-integrable non-negative function for $i = 1, 2, \ldots n$, $D = \left[0, \int_X v_1\, dm\right] \times \cdots \times \left[0, \int_X v_n\, dm\right] \subset \mathbb{R}^n$, $u : D \to \mathbb{R}$ be a totally differentiable function of bounded Arzelá variation [1] such that $u(0, \ldots, 0) = 0$, and $\mu$ be the set function on $(X, \mathscr{X})$ defined by

$$\mu(A) = u\left(\int_A v_1\, dm, \ldots, \int_A v_n\, dm\right), \quad A \in \mathscr{X}.$$

Then, for each $i = 1, 2, \ldots n$, since $v_i$ is non-negative, if $A, B \in \mathscr{X}$ and $A \subset B$, then

$$\int_A v_i\, dm \leq \int_B v_i\, dm.$$

Therefor, since $V(\mu) \leq AV(u) < \infty$, where $AV(u)$ is the Arzelá variation [1] of $u$, it follows that $\mu$ is of bounded variation. By Example 1 we have that

$\mu \in \overline{D}_{A_0}(X, \mathscr{X}, m) \cap \underline{D}_X(X, \mathscr{X}, m)$ and that

$$\frac{d\mu}{dm_+}(A_0) = \sum_{i=1}^{n} u_i \left( \int_{A_0} v_1 \, dm, \ldots, \int_{A_0} v_n \, dm \right) v_i 1_{A_0^c},$$

$$\frac{d\mu}{dm_-}(X) = \sum_{i=1}^{n} u_i \left( \int_{X} v_1 \, dm, \ldots, \int_{X} v_n \, dm \right) v_i,$$

where $u_i$ is the partial derivative of $u$ with respect to the $i$-th variable.

Assume that $f$ is essentially bounded, $f \geq 0$ on $A_0$, and $1_{A_0} \sim f$. Then $f$ is $\mu$-Choquet integrable. Let $\nu$ be the set function on $(X, \mathscr{X})$ defined by

$$\nu(A) = (\mathrm{C}) \int_A f \, d\mu, \quad A \in \mathscr{X}.$$

Then $\nu$ is upper differentiable at $A_0$ with respect to $m$ and

$$\frac{d\nu}{dm_+}(A_0) = \sum_{i=1}^{n} K_i v_i 1_{A_0^c},$$

where

$$K_i = f^+ u_i \left( \int_{A_0} v_1 \, dm, \ldots, \int_{A_0} v_n \, dm \right) - f^- u_i \left( \int_{X} v_1 \, dm, \ldots, \int_{X} v_n \, dm \right).$$

## 4 Conclusions

In this paper, we have defined upper and lower derivatives of set functions, and shown two theorems on the upper derivative of a set function represented as a Choquet indefinite integral. These theorems can be regarded as generalizations of the existing result in measure theory.

## References

1. Golubov, B.I.: Arzelá variation. In: Hazewinkel, M. (ed.) Encyclopaedia of Mathematics, vol. 1, p. 258. Kluwer, Dordrecht (1988)
2. Grabisch, M., Murofushi, T., Sugeno, M. (eds.): Fuzzy Measures and Integrals: Theory and Applications. Physica-Verlag, Heidelberg (2000)
3. Halmos, P.R.: Measure Theory. Springer, New York (1974)
4. Morris, R.J.T.: Optimization Problems Involving Set Functions. Ph. D. dissertation, University of California, Los Angels (1978)
5. Murofushi, T., Sugeno, M., Machida, M.: Non-monotonic fuzzy measures and the Choquet integral. Fuzzy Sets and Systems 64, 73–86 (1994)

# On Regularity for Non-additive Measure

Toshikazu Watanabe and Tamaki Tanaka

**Abstract.** In this paper, we give a certain result on the regularity for non-additive Borel measure under the conditions of weakly null additivity, a continuity from above, and a certain additional continuity.

**Keywords:** Non-additive measure, Regularity, Weakly null additive, Continuity.

## 1 Introduction

The regularity of measure is one of the important properties in classical measure theory. In [12], Wu and Ha generalize the regularity from a classical measure space to a finite autocontinuous fuzzy measure space and prove Lusin's theorem for the measure. Jiang and Suzuki [3] extent the result of [12] to those for a $\sigma$-finite fuzzy measure space. In [9], Song and Li investigate the regularity of null additive fuzzy measure on a metric space and prove that Lusin's theorem remains valid on fuzzy measure space under the null additivity condition. In [6], Li and Yasuda extent the regularity and corresponding theorem in the case of fuzzy Borel measures on a metric space under the weakly null additivity condition. Discussion for the regularity of fuzzy measures, see Pap [8], Jiang et al. [4], and Wu and Wu [13]. For real valued non-additive measures, see [2, 8, 11].

In this paper, in Section 3, we prove the regularity for real valued non-additive measure on a metric space in the case where the measure is weakly null-additive Borel measure and continuous from above together with a

Toshikazu Watanabe and Tamaki Tanaka
Graduate School of Science and Technology, Niigata University, 8050,
Ikarashi 2-no-cho, Nishi-ku, Niigata, 950–2181, Japan
e-mail: `wa-toshi@math.sc.niigata-u.ac.jp`, `tamaki@math.sc.niigata-u.ac.jp`

S. Li (Eds.): Nonlinear Maths for Uncertainty and its Appli., AISC 100, pp. 69–75.
springerlink.com      © Springer-Verlag Berlin Heidelberg 2011

property suggested by Sun [10]. In Section 4, we give the version of Ego-roff's Theorem for the measure on a metric space under the conditions.

## 2   Preliminaries

Let $R$ be the set of real numbers and $N$ the set of natural numbers. Denote by $\Theta$ the set of all mappings from $N$ into $N$. Let $(X, \mathcal{F})$ be a measurable space.

**Definition 1.** A set function $\mu : \mathcal{F} \to [0, \infty]$ is called a non-additive measure if it satisfies the following two conditions.
(1) $\mu(\emptyset) = 0$.
(2) If $A, B \in \mathcal{F}$ and $A \subset B$, then $\mu(A) \leq \mu(B)$.

In this paper, we always assume that $\mu$ is a finite measure on $\mathcal{F}$, that is, $\mu(X) < \infty$.

**Definition 2.** (1) $\mu$ is called continuous from above if $\lim_{n \to \infty} \mu(A_n) = \mu(A)$ whenever $\{A_n\} \subset \mathcal{F}$ and $A \in \mathcal{F}$ satisfy $A_n \searrow A$ and there exists $n_0$ such that $\mu(A_{n_0}) < \infty$.
(2) $\mu$ is called continuous from below if $\lim_{n \to \infty} \mu(A_n) = \mu(A)$ whenever $\{A_n\} \subset \mathcal{F}$ and $A \in \mathcal{F}$ satisfy $A_n \nearrow A$.
(3) $\mu$ is called a fuzzy measure if it is continuous from above and below.
(4) $\mu$ is called weakly null-additive if $\mu(A \cup B) = 0$ whenever $A, B \in \mathcal{F}$ and $\mu(A) = \mu(B) = 0$; see [11].
(5) $\mu$ is called strongly order continuous if it is continuous from above at measurable sets of measure 0, that is, for any $\{A_n\} \subset \mathcal{F}$ and $A \in \mathcal{F}$ with $A_n \searrow A$ and $\mu(A) = 0$, it holds that $\lim_{n \to \infty} \mu(A_n) = 0$.
(6) $\mu$ has property (S) if for any sequence $\{A_n\} \subset \mathcal{F}$ with $\lim_{n \to \infty} \mu(A_n) = 0$, there exists a subsequence $\{A_{n_k}\}$ such that $\mu(\cap_{i=1}^{\infty} \cup_{k=i}^{\infty} A_{n_k}) = 0$; see [10].

**Definition 3.** Let $\{f_n\}$ be a sequence of $\mathcal{F}$-measurable real valued functions on $X$ and $f$ also such a function.
(1) $\{f_n\}$ is called convergent $\mu$-a.e. to $f$ if there exists an $A \in \mathcal{F}$ with $\mu(A) = 0$ such that $\{f_n\}$ converges to $f$ on $X \setminus A$.
(2) $\{f_n\}$ is called $\mu$-almost uniformly convergent to $f$ if there exist a decreasing net $\{B_\gamma \mid \gamma \in \Gamma\} \subset \mathcal{F}$ such that for any $\varepsilon > 0$, there exists a $\gamma \in \Gamma$ such that $\mu(B_\gamma) < \varepsilon$ and $\{f_n\}$ converges to $f$ uniformly on each subset $X \setminus B_\gamma$.

For the weakly null-additivity, we give the following Lemma.

**Lemma 1.** *If $\mu$ is strongly order continuous and has property (S), then the following two conditions are equivalent:*

(i) $\mu$ *is weakly null-additive.*

(ii) *For any* $\varepsilon > 0$ *and double sequence* $\{A_{m,n}\} \subset \mathcal{F}$ *satisfying that* $A_{m,n} \searrow D_m$ *as* $n \to \infty$ *and* $\mu(D_m) = 0$ *for each* $m \in N$, *there exists a* $\theta \in \Theta$ *such that* $\mu\left(\cup_{m=1}^{\infty} A_{m,\theta(m)}\right) < \varepsilon$.

*Proof.* (i)$\to$ (ii) Let $\{A_{m,n}\}$ be a double sequence such that $A_{m,n} \searrow D_m$ as $n \to \infty$ and $\mu(D_m) = 0$ for each $m \in N$. Put $B_{m,n} = \cup_{j=1}^{m} A_{j,n}$ and $F_m = \cup_{j=1}^{m} D_j$, then $\{B_{m,n}\}$ is increasing for each $n \in N$ and $B_{m,n} \searrow F_m$ as $n \to \infty$. Since $\mu$ is weakly null additive, $\mu(F_m) = 0$. Since $\mu$ is strongly order continuous, for any integer $m$, there exists $n_0(m)$ such that $\mu(B_{m,n_0(m)}) < \frac{1}{m}$. Then $\lim_{m\to\infty} \mu(B_{m,n_0(m)}) = 0$. By property (S), there exists a strictly increasing sequence $\{m_i\} \subset N$ such that $\mu(\cap_{j=1}^{\infty} \cup_{i=j}^{\infty} B_{m_i,n_0(m_i)}) = 0$. For any $\varepsilon > 0$, since $\mu$ is strongly order continuous, there exists a $j_0 \in N$ such that $\mu(\cup_{i=j_0}^{\infty} B_{m_i,n_0(m_i)}) < \varepsilon$. Define $\theta \in \Theta$ such that $\theta(m) = n_0(m_{j_0})$ if $1 \le m \le m_{j_0}$ and $\theta(m) = n_0(m_i)$ if $m_{i-1} < m \le m_i$ for some $i > j_0$. Since $\{B_{m,n}\}$ is increasing for each $n \in N$, we have $\cup_{i=j_0}^{\infty} B_{m_i,n_0(m_i)} = \cup_{m=1}^{\infty} B_{m,\theta(m)}$. Since $\cup_{m=1}^{\infty} A_{m,\theta(m)} \subset \cup_{m=1}^{\infty} B_{m,\theta(m)}$, we have $\mu(\cup_{m=1}^{\infty} A_{m,\theta(m)}) \le \mu(\cup_{m=1}^{\infty} B_{m,\theta(m)})$. Thus, (ii) holds.

(ii)$\to$(i) Let $F, G \in \mathcal{F}$ and $\mu(F) = \mu(G) = 0$. Define a double sequence $\{A_{m,n}\} \subset \mathcal{F}$ such that $A_{1,n} = F$, $A_{2,n} = G$ and $A_{m,n} = \emptyset$ ($m \ge 3$) for any $n \in N$. Let $D_1 = F$, $D_2 = G$ and $D_m = \emptyset$ ($m \ge 3$) for any $n \in N$. By assumption, for any $\varepsilon > 0$ there exists a $\theta \in \Theta$ such that $\mu(\cup_{m=1}^{\infty} A_{m,\theta(m)}) < \varepsilon$. Since $\cup_{m=1}^{\infty} A_{m,\theta(m)} = F \cup G$, we have $\mu(F \cup G) < \varepsilon$. Then we have $\mu(F \cup G) = 0$. $\qquad\square$

## 3 Regularity of Measure

Let $X$ be a Hausdorff space. Denote by $\mathcal{B}(X)$ the $\sigma$-field of all Borel subsets of $X$, that is, the $\sigma$-field generated by the open subsets of $X$. A non-additive measure defined on $\mathcal{B}(X)$ is called a non-additive Borel measure on $X$. In Theorem 1, we prove that the regularity is enjoyed by non-additive Borel measures which is continuous from above with weakly null-additivity and has property (S). The proof is similar to that of [6, Theorem 1] and given here for completeness.

**Definition 4 ( [12]).** Let $\mu$ be a non-additive Borel measure on $X$. $\mu$ is called regular if for any $\varepsilon > 0$ and $A \in \mathcal{B}(X)$, there exist a closed set $F_\varepsilon$ and an open set $G_\varepsilon$ such that $F_\varepsilon \subset A \subset G_\varepsilon$ and $\mu(G_\varepsilon \setminus F_\varepsilon) < \varepsilon$.

**Theorem 1.** *Let* $X$ *be a metric space and* $\mathcal{B}(X)$ *a* $\sigma$-*field of all Borel subsets of* $X$. *Let* $\mu$ *be a non-additive Borel measure on* $X$ *which is weakly null additive, continuous from above and has property* (S). *Then* $\mu$ *is regular.*

*Proof.* Let $\mu$ be a non-additive Borel measure. Denote by $\mathcal{E}$ the family of Borel subsets $A$ of $X$ with the property that for any $\varepsilon > 0$, there exist a closed set $F_\varepsilon$ and an open set $G_\varepsilon$ such that

$$F_\varepsilon \subset A \subset G_\varepsilon \text{ and } \mu(G_\varepsilon \setminus F_\varepsilon) < \varepsilon.$$

We first show that $\mathcal{E}$ is a $\sigma$-field. It is obvious that $\mathcal{E}$ is closed for complementation and contains $\emptyset$ and $X$. We show that $\mathcal{E}$ is closed for countable unions. Let $\{A_m\}$ be a sequence of $\mathcal{E}$ and put $A = \cup_{m=1}^\infty A_m$ on $X$. Then for each $m \in N$, there exist double sequences $\{F_{m,n}\}$ of closed sets and $\{G_{m,n}\}$ of open sets such that

$$F_{m,n} \subset A_m \subset G_{m,n} \text{ and } \mu(G_{m,n} \setminus F_{m,n}) < \frac{1}{n} \text{ for all } n \in N.$$

We may assume that, for each $m \in N$, $\{F_{m,n}\}$ is increasing and $\{G_{m,n}\}$ is decreasing without loss of generality. For each $m \in N$, put $D_m = \cap_{n=1}^\infty (G_{m,n} \setminus F_{m,n})$. Since $\mu$ is continuous from above and

$$(G_{m,n} \setminus F_{m,n}) \searrow D_m \text{ as } n \to \infty,$$

we have $\mu(D_m) = \lim_{n\to\infty} \mu(G_{m,n} \setminus F_{m,n}) = 0$. As $\mu$ is weakly null additive, by Lemma 1, for any $\varepsilon > 0$ there exists a $\theta \in \Theta$ such that

$$\mu\left( \bigcup_{m=1}^\infty (G_{m,\theta(m)} \setminus F_{m,\theta(m)}) \right) < \frac{\varepsilon}{2},$$

and then

$$\mu\left( \bigcup_{m=1}^\infty G_{m,\theta(m)} \setminus \bigcup_{m=1}^\infty F_{m,\theta(m)} \right) < \frac{\varepsilon}{2}.$$

Since

$$\left( \bigcup_{m=1}^\infty G_{m,\theta(m)} \setminus \bigcup_{m=1}^N F_{m,\theta(m)} \right) \searrow \left( \bigcup_{m=1}^\infty G_{m,\theta(m)} \setminus \bigcup_{m=1}^\infty F_{m,\theta(m)} \right)$$

as $N \to \infty$, by the continuity from above and monotonicity of $\mu$, there exists an $N_0 \in N$ such that

$$\mu\left( \bigcup_{m=1}^\infty G_{m,\theta(m)} \setminus \bigcup_{m=1}^{N_0} F_{m,\theta(m)} \right) - \mu\left( \bigcup_{m=1}^\infty G_{m,\theta(m)} \setminus \bigcup_{m=1}^\infty F_{m,\theta(m)} \right) < \frac{\varepsilon}{2}.$$

Then we have

$$\mu\left( \bigcup_{m=1}^\infty G_{m,\theta(m)} \setminus \bigcup_{m=1}^{N_0} F_{m,\theta(m)} \right) < \frac{\varepsilon}{2} + \frac{\varepsilon}{2} = \varepsilon.$$

Denote $F_\varepsilon = \cup_{m=1}^{N_0} F_{m,\theta(m)}$ and $G_\varepsilon = \cup_{m=1}^\infty G_{m,\theta(m)}$, then $F_\varepsilon$ is closed, $G_\varepsilon$ is open and we have

$$F_\varepsilon \subset A \subset G_\varepsilon \text{ and } \mu(G_\varepsilon \smallsetminus F_\varepsilon) < \varepsilon.$$

Therefore $A \in \mathcal{E}$. Thus we have proved that $\mathcal{E}$ is a $\sigma$-field.

Next we verify that $\mathcal{E}$ contains all closed subsets of $X$. Let $F$ be closed in $X$. Since $X$ is a metric space, one can find a sequence $\{G_n\}$ of open subsets of $X$ such that $G_n \searrow F$, and hence $\lim_{n\to\infty} \mu(G_n \smallsetminus F) = 0$ by the continuity from above. Thus, we have $F \in \mathcal{E}$. Consequently, $\mathcal{E}$ is a $\sigma$-field which contains all closed subsets of $X$, so that it also contains all Borel subsets of $X$, that is, $\mathcal{E} \subset \mathcal{B}(X)$. Therefore $\mu$ is regular. $\qquad\square$

*Example 1.* Let $X = [0,1]$ be a metric space with the metric $d(x,y) = |x-y|$, $\mathcal{B}(X)$ a Borel measure of $X$ and $m$ the Lebesgue measure on $\mathcal{B}(X)$. Define

$$\mu(A) = \begin{cases} a \cdot m(A) & \text{if } m(A) < 1, \\ 1 & \text{if } m(A) = 1, \end{cases}$$

where $0 < a < 1$. Then $\mu$ is a non-additive measure. It is easy to see that $\mu$ is continuous from above. In fact, let $\{A_n\} \subset \mathcal{B}(X)$ a sequence with $A_n \searrow A$ where $A \in \mathcal{B}(X)$. We consider the following cases: (i) $m(A) = 1$, (ii) $m(A) < 1$. In cases (i), since $m(A_n) = 1$, the assertion is clear. In cases (ii), when $m(A_n) < 1$ the assertion is clear, however, when $m(A_n) = 1$ it is impossible. Thus $\mu$ is continious from above. Since $m$ is the Lebesgue measure, weakly null additivity and property (S) of $\mu$ hold. $\mu$ is not continuous from below. In fact, if we take $A_n = \left[0, 1 - \frac{1}{n}\right] \cup \{1\}$, $n \in N$, then $A_n \nearrow X$. Nevertheless, we have $\mu(A_n) = a \cdot m(A_n) = a \cdot \left(1 - \frac{1}{n}\right) \nearrow a < 1 = \mu(X)$.

## 4   Egoroff's Theorem

In this section, we prove a version of Egoroff's theorem for a measure which is weakly null-additive, continuous from above and has property (S) defined on a metric space. The proof is similar to [6, Proposition 4] and given here for completeness. Let us begin with giving the following result; see [1].

**Theorem 2.** *Let $\mu$ be a non-additive measure which is strongly order continuous and has property* (S). *Let $\{f_n\}$ be a sequence of $\mathcal{F}$-measurable real valued functions on $X$ and $f$ also such a function. If $\{f_n\}$ converges $\mu$-a.e. to $f$, then $\{f_n\}$ converges $\mu$-almost uniformly to $f$.*

**Theorem 3.** *Let $\mu$ be a non-additive Borel measure which is strongly order continuous and has property* (S). *Let $\{f_n\}$ be a sequence of Borel measurable real valued functions on $X$ and $f$ also such a function. If $\{f_n\}$ converges $\mu$-a.e. to $f$, then there exists an increasing sequence $\{A_m\} \subset \mathcal{B}(X)$ such that $\mu(X \smallsetminus \cup_{m=1}^\infty A_m) = 0$ and $\{f_n\}$ converges to $f$ uniformly on $A_m$ for each $m \in N$.*

*Proof.* Since $\{f_n\}$ converges $\mu$-a.e. to $f$, by Theorem 2, for any $\varepsilon > 0$, there exists a decreasing net $\{B_\gamma \mid \gamma \in \Gamma\}$ such that $\mu(B_\gamma) < \varepsilon$ and that $\{f_n\}$ converges to $f$ uniformly on each set $X \setminus B_\gamma$. For any $m \in N$, there exists a $\{\gamma_m\}$ such that $\mu(B_{\gamma_m}) < \varepsilon$. Put $A_m = X \setminus \cap_{i=1}^m B_{\gamma_i}$ for each $m \in N$. The proof is complete. $\qquad\qquad\square$

**Theorem 4.** *Let $X$ be a metric space and $\mu$ a non-additive Borel measure which is weakly null additive, continuous from above and has property* (S). *Let $\{f_n\}$ be a sequence of Borel measurable real valued functions on $X$ and $f$ also such a function. If $\{f_n\}$ converges $\mu$-a.e. to $f$, then for any $\varepsilon > 0$, there exists a closed set $F_\varepsilon$ such that $\mu(X \setminus F_\varepsilon) < \varepsilon$ and $\{f_n\}$ converges to $f$ uniformly on each $F_\varepsilon$.*

*Proof.* Since $\{f_n\}$ converges $\mu$-a.e. to $f$, by Theorem 3, there exists an increasing sequence $\{A_m\} \subset \mathcal{B}(X)$ such that $\{f_n\}$ converges to $f$ uniformly on $A_m$ for each $m \in N$ and $\mu(X \setminus \cup_{m=1}^\infty A_m) = 0$. By Theorem 1, $\mu$ is regular. Then for each $m \in N$, there exists an increasing sequence $\{F_{m,n}\}$ of closed sets such that $F_{m,n} \subset A_m$ and $\mu(A_m \setminus F_{m,n}) < \frac{1}{n}$ for any $n \in N$. Without loss of generality, we can assume that for each $m \in N$, $\{A_m \setminus F_{m,n}\}$ is decreasing as $n \to \infty$. Then we have

$$(A_m \setminus F_{m,n}) \searrow \bigcap_{n=1}^\infty (A_m \setminus F_{m,n}) \text{ as } n \to \infty.$$

Put $X_{m,n} = (X \setminus \cup_{m=1}^\infty A_m) \cup (A_m \setminus F_{m,n})$ and $D_m = \cap_{n=1}^\infty X_{m,n}$. Then for each $m \in N$, $X_{m,n} \searrow D_m$ as $n \to \infty$. Since

$$\mu\left(\bigcap_{n=1}^\infty (A_m \setminus F_{m,n})\right) \le \mu(A_m \setminus F_{m,n}),$$

we have

$$\mu\left(\bigcap_{n=1}^\infty (A_m \setminus F_{m,n})\right) = \lim_{n \to \infty} \mu(A_m \setminus F_{m,n}) = 0.$$

By the weakly null additivity of $\mu$, we have $\mu(D_m) = 0$ for any $m \in N$. For any $\varepsilon > 0$, by Lemma 1, there exists a $\theta \in \Theta$ such that $\mu\left(\cup_{m=1}^\infty X_{m,\theta(m)}\right) < \frac{\varepsilon}{2}$. Since $X \setminus \cup_{m=1}^\infty F_{m,\theta(m)} \subset \cup_{m=1}^\infty X_{m,\theta(m)}$, we have

$$\mu\left(X \setminus \bigcup_{m=1}^\infty F_{m,\theta(m)}\right) < \frac{\varepsilon}{2}.$$

On the other hand, since $\left(X \setminus \cup_{m=1}^N F_{m,\theta(m)}\right) \searrow \left(X \setminus \cup_{m=1}^\infty F_{m,\theta(m)}\right)$ as $N \to \infty$ and $\mu$ is continuous from above, there exists an $N_0 \in N$ such that

$$\mu\left(X \setminus \cup_{m=1}^{N_0} F_{m,\theta(m)}\right) - \mu\left(X \setminus \cup_{m=1}^\infty F_{m,\theta(m)}\right) < \frac{\varepsilon}{2}.$$

Then we have

$$\mu \left( X \smallsetminus \bigcup_{m=1}^{N_0} F_{m,\theta(m)} \right) < \frac{\varepsilon}{2} + \frac{\varepsilon}{2} = \varepsilon.$$

Denote $F_\varepsilon = \cup_{m=1}^{N_0} F_{m,\theta(m)}$, then $F_\varepsilon$ is a closed set, $\mu(X \smallsetminus F_\varepsilon) < \varepsilon$ and $F_\varepsilon \subset \cup_{m=1}^{N} A_m$. It is easy to see that $\{f_n\}$ converges to $f$ uniformly on $F_\varepsilon$. $\qquad \square$

**Acknowledgements.** The authors would like to express their hearty thanks to referee.

# References

1. Asahina, S., Uchino, K., Murofushi, T.: Relationship among continuity conditions and null-additivity conditions in non-additive measure theory. Fuzzy Sets and Systems 157, 691–698 (2006)
2. Denneberg, D.: Non-Additive Measure and Integral, 2nd edn. Kluwer Academic Publishers, Dordrecht (1997)
3. Jiang, Q., Suzuki, H.: Fuzzy measures on metric spaces. Fuzzy Sets and Systems 83, 99–106 (1996)
4. Jiang, Q., Wang, S., Ziou, D.: A further investigation for fuzzy measures on metric spaces. Fuzzy Sets and Systems 105, 293–297 (1999)
5. Kawabe, J.: Regularity and Lusin's theorem for Riesz space-valued fuzzy measures. Fuzzy Sets and Systems 158, 895–903 (2007)
6. Li, J., Yasuda, M.: Lusin's theorems on fuzzy measure spaces. Fuzzy Sets and Systems 146, 121–133 (2004)
7. Li, J., Yasuda, M., Jiang, Q., Suzuki, H., Wang, Z., Klir, G.J.: Convergence of sequence of measurable functions on fuzzy measure spaces. Fuzzy Sets and Systems 87, 317–323 (1997)
8. Pap, E.: Null-Additive Set Functions. Kluwer Academic Publishers, Dordrecht (1995)
9. Song, J., Li, J.: Regularity of null-additive fuzzy measure on metric spaces. International Journal of General Systems 32, 271–279 (2003)
10. Sun, Q.: Property (S) of fuzzy measure and Riesz's theorem. Fuzzy Sets and Systems 62, 117–119 (1994)
11. Wang, Z., Klir, G.J.: Fuzzy Measure Theory. Plenum Press, New York (1992)
12. Wu, C., Ha, M.: On the regularity of the fuzzy measure on metric fuzzy measure spaces. Fuzzy Sets and Systems 66, 373–379 (1994)
13. Wu, J., Wu, C.: Fuzzy regular measures on topological spaces. Fuzzy Sets and Systems 119, 529–533 (2001)

# Autocontinuity from below of Set Functions and Convergence in Measure

Jun Li, Masami Yasuda, and Ling Zhou

**Abstract.** In this note, the concepts of strong autocontinuity from below and strong converse autocontinuity from below of set function are introduced. By using four types of autocontinuity from below of monotone measure, the relationship between convergence in measure and pseudo-convergence in measure for sequence of measurable function are discussed.

**Keywords:** Monotone measure, Autocontinuity from below, Convergence in measure, Pseudo-convergence in measure.

## 1 Introduction

In non-additive measure theory, there are several different kinds of convergence for sequence of measurable functions, such as almost everywhere convergence, pseudo-almost everywhere convergence, convergence in measure, and convergence pseudo-in measure. The implication relationship between such convergence concepts are closely related to the structural characteristics of set functions. In this direction there are a lot of results ( [5, 7, 2, 6, 3, 10, 4, 8, 9, 11, 12, 14, 15]).

Jun Li
School of Science, Communication University of China, Beijing 100024, China
e-mail: lijun@cuc.edu.cn

Masami Yasuda
Department of Mathematics & Informatics, Faculty of Science, Chiba University,
Chiba 263-8522, Japan
e-mail: yasuda@math.s.chiba-u.ac.jp

Ling Zhou
College of Applied Sciences, Xinjiang Agricultural University, Xinjiang,
Urumqi 830052, China
e-mail: zhoulin5703@163.com

S. Li (Eds.): Nonlinear Maths for Uncertainty and its Appli., AISC 100, pp. 77–83.
springerlink.com
© Springer-Verlag Berlin Heidelberg 2011

In this note, we further discuss the relationship between convergence in measure and convergence pseudo-in measure for sequence of measurable functions. We shall introduce the concepts of strong autocontinuity from below and strong converse autocontinuity from below of a set function. By using the two types of autocontinuity from below of monotone measures, we investigate the inheriting of convergence in measure and convergence pseudo-in measure for sequence of measurable function under the common addition operation "+" and logic addition operation "∨". The implication relationship between convergence in measure and pseudo-convergence in measure are shown by using autocontinuity from below and converse autocontinuity from below, respectively.

## 2   Preliminaries

Let $X$ be a non-empty set, $\mathcal{F}$ a $\sigma$-algebra of subsets of $X$, and $(X, \mathcal{F})$ denotes the measurable space.

**Definition 1.** ( [9, 15]) Set function $\mu : \mathcal{F} \to [0, +\infty]$ is called a *monotone measure* on $(X, \mathcal{F})$ iff it satisfies the following requirements:

(1)  $\mu(\emptyset) = 0$;                                              (vanishing at $\emptyset$)
(2)  $A \subset B$ and $A, B \in \mathcal{F} \implies \mu(A) \leq \mu(B)$.           (monotonicity)

When $\mu$ is a monotone measure, the triple $(X, \mathcal{F}, \mu)$ is called a monotone measure space ( [9, 15]).

In some literature, a set function $\mu$ satisfying the conditions (1) and (2) of Definition 1 is called a fuzzy measure or a non-additive measure .

In this paper, all the considered sets are supposed to belong to $\mathcal{F}$ and $\mu$ is supposed to be a finite monotone measure, i.e., $\mu(X) < \infty$. All concepts and symbols not defined may be found in [9, 15].

**Definition 2.** ( [1]) A set function $\mu : \mathcal{F} \to [0, +\infty)$ is said to have *pseudometric generating property* (for short *p.g.p*), if for any $\{E_n\} \subset \mathcal{F}$ and $\{F_n\} \subset \mathcal{F}$,

$$\mu(E_n) \vee \mu(F_n) \to 0 \implies \mu(E_n \cup F_n) \to 0.$$

*Note*: The concept of pseudometric generated property goes back to Dobrakov and Farkova in seventies, and this was related to Frechet-Nikodym topology [1, 9].

Let $\mathbf{F}$ be the class of all finite real-valued measurable functions on $(X, \mathcal{F}, \mu)$, and let $A \in \mathcal{F}, f \in \mathbf{F}, f_n \in \mathbf{F}$ $(n = 1, 2, \ldots)$ and $\{f_n\}$ denote a sequence of measurable functions. We say that $\{f_n\}$ *converges in measure $\mu$ to*

$f$ on $A$, and denote it by $f_n \xrightarrow[A]{\mu} f$, if for any given $\sigma > 0$, $\lim\limits_{n \to +\infty} \mu(\{|f_n - f| \geq \sigma\} \cap A) = 0$; $\{f_n\}$ converges pseudo-in measure $\mu$ to $f$ on $A$, and denote it by $f_n \xrightarrow[A]{p,\mu} f$, if for any given $\sigma > 0$, $\lim\limits_{n \to +\infty} \mu(\{|f_n - f| < \sigma\} \cap A) = \mu(A)$; $\{f_n\}$ converges pseudo-in measure $\mu$ to $f$ in $A$, and denote it by $f_n \xrightarrow{p,\mu} f$ in $A$, if $f_n \xrightarrow{p,\mu} f$ on $C$ for all $C \in A \cap \mathcal{F}$.

## 3   Autocontinuity of Set Function

In [14] Wang introduced the concepts of autocontinuity from below and converse-autocontinuity from below of set function, and discussed the convergence for sequence of measurable functions by using the structure of set functions. Now we shall introduce the concepts of strong autocontinuity from below and strong converse-autocontinuity from below for set functions and show their properties.

**Definition 3.** ( [9, 14, 15]) Let $(X, \mathcal{F}, \mu)$ be a monotone measure space.
(1)  $\mu$ is said to be *autocontinuous from below* and denote it by *autoc.↑*, if for any $E \in \mathcal{F}, \{F_n\} \subset \mathcal{F}$,

$$\mu(F_n) \to 0 \implies \mu(E - F_n) \to \mu(E);$$

(2)  $\mu$ is said to be *converse-autocontinuous from below* and denote it by *c.autoc.↑*, if for any $A \in \mathcal{F}, \{B_n\} \subset A \cap \mathcal{F}$,

$$\mu(B_n) \to \mu(A) \implies \mu(A - B_n) \to 0.$$

**Definition 4.** Let $(X, \mathcal{F}, \mu)$ be a monotone measure space.
(1)  $\mu$ is said to be *strong autocontinuous from below* and denote it by *s.autoc.↑*, if

$$\mu(E_n) \vee \mu(F_n) \to 0 \implies \mu(A - E_n \cup F_n) \to \mu(A),$$

for any $A \in \mathcal{F}, \{E_n\} \subset \mathcal{F}$ and $\{F_n\} \subset \mathcal{F}$;
(2)  $\mu$ is said to be *strong converse-autocontinuous from below* and denote it by *s.c.autoc.↑*, if

$$\mu(A - E_n) \wedge \mu(A - F_n) \to \mu(A) \implies \mu(E_n \cup F_n) \to 0,$$

for any $A \in \mathcal{F}, \{E_n\} \subset A \cap \mathcal{F}$ and $\{F_n\} \subset A \cap \mathcal{F}$.

**Proposition 1.** *If $\mu$ is s.autoc.↑ (resp. s.c.autoc.↑), then it is autoc.↑ (resp. c.autoc.↑).*

**Proposition 2.** *If $\mu$ is autoc.↑ and has p.g.p, then it is s.autoc.↑.*

**Proposition 3.** *If $\mu$ is c.autoc.↑ and has p.g.p, then it is s.c.autoc.↑.*

## 4  Convergence in Measure

In this section, we study the application relationship between convergence in measure and convergence pseudo-in measure on monotone measure spaces.

The first conclusion of the following theorem due to Wang [15].

**Theorem 1.**  *Let $\mu$ be a monotone measure. Then,*

(1) $\mu$ *is autoc.↑ iff* $f_n \xrightarrow[A]{p.\mu} f$ *whenever* $f_n \xrightarrow[A]{\mu} f$, $\forall A \in \mathcal{F}, f, f_n \in \mathbf{F}$;

(2) $\mu$ *is c.autoc↑, iff* $f_n \xrightarrow[A]{\mu} f$ *whenever* $f_n \xrightarrow[A]{p.\mu} f$, $\forall A \in \mathcal{F}, f, f_n \in \mathbf{F}$.

*Proof.* We only prove (2). Let $\mu$ be c.autoco↑. If $f_n \xrightarrow[A]{p.\mu} f$, then for any given $\sigma > 0$, we have

$$\lim_{n \to +\infty} \mu(\{|f_n - f| < \sigma\} \cap A) = \mu(A)$$

and therefore, using the converse-autocontinuity from below of $\mu$, we have

$$\lim_{n \to +\infty} \mu(\{|f_n - f| \geq \sigma\} \cap A) = \lim_{n \to +\infty} \mu(A - \{|f_n - f| < \sigma\})$$
$$= 0.$$

So $f_n \xrightarrow[A]{\mu} f$.

Conversly, for any $A \in \mathcal{F}, \{B_n\} \subset A \cap \mathcal{F}$, and $\mu(B_n) \to \mu(A)$, we define measurable function sequences $\{f_n\}$ by

$$f_n = \chi_{B_n} = \begin{cases} 0 \text{ if } x \notin B_n \\ 1 \text{ if } x \in B_n, \end{cases}$$

n = 1, 2,..., and denote $f \equiv 1$. It is easy to see that $f_n \xrightarrow[A]{p.\mu} f$. If it implies $f_n \xrightarrow[A]{\mu} f$, then for $\sigma = \frac{1}{2}$, we have

$$\lim_{n \to +\infty} \mu(\{|f_n - f| \geq \frac{1}{2}\} \cap A) = 0.$$

As

$$\{|f_n - f| \geq \frac{1}{2}\} \cap A = \{1 - \chi_{B_n} \geq \frac{1}{2}\} \cap A = A - B_n.$$

So $\lim_{n \to +\infty} \mu(A - B_n) = 0$. This shows that $\mu$ is c.autoc↑. $\qquad\square$

The following theorems describe the inheriting of convergence in measure and convergence pseudo-in measure for sequence of measurable function under the common addition operation.

**Theorem 2.** *Let $\mu$ be a monotone measure.*

*(1) If $\mu$ is s.autoc.$\uparrow$, then $f_n \xrightarrow{\mu} f$ and $g_n \xrightarrow{\mu} g$ on $A$ imply*

$$\alpha f_n + \beta g_n \xrightarrow[D]{p.\mu} \alpha f + \beta g,$$

*for any $D \in A \cap \mathcal{F}, \alpha, \beta \in R^1$.*

*(2) If $\mu$ is s.c.autoc.$\uparrow$, then $f_n \xrightarrow[A]{p.\mu} f$ and $g_n \xrightarrow[A]{p.\mu} g$ imply*

$$\alpha f_n + \beta g_n \xrightarrow[A]{\mu} \alpha f + \beta g,$$

*for any $A \in \mathcal{F}, \alpha, \beta \in R^1$.*

*Proof.* It is similar to the proof of Theorem 1.    □

The following Theorem 3 and 4 describe respectively the characteristics of strong autocontinuity from below and strong converse-autocontinuity from below of set functions.

**Theorem 3.** *The following statements are equivalent:*

*(1)  $\mu$ is s.autoc.$\uparrow$;*

*(2)  $f_n + g_n \xrightarrow[A]{p.\mu} 0$ whenever $f_n \xrightarrow[A]{\mu} 0$ and $g_n \xrightarrow[A]{\mu} 0, \forall A \in \mathcal{F}$;*

*(3)  $f_n \vee g_n \xrightarrow[A]{p.\mu} 0$ whenever $f_n \xrightarrow[A]{\mu} 0$ and $g_n \xrightarrow[A]{\mu} 0, \forall A \in \mathcal{F}$.*

*Proof.* $(1) \implies (2)$. It follows directly from Theorem 2 above.

$(2) \implies (3)$. For any $A \in \mathcal{F}$, if $f_n \xrightarrow[A]{\mu} 0$ and $g_n \xrightarrow[A]{\mu} 0$, then $|f_n| \xrightarrow[A]{\mu} 0$ and $|g_n| \xrightarrow[A]{\mu} 0$. By condition (2), we have $|f_n| + |g_n| \xrightarrow{p.\mu} 0$ on $A$, therefore, for any $\sigma > 0$,

$$\lim_{n \to +\infty} \mu(\{|f_n| + |g_n| < \sigma\} \cap A) = \mu(A).$$

Noting that $|f_n \vee g_n| \leq |f_n| + |g_n|$, we get

$$\{|f_n| + |g_n| < \sigma\} \cap A \subseteq \{|f_n \vee g_n| < \sigma\} \cap A \subseteq A.$$

So

$$\lim_{n \to +\infty} \mu(\{|f_n \vee g_n| < \sigma\} \cap A) = \mu(A).$$

This shows $f_n \vee g_n \xrightarrow{p.\mu} 0$ on $A$.

(3)$\Longrightarrow$(1). For any $\{E_n\} \subset \mathcal{F}, \{F_n\} \subset \mathcal{F}$ with $\lim_{n\to\infty} \mu(E_n) \vee \mu(F_n) = 0$, we define measurable function sequences $\{f_n\}$ and $\{g_n\}$ by

$$f_n = \chi_{E_n} = \begin{cases} 0 \text{ if } x \notin E_n \\ 1 \text{ if } x \in E_n \end{cases}$$

and

$$g_n = \chi_{F_n} = \begin{cases} 0 \text{ if } x \notin F_n \\ 1 \text{ if } x \in F_n, \end{cases}$$

$n = 1, 2,...$, then $f_n \xrightarrow{\mu} 0$ on $A$ and $g_n \xrightarrow{\mu} 0$ on $A$. Thus, $f_n \vee g_n \xrightarrow{\mu} 0$ on $A$. Therefore for $\sigma = \frac{1}{2}$, we have

$$\lim_{n\to+\infty} \mu(\{f_n \vee g_n < \frac{1}{2}\} \cap A) = \mu(A).$$

Noting $f_n \vee g_n = \chi_{E_n} \vee \chi_{F_n} = \chi_{E_n \cup F_n}$, and

$$\{\chi_{E_n} \vee \chi_{F_n} < \frac{1}{2}\} \cap A = A - \{\chi_{E_n} \vee \chi_{F_n} \geq \frac{1}{2}\} = A - E_n \cup F_n.$$

So

$$\lim_{n\to+\infty} \mu(A - E_n \cup F_n) = \mu(A).$$

That is, $\mu$ is s.autoc.$\uparrow$. □

**Theorem 4.** *The following statements are equivalent:*
*(1)* $\mu$ *is s.c.autoc.$\uparrow$;*
*(2)* $f_n + g_n \xrightarrow[A]{\mu} 0$ *whenever* $f_n \xrightarrow[A]{p.\mu} 0$ *and* $g_n \xrightarrow[A]{p.\mu} 0$, $\forall A \in \mathcal{F}$;
*(3)* $f_n \vee g_n \xrightarrow[A]{\mu} 0$ *whenever* $f_n \xrightarrow[A]{p.\mu} 0$ *and* $g_n \xrightarrow[A]{p.\mu} 0$, $\forall A \in \mathcal{F}$.

*Proof.* It is similar to the proof of Theorem 3. □

**Acknowledgements.** The first author was supported by NSFC Grant No. 70771010. The second author was supported by JSPS Grant No.22540112.

# References

1. Dobrakov, I., Farkova, J.: On submeasures II. Math. Slovoca 30, 65–81 (1980)
2. Li, J.: Order continuous of monotone set function and convergence of measurable functions sequence. Applied Mathematics and Computation 135(2-3), 211–218 (2003)
3. Liu, Y., Liu, B.: The relationship between structural characteristics of fuzzy measure and convergences of sequences of measurable functions. Fuzzy Sets and Systems 120, 511–516 (2001)

4. Li, J., Mesiar, R., Zhang, Q.: Absolute continuity of monotone measure and convergence in measure. Communications in Computer and Information Science 80, 500–504 (2010)
5. Li, J., Yasuda, M., Jiang, Q., Suzuki, H., Wang, Z., Klir, G.J.: Convergence of sequence of measurable functions on fuzzy measure space. Fuzzy Sets and Systems 87, 385–387 (1997)
6. Li, J., Yasuda, M.: On Egoroff's theorem on finite monotone non-additive measure space. Fuzzy Sets and Systems 153, 71–78 (2005)
7. Li, J., Zhang, Q.: Asymptotic structural characteristics of monotone measure and convergence in monotone measure. The Journal of Fuzzy Mathematics 9(2), 447–459 (2001)
8. Murofushi, T., Uchino, K., Asahina, S.: Conditions for Egoroff's theorem in non-additive measure theory. Fuzzy Sets and Systems 146, 135–146 (2004)
9. Pap, E.: Null-additive Set Functions. Kluwer Academic Press, Dordrecht (1995)
10. Song, J., Li, J.: Lebesgue theorems in non-additive measure theory. Fuzzy Sets and Systems 149, 543–548 (2005)
11. Sun, Q.: Property (S) of fuzzy measure and Riesz's theorem. Fuzzy Sets and Systems 62, 117–119 (1994)
12. Takahashi, M., Murofushi, T.: Relationship between convergence concepts in fuzzy measure theory. In: Proc. 11th IFSA World Congress, vol. I, pp. 467–473 (2005)
13. Uchino, K., Murofushi, T.: Relations between mathematical properties of fuzzy measures. In: Proc. 10th IFSA World Congress, pp. 27–30 (2003)
14. Wang, Z.: Asymptotic structural characteristics of fuzzy measure and their applications. Fuzzy Sets and Systems 16, 277–290 (1985)
15. Wang, Z., Klir, G.J.: Generalized Measure Theory. Springer, Boston (2009)

# On Lusin's Theorem for Non-additive Measure

Tamaki Tanaka and Toshikazu Watanabe

**Abstract.** In this paper, we prove Lusin's theorem remains valid for non-additive Borel measure under the conditions of weakly null additivity, continuity from above and a certain additional continuity.

**Keywords:** Non-additive measure, Lusin's theorem, Weakly null additive, Continuity.

## 1 Introduction

Lusin's theorem is one of the most fundamental theorems in classical measure theory and does not hold in non-additive measure theory without additional conditions. In [13], Wu and Ha generalize Lusin's theorem from a classical measure space to a finite autocontinuous fuzzy measure space. Jiang and Suzuki [2] extent the result of [13] to a $\sigma$-finite fuzzy measure space. In [9], Song and Li investigate the regularity of null additive fuzzy measure on a metric space and show Lusin's theorem remains valid on fuzzy measure space under the null additivity condition. In [5], Li and Yasuda extent the theorem into the Borel measures on a metric space under the weakly null additivity condition. For the regularity fuzzy measures, see Pap [7], Jiang et al. [3], and Wu and Wu [14]. For real valued non-additive measures, see [1,7,11].

In [12], we investigate the regularity of non-additive Borel measures on a metric space in the case where the measure is weakly null-additive and continuous from above together with a certain property suggested by Sun [10]. In this paper, we prove Lusin's theorem remains valid for the measure on a metric space under the conditions.

Tamaki Tanaka and Toshikazu Watanabe
Graduate School of Science and Technology, Niigata University, 8050,
Ikarashi 2-no-cho, Nishi-ku, Niigata, 950–2181, Japan
e-mail: tamaki@math.sc.niigata-u.ac.jp, wa-toshi@math.sc.niigata-u.ac.jp

S. Li (Eds.): Nonlinear Maths for Uncertainty and its Appli., AISC 100, pp. 85–92.
springerlink.com                           © Springer-Verlag Berlin Heidelberg 2011

## 2  Preliminaries

Let $R$ be the set of real numbers and $N$ the set of natural numbers. Denote by $\Theta$ the set of all mappings from $N$ into $N$. Let $(X, \mathcal{F})$ be a measurable space.

**Definition 1.** A set function $\mu : \mathcal{F} \to [0, \infty]$ is called a non-additive measure if it satisfies the following two conditions.
(1) $\mu(\emptyset) = 0$.
(2) If $A, B \in \mathcal{F}$ and $A \subset B$, then $\mu(A) \leq \mu(B)$.

In this paper, we always assume that $\mu$ is a finite measure on $\mathcal{F}$, that is, $\mu(X) < \infty$.

**Definition 2.** (1) $\mu$ is called continuous from above if $\lim_{n\to\infty} \mu(A_n) = \mu(A)$ whenever $\{A_n\} \subset \mathcal{F}$ and $A \in \mathcal{F}$ satisfy $A_n \searrow A$ and there exists $n_0$ such that $\mu(A_{n_0}) < \infty$.
(2) $\mu$ is called continuous from below if $\lim_{n\to\infty} \mu(A_n) = \mu(A)$ whenever $\{A_n\} \subset \mathcal{F}$ and $A \in \mathcal{F}$ satisfy $A_n \nearrow A$.
(3) $\mu$ is called a fuzzy measure if it is continuous from above and below.
(4) $\mu$ is called weakly null-additive if $\mu(A \cup B) = 0$ whenever $A, B \in \mathcal{F}$ and $\mu(A) = \mu(B) = 0$; see [11].
(5) $\mu$ is called strongly order continuous if it is continuous from above at measurable sets of measure 0, that is, for any $\{A_n\} \subset \mathcal{F}$ and $A \in \mathcal{F}$ with $A_n \searrow A$ and $\mu(A) = 0$, it holds that $\lim_{n\to\infty} \mu(A_n) = 0$.
(6) $\mu$ has property (S) if for any sequence $\{A_n\} \subset \mathcal{F}$ with $\lim_{n\to\infty} \mu(A_n) = 0$, there exists a subsequence $\{A_{n_k}\}$ such that $\mu(\cap_{i=1}^{\infty} \cup_{k=i}^{\infty} A_{n_k}) = 0$; see [10].

**Definition 3.** Let $\{f_n\}$ be a sequence of $\mathcal{F}$-measurable real valued functions on $X$ and $f$ also such a function.
(1) $\{f_n\}$ is called convergent $\mu$-a.e. to $f$ if there exists an $A \in \mathcal{F}$ with $\mu(A) = 0$ such that $\{f_n\}$ converges to $f$ on $X \setminus A$.

## 3  Regularity of Measure

In [12], first, for the weakly null-additivity, we give the following Lemma. Second, we also showed the regularity of non-additive Borel measures on a metric space (Theorem 1). Finally, we showed a version of Egoroff's theorem for the measures on a metric space (Theorem 2).

**Lemma 1.** *If $\mu$ is strongly order continuous and has property (S), then the following two conditions are equivalent:*
(i) *$\mu$ is weakly null-additive.*
(ii) *For any $\varepsilon > 0$ and double sequence $\{A_{m,n}\} \subset \mathcal{F}$ satisfying that $A_{m,n} \searrow D_m$ as $n \to \infty$ and $\mu(D_m) = 0$ for each $m \in N$, there exists a $\theta \in \Theta$ such that $\mu\left(\cup_{m=1}^{\infty} A_{m,\theta(m)}\right) < \varepsilon$.*

Let $X$ be a Hausdorff space. Denote by $\mathcal{B}(X)$ the $\sigma$-field of all Borel subsets of $X$, that is, the $\sigma$-field generated by the open subsets of $X$. A non-additive measure defined on $\mathcal{B}(X)$ is called a non-additive Borel measure on $X$.

**Definition 4 ( [13]).** Let $\mu$ be a non-additive Borel measure on $X$. $\mu$ is called regular if for any $\varepsilon > 0$ and $A \in \mathcal{B}(X)$, there exist a closed set $F_\varepsilon$ and an open set $G_\varepsilon$ such that $F_\varepsilon \subset A \subset G_\varepsilon$ and $\mu(G_\varepsilon \setminus F_\varepsilon) < \varepsilon$.

**Theorem 1.** *Let $X$ be a metric space and $\mathcal{B}(X)$ a $\sigma$-field of all Borel subsets of $X$. Let $\mu$ be a non-additive Borel measure on $X$ which is weakly null additive, continuous from above and has property (S). Then $\mu$ is regular.*

**Theorem 2.** *Let $X$ be a metric space and $\mu$ a non-additive Borel measure which is weakly null additive, continuous from above and has property (S). Let $\{f_n\}$ be a sequence of Borel measurable real valued functions on $X$ and $f$ also such a function. If $\{f_n\}$ converges $\mu$-a.e. to $f$, then for any $\varepsilon > 0$, there exists a closed set $F_\varepsilon$ such that $\mu(X \setminus F_\varepsilon) < \varepsilon$ and $\{f_n\}$ converges to $f$ uniformly on each $F_\varepsilon$.*

## 4 Lusin's Theorem

In this section, we shall further generalize well-known Lusin's theorem in classical measure theory to that of non-additive measure spaces by using the results obtained in Sections 3. The proof is similar to that of [5, Theorem 4] and given here for completeness. For real valued fuzzy measure case, see [5], and for Riesz space-valued fuzzy measure case, see [4].

**Theorem 3.** *Let $X$ be a metric space and $\mu$ a non-additive Borel measure which is weakly null additive, continuous from above and has property (S). Let $f$ be a Borel measurable real valued function on $X$. Then for any $\varepsilon > 0$, there exists a closed set $F_\varepsilon$ such that $\mu(X \setminus F_\varepsilon) < \varepsilon$ and $f$ is continuous on each $F_\varepsilon$.*

*Proof.* We prove the theorem stepwise in the following two situations.

(a) Suppose that $f$ is a simple function, that is, $f(x) = \sum_{m=1}^{s} a_m \chi_{A_m}(x)$ $(x \in X)$, where $a_m \in R$ $(m = 1, 2, \ldots, s)$, $\chi_{A_m}(x)$ is the characteristic function of the Borel set $A_m$ and $X = \sum_{m=1}^{s} A_m$ (a disjoint finite union). By Theorem 1, $\mu$ is regular. Then for each $m \in N$, there exists a sequence $\{F_{m,n}\}$ of closed sets such that $F_{m,n} \subset A_m$ and $\mu(A_m \setminus F_{m,n}) < \frac{1}{n}$ for any $n \in N$. We may assume that $\{F_{m,n}\}$ is increasing in $n$ for each $m$, without any loss of generality. Put $B_{m,n} = A_m \setminus F_{m,n}$ if $m = 1, \ldots, s$ and $B_{m,n} = \emptyset$ if $m > s$, and put $D_m = \cap_{n=1}^{\infty} B_{m,n}$. We have $\mu(D_m) = 0$. By Lemma 1, for any $\varepsilon > 0$, there exists a $\theta \in \Theta$ such that $\mu\left(\bigcup_{m=1}^{\infty} \left(A_m \setminus F_{m,\theta(m)}\right)\right) < \varepsilon$. Put $F_\varepsilon = \cup_{m=1}^{s} F_{m,\theta(m)}$, then $f$ is continuous on the closed subset $F_\varepsilon$ of $X$ and we have

$$\mu\left(X \setminus F_\varepsilon\right) = \mu\left(\bigcup_{m=1}^{s} A_m \setminus \bigcup_{m=1}^{s} F_{m,\,\theta(m)}\right) \le \mu\left(\bigcup_{m=1}^{s} \left(A_m \setminus F_{m,\,\theta(m)}\right)\right).$$

Therefore we have $\mu\left(X \setminus F_\varepsilon\right) < \varepsilon$.

(b) Let $f$ be a Borel measurable real valued function. Then there exists a sequence $\{\phi_m\}$ of simple functions such that $\phi_m \to f$ as $m \to \infty$ on $X$. By the result obtained in (a), for each simple function $\phi_m$ and every $n \in N$, there exists a closed set $X_{m,n} \subset X$ such that $\phi_m$ is continuous on $X_{m,n}$ and $\mu\left(X \setminus X_{m,n}\right) < \frac{1}{n}$. Without loss of generality, we can assume that the sequence $\{X_{m,n}\}$ of closed sets is increasing with respect to $n$ for each $m$ (otherwise, we can take $\cup_{i=1}^{n} X_{m,i}$ instead of $X_{m,n}$ and noting that $\phi_m$ is a simple function, it remains continuous on $\cup_{i=1}^{n} X_{m,i}$). Since

$$\left(X \setminus X_{m,n}\right) \searrow \bigcap_{n=1}^{\infty} \left(X \setminus X_{m,n}\right) \text{ as } n \to \infty,$$

we have $\mu\left(\cap_{n=1}^{\infty}\left(X \setminus X_{m,n}\right)\right) = \lim_{n \to \infty} \mu\left(X \setminus X_{m,n}\right) = 0$. By using Lemma 1, for any $n \in N$, there exists a sequence $\{\tau_n\} \subset \Theta$ such that

$$\mu\left(\bigcup_{m=1}^{\infty} \left(X \setminus X_{m,\,\tau_n(m)}\right)\right) < \frac{1}{n},$$

that is, $\mu\left(X \setminus \cap_{m=1}^{\infty} X_{m,\,\tau_n(m)}\right) < \frac{1}{n}$. Since the double sequence $\{X \setminus X_{m,n}\}$ is decreasing in $n \in N$ for each $m \in N$, without any loss of generality, we may assume that for fixed $m \in N$, $\tau_1(m) < \tau_2(m) < \cdots < \tau_n(m) < \cdots$. Put $H_n = \cap_{m=1}^{\infty} X_{m,\,\tau_n(m)}$, then we have a sequence $\{H_n\}$ of closed sets satisfying $H_1 \subset H_2 \subset \cdots$. Since

$$\left(X \setminus H_n\right) \searrow \left(X \setminus \bigcup_{n=1}^{\infty} H_n\right) \text{ as } n \to \infty,$$

we have

$$\mu\left(X \setminus \bigcup_{n=1}^{\infty} H_n\right) = \lim_{n \to \infty} \mu\left(X \setminus H_n\right) = 0.$$

Noting that $\phi_m$ is continuous on $X_{m,n}$ and $H_n \subset X_{m,\,\tau_n(m)}$, $\phi_m$ is continuous on $H_n$ for every $m \in N$.

On the other hand, since $\phi_m \to f$ as $m \to \infty$ on $X$, by Theorem 2, there exists a sequence $\{K_n\}$ of closed sets such that $\mu\left(X \setminus K_n\right) < \frac{1}{n}$ and $\{\phi_m\}$ converges to $f$ uniformly on $K_n$ for every $n \in N$. We may assume that $\{K_n\}$ is increasing in $n$ for each $m$, without any loss of generality. Since $\left(X \setminus K_n\right) \searrow \left(X \setminus \cup_{n=1}^{\infty} K_n\right)$ as $n \to \infty$, we have $\mu\left(X \setminus \cup_{n=1}^{\infty} K_n\right) = \lim_{n \to \infty} \mu\left(X \setminus K_n\right) = 0$. Then $\{\phi_m\}$ converges to $f$ uniformly on $K_n$ for every $n \in N$. Consider the sequence $\{(X \setminus H_n) \cup (X \setminus K_n)\}$, then we have

$$(X \smallsetminus H_n) \cup (X \smallsetminus K_n) \searrow \left( X \smallsetminus \bigcup_{n=1}^{\infty} H_n \right) \cup \left( X \smallsetminus \bigcup_{n=1}^{\infty} K_n \right) \text{ as } n \to \infty.$$

Since $\mu$ is weakly null-additive, we have

$$\mu \left( \left( X \smallsetminus \bigcup_{n=1}^{\infty} H_n \right) \cup \left( X \smallsetminus \bigcup_{n=1}^{\infty} K_n \right) \right) = 0.$$

Moreover, since $\mu$ is continuous from above, for any $\varepsilon > 0$, there exists an $n_0$ such that $\mu \left( (X \smallsetminus H_{n_0}) \cup (X \smallsetminus K_{n_0}) \right) < \varepsilon$. Put $F_\varepsilon = H_{n_0} \cap X_{n_0}$, then $F_\varepsilon$ is a closed set and $\mu (X \smallsetminus F_\varepsilon) < \varepsilon$. We show that $f$ is continuous on $F_\varepsilon$. In fact, $F_\varepsilon \subset H_{n_0}$ and $\phi_m$ is continuous on $H_{n_0}$, therefore $\phi_m$ is continuous on $F_\varepsilon$ for each $m \in N$. Noting that $\{\phi_m\}$ converges to $f$ uniformly on $F_\varepsilon$, $f$ is continuous on $F_\varepsilon$. □

## 5  Applications of Lusin's Theorem

In [5], Li and Yasuda give applications Lusin's Theorem for fuzzy measure. In this section, we prove the same results when the measure is weakly non-additive Borel, continuous from above and has property $(S)$ on metric spaces. The proof is the same as that of [5] and is given for completeness.

Let $\mu$ be a non-additive measure on $\mathcal{B}(X)$ and $f$ a non-additive real valued measurable function on $X$. We define the Sugeno Integral of $f$ on $X$ with respect to $\mu$, denoted by $(S) \int f d\mu$, as follows:

$$(S) \int f d\mu = \sup_{0 \le \alpha < +\infty} [\alpha \wedge \mu(\{x \in X \mid f(x) \ge \alpha\})].$$

The Choquet integral of $f$ on $X$ with respect to $\mu$, denoted by $(C) \int f d\mu$, is defined by

$$(C) \int f d\mu = \int_0^{\infty} \mu (\{x \in X \mid f(x) > t\}) \, dt,$$

where the right side integral is Lebesgue integral.

We say that a sequence $\{f_n\}$ of measurable function converges to $f$ in non-additive measure $\mu$, and denote it by $f_n \overset{\mu}{\to} f$, if for any $\varepsilon > 0$, $\lim_{n\to\infty} \mu (\{x \in X \mid |f_n(x) - f(x)| \ge \varepsilon\}) = 0$.

**Theorem 4.** *Let $\mu$ be a non-additive measure on $\mathcal{B}(X)$ which is weakly null-additive, continuous from above and has property $(S)$ and $f$ a non-additive real valued measurable function on $\mathcal{B}(X)$, then there exists a sequence $\{\phi_n\}$ of continuous function on $X$ such that $\phi_n \overset{\mu}{\to} f$. Moreover, if $|f| \le M$, then $|\phi| \le M$, $n \in N$.*

*Proof.* For every $n \in N$, by Theorem 3, there exists closed subset $F_n$ of $X$ such that $f$ is continuous on $F_n$ and $\mu(X \smallsetminus F_n) < \frac{1}{n}$. By Tietze's extension Theorem [8], for every $n \in N$, there exists continuous function $\phi_n$ on $X$ such that $\phi_n(x) = f(x)$ for $x \in F_n$ and if $|f| \leq M$, then $|\phi_n| \leq M$. We show that $\{\phi_n\}$ converges to $f$ in the non-additive measure. In fact, for any $\varepsilon > 0$, we have $\{x \in X \mid |\phi_n(x) - f(x)| \geq \varepsilon\} \subset X \smallsetminus F_n$ and thus $\mu(\{x \in X \mid |\phi_n(x) - f(x)| \geq \varepsilon\}) \leq \mu(X \smallsetminus F_n) < \frac{1}{n}$, for $n \in N$. Therefore we have $\lim_{n \to \infty} \mu(\{x \in X \mid |\phi_n(x) - f(x)| \geq \varepsilon\}) = 0$. $\qquad\square$

**Theorem 5.** *Let $\mu$ be a non-additive measure on $\mathcal{B}(X)$ which is weakly null-additive, continuous from above and has property $(S)$ and $f$ a non-additive real valued measurable function on $X$, then there exists a sequence $\{\phi_n\}$ of continuous function such that*

$$(S) \lim_{n \to \infty} \int |\phi_n - f| d\mu = 0.$$

*Moreover, if $|f| \leq M$, then $|\phi_n| \leq M$, $n \in N$ and*

$$(C) \lim_{n \to \infty} \int |\phi_n - f| d\mu = 0.$$

*Proof.* By Theorem 4, there exists a sequence $\{\phi_n\}$ of continuous function such that $\phi_n \overset{\mu}{\to} f$. By Theorem 7.4 in [11], we have $\lim_{n \to \infty}(S) \int |\phi_n - f| d\mu = 0$. By Theorem 4, if $|f| \leq M$, then $|\phi_n| \leq M$ $n \in N$. Put

$$g_n(t) = \mu(\{x \in X \mid |\phi_n(x) - f(x)| > t\}), \, t \in [0, \infty)$$

since $\phi_n \overset{\mu}{\to} f$, we have $g_n(t) \to 0$ $\mu$-a.e on $[0, \infty)$ as $n \to \infty$. Note that $|g_n(t)| \leq \mu(X) < \infty$, we have $g_n(t) = 0$ for any $t > 2M$, $n \in N$. Applying the Bounded Convergence theorem in Lebesgue integral theory [8] to the function sequence $\{g_n(t)\}$, we have

$$\int_0^\infty g_n(t)dt = \int_0^{2M} g_n(t)dt \to 0 \text{ as } n \to \infty.$$

Then we have $\lim_{n \to \infty}(C) \int |\phi_n - f| d\mu = 0$. $\qquad\square$

In the following argument, let $X = R^1$. Then the following theorems are also proved, see [5].

**Theorem 6.** *Let $\mu$ be a non-additive measure on $\mathcal{B}(X)$ which is weakly null-additive, continuous from above and has property $(S)$ and $f$ a non-additive real valued measurable function on $[a, b]$, then there exists a sequence $\{P_n\}$ of polynomials on $[a, b]$ such that $P_n \overset{\mu}{\to} f$ on $[a, b]$. Moreover, if $|f| \leq M$, then $|P_n| \leq M + 1$, $n \in N$.*

*Proof.* Consider the problem on the reduced non-additive measure space $([a,b], [a,b] \cap \mathcal{B}(X), \mu)$. By Theorem 4, there exists a sequence $\{\phi_n\}$ of continuous function on $[a,b]$ such that $\phi_n \xrightarrow{\mu} f$ on $[a,b]$. There exists a subsequence $\{\phi_{n_k}\}$ of $\{\phi_n\}$ such that $\mu(\{x \in X \mid |\phi_{n_k}(x) - f(x)| \geq \frac{1}{2k}\}) < \frac{1}{k}$ for any $k \in N$. Since $\phi_{n_k}$ is continuous on $[a,b]$, by using Weierstrass's theorem [8], for any $k \in N$, there exists a polynomial $P_k$ on $[a,b]$ such that $|P_k(x) - \phi_{n_k}(x)| < \frac{1}{2k}$ for any $x \in [a,b]$. Then for any $k \in N$, we have $\{x \in X \mid |P_k(x) - \phi_{n_k}(x)| \geq \frac{1}{2k}\} = \emptyset$. Noting that $\{x \in X \mid |P_k(x) - \phi_{n_k}(x)| \geq \frac{1}{k}\} \subset A_k \cup B_k$, where $A_k = \{x \in X \mid |P_k(x) - \phi_{n_k}(x)| \geq \frac{1}{2k}\}$ and $B_k = \{x \in X \mid |\phi_{n_k}(x) - f(x)| \geq \frac{1}{2k}\}$. Since $A_k \cup B_k = \{x \in X \mid |\phi_{n_k}(x) - f(x)| \geq \frac{1}{2k}\}$, we have $\mu(\{x \in X \mid |P_k(x) - f(x)| \geq \frac{1}{k}\}) < \frac{1}{k}$. Moreover we prove that $P_n \xrightarrow{\mu} f$ on $[a,b]$. In fact, any $\varepsilon > 0$, there exists an $n_0$ such that $\frac{1}{n_0} < \varepsilon$, then for any $n \geq n_0$, we have $\{x \in X \mid |P_n(x) - f(x)| \geq \varepsilon\} \subset \{x \in X \mid |P_n(x) - f(x)| \geq \frac{1}{n}\}$, and $\mu(\{x \in X \mid |P_n(x) - f(x)| \geq \varepsilon\}) \leq \mu(\{x \in X \mid |P_n(x) - f(x)| \geq \frac{1}{n}\}) < \frac{1}{n}$. Therefore $P_n \xrightarrow{\mu} f$. From above, if $|f| \leq M$, then $|\phi_{n_k}| \leq M$. Since for every $P_k$, $|P_k(x) - \phi_{n_k}(x)| < \frac{1}{2k}$ for all $x \in [a,b]$, we have $|P_n| \leq M+1$, $n \in N$. $\square$

We can prove the following result in a similar way to the proof of Theorem 5.

**Theorem 7.** *Let $\mu$ be a non-additive measure on $\mathcal{B}(X)$ which is weakly null-additive, continuous from above and has property $(S)$ and $f$ a non-additive real valued measurable function on $[a,b]$, then there exists a sequence $\{P_n\}$ of polynomials on $[a,b]$ such that*

$$(S) \lim_{n \to \infty} \int |P_n - f| d\mu - 0.$$

*Moreover, if $|f| \leq M$, then $|P_n| \leq M+1$, $n \in N$ and*

$$(C) \lim_{n \to \infty} \int |P_n - f| d\mu = 0.$$

Similarly, we can prove the following.

**Theorem 8.** *Let $\mu$ be a non-additive measure on $\mathcal{B}(X)$ which is weakly null-additive, continuous from above and has property $(S)$ and $f$ a non-additive real valued measurable function on $[a,b]$, then there exists a sequence $\{s_n\}$ of step functions on $[a,b]$ such that $s_n \xrightarrow{\mu} f$ and*

$$(S) \lim_{n \to \infty} \int |s_n - f| d\mu = 0.$$

*Moreover, if $|f|$ is Choquet integrable, then so is $|s_n|$ and*

$$(C) \lim_{n \to \infty} \int |s_n - f| d\mu = 0.$$

**Acknowledgements.** The authors would like to express their hearty thanks to referee.

# References

1. Denneberg, D.: Non-Additive Measure and Integral, 2nd edn. Kluwer Academic Publishers, Dordrecht (1997)
2. Jiang, Q., Suzuki, H.: Fuzzy measures on metric spaces. Fuzzy Sets and Systems 83, 99–106 (1996)
3. Jiang, Q., Wang, S., Ziou, D.: A further investigation for fuzzy measures on metric spaces. Fuzzy Sets and Systems 105, 293–297 (1999)
4. Kawabe, J.: Regularity and Lusin's theorem for Riesz space-valued fuzzy measures. Fuzzy Sets and Systems 158, 895–903 (2007)
5. Li, J., Yasuda, M.: Lusin's theorems on fuzzy measure spaces. Fuzzy Sets and Systems 146, 121–133 (2004)
6. Li, J., Yasuda, M., Jiang, Q., Suzuki, H., Wang, Z., Klir, G.J.: Convergence of sequence of measurable functions on fuzzy measure spaces. Fuzzy Sets and Systems 87, 317–323 (1997)
7. Pap, E.: Null-Additive Set Functions. Kluwer Academic Publishers, Dordrecht (1995)
8. Royden, H.L.: Real Analysis. Macmillan, New York (1966)
9. Song, J., Li, J.: Regularity of null-additive fuzzy measure on metric spaces. International Journal of General Systems 32, 271–279 (2003)
10. Sun, Q.: Property (S) of fuzzy measure and Riesz's theorem. Fuzzy Sets and Systems 62, 117–119 (1994)
11. Wang, Z., Klir, G.J.: Fuzzy Measure Theory. Plenum Press, New York (1992)
12. Watanabe, T., Tanaka, T.: On regularity for non-additive measure (preprint)
13. Wu, C., Ha, M.: On the regularity of the fuzzy measure on metric fuzzy measure spaces. Fuzzy Sets and Systems 66, 373–379 (1994)
14. Wu, J., Wu, C.: Fuzzy regular measures on topological spaces. Fuzzy Sets and Systems 119, 529–533 (2001)

# Multiple-Output Choquet Integral Models and Their Applications in Classification Methods

Eiichiro Takahagi

**Abstract.** Two types of the multiple-output Choquet integral models are defined. Vector-valued Choquet integral models are vector-valued functions calculated by $m$ times Choquet integral calculations with respect to the $m$-th fuzzy measure of a fuzzy measure vector. Logical set-function-valued Choquet integral models are set-function-valued functions that can be used in classification. The set function shows singleton, overlap, and unclassifiable degrees. The sum value of the set function is equal to 1. A method for transformation from vector-valued Choquet integral models to logical set-function-valued Choquet integral models is proposed.

**Keywords:** Choquet Integral, Vector-valued function, Classification, Overlap and unclassifiable degrees.

## 1 Introduction

Choquet integral [1] models are useful comprehensive models [3]. In [6], Choquet integral models are extended to multiple-output models; thus, for a input vector $(x_1, \ldots, x_n)$ and fuzzy measures $\mu_1, \ldots, \mu_m$, by calculating $m$ times Choquet integrals, $m$ dimensional output vector $(y_1, \ldots, y_m)$ is obtained. The properties of the output vector are dependent on the restrictions on the fuzzy measures. For example, if $\sum \mu_j(A) = 1, \forall A$, then $\sum y_j = 1$. Because of this property, Choquet integral models can be used in classification. In sections 2 and 3, we present the definitions and properties of the multi-output Choquet integral models. In section 4, we propose a logical set-function-valued Choquet integral model and a method for the transformation from vector-valued Choquet integral models to set-function-valued Choquet integral models.

Eiichiro Takahagi
School of Commerce, Senshu University, 2-1-1, Tamaku, Kawasaki, 214-8580, Japan
e-mail: takahagi@isc.senshu-u.ac.jp

S. Li (Eds.): Nonlinear Maths for Uncertainty and its Appli., AISC 100, pp. 93–100.
springerlink.com &copy; Springer-Verlag Berlin Heidelberg 2011

## 2  Definitions

### 2.1  Choquet Integral

**Definition 1.** $X = \{1, \ldots, n\}$ is the set of evaluation items ($n$: number of evaluation items), $x_i \in \mathbb{R}^+$ is the input value of the $i^{\text{th}}$ item, and $y$ is the comprehensive evaluation value.

**Definition 2.** A non-monotone fuzzy measure $\mu$ is defined as

$$\mu : 2^X \to \mathbb{R} \ , \ \mu(\emptyset) = 0. \tag{1}$$

**Definition 3.** The Choquet integral with respect to $\mu$ is defined as

$$y = f_\mu^C(x_1, \ldots, x_n) \equiv \sum_{i=1}^n [x_{\sigma(i)} - x_{\sigma(i+1)}]\mu(\{\sigma(1), \ldots, \sigma(i)\}), \tag{2}$$

where $\sigma$ is a permutation such that $x_{\sigma(1)} \geq \ldots \geq x_{\sigma(n)}$ and $X = \{\sigma(1), \ldots, \sigma(n)\}$, and let $x_{\sigma(n+1)} = 0$.

### 2.2  Logical Choquet Integral

The logical Choquet integral [5] was introduced to deal with fuzzy values (the interval $[0, 1]$) and the fuzzy switching functions. The input and output values, the domain of fuzzy measures (set functions), and the integration range are the interval $[0, 1]$.

**Definition 4.** $x_i^\sharp \in [0, 1]$ is the input value of the $i^{\text{th}}$ item, and $y^\sharp$ is the comprehensive evaluation value.

**Definition 5.** The extended fuzzy measure $\mu^\sharp$ is a set function to the interval $[0, 1]$, that is

$$\mu^\sharp : 2^X \to [0, 1]. \tag{3}$$

**Definition 6.** The extended Choquet integral (the Choquet integral with respect to set functions) is defined as

$$y^\sharp = f_{\mu^\sharp}^{EC}(x_1^\sharp, \ldots, x_n^\sharp) \equiv \sum_{i=0}^n [x_{\sigma(i)}^\sharp - x_{\sigma(i+1)}^\sharp]\mu^\sharp(\{\sigma(1), \ldots, \sigma(i)\}), \tag{4}$$

where $x_{\sigma(0)}^\sharp = 1$, $x_{\sigma(n+1)}^\sharp = 0$ and $\{\sigma(1), \ldots, \sigma(i)\} = \emptyset$ when $i = 0$.

The extended Choquet integral can be calculated by using the Choquet integral as follows:

$$f_{\mu^\sharp}^{EC}(x_1^\sharp, \ldots, x_n^\sharp) = f_\mu^C(x_1^\sharp, \ldots, x_n^\sharp) + \mu^\sharp(\emptyset), \tag{5}$$

where $\mu(A) = \mu^\sharp(A) - \mu^\sharp(\emptyset), \forall A \in 2^X$.

## 2.3 Vector-Valued Choquet Integral Model

Vector-valued Choquet integral model [6] is an extension of the product of a matrix and a vector. Let us input the vectors $\mathbf{x} = (x_1, \ldots, x_n)$ and $\mathbf{x}^\sharp = (x_1^\sharp, \ldots, x_n^\sharp)$.

**Definition 7.** Vector-valued Choquet integral models are vector-valued functions calculated by $m$ times Choquet integral calculations with respect to the $m^{\text{th}}$ fuzzy measure vector.

$$\mathbf{y} = f_{\boldsymbol{\mu}}(\mathbf{x}), \quad \text{where } y_j = f_{\mu_j}^C(x_1, \ldots, x_n), j = 1, \ldots, m. \tag{6}$$

**Definition 8.** Logical vector-valued Choquet integral models are defined as

$$\mathbf{y}^\sharp = f_{\boldsymbol{\mu}^\sharp}(\mathbf{x}^\sharp), \quad \text{where } y_j^\sharp = f_{\mu_j^\sharp}^{EC}(x_1^\sharp, \ldots, x_n^\sharp), j = 1, \ldots, m. \tag{7}$$

# 3 Properties of the Vector-Valued Choquet Integral

*Property 1.* If $\sum_{j=1}^m \mu_j(A) = | A |, \forall A \in 2^X$, then for any $\mathbf{x} \in \mathbb{R}^{+n}, \mathbf{y} = f_{\boldsymbol{\mu}}(\mathbf{x})$ satisfies the property, $\sum_{j=1}^m y_j = \sum_{i=1}^n x_i$, where $| A |$ is the number of elements of the set $A$.

*Property 2.* If

$$\sum_{j=1}^m \mu_j^\sharp(A) = 1, \forall A \in 2^X \tag{8}$$

for any $\mathbf{x}^\sharp \in [0,1]^n$, $\mathbf{y}^\sharp = f_{\boldsymbol{\mu}^\sharp}^{EC}(\mathbf{x}^\sharp)$ satisfies the properties, $y_j^\sharp \in [0,1], j = 1, \ldots, m$ and

$$\sum_{j=1}^m y_j^\sharp = 1. \tag{9}$$

# 4 Classification

Fuzzy integral models are used in the classification method such as [4]. In this section, we propose a normalised outputs model by using the Choquet integral model.

## 4.1 Classification by the Vector-Valued Choquet Integral

It is useful to apply the logical vector-valued Choquet integral to the classification models. The classification rules are given by the extended fuzzy measures derived from the fuzzy switching functions, linear functions, etc. Let $Y = \{1, \ldots, m\}$ be a set of classes. The fuzzy measure values are assigned from the functions $g_j : \{0, 1\}^n \to [0, 1], j \in Y$. The fuzzy switching functions that are the linear functions shown in section 5 are examples of $g_j$.

$$\mu_j^\sharp(A) = g_j(\mathbf{x}^{\sharp A}), \quad \text{where } x_i^{\sharp A} = \begin{cases} 1 & \text{if } i \in A \\ 0 & \text{otherwise.} \end{cases}$$

Applying the vector-valued Choquet integral $\mathbf{y}^\sharp = f_{\boldsymbol{\mu}^\sharp}^{EC}(\mathbf{x}^\sharp)$, $\mathbf{y}^\sharp \in [0, 1]^m$ is the degree of belogingness of each class. Choquet integral is a linear interpolation function among binary inputs [2].

## 4.2  Logical Set-Function-Valued Choquet Integral Model

If the fuzzy measures satisfy the equation (8), then the output vector $\mathbf{y}^\sharp$ satisfies $\sum_j y_j^\sharp = 1$. As the conditions are not always satisfied when the fuzzy measures are identified individually, a logical set-function-valued Choquet integral model that includes overlap classification and unclassified degrees is proposed. The rules of the classification are $2^m$ fuzzy measures,

$$\nu_B^\sharp : 2^X \to [0, 1], \ \forall B \in 2^Y, \tag{10}$$

and $z_B^\sharp$ are calculated by using the extended Choquet integrals as follows:

$$z_B^\sharp = f_{\nu_B^\sharp}^{EC}(x_1^\sharp, \ldots, x_n^\sharp), \ \forall B \in 2^Y. \tag{11}$$

**Definition 9.** The function for the transformation from $\mu_j^\sharp(A)$ ($j \in Y$) to $\nu_B^\sharp$ ($B \in 2^Y$), $\forall A \in 2^X$ is defined as follows:

$$\nu_B^\sharp(A) = \begin{cases} \sum_{B \subseteq C} [(-1)^{|C \setminus B|} \min_{j \in C} \mu_j^\sharp(A)] & \text{if } B \neq \emptyset \\ 1 - \sum_{B \in (2^Y \setminus \emptyset)} \nu_B^\sharp(A) & \text{otherwise.} \end{cases} \tag{12}$$

The sum of the transformed $\nu_B^\sharp$ is equal to 1 even if $\mu^\sharp$ does not satisfy the equation (8).

*Property 3.* For any $\mu_1^\sharp, \ldots, \mu_m^\sharp$, the transformed $\nu_B^\sharp$ represented by equation (12) satisfies

$$\sum_{B \in 2^Y} \nu_B^\sharp(A) = 1, \ \forall A \in 2^X. \tag{13}$$

*Property 4.* For any $\mu_1^\sharp, \ldots, \mu_m^\sharp$ and $x_1^\sharp, \ldots, x_n^\sharp$, using the transformed $\nu_B^\sharp$ by equation (12),

$$\sum_{B \in 2^Y} z_B^\sharp = \sum_{B \in 2^Y} f_{\nu_B^\sharp}^{EC}(x_1^\sharp, \ldots, x_n^\sharp) = 1. \tag{14}$$

**Definition 10.** The whole fuzzy measure $\nu_B^\diamond$ is defined as follows:

$$\nu_B^\diamond(A) = \begin{cases} \displaystyle\sum_{C \subseteq B, C \neq \emptyset} \nu_C^\sharp(A) & \text{if } B \in (2^Y \setminus \emptyset) \\ \nu_\emptyset^\sharp(A) & \text{otherwise .} \end{cases} \tag{15}$$

If $\mid B \mid = 1$, then $z_B^\sharp$ shows the singleton degree of the class. For example, $z_{\{2\}}^\sharp$ shows the singleton degree of class 2; this singleton degree of class 2 does not include the overlap degree of both classes 1 and 2. If $\mid B \mid > 1$, then $z_B^\sharp$ shows the overlap degree of $B$. For example, $z_{\{1,3\}}^\sharp$ shows the overlap degree that belongs to both classes 1 and 3. The whole degree which belongs to classes 1 and 3 is $z_{\{1,3\}}^\diamond = f_{\nu_{\{1,3\}}^\diamond}^{EC}(x_1^\sharp, \ldots, x_n^\sharp) = z_{\{1,3\}}^\sharp + z_{\{1\}}^\sharp + z_{\{3\}}^\sharp$. $z_\emptyset^\sharp$ shows the degree that does not belong to any of the classes.

*Property 5.* For any $(\mu_1^\sharp, \ldots, \mu_m^\sharp)$ and $(x_1^\sharp, \ldots, x_n^\sharp)$, the transformed $\nu_B^\sharp, \forall B \in 2^Y$ by equation (12) and $z_B^\sharp = f_{\nu_B^\sharp}^{EC}(x_1^\sharp, \ldots, x_n^\sharp), \forall B \in 2^Y$ have the following properties:

$$\mu_i^\sharp(A) = \sum_{i \in B} \nu_B^\sharp(A), \forall A \in 2^X \tag{16}$$

$$y_i^\sharp = \sum_{i \in B} z_B^\sharp \tag{17}$$

$$\nu_\emptyset^\sharp(A) = 1 - \max_{j=1,\ldots,m} \mu_j^\sharp(A), \forall A \in 2^X \tag{18}$$

$$z_\emptyset^\sharp = 1 - \max_{j=1,\ldots,m} y_j^\sharp \tag{19}$$

$$\nu_B^\sharp(A) \in [0,1], \forall A \in 2^X, \forall B \in 2^Y \tag{20}$$

$$z_B^\sharp \in [0,1], \forall B \in 2^Y \tag{21}$$

## 5  Numerical Examples

### 5.1  *Linear Functions*

First example is a recommendation system for selecting either arts course or science course in high schools. Input values are the language score ($x_1^\sharp$) and the mathematics score ($x_2^\sharp$). Output values are the arts course ($y_1^\sharp$) and the science course ($y_2^\sharp$). The classification functions are represented by the following linear functions:

$$g_1(x_1, x_2) = 0.5x_1 + 0.4x_2 + 0.1 \tag{22}$$
$$g_2(x_1, x_2) = \quad 0.3x_1 + 0.7x_2. \tag{23}$$

**Table 1** Fuzzy Measures corresponding to equations (22) and (23)

| A (Sets) | $\mu_1^\sharp(A)$ | $\mu_2^\sharp(A)$ | $\nu_{\{1\}}^\sharp(A)$ | $\nu_{\{2\}}^\sharp(A)$ | $\nu_{\{1,2\}}^\sharp(A)$ | $\nu_\emptyset^\sharp(A)$ | $\nu_{\{1,2\}}^\circ(A)$ |
|---|---|---|---|---|---|---|---|
| $\emptyset$ | 0.1 | 0 | 0.1 | 0 | 0 | 0.9 | 0.1 |
| $\{1\}$ | 0.6 | 0.3 | 0.3 | 0 | 0.3 | 0.4 | 0.6 |
| $\{2\}$ | 0.5 | 0.7 | 0 | 0.2 | 0.5 | 0.3 | 0.7 |
| $\{1,2\}$ | 1.0 | 1.0 | 0 | 0 | 1.0 | 0 | 1.0 |

By using the equation (10),(12), and (15), we obtain the values listed in table 1; these values satisfy the properties 3, 4, and 5. Figure 1 is the graph when $x_1 = 0.4$. As $x_2$ increases, $z_{\{1\}}^\sharp$ decreases because $z_{\{1\}}^\sharp$ is the sigleton degree of the art course, and the mathematics weights of art course (equation (22)) is lower than science course (equation (23)).

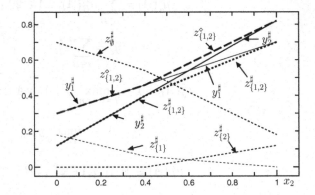

**Fig. 1** Outputs corresponding to the Art and Science model ($x_1^\sharp = 0.4$)

## 5.2 Fuzzy Switching Function

The following are fuzzy switching functions with $n = 4$ and $m = 2$ and they are defiend as

$$g_1(x_1, x_2, x_3, x_4) = (0.8 \wedge x_1 \wedge x_2) \vee (0.6 \wedge x_2 \wedge x_4) \vee (x_1 \wedge x_2 \wedge x_4) \tag{24}$$
$$g_2(x_1, x_2, x_3, x_4) = (0.7 \wedge x_1 \wedge x_3) \vee (0.8 \wedge x_3 \wedge x_4) \vee (x_1 \wedge x_3 \wedge x_4) \tag{25}$$

Those functions indicate the following:

- If $x_1$, $x_2$, and $x_4$ are fulfilled, then the object is classified as a class 1 object with degree 1.
- If $x_1$ and $x_2$, or $x_2$ and $x_4$ are fulfilled, then the object is classified as a class 1 object with a certain degree.
- If $x_1$, $x_3$, and $x_4$ are fulfilled, then the object is classified as a class 2 object with degree 1.
- If $x_1$ and $x_3$, or $x_3$ and $x_4$ are fulfilled, then the object is classified as a class 2 object with a certain degree.

**Table 2** Fuzzy Measures corresponding to equations (24) and (25)

| $A$ (Sets) | $\mu_1^\sharp(A)$ | $\mu_2^\sharp(A)$ | $\nu_{\{1\}}^\sharp(A)$ | $\nu_{\{2\}}^\sharp(A)$ | $\nu_{\{1,2\}}^\sharp(A)$ | $\nu_{\emptyset}^\sharp(A)$ | $\nu_{\{1,2\}}^\circ(A)$ |
|---|---|---|---|---|---|---|---|
| $\{\}$ | 0 | 0 | 0 | 0 | 0 | 1 | 0 |
| $\{1\}$ | 0 | 0 | 0 | 0 | 0 | 1 | 0 |
| $\{2\}$ | 0 | 0 | 0 | 0 | 0 | 1 | 0 |
| $\{1,2\}$ | 0.8 | 0 | 0.8 | 0 | 0 | 0.2 | 0.8 |
| $\{3\}$ | 0 | 0 | 0 | 0 | 0 | 1 | 0 |
| $\{1,3\}$ | 0 | 0.7 | 0 | 0.7 | 0 | 0.3 | 0.7 |
| $\{2,3\}$ | 0 | 0 | 0 | 0 | 0 | 1 | 0 |
| $\{1,2,3\}$ | 0.8 | 0.7 | 0.1 | 0 | 0.7 | 0.2 | 0.8 |
| $\{4\}$ | 0 | 0 | 0 | 0 | 0 | 1 | 0 |
| $\{1,4\}$ | 0 | 0 | 0 | 0 | 0 | 1 | 0 |
| $\{2,4\}$ | 0.6 | 0 | 0.6 | 0 | 0 | 0.4 | 0.6 |
| $\{1,2,4\}$ | 1 | 0 | 1 | 0 | 0 | 0 | 1 |
| $\{3,4\}$ | 0 | 0.8 | 0 | 0.8 | 0 | 0.2 | 0.8 |
| $\{1,3,4\}$ | 0 | 1 | 0 | 1 | 0 | 0 | 1 |
| $\{2,3,4\}$ | 0.6 | 0.8 | 0 | 0.2 | 0.6 | 0.2 | 0.8 |
| $X$ | 1 | 1 | 0 | 0 | 1 | 0 | 1 |

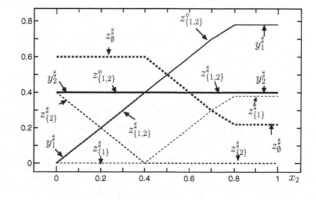

**Fig. 2** Outputs corresponding to the fuzzy switching functions ($x_1 = 0.8$, $x_3 = 0.4$, and $x_4 = 0.7$)

Table 2 shows the fuzzy measures of the fuzzy switching functions. Figure 2 shows graphs for the case in which $x_1 = 0.8$, $x_3 = 0.4$, and $x_4 = 0.7$. When $0 \le x_2 \le 0.4$, that is $x_2 \le x_3$, the rule $g_1$ is given priority over $g_2$. Therefore, $z_{\{1\}}^\# = 0$.

## 6  Conclusions

We show the method for the transformation from vector-valued classification functions by Choquet integrals to set-function-valued functions, which can express singleton, overlap, or unclassifiable degrees. However, as the examples were artificial models, real applications have to be developed.

## References

1. Choquet, G.: Theory of capacities. Annales de l'Institut Fourier 5, 131–295 (1954)
2. Grabisch, M.: The Choquet integral as a linear interpolator. In: 10th Int. Conf. on Information Processing and Management of Uncertainty in Knowledge-Based Systems (IPMU 2004), Perugia, Italy, pp. 373–378 (2004)
3. Murofushi, T., Sugeno, M.: A theory of fuzzy measures: representations, the Choquet integral, and null sets. J. Math. Anal. Appl. 159-2, 532–549 (1991)
4. Tahani, H., Keller, J.M.: Information fusion in computer vision using the fuzzy integral. IEEE Trans. on Systems, Man and Cybernetics 20-3, 733–741 (1990)
5. Takahagi, E.: Fuzzy integral based fuzzy switching functions. In: Peters, J.F., Skowron, A., Dubois, D., Grzymała-Busse, J.W., Inuiguchi, M., Polkowski, L. (eds.) Transactions on Rough Sets II. LNCS, vol. 3135, pp. 129–150. Springer, Heidelberg (2004)
6. Takahagi, E.: A Choquet integral model with multiple outputs and its application to classifications. Journal of Japan Society for Fuzzy Theory and Intelligent Informatics 22, 481–484 (2010)

# On Convergence Theorems of Set-Valued Choquet Integrals

Hongxia Wang and Shoumei Li

**Abstract.** The article aims at discussing the Choquet integrals of set-valued random variables with respect to capacities. We firstly state representation theorems and subadditive property of set-valued Choquet integrals. Then we mainly prove Fatou's Lemmas, Lesbesgue dominated convergence theorem and monotone convergence theorems of set-valued Choquet integrals under the weaker conditions than that in previous works.

**Keywords:** Set-valued Choquet integral, Set-valued random variable, Capacity, Kuratowski convergence.

## 1 Introduction

It is well known that classical probability theory and statistical methods are powerful tools for dealing with stochastic phenomena with many applications. However, there are many uncertain phenomena which can not be easily modeled by using classical probability theory in finance and economics. The famous counterexamples are the Allais paradox [1] and the Ellsberg paradox [5]. In 1953, Choquet introduced concepts of capacities and the Choquet integral [3]. Capacity is non-additive measure and the Choquet integral is one kind of nonlinear expectations. Many papers developed the Choquet theory and its applications. We would like to thank the excellent overview paper written by Wang and Yan [14]. Here we specially mention that Schmeidler introduced the Choquet expected utility (CEU) [13]. Under the framework of CEU theory, Wang and Yan gave the solutions to Allais' paradox and Ellsberg's paradox. On the other hand, there is also another kind of uncertain phenomena, which can be described by a set-valued random variable

Hongxia Wang and Shoumei Li
Department of Applied Mathematics, Beijing University of Technology, Beijing, 100124, P.R. China
e-mail: xiahongwang@emails.bjut.edu.cn, lisma@bjut.edu.cn

S. Li (Eds.): Nonlinear Maths for Uncertainty and its Appli., AISC 100, pp. 101–108.
springerlink.com                    © Springer-Verlag Berlin Heidelberg 2011

(also called random sets, multifunctions, correspondences in literature) and
the Aumann integral (cf. [2]). Aumann and others used multifunctions to
discuss the competitive equilibria problem in economics. After that, the theory of set-valued random variables with its applications was developed very
deeply and extensively (e.g. [6,10,11]).

There are many complex systems in which we have to deal with two uncertainty phenomena at the same time. Jiang and Kwon [7] introduced the
concept of set-valued Choquet integrals and discussed some properties of
this kind of integrals. Jang et al. studied some convergence theorems of a
sequence of set-valued Choquet integrals in [8,9]. Zhang et al. revised some
mistakes in above papers and proved convergence theorems for set-valued
Choquet integrals with respect to an $m$-continuous and continuous capacity
in the Kuratowski sense in [15]. But it is necessary to point out that we
may weaken the sufficient conditions for some convergence results in [8,9,15],
which is our main purpose.

This paper is organized as follows. Section 2 is for preliminaries and notions
of capacity, Choquet integral, set-valued random variable, and set-valued
Choquet integral. In Section 3, we shall discuss representation theorems and
subadditive property, and then mainly build the convergence theorems of set-
valued Choquet integrals under the weaker conditions than that in previous
works.

## 2   Preliminaries and Notations

Throughout this paper, assume that $(\Omega, \mathcal{F})$ is a measurable space, $\mathbb{R}$ ($\mathbb{R}^+$)
is the set of (non-negative) real numbers, $\mathbf{K}(\mathbb{R}^+)$ is the class of nonempty
closed subsets of $\mathbb{R}^+$ and $\mathcal{P}_0(\mathbb{R}^+)$ is the family of all nonempty subsets of $\mathbb{R}^+$.
$\mathbb{N}$ is the set of natural numbers.

We first recall some concepts and some elementary results of capacity and
the Choquet integral.

**Definition 1.** [3] A set function $\mu : \mathcal{F} \to [0,1]$ is called a capacity if it
satisfies: (1) $\mu(\emptyset) = 0$, $\mu(\Omega) = 1$; (2) $\mu(A) \leq \mu(B)$ for any $A \subseteq B$ and
$A, B \in \mathcal{F}$. A triplet $(\Omega, \mathcal{F}, \mu)$ is called a capacity space.

The concepts of a capacity $\mu$ on $\mathcal{F}$ is continuous from below (above) and
continuous are the same as that in classical probability. A capacity $\mu$ is called
*concave (or submodular)* if $\mu(A \cup B) + \mu(A \cap B) \leq \mu(A) + \mu(B)$, $A, B \in \mathcal{F}$.

Assume that $(\Omega, \mathcal{F}, \mu)$ is a capacity space. If $B \subset \Omega$, exists $A \in \mathcal{F}$, such
that $B \subset A$ and $\mu(A) = 0$, then $B$ is called a $\mu$-nullset. A property depending
on $\omega \in \Omega$ is said to hold almost everywhere with respect to $\mu$, abbreviated
a.e.$[\mu]$, if there is a $\mu$-nullset $N$ such that the property is valid outside $N$.

**Definition 2.** [3] The Choquet integral of a measurable function $f : \Omega \to$
$\mathbb{R}^+$ with respect to capacity $\mu$ on $A \in \Omega$ is defined by

$$(C) \int_A f d\mu = \int_0^{+\infty} \mu((f \geq t) \cap A) dt,$$

where the integral in the right hand is taken in the sense of Riemann. Instead of $(C) \int_\Omega f d\mu$, we shall write $(C) \int f d\mu$. If $(C) \int f d\mu < \infty$, we say that $f$ is Choquet integrable.

**Definition 3.** Let $\mu$ be a capacity on $(\Omega, \mathcal{F})$. We define

$$L_C(\mu) = \left\{ f : \Omega \to \mathbb{R}^+ \text{ is measurable} : (C) \int f d\mu < \infty \right\}.$$

The following two lemmas come from the article [4, 12].

**Lemma 1.** *(Fatou's lemmas) For given a sequence of non-negative measurable functions $\{f_n\}$, (1) if $\mu$ is continuous from below, then*

$$(C) \int \underline{\lim}_{n \to \infty} f_n d\mu \leq \underline{\lim}_{n \to \infty} (C) \int f_n d\mu;$$

*(2) if $\mu$ be continuous from above and there exists $g \in L_C(\mu)$, such that $f_n \leq g$ for $n \geq 1$, then*

$$\overline{\lim}_{n \to \infty} (C) \int f_n d\mu \leq (C) \int \overline{\lim}_{n \to \infty} f_n d\mu.$$

**Lemma 2.** *(Lebesgue dominated convergence theorem) Let $\mu$ be continuous. If $f_n \to f$ a.c.[$\mu$], and there exists $g \in L_C(\mu)$, such that $f_n \leq g$ for $n \geq 1$, then $(C) \int f_n d\mu \to (C) \int f d\mu$.*

In the following subsection, we shall list some preliminaries about set-valued random variables and set-valued Choquet integrals.

Let $F : \Omega \to \mathcal{P}_0(\mathbb{R}^+)$ be a mapping, called a set-valued mapping. The set $G(F) = \{(\omega, x) \in \Omega \times \mathbb{R}^+ : x \in F(\omega)\}$, is called the graph of $F$, and the set $F^{-1}(A) = \{\omega \in \Omega : F(\omega) \cap A \neq \emptyset\}$, $A \subset \mathbb{R}^+$, the inverse image of $F$.

**Definition 4.** [11] A set-valued mapping $F : \Omega \to \mathbf{K}(\mathbb{R}^+)$ is called measurable if, for each open subset $O \subset \mathbb{R}^+$, $F^{-1}(O) \in \mathcal{F}$. A measurable set-valued mapping is also called a set-valued random variable.

**Theorem 1.** *[11] Let $F : \Omega \to \mathbf{K}(\mathbb{R}^+)$ be a set-valued mapping. If $\mathcal{F}$ is complete with respect to some $\sigma$-finite measure, then the following conditions are equivalent:*

*(1) for each Borel set $B \subset \mathcal{B}(\mathbb{R}^+)$, $F^{-1}(B) \in \mathcal{F}$, where $\mathcal{B}(\mathbb{R}^+)$ is the Borel $\sigma$-field of $\mathbb{R}^+$;*
*(2) for each closed set $C \subset \mathbb{R}^+$, $F^{-1}(C) \in \mathcal{F}$;*

*(3) for each open set $O \subset \mathbb{R}^+$, $F^{-1}(O) \in \mathcal{F}$;*
*(4) $G(F)$ is $\mathcal{F} \times \mathcal{B}(\mathbb{R}^+)$-measurable.*

**Definition 5.** [11] A function $f : \Omega \to \mathbb{R}^+$ is called a selection for a set-valued mapping $F : \Omega \to \mathbf{K}(\mathbb{R}^+)$ if $f(\omega) \in F(\omega)$ for all $\omega \in \Omega$.

**Theorem 2.** *[11] Assume that $F : \Omega \to \mathbf{K}(\mathbb{R}^+)$ is a set-valued mapping. Then the following statements are equivalent:*
*(1) $F$ is a set-valued random variable;*
*(2) there exists a countable family $\{f_n : n \in \mathbb{N}\}$ of measurable selections of $F$ such that*

$$F(\omega) = \mathrm{cl}\{f_n(\omega) : n \in \mathbb{N}\}, \ for \ all \ \omega \in \Omega.$$

Now we introduce the concept of set-valued Choquet integrals.

**Definition 6.** Let $F : \Omega \to \mathbf{K}(\mathbb{R}^+)$ be a set-valued random variable, and $\mu$ be a capacity on $(\Omega, \mathcal{F})$. We define the family of Choquet integrable selections of $F$ a.e.$[\mu]$ as

$$S_C(F) = \Big\{ f \in L_C(\mu) : f(\omega) \in F(\omega) \ \ a.e.[\mu] \Big\}.$$

It is easy to show that $S_C(F)$ is a closed subset of $L_C(\mu)$.

**Definition 7.** [7] Let $F : \Omega \to \mathbf{K}(\mathbb{R}^+)$ be a set-valued random variable. The Choquet integral of $F$ with respect to capacity $\mu$ on $A \in \mathcal{F}$ is defined by

$$(C) \int_A F d\mu = \Big\{ (C) \int_A f d\mu : f \in S_C(F) \Big\}.$$

*Remark 1.* (1) Instead of $(C) \displaystyle\int_\Omega F d\mu$, we shall write $(C) \displaystyle\int F d\mu$.

(2) A set-valued random variable $F$ is said to be integrable with respect to $\mu$ if $(C) \displaystyle\int F d\mu \neq \emptyset$.

(3) $(C) \displaystyle\int F d\mu$ is closed [15].

**Definition 8.** Let $F : \Omega \to \mathbf{K}(\mathbb{R}^+)$ be a set-valued random variable. $F$ is called Choquet integrably bounded if $\|F(\omega)\|_{\mathbf{K}} \in L_C(\mu)$, where

$$\|F(\omega)\|_{\mathbf{K}} = \sup \Big\{ |x| : x \in F(\omega) \Big\}, \ \text{for all} \ \omega \in \Omega.$$

*Remark 2.* (1) This definition is different from Definition 2.5 in [7]: a set-valued function $F$ is said to be Choquet integrably bounded if there is $g \in L_C(\mu)$ such that

$$\|F(\omega)\|_{\mathbf{K}} \leq g(\omega).$$

In Definition 9, we directly take $g = \|F\|_{\mathbf{K}}$ since $\|F\|_{\mathbf{K}}$ is measurable from Theorem 2.

(2) It is easy to prove that $S_C(F)$ is not empty if $F$ is Choquet integrably bounded.

**Definition 9.** [11] Let $A_n \subset \mathbf{K}(\mathbb{R}^+)$, we write

$$\liminf_{n \to \infty} A_n = \{x \in \mathbb{R}^+ : x = \lim_{n \to \infty} x_n, \ x_n \in A_n, \ n \geq 1\},$$

$$\limsup_{n \to \infty} A_n = \{x \in \mathbb{R}^+ : x = \lim_{k \to \infty} x_{n_k}, \ x_{n_k} \in A_{n_k}, \ k \geq 1\}.$$

If $\liminf\limits_{n \to \infty} A_n = \limsup\limits_{n \to \infty} A_n = A$, then $\{A_n\}$ is said to be the Kuratowski convergent to $A$ and it is simply noted with $A_n \to A, n \to \infty$.

## 3  Main Results

Throughout this section, let $(\Omega, \mathcal{F})$ be a measurable space and $\mu$ be a capacity on $\Omega$. The family of all set-valued random variables is denoted by $\mathcal{U}[\Omega; \mathbf{K}(\mathbb{R}^+)]$. Since the page limitation, we have to omit some proofs in this section. We firstly present representation theorems of set-valued random variables.

**Theorem 3.** *Let $F \in \mathcal{U}[\Omega; \mathbf{K}(\mathbb{R}^+)]$. If $F$ is Choquet integrably bounded, then there exists a sequence $\{f_n : n \in \mathbb{N}\} \subset S_C(F)$ such that $F(\omega) = \mathrm{cl}\{f_n(\omega) : n \in \mathbb{N}\}$ for all $\omega \in \Omega$.*

**Theorem 4.** *Let $F \in \mathcal{U}[\Omega; \mathbf{K}(\mathbb{R}^+)]$. If $S_C(F) \neq \emptyset$ and $\mu$ is submodular, then there exists a sequence $\{f_n : n \in \mathbb{N}\} \subset S_C(F)$ such that $F(\omega) = \mathrm{cl}\{f_n(\omega) : n \in \mathbb{N}\}$ for all $\omega \in \Omega$.*

The next theorem shows that the subadditive of Choquet integral can be extended to the case of set-valued Choquet integral.

**Theorem 5.** *Assume that the capacity $\mu$ is submodular and $F_1, F_2 \in \mathcal{U}[\Omega; \mathbf{K}(\mathbb{R}^+)]$ are Choquet integrably bounded. Then we have*

$$(C) \int (F_1 + F_2) d\mu \leq (C) \int F_1 d\mu + (C) \int F_2 d\mu.$$

Now we shall mainly establish convergence theorems of Choquet integrals for sequences of set-valued random variables.

In the following convergence theorems, we always assume that $\mu$ is a continuous capacity and $F_n \in \mathcal{U}[\Omega; \mathbf{K}(\mathbb{R}^+)]$.

For any $A \in \mathcal{P}_0(\mathbb{R})$ and $x \in \mathbb{R}$ the distance between $x$ and $A$ is defined by $d(x, A) = \inf_{y \in A} |y - x|$. Now we prove the following Fatou's Lemmas.

**Theorem 6.** *Let $\mu$ be submodular. Assume that $F_n$ are Choquet integrably bounded and there exists a positive function $g \in L_C(\mu)$, such that $d(0, F_n) \leq g$ a.e.[$\mu$] for all $n \geq 1$. If $F = \liminf\limits_{n\to\infty} F_n$ a.e.[$\mu$] and $S_C(F) \neq \emptyset$ then*

$$(C) \int F d\mu \subset \liminf_{n\to\infty} (C) \int F_n d\mu.$$

*Proof.* For each $f \in S_C(F)$ and $n \geq 1$, define $G_n : \Omega \to \mathbf{K}(\mathbb{R}^+)$ by

$$G_n(\omega) = \left\{ x \in F_n(\omega) : |f(\omega) - x| \leq d(f(\omega), F_n(\omega)) + \frac{1}{n} \right\}, \quad \omega \in \Omega.$$

Since $d(x, F_n(\omega))$ is continuous with respect to $x \in \mathbb{R}^+$ and is measurable with respect to $\omega \in \Omega$, $d(x, F_n(\omega))$ is $\mathcal{F} \times \mathcal{B}(\mathbb{R}^+)$-measurable. Thus $d(f(\omega), F_n(\omega))$ is measurable. Furthermore,

$$\phi_n(x, \omega) = |f(\omega) - x| - d(f(\omega), F_n(\omega))$$

is continuous with respect to $x \in \mathbb{R}^+$ and is measurable with respect to $\omega \in \Omega$. Then $\phi_n(x, \omega)$ is $\mathcal{F} \times \mathcal{B}(\mathbb{R}^+)$-measurable. Hence $G(G_n) \in \mathcal{F} \times \mathcal{B}(\mathbb{R}^+)$. By using Theorem 1 and Theorem 2, there is an $\mathcal{F}$-measurable function $f_n$ such that $f_n(\omega) \in G_n(\omega)$ for all $\omega \in \Omega$. Since $f_n \leq \|F\|_{\mathbf{K}} \in L_C(\mu)$ a.e.[$\mu$], we have $f_n$ is Choquet integrable, i.e., $f_n \in S_C(F_n)$. Since $f(\omega) \in F(\omega) = \liminf\limits_{n\to\infty} F_n(\omega)$ a.e.[$\mu$] implies $d(f(\omega), F_n(\omega)) \to 0$ a.e.[$\mu$] as $n \to \infty$, and

$$|f(\omega) - f_n(\omega)| \leq d(f(\omega), F_n(\omega)) + \frac{1}{n} \quad a.e.[\mu],$$

we have $|f(\omega) - f_n(\omega)| \to 0$ a.e.[$\mu$] as $n \to \infty$. Since

$$0 \leq f_n(\omega) \leq |f(\omega) - f_n(\omega)| + f(\omega) \leq d(f(\omega), F_n(\omega)) + f(\omega) + \frac{1}{n}$$
$$\leq d(0, F_n(\omega)) + 2f(\omega) + \frac{1}{n} \leq g(\omega) + 2f(\omega) + \frac{1}{n} \quad a.e.[\mu],$$

and $\mu$ is submodular which implies $(C) \int (g + 2f + \frac{1}{n}) d\mu \leq (C) \int g d\mu + 2(C) \int f d\mu + \frac{1}{n} < \infty$, from Lemma 2, it follows that

$$d\left( (C) \int f d\mu, (C) \int F_n d\mu \right) \leq \left| (C) \int f d\mu - (C) \int f_n d\mu \right| \to 0.$$

Hence, $(C) \int f d\mu \in \liminf\limits_{n\to\infty} (C) \int F_n d\mu$, which implies the conclusion of the Theorem. $\qquad\square$

**Theorem 7.** *Assume that there exists* $g \in L_C(\mu)$, *such that* $\|F_n\|_{\mathbf{K}} \leq g$ *a.e.*$[\mu]$ *for all* $n \geq 1$. *If* $F = \liminf\limits_{n \to \infty} F_n$ *a.e.*$[\mu]$, *then*

$$(C) \int F d\mu \subset \liminf_{n \to \infty} (C) \int F_n d\mu.$$

*Remark 3.* The sufficient condition of Theorem 3.13 (Fatou's lemma) in [15] is: $\mu$ is continuous, $m$-continuous and there exists a Choquet integrable function $g$, such that $\|F_n\|_{\mathbf{K}} \leq g$ for $n \geq 1$. Obviously, the sufficient condition of the above Theorem is weaker.

**Theorem 8.** *Assume that there exists* $g \in L_C(\mu)$ *such that* $\|F_n\|_{\mathbf{K}} \leq g$ *a.e.*$[\mu]$ *for all* $n \geq 1$. *If* $F = \limsup\limits_{n \to \infty} F_n$ *a.e.*$[\mu]$ *then*

$$\limsup_{n \to \infty} (C) \int F_n d\mu \subset (C) \int F d\mu.$$

*Proof.* Take $y \in \limsup\limits_{n \to \infty} (C) \int F_n d\mu$. Then there exists a sequence $\{y_{n_j} : j \in \mathbb{N}\}$ with limit $y$, such that $y_{n_j} = (C) \int f_{n_j} d\mu$ with $f_{n_j} \in S_C(F_{n_j}), j \in \mathbb{N}$.

Since

$$\limsup_{j \to \infty} F_{n_j} \subseteq \limsup_{n \to \infty} F_n = F \ a.e.[\mu],$$

and $f_{n_j} \in S_U(F_{n_j}), j \subset \mathbb{N}$, we have $\overline{\lim}_{j \to \infty} f_{n_j}(\omega) \in F(\omega)$ a.e.$[\mu]$. Hence, there exists a subsequence $\{n_{j_k} : k \in \mathbb{N}\}$ of $\{n_j : j \in \mathbb{N}\}$, such that $\lim\limits_{k \to \infty} f_{n_{j_k}}(\omega) \in F(\omega)$ a.e.$[\mu]$ due to the concept of upper limit. Define $f(\omega) = \lim\limits_{k \to \infty} f_{n_{j_k}}(\omega)$. Since $f_{n_{j_k}} \leq \|F_{n_{j_k}}\|_{\mathbf{K}} \leq g, k \in \mathbb{N}$, and $g \in L_C(\mu)$, we have

$$(C) \int f d\mu = \lim_{k \to \infty} (C) \int f_{n_{j_k}} d\mu = y$$

by using Lemma 2. So $y \in (C) \int F d\mu$, which completes the proof. $\square$

From above two theorems, we have the following Lebesgue dominated convergence theorem of set-valued Choquet integrals in the sense of Kuratowski convergence.

**Theorem 9.** *Assume that there exists* $g \in L_C(\mu)$, *such that* $\|F_n\|_{\mathbf{K}} \leq g$ *a.e.*$[\mu]$ *for all* $n \geq 1$. *If* $F_n \to F(n \to \infty)$ *a.e.*$[\mu]$ *then*

$$\lim_{n \to \infty} (C) \int F_n d\mu = (C) \int F d\mu.$$

Finally, we shall discuss the monotone convergence theorems of set-valued Choquet integrals in the sense of Kuratowski convergence.

**Theorem 10.** *Assume that $F_{n+1} \supset F_n$ for all $n \in \mathbb{N}$, and $F = \bigcup\limits_{n=1}^{\infty} F_n$. If $S_C(F) \neq \emptyset$ and $F$ is Choquet integrably bounded, then*

$$\lim_{n \to \infty} (C) \int F_n d\mu = (C) \int F d\mu.$$

**Theorem 11.** *Assume that $F_{n+1} \subset F_n$ for all $n \in \mathbb{N}$, and $F = \bigcap\limits_{n=1}^{\infty} F_n$. If $S_C(F) \neq \emptyset$ and $F_1$ is Choquet integrably bounded, then*

$$\lim_{n \to \infty} (C) \int F_n d\mu = (C) \int F d\mu.$$

**Acknowledgements.** This paper is supported by PHR (No. 201006102) and Beijing Natural Science Foundation (Stochastic Analysis with uncertainty and applications in finance). We would like to thank referees for their valuable suggestions.

# References

1. Allais, M.: Le comportement de l'homme rationnel devant le risque: Critique des postulates et axiomes de l'ecole americaine. Econometrica 21, 503–546 (1953)
2. Aumann, R.J.: Integrals of set-valued functions. J. Math. Appl. 12, 1–12 (1965)
3. Choquet, G.: Theory of capacities. Ann. Inst. Fourier 5, 131–295 (1953)
4. Denneberg, D.: Non-Additive Measure and Integral. Kluwer Academic Publishers, Boston (1994)
5. Ellsberg, D.: Risk, ambiguity, and the Savage axioms. Quart. J. Econom. 75, 643–669 (1961)
6. Hiai, F., Umegaki, H.: Integrals, conditional expectations and martingales of multivalued functions. J. Multiva. Anal. 7, 149–182 (1977)
7. Jiang, L.C., Kwon, J.S.: On the representation of Choquet integrals of set-valued functions, and null sets. Fuzzy Sets and Systems 112, 233–239 (2000)
8. Jiang, L.C., Kim, Y.K., Jeon, J.D.: On set-valued Choquet integrals and convergence theorems. Advanced Studies in Contemporary Mathematics 6, 63–76 (2003)
9. Jiang, L.C., Kim, Y.K., Jeon, J.D.: On set-valued Choquet integrals and convergence theorems (II). Bull. Korean Math. Soc. 40, 139–147 (2003)
10. Klein, E., Thompson, A.: Theory of Correspondence. Wiley, New York (1984)
11. Li, S., Ogura, Y., Kreinovich, V.: Limit theorems and applications of set-valued and fuzzy set-valued random variables. Kluwer Academic Publishers, Netherlands (2002)
12. Pap, E.: Null-Additive Set-Functions. Kluwer, Dordrecht (1994)
13. Schmeidler, D.: Subjective probability and expected utility without additivity. Econometrica 57, 571–587 (1989)
14. Wang, Z., Yan, J.A.: A selective overview of applications of Choquet integrals (2006) (manuscript)
15. Zhang, D.L., Guo, C.M., Liu, S.Y.: Set-valued Choquet integrals revisited. Fuzzy Sets and Systems 147, 475–485 (2004)

# On Nonlinear Correlation of Random Elements

Hung T. Nguyen

**Abstract.** In view of imprecise data in a variety of situations, we proceed to investigate dependence structure of random closed sets. Inspired by the modeling and quantifying of nonlinear dependence structures of random vectors, using copulas, we look at the extension of copula connection to the case of infinitely separable metric spaces with applications to the space of closed sets of a Hausdorff, locally compact, second countable space.

**Keywords:** Copulas, Correlation, Dependence structures, Random closed sets.

## 1 Introduction

In view of interests in financial risk management, there is a need to take a closer look at the problem of modeling and quantifying dependence structures among random variables. *Traditionally*, this problem is very simple. When two random variables $X$ and $Y$ are not independent, one quantifies their dependence by using Pearson's correlation coefficient. Even it was spelled out that Pearson's correlation coefficient is a measure of *linear dependence* between the variables, there was no comments on how to model and quantify other (nonlinear) dependence structures. Of course, linearity is always considered as a first approximation to nonlinearity. It is the field of financial economics which has triggered a reexamination of Pearson linear correlation analysis. First, there are random variables which have infinite variances (namely those with heavy-tailed distributions). Pearson's correlation cannot

Hung T. Nguyen
Department of Mathematical Sciences, New Mexico State University, Las Cruces, NM 88003, USA
e-mail: hunguyen@nmsu.edu

S. Li (Eds.): Nonlinear Maths for Uncertainty and its Appli., AISC 100, pp. 109–115.
springerlink.com      © Springer-Verlag Berlin Heidelberg 2011

be even defined for such variables. Second, it is well known that Pearson's correlation coefficient is not invariant with changes of scales. But most importantly, Pearson's correlation coefficient is only the degree to which how removed the dependence of two variables from being *linear*. In other words, one only concerned with *linear correlation*. On the other hand, linearity dependence and its quantification are not sufficient to specify the multivariate model governing the random vector of interest, given its marginal distributions. This is so because dependence structures are of various forms. The joint distribution of a random vector contains all information about its behavior, including its dependence structure. How to extract the dependence structure of a random vector from the knowledge of its joint distribution? The answers to this question and to "how to relate marginal distributions to the joint distribution?" are fortunately given by Sklar's theory of copulas. This has opened the door to investigating *nonlinear dependence structures and their quantifications (correlations)*.

After describing the state-of-the-art of modeling and quantifying nonlinear dependence structures of random vectors using copulas, as well as discussing open research issues, we embark on the problem of nonlinear correlation of random closed sets.

## 2  Why Nonlinear Correlation?

Let $(\Omega, \mathcal{A})$ and $(S, \mathcal{S})$ be two measurable spaces. A map $X : \Omega \to S$ is called a random element when $X^{-1}(\mathcal{S}) \subseteq \mathcal{A}$. If $P$ is a probability measure on $\mathcal{A}$, then the (probability) law of $X$ is the probability measure $P_X = PX^{-1}$ on $\mathcal{S}$. For two random elements $X$ and $Y$, defined on $(\Omega, \mathcal{A}, P)$ with values in $(S, \mathcal{S})$, $(T, \mathcal{T})$, respectively, the law of the random element $Z = (X, Y)$ is a probability measure $P_Z$ on $\mathcal{S} \otimes \mathcal{T}$ such that

$$P_X = P_Z \circ \pi_S^{-1}, P_Y = P_Z \circ \pi_T^{-1}$$

where $\pi_S^{-1}, \pi_T^{-1}$ denote the projections from $S \times T$ onto $S, T$, respectively, in other words, $P_X$ and $P_Y$ are marginal laws. $X$ and $Y$ are said to be independent when $P_Z = P_X \otimes P_Y$ (product measure), i.e., for any $A \in \mathcal{S} \otimes \mathcal{T}$,

$$P_Z(A) = \int_{S \times T} 1_A(s, t) dP_X(s) dP_Y(t)$$

We are concerned with the case where $X$ and $Y$ are not independent (they are dependent). Note that, this is somewhat similar to the concern about noncompactness of sets in infinitely dimensional metric spaces in metric fixed point theory (see e.g. [1]) where the quantification of noncompactness by some measures of noncompactness, such as Kuratowski's one, is needed to investigate sufficient conditions for the fixed point property of non-expansive mappings. Here, we wish to model and quantify dependence structures in multivariate models for statistical decision-making.

Now the dependence between $X$ and $Y$ can be of various different forms. For example, for random variables, i.e. when $S = T = \overline{\mathbb{R}}$, if $Y = aX + b$, then they are linearly dependent; if $Y = X^2$, then they are nonlinearly dependent. When focusing on linear dependence (as a first approximation) the usual approach is this. Assuming that both $X$ and $Y$ have finite variances, one quantifies its strength (i.e., measures how far the dependence is removed from linearity) by using Pearson's correlation coefficient. Clearly Pearson's correlation cannot detect nonlinear dependence. One well-known drawback of Pearson's correlation it that it is not invariant with respect to a change of scale. As far as financial economics is concerned, Pearson's correlation is not defined since financial variables usually have heavy-tailed distributions, and hence have infinite variances (see e.g., [9]).

For other random elements, as observed data, such as random closed sets of $\overline{\mathbb{R}}^d$, the space of closed sets is nonlinear. As such, we need to model and quantify nonlinear dependence.

## 3 Entering Copulas

For the case of random vectors, i.e., when $S = \overline{\mathbb{R}}^d$ with its Borel $\sigma$-field $\mathcal{B}\left(\overline{\mathbb{R}}^d\right)$, the modeling of dependence structures is solved by copulas (see [2, 5, 11]). Specifically, in view of Lebesgue-Stieltjes theorem, it suffices to look at multivariate distribution functions instead of probability measures. If $Z = (X, Y)$ has joint distribution $H$ and marginals $F, G$, then $C : [0,1]^2 \to [0,1]$, where $C(u, v) = H(F^{-1}(u), G^{-1}(v))$ is a copula such that, for any $(x, y) \in \overline{\mathbb{R}}^2$, $H(x, y) = C(F(x), G(y))$, assuming, for simplicity that $F$ and $G$ are continuous.

The copula $C$ of $(X, Y)$ models their dependence structure. For example, $X$ and $Y$ are comonotonic if and only if $C(u, v) = u \wedge v$; $X$ and $Y$ are counter-comonotonic if and only if $C(u, v) = \max\{u + v - 1, 0\}$. Of course, $C(u, v) = uv$ if and only if $X$ and $Y$ are independent.

The problem of quantifying a given dependence structure represented by a copula is rather delicate. Let $\mathcal{C}$ be the space of all bivariate copulas. A (copula-based) correlation measure is a functional $\varphi : \mathcal{C} \to [-1, 1]$. Just like modeling financial risk, the best we can do is to list desirable properties a correlation measure should possess. For example, $\kappa(X, Y) = \varphi(C)$ should satisfies: $\kappa(X, Y)$ is defined for every pair of $X$ and $Y$ (regardless of their finite or infinite variances), symmetric, is zero when $X$ and $Y$ are independent.

But, whatever we associate a measure, called a correlation measure, to a dependence structure, i.e., to a given copula $C$, that correlation measure should be a function of $C$ alone, since copulas are invariant with respect to strictly increasing transformations. The Pearson's correlation is not scale invariant since it is a function of $C$ and the marginal distributions.

While the copula of $(X, Y)$ represents their dependence structure, one usually looks at some specific way that $X$ and $Y$ are related to each other, for example the classical Kandell $\tau$ rank correlation is based on the concept of concordance, whereas the Spearman rho $\rho$ measures how far $C$ is removed from independence. These copula-based correlations are, respectively

$$\tau(X, Y) = \int_0^1 \int_0^1 C(u, v) dC(u, v) - 1,$$

$$\rho(X, Y) = 12 \int_0^1 \int_0^1 [C(u, v) - uv] du dv$$

One could ask: what is a copula-based correlation measuring how far $C$ is removed from linearity?

A useful local correlation is the tail dependence correlation

$$\lambda(X, Y) = \lim_{\alpha \nearrow 1} \frac{1 - 2\alpha + C(\alpha, \alpha)}{1 - \alpha}$$

The copula connection between distributions $H, F$ and $G$ is written in terms of their associated probability laws as, for any $x, y$ in $\overline{\mathbb{R}}$,

$$dH(([-\infty, x] \times [-\infty, y]) = C(dF([-\infty, x], dG([-\infty, y]))$$

noting that $\sigma\{[-\infty, x] : x \in \overline{\mathbb{R}}\} = \mathcal{B}(\overline{\mathbb{R}})$, and $dF, dG$ are nonatomic probability measures.

$X$ and $Y$ are independent if and only if $C(u, v) = uv$, so that, for any $A, B$ in $\mathcal{B}(\overline{\mathbb{R}})$,

$$dH(A \times B) = C(dF(A), dG(B))$$

## 4   Copulas for Probability Measures

In infinitely dimensional (polish) spaces, such as $C([0, 1])$, $D([0, 1])$, we have to work directly with probability laws on them, since there is no counterpart of distribution functions as opposed to Euclidean spaces. Since copulas are essential for modeling and quantifying dependence structures of random vectors, we would like to know whether there are some copula connection between probability measures. To our knowledge, the only work in this direction is [8] in which a formal generalization of

$$dH(([-\infty, x] \times [-\infty, y]) = C(dF([-\infty, x], dG([-\infty, y]))$$

is obtained for general polish spaces. Note that, without referring to Sklar's copulas, Strassen [12] investigated the existence of a joint probability measure with given marginals and bounds.

The setting is this. For two random elements $X$ and $Y$, defined on $(\Omega, \mathcal{A}, P)$ with values in $(S, \mathcal{S})$, $(T, \mathcal{T})$, respectively, the law of the random element $Z = (X, Y)$ is a probability measure $P_Z$ on $\mathcal{S} \otimes \mathcal{T}$ such that $P_X = P_Z \circ \pi_S^{-1}$, $P_Y = P_Z \circ \pi_T^{-1}$, where $\pi_S^{-1}$, $\pi_T^{-1}$ denote the projections from $S \times T$ onto $S, T$, respectively, in other words, $P_X$ and $P_Y$ are marginal laws. We seek some copula connection between $P_Z, P_X$ and $P_Y$.

The main result of Scarsini [10] is this. Let $\mathbb{A} \subseteq \mathcal{S}, \mathbb{B} \subseteq \mathcal{T}$ be totally ordered by set inclusion, and suppose $P_X(\mathcal{S}) = P_Y(\mathcal{T}) = [0, 1]$. Then there exists a copula $C$, depending upon $\mathbb{A}$ and $\mathbb{B}$, such that, for any $A \in \mathbb{A}$, $B \in \mathbb{B}$,

$$P_Z(A \times B) = C(P_X(A), P_Y(B))$$

There was no discussions about dependence modeling, let alone correlation analysis, in this infinitely dimensional abstract setting.

## 5  The Case of Random Closed Sets

Here is an important example of a type of random elements with values in infinitely dimensional polish spaces. Coarse data are sets rather than points in euclidean spaces. Now random vectors can be viewed as random elements taking singletons $\{x\}$ as values, where the $\{x\}$ are closed sets of $\overline{\mathbb{R}}^d$. Thus, random closed subsets of $\overline{\mathbb{R}}^d$ are natural generalizations of values of random vectors. Closed subsets of $\overline{\mathbb{R}}^d$ (or, more generally, of a Hausdorff, locally compact and second countable space) are bona fide values of random elements, thanks to Matheron [3], see also [3, 6].

The space $\mathcal{F}$ of all closed subsets of $\mathbb{R}^d$, equiped with the hit-or-miss topology (see [3, 4, 6]) is compact, Hausdorff and second countable, and as such, it is metrizable and separable. A metric compatible with this topology is the stereographic distance. Specifically, using one-point compactification of $\mathbb{R}^d$ as the sphere $D^d \subseteq \mathbb{R}^{d+1}$, and the Euclidean metric $\rho$ on $\mathbb{R}^{d+1}$, the stereographic Hausdorff distance on $\mathcal{F}$ is

$$H_\rho(A, B) = \rho(A' \cup NP, B' \cup NP)$$

where $\rho$ is the Hausdorff distance on compact sets of $D^d$, and $A', B'$ are stereographic projections of $A, B$ on $D^d$, and $NP$ stands for "North pole" of $D^d$.

*Remark.* The space $\mathcal{F}$ with the set inclusion is a partially ordered set. In fact, it is a continuous lattice, and its hit-or-miss topology is precisely the Lawson topology (see e.g., [7]). The framework of continuous lattices is useful for extending closed sets to upper semi-continuous functions.

A random closed set is a map $X : \Omega \to \mathcal{F}$, such that $X^{-1}(\mathcal{B}(\mathcal{F})) \subseteq \mathcal{A}$, where $\mathcal{B}(\mathcal{F})$ denotes the Borel $\sigma$-field respect to the hit-or-miss topology. Its law is a probability measure on $\mathcal{B}(\mathcal{F})$. Unlike function spaces like $C([0,1]), D([0,1])$, there is a counterpart of Lebesgue-Stieltjes theorem, namely probability measures on $\mathcal{B}(\mathcal{F})$ are in a bijective correspondence with capacity functionals (playing the role of distribution functions of random vectors). Specifically, let $\mathcal{K}$ be the class of compact subsets of $\mathbb{R}^d$. A capacity functional is a map $F : \mathcal{K} \to [0,1]$ such that

(i) $F(\varnothing) = 0$,

(ii) if $K_n \searrow K$ then $F(K_n) \searrow F(K)$,

(iii) $F$ is alternating of infinite order, i.e.,

$$F\left(\bigcap_{i=1}^{n} K_i\right) \leq \sum_{\varnothing \neq I \subseteq \{1,2,\dots,n\}} (-1)^{|I|+1} F\left(\bigcup_{i \in I} K_i\right).$$

The Choquet theorem is this. If $F$ is a capacity functional, then there exists a unique probability measure $Q$ on $\mathcal{B}(\mathcal{F})$ such that $Q(A \in \mathcal{F} : A \cap K \neq \varnothing) = F(K)$, for any $K \in \mathcal{K}$.

*Remark.* For a random set with values as subsets of a finite set, the dual of its capacity functional, i.e., $A \to 1 - F(A^c)$ is its distribution function (also called a belief function) where $\subseteq$ replaces $\leq$. In general, the Choquet theorem is not valid in non-locally compact Polish spaces, see [8].

In view of Choquet theorem, copula connections among probability measures on $\mathcal{B}(\mathcal{F})$ could be investigated at the level of their associated capacity functionals. But a direct application of Scarsini's result is also possible.

Unlike the case of random vectors (viewed as random closed sets taking singleton sets as values), dependence structures among general random closed sets could be broken down further in terms of factors such as shape, location and size. For example, suppose we are interested in the dependence with respect to size of two random closed sets $X$ and $Y$, we proceed as follows. Let $L$ denote the Lebesgue measure on $\mathbb{R}^d$. Then a functional $\Phi : \mathcal{F} \to \overline{\mathbb{R}}^+$ is considered, where $\Phi(A) = L(A)$.

In general, let $\Phi : \mathcal{F} \to \overline{\mathbb{R}}$ be a continuous functional. Let $A_x = \Phi^{-1}([-\infty, x])$, and $\mathbb{A} = \{A_x : x \in \overline{\mathbb{R}}\}$. Then $\mathbb{A}$ is a chain. Assuming that $P_X(\mathcal{B}(\mathcal{F})) = P_Y(\mathcal{B}(\mathcal{F})) = [0,1]$, there exists a unique copula $C_\Phi$ such that, for any $x, y \in \overline{\mathbb{R}}$,

$$P_{(X,Y)}(A_x \times A_y) = C_\Phi(P_X(A_x), P_Y(A_y))$$

from which one can proceed to define various concepts of copula-based correlation for random closed sets. It is clear that further studies are needed to complete this program.

# References

1. Ayerbe Toledano, J.M., Dominguez Benavides, T., Lopez Acedo, T.: Measures of Noncompactness in Metric Fixed Point Theory. Birkhauser-Verlag, Basel (1997)
2. Joe, H.: Multivariate Models and Dependence Concepts. Chapman and Hall/CRC (1997)
3. Matheron, G.: Random Sets and Integral Geometry. J. Wiley, Chichester (1975)
4. Molchanov, I.: Theory of Random Sets. Springer, Heidelberg (2005)
5. Nelsen, R.B.: An Introduction to Copulas. Springer, Heidelberg (2006)
6. Nguyen, H.T.: An Introduction to Random Sets. Chapman and Hall/CRC (2006)
7. Nguyen, H.T., Tran, H.: On a continuous lattice approach to modeling of coarse data in system analysis. Journal of Uncertain Systems 1(1), 62–73 (2007)
8. Nguyen, H.T., Nguyen, N.T.: A negative version of Choquet theorem for Polish spaces. East-West Journal of Mathematics 1, 61–71 (1998)
9. Resnick, S.I.: Heavy-Tail Phenomena, Probabilistic and Statistical Modeling. Springer, Heidelberg (2007)
10. Scarsini, M.: Copulae of probability measures on product spaces. J. Multivariate Anal. 31, 201–219 (1989)
11. Sklar, A.: Fonctions de repartition a n dimensions et leur marges. Publ. Inst. Statist. Univ. Paris 8, 229–231 (1959)
12. Strassen, V.: The existence of probability measures with given marginals. Ann. Math. Statist. 36, 423–439 (1965)

# On Fuzzy Stochastic Integral Equations–A Martingale Problem Approach

Mariusz Michta

**Abstract.** In the paper we consider fuzzy stochastic integral equations using the methods of stochastic inclusions. The idea is to consider an associated martingale problem and its solutions in order to obtain a solution to the fuzzy stochastic equation.

**Keywords:** Fuzzy-valued integral, Fuzzy-stochastic equation.

## 1 Introduction

Fuzzy stochastic differential equations were studied recently by Kim in [7], Ogura in [14], and Malinowski and Michta in [10] and [11]. In this note we describe a new and different method used to the study the notion of fuzzy stochastic differential or integral equations proposed recently in [12]. We study the existence of solutions of a fuzzy stochastic differential equation driven by the Brownian motion under weaker conditions than Lipschitz continuity imposed on the right-hand side and considered earlier in the cited papers. In our approach, we use a new and different method applied to stochastic fuzzy systems. We interpret the fuzzy stochastic equation as a family of stochastic differential inclusions. The idea implemented in this note is to solve those inclusions via the appropriately defined martingale problem and then apply the theorem of Negoita and Ralescu. In the deterministic case, our approach corresponds to ideas and comments presented in [1], [5] and [2], where two different approaches to the fuzzy differential equation were presented and compared. Moreover, the idea presented here enables us to overcome (in different way than recently proposed in [10]) difficulties with a possible unboundedness of a set-valued and (consequently) of a fuzzy

Mariusz Michta
Institute of Mathematics and Informatics, Opole University,
Oleska 48, Opole, Poland
e-mail: m.michta@wmie.uz.zgora.pl

S. Li (Eds.): Nonlinear Maths for Uncertainty and its Appli., AISC 100, pp. 117–124.
springerlink.com                    © Springer-Verlag Berlin Heidelberg 2011

stochastic integral, which seems to be an open problem. Consequently, we are able to consider fuzzy stochastic equations not only with a single-valued but also with a fuzzy-valued integrands appearing in diffusion terms.

## 2 Set-Valued Trajectory Stochastic Integral

Let $(\Omega, \mathbf{F}, \{\mathbf{F}_t\}_{t \geq 0}, P)$ be a complete filtered probability space satisfying the usual hypothesis, i.e., $\{\mathbf{F}_t\}_{t \geq 0}$ is an increasing and right continuous family of sub-$\sigma$-fields of $\mathbf{F}$ and $\mathbf{F}_0$ contains all $P$-null sets. Let $\mathcal{P}$ denote the smallest $\sigma$-field on $R_+ \times \Omega$ with respect to which every left-continuous and $\{\mathbf{F}_t\}_{t \geq 0}$-adapted process is measurable. An $R^d$-valued stochastic process $X$ is said to be predictable if $X$ is $\mathcal{P}$-measurable. One has $\mathcal{P} \subset \beta \otimes \mathbf{F}$, where $\beta$ denotes the Borel $\sigma$-field on $R_+$. Let $Z$ be an $\{\mathbf{F}_t\}_{t \geq 0}$-adapted and càdlàg (right hand continuous with left limits) semimartingale with values in $R^1$, i.e., $Z = M + A$, $Z_0 = 0$ where $M$ is an $\{\mathbf{F}_t\}_{t \geq 0}$-adapted local martingale and $A$ is an $\{\mathbf{F}_t\}_{t \geq 0}$-adapted, càdlàg process with finite variation on compact intervals in $R_+$ (see [15] for details). By $\mathcal{H}^2$ we denote the space of $\{\mathbf{F}_t\}_{t \geq 0}$-adapted semimartingales with a finite $\mathcal{H}^2$-norm:

$$||Z||_{\mathcal{H}^2} := ||[M, M]_\infty^{1/2}||_{\mathrm{L}^2} + || \left( \int_0^\infty |dA_t| \right) ||_{\mathrm{L}^2} < \infty,$$

where $[M, M]$ denotes the quadratic variation process for a local martingale $M$, while $|A|. := \int_0^\cdot |dA_s|$ is the total variation of the random measure induced by the paths of the process $A$. Let $\mu_M$ denote the Doléans-Dade measure for the martingale $M$ (see [3]). Then for all $f \in L^2(R_+ \times \Omega, \mathcal{P}, \mu_M)$ one has

$$\int_{[0,t] \times \Omega} |f|^2 d\mu_M = E \left( \int_0^t |f_s|^2 d[M, M]_s \right) = E| \int_0^t f_s dM_s|^2,$$

for $t \geq 0$. Let us also define a random measure on $R_+$

$$\gamma(\omega, dt) := |A(\omega)|_\infty |dA_t(\omega)|$$

and a measure associated with the process $A$ by the formula:

$$\nu_A(C) := \int_\Omega \int_0^\infty I_C(\omega, t) \gamma(\omega, dt) P(d\omega)$$

for every $C \in \mathcal{P}$. Then we have

$$\nu_A(R_+ \times \Omega) = E \left( \int_0^\infty |dA_s| \right)^2.$$

Hence $\nu_A$ is a finite measure on $\mathcal{P}$. Finally, we define a finite measure $\mu_Z$ associated with $Z \in \mathcal{H}^2$ by $\mu_Z := \mu_M + \nu_A$. Let us denote $L^2_\mathcal{P}(\mu_Z) := L^2(R_+ \times \Omega, \mathcal{P}, \mu_Z)$. Particularly, when $Z = W$ is a standard Wiener process we have $d\mu_W = dt \times dP$ and we may concentrate on the class of square integrable and nonanticipating processes.

By $Cl(R^d)$ $(Cl_b(R^d))$ we denote the family of all nonempty, closed (nonempty, closed and bounded) subsets of $R^d$. Similarly, $Comp(R^d)$ (resp. $Conv(R^d)$) is the family of all nonempty and compact (resp. compact and convex) subsets of $R^d$, endowed with the Hausdorff metric $H_{R^d}$. Let $F = (F(t))_{t \geq 0}$ be a set-valued stochastic process with values in $Cl(R^d)$. We call $F$ to be $\{\mathbf{F}_t\}_{t \geq 0}$-adapted if $F(t)$ is $\mathbf{F}_t$-measurable for each $t \in R_+$. It is predictable if $F$ is $\mathcal{P}$-measurable. Let us define the set

$$S_{\mathcal{P}}^2(F, \mu_Z) := \{f \in L_{\mathcal{P}}^2(\mu_Z) : f \in F \ \mu_Z \ a.e.\}.$$

We say that $F$ is $L_{\mathcal{P}}^2(\mu_Z)$-integrably bounded if $\|F\| \in L_{\mathcal{P}}^2(\mu_Z)$, where $\|F\| := H_{R^d}(F, \{0\})$. By Kuratowski and Ryll-Nardzewski Selection Theorem (see e.g. [8] ) it follows that $S_{\mathcal{P}}^2(F, \mu_Z) \neq \emptyset$ and for every $f \in S_{\mathcal{P}}^2(F, \mu_Z)$ the Itô stochastic integral $\int f_s dZ_s$ exists. Hence for every $t \geq 0$, we define the set

$$\int_0^t F_s dZ_s := \left\{ \int_0^t f_s dZ_s : f \in S_{\mathcal{P}}^2(F, \mu_Z) \right\}$$

which is called the set-valued trajectory stochastic integral of $F$ with respect to semimartingale $Z$. By [12] we have the following result needed in the sequel.

**Theorem 1.** *For each $n \geq 1$, let $F_n : R_+ \times \Omega \to Comp(R^d)$ be a predictable multivalued mapping such that $F_1$ is $L_{\mathcal{P}}^2(\mu_Z)$- integrably bounded and $F_1 \supset F_2 \supset ... \supset F$ $\mu_Z$-a.e. and let $F := \bigcap_{n \geq 1} F_n$ $\mu_Z$-a.e.. Then for every $t \geq 0$ it holds $\int_0^t F_s dZ_s = \bigcap_{n \geq 1} \int_0^t (F_n)_s dZ_s$.*

## 3   Fuzzy Trajectory Stochastic Integral

Let $\mathcal{X}$ be a given metric space with a Borel $\sigma$-field $\beta(\mathcal{X})$ and as before, let $(\Omega, \mathbf{F}, \{\mathbf{F}_t\}_{t \geq 0}, P)$ be a given filtered probability space. We also let $Z$ to be a given $\mathcal{H}^2$-semimartingale.

**Definition 1.** By a fuzzy set $u \in \mathcal{F}(\mathcal{X})$ we mean a function $u : \mathcal{X} \to [0, 1]$ for which the $\alpha$-level set $[u]^\alpha := \{x \in \mathcal{X} : u(x) \geq \alpha\} \in Cl_b(\mathcal{X})$ for all $\alpha \in (0, 1]$. The support of $u$ is defined by $[u]^0 := cl\{x \in \mathcal{X} : u(x) > 0\}$.

We consider also

$$\mathcal{F}_{Comp}(\mathcal{X}) = \{u \in \mathcal{F}(\mathcal{X}) : [u]^\alpha \in Comp(\mathcal{X}), \alpha \in [0, 1]\}$$

and

$$\mathcal{F}_{Conv}(\mathcal{X}) = \{u \in \mathcal{F}(\mathcal{X}) : [u]^\alpha \in Conv(\mathcal{X}), \alpha \in [0, 1]\}$$

in the case when $\mathcal{X}$ is a linear normed space. In what follows, we will consider the case $\mathcal{X} = R^d$. The following metric in $\mathcal{F}_{Comp}(R^d)$ is often used (see e.g. [9]):

$$d_\infty(u,v) := \sup_{\alpha \in [0,1]} H_{R^d}([u]^\alpha, [v]^\alpha) \text{ for } u, v \in \mathcal{F}_{Comp}(R^d).$$

By a fuzzy random variable we mean a function $u : \Omega \to \mathcal{F}(R^d)$ such that $[u(\cdot)]^\alpha : \Omega \to Cl_b(R^d)$ is an **F**-measurable set-valued mapping for every $\alpha \in [0,1]$. The fuzzy-valued random function $f : R_+ \times \Omega \to \mathcal{F}(R^d)$ is said to be predictable (resp. nonanticipating) if the set-valued mapping $[f]^\alpha : R_+ \times \Omega \to Cl_b(R^d)$ is $\mathcal{P}$ (resp. $\mathcal{N}$) measurable for every $\alpha \in [0,1]$. The family $(u_t, t \in R_+)$ of fuzzy random variables is called a fuzzy stochastic process. It is called $(\mathbf{F}_t)_{t \geq 0}$-adapted if $u_t$ is an $\mathbf{F}_t$-measurable fuzzy random variable for every $t \in R_+$. We recall the following version of the theorem of Negoita and Ralescu.

**Theorem 2.** ( [13]) Let $Y$ be a set and let $\{Y_\alpha, \alpha \in [0,1]\}$ be a family of subsets of $Y$ such that

a) $Y_0 = Y$,
b) $\alpha_1 \leq \alpha_2 \Rightarrow Y_{\alpha_1} \supset Y_{\alpha_2}$,
c) $\alpha_n \nearrow \alpha \Rightarrow Y_\alpha = \bigcap_{n=1}^\infty Y_{\alpha_n}$.

   Then the function $\phi : Y \to [0,1]$ defined by $\phi(x) = \sup\{\alpha \in [0,1] : x \in Y_\alpha\}$ has the property that $\{x \in Y : \phi(x) \geq \alpha\} = Y_\alpha$ for any $\alpha \in [0,1]$.

For $f : R_+ \times \Omega \to \mathcal{F}(R^d)$ being a predictable fuzzy random function it is called to be $L^2_\mathcal{P}(\mu_Z)$-integrably bounded if $\||[f]^0\| \in L^2_\mathcal{P}(\mu_Z)$. For such a predictable fuzzy random function $f$ let us consider the trajectory set-valued stochastic integral $Y_\alpha(t) := \int_0^t [f]_s^\alpha dZ_s$ for any $t \in R_+$ and every $\alpha \in [0,1]$. Then by Theorem 1 and Theorem 2 ( [13]), for every fixed $t \in R_+$ there exists a fuzzy set $X(f,Z)_t \in \mathcal{F}(L^2(\Omega, \mathbf{F}_t, P, R^d))$ such that $[X(f,Z)_t]^\alpha = \int_0^t [f]_s^\alpha dZ_s$ for every $t \in R_+$ and every $\alpha \in [0,1]$. Having the family of just described fuzzy sets $\{X(f,Z)_t, t \in R_+\}$, one can introduce ( [12]):

**Definition 2.** By a fuzzy trajectory stochastic integral of the predictable and $L^2_\mathcal{P}(\mu_Z)$-integrably bounded fuzzy random function $f$ with respect to the semimartingale $Z$ we mean the family of fuzzy sets $\{X(f,Z)_t, t \in R_+\}$ described above. We denote it by $X(f,Z)_t := (\mathcal{F})\int_0^t f dZ$ for $t \in R_+$.

## 4   Fuzzy Stochastic Differential Equation

Let us assume now $Z = W$, where $W$ is an $m$-dimensional Wiener process defined on some filtered probability space $(\Omega, \mathbf{F}, \{\mathbf{F}_t\}_{t \in [0,T]}, P)$. By $R^{d \times m}$ we denote the space of all $d \times m$ matrices $(g_{ij})_{d \times k}$ with real elements, equipped with the norm: $\|(g_{ij})_{d \times m}\| = \max_{1 \leq i \leq d, 1 \leq j \leq m} |g_{ij}|$. By $\mathcal{F}_t^x$ we denote a $\sigma$-field generated by the process $x$ to the time $t$, i.e. $\mathcal{F}_t^x = \sigma\{x(s); s \leq t\}$. By $P^x$ we denote the distribution (probability law) of the process $x$ under the probability $P$. We consider $\beta([0,T]) \times \beta(R^d)$-measurable fuzzy valued functions $f : [0,T] \times R^d \to \mathcal{F}_{Conv}(R^d)$ and $g : [0,T] \times R^d \to \mathcal{F}_{Conv}(R^{d \times m})$. By the fuzzy stochastic equation we mean the formal relation

$$dx(t) = f(t, x(t))dt + g(t, x(t)dW_t, \ t \in [0,T] \tag{1}$$
$$x(0) = x_0 \in \mathcal{F}_{Conv}(R^d)$$

which is interpreted as a family of stochastic fuzzy integral inclusions

$$x(t) - x(s) \in \left[(F)\int_s^t f(\tau, x(\tau))d\tau\right]^\alpha + \left[(F)\int_s^t g(\tau, x(\tau))dW_\tau\right]^\alpha,$$
$$x(0) \in [x_0]^\alpha,$$

where $0 \leq s \leq t \leq T$. Or equivalently as

$$x(t) - x(s) \in \int_s^t [f(\tau, x(\tau)]^\alpha \, d\tau + \int_s^t [g(\tau, x(\tau)]^\alpha \, dW_\tau, \ 0 \leq s \leq t \leq T, \tag{1$^\alpha$}$$
$$x(0) \in [x_0]^\alpha,$$

for $\alpha \in [0,1]$. The fuzzy trajectory stochastic integrals above are defined as in Definition 2, while the set-valued trajectory stochastic integrals in $(1^\alpha)$ are their $\alpha$-level sets taking as a semimartingale $Z_t = t$ or $Z_t = W_t$, provided they are nonempty sets. The stochastic inclusion $(1^\alpha)$ can only have any significance as a replacement for the fuzzy stochastic differential equation (1) if the solutions of $(1^\alpha)$ generate fuzzy sets. Following [12] this is true for distributions (probability laws) of the solution processes to $(1^\alpha)$. For this aim we describe first the set of solutions to the stochastic inclusion $(1^\alpha)$ for a fixed $\alpha \in [0,1]$.

**Definition 3.** By a solution to the stochastic inclusion $(1^\alpha)$ we mean a $d$-dimensional, continuous stochastic process $x$ defined on some probability space $(\Omega, \mathbf{F}, P)$ with the filtration $(\mathbf{F}_t^x)_{t \in [0,T]}$, an $(\mathbf{F}_t^x)_{t \in [0,T]}$-Wiener process $W$ and stochastic processes $u \in S_{\mathcal{N}^x}^2 ([f \circ x]^\alpha, \mu_W)$ and $v \in S_{\mathcal{N}^x}^2 ([g \circ x]^\alpha, \mu_W)$ such that:

$$x(t) = x(0) + \int_0^t u_s ds + \int_0^t v_s dW_s, \ t \in [0,T] \tag{2$^\alpha$}$$
$$x(0) \in [x_0]^\alpha$$

where $\mathcal{N}^x$ denotes here a $\sigma$-field of nonanticipating subsets in $[0,T] \times \Omega$, generated by the filtration $(\mathbf{F}_t^x)_{t \in [0,T]}$.

The solution defined above is thought as a system $(\Omega, \mathbf{F}, P, W, \{\mathbf{F}_t^x\}_{t \in [0,T]}, x)$ in which all elements can depend on the fixed level index $\alpha \in [0, 1]$. To avoid such a dependence one can use an equivalent approach based on martingale problems on the path space (canonical space) (see [12] for details). This allows identification of weak solutions with their distributions which is crucial for the notion of fuzzy solution to the equation (1). For these aims let $C :=$ $C([0, T], R^d)$ be the space of continuous, $R^d$-valued functions with the Borel $\sigma$-field $\beta(C)$. Define a coordinate process (canonical projections) $\pi_t : C \to R^d$, $\pi_t(x) = x(t)$, and its natural filtration $(\mathcal{A}_t)_{t \in [0,T]}, \mathcal{A}_t := \sigma\{\pi_s : s \leq t\}, t \in$ $[0, T]$. Let us take its right-continuous version $(\mathcal{A}_t^+)_{t \in [0,T]}, \mathcal{A}_t^+ := \mathcal{A}_{t+} =$ $\bigcap_{s > t} \mathcal{A}_s$. Let $a : [0, T] \times C \to R^d$, $b : [0, T] \times C \to R^{d \times m}$ be $\beta([0, T]) \times \beta(C)$-measurable functions. By $C_b^2(R^d)$ we denote the space of all bounded and twice continuously differentiable functions $z : R^d \to R^1$. For $z \in C_b^2(R^d)$ and $y \in C$ we let:

$$(\mathcal{L}_t z)(y) := \frac{1}{2} \sum_{i=1}^d \sum_{k=1}^d \gamma_{ik}(t, y) \frac{\partial^2 z(y(t))}{\partial x_i \partial x_k} + \sum_{i=1}^d a_i(t, y) \frac{\partial z(y(t))}{\partial x_i},$$

where $\gamma_{ik}(t, y) = \sum_{j=1}^m b_{ij}(t, y) b_{kj}(t, y)$; $1 \leq i, k \leq d$.
Let $\mathcal{M}(C)$ denote the set of all probability measures on $(C, \beta(C))$.

**Definition 4.** Let $\alpha \in [0, 1]$. A probability measure $Q \in \mathcal{M}(C)$ is said to be a solution to the local martingale problem of $(1^\alpha)$ if it satisfies:

   $i)$ $Q\{\pi_0 \in [x_0]^\alpha\} = 1$

   $ii)$ there exist measurable mappings $a : [0, T] \times C \to R^d$, and $b : [0, T] \times C \to$ $R^{d \times m}$, such that $a(t, y) \in [f(t, y(t))]^\alpha$, $b(t, y) \in [g(t, y(t))]^\alpha$ $dt \times dQ$-a.e., and for every $z \in C_b^2(R^d)$ the process $(M_t^z)$ ( on $(C, \beta(C), Q)$ ):

$$M_t^z := z \circ \pi_t - z \circ \pi_0 - \int_0^t (\mathcal{L}_s z) ds : t \in [0, T]$$

is a $(\mathcal{A}_t^+, Q)$-local martingale.

Let $\mathcal{R}^\alpha(f, g, [x_0]^\alpha)$ denote the set of those measures $Q \in \mathcal{M}(C)$, which are solutions to the local martingale problem of $(1^\alpha)$. The space $\mathcal{M}(C)$ and the set $\mathcal{R}^\alpha(f, g, [x_0]^\alpha)$ can be equipped with a topology of weak convergence of probability measures. We have the following connection between the solutions to the stochastic inclusion $(1^\alpha)$ and solutions to the martingale problem described above (see [12]).

**Proposition 1.** Let $f : [0, T] \times R^d \to \mathcal{F}_{Conv}(R^d)$ and $g : [0, T] \times R^d \to$ $\mathcal{F}_{Conv}(R^{d \times m})$ be $\beta([0, T]) \times \beta(R^d)$-measurable fuzzy valued functions. Then for every $\alpha \in [0, 1]$ and $x_0 \in \mathcal{F}_{Conv}(R^d)$ there exists a weak solution to the stochastic inclusion $(1^\alpha)$ if and only if $\mathcal{R}^\alpha(f, g, [x_0]^\alpha) \neq \emptyset$.

Further analysis in this section makes use of the following notions.

**Definition 5.** A set $U \subset R^{d \times m}$ is said to satisfy a (DC) property (diagonal convexity) if the set $D(U) := \{uu^T : u \in U\}$ is a convex subset in $R^{d \times d}$, where $u^T$ denotes the transposition for $u$. Consequently, a fuzzy valued function $g : [0, T] \times R^d \to \mathcal{F}_{Conv}(R^{d \times m})$ is said to satisfy a (DC) property if the set $[g(t, x)]^\alpha \subset R^{d \times m}$ satisfies such property for every $(t, x) \in [0, T] \times R^d$ and $\alpha \in [0, 1]$.

Having the characterization given in Proposition 3 one can state the notion of a fuzzy solution to equation (1).

**Definition 6.** By the fuzzy solution of the fuzzy stochastic differential equation (1) we mean a fuzzy set $X(x_0, f, g) \in \mathcal{F}_{Comp}(\mathcal{M}(C))$ such that $[X(x_0, f, g)]^\alpha = \mathcal{R}^\alpha(f, g, [x_0]^\alpha)$ for every $\alpha \in [0, 1]$.

The main result of this note comes from [12], where we refer the reader for the proof.

**Theorem 3.** Let $f : [0, T] \times R^d \to \mathcal{F}_{Conv}(R^d)$, and $g : [0, T] \times R^d \to \mathcal{F}_{Conv}(R^{d \times m})$ be $\beta([0, T]) \times \beta(R^d)$-measurable and bounded fuzzy valued functions such that $f(t, \cdot) : R^d \to (\mathcal{F}_{Conv}(R^d), d_\infty)$ and $g(t, \cdot) : R^d \to (\mathcal{F}_{Conv}(R^{d \times m}), d_\infty)$ are continuous for each fixed $t \in [0, T]$. Assume also that $g$ satisfies a (DC) property and let $x_0 \in \mathcal{F}_{Conv}(R^d)$. Then there exists a fuzzy solution of the fuzzy stochastic equation (1).

## 5 Applications

This meaning of the stochastic fuzzy equation described above reflects the idea used earlier in the deterministic case by Hullermeier in [5] and next by Agarwal, O' Regan, Lakshmikantham in [1] (and others), which was an alternative approach to the notion of the fuzzy differential equation initiated by Kaleva in [6]. In [5] and [1] the authors considered fuzzy differential equation

$$dx(t) = f(t, x(t))dt, \quad x(0) = x_0; \tag{3}$$

interpreted as a family of integral inclusions:

$$x(t) - x(s) \in \int_s^t [f(\tau, x(\tau))]^\alpha \, d\tau, \quad x(0) \in [x_0]^\alpha;$$

for $\alpha \in [0, 1]$. The notion of a fuzzy solution of fuzzy differential equation was described as follows: for a fixed $\alpha \in [0, 1]$, let $S(\alpha)$ be a set of all solutions of the inclusion above ( provided it is nonempty), i.e.

$$S(\alpha) := \left\{ x \in C([0, T], R^d) : x(t) - x(s) \in \int_s^t [f(\tau, x(\tau))]^\alpha \, d\tau, 0 \le s \le t \le T, x(0) \in [x_0]^\alpha \right\}.$$

Then under appropriate conditions $S(\alpha)$ is nonempty and compact subset of $C([0,T], R^d)$ for every $\alpha \in [0,1]$. Moreover if the family $(S(\alpha) : \alpha \in [0,1])$ satisfies conditions of Negoita and Ralescu type theorem, therefore there exists a fuzzy set (let say) $X(f,x_0) \in \mathcal{F}_{Comp}(C([0,T], R^d))$ such that $[X(f,x_0)]^\alpha = S(\alpha)$ for every $\alpha \in [0,1]$. In [1] just a fuzzy set is called a fuzzy solution to the fuzzy differential equation (3). Taking the diffusion term $g = \theta$ (i.e., zero in a fuzzy sense) in the fuzzy stochastic equation (1), we obtain $\mathcal{R}^\alpha(f, \theta, [x_0]^\alpha) = \{\delta_x : x \in S(\alpha)\}$ and consequently $[X(f, \theta, x_0)]^\alpha = \{\delta_x : x \in [X(f,x_0)]^\alpha\}$ for every $\alpha \in [0,1]$. In this sense the martingale approach in the stochastic case generalizes the deterministic case and the fuzzy stochastic equation (1) generalizes the fuzzy differential equation (3) above.

# References

1. Agarwal, R.P., O'Regan, D., Lakshmikantham, V.: A stacking theorem approach for fuzzy differential equations. Nonlinear Anal. 55, 299–312 (2003)
2. Bhaskar, T.G., Lakshmikantham, V., Devi, V.: Revisiting fuzzy differential equations. Nonlinear Anal. 58, 351–358 (2004)
3. Chung, K.L., Williams, R.J.: Introduction to Stochastic Integration. Birkhauser, Boston (1983)
4. Hiai, F., Umegaki, H.: Integrals, conditional expectations and martingales for multivalued functions. J. Multivar. Anal. 7, 147–182 (1977)
5. Hullermeier, E.: An approach to modelling and simulation of uncertain dynamical systems. Int. J. Uncertainty Fuzziness Knowledge Based Systems 5, 117–137 (1997)
6. Kaleva, O.: Fuzzy differential equations. Fuzzy Sets and Systems 24, 301–317 (1987)
7. Kim, J.H.: On fuzzy stochastic differential equations. J. Korean Math. Soc. 42, 153–169 (2005)
8. Kisielewicz, M.: Differential Inclusions and Optimal Control. Kluwer Acad. Publ., Dordrecht (1991)
9. Lakshmikantham, V., Mohapatra, R.N.: Theory of Fuzzy Differential Equations and Inclusions. Taylor and Francis Publishers, London (2003)
10. Malinowski, M.T., Michta, M.: Fuzzy stochastic integral equations. Dynam. Systems Appl. 19, 473–494 (2010)
11. Malinowski, M.T., Michta, M.: Stochastic fuzzy differential equations with an application. Kybernetika 47, 123–143 (2011)
12. Michta, M.: On set-valued stochastic integrals and fuzzy stochastic equations. Fuzzy Sets and Systems (2011), doi:10.1016/j.fss.2011.01.007
13. Negoita, C.V., Ralescu, D.A.: Applications of Fuzzy Sets to System Analysis. Wiley, New York (1975)
14. Ogura, Y.: On stochastic differential equations with fuzzy set coefficients. In: Soft Methods for Handling Variability and Imprecision, pp. 263–270. Springer, Heidelberg (2008)
15. Protter, P.: Stochastic Integration and Differential Equations: A New Approach. Springer, New York (1990)

# Set-Valued Stochastic Integrals with Respect to Poisson Processes in a Banach Space

Jinping Zhang and Itaru Mitoma

**Abstract.** In a separable Banach space $\mathfrak{X}$, after studying $\mathfrak{X}$-valued stochastic integrals with respect to Poisson random measure $N(dsdz)$ and the compensated Poisson random measure $\tilde{N}(dsdz)$ generated by stationary Poisson stochastic process $\mathbf{P}$, we prove that if the characteristic measure $\nu$ of $\mathbf{P}$ is finite, the stochastic integrals (denoted by $\{J_t(F)\}$ and $\{I_t(F)\}$ separately) for set-valued stochastic process $\{F(t)\}$ are integrably bounded and convex a.s. Furthermore, the set-valued integral $\{I_t(F)\}$ with respect to compensated Poisson random measure is a right continuous (under Hausdorff metric) set-valued martingale.

**Keywords:** Poisson Random Measure, Compensated Poisson Random Measure, Set-Valued Stochastic Integral.

## 1 Introduction

Recently, stochastic integrals for set-valued stochastic processes with respect to Brownian motion and martingale have been received much attention, e.g. see [5, 6, 9, 11, 13, 16]. Correspondingly, the set-valued differential equations are studied, e.g. see [11, 12, 14, 15, 17]. Michta [10] extended the integrator to a larger class: semimartingales. But the integrable boundedness of the corresponding set-valued stochastic integrals are not obtained since the semimartingale may not be of finite variation.

Jinping Zhang
Department of Mathematics and Physics, North China Electric Power University,
Beijing 102206, China
e-mail: zhangjinping@ncepu.edu.cn

Itaru Mitoma
Department of Mathematics, Saga University, Saga 840-8502, Japan
e-mail: mitoma@ms.saga-u.ac.jp

S. Li (Eds.): Nonlinear Maths for Uncertainty and its Appli., AISC 100, pp. 125–134.
springerlink.com                                  © Springer-Verlag Berlin Heidelberg 2011

Poisson stochastic processes are special but they play an important role both on random mathematics (cf. [4,2,7]) and on applied fields for example in financial mathematics [7]. If the characteristic measure $\nu$ of stationary Poisson process of $\mathbf{P}$ is finite, then both the Poisson random measure $N(dsdz)$ (where $z \in Z$, the state space of $\mathbf{P}$) and compensated Poisson random measure $\tilde{N}(dsdz)$ are of finite variation. We will prove that the stochastic integrals for set-valued $\mathscr{S}$-predictable (see Definition 1) process with respect to $N(dsdz)$ and $\tilde{N}(dsdz)$ are integrably bounded, convex a.s. Furthermore, when the $\sigma$-algebra $\mathcal{F}$ is separable, the set-valued stochastic integrals with respect to compensated Poisson random measure is a right continuous (under Hausdorff metric) set-valued martingale. The maximal inequality is studied in another paper [18].

This paper is organized as follows: Section 3 is on the notations and preliminaries on set-valued theory. In Section 3, at first we study the stochastic integrals for $\mathfrak{X}$-valued $\mathscr{S}$-predictable process with respect to $N(dsdz)$ and $\tilde{N}(dsdz)$. Then we study the stochastic integrals for set-valued $\mathscr{S}$-predictable processes with respect to $N(dsdz)$ and $\tilde{N}(dsdz)$.

Note: since the limitation of pages, in this paper we only list our results without proof. The detailed proofs are given in [18].

## 2  Preliminaries

Let $(\Omega, \mathcal{F}, P)$ be a complete probability space, $\{\mathcal{F}_t\}_{t \geq 0}$ a filtration satisfying the usual conditions, which means $\mathcal{F}_0$ includes all $P$-null sets in $\mathcal{F}$, the filtration is non-decreasing and right continuous. Let $\mathcal{B}(E)$ be the Borel field of a topological space $E$, $(\mathfrak{X}, \| \cdot \|)$ a separable Banach space equipped with the norm $\| \cdot \|$ and $\mathbf{K}(\mathfrak{X})$ (resp. $\mathbf{K_b}(\mathfrak{X})$) the family of all nonempty closed (resp. bounded closed) subsets of $\mathfrak{X}$. Let $1 \leq p < +\infty$ and $L^p(\Omega, \mathcal{F}, P; \mathfrak{X})$ (denoted briefly by $L^p(\Omega; \mathfrak{X})$) be the Banach space of equivalent class of $\mathfrak{X}$-valued $\mathcal{F}$-measurable functions $f : \Omega \to \mathfrak{X}$ such that the norm $\|f\|_p = \left\{ \int_\Omega \|f(\omega)\|^p dP \right\}^{1/p}$ is finite. An $\mathfrak{X}$-valued function $f$ is called $L^p$-integrable if $f \in L^p(\Omega; \mathfrak{X})$.

A set-valued function $F : \Omega \to \mathbf{K}(\mathfrak{X})$ is said to be *measurable* if for any open set $O \subset \mathfrak{X}$, the inverse $F^{-1}(O) := \{\omega \in \Omega : F(\omega) \cap O \neq \emptyset\}$ belongs to $\mathcal{F}$. Such a function $F$ is called a *set-valued random variable*. Let $\mathcal{M}(\Omega, \mathcal{F}, P; \mathbf{K}(\mathfrak{X}))$ be the family of all set-valued random variables, which is briefly denoted by $\mathcal{M}(\Omega; \mathbf{K}(\mathfrak{X}))$.

For $A, B \in 2^{\mathfrak{X}}$ (the power set of $\mathfrak{X}$), $H(A, B) \geq 0$ is defined by

$$H(A, B) := \max\{\sup_{x \in A} \inf_{y \in B} \|x - y\|, \sup_{y \in B} \inf_{x \in A} \|x - y\|\}.$$

$H(A, B)$ for $A, B \in \mathbf{K}_b(\mathfrak{X})$ is called the *Hausdorff metric*. It is well-known that $\mathbf{K}_\mathbf{b}(\mathfrak{X})$ equipped with the $H$-metric denoted by $((\mathbf{K}_b(\mathfrak{X}), H))$ is a complete metric space (cf. [8]).

For $F \in \mathcal{M}(\Omega, \mathbf{K}(\mathfrak{X}))$, the family of all $L^p$-integrable selections is defined by

$$S_F^p(\mathcal{F}) := \{f \in L^p(\Omega, \mathcal{F}, P; \mathfrak{X}) : f(\omega) \in F(\omega) \ a.s.\}.$$

In the following, $S_F^p(\mathcal{F})$ is denoted briefly by $S_F^p$. If $S_F^p$ is nonempty, $F$ is said to be $L^p$-*integrable*. $F$ is called $L^p$-*integrably bounded* if there exits a function $h \in L^p(\Omega, \mathcal{F}, P; \mathbb{R})$ such that $\|x\| \leq h(\omega)$ for any $x$ and $\omega$ with $x \in F(\omega)$. It is equivalent to that $\|F\|_\mathbf{K} \in L^p(\Omega; \mathbb{R})$, where $\|F(\omega)\|_\mathbf{K} := \sup_{a \in F(\omega)} \|a\|$. The family of all measurable $\mathbf{K}(\mathfrak{X})$-valued $L^p$-integrably bounded functions is denoted by $L^p(\Omega, \mathcal{F}, P; \mathbf{K}(\mathfrak{X}))$. Write it for brevity as $L^p(\Omega; \mathbf{K}(\mathfrak{X}))$.

**Proposition 1.** ( [3]) *Let* $F_1, F_2 \in \mathcal{M}(\Omega; \mathfrak{X})$ *and* $F(\omega) = cl(F_1(\omega) + F_2(\omega))$ *for all* $\omega \in \Omega$. *Then* $F \in \mathcal{M}(\Omega; \mathfrak{X})$.

Let $\mathbb{R}_+$ be the set of all nonnegative real numbers and $\mathcal{B}_+ := \mathcal{B}(\mathbb{R}_+)$. $\mathbb{N}$ denotes the set of natural numbers. An $\mathfrak{X}$-valued stochastic process $f = \{f_t : t \geq 0\}$ (or denoted by $f = \{f(t) : t \geq 0\}$ )is defined as a function $f : \mathbb{R}_+ \times \Omega \longrightarrow \mathfrak{X}$ with the $\mathcal{F}$-measurable section $f_t$, for $t \geq 0$. We say $f$ is *measurable* if $f$ is $\mathcal{B}_+ \otimes \mathcal{F}$-measurable. The process $f = \{f_t : t \geq 0\}$ is called $\mathcal{F}_t$-*adapted* if $f_t$ is $\mathcal{F}_t$-measurable for every $t \geq 0$. $f = \{f_t : t \geq 0\}$ is called *predictable* is it is $\mathcal{P}$-measurable, where $\mathcal{P}$ is the $\sigma$-algebra generated by all left continuous and $\mathcal{F}_t$-adapted stochastic processes.

In a fashion similar to the $\mathfrak{X}$-valued stochastic process, a *set-valued stochastic process* $F = \{F_t : t \geq 0\}$ is defined as a set-valued function $F : \mathbb{R}_+ \times \Omega \longrightarrow \mathbf{K}(\mathfrak{X})$ with $\mathcal{F}$-measurable section $F_t$ for $t \geq 0$. It is called *measurable* if it is $\mathcal{B}_+ \otimes \mathcal{F}$-measurable, and $\mathcal{F}_t$-*adapted* if for any fixed $t$, $F_t(\cdot)$ is $\mathcal{F}_t$-measurable. $F = \{F_t : t \geq 0\}$ is called *predictable* if it is $\mathcal{P}$-measurable.

# 3 Set-Valued Stochastic Integrals with Respect to Poisson Process

In this section, at first we will study the stochastic integrals with respect to the Poisson random measure and compensated Poisson random measure for $\mathfrak{X}$-valued stochastic processes. Then we study the corresponding stochastic integrals for set-valued stochastic processes.

## 3.1 Single Valued Stochastic Integrals w.r.t. Poisson Process

Let $(\mathfrak{X}, \| \cdot \|)$ be a separable Banach space and $Z$ an another separable Banach space with $\sigma$-algebra $\mathcal{B}(Z)$. A *point function* $\mathbf{p}$ on $Z$ means a mapping

$\mathbf{p} : \mathbf{D_p} \to Z$, where the domain $\mathbf{D_p}$ is a countable subset of $[0, T]$. $\mathbf{p}$ defines a counting measure $N_\mathbf{p}(dtdz)$ on $[0, T] \times Z$ (with the product $\sigma$-algebra $\mathcal{B}([0, T]) \otimes \mathcal{B}(Z)$) by

$$N_\mathbf{p}((0, t], U) := \#\{\tau \in \mathbf{D_p} : \tau \le t, \mathbf{p}(\tau) \in U\}, \tag{1}$$
$$t \in (0, T], \ U \in \mathcal{B}(Z).$$

The for $0 \le s < t \le T$,

$$N_\mathbf{p}((s, t], U) := N_\mathbf{p}((0, t], U) - N_\mathbf{p}((0, s], U). \tag{2}$$

In the following, we also write $N_\mathbf{p}((0, t], U)$ as $N_\mathbf{p}(t, U)$.

A *point process* is obtained by randomizing the notion of point functions. If there is a continuous $\mathcal{F}_t$-adapted increasing process $\hat{N}_\mathbf{p}$ such that for $U \in \mathcal{B}(Z)$ and $t \in [0, T]$, $\tilde{N}_\mathbf{p}(t, U) := N_\mathbf{p}(t, U) - \hat{N}_\mathbf{p}(t, U)$ is an $\mathcal{F}_t$-martingale, then the random measure $\{\hat{N}_\mathbf{p}(t, U)\}$ is called the *compensator* of the point process $\mathbf{p}$ (or $\{N_\mathbf{p}(t, U)\}$) and the process $\{\tilde{N}_\mathbf{p}(t, U)\}$ is called the *compensated* point process.

A point process $\mathbf{p}$ is called *Poisson Process* if $N_\mathbf{p}(dtdz)$ is a Poisson random measure on $[0, T] \times Z$. A Poisson point process is stationary if and only if its intensity measure $\nu_\mathbf{p}(dtdz) = E[N_\mathbf{p}(dtdz)]$ is of the form

$$\nu_\mathbf{p}(dtdz) = \nu(dz)dt \tag{3}$$

for some measure $\nu(dz)$ on $(Z, \mathcal{B}(Z))$. $\nu(dz)$ is called the *characteristic measure of* $\mathbf{p}$.

Let $\nu$ be a $\sigma$- finite measure on $(Z, \mathcal{B}(Z))$, (i.e. there exists $U_i \in \mathcal{B}(Z), i \in \mathbb{N}$, pairwise disjoint such that $\nu(U_i) < \infty$ for all $i \in \mathbb{N}$ and $Z = \cup_{i=1}^{\infty} U_i$), and $\mathbf{p} = (\mathbf{p}_t)$ the $\mathcal{F}_t$-adapted stationary Poisson point process on $Z$ with the characteristic measure $\nu$ such that the compensator $\hat{N}_\mathbf{p}(t, U) = E[N_\mathbf{p}(t, U)] = t\nu(U)$ (non-random).

For convenience, from now on, we will omit the subscript $\mathbf{p}$ appeared in the above notations.

**Definition 1.** An $\mathfrak{X}$-valued function defined on $[0, T] \times Z \times \Omega$ is called $\mathscr{S}$-predictable if the mapping $(t, z, \omega) \to f(t, z, \omega)$ is $\mathscr{S}/\mathcal{B}(\mathfrak{X})$-measurable where $\mathscr{S}$ is the smallest $\sigma$-algebra on $[0, T] \times Z \times \Omega$ with respect to all $g$ having the following properties are measurable:
   (i) for each $t \in [0, T]$, $(z, \omega) \to g(t, z, \omega)$ is $\mathcal{B}(Z) \otimes \mathcal{F}_t$-measurable;
   (ii) for each $(z, \omega)$, $t \to (t, z, \omega)$ is left continuous.

*Remark 1.* $\mathscr{S} = \mathcal{P} \otimes \mathcal{B}(Z)$, where $\mathcal{P}$ denotes the $\sigma$-field on $[0, t] \times \Omega$ generated by all left continuous and $\mathcal{F}_t$-adapted processes.

Set

$$\mathscr{L} = \Big\{ f(t,z,\omega) : f \text{ is } \mathscr{S}-predictable \text{ and}$$

$$E\Big[ \int_0^T \int_Z \|f(t,z,\omega)\|^2 \nu(dz)dt \Big] < \infty \Big\}$$

equipped with the norm

$$\|f\|_{\mathscr{L}} := \Big( E\Big[ \int_0^T \int_Z \|f(t,z,\omega)\|^2 \nu(dz)dt \Big] \Big)^{1/2}.$$

Similar to the Lemma 2.2 in [7], we have the following result:

**Lemma 1.** *Let* $f(z)$ *be an* $\mathcal{F}_s \otimes \mathcal{B}(Z)$-*measurable random variable taking values in* $\mathfrak{X}$ *such that* $E\big[ \int_Z \|f(z)\|^2 \nu(dz) \big] < \infty$. *Then it holds for any* $s < t \leq T$,

$$E\Big[ \Big( \int_Z f(z)\tilde{N}((s,t],dz) \Big) \mid \mathcal{F}_s \Big] = 0 \text{ a.s.} \tag{4}$$

$$E\Big[ \Big\| \int_Z f(z)\tilde{N}((s,t],dz) \Big\|^2 \mid \mathcal{F}_s \Big] = (t-s) \int_Z \|f(z)\|^2 \nu(dz) \text{ a.s.} \tag{5}$$

$$E\Big[ \Big( \int_Z f(z)N((s,t],dz) \Big) \mid \mathcal{F}_s \Big] = (t-s) \int_Z f(z)\nu(dz) \text{ a.s.} \tag{6}$$

*If* $\nu$ *is of finite measure, then there exists a constant* $C$ *(independent of* $t$*) such that*

$$E\Big[ \Big\| \int_Z f(z)N((s,t],dz) \Big\|^2 \mid \mathcal{F}_s \Big] \leq C \int_Z \|f(z)\|^2 \nu(dz) \text{ a.s.} \tag{7}$$

From now on, we suppose $\nu$ is finite in the measurable space $(Z, \mathcal{B}(Z))$.

Let $\mathbb{S}$ be the subspace of those $f \in \mathscr{L}$ for which there exists a partition $0 = t_0 < t_1 < \cdots < t_n = T$ of of $[0,T]$ such that

$$f(t,z,\omega) = f(0,z,\omega)\chi_{\{0\}}(t) + \sum_{i=1}^n \chi_{(t_{i-1},t_i]}(t)f(t_{i-1},z,\omega).$$

**Lemma 2.** $\mathbb{S}$ *is dense in* $\mathscr{L}$ *with respect to the norm* $\|\cdot\|_{\mathscr{L}}$.

Let $f$ be in $\mathbb{S}$ and

$$f(t,z,\omega) = f(0,z,\omega)\chi_{\{0\}}(t) + \sum_{i=1}^n \chi_{(t_{i-1},t_i]}(t)f(t_{i-1},z,\omega),$$

where $0 = t_0 < t_1 < \cdots < t_n = T$ is a partition of $[0,T]$. Define

$$J_T(f) = \int_{0+}^T \int_Z f(s,z,\omega)N(dzdt) := \sum_{i=1}^n \int_Z f(t_{i-1},z,\omega)N((t_{i-1},t_i],dz),$$

$$\tag{8}$$

and

$$I_T(f) = \int_{0+}^{T} \int_Z f(s,z,\omega)\tilde{N}(dzdt) := \sum_{i=1}^{n} \int_Z f(t_{i-1},z,\omega)\tilde{N}((t_{i-1},t_i],dz).$$
(9)

For any integer $0 \le k \le n$, let

$$M_k = \sum_{i=1}^{k} \int_Z f_{t_{i-1}} \tilde{N}((t_{i-1},t_i],dz)$$

then $M_k$ is $\mathcal{F}_{t_k}$-measurable, $E[M_k] = 0$, $E[I_T(f)] = E[M_n] = 0$ and

$$E[M_k|\mathcal{F}_{t_{k-1}}] = E[(M_{k-1} + \int_Z f_{t_{k-1}}\tilde{N}((t_{i-1},t_i],dz)|\mathcal{F}_{t_{k-1}}]$$

$$= M_{k-1} + \int_Z f_{t_{k-1}}E[\tilde{N}((t_{i-1},t_i],dz)] = M_{k-1}.$$

That is to say $\{M_k, \mathcal{F}_{t_k} : 1 \le k \le n\}$ is an $\mathfrak{X}$-valued martingale.

For any $t \in (0,T]$, define

$$J_t(f) = \int_{0+}^{t} \int_Z f(s,z,\omega)N(dzds) := \sum_{i=1}^{n} \int_Z f(t_{i-1},z,\omega)N((t_{i-1}\wedge t, t_i \wedge t],dz),$$
(10)

and

$$I_t(f) = \int_{0+}^{T} \int_Z f(s,z,\omega)\tilde{N}(dzds) := \sum_{i=1}^{n} \int_Z f(t_{i-1},z,\omega)\tilde{N}((t_{i-1}\wedge t, t_i \wedge t],dz).$$
(11)

By Lemma 2, for any $f \in \mathcal{L}$, there exist a sequence $\{f^n : n \in \mathbb{N}\}$ in $\mathbb{S}$ such that $\{f^n\}$ converges to $f$ with respect to $\|\cdot\|_{\mathscr{L}}$ and $\left\{\int_{0+}^{t} \int_Z f^n N(dzds)\right\}$ converges to a limit in $L^2$-sense. The limit is denoted by

$$I_t(f) := \int_{0+}^{t} \int_Z f(s,z,\omega)\tilde{N}(dzds),$$

which is called *stochastic integral of $f$* with respect to compensated Poisson random measure $\tilde{N}_t$. Similarly, we can define the *stochastic integral of $f$* with respect to Poisson random measure $N$, denoted by $J_t(f) = \int_{0+}^{t} \int_Z f(s,z,\omega)N(dzds)$. Similarly, for any $0 < s < t < T$, the integrals $\int_s^t \int_Z f(\tau,z,\omega)\tilde{N}_{\mathbf{p}}(dzd\tau)$ and $\int_s^t \int_Z f(\tau,z,\omega)N(dzd\tau)$ can be defined.

*Remark 2.* When the measure $\nu$ is finite, for any $A \in \mathcal{B}(Z)$, both $\{N(t,A)\}$ and $\{\tilde{N}(t,A)\}$ are of finite variation processes. The stochastic integrals coincide with Lebesgue-Stieltjes integrals.

**Proposition 2.** *For any $f \in \mathbb{S}$, both $\{I_t(f)\}$ and $\{J_t(f)\}$ are $\mathcal{F}_t$-adapted square-integrable processes. Moreover, $\{I_t(f)\}$ is an $\mathfrak{X}$-valued right continuous martingale with mean zero.*

**Definition 2.** ( [1]) A Banach space $(\mathfrak{X}, \|\cdot\|)$ is called M-type 2 if and only if there exists a constant $C_{\mathfrak{X}} > 0$ such that for any $\mathfrak{X}$-valued martingale $\{\mathbf{M_k}\}$, it holds that

$$\sup_k E[\|\mathbf{M}_k\|^2] \leq C_{\mathfrak{X}} \sum_k E[\|\mathbf{M}_k - \mathbf{M}_{k-1}\|^2]. \tag{12}$$

**Theorem 1.** *Let $\mathfrak{X}$ be of M-type 2 and $(Z, \mathcal{B}(Z))$ a separable Banach space with finite measure $\nu$. Let $N_{\mathbf{p}}$ be stationary Poisson process with characteristic measure $\nu$. Let $f$ be in $\mathscr{L}$. Then there exists a constant $C$ such that*

$$E\left[\sup_{0<s\leq t} \left\| \int_{0+}^s \int_Z f(\tau, z, \omega)\tilde{N}(dzd\tau)\right\|^2\right] \leq C\int_0^t \int_Z E[\|f(s, z, \omega)\|^2]\nu(dz)ds, \tag{13}$$

*and*

$$E\left[\sup_{0<s\leq t} \left\| \int_{0+}^s \int_Z f(\tau, z, \omega)N(dzd\tau)\right\|^2\right] \leq C\int_0^t \int_Z E[\|f(s, z, \omega)\|^2]\nu(dz)ds, \tag{14}$$

*where $C$ depends on the constant $C_{\mathfrak{X}}$ in Definition 1.*

## 3.2 Set-Valued Stochastic Integrals w.r.t. Poisson Process

Now we study the set-valued stochastic integration with respect to the Poisson random measure and the compensated Poisson random measure in an M-type 2 separable Banach space $\mathfrak{X}$.

A set-valued stochastic process $F = \{F_t\} : Z \times [0, T] \times \Omega \to \mathbf{K}(\mathfrak{X})$ is called $\mathscr{S}$-predictable if $F(z, t, \omega)$ is $\mathscr{S}/\sigma(\mathcal{C})$-measurable.

Set

$$\mathscr{M} = \Big\{ F(t, z, \omega) : F \text{ is } \mathscr{S}-predictable \text{ and}$$

$$E\Big[\int_0^T \int_Z \|F(t, z, \omega)\|_{\mathbf{K}}^2 \nu(dz)dt\Big] < \infty\Big\}$$

Given a set-valued stochastic process $F \in \mathscr{M}$, the $\mathfrak{X}$-valued stochastic process $f \in \mathscr{S}$ is called $\mathscr{S}$-selection if $f(z, t, \omega) \in F(z, t, \omega)$ for a.e. $(z, t, \omega)$. The family of all $\mathscr{L}$-selections is denoted by $S(F)$, that is

$$S(F) = \{f \in \mathscr{L} : f(z, t, \omega) \in F(z, t, \omega) \text{ for a.e. } (z, t, \omega)\}.$$

Set

$$\tilde{\Gamma}_t := \{\int_{0+}^t \int_Z f(z,s,\omega)\tilde{N}(ds,dz) : (f(t))_{t\in[0,T]} \in S(F)\},$$

$$\Gamma_t := \{\int_{0+}^t \int_Z f(z,s,\omega)N(ds,dz) : (f(t))_{t\in[0,T]} \in S(F)\},$$

*Remark 3.* It is easy to see for any $t \in [0,T]$, $\tilde{\Gamma}_t$ and $\Gamma_t$ are the subsets of $L^2[\Omega, \mathcal{F}_t, P; \mathfrak{X}]$. Furthermore, if $\{F_t, \mathcal{F}_t : t \in [0,T]\}$ is convex, then $\tilde{\Gamma}_t$ and $\Gamma_t$ so do.

Let $de\tilde{\Gamma}_t$ (resp. $de\Gamma_t$) denote the decomposable set of $\tilde{\Gamma}_t$ (resp. $\Gamma_t$) with respect to $\sigma$-algebra $\mathcal{F}_t$, $\overline{de}\tilde{\Gamma}_t$ (resp. $\overline{de}\Gamma_t$)the decomposable closed hull of $\tilde{\Gamma}_t$ (resp. $\Gamma_t$)with respect to $\mathcal{F}_t$, where the closure is taken in $L^1(\Omega, \mathfrak{X})$. That is to say, for any $g \in \overline{de}\tilde{\Gamma}_t$ (resp. $\overline{de}\Gamma_t$)and any given $\epsilon > 0$, there exists a finite $\mathcal{F}_t$-measurable partition $\{A_1, ..., A_m\}$ of $\Omega$ and $(f^1(t))_{t\in[0,T]}, ..., (f^m(t))_{t\in[0,T]} \in S(F)$ such that

$$\|g - \sum_{k=1}^m \chi_{A_k} \int_{0+}^t \int_Z f^k(s)\tilde{N}(dsdz)\|_{L^1} < \epsilon.$$

$$(resp. \|g - \sum_{k=1}^m \chi_{A_k} \int_{0+}^t \int_Z f^k(s)N(dsdz)\|_{L^1} < \epsilon)$$

Similar to Theorem 4.1 in [16], we have

**Theorem 2.** *Let* $\{F_t, \mathcal{F}_t : t \in [0,T]\} \in \mathcal{M}$, *then for any* $t \in [0,T]$, $\overline{de}\Gamma_t \subset L^1(\Omega, \mathcal{F}_t, P; \mathfrak{X})$. *Moreover, there exists a set-valued random variable* $J_t(F) \in \mathcal{M}(\Omega, \mathcal{F}_t, P; \mathbf{K}(\mathfrak{X}))$ *such that* $S^1_{J_t(F)}(\mathcal{F}_t) = \overline{de}\Gamma_t$. *Similarly, there exists a set-valued random variable* $I_t(F) \in \mathcal{M}(\Omega, \mathcal{F}_t, P; \mathbf{K}(\mathfrak{X}))$ *such that* $S^1_{I_t(F)}(\mathcal{F}_t) = \overline{de}\tilde{\Gamma}_t$.

**Definition 3.** Set-valued stochastic processes $(J_t(F))_{t\in[0,T]}$ and $(I_t(F))_{t\in[0,T]}$ defined as above are called stochastic integral of $\{F_t, \mathcal{F}_t : t \in [0,T]\} \in \mathcal{M}$ with respect to Poisson random measure $N(dsdz)$ and compensated random measure $\tilde{N}(dsdz)$ respectively. For each $t$, we denote $I_t(F) = \int_{0+}^t \int_Z F_s\tilde{N}(dsdz)$, $J_t(F) = \int_{0+}^t \int_Z F_sN(dsdz)$. Similarly, for $0 < s < t$, we also can define the set-valued random variable $I_{s,t}(F) = \int_s^t \int_Z F_u\tilde{N}(dsdz)$, $J_{s,t}(F) = \int_s^t \int_Z F_uN(dsdz)$.

**Theorem 3.** *Let* $\{F_t, \mathcal{F}_t : t \in [0,T]\} \in \mathcal{M}$. *Then for any* $t \in [0,T]$, *the stochastic integrals* $I_t(F)$ *and* $J_t(F)$ *are integrably bounded and convex a.s.*

Since for $f \in \mathcal{L}$, the integral process $\{I_t(f)\}$ with respect to compensated random measure is an $\mathfrak{X}$-valued martingale, then we have

**Theorem 4.** *Let* $\{F_t, \mathcal{F}_t : t \in [0,T]\} \in \mathcal{M}$, *then the stochastic integral* $\{I_t(F), \mathcal{F}_t : t \in [0,T]\}$ *is a set-valued submartingale.*

**Lemma 3.** *If* $\mathcal{F}$ *is separable with respect to probability measure* $P$, *for* $\{F_t, \mathcal{F}_t : t \in [0,T]\} \in \mathcal{M}$, *there exists a sequence* $\{f^n : n \in \mathbb{N}\} \subset S(F)$, *such that for every* $t \in [0,T]$,

$$S^1_{I_t(F)} = \overline{de}\{\int_{0+}^t \int_Z f^n_s \tilde{N}(dsdz) : n \in \mathbb{N}\}, \qquad (15)$$

$$S^1_{J_t(F)} = \overline{de}\{\int_{0+}^t \int_Z f^n_s N(dsdz) : n \in \mathbb{N}\}, \qquad (16)$$

*where the closure is taken in* $L^1$, *decomposability is with respect to* $\mathcal{F}_t$.

**Theorem 5.** *(Castaing representation of set-valued stochastic integrals)* *Assume* $\mathcal{F}$ *is separable with respect to the probability measure* $P$. *Then for a set-valued stochastic process* $\{F_t, \mathcal{F}_t : t \in [0,T]\} \in \mathcal{M}$, *there exists a sequence* $\{(f^i_t)_{t \in [0,T]} : i = 1,2,...\} \subset S(F)$ *such that for each* $t \in [0,T]$, $F(t,z,\omega) = cl\{(f^i_t(z,\omega)) : i = 1,2,...\}$ *a.s., and*

$$I_t(F)(\omega) = cl\{\int_{0+}^t \int_Z f^i_s(z,\omega)\tilde{N}(dsdz)(\omega) : i = 1,2,...\} \text{ a.s.}$$

*and*

$$J_t(F)(\omega) = cl\{\int_{0+}^t \int_Z f^i_s(z,\omega)N(dsdz)(\omega) : i = 1,2,...\} \text{ a.s.}$$

**Theorem 6.** *Assume* $\mathcal{F}$ *is separable with respect to* $P$, *a set-valued stochastic process* $\{F_t, \mathcal{F}_t : t \in [0,T]\} \in \mathcal{M}$. *Then the following holds*

$$I_t(F)(\omega) = cl\{I_{t_1}(F)(\omega) + \int_{t_1}^t F_s(\omega)\tilde{N}(dsdz)(\omega)\}, a.s.$$

$$J_t(F)(\omega) = cl\{J_{t_1}(F)(\omega) + \int_{t_1}^t F_s(\omega)N(dsdz)(\omega)\}, a.s.$$

*where the closure is taken in* $\mathfrak{X}$.

**Theorem 7.** *Assume* $\mathcal{F}$ *is separable with respect to* $P$, *a set-valued stochastic process* $\{F_t, \mathcal{F}_t : t \in [0,T]\} \in \mathcal{M}$. *Then* $\{I_t(F)\}$ *is a set-valued right continuous (with respect to Hausdorff metric) martingale.*

**Acknowledgements.** The authors would like to express their gratitude to referees for valuable suggestions. This paper is partly supported by The Project Sponsored by SRF for ROCS, SEM and The Fundamental Research Funds for the Central Universities (No. 10QL25).

# References

1. Brzeźniak, Z., Carroll, A.: Approximations of the Wong-Zakai differential equations in M-type 2 Banach spaces with applications to loop spaces. Séminaire de Probabilitiés XXXVII, 251–289 (2003)
2. Dettweiler, E.: A characterization of the Banach spaces of type $p$ by Lévy measures. Math. Z. 157, 121–130 (1977)
3. Hiai, F., Umegaki, H.: Integrals, conditional expectations and martingales of multivalued functions. J. Multivar. Anal. 7, 149–182 (1977)
4. Ikeda, N., Watanabe, S.: Stochastic Differential Equations and Diffusion Processes. North-Holland publishing company, Amsterdam (1981)
5. Jung, E.J., Kim, J.H.: On set-valued stochastic integrals. Stoch. Anal. Appl. 21(2), 401–418 (2003)
6. Kim, B.K., Kim, J.H.: Stochastic integrals of set-valued processes and fuzzy processes. Journal of Mathematical Analysis and Applications 236, 480–502 (1999)
7. Kunita, H.: Itô's stochastic caculus: Its surprising power for applications. Stochastic Processes and Their Applications 120, 622–652 (2010)
8. Li, S., Ogura, Y., Kreinovich, V.: Limit Theorems and Applications of Set-Valued and Fuzzy Sets-Valued Random Variables. Kluwer Academic Publishers, Dordrecht (2002)
9. Li, S., Ren, A.: Representation theorems, set-valued and fuzzy set-valued Itô Integral. Fuzzy Sets and Systems 158, 949–962 (2007)
10. Michta, M.: On set-valued stochastic integrals and fuzzy stochastic equations. Fuzzy Sets and Systems (2011), doi:10.10T6/j.fss.2011.01.007
11. Mitoma, I., Okazaki, Y., Zhang, J.: Set-valued stochastic differential equations in M-type 2 Banach space. Communications on Stochastic Analysis 4(2), 215–237 (2010)
12. Zhang, J.: Integrals and Stochastic Differential Equations for Set-Valued Stochastic Processes. PhD Thesis. Saga University, Saga (2009)
13. Zhang, J.: Set-valued stochastic integrals with respect to a real valued martingale. In: Soft Method for Handling Variability and Imprecision ASC 48. Springer, Heidelberg (2008)
14. Zhang, J.: Stochastic differential equations for set-valued stochastic processes. In: Information and Mathematics of Non-Additivity and Non-Extensivity, RIMS Kokyouroku 1630. Kyoto University, Kyoto (2009)
15. Zhang, J., Li, S., Mitoma, I., Okazaki, Y.: On the solution of set-valued stochastic differential equations in M-type 2 Banach space. Tohoku Math. J. 61, 417–440 (2009)
16. Zhang, J., Li, S., Mitoma, I., Okazaki, Y.: On Set-Valued Stochastic Integrals in an M-type 2 Banach Space. J. Math. Anal. Appl. 350, 216–233 (2009)
17. Zhang, J., Ogura, Y.: On set-valued stochastic differential equation of jump type. In: Proceedings of The 10th International Conference on Intelligent Technologies (2009)
18. Zhang, J., Mitoma, I.: On set-valued stochastic differential equations driven by Poisson processes in a Banach space (Preprint)

# The Convergence Theorems of Set-Valued Pramart in a Banach Space

Li Guan and Shoumei Li

**Abstract.** Martingale theory plays an important role in probability theory and applications such as mathematical finance, system control and so on. Classical martingale theory has been extended to more general cases, i.e. the theory of set-valued martingales and fuzzy set-valued martingales. In this paper, we shall introduce the concept of set-valued asymptotic martingale in probability (pramart for short) in a Banach space and discuss its some properties. Then we shall prove two convergence theorems of set-valued pramart in the sense of $\triangle$ and Hausdorff metric in probability respectively.

**Keywords.** Set-valued random variables, Hausdorff metric, Set-valued pramart.

## 1 Introduction

It is well known that the classical martingale is one of the most important stochastic processes, and martingale theory is one necessary tool especially in stochastic analysis and applications, for example, mathematical finance [28]. By the development of stopping time techniques, it is allowed to generate the concepts of martingale. The outcome of this effort was the introduction and detailed study of vector-valued asymptotic martingales (amarts for short), uniform amarts (e.g. cf. [9], [10]) and amart in probability (pramart for short)(cf. [1], [24]). Amart and pramart theory is important because it is not only the extension of classical martingale theory but also includes many classical limit theories. For example, in the finite Euclidean space, each uniformly integrable sequence of random variables is a pramart. It is a more

Li Guan and Shoumei Li
Department of Applied Mathematics, Beijing University of Technology,
100 Pingleyuan, Chaoyang District, Beijing, 100124, P.R. China
e-mail: guanli@bjut.edu.cn, lisma@bjut.edu.cn

S. Li (Eds.): Nonlinear Maths for Uncertainty and its Appli., AISC 100, pp. 135–142.
springerlink.com                                  © Springer-Verlag Berlin Heidelberg 2011

general process than martingale and amart. So it is interesting to study convergences of pramarts. From mathematical view, pramart theory should be considered in more general spaces such as Banach spaces, general topological spaces, hyperspaces and function spaces of Banach spaces. A hyperspace is a family of subsets of some basic Banach space $\mathfrak{X}$, which means that the elements of hyperspace are subsets of $\mathfrak{X}$. Usually, a hyperspace is not a linear space. A pramart in a hyperspace is a set-valued pramart of $\mathfrak{X}$.

In the past 40 years, the theory of set-valued random variables (or random sets, set-valued functions, multivalued functions, in literature) has been developed quite extensively (cf. [2], [3], [13], [14], [15], [22], [34] etc.). Many results have been obtained for set-valued martingale theory. For examples, the representation theorem of set-valued martingales was proved by using martingale selections by Luu [23]; The convergence theorems of martingales, submartingales and supermartingales under various settings were obtained by many authors, such as Hess [12] , Korvin and Kleyle [16], Li and Ogura [19], [20], Papageorgiou [29], [30], Wang and Xue [36], etc.. Papageorgiou [31] also proved the convergence of set-valued uniform amart in the sense of Kuratowski-Mosco, and discussed the weak convergence of set-valued amart. In [17], we provided an optional sampling theorem, a quasi-Riesz decomposition theorem and a representation theorem for set-valued amarts.

For set-valued pramart, it is divided into two concepts, i.e. superpramart and subpramart. In [35], Zhang *et al.* proved some convergence theorems of weakly compact convex set-valued superpramart in the sense of Kuratowski-Mosco convergence. In [1] Ahmed got the convergence theorems of vector-valued and set-valued pramart in the sense of Mosco convergence. The discussions for set-valued pramart are not enough and the results are less. The aim of this paper is to obtain more properties and more convergence theorems for set-valued pramart in a Banach space.

This paper is organized as follows. In section 2, we shall briefly introduce some concepts and notations on set-valued random variables. In section 3, we shall introduce the concepts of set-valued pramart and prove the properties of set-valued pramart. In section 4, we shall state our main results in the form of convergence theorems. In view of the limitation of pages, we omit the proof in this paper.

## 2   Preliminaries on Set-Valued Random Variables

Throughout this paper, we assume that $(\Omega, \mathcal{A}, \mu)$ is a nonatomic complete probability space, $(\mathfrak{X}, \| \cdot \|)$ is a real separable Banach space, $\mathbb{N}$ is the set of nature numbers, $\mathbf{K}_k(\mathfrak{X})$ is the family of all nonempty compact subsets of $\mathfrak{X}$, and $\mathbf{K}_{kc}(\mathfrak{X})$ is the family of all nonempty compact convex subsets of $\mathfrak{X}$.

Let $A$ and $B$ be two nonempty subsets of $\mathfrak{X}$ and let $\lambda \in \mathbb{R}$, the set of all real numbers. We define addition and scalar multiplication as

$$A + B = \{a + b : a \in A, b \in B\},$$

$$\lambda A = \{\lambda a : a \in A\}.$$

The Hausdorff metric on $\mathbf{K}_k(\mathfrak{X})$ is defined by

$$d_H(A, B) = \max\{\sup_{a \in A} \inf_{b \in B} \|a - b\|, \ \sup_{b \in B} \inf_{a \in A} \|a - b\|\},$$

for $A$, $B \in \mathbf{K}_k(\mathfrak{X})$. For an $A$ in $\mathbf{K}_k(\mathfrak{X})$, let $\|A\|_{\mathbf{K}} = d_H(\{0\}, A)$. The metric space $(\mathbf{K}_k(\mathfrak{X}), d_H)$ is complete and separable, and $\mathbf{K}_{kc}(\mathfrak{X})$ is a closed subset of $(\mathbf{K}_k(\mathfrak{X}), d_H)$ (cf. [22], Theorems 1.1.2 and 1.1.3). For more general hyperspaces, more topological properties of hyperspaces, readers may refer to a good book [5].

For each $A \in \mathbf{K}_{kc}(\mathfrak{X})$, define the support function by

$$s(x^*, A) = \sup_{a \in A} < x^*, a >, \quad x^* \in \mathfrak{X}^*,$$

where $\mathfrak{X}^*$ is the dual space of $\mathfrak{X}$.

A set-valued mapping $F : \Omega \to \mathbf{K}_k(\mathfrak{X})$ is called *a set-valued random variable (or a random set, or a multifunction)* if, for each open subset $O$ of $\mathfrak{X}$, $F^{-1}(O) = \{\omega \in \Omega : F(\omega) \cap O \neq \emptyset\} \in \mathcal{A}$.

In fact, set-valued random variables can be defined as a mapping from $\Omega$ to the family of all closed subsets of $\mathfrak{X}$. Since our main results shall be only related to compact set-valued random variables, we limit the definition above in the compact case. Concerning its equivalent definitions, please refer to [6], [13] [22] and [35].

A set-valued random variable $F$ is called *integrably bounded* (cf. [13] or [22]) if $\int_\Omega \|F(\omega)\|_{\mathbf{K}} d\mu < \infty$.

Let $L^1[\Omega, \mathcal{A}, \mu; \mathbf{K}_k(\mathfrak{X})]$ denote the space of all integrably bounded random variables, and $L^1[\Omega, \mathcal{A}, \mu; \mathbf{K}_{kc}(\mathfrak{X})]$ denote the space of all integraly bounded random variables taking values in $\mathbf{K}_{kc}(\mathfrak{X})$. For $F, G \in L^1[\Omega, \mathcal{A}, \mu; \mathbf{K}_k(\mathfrak{X})]$, $F = G$ if and only if $F(\omega) = G(\omega)$ a.e.$(\mu)$.

We define the following metric for $F, G$

$$\triangle(F, G) = \int_\Omega d_H(F, G) d\mu,$$

then $(L^1[\Omega, \mathcal{A}, \mu; \mathbf{K}_k(\mathfrak{X})], \triangle)$ is a complete metric space (cf. [13], [22]).

For each set-valued random variable $F$, *the expectation of $F$*, denoted by $E[F]$, is defined as

$$E[F] = \Big\{ \int_\Omega f d\mu : f \in S_F \Big\},$$

where $\int_\Omega f d\mu$ is the usual Bochner integral in $L^1[\Omega, \mathfrak{X}]$, the family of integrable $\mathfrak{X}$-valued random variables, and $S_F = \{f \in L^1[\Omega; \mathfrak{X}] : f(\omega) \in F(\omega), a.e.(\mu)\}$. This integral was first introduced by Aumann [3], called Aumann integral in literature.

Let $\mathcal{A}_0$ be a sub-$\sigma$-field of $\mathcal{A}$ and let $S_F(\mathcal{A}_0)$ denote the set of all $\mathcal{A}_0$-measurable mappings in $S_F$. The *conditional expectation* $E[F|\mathcal{A}_0]$ of an $F \in L^1[\Omega, \mathcal{A}, \mu; \mathbf{K}_{kc}(\mathfrak{X})]$ is determined as a $\mathcal{A}_0$-measurable element of $L^1[\Omega, \mathcal{A}, \mu; \mathbf{K}_{kc}(\mathfrak{X})]$ by

$$S_{E[F|\mathcal{A}_0]}(\mathcal{A}_0) = cl\{E(f|\mathcal{A}_0) : f \in S_F\},$$

where the closure is taken in the $L^1[\Omega, \mathfrak{X}]$. This definition was first introduced by Hiai and Umegaki in [13].

In the following, we assume that $\{\mathcal{A}_n : n \geq 1\}$ is an increasing sequence of sub-$\sigma$-fields of $\mathcal{A}$ such that $\mathcal{A}_\infty = \sigma(\bigcup_{n \geq 1} \mathcal{A}_n)$.

$\{F_n, \mathcal{A}_n : n \in \mathbb{N}\}$ is called a *set-valued martingale (submartingale, super-martingale resp.)* if

(1) $F_n \in L^1[\Omega, \mathcal{A}_n, \mu; \mathbf{K}_{kc}(\mathfrak{X})]$ for all $n \in \mathbb{N}$ ,

(2) for any $n \in \mathbb{N}$, $F_n = (\subseteq, \supseteq$ resp.$)E[F_{n+1}|\mathcal{A}_n]$ a.e.$(\mu)$.

The following concept will be used later.

**Definition 1.** A Banach space $\mathfrak{X}$ is said to have the Radon Nikodym property (RNP) with respect to the finite measure space $(\Omega, \mathcal{A}, \mu)$, if for each $\mu$-continuous $\mathfrak{X}$-valued measure $m : \mathcal{A} \to \mathfrak{X}$ of bounded variation, there exists an integrable function $f : \Omega \to \mathfrak{X}$ such that

$$m(A) = \int_A f d\mu, \quad \text{for all } A \in \mathcal{A}.$$

It is known that every separable dual space of a separable Banach space and every reflexive space have the RNP [7].

## 3 Properties of Set-Valued Pramarts

A function $\tau : \Omega \to \mathbb{N} \bigcup \{\infty\}$ is said to be a stopping time with respect to $\{\mathcal{A}_n : n \in \mathbb{N}\}$, if for each $n \geq 1$,

$$\{\tau = n\} =: \{\omega \in \Omega : \tau(\omega) = n\} \in \mathcal{A}_n.$$

The set of all stopping times is denoted by $T^*$. And we say that $\tau_1 \leq \tau_2$ if and only if

$$\tau_1(\omega) \leq \tau_2(\omega), \quad \text{for all } \omega \in \Omega.$$

Let $T$ denote the set of all bounded stopping times. For any given $\sigma \in T$, denote $T(\sigma) = \{\tau \in T : \sigma \leq \tau\}$. Given $\tau \in T$, define

$$\mathcal{A}_\tau = \{A \in \mathcal{A} : A \cap \{\tau = n\} \in \mathcal{A}_n, n \geq 1\}.$$

Then $\mathcal{A}_\tau$ is a sub-$\sigma$-field of $\mathcal{A}$. If $X_n \in L^1[\Omega, \mathcal{A}_n, \mu; \mathbf{F}_{kc}(\mathfrak{X})]$ for any $n \in \mathbb{N}$, define

$$X_\tau(\omega) = X_{\tau(\omega)}(\omega) \quad \text{for all } \omega \in \Omega,$$

then $X_\tau : \Omega \to \mathbf{F}_{kc}(\mathfrak{X})$ is $\mathcal{A}_\tau$-measurable.

Now we give the definition of a set-valued pramart.

**Definition 2.** An adapted set-valued process $\{F_n, \mathcal{A}_n : n \geq 1\}$ is called a set-valued pramart, if for any given $\delta > 0$, we have

$$\lim_{\sigma \in T} \sup_{\tau \in T(\sigma)} \mu\{d_H(F_\sigma, E[F_\tau | \mathcal{A}_\sigma]) > \delta\} = 0.$$

The concept of set-valued pramart is the extension of both concepts of real-valued pramart and set-valued martingale. And we obviously can know that every set-valued uniform amart is a set-valued pramart. We also have the following result.

**Theorem 1.** If $\{F_n, \mathcal{A}_n : n \geq 1\}$ is set-valued pramart, then for any $x^* \in \mathfrak{X}^*$, $\{s(x^*, F_n), \mathcal{A}_n : n \geq 1\}$ is a real-valued pramart.

**Theorem 2.** If $\{F_n : n \geq 1\}$ and $\{G_n : n \geq 1\}$ are set-valued pramarts, then
(1) $\{F_n \cap G_n, \mathcal{A}_n : n \geq 1\}$ is a set-valued pramart.
(2) $\{F_n \cup G_n, \mathcal{A}_n : n \geq 1\}$ is a set-valued pramart.

**Theorem 3.** If $\{F_n, \mathcal{A}_n : n \geq 1\}$ is a set-valued pramart, then $\{F_{\rho \wedge n}, \mathcal{A}_{\rho \wedge n} : n \geq 1\}$ is a set-valued pramart.

## 4 Convergence Theorems of Set-Valued Pramarts

In this section, we will give two convergence theorems of set-valued pramarts in the sense of Hausdorff metric $d_H$ in probability.

**Theorem 4.** Assume that $\{F_n, \mathcal{A}_n : n \geq 1\} \subset L^1[\Omega, \mathcal{A}, \mu; \mathbf{K}_{kc}(\mathfrak{X})]$ is a set-valued pramart, then there exists a set-valued martingale $\{G_n, \mathcal{A}_n : n \geq 1\} \subset L^1[\Omega, \mathcal{A}, \mu; \mathbf{K}_{kc}(\mathfrak{X})]$ such that for any $\delta > 0$, there is

$$\lim_\sigma \mu\{\triangle(F_\sigma, G_\sigma) > \delta\} = 0.$$

**Lemma 1.** Let $\{A_n, B_n; n \geq 1\} \subset \mathbf{K}_{bc}(\mathfrak{X})$, and for any $\varepsilon > 0$,

$$\lim_{n \to \infty} \mu\left\{d_H(A_n, B_n) > \frac{\varepsilon}{2}\right\} = 0.$$

(i) If there exists $B \in \mathbf{K}_{bc}(\mathfrak{X})$ such that $\lim\limits_{n\to\infty} d_H(B_n, B) = 0$, then for any $\varepsilon > 0$, we have

$$\lim_{n\to\infty} \mu\Big\{ d_H(A_n, B) > \varepsilon \Big\} = 0$$

(ii) If there exists $B \in \mathbf{K}_{bc}(\mathfrak{X})$ such that $\lim\limits_{n\to\infty} \|B_n\|_{\mathbf{K}} = \|B\|_{\mathbf{K}}$, then we have

$$\mu\Big\{ \big| \|A_n\|_{\mathbf{K}} - \|B\|_{\mathbf{K}} \big| > \varepsilon \Big\} \to 0, \quad (n \to \infty).$$

**Theorem 5.** *Let $\mathfrak{X}$ be a finite dimensional space, $\{F_n : n \geq 1\} \subset L^1[\Omega, \mathcal{A}, \mu; \mathbf{K}_{bc}(\mathfrak{X})]$ be a set-valued pramart, and $\sup\limits_{n} E[\|F_n\|_{\mathbf{K}}] < \infty$, if one of the following conditions is satisfied,*
    *(i) $\mathfrak{X}$ has Radon Nikodym property and $\mathfrak{X}^*$ is separable;*
    *(ii) there exists weak compact convex set-valued $G$, such that*

$$E[F_m | \mathcal{A}_n] \subset G, \quad a.e., m \geq n \geq 1;$$

*then there exists $F \in L^1[\Omega, \mathcal{A}, \mu; \mathbf{K}_{bc}(\mathfrak{X})]$, such that*

$$\mu\Big\{ d_H(F_n, F) > \varepsilon \Big\} \to 0, \quad n \to \infty.$$

$$\mu\Big\{ \big| \|F_n\|_{\mathbf{K}} - \|F\|_{\mathbf{K}} \big| > \varepsilon \Big\} \to 0, \quad n \to \infty.$$

**Acknowledgements.** The authors would like to thank research supported by Beijing Natural Science Foundation(Stochastic Analysis with uncertainty and applications in finance) , PHR (No. 201006102), Fund of Yuan of BJUT(No.97006013200701), Start Fund of Doctor(No.X0006013200901), Fund of Scientific Research, Basic Research Foundation of Natural Science of BJUT(No.00600054K2002).

# References

1. Ahmed, C.-D.: On almost sure convergence of vector valued Pramart and multivalued pramarts. Journal of Convex Analysis 3(2), 245–254 (1996)
2. Artstein, Z., Vitale, R.A.: A strong law of large numbers for random compact sets. Ann. Probab. 3, 879–882 (1975)
3. Aumann, R.: Integrals of set valued functions. J. Math. Anal. Appl. 12, 1–12 (1965)
4. Bagchi, S.: On a.s. convergence of classes of multivalued asymptotic martingales. Ann. Inst. H. Poincaré, Probabilités et Statistiques 21, 313–321 (1985)
5. Beer, G.: Topologies on Closed and Closed Convex Sets. Mathematics and Its Applications. Kluwer Academic Publishers, Dordrecht (1993)
6. Castaing, C., Valadier, M.: Convex Analysis and Measurable Multifunctions. Lect. Notes in Math., vol. 580. Springer, New York (1977)
7. Chatterji, S.D.: Martingale convergence and the Radom-Nikodym theorem in Banach spaces. Math. Scand. 22, 21–41 (1968)

8. Colubi, A., López-Díaz, M., Domínguez-Menchero, J.S., Gil, M.A.: A generalized strong law of large numbers. Probab. Theory and Rel. Fields 114, 401–417 (1999)
9. Edgar, G.A., Sucheston, L.: Amarts: A class of asympotic martingales A. Discrete parameter. J. Multivariate Anal. 6, 193–221 (1976)
10. Egghe, L.: Stopping Time Techniques for Analysis and Probabilists. Cambridge University Press, Cambridge (1984)
11. Feng, Y.: Strong law of large numbers for stationary sequences of random upper semicontinuous functions. Stoch. Anal. Appl. 22, 1067–1083 (2004)
12. Hess, C.: Measurability and integrability of the weak upper limit of a sequence of multifunctions. J. Math. Anal. Appl. 153, 226–249 (1983)
13. Hiai, F., Umegaki, H.: Integrals, conditional expectations and martingales of multivalued functions. J. Multivar. Anal. 7, 149–182 (1977)
14. Jung, E.J., Kim, J.H.: On set-valued stochastic integrals. Stoch. Anal. Appl. 21(2), 401–418 (2003)
15. Klein, E., Thompson, A.C.: Theory of Correspondences Including Applications to Mathematical Economics. John Wiley & Sons, Chichester (1984)
16. de Korvin, A., Kleyle, R.: A convergence theorem for convex set valued supermartingales. Stoch. Anal. Appl. 3, 433–445 (1985)
17. Li, S., Guan, L.: Decomposition and representation theorem of set-valued amarts. International Journal of Approximation Reasoning 46, 35–46 (2007)
18. Li, S., Ogura, Y.: Fuzzy random variables, conditional expectations and fuzzy martingales. J. Fuzzy Math. 4, 905–927 (1996)
19. Li, S., Ogura, Y.: Convergence of set valued sub and supermartingales in the Kuratowski-Mosco sense. Ann. Probab. 26, 1384–1402 (1998)
20. Li, S., Ogura, Y.: Convergence of set valued and fuzzy valued martingales. Fuzzy Sets and Syst. 101, 453–461 (1999)
21. Li, S., Ogura, Y.: A convergence theorem of fuzzy valued martingale in the extended Hausdorff metric $H_\infty$. Fuzzy Sets and Syst. 135, 391–399 (2003)
22. Li, S., Ogura, Y., Kreinovich, V.: Limit Theorems and Applications of Set-Valued and Fuzzy Set-Valued Random Variables. Kluwer Academic Publishers, Dordrecht (2002)
23. Luu, D.Q.: Representations and regularity of multivalued martingales. Acta Math. Vietn. 6, 29–40 (1981)
24. Luu, D.Q.: On convergence of vector-valued weak amarts and Pramarts. Vietnam Journal of Mathematics 34(2), 179–187 (2006)
25. Meyer, P.A. (ed.): Le retournement du temps, d'après Chung et Walsh, In Seminaire de Probabilités. Lecture Notes in Math., vol. 191. Springer, Berlin (1971)
26. Molchanov, I.: Theory of Random Sets. Springer, London (2005)
27. Molchanov, I.: On strong laws of large numbers for random upper semicontinuous functions. J. Math. Anal. Appl. 235, 249–355 (1999)
28. Musiela, M., Rutkowski, M.: Martingale Methods in Financial Modelling. Springer, Heidelberg (1997)
29. Papageorgiou, N.S.: On the theory of Banach space valued multifunctions. 1. integration and conditional expectation. J. Multiva. Anal. 17, 185–206 (1985)
30. Papageorgiou, N.S.: A convergence theorem for set valued multifunctions. 2. Set valued martingales and set valued measures. J. Multiva. Anal. 17, 207–227 (1987)

31. Papageorgiou, N.S.: On the conditional expectation and convergence properties of random sets. Trans. Amer. Math. Soc. 347, 2495–2515 (1995)
32. Puri, M.L., Ralescu, D.A.: Fuzzy random variables. J. Math. Anal. Appl. 114, 409–422 (1986)
33. Puri, M.L., Ralescu, D.A.: Convergence theorem for fuzzy martingales. J. Math. Anal. Appl. 160, 107–121 (1991)
34. Taylor, R.L., Inoue, H.: Convergence of weighted sums of random sets. Stoch. Anal. Appl. 3, 379–396 (1985)
35. Zhang, W., Li, S., Wang, Z., Gao, Y.: Set-Valued Stochastic Processes. Science Publisher (2007) (in Chinese)
36. Wang, Z., Xue, X.: On convergence and closedness of multivalued martingales. Trans. Amer. Math. Soc. 341, 807–827 (1994)

# Fuzzy Stochastic Integral Equations Driven by Martingales

Marek T. Malinowski and Mariusz Michta

**Abstract.** Exploiting the properties of set-valued stochastic trajectory integrals we consider a notion of fuzzy stochastic Lebesgue–Stieltjes trajectory integral and a notion of fuzzy stochastic trajectory integral with respect to martingale. Then we use these integrals in a formulation of fuzzy stochastic integral equations. We investigate the existence and uniqueness of solution to such the equations.

**Keywords:** Fuzzy stochastic integrals, Fuzzy stochastic integral equation.

## 1 Introduction

The notions of set-valued stochastic integrals have been introduced by Kisielewicz in [7] and since then they were successfully used in the topic of stochastic differential inclusions. The Kisielewicz approach utilizes Aumann's concept of set-valued integral and leads to set-valued stochastic integral as a subset of the space of square integrable random vectors. We call this type of integral a set-valued stochastic trajectory integral. Further studies treated on set-valued stochastic integrals are contained in [5, 9, 11, 25, 26]. In [5] the definition of set-valued stochastic integral has been modified in

Marek T. Malinowski
Faculty of Mathematics, Computer Science and Econometrics, University of Zielona Góra, Szafrana 4a, 65-516 Zielona Góra, Poland
e-mail: m.malinowski@wmie.uz.zgora.pl

Mariusz Michta
Faculty of Mathematics, Computer Science and Econometrics, University of Zielona Góra, Szafrana 4a, 65-516 Zielona Góra, Poland
Institute of Mathematics and Informatics, Opole University, Oleska 48, 45-052 Opole, Poland
e-mail: m.michta@wmie.uz.zgora.pl

S. Li (Eds.): Nonlinear Maths for Uncertainty and its Appli., AISC 100, pp. 143–150.
springerlink.com © Springer-Verlag Berlin Heidelberg 2011

order to describe it as a set-valued adapted stochastic process. The definition from [5] was then modified in [11] by considering the predictable set-valued processes as a set-valued random variable in the product space $I\!R_+ \times \Omega$. In [26] the authors considered set-valued stochastic integrals (also as the set-valued stochastic processes) with respect to a real-valued Brownian motion, whereas in [25] with respect to a real-valued, continuous, square integrable martingale. There are also the papers with the studies and applications of the set-valued integrals with set-valued integrators [18, 19] and recently [9].

Incorporating the notions of set-valued stochastic integrals, the set-valued differential equations - a new subject ( [8, 10, 14, 15, 21, 27]) - generalize the theory of classical stochastic differential equations. An other approach can be found in [22]. In the papers [8,10,21,27] such the set-valued equations contain the diffusion part (driven by a Brownian motion) being a single-valued Itô's integral.

This note is an abbreviated version of the manuscript [17], where we use the Kisielewicz concept of set-valued stochastic trajectory integral. We consider such integral driven by a martingale. It allows us to define a fuzzy stochastic trajectory integrals and study fuzzy-set-valued stochastic integral equations with (essentially) set-valued stochastic integrals. The stochastic fuzzy integral equations can be adequate in modelling of the dynamics of real phenomena which are subjected to two kinds of uncertainties: randomness and fuzziness, simultaneously. Some results in this area are contained in [3, 4, 6, 12, 13, 15, 16, 23].

## 2  Preliminaries

Let $\mathcal{X}$ be a separable, reflexive Banach space, $\mathcal{K}_c^b(\mathcal{X})$ the family of all nonempty closed, bounded and convex subsets of $\mathcal{X}$. The Hausdorff metric $H_{\mathcal{X}}$ in $\mathcal{K}_c^b(\mathcal{X})$ is defined by

$$H_{\mathcal{X}}(A, B) = \max\Big\{\sup_{a \in A} \mathrm{dist}_{\mathcal{X}}(a, B), \sup_{b \in B} \mathrm{dist}_{\mathcal{X}}(b, A)\Big\},$$

where $\mathrm{dist}_{\mathcal{X}}(a, B) = \inf_{b \in B} \|a - b\|_{\mathcal{X}}$ and $\| \cdot \|_{\mathcal{X}}$ denotes a norm in $\mathcal{X}$.

It is known that $(\mathcal{K}_c^b(\mathcal{X}), H_{\mathcal{X}})$ is a complete, separable metric space.

Let $(U, \mathcal{U}, \mu)$ be a measure space. A set-valued mapping (multifunction) $F\colon U \to \mathcal{K}_c^b(\mathcal{X})$ is said to be $\mathcal{U}$-measurable (or measurable, for short) if it satisfies: $\{u \in U : F(u) \cap C \neq \emptyset\} \in \mathcal{U}$ for every closed set $C \subset \mathcal{X}$.

A measurable multifunction $F\colon U \to \mathcal{K}_c^b(\mathcal{X})$ is said to be $L_{\mathcal{U}}^p(\mu)$-integrably bounded $(p \geq 1)$, if there exists $h \in L^p(U, \mathcal{U}, \mu; I\!R_+)$ such that the inequality $\||F\||_{\mathcal{X}} \leq h$ holds $\mu$-a.e., where $\||A\||_{\mathcal{X}} = H_{\mathcal{X}}(A, \{0\}) = \sup_{a \in A} \|a\|_{\mathcal{X}}$ for $A \in \mathcal{K}^b(\mathcal{X})$, and $I\!R_+ = [0, \infty)$.

Let $\mathcal{M}$ be a set of $\mathcal{U}$-measurable mappings $f\colon U \to \mathcal{X}$. The set $\mathcal{M}$ is said to be decomposable if for every $f_1, f_2 \in \mathcal{M}$ and every $A \in \mathcal{U}$ it holds $f_1 1_A + f_2 1_{U \setminus A} \in \mathcal{M}$.

Denote $I = [0, T]$, where $T < \infty$. Let $(\Omega, \mathcal{A}, \{\mathcal{A}_t\}_{t \in I}, P)$ be a complete filtered probability space satisfying the usual hypotheses.

At this moment we put $\mathcal{X} = \mathbb{R}^d$, $U = I \times \Omega$, $\mathcal{U} = \mathcal{P}$, where $\mathcal{P}$ denotes the $\sigma$-algebra of the predictable elements in $I \times \Omega$. A stochastic process $f \colon I \times \Omega \to \mathbb{R}^d$ is called predictable if $f$ is $\mathcal{P}$-measurable.

A mapping $F \colon I \times \Omega \to \mathcal{K}_c^b(\mathbb{R}^d)$ is said to be a set-valued stochastic process, if for every $t \in I$ the mapping $F(t, \cdot) \colon \Omega \to \mathcal{K}_c^b(\mathbb{R}^d)$ is an $\mathcal{A}$-measurable multifunction. If for every fixed $t \in I$ the mapping $F(t, \cdot) \colon \Omega \to \mathcal{K}_c^b(\mathbb{R}^d)$ is an $\mathcal{A}_t$-measurable multifunction then $F$ is called $\{\mathcal{A}_t\}$-adapted. A set-valued stochastic process $F$ is predictable if it is a $\mathcal{P}$-measurable multifunction.

By a fuzzy set $u$ of the space $\mathcal{X}$ we mean a function $u \colon \mathcal{X} \to [0, 1]$. We denote this fact as $u \in \mathcal{F}(\mathcal{X})$. For $\alpha \in (0, 1]$ denote $[u]^\alpha := \{x \in \mathcal{X} : u(x) \geq \alpha\}$ and let $[u]^0 := \mathrm{cl}_{\mathcal{X}}\{x \in \mathcal{X} : u(x) > 0\}$, where $\mathrm{cl}_{\mathcal{X}}$ denotes the closure in $(\mathcal{X}, \|\cdot\|_{\mathcal{X}})$. The sets $[u]^\alpha$ are called the $\alpha$-level sets of fuzzy set $u$, and 0-level set $[u]^0$ is called the support of $u$.

Denote $\mathcal{F}_c^b(\mathcal{X}) = \{u \in \mathcal{F}(\mathcal{X}) : [u]^\alpha \in \mathcal{K}_c^b(\mathcal{X}) \text{ for every } \alpha \in [0, 1]\}$. In this set we consider two metrics: the generalized Hausdorff metric

$$D_{\mathcal{X}}(u, v) := \sup_{\alpha \in [0,1]} H_{\mathcal{X}}([u]^\alpha, [v]^\alpha),$$

and the Skorohod metric

$$D_S^{\mathcal{X}}(u, v) := \inf_{\lambda \in \Lambda} \max\left\{ \sup_{t \in [0,1]} |\lambda(t) - t|, \sup_{t \in [0,1]} H_{\mathcal{X}}(x_u(t), x_v(\lambda(t))) \right\},$$

where $\Lambda$ denotes the set of strictly increasing continuous functions $\lambda \colon [0, 1] \to [0, 1]$ such that $\lambda(0) = 0$, $\lambda(1) = 1$, and $x_u, x_v \colon [0, 1] \to \mathcal{K}^b(\mathcal{X})$ are the càdlàg representations for the fuzzy sets $u, v \in \mathcal{F}^b(\mathcal{X})$, see [1] for details. The space $(\mathcal{F}^b(\mathcal{X}), D_{\mathcal{X}})$ is complete and non-separable, and the space $(\mathcal{F}^b(\mathcal{X}), D_S^{\mathcal{X}})$ is Polish.

For our aims we will consider two cases of $\mathcal{X}$. Namely we will take $\mathcal{X} = \mathbb{R}^d$ or $\mathcal{X} = L^2$, where $L^2 = L^2(\Omega, \mathcal{A}, P; \mathbb{R}^d)$ and we assume that $\sigma$-algebra $\mathcal{A}$ is separable with respect to probability measure $P$.

**Definition 1.** (Puri–Ralescu [24]). By a fuzzy random variable we mean a function $\mathfrak{u} \colon \Omega \to \mathcal{F}_c^b(\mathcal{X})$ such that $[\mathfrak{u}(\cdot)]^\alpha \colon \Omega \to \mathcal{K}_c^b(\mathcal{X})$ is an $\mathcal{A}$-measurable multifunction for every $\alpha \in (0, 1]$.

This definition is one of the possible to be considered for fuzzy random variables, and because of our further aims we recall some facts about measurability concepts for fuzzy-set-valued mappings. Generally, having a metric $d$ in the set $\mathcal{F}_c^b(\mathcal{X})$ one can consider $\sigma$-algebra $\mathcal{B}_d$ generated by the topology induced by $d$. Then a fuzzy random variable $\mathfrak{u}$ can be viewed as a measurable (in the classical sense) mapping between two measurable spaces, namely $(\Omega, \mathcal{A})$ and $(\mathcal{F}_c^b(\mathcal{X}), \mathcal{B}_d)$. Using the classical notation, we write this as: $\mathfrak{u}$ is

$A|\mathcal{B}_d$-measurable. It is known (see [1]) that for a mapping $\mathfrak{u}\colon \Omega \to \mathcal{F}_c^b(\mathcal{X})$, where $(\Omega, \mathcal{A}, P)$ is a given probability space, it holds:

- $\mathfrak{u}$ is the fuzzy random variable if and only if $\mathfrak{u}$ is $A|\mathcal{B}_{D_{\tilde{S}}}$-measurable,
- if $\mathfrak{u}$ is $A|\mathcal{B}_{D_{\mathcal{X}}}$-measurable, then it is the fuzzy random variable; the opposite implication is not true.

A fuzzy-set-valued mapping $\mathfrak{f}\colon I \times \Omega \to \mathcal{F}_c^b(\mathcal{X})$ is called a fuzzy stochastic process if $\mathfrak{f}(t, \cdot)\colon \Omega \to \mathcal{F}_c^b(\mathcal{X})$ is a fuzzy random variable for every $t \in I$.

The fuzzy stochastic process $\mathfrak{f}\colon I \times \Omega \to \mathcal{F}_c^b(\mathcal{X})$ is said to be predictable if the set-valued mapping $[\mathfrak{f}]^\alpha\colon I \times \Omega \to \mathcal{K}_c^b(\mathcal{X})$ is $\mathcal{P}$-measurable for every $\alpha \in (0, 1]$.

Let $\mathfrak{f}\colon I \times \Omega \to \mathcal{F}_c^b(\mathcal{X})$ be a predictable fuzzy stochastic process. The process $\mathfrak{f}$ is said to be $L_{\mathcal{P}}^2(\mu)$-integrably bounded, if $\| \|[\mathfrak{f}]^0\| \| \in L_{\mathcal{P}}^2(\mu)$.

## 3　Fuzzy Stochastic Lebesgue–Stieltjes Trajectory Integral

Let $A\colon I \times \Omega \to \mathbb{R}$ be an $\{\mathcal{A}_t\}$-adapted stochastic process. We will assume that its trajectories are continuous and of finite variation on $I$, $A(0, \cdot) = 0$ $P$-a.e.

Let $|A(\omega)|_t$ denotes the total variation of the random measure induced by the trajectories of the process $A$. We will assume that $\mathbb{E}|A|_T^2 < \infty$. Denote by $\Gamma$ a random measure on $I$ which is defined as follows: $\Gamma_{A(\cdot, \omega)}(dt) = |A(\omega)|_T d|A(\omega)|_t$. Finally, we define the measure $\nu_A$ on $(I \times \Omega, \mathcal{P})$ by

$$\nu_A(C) := \int\limits_{I \times \Omega} \mathbf{1}_C(t, \omega) \Gamma_{A(\cdot, \omega)}(dt) P(d\omega) \quad \text{for} \quad C \in \mathcal{P}.$$

Denote $L_{\mathcal{P}}^2(\nu_A) := L^2(I \times \Omega, \mathcal{P}, \nu_A; \mathbb{R}^d)$. Let $F\colon I \times \Omega \to \mathcal{K}^b(\mathbb{R}^d)$ be a predictable set-valued stochastic process. Let us define the set $S_{\mathcal{P}}^2(F, \nu_A) := \{f \in L_{\mathcal{P}}^2(\nu_A) : f \in F, \ \nu_A\text{-a.e.}\}$. Notice that $S_{\mathcal{P}}^2(F, \nu_A) \neq \emptyset$.

**Definition 2.** For a predictable and $L_{\mathcal{P}}^2(\nu_A)$-integrably bounded set-valued stochastic process $F\colon I \times \Omega \to \mathcal{K}_c^b(\mathbb{R}^d)$ and for $\tau, t \in I$, $\tau < t$ the set-valued stochastic Lebesgue–Stieltjes trajectory integral (over interval $[\tau, t]$) of $F$ with respect to the process $A$ is the set

$$(S) \int_\tau^t F(s) dA_s := \left\{ \int_\tau^t f(s) dA_s : f \in S_{\mathcal{P}}^2(F, \nu_A) \right\}.$$

In the rest of the paper, for the sake of convenience, we will write $L^2$ instead of $L^2(\Omega, \mathcal{A}, P; \mathbb{R}^d)$ and $L_t^2$ instead of $L^2(\Omega, \mathcal{A}_t, P; \mathbb{R}^d)$.

**Theorem 1.** *Let $F\colon I \times \Omega \to \mathcal{K}_c^b(\mathbb{R}^d)$ be a predictable and $L_{\mathcal{P}}^2(\nu_A)$-integrably bounded set-valued stochastic process. Then*

(a) $S_{\mathcal{P}}^2(F, \nu_A)$ is nonempty, bounded, closed, convex, weakly compact and decomposable subset of $L_{\mathcal{P}}^2(\nu_A)$,

(b) $(S) \int_\tau^t F(s) dA_s$ is nonempty, bounded, closed, convex and weakly compact subset of $L_t^2$ for every $\tau, t \in I$, $\tau < t$.

**Theorem 2.** For each $n \in \mathbb{N}$, let $F_n : I \times \Omega \to \mathcal{K}^b(\mathbb{R}^d)$ be a predictable set-valued stochastic process such that $F_1$ is $L_{\mathcal{P}}^2(\nu_A)$-integrably bounded and $F_1 \supset F_2 \supset \ldots \supset F$ $\nu_A$-a.e., where $F := \bigcap_{n=1}^\infty F_n$ $\nu_A$-a.e. Then for every $\tau, t \in I$, $\tau < t$ it holds $(S) \int_\tau^t F(s) dA_s = \bigcap_{n=1}^\infty (S) \int_\tau^t F_n(s) dA_s$.

Using the Representation Theorem of Negoita–Ralescu we obtain:

**Theorem 3.** Assume that $\mathfrak{f} : I \times \Omega \to \mathcal{F}_c^b(\mathbb{R}^d)$ is a predictable and $L_{\mathcal{P}}^2(\nu_A)$-integrably bounded fuzzy stochastic process. Then for every $\tau, t \in I$, $\tau < t$ there exists a unique fuzzy set in $\mathcal{F}_c^b(L^2)$ denoted by $(F) \int_\tau^t \mathfrak{f}(s) dA_s$ such that for every $\alpha \in (0, 1]$ it holds $\left[ (F) \int_\tau^t \mathfrak{f}(s) dA_s \right]^\alpha = (S) \int_\tau^t [\mathfrak{f}(s)]^\alpha dA_s$, and $\left[ (F) \int_\tau^t \mathfrak{f}(s) dA_s \right]^0 \subset (S) \int_\tau^t [\mathfrak{f}(s)]^0 dA_s$.

For a fuzzy stochastic trajectory integral $(F) \int_\tau^t \mathfrak{f}(s) dA_s$ defined in Theorem 3 the following properties hold true.

**Theorem 4.** Let $\mathfrak{f}_1, \mathfrak{f}_2 : I \times \Omega \to \mathcal{F}_c^b(\mathbb{R}^d)$ be the predictable and $L_{\mathcal{P}}^2(\nu_A)$-integrably bounded fuzzy stochastic processes. Then

(a) for every $\tau, a, t \in I$, $\tau \leq a \leq t$ it holds $(F) \int_\tau^t \mathfrak{f}_1(s) dA_s = (F) \int_\tau^a \mathfrak{f}_1(s) dA_s + (F) \int_a^t \mathfrak{f}_1(s) dA_s$,

(b) for every $\tau, t \in I$, $\tau < t$ it holds $D_{L^2}^2 \left( (F) \int_\tau^t \mathfrak{f}_1(s) dA_s, (F) \int_\tau^t \mathfrak{f}_2(s) dA_s \right)$
$\leq 2 \int_{[\tau, t] \times \Omega} D_{\mathbb{R}^d}^2(\mathfrak{f}_1, \mathfrak{f}_2) d\nu_A$,

(c) for every $\tau \in [0, T)$ the mapping $[\tau, T] \ni t \mapsto (F) \int_\tau^t \mathfrak{f}_1(s) dA_s \in \mathcal{F}_c^b(L^2)$ is continuous with respect to the metric $D_{L^2}$.

## 4  Fuzzy Stochastic Trajectory Integral with Respect to Martingales

In this section we consider the notion of fuzzy stochastic trajectory integral, where the integrator is a martingale.

Let $M : I \times \Omega \to \mathbb{R}$ be a square-integrable $\{\mathcal{A}_t\}$-martingale with continuous trajectories, $M(0, \cdot) = 0$ $P$-a.e. The martingale $M$ generates a measure $\mu_M$ defined on measurable space $(I \times \Omega, \mathcal{P})$ which is called the Doléans-Dade measure (see e.g. [2]).

Denote $L_{\mathcal{P}}^2(\mu_M) := L^2(I \times \Omega, \mathcal{P}, \mu_M; \mathbb{R}^d)$. Let $G : I \times \Omega \to \mathcal{K}_c^b(\mathbb{R}^d)$ be a predictable and $L_{\mathcal{P}}^2(\mu_M)$-integrably bounded set-valued stochastic process. Then the set $S_{\mathcal{P}}^2(G, \mu_M) := \{g \in L_{\mathcal{P}}^2(\mu_M) : g \in G, \mu_M\text{-a.e.}\}$ is nonempty.

**Definition 3.** For a predictable and $L^2_{\mathcal{P}}(\mu_M)$-integrably bounded set-valued stochastic process $G\colon I \times \Omega \to \mathcal{K}^b_c(\mathbb{R}^d)$ and for $\tau, t \in I$, $\tau < t$ the set-valued stochastic trajectory integral (over interval $[\tau, t]$) of $G$ with respect to the continuous, square-integrable martingale $M$ is the set

$$(S) \int_\tau^t G(s)dM_s := \left\{ \int_\tau^t g(s)dM_s : g \in S^2_{\mathcal{P}}(G, \mu_M) \right\}.$$

**Theorem 5.** Let $G\colon I \times \Omega \to \mathcal{K}^b_c(\mathbb{R}^d)$ be a predictable and $L^2_{\mathcal{P}}(\mu_M)$-integrably bounded set-valued stochastic process. Then

(a) $S^2_{\mathcal{P}}(G, \mu_M)$ is nonempty, bounded, closed, convex, weakly compact and decomposable subset of $L^2_{\mathcal{P}}(\mu_M)$,

(b) $(S) \int_\tau^t G(s)dM_s$ is nonempty, bounded, closed, convex and weakly compact subset of $L^2_t$, for every $\tau, t \in I$, $\tau < t$.

**Theorem 6.** Let $G_n\colon I \times \Omega \to \mathcal{K}^b_c(\mathbb{R}^d)$, $n \in \mathbb{N}$, be the predictable set-valued stochastic processes such that $G_1$ is $L^2_{\mathcal{P}}(\mu_M)$-integrably bounded and $G_1 \supset G_2 \supset \ldots \supset G$ $\mu_M$-a.e., where $G := \bigcap_{n=1}^\infty G_n$ $\mu_M$-a.e. Then for every $\tau, t \in I$, $\tau < t$ it holds $(S) \int_\tau^t G(s)dM_s = \bigcap_{n=1}^\infty (S) \int_\tau^t G_n(s)dM_s$.

**Theorem 7.** Assume that $\mathfrak{g}\colon I \times \Omega \to \mathcal{F}^b_c(\mathbb{R}^d)$ is a predictable and $L^2_{\mathcal{P}}(\mu_M)$-integrably bounded fuzzy stochastic process. Then for every $\tau, t \in I$, $\tau < t$ there exists a unique fuzzy set in $\mathcal{F}^b_c(L^2)$ denoted by $(F) \int_\tau^t \mathfrak{g}(s)dM_s$ such that for every $\alpha \in (0, 1]$ it holds $\left[ (F) \int_\tau^t \mathfrak{g}(s)dM_s \right]^\alpha = (S) \int_\tau^t [\mathfrak{g}(s)]^\alpha dM_s$, and $\left[ (F) \int_\tau^t \mathfrak{g}(s)dM_s \right]^0 \subset (S) \int_\tau^t [\mathfrak{g}(s)]^0 dM_s$.

For a fuzzy stochastic trajectory integral $(F) \int_\tau^t \mathfrak{g}(s)dM_s$ defined in Theorem 7 the following properties hold true.

**Theorem 8.** Let $\mathfrak{g}_1, \mathfrak{g}_2\colon I \times \Omega \to \mathcal{F}^b_c(\mathbb{R}^d)$ be the predictable and $L^2_{\mathcal{P}}(\mu_M)$-integrably bounded fuzzy stochastic processes. Then

(a) for every $\tau, a, t \in I$, $\tau \le a \le t$ it holds $(F) \int_\tau^t \mathfrak{g}_1(s)dM_s = (F) \int_\tau^a \mathfrak{g}_1(s)dM_s + (F) \int_a^t \mathfrak{g}_1(s)dM_s$,

(b) for every $\tau, t \in I$, $\tau < t$ it holds $D^2_{L^2}\left( (F) \int_\tau^t \mathfrak{g}_1(s)dM_s, (F) \int_\tau^t \mathfrak{g}_2(s)dM_s \right)$
$\le \int_{[\tau,t] \times \Omega} D^2_{\mathbb{R}^d}(\mathfrak{g}_1, \mathfrak{g}_2)d\mu_M$,

(c) for every $\tau \in [0, T)$ the mapping $[\tau, T] \ni t \mapsto (F) \int_\tau^t \mathfrak{g}_1(s)dM_s \in \mathcal{F}^b_c(L^2)$ is continuous with respect to the metric $D_{L^2}$.

## 5   Solutions of Fuzzy Stochastic Integral Equations

Now we consider the fuzzy stochastic integral equations with the fuzzy stochastic trajectory integrals defined in the preceding sections.

Let $\mathfrak{f}, \mathfrak{g}: I \times \Omega \times \mathcal{F}_c^b(L^2) \to \mathcal{F}_c^b(I\!R^d)$ and let $X_0 \in \mathcal{F}_c^b(L_0^2)$. By a fuzzy stochastic integral equation we mean the following relation in the metric space $(\mathcal{F}_c^b(L^2), D_{L^2})$:

$$X(t) = X_0 + (F) \int_0^t \mathfrak{f}(s, X(s))dA_s + (F) \int_0^t \mathfrak{g}(s, X(s))dM_s \text{ for } t \in I. \quad (1)$$

**Definition 4.** By a solution to (1) we mean a $D_{L^2}$-continuous mapping $X: I \to \mathcal{F}_c^b(L^2)$ that satisfies (1). A solution $X: I \to \mathcal{F}_c^b(L^2)$ to (1) is unique if $X(t) = Y(t)$ for every $t \in I$, where $Y: I \to \mathcal{F}_c^b(L^2)$ is any solution to (1).

Assume that $\mathfrak{f}, \mathfrak{g}: I \times \Omega \times \mathcal{F}_c^b(L^2) \to \mathcal{F}_c^b(I\!R^d)$ satisfy:

(f1)    the mappings $\mathfrak{f}, \mathfrak{g}: I \times \Omega \times \mathcal{F}_c^b(L^2) \to \mathcal{F}_c^b(I\!R^d)$ are $\mathcal{P} \otimes \mathcal{B}_{D_S^{L^2}} | \mathcal{B}_{D_S^{I\!R^d}}$-measurable,

(f2)    there exists a constant $K > 0$ such that

$$\max\{D_{I\!R^d}(\mathfrak{f}(t, \omega, u), \mathfrak{f}(t, \omega, v)), D_{I\!R^d}(\mathfrak{g}(t, \omega, u), \mathfrak{g}(t, \omega, v))\} \leq K D_{L^2}(u, v),$$

for every $(t, \omega) \in I \times \Omega$, and every $u, v \in \mathcal{F}_c^b(L^2)$,

(f3)    there exists a constant $C > 0$ such that

$$\max\{D_{I\!R^d}(\mathfrak{f}(t, \omega, u), \hat{\theta}), D_{I\!R^d}(\mathfrak{g}(t, \omega, u), \hat{\theta})\} \leq C(1 + D_{L^2}(u, \hat{\Theta})),$$

for every $(t, \omega) \in I \times \Omega$, and every $u \in \mathcal{F}_c^b(L^2)$,

(f4)    there exists a constant $Q > 0$ such that $2K^2 \sup_{t \in I} e^{-Qt} \left(2 \int_{[0,t] \times \Omega} e^{Qs} d\nu_A \right.$

$\left. + \int_{[0,t] \times \Omega} e^{Qs} d\mu_M \right) < 1.$

The description of the symbols $\hat{\theta}, \hat{\Theta}$ appearing in (f3) is as follows: let $\theta, \Theta$ denote the zero elements in $I\!R^d$ and $L^2$, respectively, the symbols $\hat{\theta}, \hat{\Theta}$ are their fuzzy counterparts, i.e. $\hat{\theta} \in \mathcal{F}_c^b(I\!R^d)$ and $[\hat{\theta}]^\alpha = \{\theta\}$ for every $\alpha \in [0, 1]$, also $\hat{\Theta} \in \mathcal{F}_c^b(L^2)$ and $[\hat{\Theta}]^\alpha = \{\Theta\}$ for every $\alpha \in [0, 1]$.

**Theorem 9.** Let $X_0 \in \mathcal{F}_c^b(L_0^2)$, and $\mathfrak{f}, \mathfrak{g}: I \times \Omega \times \mathcal{F}_c^b(L^2) \to \mathcal{F}_c^b(I\!R^d)$ satisfy the conditions (f1)-(f4). Then the equation (1) has a unique solution.

# References

1. Colubi, A., Domínguez-Menchero, J.S., López-Díaz, M., Ralescu, D.A.: A $D_E[0, 1]$ representation of random upper semicontinuous functions. Proc. Amer. Math. Soc. 130, 3237–3242 (2002)
2. Chung, K.L., Williams, R.J.: Introduction to Stochastic Integration. Birkhäuser, Boston (1983)
3. Fei, W.: Existence and uniqueness of solution for fuzzy random differential equations with non-Lipschitz coefficients. Inform. Sci. 177, 4329–4337 (2007)
4. Feng, Y.: Fuzzy stochastic differential systems. Fuzzy Sets Syst. 115, 351–363 (2000)

5. Jung, E.J., Kim, J.H.: On set-valued stochastic integrals. Stoch. Anal. Appl. 21, 401–418 (2003)
6. Kim, J.H.: On fuzzy stochastic differential equations. J. Korean Math. Soc. 42, 153–169 (2005)
7. Kisielewicz, M.: Properties of solution set of stochastic inclusions. J. Appl. Math. Stochastic Anal. 6, 217–236 (1993)
8. Li, J., Li, S.: Ito type set-valued stochastic differential equation. J. Uncertain Syst. 3, 53–63 (2009)
9. Li, S., Li, J., Li, X.: Stochastic integral with respect to set-valued square integrable martingales. J. Math. Anal. Appl. 370, 659–671 (2010)
10. Li, J., Li, J., Li, S., Ogura, Y.: Strong solutions of Itô type set-valued stochastic differential equations. Acta Math. Sinica Engl. Ser. 26, 1739–1748 (2010)
11. Li, S., Ren, A.: Representation theorems, set-valued and fuzzy set-valued Ito integral. Fuzzy Sets Syst. 158, 949–962 (2007)
12. Malinowski, M.T.: On random fuzzy differential equations. Fuzzy Sets Syst. 160, 3152–3165 (2009)
13. Malinowski, M.T.: Existence theorems for solutions to random fuzzy differential equations. Nonlinear Anal. TMA 73, 1515–1532 (2010)
14. Malinowski, M.T., Michta, M.: Stochastic set differential equations. Nonlinear Anal. TMA 72, 1247–1256 (2010)
15. Malinowski, M.T., Michta, M.: Fuzzy stochastic integral equations. Dynam. Systems Appl. 19, 473–494 (2010)
16. Malinowski, M.T., Michta, M.: Stochastic fuzzy differential equations with an application. Kybernetika 47, 123–143 (2011)
17. Malinowski, M.T., Michta, M.: Set-valued stochastic integral equations driven by martingales (Preprint)
18. Michta, M.: Stochastic inclusions with multivalued integrators. Stoch. Anal. Appl. 20, 847–862 (2002)
19. Michta, M., Motyl, J.: Weak solutions to stochastic inclusions with multivalued integrator. Dynam. Systems Appl. 14, 323–334 (2005)
20. Michta, M.: On set-valued stochastic integrals and fuzzy stochastic equations. Fuzzy Sets Syst. (2011), doi:10.1016/j.fss.2011.01.07
21. Mitoma, I., Okazaki, Y., Zhang, J.: Set-valued stochastic differential equation in M-type 2 Banach space. Comm. Stoch. Anal. 4, 215–237 (2010)
22. Ogura, Y.: On stochastic differential equations with set coefficients and the Black & Scholes model. In: Proceedings of the 8th International Conference on Intellgent Technologies (2007)
23. Ogura, Y.: On stochastic differential equations with fuzzy set coefficients. In: Dubois, D., et al. (eds.) Soft Methods for Handling Variability and Imprecision, ASC 48. Springer, Berlin (2008)
24. Puri, M.L., Ralescu, D.A.: Fuzzy random variables. J. Math. Anal. Appl. 91, 552–558 (1983)
25. Zhang, J.: Set-valued stochastic integrals with respect to a real valued martingale. In: Dubois, D., et al. (eds.) Soft Methods for Handling Variability and Imprecision. Springer, Berlin (2008)
26. Zhang, J., Li, S., Mitoma, I., Okazaki, Y.: On set-valued stochastic integrals in an M-type 2 Banach space. J. Math. Anal. Appl. 350, 216–233 (2009)
27. Zhang, J., Li, S., Mitoma, I., Okazaki, Y.: On the solutions of set-valued stochastic differential equations in M-type 2 Banach spaces. Tohoku Math. J. 61, 417–440 (2009)

# Moderate Deviations of Random Sets and Random Upper Semicontinuous Functions

Xia Wang

**Abstract.** In this paper, we obtain moderate deviations of random sets which take values of bounded closed convex sets on the underling separable Banach space with respect to the Hausdorff distance $d_H$. We also get moderate deviations for random upper semicontinuous functions whose values are of bounded closed convex levels on the underling separable Banach space in the sense of the uniform Hausdorff distance $d_H^\infty$. The main tool is the work of Wu on the moderate deviations for empirical processes [15].

**Keywords:** Random sets, Random upper semicontinuous functions, Large deviations, Moderate deviations.

## 1 Introduction

The theory of large deviation principle (LDP) and moderate deviation principle (MDP) deals with the asymptotic estimation of probabilities of rare events and provides exponential bound on probability of such events. Some authors have discussed LDP and MDP on random sets and random upper semicontinuous functions. In 1999, Cerf [2] proved LDP for sums i.i.d. compact random sets in a separable type $p$ Banach space with respect to the Hausdorff distance $d_H$, which is called Cramér type LDP. In 2006, Terán obtained Cramér type LDP of random upper semicontinuous functions whose level sets are compact [11], and Bolthausen type LDP of random upper semicontinuous functions whose level sets are compact convex [12] on a separable Banach space in the sense of the uniform Hausdorff distance $d_H^\infty$. In 2009, Ogura and Setokuchi [9] proved a Cramér type LDP for random upper

Xia Wang
College of Applied Sciences, Beijing University of Technology,
Beijing 100124, P.R. China
e-mail: wangxia@bjut.edu.cn

S. Li (Eds.): Nonlinear Maths for Uncertainty and its Appli., AISC 100, pp. 151–159.

semicontiunous functions on the underling separable Banach space with respect to the metric $d_Q$ (see [9] for the notation) in a different method, which is weaker than the uniform Hausdorff distance $d_H^\infty$. In 2010, Ogura, Li and Wang [8] also discussed LDP for random upper semicontinuous functions whose underlying space is d-dimensional Euclidean space $\mathbb{R}^d$ under various topologies for compact covex random sets and random upper semicontinuous functions, Wang [13] considered LDP joint behavior of a family of random compact sets indexed by $t$, which was called sample path or functional LDP in some textbooks and literatures, Wang and Li [14]obtained LDP for bounded closed convex random sets and related random upper semicontiunous functions. About the aspect on the MDP of random sets and random upper semicontiunous functions, Ogura, Li and Wang [8]first discussed a MDP of random compact covex sets with respect to Hausdorff distance $d_H$ and a MDP of random upper semicontinuous functions whose level sets are compact convex with respect to the distance $d_p^0$ on the underling separable Banach space, in their paper, they embedded them into a separable Banach space respectively by support function and support process, and then used the classical result of MDP on the separable Banach space(see Chen [3]) and obtained their result. However, previous work about MDP was restricted to compact convex random sets and compact convex random upper semicontinuous functions. In this paper, we will obtain MDP for bounded closed convex random sets with respect to the Hausdorff distance $d_H$ and related random upper semicontiunous functions in the sense of the uniform Hausdorff distance $d_H^\infty$. We also embed them into a Banach space respectively by support and support process, but they are not separable, and we can't again use the classical result of MDP [3], then we continue to embed those Banach spaces to Wu's spaces (not necessarily separable), where Wu [15] has obtained MDP, so the main tool is the work of Wu on the moderate deviations for empirical processes [15].

The paper is structured as follows. Section 2 will give some preliminaries about bounded closed convex random sets and random upper semicontinuous functions. In section 3, we will give moderate deviations of random sets which take values of bounded closed convex sets on the underling separable Banach space with respect to the Hausdorff distance $d_H$, and prove that of random upper semicontiunous functions whose values are of bounded closed convex levels on the underling separable Banach space in the sense of the uniform Hausdorff distance $d_H^\infty$.

## 2   Preliminaries

Throughout this paper, we assume that $(\Omega, \mathcal{A}, P)$ is a complete probability space, $(\mathfrak{X}, \| \cdot \|_{\mathfrak{X}})$ is a real separable Banach space with its dual space $\mathfrak{X}^*$, which is separable with respect to usual norm $\| \cdot \|_{\mathfrak{X}^*}$. $\mathcal{K}(\mathfrak{X})$ is the family of all non-empty closed subsets of $\mathfrak{X}$, $\mathcal{K}_b(\mathfrak{X})$(resp. $\mathcal{K}_{bc}(\mathfrak{X})$) is the family of all non-empty bounded closed (resp. bounded closed convex) subsets of $\mathfrak{X}$.

Let $A$ and $B$ be two non-empty subsets of $\mathfrak{X}$ and let $\lambda \in \mathbb{R}$, we can define addition and scalar multiplication by $A + B = cl\{a+b : a \in A, \ b \in B\}, \lambda A = \{\lambda a : \ a \in A\}$, where $clA$ is the closure of set $A$ taken in $\mathfrak{X}$. The Hausdorff distance on $\mathcal{K}_b(\mathfrak{X})$ is defined by

$$d_H(A, B) = \max\Big\{ \sup_{a \in A} \inf_{b \in B} \|a - b\|_{\mathfrak{X}}, \sup_{b \in B} \inf_{a \in A} \|a - b\|_{\mathfrak{X}}\Big\}.$$

In particular, we denote $\|A\|_{\mathcal{K}} = d_H(\{0\}, A) = \sup_{a \in A}\{\|a\|_{\mathfrak{X}}\}$. Then $(\mathcal{K}_b(\mathfrak{X}), d_H)$ is a complete metric space (see Li, Ogura and Kreinovich [6, p.5 Theorem 1.1.2].

$X$ is called bounded closed convex random sets, if it is a measurable mapping from the space $(\Omega, \mathcal{A}, P)$ to the space $(\mathcal{K}_{bc}(\mathfrak{X}), \mathfrak{B}(\mathcal{K}_{bc}(\mathfrak{X})))$, where $\mathfrak{B}(\mathcal{K}_{bc}(\mathfrak{X}))$ is the Borel $\sigma$-field of $\mathcal{K}_{bc}(\mathfrak{X})$ generated by the Hausdorff distance $d_H$. The expectation of $X$ denoted by $E[X]$, is defined by $E[X] = cl\{\int_\Omega \xi dP : \xi \in S_X\}$, where $\int_\Omega \xi dP$ is the usual Bochner integral in $L^1[\Omega; \mathfrak{X}]$ (the family of integral $\mathfrak{X}$-valued random variables), and $S_X = \{\xi \in L^1[\Omega; \mathfrak{X}] : \xi(\omega) \in X(\omega), a.e. \ P\}$. We call $E[X]$ Auman integral (see Auman [1]).

Let $S^*$ be unit sphere of $\mathfrak{X}^*$ with strong topology whose related strong distance is denoted by $d_s^*$. Since we assume the dual space $\mathfrak{X}^*$ is a separable Banach space, the unit sphere $S^*$ is also separable. Let $D_1 = \{x_1^*, x_2^*, \cdots\}$ be the countable dense subset in the unit sphere $S^*$. Denote by $C(S^*, d_s^*)$ be space of all continuous functions on $S^*$ with the strong topology with the uniform norm $\|\cdot\|_{C(S^*)}(\|f\|_{C(S^*)} = \sup\{|f(x^*)| : x^* \in S^*\}$, for $f \in C(S^*, d_s^*)$, in fact $\|f\|_{C(S^*)} = \sup\{|f(x^*)| : x^* \in D_1\})$. We know that $C(S^*, d_s^*)$ is a Banach space, and in general it is not separable.

For each $A \in \mathcal{K}_{bc}(\mathfrak{X})$, we define its support function $s(A) : S^* \to \mathbb{R}$ as

$$s(A)(x^*) = \sup\{x^*(x) : \ x \in A\}, \quad x^* \in S^*.$$

The mapping $s : \mathcal{K}_{bc}(\mathfrak{X}) \to C(S^*, d_s^*)$ has the following properties: for any $A_1, A_2 \in \mathcal{K}_{bc}(\mathfrak{X})$ and $\lambda \in \mathbb{R}^+ = [0, \infty),(1)s(A_1 + A_2) = s(A_1) + s(A_2), (2)s(\lambda A_1) = \lambda s(A_1), (3)d_H(A_1, A_2) = \|s(A_1) - s(A_2)\|_{C(S^*)}$.

In fact, the mapping $s$ is an isometric embedding of $(\mathcal{K}_{bc}(\mathfrak{X}), d_H)$ into a closed convex cone of the Banach space $(C(S^*, d_s^*), \|\cdot\|_{C(S^*)})$ (see Li, Ogura and Kreinovich [6, p.11 Theorem 1.1.12]).

In the following, we introduce the definition of a random upper semicontinuous function. Let $I = [0, 1], I_{0+} = (0, 1]$. Let $\mathcal{F}_b(\mathfrak{X})$ denote the family of all functions $u : \mathfrak{X} \to I$ satisfying the conditions: (1) the 1-level set $[u]_1 = \{x \in \mathfrak{X} : u(x) = 1\} \neq \emptyset$, (2) each $u$ is upper semicontinuous, i.e. for each $\alpha \in I_{0+}$, the $\alpha$ level set $[u]_\alpha = \{x \in \mathfrak{X} : u(x) \geq \alpha\}$ is a closed subset of $\mathfrak{X}$, (3) the support set $[u]_0 = cl\{x \in \mathfrak{X} : u(x) > 0\}$ is bounded.

Let $\mathcal{F}_{bc}(\mathfrak{X})$(resp. $\mathcal{F}_c(\mathfrak{X})$) be the family of all bounded closed convex (resp. convex) upper semicontinuous functions. It is known that $u$ is convex in the above sense if and only if, for any $\alpha \in I$, $[u]_\alpha \in \mathcal{F}_c(\mathfrak{X})$(see Chen [3, Theorem 3.2.1]).

For any two upper semicontinuous functions $u_1, u_2$, define

$$(u_1 + u_2)(x) = \sup_{x_1 + x_2 = x} \min\{u_1(x_1), u_2(x_2)\} \quad \text{for any} \quad x \in \mathfrak{X}.$$

Similarly, for any upper semicontinuous function $u$ and for any $\lambda \geq 0$ and $x \in \mathfrak{X}$, define

$$(\lambda u)(x) = \begin{cases} u(\frac{x}{\lambda}), & \text{if } \lambda \neq 0, \\ I_0(x), & \text{if } \lambda = 0, \end{cases}$$

where $I_0$ is the indicator function of 0. It is known that for any $\alpha \in [0,1]$, $[u_1 + u_2]_\alpha = [u_1]_\alpha + [u_2]_\alpha$, $[\lambda u]_\alpha = \lambda[u]_\alpha$.

The following distance is the uniform Hausdorff distance which is extension of the Hausdorff distance $d_H$ : for $u, v \in \mathcal{F}_b(\mathfrak{X})$, $d_H^\infty(u, v) = \sup_{\alpha \in I} d_H([u]_\alpha, [v]_\alpha)$, this distance is the strongest one considered in the literatures. The space $(\mathcal{F}_{bc}(\mathfrak{X}), d_H^\infty)$ is complete. We denote $\|u\|_{\mathcal{F}} = d_H^\infty(u, I_{\{0\}}) = \|u_0\|_{\mathcal{K}}$.

$X$ is called a random upper semicontinuous function (or random fuzzy set or fuzzy set-valued random variable), if it is a measurable mapping $X : (\Omega, \mathcal{A}, P) \to (\mathcal{F}_{bc}(\mathfrak{X}), \mathfrak{B}(\mathcal{F}_{bc}(\mathfrak{X})))$ (where $\mathfrak{B}(\mathcal{F}_{bc}(\mathfrak{X}))$ is the Borel $\sigma$-field of $\mathcal{F}_{bc}(\mathfrak{X})$ generated by the uniform Hausdorff distance $d_H^\infty$). It is well known that the level mappings $L_\alpha : U \mapsto [U]_\alpha (\alpha \in I)$ are continuous from the space $(\mathcal{F}_{bc}(\mathfrak{X}), d_H^\infty)$ to the space $(\mathcal{K}_{bc}(\mathfrak{X}), d_H)$, so if $X$ is a random upper semicontinuous function, then $[X]_\alpha$ is a bounded closed convex random set for any $\alpha \in I$. The expectation of an $\mathcal{F}_{bc}(\mathfrak{X})$-valued random variable $X$, denoted by $E[X]$, is an element in $\mathcal{F}_{bc}(\mathfrak{X})$ such that for every $\alpha \in I$, $[E[X]]_\alpha = cl \int_\Omega [X]_\alpha dP = cl\{E\xi : \xi \in S_{[X]_\alpha}\}$.

Let $D(I, C(S^*, d_s^*)) = \{f : I \to C(S^*, d_s^*) \text{ is left continuous at } I_{0+}, \text{ right continuous at } 0 \text{ and bounded, and } f \text{ has right limit in } (0,1)\}$. Then it is a Banach space with respect to the norm $\|f\|_D = \sup_{\alpha \in I} \|f(\alpha)\|_{C(S^*)}$ ( see Li Ogura and Nguyen [7, Lemma 3.1]), and it is not separable.

For any $u \in \mathcal{F}_{bc}(\mathfrak{X})$, The support process of $u$ is defined to be the process

$$j(u)(\alpha, x^*) = s([u]_\alpha)(x^*) = \sup_{x \in [u]_\alpha} \{x^*(x)\}, \quad (\alpha, x^*) \in I \times S^*.$$

The mapping $j : \mathcal{F}_{bc}(\mathfrak{X}) \to D(I, C(S^*, d_s^*))$ has the following properties: (1) $j(u + v) = j(u) + j(v)$, for any $u, v \in \mathcal{F}_{bc}(\mathfrak{X})$, (2) $j(\lambda u) = \lambda j(u)$, $\lambda \geq 0$, for any $u \in \mathcal{F}_{bc}(\mathfrak{X})$, (3)$\|j(u) - j(v)\|_D = d_H^\infty(u, v)$, for any $u, v \in \mathcal{F}_{bc}(\mathfrak{X})$.

In fact, the mapping $j$ is an isometrically embedding of $(\mathcal{F}_{bc}(\mathfrak{X}), d_H^\infty)$ into a closed convex cone of the Banach space $(D(I, C(S^*, d_s^*)), \|\cdot\|_D)$, which is not separable.

Our ideal for obtaining our main results is that we first embed respectively the space $(\mathcal{K}_{bc}(\mathfrak{X}), d_H)$ into a closed convex cone of the Banach space $(C(S^*, d_s^*), \|\cdot\|_{C(S^*)})$ by support function, and the space $(\mathcal{F}_{bc}(\mathfrak{X}), d_H^\infty)$ into a closed convex cone of the Banach space $(D(I, C(S^*, d_s^*)), \|\cdot\|_D)$ by support process, but both of them are not separable, we can't again use the classical result of MDP [3], where the Banach space is needed to be separable, then we respectively continue to embed those Banach space into Wu's spaces (not necessarily separable) by linear and isometric mappings $g_1$ and $g_2$(see the following notations), where Wu [15] has obtained LDP, so we can obtain our main results. In the following, we will introduce some notations that we need corresponding to Wu's paper [15].

Since $D_1 = \{x_1^*, x_2^*, \cdots\}$ is countable dense in the unit sphere $S^*$, $\widetilde{D_1} = \{\widetilde{x_1^*}, \widetilde{x_2^*}, \cdots\}$ is a subset of the unit ball of the dual space of $(C(S^*, d_s^*), \|\cdot\|_{C(S^*)})$, where $\widetilde{x_i^*}(f) = f(x_i^*)$, for any $i \in \mathbb{N}, f \in C(S^*, d_s^*)$. Let $\ell_\infty(\widetilde{D_1})$ be the space of all bounded real function on $\widetilde{D_1}$ with supnorm $\|F\|_{\ell_\infty(\widetilde{D_1})} = \sup_{\nu \in \widetilde{D_1}} |F(\nu)|$. This is a nonseparable Banach space.

Denote $M_b(C(S^*, d_s^*), \|\cdot\|_{C(S^*)})$ be space of sighed measures of finite variations on $(C(S^*, d_s^*), \|\cdot\|_{C(S^*)})$. For every $\nu \in M_b(C(S^*, d_s^*), \|\cdot\|_{C(S^*)})$, we can define an element $\nu^{\widetilde{D_1}}$ in $\ell_\infty(\widetilde{D_1})$ as $\nu^{\widetilde{D_1}}(\widetilde{x_i^*}) = \nu(\widetilde{x_i^*}) = \int_{C(S^*, d_s^*)} \widetilde{x_i^*} d\nu$, for all $\widetilde{x_i^*} \in \widetilde{D_1}$. In particular, denote the mapping $g_1 : C(S^*, d_s^*) \to \ell_\infty(\widetilde{D_1})$ given by

$$g_1(f) = \delta_f^{\widetilde{D_1}}, \quad \delta_f^{\widetilde{D_1}}(\widetilde{x_i^*}) = \delta_f(\widetilde{x_i^*}) = \int_{C(S^*, d_s^*)} \widetilde{x_i^*} d\delta_f = \widetilde{x_i^*}(f) = f(x_i^*),$$

for all $\widetilde{x_i^*} \in \widetilde{D_1}, \delta_f$ is the Dirac measure concentrated at $f$. In fact, the mapping $g_1$ is linear and isometric from the Banach space $C(S^*, d_s^*)$ to $\ell_\infty(\widetilde{D_1})$, i.e.

(1)$g_1(\alpha f + \beta h) = \alpha g_1(f) + \beta g_1(h)$ for any $f, h \in C(S^*, d_s^*), \alpha, \beta \in \mathbb{R}$,

(2)$\|f - h\|_{C(S^*)} = \sup_{x^* \in S^*} |\delta_f^{\widetilde{D_1}}(\widetilde{x^*}) - \delta_h^{\widetilde{D_1}}(\widetilde{x^*})| = \|g_1(f) - g_1(h)\|_{\ell_\infty(\widetilde{D_1})}.$

Let $Q_0$ be all rational numbers in the interval I, $D_2 = Q_0 \times D_1$, $\widetilde{D_2} = \{\widetilde{(\alpha, x^*)} : (\alpha, x^*) \in D_2\}$ is a subset of the unit ball of the dual space of $(D(I, C(S^*, d_s^*)), \|\cdot\|_D)$, where $\widetilde{(\alpha, x^*)}(f) = f(\alpha, x^*)$, for any $f \in D(I, C(S^*, d_s^*))$. Let $\ell_\infty(\widetilde{D_2})$ be the space of all bounded real function on $\widetilde{D_2}$ with supnorm $\|F\|_{\ell_\infty(\widetilde{D_2})} = \sup_{\nu \in \widetilde{D_2}} |F(\nu)|$. This is a nonseparable Banach space.

Denote $M_b(D(I, C(S^*, d_s^*)), \|\cdot\|_D)$ be space of sighed measures of finite variations on $(D(I, C(S^*, d_s^*)), \|\cdot\|_D)$. For every $\nu \in M_b(D(I, C(S^*, d_s^*)), \|\cdot\|_D)$, we can also define an element $\nu^{\widetilde{D_2}}$ in $\ell_\infty(\widetilde{D_2})$ as $\nu^{\widetilde{D_2}}((\widetilde{\alpha, x_i^*})) = \nu((\widetilde{\alpha, x_i^*})) = \int_{D(I, C(S^*, d_s^*))} (\widetilde{\alpha, x^*}) d\nu$, for all $(\widetilde{\alpha, x_i^*}) \in \widetilde{D_2}$. In particular, we define another mapping $g_2 : D(I, C(S^*, d_s^*)) \to \ell_\infty(\widetilde{D_2})$ given by

$$g_2(f) = \delta_f^{\widetilde{D_2}},$$

$$\delta_f^{\widetilde{D_2}}((\widetilde{\alpha, x^*})) = \delta_f((\widetilde{\alpha, x^*})) = \int_{D(I, C(S^*, d_s^*))} (\widetilde{\alpha, x^*}) d\delta_f = (\widetilde{\alpha, x^*})(f) = f(\alpha, x^*),$$

for all $(\widetilde{\alpha, x^*}), f \in \widetilde{D_2}$. In fact, the mapping $g_2$ is also linear and isometric from Banach space $D(I, C(S^*, d_s^*))$ to $\ell_\infty(\widetilde{D_2})$, i.e.

(1) $g_2(\alpha f + \beta h) = \alpha g_2(f) + \beta g_2(h)$ for any $f, h \in D(I, C(S^*, d_s^*)), \alpha, \beta \in \mathbb{R}$,

(2) $\|f - h\|_D = \sup_{\alpha \in I} \sup_{x^* \in S^*} |f(\alpha, x^*) - h(\alpha, x^*)| = \sup_{(\alpha, x^*) \in D_2} |f(\alpha, x^*) - h(\alpha, x^*)|$

$$= \sup_{(\widetilde{\alpha, x^*}) \in \widetilde{D_2}} |\delta_f^{\widetilde{D_2}}((\widetilde{\alpha, x^*})) - \delta_h^{\widetilde{D_2}}((\widetilde{\alpha, x^*}))| = \|g_2(f) - g_2(h)\|_{\ell_\infty(\widetilde{D_2})}.$$

Wu [15] has obtained MDP on the spaces $(\ell_\infty(\widetilde{D_1}), \|\cdot\|_{\ell_\infty(\widetilde{D_1})})$ and $(\ell_\infty(\widetilde{D_2}), \|\cdot\|_{\ell_\infty(\widetilde{D_2})})$.

## 3  Main Results and Proofs

Before giving MDP for random sets and random upper semicontinuous functions, we define rate functions and LDP. We refer to the books of Dembo and Zeitouni [4] and Deuschel and Stroock [5] for the general theory on large deviations (also see Yan, Peng, Fang and Wu [16]).

Let $E$ be a regular Hausdorff topological and $\{\mu_\lambda : \lambda > 0\}$ be a family of probability measures on $(E, \mathcal{E})$, where $\mathcal{E}$ is the Borel $\sigma$-algebra. A *rate function* is a lower semicontinuous mapping $I : E \to [0, \infty]$. A *good rate function* is a rate function such that the level sets $\{x : I(x) \le \alpha\}$ are compact subset of $E$. let $b_\lambda$ be a positive function on $(0, +\infty)$ satisfying $\lim_{\lambda \to +\infty} b_\lambda \to +\infty$. A family of probability measures $\{\mu_\lambda : \lambda > 0\}$ on the measurable space $(E, \mathcal{E})$ is said to satisfy the *LDP* with speed $\frac{1}{b_\lambda}$ and with the rate function $I$ if, for all open set $V \subset \mathcal{E}$, $\liminf_{\lambda \to \infty} \frac{1}{b_\lambda} \ln \mu_\lambda(V) \ge -\inf_{x \in V} I(x)$, for all closed set $U \subset \mathcal{E}$, $\limsup_{\lambda \to \infty} \frac{1}{b_\lambda} \ln \mu_\lambda(U) \le -\inf_{x \in U} I(x)$,

In the following, we give our main two results. We first present MDP for $(\mathcal{K}_{bc}(\mathfrak{X}), d_H)$-valued *i.i.d.* random variables.

Suppose $\{a(n)\}_{n\geq 1}$ be a real number sequence satisfying: $a(n)/n \to 0$, $a(n)/\sqrt{n} \to \infty$.

**Theorem 1.** *Let $X, X_1, \ldots, X_n$ be $\mathcal{K}_{bc}(\mathfrak{X})$-valued i.i.d. random variables satisfying $E(s(X)(x^*))^2 < \infty$, $\forall x^* \in D_1$, and (1) The space $(\widetilde{D}_1, d_2^{(1)})$ is totally bounded, (2) $d_H\left(\frac{X_1+\cdots+X_n}{a(n)}, \frac{nE[X]}{a(n)}\right) \xrightarrow{P} 0$, (3) There exists $M > 0$, such that, for all $\varepsilon > 0$, $\limsup_{n\to\infty} \frac{n}{a^2(n)} \log(nP\{\|X\|_\mathcal{K} > \varepsilon\}) \leq -\varepsilon^2/M$.*
*Then for any open set $U \subset \mathbb{R}$,*

$$\liminf_{n\to\infty} \frac{n}{a^2(n)} \log P\left\{d_H\left(\frac{X_1+\cdots+X_n}{a(n)}, \frac{nE[X]}{a(n)}\right) \in U\right\}$$
$$\geq -\inf_{x\in U}\{\inf\{I^{(1)}_{\ell_\infty(\widetilde{D}_1)}(F) : \|F\|_{\ell_\infty(\widetilde{D}_1)} = x\}\},$$

*any for any closed set $V \subset \mathbb{R}$,*

$$\limsup_{n\to\infty} \frac{n}{a^2(n)} \log P\left\{d_H\left(\frac{X_1+\cdots+X_n}{a(N)}, \frac{nE[X]}{a(n)}\right) \in V\right\}$$
$$\leq -\inf_{x\in V}\{\inf\{I^{(1)}_{\ell_\infty(\widetilde{D}_1)}(F) : \|F\|_{\ell_\infty(\widetilde{D}_1)} = x\}\},$$

*where*

$$I^{(1)}_{\ell_\infty(\widetilde{D}_1)}(F) = \inf\{I(\nu) : \nu \in M_b(C(S^*, d_s^*), \|\cdot\|_{C(S^*)}), \nu^{\widetilde{D}_1} = F \text{ on } \widetilde{D}_1\}$$

*and $I(\nu) = \frac{1}{2}\int_{C(S^*,d_s^*)}\left(\frac{d\nu}{d(P\circ s(X)^{-1})}\right)^2 d(P\circ s(X)^{-1})$, if $\nu \ll d(P\circ s(X)^{-1})$, $\nu(C(S^*, d_s^*)) = 0$. Otherwise, $I(\nu) = +\infty$.*

The type of the large deviations above is usually called the moderate deviations (MDP). We omit the proof of Theorem 1 because its key step is included in the proof of Theorem 2 below.    □

In the following, we give MDP for $(\mathcal{F}_{bc}(\mathfrak{X}), d_H^\infty)$-valued i.i.d. random variables.

**Theorem 2.** *Let $X, X_1, \ldots, X_n$ be $\mathcal{F}_{bc}(\mathfrak{X})$-valued i.i.d. random variables satisfying $E(j(X)(\alpha, x^*))^2 < \infty$, $\forall (\alpha, x^*) \in D_2$ and (1) The space $(\widetilde{D}_2, d_2^{(2)})$ is totally bounded, (2) $d_H^\infty\left(\frac{X_1+\cdots+X_n}{a(n)}, \frac{nE[X]}{a(n)}\right) \xrightarrow{P} 0$, (3) There exists $M > 0$, such that, for all $\varepsilon > 0$, $\limsup_{n\to\infty} \frac{n}{a^2(n)} \log(nP\{\|X\|_\mathcal{F} > \varepsilon\}) \leq -\varepsilon^2/M$.*
*Then for any open set $U \subset \mathbb{R}$,*

$$\liminf_{n\to\infty} \frac{n}{a^2(n)} \log P\left\{d_H^\infty\left(\frac{X_1+\cdots+X_n}{a(n)}, \frac{nE[X]}{a(n)}\right) \in U\right\}$$
$$\geq -\inf_{x\in U}\{\inf\{I^{(2)}_{\ell_\infty(\widetilde{D}_2)}(F) : \|F\|_{\ell_\infty(\widetilde{D}_2)} = x\}\}, \tag{1}$$

*and for any closed set $V \subset \mathbb{R}$,*

$$\limsup_{n \to \infty} \frac{n}{a^2(n)} \log P \left\{ d_H^\infty \left( \frac{X_1 + \cdots + X_n}{a(N)}, \frac{nE[X]}{a(n)} \right) \in V \right\}$$

$$\leq - \inf_{x \in V} \{ \inf \{ I_{\ell_\infty(\widetilde{D}_2)}^{(2)}(F) : \|F\|_{\ell_\infty(\widetilde{D}_2)} = x \} \}, \tag{2}$$

*where*

$$I_{\ell_\infty(\widetilde{D}_2)}^{(2)}(F)$$

$$= \inf \{ I(\nu) : \quad \nu \in M_b(D(I, C(S^*, d_s^*))), \| \cdot \|_D), \nu^{\widetilde{D}_2} = F \text{ on } \widetilde{D}_2 \} \tag{3}$$

*and* $I(\nu) = \frac{1}{2} \int_{D(I,C(S^*,d_s^*))} (\frac{d\nu}{d(P \circ (j(X))^{-1})})^2 d(P \circ (j(X))^{-1})$, *if* $\nu \ll d(P \circ (j(X))^{-1}), \nu(D(I, C(S^*, d_s^*))) = 0$. *Otherwise* $I(\nu) = +\infty$.

*Proof.* For each $\alpha \in I$ and $i \in \mathbb{N}$, random sets $[X_i]_\alpha \in \mathcal{F}_{bc}(\mathfrak{X})$, and also $\text{cl}E[[X_i]_\alpha] \in \mathcal{F}_{bc}(\mathfrak{X})$. Further $s(\text{cl}E[[X_i]_\alpha]) = s(E([X_i]_\alpha) = E[s([X_i]_\alpha]$ (cf.Li, Ogura and Kreinovich [6, p.46, Theorem 2.1.12]). By the properties of linear and isometric mapping $g_2$, hence we obtain

$$d_H^\infty \left( \frac{X_1 + \cdots + X_n}{a(n)}, \frac{nE[X]}{a(n)} \right)$$

$$= \sup_{\alpha \in I} d_H \left( \frac{[X_1]_\alpha + \cdots + [X_n]_\alpha}{a(n)}, \frac{n\text{cl}E[[X]_\alpha]}{a(n)} \right)$$

$$= \sup_{\alpha \in I} \| \frac{s([X_1]_\alpha) + \cdots + s([X_n]_\alpha)}{a(n)} - \frac{ns(\text{cl}E[[X]_\alpha])}{a(n)} \|_{C(S^*)}$$

$$= \sup_{\alpha \in I} \| \frac{s([X_1]_\alpha) + \cdots + s([X_n]_\alpha)}{a(n)}, \frac{ns(E[[X]_\alpha])}{a(n)} \|_{C(S^*)}$$

$$= \| \frac{j(X_1) + \cdots + j(X_n)}{a(n)} - \frac{nE[j(X)]}{a(n)} \|_D$$

$$= \| g_2 \left( \frac{j(X_1) + \cdots + j(X_n)}{a(n)} \right) - g_2 \left( \frac{nE[j(X)]}{a(n)} \right) \|_{\ell_\infty(\widetilde{D}_2)}$$

$$= \| \left( \delta_{\frac{j(X_1) + \cdots + j(X_n)}{a(n)}} - \frac{n}{a(n)} P \circ (j(X))^{-1} \right)^{\widetilde{D}_2} \|_{\ell_\infty(\widetilde{D}_2)}.$$

In view of the conditions of this Theorem and the above equation, we have $\{ j(X), j(X_n) : n \in \mathbb{N} \}$ are $(D(I, C(S^*, d_s^*)), \| \cdot \|_D)$-valued i.i.d. random variables satisfying $\widetilde{D}_2 \subset L^2(D(I, C(S^*, d_s^*), P \circ (j(X))^{-1})$ and $(i)$The space $(\widetilde{D}_2, d_2^{(2)})$ is totally bounded, $(ii) \left( \delta_{\frac{j(X_1) + \cdots + j(X_n)}{a(n)}} - \frac{n}{a(n)} P \circ (j(X))^{-1} \right)^{\widetilde{D}_2}$

$\overset{P}{\longrightarrow} 0$, in $\ell_\infty(\widetilde{D}_2)$, $(iii)$There exists $M > 0$, such that, for all $\varepsilon > 0$,

$$\limsup_{n \to \infty} \frac{n}{a^2(n)} \log(nP \left\{ \| \delta_{j(X)}^{\widetilde{D}_2} \|_{\ell_\infty(\widetilde{D}_2)} > \varepsilon \right\}) \leq -\varepsilon^2/M.$$

By Theorem 5 in Wu [15], we know $\{P \circ ((\frac{1}{a(n)} \sum_{i=1}^{n} \delta_{j(X_i)} - \frac{n}{a(n)} P \circ$
$(j(X)))^{-1})^{\widetilde{D_2}})^{-1}) : n \in \mathbb{N}\}$ as $n \to \infty$ satisfy the MDP in $\ell_\infty(\widetilde{D_2})$ with speed
$\frac{n}{a^2(n)}$ and with the rate function given in (2). Since the mapping $\| \cdot \|_{\ell_\infty(\widetilde{D_2})}$
from $(D(I, C(S^*, d_s^*)), \| \cdot \|_D)$ to $\mathbb{R}$ is continuous, and due to the contraction
priciple [4, p.126, Theorem 4.2], then we easily obtain (1) and (5) in Theorem
2. So we complete the proof of Theorem 2.

**Acknowledgements.** This research is partially supported by Basic Research
Foundation of Natural Science of BJUT (No. 00600054K2002) and supported by
PHR (No. 201006102) and Beijing Natural Science Foundation(Stochastic Analysis
with uncertainty and applications in finance).

# References

1. Auman, A.: Integrals of set valued functions. J. Math. Anal. Appl. 12, 1–12 (1965)
2. Cerf, R.: Large deviations for sums of i.i.d. random compact sets. Pro. Amer.
   Math. Soc. 127, 2431–2436 (1999)
3. Chen, Y.: Fuzzy systems and mathematics. Huazhong institute press of Science
   and Technology, Wuhan (1984) (in Chinese)
4. Dembo, A., Zeitouni, O.: Large Deviations Techniques and applications, 2nd
   edn. Springer, Heidelberg (1998)
5. Deuschel, J.D., Strook, D.W.: Large Deviations. Academin Press, Inc., Boston
   (1989)
6. Li, S., Ogura, Y., Kreinovich, V.: Limit Theorems and Applications of Set-
   valued and Fuzzy-valued Random Variables. Kluwer Academic Publishers,
   Dordrecht (2002)
7. Li, S., Ogura, Y., Nguyen, H.T.: Gaussian processes and martingales for fuzzy
   valued random variables with continuous parameter. Information Sciences 133,
   7–21 (2001)
8. Ogura, Y., Li, S., Wang, X.: Large and moderate deviations of random upper
   semicontinuous functions. Stoch. Anal. Appl. 28, 350–376 (2010)
9. Ogura, Y., Setokuchi, T.: Large deviations for random upper semicontinuous
   functions. Tohoku. Math. J. 61, 213–223 (2009)
10. Puri, M.L., Ralescu, D.A.: Fuzzy random variables. J. Math. Anal. Appl. 114,
    406–422 (1986)
11. Teran, P.: A large deviation principle for random upper semicontimuous func-
    tions. Pro. Amer. Math. Soc. 134, 571–580 (2006)
12. Teran, P.: On Borel measurability and large deviations for fuzzy random vari-
    ables. Fuzzy Sets and Systems 157, 2558–2568 (2006)
13. Wang, X.: Sample path large deviations for random compact sets. International
    Journal of Intelligent Technologies and Applied Statistic 3(3), 323–339 (2010)
14. Wang, X.: Large deviations of random compact sets and random upper semi-
    continuous functions. In: Borgelt, C., et al. (eds.) Combining Soft Computing
    and Statistical Methods in Data Analysis, pp. 627–634. Springer, Berlin (2010)
15. Wu, L.M.: Large deviations, moderate deviations and LIL for empirical pro-
    cesses. Ann. Probab. 22, 17–27 (1994)
16. Yan, J.A., Peng, S.G., Fang, S.Z., Wu, L.M.: Several Topics in Stochstic Anal-
    ysis. Academic Press of China, Beijing (1997)

# Solution of Random Fuzzy Differential Equation

Jungang Li and Jinting Wang

**Abstract.** Random Fuzzy Differential Equation(RFDE) describes the phenomena not only with randomness but also with fuzziness. It is widely used in fuzzy control and artificial intelligence etc. In this paper, we shall discuss RFDE as follows:

$$d\tilde{F}(t) = \tilde{f}(t, \tilde{F}(t))dt + g(t, \tilde{F}(t))dB_t,$$

where $\tilde{f}(t, \tilde{F}(t))dt$ is related to fuzzy set-valued stochastic Lebesgue integral, $g(t, \tilde{F}(t))dB_t$ is related to Itô integral. Firstly we shall give some basic results about set-valued and fuzzy set-valued stochastic processes. Secondly, we shall discuss the Lebesgue integral of a fuzzy set-valued stochastic process with respect to time $t$, especially the Lebesgue integral is a fuzzy set-valued stochastic process. Finally by martingale moment inequality, we shall prove a theorem of existence and uniqueness of solution of random fuzzy differential equation.

**Keywords:** Fuzzy set-valued stochastic process, Random fuzzy differential equation, Fuzzy set-valued Lebesgue integral, Level-set process.

## 1  Introduction

Itô type stochastic differential equations have been widely used in the stochastic control (e.g. [5]) and financial mathematics (e.g. [17]). Random fuzzy differential equations(RFDEs) deal with the real phenomena not only with randomness but also with fuzziness. Puri and Ralescu introduced fuzzy set-valued random variable in [16], and gave the concept of differentiability by

Jungang Li and Jinting Wang
Department of Mathematics, Beijing Jiaotong University, Beijing 100044, P.R. China
e-mail: lijg@bjtu.edu.cn, jtwang@bjtu.edu.cn

S. Li (Eds.): Nonlinear Maths for Uncertainty and its Appli., AISC 100, pp. 161–168.
springerlink.com        © Springer-Verlag Berlin Heidelberg 2011

Hukuhara difference in [15]. Li *et al* discussed Lebesgue integral of a set-valued stochastic process with repect to time $t$ and Lebesgue integral of a fuzzy set-valued stochastic process with repect to time $t$ in [6] [7] [9] [10].

There are some nice papers on RFDEs. Feng [3] [4] studied mean-square fuzzy stochastic differential systems by mean-square derivative introduced in [2]. In [1], Fei proved the existence and uniqueness of the solution for RFDEs with non-Lipschitz coefficients. Malinowski proved the existence and uniqueness of the solution to RFDEs with global Lipschits-type condition in [13] and discussed local solution and global solution to RFDEs in [14]. Since the existence of Hukuhara difference is a difficult problem (c.f. [9]), we shall discuss random fuzzy differential equation with level sets and selections by martingale moment inequality in this paper.

We organize our paper as follows: in section 2, we shall introduce some necessary notations, definitions and results about set-valued stochastic processes and fuzzy set-valued stochastic processes. In section 3, we shall give Lebesgue integral of a fuzzy set-valued stochastic process with respect to time $t$ and discuss its properties, especially the integral is a fuzzy set-valued stochastic process. Finally, we prove the theorem of existence and uniqueness of solution to RFDE.

## 2   Fuzzy Set-Valued Stochastic Processes

Throughout this paper, assume that $(\Omega, \mathcal{A}, \mu)$ is a complete atomless probability space, $I = [0, T]$, the $\sigma$-field filtration $\{\mathcal{A}_t : t \in I\}$ satisfies the usual conditions (i.e. containing all null sets, non-decreasing and right continuous). We assume that $\mathcal{A}$ is $\mu$-separable as for the almost everywhere problem (cf. [9]). $R$ is the set of all real numbers, $N$ is the set of all natural numbers, $R^d$ is the $d$-dimensional Euclidean space with usual norm $\| \cdot \|$, $\mathcal{B}(E)$ is the Borel field of the metric space $E$. Let $f = \{f(t), \mathcal{A}_t : t \in I\}$ be a $R^d$-valued adapted stochastic process. It is said that $f$ is progressively measurable if for any $t \in I$, the mapping $(s, \omega) \mapsto f(s, \omega)$ from $[0, t] \times \Omega$ to $R^d$ is $\mathcal{B}([0, t]) \times \mathcal{A}_t$-measurable. Each right continuous (left continuous) adapted process is progressively measurable. Assume that $\mathcal{L}^p(R^d)$ denotes the set of $R^d$-valued stochastic processes $f = \{f(t), \mathcal{A}_t : t \in I\}$ such that $f$ satisfying (a) $f$ is progressively measurable; and (b)

$$|||f|||_p = \left[ E\left( \int_0^T \|f(t, \omega)\|^p ds \right) \right]^{1/p} < \infty.$$

Let $f, f' \in \mathcal{L}^p(R^d)$, $f = f'$ if and only if $|||f - f'|||_p = 0$. Then $(\mathcal{L}^p(R^d), |||\cdot|||_p)$ is complete. Now we review notation and concepts of set-valued stochastic processes. Assume that $\mathbf{K}(R^d)$ is the family of all nonempty, closed subsets of $R^d$, and $\mathbf{K}_c(R^d)$ (*resp.* $\mathbf{K}_k(R^d)$, $\mathbf{K}_{kc}(R^d)$) is the family of all nonempty closed convex (*resp.* compact, compact convex) subsets of $R^d$.

Let $\mathbf{F}(R^d)$ denote the family of all functions $\nu : R^d \to [0,1]$ which satisfy the following two conditions:

(1) each $\nu$ is an upper semicontinuous function, i.e., for each $\alpha \in (0,1]$ the level set $\nu_\alpha = \{x \in R^d : \nu(x) \geq \alpha\}$ is a closed subset of $R^d$,

(2) the level set $\nu_1 = \{x \in R^d : \nu(x) = 1\} \neq \emptyset$.

Let $\mathbf{F}_k(R^d)$ denote the family of all functions $\nu$ in $\mathbf{F}(R^d)$ with property (3) their support sets $\nu_{0+} = \mathrm{cl}\{x \in R^d : \nu(x) > 0\} \in \mathbf{K}_k(R^d)$.

A *fuzzy set-valued random variable ( or fuzzy random variable, fuzzy random set)* is a function $\tilde{X} : \Omega \to \mathbf{F}(R^d)$, such that its level set

$$\tilde{X}_\alpha(\omega) = \{x \in R^d : \tilde{X}(\omega)(x) \geq \alpha\}$$

is a set-valued random variable for every $\alpha \in (0,1]$.

**Definition 1.** $\tilde{F} = \{\tilde{F}(t) : t \in I\}$ is called a *fuzzy set-valued stochastic process* if $\tilde{F} : I \times \Omega \to \mathbf{F}(R^d)$ is such that for any fixed $t \in I$, $\tilde{F}(t, \cdot)$ is a fuzzy set-valued random variable. For any $\alpha \in (0,1]$, $\tilde{F}_\alpha = \{\tilde{F}_\alpha(t) : t \in I\}$ is a set-valued stochastic process called *$\alpha$-level set process*.

**Definition 2.** A set-valued stochastic process $F = \{F(t) : t \in I\}$ is called *progressively measurable*, if for any $A \in \mathcal{B}(R^d)$ and any $t \in I$, $\{(s, \omega) \in [0,t] \times \Omega : F(s, \omega) \cap A \neq \emptyset\} \in \mathcal{B}([0,t]) \times \mathcal{A}_t$. $F$ is called $\mathcal{L}^1$-bounded, if the real stochastic process $\{\|F(t)\|_{\mathbf{K}}, \mathcal{A}_t : t \in I\} \in \mathcal{L}^1(R)$.

**Definition 3.** A $R^d$-valued progressively process $\{f(t), \mathcal{A}_t : t \in I\} \in \mathcal{L}^1(R^d)$ is called an $\mathcal{L}^1$-selection of $F = \{F(t), \mathcal{A}_t : t \in I\}$ if $f(t, \omega) \in F(t, \omega)$ for a.e. $(t, \omega) \in I \times \Omega$.

Let $S^1(\{F(\cdot)\})$ or $S^1(F)$ denote the family of all $\mathcal{L}^1$-selections of $F = \{F(t), \mathcal{A}_t : t \in I\}$, i.e.

$$S^1(F) = \left\{ \{f(t) : t \in I\} \in \mathcal{L}^1(R^d) : f(t, \omega) \in F(t, \omega), \text{ for a.e. } (t, \omega) \in I \times \Omega \right\}.$$

Let $\mathcal{L}^1(\mathbf{K}(R^d))$ denote the set of all $\mathcal{L}^1$-bounded progressively measurable $\mathbf{K}(R^d)$-valued stochastic processes. Similarly, we have notations $\mathcal{L}^1(\mathbf{K}_c(R^d))$, $\mathcal{L}^1(\mathbf{K}_k(R^d))$ and $\mathcal{L}^1(\mathbf{K}_{kc}(R^d))$.

**Definition 4.** A fuzzy set-valued stochastic process $\tilde{F} = \{\tilde{F}(t) : t \in I\}$ is called *progressively measurable*, if for any $\alpha \in (0,1]$, $\tilde{F}_\alpha = \{\tilde{F}_\alpha(t) : t \in I\}$ is a progressively measurable set-valued stochastic process.

**Definition 5.** A fuzzy set-valued stochastic process $\tilde{F} = \{\tilde{F}(t) : t \in I\}$ is called *integrably bounded*, if $\tilde{F}_{0+} = \{\tilde{F}_{0+}(t) : t \in I\}$ is $\mathcal{L}^1$-bounded. Let $\mathcal{L}^1[\Omega, \mathcal{A}, \mu; \mathbf{F}_k(R^d)]$ be the set of all integrably bounded progressively measurable fuzzy set-valued stochastic processes and similarly $\mathcal{L}^1[\Omega, \mathcal{A}, \mu; \mathbf{F}_{kc}(R^d)]$.

Concerning more definitions and more results of set-valued and fuzzy set-valued random variables or stochastic processes, readers could refer to the book [11].

## 3 Fuzzy Set-Valued Lebesgue Integral and Random Fuzzy Differential Equation

Now we introduce Aumann type Lebesgue integral of a fuzzy set-valued stochastic differential equation, for more details we could refer to [10]. Here, we use $\mathcal{L}^1(\mathbf{F}_k(R^d))$. In fact, the following holds for $\mathcal{L}^p(\mathbf{F}_k(R^d))$, $p \geq 1$.

**Definition 6.** Let a fuzzy set-valued stochastic process $\tilde{F} = \{\tilde{F}(t) : t \in I\} \in \mathcal{L}^1(\mathbf{F}_k(R^d))$. For any $\alpha \in (0,1]$, $t \in I$, $\omega \in \Omega$, define

$$(A) \int_0^t \tilde{F}_\alpha(s,\omega)ds := \left\{ \int_0^t f_\alpha(s,\omega)ds : f_\alpha \in S^1(\tilde{F}_\alpha) \right\},$$

where $\int_0^t f_\alpha(s,\omega)ds$ is the Lebesgue integral. $(A) \int_0^t \tilde{F}(s,\omega)ds$ is called the Aumann type Lebesgue integral of the fuzzy set-valued stochastic process $\tilde{F}$ with respect to time $t$. For any $0 \leq u < t < T$,

$$(A) \int_u^t \tilde{F}(s,\omega)ds := (A) \int_0^t I_{[u,t]}(s)\tilde{F}(s,\omega)ds.$$

**Theorem 1.** Let a fuzzy set-valued stochastic process $\tilde{F} \in \mathcal{L}^1(\mathbf{F}_k(R^d))$. Then for any $\alpha \in (0,1]$, the set-valued mapping $L_t(\tilde{F}_\alpha) : \Omega \to \mathbf{K}_{kc}(R^d)$ defined by

$$L_t(\tilde{F}_\alpha)(\omega) = (A) \int_0^t \tilde{F}_\alpha(s,\omega)ds$$

is measurable, i.e. $L_t(\tilde{F}_\alpha)$ is a set-valued random variable, and

$$L_t(\tilde{F}_\alpha)(\omega) = (A) \int_0^t \mathrm{co}\tilde{F}_\alpha(s,\omega)ds.$$

**Remark.** We are interested in the set of all selections of the integral stochastic process $L(\tilde{F}_\alpha)$. For any fixed $t \in I$, $I_t(f_\alpha)(\omega) =: \int_0^t f_\alpha(s,\omega)ds$ is an $\mathcal{A}_t$-measurable function with respect to $\omega$ for any given $f_\alpha \in S^1(\tilde{F}_\alpha)$. Thus $I_t(f_\alpha)(\cdot) =: \int_0^t f_\alpha(s,\cdot)ds$ is a selection of $L_t(\tilde{F}_\alpha)$. As a matter of fact, we have the following Theorem.

**Theorem 2.** Assume that a fuzzy set-valued stochastic process $\tilde{F} \in \mathcal{L}^1(\mathbf{F}_k(R^d))$ and continue to use above notations. Then we have that for any $\alpha \in (0,1]$, $\{I_t(f_\alpha) : f_\alpha \in S^1(\tilde{F}_\alpha)\}$ is closed in $L^1[\Omega, \mathcal{A}_t, \mu; R^d]$.

**Theorem 3.** For any $\tilde{F}$, $\tilde{G} \in \mathcal{L}^1(\mathbf{F}_k(R^d))$, $\alpha \in (0,1]$, we have

$$d_H^2(L_t(\tilde{F}_\alpha)(\omega), L_t(\tilde{G}_\alpha)(\omega)) \leq t \int_0^t d_H^2(\tilde{F}_\alpha(s,\omega), \tilde{G}_\alpha(s,\omega))ds.$$

**Theorem 4.** For any fuzzy set-valued stochastic process $\tilde{F} \in \mathcal{L}^1(\mathbf{F}_k(R^d))$, there exists a unique fuzzy set-valued random variable $\tilde{L}(t)$ which belongs to

$\mathcal{L}^1[\Omega, \mathcal{A}_t, \mu; \mathbf{F}_k(R^d)]$ *such that for all* $t \in I$, $\alpha \in [0,1]$,

$$\left\{ x \in R^d : \tilde{L}(t, \omega) \geq \alpha \right\} = \int_0^t \tilde{F}_\alpha(s, \omega) ds, \quad a.e.$$

From theorem 4, the Aumann type Lebesgue integral of fuzzy set-valued stochastic process $\tilde{F}$ with respect time $t$ is the fuzzy set-valued stochastic process $\tilde{L} = \{\tilde{L}(t), t \in I\}$. Now we introduce its application in random fuzzy differential equation.

We consider the following fuzzy set-valued stochastic differential equation

$$d\tilde{F}(t) = \tilde{f}(t, \tilde{F}(t))dt + g(t, \tilde{F}(t))dB_t, \qquad (1)$$

where the fuzzy set-valued random variable $\tilde{F}(t) \in L^2(\mathbf{F}_k(R^d))$ with initial condition $\tilde{F}(0)$ being an $L^2$-bounded fuzzy set-valued random variable, for any $\alpha \in [0, 1]$, $\tilde{f}_\alpha : I \times \mathbf{K}_k(R^d) \to \mathbf{K}_k(R^d)$ is measurable, $g_\alpha : I \times \mathbf{K}_k(R^d) \to R^d \otimes R^m$ is measurable, $\tilde{f}(t) \in \mathrm{L}^2(\mathbf{F}_k(R^d))$, $g : I \times \mathbf{F}_k(R^d) \to R^d \otimes R^m$, $g \in \mathrm{L}^2(R^d)$, $B_t$ is an $m$-dimensional Brown motion. Equation (1) is equivalent to the integral form:

$$\tilde{F}(t) = \tilde{F}(0) + (A) \int_0^t \tilde{f}(s, \tilde{F}(s))ds + \int_0^t g(s, \tilde{F}(s))dB_s, \qquad (2)$$

or the level set integral equation: for any $\alpha \in (0, 1]$,

$$\tilde{F}_\alpha(t) = \tilde{F}_\alpha(0) + (A) \int_0^t \tilde{f}_\alpha(s, F_\alpha(s))ds + \int_0^t g_\alpha(s, F_\alpha(s))dB_s.$$

**Theorem 5 (Existence and uniqueness theorem).** *For* $\alpha \in (0, 1]$, $\tilde{F}_\alpha \in \mathbf{K}_k(R^d)$, $t \in I$, *assume that* $\tilde{f}_\alpha(t, \tilde{F}_\alpha)$, $g_\alpha(t, \tilde{F}_\alpha)$ *satisfy the following conditions:*

*(i) linear growth condition:*

$$\|\tilde{f}_\alpha(t, \tilde{F}_\alpha)\|_{\mathbf{K}}^2 + \|g_\alpha(t, \tilde{F}_\alpha)\|^2 \leq K^2(1 + \|\tilde{F}_\alpha\|_{\mathbf{K}}^2);$$

*(ii) Lipschitz continuous condition:*

$$d_H(\tilde{f}_\alpha(t, \tilde{F}_{1,\alpha}), \tilde{f}_\alpha(t, \tilde{F}_{2,\alpha})) + \|g_\alpha(t, \tilde{F}_{1,\alpha}) - g_\alpha(t, \tilde{F}_{2,\alpha})\| \leq K d_H(\tilde{F}_{1,\alpha}, \tilde{F}_{2,\alpha});$$

*where* $K$ *is a positive constant. Then for any given initial* $L^2$*-bounded fuzzy set-valued random variable* $\tilde{F}(0)$, *there is a solution to the equation (1), and the solution is unique.*

*Proof. Step 1* We first note that, for any $\alpha \in (0, 1]$, each $t \in I$,

$$E[\|\tilde{f}_\alpha(t, \tilde{F}_\alpha(t))\|_{\mathbf{K}}^2] + E[\|g(t, \tilde{F}_\alpha(t))\|^2] \leq 4K^2(1 + E[\|\tilde{F}_\alpha(t)\|_{\mathbf{K}}^2]). \qquad (3)$$

Since $\tilde{F}_\alpha(s, \omega)$ is progressively measurable, then $\tilde{f}_\alpha(t, \tilde{F}_\alpha(t))$ and $g(t, \tilde{F}_\alpha(t))$ are also progressively measurable. Hence, $\tilde{f}_\alpha(t, \tilde{F}_\alpha(t)) \in \mathcal{L}^2(\mathbf{K}_k(R^d))$ and $g(t, \tilde{F}_\alpha(t)) \in L^2(R^d \otimes R^m)$.

To use successively approximation method, we define, for each $t \in I$,

$$\tilde{F}_{0,\alpha}(t) = \tilde{F}_\alpha(0),$$

$$\tilde{F}_{n+1,\alpha}(t) = \tilde{F}_\alpha(0) + \int_0^t \tilde{f}_\alpha(s, \tilde{F}_{n,\alpha}(s))ds + \int_0^t g(s, \tilde{F}_{n,\alpha}(s))dB_s, \quad n \geq 0.$$

We will show that $\tilde{F}_{n,\alpha} = \tilde{F}_{n,\alpha}(s, \omega)$ belongs to $\mathcal{L}^2(\mathbf{K}_k(R^d))$. This assertion is obvious for $n = 0$. Assume that $\tilde{F}_{n,\alpha} \in \mathcal{L}^2(\mathbf{K}_k(R^d))$. Then $\tilde{f}_\alpha(t, \tilde{F}_{n,\alpha}(t)) \in \mathcal{L}^2(\mathbf{K}_k(R^d))$ and $g(t, \tilde{F}_{n,\alpha}(t)) \in \mathcal{L}^2(R^d \otimes R^m)$ by (3). We further have

$$E[\|\tilde{F}_{n+1,\alpha}(t)\|_{\mathbf{K}}^2] \leq 3E\left[\|\tilde{F}_\alpha(0)\|_{\mathbf{K}}^2 + \left\|\int_0^t \tilde{f}_\alpha(s, \tilde{F}_{n,\alpha}(s))ds\right\|_{\mathbf{K}}^2 \right.$$
$$\left. + \left\|\int_0^t g(s, \tilde{F}_{n,\alpha}(s))dB_s\right\|^2\right].$$

Since $E[\|\int_0^t g(s, \tilde{F}_{n,\alpha}(s))dB_s\|^2] = E[\int_0^t \|g(s, \tilde{F}_{n,\alpha}(s))\|^2 ds]$, the equation above with (3) implies $\tilde{F}_{n+1,\alpha} \in \mathcal{L}^2(\mathbf{K}_k(R^d))$. Hence $\tilde{F}_{n,\alpha} \in \mathcal{L}^2(\mathbf{K}_k(R^d))$ for all $n \geq 0$ by induction.

*Step 2* In this step we shall prove the existence of a solution to equation (2). By Theorem 3 together with the inequality

$$d_H(a_1 + A_1 + B_1, a_2 + A_2 + B_2) \leq \|a_1 - a_2\| + d_H(A_1, A_2) + d_H(B_1, B_2),$$

for $a_i \in R^d$, $A_i, B_i \in \mathbf{K}_k(R^d)$, we have

$$d_H(\tilde{F}_{n,\alpha}(t), \tilde{F}_{n+1,\alpha}(t)) \leq \int_0^t d_H(\tilde{f}(s, \tilde{F}_{n-1,\alpha}(s)), \tilde{f}(s, \tilde{F}_{n,\alpha}(s)))ds$$
$$+ \left\|\int_0^t (g(s, \tilde{F}_{n-1,\alpha}(s)) - g(s, \tilde{F}_{n,\alpha}(s)))dB_s\right\|.$$

By the martingale moment inequality, it holds

$$E\left[\sup_{u \in [0, t]} \left\|\int_0^u (g(s, \tilde{F}_{n-1,\alpha}(s)) - g(s, \tilde{F}_{n,\alpha}(s)))dB_s\right\|^2\right]$$
$$\leq 4E\left[\int_0^t \|g(s, \tilde{F}_{n-1,\alpha}(s)) - g(s, \tilde{F}_{n,\alpha}(s))\|^2 ds\right].$$

By the two inequalities above and condition (ii), we have

$$\Delta_n(t) \leq 2E\Big[T \int_0^t d_H^2(\tilde{f}(s, \tilde{F}_{n-1,\alpha}(s)), \tilde{f}(s, \tilde{F}_{n,\alpha}(s)))ds$$

$$+ 4\int_0^t \|g(s, \tilde{F}_{n-1,\alpha}(s)) - g(s, \tilde{F}_{n,\alpha}(s))\|^2 ds\Big]$$

$$\leq 2(T \vee 4)K^2 E\Big[\int_0^t d_H^2(\tilde{F}_{n-1,\alpha}(s), \tilde{F}_{n,\alpha}(s))ds\Big]$$

$$\leq 2(T \vee 4)K^2 \int_0^t \Delta_{n-1}(s)ds,$$

where

$$\Delta_n(t) = E\Big[\sup_{s\in[0,\,t]} d_H^2(\tilde{F}_{n,\alpha}(s, \omega), \tilde{F}_{n+1,\alpha}(s, \omega))\Big].$$

Hence, we have

$$\Delta_n(T) \leq c^n \int_0^T \int_0^{s_1} \cdots \int_0^{s_{n-1}} \Delta_1(s_n)ds_n ds_{n-1} \cdots ds_1 \leq \frac{(cT)^n}{n!}\Delta_1(T),$$

for some constants $c > 0$. Therefore, one has

$$\sum_{n=1}^{\infty} \Delta_n(T) < \infty, \quad \text{or} \quad \sum_{n=1}^{\infty} \sup_{s\in[0,\,T]} d_H^2(\tilde{F}_{n,\alpha}(s, \omega), \tilde{F}_{\alpha}(s, \omega)) < \infty, \quad a.e.$$

This ensures the existence $\tilde{F}_{\alpha}(t, \omega) \in \mathbf{K}_k(R^d)$ such that

$$\lim_{n\to\infty} \sup_{s\in[0,\,T]} d_H(\tilde{F}_{n,\alpha}(s, \omega), \tilde{F}_{\alpha}(s, \omega)) = 0, \qquad a.e.$$

Since $[0,1]$ is separable, there is a solution $\tilde{F}$ which is a fuzzy set-valued stochastic process satisfying equation (2).

*Step 3* In this step we shall prove the uniqueness of the solutions. Let $\tilde{F}$ and $\tilde{G}$ be two solutions to equation (2), and for any $\alpha \in (0, 1]$, denote

$$\Delta(t) = E[\sup_{s\in[0,\,t]} d_H^2(\tilde{F}_{\alpha}(s, \omega), \tilde{G}_{\alpha}(s, \omega))].$$

Then, through the same way as above, we have

$$\Delta(t) \leq 2(T \vee 4)K^2 \int_0^t \Delta(s)ds,$$

which implies

$$\Delta(T) \leq \frac{(cT)^n}{n!}\Delta(T).$$

Letting $n \to \infty$, we obtain $\Delta(T) = 0$ and the uniqueness follows. $\square$

**Acknowledgements.** We would like to thank the referees for their valuable comments. This work was supported by National Science Foundation of China under Grant nos. 10871020 and 11010301053.

# References

1. Fei, W.: Existence and uniqueness of solution for fuzzy random differential equations with non-Lipschitz coefficients. Infor. Sci. 177, 4329–4337 (2007)
2. Feng, Y.: Mean-square integral and differential of fuzzy stochastic processes. Fuzzy Sets and Syst. 102, 271–280 (1999)
3. Feng, Y.: Fuzzy stochastic differential systems. Fuzzy Sets and Syst. 115, 351–363 (2000)
4. Feng, Y.: The solutions of linear fuzzy stochastic differential systems. Fuzzy Sets and Syst. 140, 341–354 (2003)
5. Ikeda, N., Watanabe, S.: Stochastic Differential Equations and Diffusion Processes. North-Holland, Kodansha (1981)
6. Li, J., Li, S.: Set-valued stochastic Lebesgue integral and representation theorems. Int. J. Comput. Intell. Syst. 1, 177–187 (2008)
7. Li, J., Li, S.: Aumann type set-valued Lebesgue integral and representation theorem. Int. J. Comput. Intell. Syst. 2, 83–90 (2009)
8. Li, J., Li, S., Xie, Y.: The space of fuzzy set-valued square integrable martingales. In: IEEE Int. Conf. Fuzzy Syst., Korea, pp. 872–876 (2009)
9. Li, J., Li, S., Ogura, Y.: Strong solution of Itô type set-valued stochastic differential equation. Acta Mathematica Sinica 26, 1739–1748 (2010)
10. Li, J., Wang, J.: Fuzzy set-valued stochastic Lebesgue integral (unpublished)
11. Li, S., Ogura, Y., Kreinovich, V.: Limit Theorems and Applications of Set-Valued and Fuzzy Set-Valued Random Variables. Kluwer Academic Publishers, Dordrecht (2002)
12. Li, S., Ren, A.: Representation theorems, set-valued and fuzzy set-valued Itô integral. Fuzzy Sets and Syst. 158, 949–962 (2007)
13. Malinowski, M.: On random fuzzy differential equation. Fuzzy Sets and Syst. 160, 3152–3165 (2009)
14. Malinowski, M.: Existence theorems for solutions to random fuzzy differential equations. Nonlinear Analysis 73, 1515–1532 (2010)
15. Puri, M., Ralescu, D.: Differentials of fuzzy functions. J. Math. Anal. Appl. 91, 552–558 (1983)
16. Puri, M., Ralescu, D.: Fuzzy random variables. J. Math. Anal. Appl. 114, 409–422 (1986)
17. Steele, J.: Stochastic Calculus and Financial Applications. Springer, New York (2001)

# Completely Monotone Outer Approximations of Lower Probabilities on Finite Possibility Spaces

Erik Quaeghebeur

**Abstract.** Drawing inferences from general lower probabilities on finite possibility spaces usually involves solving linear programming problems. For some applications this may be too computationally demanding. Some special classes of lower probabilities allow for using computationally less demanding techniques. One such class is formed by the completely monotone lower probabilities, for which inferences can be drawn efficiently once their Möbius transform has been calculated. One option is therefore to draw approximate inferences by using a completely monotone approximation to a general lower probability; this must be an outer approximation to avoid drawing inferences that are not implied by the approximated lower probability. In this paper, we discuss existing and new algorithms for performing this approximation, discuss their relative strengths and weaknesses, and illustrate how each one works and performs.

**Keywords:** lower probabilities, Outer approximation, Complete monotonicity, Belief functions, Möbius transform.

## 1 Introduction

In the theory of coherent lower previsions—or, more colloquially, of imprecise probabilities—the procedure of natural extension is the basic technique for drawing inferences [11, §3.1]. In a finitary setting, i.e., one with a finite possibility space $\Omega$ and in which the lower prevision $\underline{P}$ is assessed for a finite collection of gambles (random variables) $\mathcal{K} \subseteq \mathbb{R}^\Omega$, calculating the natural extension $\underline{E}_P f$ for a gamble $f$ in $\mathbb{R}^\Omega$ corresponds to solving a linear programming (LP) problem:

Erik Quaeghebeur
SYSTeMS Research Group, Ghent University, Technologiepark-Zwijnaarde 914,
9052 Gent
e-mail: Erik.Quaeghebeur@UGent.be

S. Li (Eds.): Nonlinear Maths for Uncertainty and its Appli., AISC 100, pp. 169–178.
springerlink.com

$$\underline{E}_{\underline{P}}f := \max\{\alpha \in \mathbb{R} : f - \alpha \geq \textstyle\sum_{g \in \mathcal{K}} \lambda_g \cdot (g - \underline{P}g), \lambda \in \mathbb{R}_{\geq 0}^{\mathcal{K}}\}. \qquad (1)$$

In applications where $\mathcal{K}$ is large or where the natural extension needs to be calculated for a large number of gambles, such LP problems may be too computationally demanding. But natural extension preserves dominance: if $\underline{P}_* \leq \underline{P}$ then $\underline{E}_{\underline{P}_*} \leq \underline{E}_{\underline{P}}$.

So first we restrict $\mathcal{K}$ and only consider lower previsions $\underline{P}$ that are defined on the set of (indicators of) events; i.e., we only consider lower probabilities $\underline{P}$ defined on the power set $2^\Omega$. Now, for 2-monotone lower probabilities $\underline{P}_*$, which form a subclass of the coherent lower probabilities, the natural extension $\underline{E}_{\underline{P}_*}$ can be calculated more efficiently using Choquet integration [11, see, e.g.,§3.2.4]:

$$\underline{E}_{\underline{P}_*}f = (C)\int f d\underline{P}_* := \min f + \textstyle\int_{\min f}^{\max f} \underline{P}_*\{\omega \in \Omega : f\omega \geq t\}\mathrm{d}t, \qquad (2)$$

So if we can find a 2-monotone outer approximation $\underline{P}_*$ to $\underline{P}$, i.e., such that $\underline{P}_* A \leq \underline{P}A$ for all events $A \subseteq \Omega$, we can efficiently calculate the outer approximation $\underline{E}_{\underline{P}_*}$ to $\underline{E}_{\underline{P}}$.

How can we go about this? Every coherent lower prevision $\underline{P}$ can be written as a convex combination of extreme coherent lower previsions [7] that is not-necessarily unique [6, e.g.,§2.3.3, ¶4]; in the finitary case, the set $\mathcal{E}^c(\mathcal{K})$ of extreme coherent lower previsions on $\mathcal{K}$ is finite. So $\underline{P} = \sum_{\underline{Q} \in \mathcal{E}^c(\mathcal{K})} \lambda \underline{Q} \cdot \underline{Q}$, where $\lambda : \mathcal{E}^c(\mathcal{K}) \to [0,1]$ is a function that generates coefficients of a convex $\mathcal{E}^c(\mathcal{K})$-decomposition of $\underline{P}$. The same holds for 2-monotone lower probabilities, but with a different set of extreme members $\mathcal{E}^2(\Omega)$ [8,6]. The idea is to find a $\nu : \mathcal{E}^2(\Omega) \to [0,1]$ such that $\underline{P}_* := \sum_{\underline{Q} \in \mathcal{E}^2(\Omega)} \nu \underline{Q} \cdot \underline{Q}$ is an—in some sense—good outer approximation to $\underline{P}$.

It is impractical to consider all elements of $\mathcal{E}^2(\Omega)$: finding this set is computationally very demanding and with increasing $|\Omega|$ it quickly becomes very large [8, §4]. In this paper, our strategy is to only retain the subclass $\mathcal{E}^\infty(\Omega)$ of vacuous lower probabilities: each such lower prevision essentially corresponds to an assessment that a given event $A$ of $\Omega$ occurs; the corresponding natural extension is given by $\underline{E}_A f := \min_{\omega \in A} f\omega$. Lower probabilities $\underline{P}_*$ that can be written as a convex combination of vacuous lower probabilities are called completely monotone. The decomposition of such a lower probability, i.e., the coefficient function $\nu : 2^\Omega \to [0,1]$, is unique and determines it as follows:

$$\underline{P}_*A = \textstyle\sum_{B \subseteq A} \nu B, \qquad \underline{E}_{\underline{P}_*}f = \sum_{B \subseteq \Omega} \nu B \cdot \underline{E}_B f = \sum_{B \subseteq \Omega} \nu B \cdot \min_{\omega \in B} f\omega. \qquad (3)$$

The left-hand equation is called Möbius inversion; the right-hand one is an alternative to Choquet integration for calculating the natural extension.

Mathematically, completely monotone lower probabilities coincide with the belief functions of Dempster–Shafer theory [4,9]. From this theory, we know that the coefficients of the decomposition—which we call basic belief mass assignments in this paper—can be obtained by using the Möbius transform of $\underline{P}_*$; i.e., the coefficient of $\underline{E}_A$ is

$$\nu A = \sum_{B \subseteq A}(-1)^{|A \setminus B|} \cdot \underline{P}_* B = \underline{P}_* A - \sum_{B \subset A} \nu B, \qquad (4)$$

where the last expression shows how these coefficients can be calculated recursively. Obviously, $\underline{P}_*$ must be defined for all events to calculate these coefficients; if necessary, one should extend it to all indicator functions first to obtain the lower probability on all events.

In this paper, we assume a lower probability $\underline{P}$ is given that is defined on all events. We discuss a number of algorithms that allow us to obtain a basic belief mass assignment function $\nu$ that determines—via Equation (3)—a completely monotone lower probability $\underline{P}_*$ that is an outer approximation to $\underline{P}$, i.e., $\underline{P}_* \leq \underline{P}$. The reason we focus on outer approximations is that they are conservative in the sense that they do not lead to inferences unwarranted by the approximated lower probability. For the algorithms to work, it is sufficient that $\underline{P}$ satisfies $\sum_{\omega \in \Omega} \underline{P}\{\omega\} \leq 1$, is nonnegative ($\underline{P} \geq 0$), monotone ($B \subseteq A \Rightarrow \underline{P}B \leq \underline{P}A$), and normed ($\underline{P}\emptyset = 0$ and $\underline{P}\Omega = 1$), all four of which we assume to be the case.

## 2 Completely Monotone Outer Approximation Algorithms

We are going to discuss four algorithms—one trivial new one, two from the literature, and one substantive new one—that fall into three classes: the first one creates a linear-vacuous mixture, the second one reduces the problem to an LP problem, and the last two are based on modifications of the Möbius transformation and are more heuristic in nature. (All algorithms have been implemented in Troffaes's `improb` software package/framework [10].) But before jumping into this material, we discuss a useful preprocessing step and introduce the lower probabilities that are used to illustrate (the results) of the techniques.

First the preprocessing step: we mentioned that the decomposition into extreme coherent lower previsions of a coherent lower prevision $\underline{P}$ is in general non-unique. However, the coefficients $\lambda\{\omega\}$ of the degenerate lower previsions—i.e., vacuous lower previsions relative to singletons $\{\omega\}$ of $\Omega$— are unique [7, Prop. 1], so we can write any coherent lower probability as a linear-imprecise mixture:

$$\underline{P}A = \kappa \cdot \sum_{\omega \in A} p\omega + (1-\kappa) \cdot \underline{R}A, \qquad \underline{E}_P f = \kappa \cdot \sum_{\omega \in \Omega} p\omega \cdot f\omega + (1-\kappa) \cdot \underline{E}_R f, \qquad (5)$$

where $\kappa := \sum_{\omega \in \Omega} \lambda\{\omega\}$ and $p\omega := \lambda\{\omega\}/\kappa$, which allows us to solve for the imprecise part $\underline{R}$. This is a coherent lower probability whose lower probability on singletons is zero. (The second equation then follows from [11, §3.4.1].) In case $\underline{P}$ is not coherent, $\underline{R}$ may end up with negative values, but we may infer from the zero values of $\underline{R}$ on singletons that these may be set to zero. Given that linear previsions are completely monotone, it makes sense to separate out the linear part—represented by the probability mass function $p$—and only approximate the imprecise part.

We use some example lower probabilities to illustrate the algorithms. All have $\Omega = \{a, b, c, d\}$: a cardinality of four allows the lower probabilities to be complex enough to be interesting without resulting in unending lists of numbers. (Also, for $|\Omega| < 4$, all lower probabilities are probability intervals and therefore 2-monotone [2, Prop. 5].) The first one in Table 1 is from the literature [5, Ex. 2], for which we also give the linear-imprecise decomposition; the second and third ones in Table 2 are especially chosen extreme coherent lower probabilities [8, 6, App. A] that highlight some of the algorithms' features and consist of an imprecise part only.

**Table 1** A lower probability $\underline{P}$, its linear-imprecise decomposition ($p, \underline{R}, \kappa = 0.737$), the 2-monotone probability interval outer approximation $\underline{R}_{\mathrm{PI}}$ of $\underline{R}$ (generated by efficient natural extension from $[\underline{R}\{\omega\}, \overline{R}\{\omega\} = 1 - \underline{R}(\Omega \setminus \{\omega\})]_{\omega \in \Omega}$ [2, cf. particularly Prop. 4]), the Möbius transform $\rho$ of $\underline{R}$, and six completely monotone outer approximations of $\underline{R}$: the linear-vacuous one $\underline{R}_{\mathrm{LV}}$; two optimal ones, $\underline{R}_{\mathrm{LPDS}}$ using a dual simplex solver and $\underline{R}_{\mathrm{LPCC}}$ using a criss-cross solver; two IRM-approximations, $\underline{R}_{\mathrm{IRM}}$ using the lexicographic order and $\pi\underline{R}_{\mathrm{IRM}}$ the inverse order; the IMRM-approximation $\underline{R}_{\mathrm{IMRM}}$. Negative values in the Möbius transform $\rho$ of $\underline{R}$ are highlighted with a gray background. Approximation values that differ from the approximated values of $\underline{R}$ are in boldface. The last row contains the sum-norm-differences between $\underline{R}$ and the approximations. The number of significant digits used has been chosen to facilitate comparisons and verification.

| Event | $\underline{P}$ | $p$ | $\underline{R}$ | $\rho$ | $\underline{R}_{\mathrm{LV}}$ | $\underline{R}_{\mathrm{LPDS}}$ | $\underline{R}_{\mathrm{LPCC}}$ | $\underline{R}_{\mathrm{IRM}}$ | $\pi\underline{R}_{\mathrm{IRM}}$ | $\underline{R}_{\mathrm{IMRM}}$ | $\underline{R}_{\mathrm{PI}}$ |
|---|---|---|---|---|---|---|---|---|---|---|---|
| $a$ | 0.0895 | 0.122 | | | | | | | | | |
| $b$ | 0.2743 | 0.372 | | | | | | | | | |
| $c$ | 0.2668 | 0.362 | | | | | | | | | |
| $d$ | 0.1063 | 0.144 | | | | | | | | | |
| $a\,b$ | 0.3947 | 0.117 | 0.117 | 0 | 0 | 0.117 | **0.046** | **0.091** | **0.066** | 0 | |
| $a\ \ c$ | 0.4506 | 0.358 | 0.358 | 0 | **0.196** | **0.079** | **0.185** | **0.193** | **0.211** | 0 | |
| $a\ \ \ \ d$ | 0.2959 | 0.381 | 0.381 | 0 | **0.352** | **0.352** | **0.242** | **0.249** | **0.244** | **0.129** | |
| $b\,c$ | 0.5837 | 0.162 | 0.162 | 0 | 0.162 | 0.162 | **0.074** | **0.074** | **0.082** | 0 | |
| $b\ \ d$ | 0.4835 | 0.391 | 0.391 | 0 | **0.227** | **0.110** | **0.219** | **0.216** | **0.227** | 0 | |
| $c\,d$ | 0.4079 | 0.132 | 0.132 | 0 | **0.002** | **0.119** | **0.099** | **0.051** | **0.081** | 0 | |
| $a\,b\,c$ | 0.7248 | 0.358 | −0.280 | 0 | 0.358 | 0.358 | **0.305** | 0.358 | 0.358 | 0.358 | |
| $a\,b\ \ d$ | 0.6224 | 0.579 | −0.310 | 0 | 0.579 | 0.579 | **0.507** | **0.556** | 0.579 | 0.579 | |
| $a\ \ c\,d$ | 0.6072 | 0.550 | −0.322 | 0 | 0.550 | 0.550 | **0.526** | **0.493** | 0.550 | 0.550 | |
| $b\,c\,d$ | 0.7502 | 0.391 | −0.295 | 0 | 0.391 | 0.391 | 0.391 | **0.341** | 0.391 | 0.391 | |
| $a\,b\,c\,d$ | 1 | 1 | 0.664 | 1 | 1 | 1 | 1 | 1 | 1 | 1 | |
| $\|\underline{R} - \underline{R}_*\|_1$ | | | | | 3.419 | 0.603 | 0.603 | 0.827 | 0.797 | 0.631 | 1.413 |

**Table 2** Two lower probabilities $\underline{P}$—the one on the left is permutation invariant [6, cf., e.g.,§2.2.6]—with their Möbius transforms $\mu$ and IMRM-approximations $\underline{P}_{\mathrm{IMRM}}$. On the left moreover two other completely monotone outer approximations, an optimal one $\underline{P}_{\mathrm{LP}}$ and an IRM-approximation $\underline{P}_{\mathrm{IRM}}$. On the right moreover the Möbius transform $\nu_6$ of the IMRM-approximation and two intermediate basic belief mass assignments ($\nu_3$ and $\nu_5$) used in its construction. (N.B.: for this lower probability, we have $\|\underline{P} - \underline{P}_{\mathrm{LP}}\|_1 = 1/2$, $\|\underline{P} - \underline{P}_{\mathrm{IRM}}\|_1 = 3/4$.) Other table elements and stylings have the same meaning as in Table 1.

| Event | $\underline{P}$ | $\mu$ | $\underline{P}_{\mathrm{LP}}$ | $\underline{P}_{\mathrm{IRM}}$ | $\underline{P}_{\mathrm{IMRM}}$ | $\underline{P}$ | $\mu$ | $\nu_3$ | $\nu_5$ | $\nu_6$ | $\underline{P}_{\mathrm{IMRM}}$ |
|---|---|---|---|---|---|---|---|---|---|---|---|
| a b | 1/3 | 1/3 | 0 | 1/10 | 1/6 | 1/2 | 1/2 | 1/2 | 1/4 | 1/4 | 1/4 |
| a c | 1/3 | 1/3 | 1/6 | 0.95/8 | 1/6 | 1/4 | 1/4 | 1/4 | 1/8 | 1/8 | 1/8 |
| a d | 1/3 | 1/3 | 1/3 | 1/7 | 1/6 | 1/4 | 1/4 | 1/4 | 1/8 | 1/8 | 1/8 |
| b c | 1/3 | 1/3 | 1/3 | 0.96/7 | 1/6 | 1/4 | 1/4 | 1/4 | 1/8 | 1/8 | 1/8 |
| b d | 1/3 | 1/3 | 1/6 | 0.99/6 | 1/6 | 0 | 0 | 0 | 0 | 0 | 0 |
| c d | 1/3 | 1/3 | 0 | 0.98/5 | 1/6 | 0 | 0 | 0 | 0 | 0 | 0 |
| a b c | 1/2 | -1/2 | 1/2 | 1.07/3 | 1/2 | 1/2 | -1/2 | -1/2 | 0 | 0 | 1/2 |
| a b d | 1/2 | -1/2 | 1/2 | 0.82/2 | 1/2 | 3/4 | 0 | 0 | 3/8 | 9/32 | 21/32 |
| a c d | 1/2 | -1/2 | 1/2 | 0.91/2 | 1/2 | 1/4 | -1/4 | -1/4 | 0 | 0 | 1/4 |
| b c d | 1/2 | -1/2 | 1/2 | 1/2 | 1/2 | 1/4 | 0 | 0 | 1/8 | 3/32 | 7/32 |
| a b c d | 1 | 0 | 1 | 1 | 1 | 1 | 1/2 | | -1/8 | 0 | 1 |
| $\|\underline{P} - \underline{P}_*\|_1$ | | | 1 | 99/70 | 1 | | | | | | 3/4 |

In the tables, we have also given the Möbius transform $\mu$ of each of these lower probabilities—or the transform $\rho$ of their imprecise part—by applying Equation (4). As is the case for all lower probabilities that are not completely monotone, some of the basic belief mass assignments so obtained are negative, but they still sum up to one [3, 9]. The algorithms we discuss all essentially construct a nonnegative basic belief mass assignment function which can be seen as resulting from shifting positive mass up in the poset of events ordered by inclusion (or shifting negative mass down) to compensate the negative mass assignments. This shifting is also the basic idea behind the last two algorithms.

**Linear-vacuous approximation.** The first algorithm is trivial: it consists in replacing a lower probability's imprecise part by the vacuous lower probability, which is identically zero except in $\Omega$, where it is 1. In terms of mass shifts, all mass of non-singletons is shifted up to the event poset's top $\Omega$.

Table 1 contains a—due to the triviality—not very interesting illustration.

**Approximation via optimization.** The second algorithm is based on the formulation of the problem as an optimization problem: we wish to find a nonnegative basic belief mass assignment function $\nu$ such that its Möbius inverse $\underline{P}_{\mathrm{LP}}$ minimizes some distance to the approximated lower probability $\underline{P}$.

We can force $\underline{P}_{\mathrm{LP}}$ to be an outer approximation by adding constraints that express its dominance by $\underline{P}$. By choosing the distance to be a linear function of $\nu$'s components, the optimization problem becomes an LP problem [1, §7]; we choose the sum-norm-distance:

$$\nu = \operatorname{argmin}\{\|\underline{P} - \underline{P}_{\mathrm{LP}}\|_1 : \underline{P}_{\mathrm{LP}} \leq \underline{P} \text{ and } \underline{P}_{\mathrm{LP}} \text{ is completely monotone}\} \tag{6}$$

$$= \operatorname{argmin}\{\textstyle\sum_{A \subseteq \Omega}|\underline{P}A - \underline{P}_{\mathrm{LP}}A| : \underline{P}_{\mathrm{LP}} \leq \underline{P} \text{ and } \underline{P}_{\mathrm{LP}} \text{ is completely monotone}\}$$

$$= \operatorname{argmax}\{\textstyle\sum_{A \subseteq \Omega}\underline{P}_{\mathrm{LP}}A : \underline{P}_{\mathrm{LP}} \leq \underline{P} \text{ and } \underline{P}_{\mathrm{LP}} \text{ is completely monotone}\}$$

$$= \operatorname{argmax}\{\textstyle\sum_{A \subseteq \Omega}\sum_{B \subseteq A}\nu B : \forall_{A \subseteq \Omega}(\textstyle\sum_{B \subseteq A}\nu B \leq \underline{P}A)$$

$$\text{and } \nu \geq 0, \sum_{B \subseteq \Omega}\nu B = 1\}$$

$$= \operatorname{argmax}\{\textstyle\sum_{B \subseteq \Omega} 2^{|\Omega \setminus B|}\nu B : \forall_{A \subseteq \Omega}(\textstyle\sum_{B \subseteq A}\nu B \leq \underline{P}A)$$

$$\text{and } \nu \geq 0, \sum_{B \subseteq \Omega}\nu B = 1\}, \tag{7}$$

where $2^{|\Omega \setminus B|}$ is the number of events $A$ that contain $B$. The third equality follows from taking into account the dominance constraints; the fourth from making the dependence on $\nu$ explicit using Equation (4). The linear-vacuous approximation shows that this linear program is feasible.

The results of this optimization approach are given for the lower probability in Table 1 and the one on the left in Table 2. In Table 1, two *differing* 'optimal' outer approximations are given, resulting from using different LP solvers. The optimal outer approximation given for the permutation invariant lower probability in Table 2 on the left is not permutation invariant itself. Both are due to the fact that in general there is no unique optimal solution and that solvers return the first one reached, which for the typical (non-interior-point) methods used lies on the border of the convex set of solutions.

The sum-norm distance can also be used as a quality criterion—one that obviously does not take symmetry aspects into account—for other approximation techniques. It has therefore also been calculated and included for the other approximations in Table 1 and 2; $\|\underline{P} - \underline{P}_{\mathrm{LP}}\|_1$ and $\|\underline{P} - \underline{P}_{\mathrm{LV}}\|_1$ provide lower and upper bounds.

**Iterative rescaling method.** The Iterative Rescaling Method or IRM [5] builds on the recursive Möbius transform formula in Equation (4), interrupting it to shift mass whenever negative mass assignments are encountered for some event.

**Algorithm 1.** IRM$(\Omega, \underline{P}) := \underline{P}_{\text{IRM}}$

1  *Form a sequence $A$ of length $2^{|\Omega|}$ by ordering all events in $2^{\Omega}$ by*
   *increasing cardinality and arbitrarily for events of equal cardinality and*
   *set $\nu\emptyset := \underline{P}\emptyset = 0$*
2  **for** $i := 1$ **to** $2^{|\Omega|} - 1$ **do** $\nu A_i := \underline{P}A_i - \sum_{B \subset A_i} \nu B$
3     **if** $\nu A_i < 0$ **then** $(\ell, \alpha) := \text{MassBasin}(A_i, \nu)$
5        **foreach** $B \subset A_i : |B| \geq \ell$ **do** $\nu B := \frac{\alpha + \nu A_i}{\alpha} \cdot \nu B$
7        $\nu A_i := 0$
8     **end**
9  **end**
10 **return** *the Möbius inverse $\underline{P}_{\text{IRM}}$ of $\nu$*

In the **if**-block, the negative mass $\nu A_i$ is distributed proportionally to its subevents of a for compensation lowest needed cardinality and up; i.e.,

$$\ell := \max\{k < |A_i| : \sum_{B \subset A:|B| \geq k} \nu B =: \alpha_k > \nu A_i\}, \qquad \alpha := \alpha_\ell.$$

For clarity, we have separated out the algorithm that calculates these parameters:

**Algorithm 2.** MassBasin$(A, \nu) := (\ell, \alpha)$

1  *Set $\ell := |A|$ and $\alpha := 0$*
2  **while** $\alpha < -\nu A$ **do** $\ell := \ell - 1$ *and* $\alpha := \alpha + \sum_{B \subset A:|B| = \ell} \nu D$
3  **return** *the (lowest needed) cardinality $\ell$ and the compensation mass $\alpha$*

The results of the IRM-algorithm are given for the lower probability in Table 1 and the one on the left in Table 2. In Table 1, two *differing* outer approximations are given, resulting from using different 'arbitrary' orderings of the events of equal cardinality. Using the sum-norm criterion, we see that the quality of the approximation depends on the order chosen. Also, the outer approximation given for the permutation invariant lower probability in Table 2 on the left is not permutation invariant itself, reflecting the impact of the arbitrary order.

Furthermore, it can be seen in the Tables that for events of a cardinality for which the optimization approximation is always exact, this is not so for the IRM-approximation; there only the last such event of the arbitrary order is exact. This is due to the fact that the IRM-algorithm does not backtrack to recalculate the mass assignments for an event after rescaling some of its subevents due to negative masses encountered for *subsequent* events.

**Iterative minimal rescaling method.** Inspired by the IRM, we have designed an approximation algorithm that avoids its defects mentioned above, at the cost of increased complexity. Furthermore, our algorithm is permutation-invariant, so it improves on the LP approach as well, in that regard.

The algorithm is still based on the recursive Möbius transform formula in Equation (4), but the rescaling approach is a bit more involved than with the IRM and gives rise to a higher number of these recursion calculations:

**Algorithm 3.** IMRM$(\Omega, \underline{P}) := \underline{P}_{\mathrm{IMRM}}$

1  Set $\nu\emptyset := \underline{P}\emptyset = 0$ and $k := 1$
2  **while** $k \leq |\Omega|$ **do** $\mathcal{A} := \emptyset$
3  $\quad$ **foreach** $A \subseteq \Omega : |A| = k$ **do** $\nu A := \underline{P}A - \sum_{B \subset A} \nu B$
5  $\quad$ $\mathcal{A} := \{ A \subseteq \Omega : |A| = k \wedge \nu A < 0 \}$
6  $\quad$ **if** $\mathcal{A} = \emptyset$ **then** $k := k + 1$
7  $\quad$ **else foreach** $A \in \mathcal{A}$ **do** $(\ell_A, \alpha_A) := \mathrm{MassBasin}(A, \nu)$
8  $\quad\quad$ $\ell := \min_{A \in \mathcal{A}} \ell_A$ and $\mathcal{B} := \{ A \in \mathcal{A} : \ell_A = \ell \}$
10 $\quad\quad$ **foreach** $A \in \mathcal{B}$ **do** $\beta_A := \sum_{B \subset A : |B| = \ell} \nu B$
11 $\quad\quad$ **foreach** $B \in \bigcup_{A \in \mathcal{B}} 2^A : |B| = \ell$ **do**
$\quad\quad\quad$ $\nu B := \max_{A \in \mathcal{B} : B \subset A} \frac{\alpha_A + \nu A}{\beta_A} \cdot \nu B$
12 $\quad\quad$ $k := \ell + 1$
13 $\quad$ **end**
14 **end**
15 **return** the Möbius inverse $\underline{P}_{\mathrm{IMRM}}$ of $\nu$

Per cardinality $k$, all basic belief mass assessments are calculated before doing any rescaling due to negative masses encountered for the events in $\mathcal{A}$. To limit the mass loss for events of lowest needed cardinality $\ell$—i.e., those most heavily penalized by the sum-norm criterion—, only their masses are rescaled during that iteration of the **while**-loop, which is the reason to restrict attention to $\mathcal{B}$. For a single event $A$ of $\mathcal{B}$, the mass loss for its cardinality-$\ell$ subevents is limited by only shifting that mass down which cannot be compensated higher up, which explains the scaling factor $\beta_A$ used. We avoid overcompensation of negative mass in one element of $\mathcal{B}$ due to a bigger deficit in another by using the largest scaling factor available, which leads to a minimal rescaling. This last point is what lead us to name the algorithm the Iterative Minimal Rescaling Method or IMRM. In general, the IMRM will not be as good as the optimization approach in terms of the sum-norm criterion: the mass is still shifted proportionally, which is not necessarily optimal.

The lower probability $\underline{P}_{\mathrm{IMRM}}$ obtained is indeed completely monotone, because for the basic belief mass assignment function we have $\nu\emptyset = 0$ from the start, $\nu A \geq 0$ for all events such that $|A| < k$ at the end of each iteration of the **while**-loop, and $\nu\Omega = \underline{P}\Omega - \sum_{B \subset \Omega} \nu B = 1 - \sum_{B \subset \Omega} \nu B$ at the end of the last iteration. It is an outer approximation because the recursion formula used tries to make $\underline{P}_{\mathrm{IMRM}}A$ equal to $\underline{P}A$ for all $A \subseteq \Omega$; subsequent rescalings can only lower this value.

The results of the IMRM-algorithm are given for the lower probability in Table 1 and both in Table 2. The positive impact of the algorithm's permutation invariance is especially clear for the left lower probability of Table 2. The algorithm itself is illustrated on Table 2's right side; there $\nu$ is given as it exists at the moment $\mathcal{A} = \emptyset$ is checked for the third, fifth, and sixth—final— iteration of the while loop; the impact of negative mass values on subsequent iterates is the prime point of interest here.

# 3 Conclusions

We have introduced and illustrated the linear-imprecise decomposition of lower probabilities, and the linear-vacuous and IMRM-algorithms for generating completely monotone outer approximations to lower probabilities. We have compared these to algorithms in the literature; permutation invariance is their main advantage.

One thing that still needs to be done is a complexity and parallelizability analysis to get a view of the relative computational burden of each of the algorithms discussed. Also interesting to investigate is the use of other objective functions in the optimization approach—e.g., using other norms, engendering nonlinear convex optimization problems—; this could lead to uniqueness of the solution and therewith permutation invariance. Both would allow for a more informed choice between the different possible completely monotone outer approximation algorithms.

**Acknowledgements.** Many thanks to Matthias Troffaes for making his software package improb publicly available, for assisting me in getting familiar with the code—so I could implement the algorithms described in this paper—, and for his very useful feedback. Thanks to Jim Hall and Jonathan Lawry for their helpful responses to my questions pertaining to their IRM-algorithm paper. Thanks to Enrique Miranda and a reviewer for their useful comments and suggestions.

# References

1. Baroni, P., Vicig, P.: An uncertainty interchange format with imprecise probabilities. Internat. J. Approx. Reason 40, 147–180 (2005), doi:10.1016/j.ijar.2005.03.001
2. de Campos, L.M., Huete, J.F., Moral, S.: Probability intervals: A tool for uncertain reasoning. Internat. J. Uncertain., Fuzziness Knowledge-Based Systems 2, 167–196 (1994), doi:10.1142/S0218488594000146
3. Chateauneuf, A., Jaffray, J.Y.: Some characterizations of lower probabilities and other monotone capacities through the use of Möbius inversion. Math. Social Sci. 17, 263–283 (1989), doi:10.1016/0165-4896(89)90056-5
4. Dempster, A.P.: Upper and lower probabilities induced by a multivalued mapping. Ann. Math. Stat. 38, 325–339 (1967), doi:10.1214/aoms/1177698950

5. Hall, J.W., Lawry, J.: Generation, combination and extension of random set approximations to coherent lower and upper probabilities. Reliab. Eng. Syst. Safety 85, 89–101 (2004), doi:10.1016/j.ress.2004.03.005
6. Quaeghebeur, E.: Learning from samples using coherent lower previsions. PhD thesis, Ghent University (2009), http://hdl.handle.net/1854/LU-495650
7. Quaeghebeur, E.: Characterizing the set of coherent lower previsions with a finite number of constraints or vertices. In: Spirtes, P., Grünwald, P. (eds.) Proceedings of UAI 2010, pp. 466–473. AUAI Press (2010), http://hdl.handle.net/1854/LU-984156
8. Quaeghebeur, E., De Cooman, G.: Extreme lower probabilities. Fuzzy Sets and Systems 159, 2163–2175 (2008), http://hdl.handle.net/1854/11713, doi:10.1016/j.fss.2007.11.020
9. Shafer, G.: A mathematical theory of evidence. Princeton University Press, Princeton (1976)
10. Troffaes, M., Quaeghebeur, E.: improb: A python module for working with imprecise probabilities (2011), https://github.com/equaeghe/improb, fork of https://github.com/mcmtroffaes/improb. Latest public release at http://packages.python.org/improb/
11. Walley, P.: Statistical Reasoning with Imprecise Probabilities. Chapman & Hall, Boca Raton (1991)

# Belief Networks and Local Computations

Radim Jiroušek

**Abstract.** This paper is one of many attempts to introduce graphical Markov models within Dempster-Shafer theory of evidence. Here we take full advantage of the notion of factorization, which in probability theory (almost) coincides with the notion of conditional independence. In Dempster-Shafer theory this notion can be quite easily introduced with the help of the operator of composition.

Nevertheless, the main goal of this paper goes even further. We show that if a belief network (a D-S counterpart of a Bayesian network) is to be used to support decision, one can apply all the ideas of Lauritzen and Spiegelhalter's local computations.

**Keywords:** Operator of composition, Factorization, Decomposable models, Conditioning.

## 1 Introduction

Graphical Markov models (GMM) [9], a technique which made computations with multidimensional probability distributions possible, opened doors for application of probabilistic methods to problem of practice. Here we have in mind especially application of the technique of local computations for which theoretical background was laid by Lauritzen and Spiegelhalter [10]. The basic idea can be expressed in a few words: a multidimensional distribution represented by a Bayesian network is first converted into a decomposable model which allows for efficient computation of conditional probabilities.

The goal of this paper is to show that the same ideas can be employed also within Dempster-Shafer theory of evidence [11].

Radim Jiroušek
Faculty of Management of University of Economics, and Institute of Information Theory and Automation, Academy of Sciences Czech Republic
e-mail: radim@utia.cas.cz

S. Li (Eds.): Nonlinear Maths for Uncertainty and its Appli., AISC 100, pp. 179–187.
springerlink.com                    © Springer-Verlag Berlin Heidelberg 2011

In this paper we consider a finite setting: space $\mathbf{X}_N = \mathbf{X}_1 \times \ldots \times \mathbf{X}_n$, and its subspaces (for $K \subseteq N$) $\mathbf{X}_K = \times_{i \in K} \mathbf{X}_i$. For a point $x = (x_1, \ldots, x_n) \in \mathbf{X}_N$ its projection into subspace $\mathbf{X}_K$ is denoted $x^{\downarrow K} = (x_{i, i \in K})$. Analogously, for $A \subseteq \mathbf{X}_N$, $A^{\downarrow K} = \{y \in \mathbf{X}_K : \exists x \in A, x^{\downarrow K} = y\}$. By a *join* of two sets $A \subseteq \mathbf{X}_K$ and $B \subseteq \mathbf{X}_L$ we understand a set

$$A \otimes B = \{x \in \mathbf{X}_{K \cup L} : x^{\downarrow K} \in A \ \& \ x^{\downarrow L} \in B\}.$$

Notice that if $K$ and $L$ are disjoint, then $A \otimes B = A \times B$, if $K = L$ then $A \otimes B = A \cap B$.

In view of this paper it is important to realize that for $C \subseteq \mathbf{X}_{K \cup L}$, $C \subseteq C^{\downarrow K} \otimes C^{\downarrow L}$, and that the equality $C = C^{\downarrow K} \otimes C^{\downarrow L}$ holds only for some of them.

## 2  Basic Assignments

A *basic assignment* (ba) $m$ on $\mathbf{X}_K$ ($K \subseteq N$) is a function $m : \mathcal{P}(\mathbf{X}_K) \to [0,1]$, for which

$$\sum_{\emptyset \neq A \subseteq \mathbf{X}_K} m(A) = 1.$$

If $m(A) > 0$, then $A$ is said to be a *focal element* of $m$. Recall that

$$Bel(A) = \sum_{\emptyset \neq B \subseteq A} m(B), \quad \text{and} \quad Pl(A) = \sum_{B \subseteq \mathbf{X}_K : B \cap A \neq \emptyset} m(B).$$

Having a ba $m$ on $\mathbf{X}_K$ one can consider its *marginal assignment* on $\mathbf{X}_L$ (for $L \subseteq K$), which is defined (for each $\emptyset \neq B \subseteq \mathbf{X}_L$):

$$m^{\downarrow L}(B) = \sum_{A \subseteq \mathbf{X}_K : A^{\downarrow L} = B} m(A).$$

**Definition 1 (Operator of composition).** For two arbitrary ba's $m_1$ on $\mathbf{X}_K$ and $m_2$ on $\mathbf{X}_L$ ($K \neq \emptyset \neq L$) a *composition* $m_1 \triangleright m_2$ is defined for each $C \subseteq \mathbf{X}_{K \cup L}$ by one of the following expressions:

[a]  if $m_2^{\downarrow K \cap L}(C^{\downarrow K \cap L}) > 0$ and $C = C^{\downarrow K} \otimes C^{\downarrow L}$ then

$$(m_1 \triangleright m_2)(C) = \frac{m_1(C^{\downarrow K}) \cdot m_2(C^{\downarrow L})}{m_2^{\downarrow K \cap L}(C^{\downarrow K \cap L})};$$

[b]  if $m_2^{\downarrow K \cap L}(C^{\downarrow K \cap L}) = 0$ and $C = C^{\downarrow K} \times \mathbf{X}_{L \setminus K}$ then

$$(m_1 \triangleright m_2)(C) = m_1(C^{\downarrow K});$$

[c]  in all other cases $(m_1 \triangleright m_2)(C) = 0$.

Let us stress that the operator of composition is something other than the famous Dempster's rule of combination [2]. While Dempster's rule was designed to combine different (independent) sources of information (it realizes fusion of sources), the operator of composition was designed to assemble (compose) factorizing basic assignments from their pieces. Notice that, e.g., for computation of $(m_1 \triangleright m_2)(C)$ it suffices to know only the values of $m_1$ and $m_2$ for the respective projections of set $C$, whereas computing Dempster's combination of $m_1$ and $m_2$ for set $C$ requires knowledge of, roughly speaking, the entire basic assignments $m_1$ and $m_2$. This is an indisputable (computational) advantage of the factorization considered in this paper. Unfortunately, the operator of composition is neither commutative nor associative. In [8, 7] we proved a number of properties concerning the operator of composition; the following ones are the most important for the purpose of this paper.

**Proposition 1.** *Let $m_1$ and $m_2$ be ba's defined on $\mathbf{X}_K$, $\mathbf{X}_L$, respectively. Then:*

  1. *$m_1 \triangleright m_2$ is a ba on $\mathbf{X}_{K \cup L}$;*
  2. *$(m_1 \triangleright m_2)^{\downarrow K} = m_1$;*
  3. *$m_1 \triangleright m_2 = m_2 \triangleright m_1 \iff m_1^{\downarrow K \cap L} = m_2^{\downarrow K \cap L}$.*

From Property 1 one immediately gets that for basic assignments $m_1, m_2$, $\ldots, m_r$ defined on $\mathbf{X}_{K_1}, \mathbf{X}_{K_2}, \ldots, \mathbf{X}_{K_r}$, respectively, the formula $m_1 \triangleright m_2 \triangleright \ldots \triangleright m_r$ defines a (possibly multidimensional) basic assignment defined on $\mathbf{X}_{K_1 \cup \ldots \cup K_r}$. However, to avoid ambiguity (recall that the operator is not associative) we have to say that, if not specified otherwise by parentheses, the operators will always be applied from left to right, i.e.,

$$m_1 \triangleright m_2 \triangleright \ldots \triangleright m_r = (\ldots (m_1 \triangleright m_2) \triangleright \ldots \triangleright m_{r-1}) \triangleright m_r.$$

Nevertheless, when designing the process of local computations for compositional models in D-S theory, which is intended to be an analogy to the process proposed by Lauritzen and Spiegelhalter in [10], one needs a type of associativity (see also [12]) expressed in the following assertion proved in [6].

**Proposition 2.** *Let $m_1, m_2$ and $m_3$ be ba's on $\mathbf{X}_{K_1}, \mathbf{X}_{K_2}$ and $\mathbf{X}_{K_3}$, respectively, such that $K_2 \supseteq K_1 \cap K_3$, and*

$$m_1^{\downarrow K_1 \cap K_2}(C^{\downarrow K_1 \cap K_2}) > 0 \implies m_2^{\downarrow K_1 \cap K_2}(C^{\downarrow K_1 \cap K_2}) > 0.$$
*Then $(m_1 \triangleright m_2) \triangleright m_3 = m_1 \triangleright (m_2 \triangleright m_3)$.*

### Belief Networks and Decomposable Models

In this subsection we introduce a Dempster-Shafer counterpart to GMM's. Studying properly probabilistic GMM's one can realize that it is the notion of *factorization* that makes it possible to represent multidimensional probability distributions efficiently. Focusing only on Bayesian networks one can see that they can be defined in probability theory in several different ways. Here we will proceed according to a rather theoretical approach which defines a Bayesain network as a probability distribution factorizing with respect to a given *acyclic directed graph* (DAG). The factorization guarantees that the independence structure of a probability distribution represented by a Bayesian network is in harmony with the well-known *d-separation criterion* [4, 9].

For Bayesian networks, this factorization principle can be formulated in the following way (here $pa(i)$ denotes the set of parents of a node $i$ of the considered DAG, and $fam(i) = pa(i) \cup \{i\}$): measure $\pi$ is a Bayesian network with a DAG $G = (N, E)$ if for each $i = 2, \ldots, |N|$ (assuming that this ordering of nodes is such that $k \in pa(j) \implies k < j$) marginal distribution $\pi^{\downarrow\{1,2,\ldots,i\}}$ factorizes with respect to couple $(\{1, 2, \ldots, i - 1\}, fam(i))$. And this is the definition which can be directly taken over into Dempster-Shafer theory.

**Definition 2 (Belief network).** We say that a ba $m$ is a *belief network* (BN) with a DAG $G = (N, E)$ if for each $i = 2, \ldots, |N|$ (assuming the enumeration meets the property that $k \in pa(j) \implies k < j$) marginal ba $m^{\downarrow\{1,2,\ldots,i\}}$ factorizes in the following sense: $m^{\downarrow\{1,2,\ldots,i\}} = m^{\downarrow\{1,2,\ldots,i-1\}} \triangleright m^{\downarrow fam(i)}$.

From this definition, which differs from those used in [3, 12], we immediately get the following description of a BN.

**Proposition 3 (Closed form for BN).** Let $G = (N, E)$ be a DAG, and $1, 2, \ldots, |N|$ be its nodes ordered in the way that parents are before their children. Ba $m$ is a BN with graph $G$ if and only if

$$m = m^{\downarrow fam(1)} \triangleright m^{\downarrow fam(2)} \triangleright \ldots \triangleright m^{\downarrow fam(|N|)}.$$

Taking advantage of the notion of factorization which is based on the operator of composition, we can also introduce decomposable ba's. In harmony with decomposable probability distribution, decomposable ba's are defined as those factorizing with respect to *decomposable graphs*, i.e. undirected graphs whose *cliques* (maximal sets of nodes inducing complete subgraphs) $C_1, C_2, \ldots, C_r$ can be ordered to meet the so-called *running intersection property* (RIP): for all $i = 2, \ldots, r$ there exists $j, 1 \leq j < i$, such that $K_i \cap (K_1 \cup \ldots \cup K_{i-1}) \subseteq K_j$.

**Definition 3 (Decomposable ba).** Consider a decomposable graph $G = (N, F)$ with cliques $C_1, C_2, \ldots, C_r$ and assume the cliques are ordered to meet RIP. We say that a ba $m$ is *decomposable* (Dba) with respect to $G = (N, F)$ if for each $i = 2, \ldots, r$ marginal ba $m^{\downarrow C_1 \cup \ldots \cup C_i}$ factorizes in the following sense:

$$m^{\downarrow C_1 \cup \ldots \cup C_i} = m^{\downarrow C_1 \cup \ldots \cup C_{i-1}} \triangleright m^{\downarrow C_i}.$$

Analogously to the closed form for a BN we get also closed form for Dba, which is again an immediate consequence of the definition.

**Proposition 4 (Closed form for Dba).** Let $G = (N, F)$ be decomposable with cliques $C_1, C_2, \ldots, C_r$ and assume the cliques are ordered to meet RIP. Ba $m$ is decomposable with respect to $G$ if and only if

$$m = m^{\downarrow C_1} \rhd m^{\downarrow C_2} \rhd \ldots \rhd m^{\downarrow C_r}.$$

## Conditioning

Unfortunately, there is no generally accepted way of conditioning in D-S theory. Though we do not have an ambition to fill in this gap, we need a tool which will enable us to answer questions like: *What is a belief for values of variable $X_j$ if we know that variable $X_i$ has a value $a$?* In probability theory the answer is given by conditional probability distribution $\pi(X_j | X_i = a)$. Let us study a possibility to obtain this conditional distribution with the help of the probabilistic operator of composition[1].

Define a *degenerated* one-dimensional probability distribution $\kappa_{|i;a}$ as a distribution of variable $X_i$ achieving probability 1 for value $X_i = a$, i.e.,

$$\kappa_{|i;a}(X_i = x) = \begin{cases} 1 & \text{if } x = a, \\ 0 & \text{otherwise.} \end{cases}$$

Now, compute $(\kappa_{|i;a} \rhd \pi)^{\downarrow\{j\}}$ for a probability distribution $\pi$ of variables $X_K$ with $i, j \in K$:

$$(\kappa_{|i;a} \rhd \pi)^{\downarrow\{j\}}(y) = ((\kappa_{|i;a} \rhd \pi)^{\downarrow\{j,i\}})^{\downarrow\{j\}}(y) = (\kappa_{|i;a} \rhd \pi^{\downarrow\{j,i\}})^{\downarrow\{j\}}(y)$$

$$- \sum_{x \in \mathbf{X}_i} \frac{\kappa_{|i;a}(x) \cdot \pi^{\downarrow\{j,i\}}(y, x)}{\pi^{\downarrow\{i\}}(x)} - \frac{\pi^{\downarrow\{j,i\}}(y, a)}{\pi^{\downarrow\{i\}}(a)} - \pi^{\downarrow\{j,i\}}(y|a)$$

Using an analogy, we consider in this paper that a proper answer to the above-raised question, in a situation when ba $m$ is taken into consideration, is given by $(m_{|i;a} \rhd m)^{\downarrow\{j\}}$ (or rather by the corresponding *Bel* function), where $m_{|i;a}$ is a ba on $\mathbf{X}_i$ with only one focal element $m(\{a\}) = 1$. This idea is moreover supported by the semantics of $m_{|i;a}$; this ba expresses the fact that we are *sure* that variable $X_i$ takes the value $a$. Therefore $m_{|i;a} \rhd m$ is a ba arising from $m$ by enforcing it to have a marginal for variable $X_i$ that is equal to $m_{|i;a}$ (see Property 2 of Proposition 1). In other words it describes the relationships among all variables from $X_N$ which is encoded in $m$, when we know that $X_i$ takes value $a$.

---

[1] In probability theory the operator of composition is defined for distributions $\pi(X_K)$ and $\kappa(X_L)$, for which $\pi^{\downarrow K \cap L}$ is absolutely continuous with respect to $\kappa^{\downarrow K \cap L}$, for each $x \in \mathbf{X}_{L \cup K}$ by the formula

$$(\pi \rhd \kappa)(x) = \frac{\pi(x^{\downarrow K})\kappa(x^{\downarrow L})}{\kappa^{\downarrow K \cap L}(x^{\downarrow K \cap L})}.$$

For the precise definition and its properties see [5].

# 3  Local Computations

As said in Introduction, by *local computations* we understand a realization of the ideas published by Lauritzen and Spiegelhalter [10]. They proposed to compute a conditional probability (as for example $\pi(X_d|X_i = a, X_j = b, X_k = c)$) for a distribution $\pi$ represented in a form of a Bayesian network in the following two steps.

1. Bayesian network is transformed into a decomposable model representing the same probability distribution $\pi$;
2. the required conditional distribution is computed by a process consisting of computations with the marginal distributions corresponding to the cliques of the respective decomposable graph.

This means that to get the desired conditional distribution one needs to know only the structure of the decomposable models (e.g. the respective decomposable graph) and the respective system of marginal distributions.

And it is the goal of this section to show that practically the same computational process can be realized also in D-S theory.

### Conversion of a BN into Dba

The process realizing this step can be directly taken over from probability theory [4]. If $G = (N, E)$ is a DAG of some belief network, then undirected graph $G = (N, \bar{E})$, where

$$\bar{E} = \left\{ \{i, j\} \in \binom{N}{2} : \exists k \in N \ \{i, j\} \subseteq fam(k) \right\},$$

is a so-called *moral graph* from which one can get the necessary decomposable graph $G = (V, F)$ (which will be uniquely specified by a system of its cliques $C_1, C_2, \ldots, C_r$) by any heuristic approach used for moral graph triangulation [1] (it is known that the process of looking for an optimal triangulated graph is a NP hard problem). Then it is an easy task to compute the necessary marginal ba's $m^{\downarrow C_1}, \ldots, m^{\downarrow C_r}$ when one realizes that there must exist an ordering (let it be the ordering $C_1, C_2, \ldots, C_r$) of the cliques meeting RIP and simultaneously

$$i \in pa(j) \implies f(i) \le f(j),$$

where $f(k) = \min(\ell : k \in C_\ell)$.

### Computation of Conditional ba

In comparison with the previous step, this computational process is much more complex. We have to show that having a decomposable ba $m = m^{\downarrow C_1} \rhd \ldots \rhd m^{\downarrow C_r}$ one can compute $(m_{|i;a} \rhd m)^{\downarrow\{j\}}$ locally.

For this, we take advantage of the famous fact that if $C_1, C_2, \ldots, C_r$ can be ordered to meet RIP, then for each $k \in \{1, 2, \ldots, r\}$ there exists an ordering meeting RIP for which $C_k$ is the first one. So consider any $C_k$ for which $i \in C_k$, and find the ordering meeting RIP which starts with $C_k$. Without loss of generality let it be $C_1, C_2, \ldots, C_r$ (so, $i \in C_1$).

Considering ba $m$ decomposable with respect to a graph with cliques $C_1, C_2, \ldots, C_r$, our goal is to compute

$$(m_{|i;a} \rhd m)^{\downarrow\{j\}} = \left(m_{|i;a} \rhd (m^{\downarrow C_1} \rhd m^{\downarrow C_2} \rhd \ldots \rhd m^{\downarrow C_r})\right)^{\downarrow\{j\}}.$$

However, at this moment we have to assume that $m^{\downarrow\{i\}}(\{a\})$ is positive. Under this assumption we can apply Proposition 2 $r - 1$ times getting

$$m_{|i;a} \rhd (m^{\downarrow C_1} \rhd m^{\downarrow C_2} \rhd \ldots \rhd m^{\downarrow C_r})$$
$$= m_{|i;a} \rhd (m^{\downarrow C_1} \rhd m^{\downarrow C_2} \rhd \ldots \rhd m^{\downarrow C_{r-1}}) \rhd m^{\downarrow C_r}$$
$$= \ldots = m_{|i;a} \rhd m^{\downarrow C_1} \rhd m^{\downarrow C_2} \rhd \ldots \rhd m^{\downarrow C_r},$$

from which computationally local process[2]

$$\bar{m}_1 = m_{|i;a} \rhd m^{\downarrow C_1},$$
$$\bar{m}_2 = \bar{m}_1^{\downarrow C_2 \cap C_1} \rhd m^{\downarrow C_2},$$
$$\bar{m}_3 = (\bar{m}_1 \rhd \bar{m}_2)^{\downarrow C_3 \cap (C_1 \cup C_2)} \rhd m^{\downarrow C_3},$$
$$\vdots$$
$$\bar{m}_r = (\bar{m}_1 \rhd \ldots \rhd \bar{m}_{r-1})^{\downarrow C_r \cap (C_1 \cup \ldots C_{r-1})} \rhd m^{\downarrow C_r},$$

yields a sequence $\bar{m}_1, \ldots, \bar{m}_r$, for which $m_{|i;a} \rhd m = \bar{m}_1 \rhd \ldots \rhd \bar{m}_r$, and each $\bar{m}_k = (m_{|i;a} \rhd m)^{\downarrow C_k}$. Therefore, to compute $(m_{|i;a} \rhd m)^{\downarrow\{j\}}$ it is enough to find any $k$ such that $j \in C_k$ because in this case $(m_{|i;a} \rhd m)^{\downarrow\{j\}} = \bar{m}_k^{\downarrow\{j\}}$.

This simple idea can be quite naturally generalized in the following sense. Considering a model with basic assignment $m$ and having a prior information about values of variables $X_{i_1} = a_1, \ldots, X_{i_t} = a_t$, the goal may be to compute

$$(m_{|i_1,\ldots,i_t;a_1,\ldots,a_t} \rhd m)^{\downarrow\{j\}} = (m_{|i_1;a_1} \rhd \ldots \rhd m_{|i_t;a_t} \rhd m)^{\downarrow\{j\}}.$$

It can be done easily just by repeating the described computational process as many times as the number of given values (in our case $t$). This is possible because ba $m_{|i_1;a_1} \rhd m = \bar{m}_1 \rhd \ldots \rhd \bar{m}_r$ is again decomposable and therefore ba's $\bar{m}_1, \ldots, \bar{m}_r$ can be again reordered so that the respective sequence of index sets meets RIP and index $i_2$ belongs to the first index set, and so on. However, and it is important to stress it, in this case we have to assume

---

[2] Notice that due to the assumption that $C_1, \ldots, C_r$ meets RIP, for each $k$ there exists $\ell$ such that $(\bar{m}_1 \rhd \ldots \rhd \bar{m}_{k-1})^{\downarrow C_k \cap (C_1 \cup \ldots C_{k-1})} = \bar{m}_\ell^{\downarrow C_k \cap (C_1 \cup \ldots C_{k-1})}$, which ensures locality of the described computations.

that the combination of given values, which specifies the condition, is a focal
element of ba $m$, i.e., regarding the condition specified above, we have to
assume that $m^{\downarrow\{i_1,\ldots,i_t\}}(\{a_1,\ldots,a_t\}) > 0$.

## 4 Conclusions

In the paper we have shown that with the help of the operator of composition it is possible to define BN's as a D-S counterpart of Bayesian networks.
Moreover, we have shown that under the assumption that a given condition
is a focal element of a ba represented by a BN, one can realize a process
yielding a basic assignment representing a *conditional belief*. This computational process can be performed locally, i.e., all the computations involves
only marginal distributions of the respective ba. The only weak point of the
presented approach is that it can be applied only under an additional assumption requiring that the prior information specifying the condition is a
focal element of the ba represented by the given BN.

**Acknowledgements.** This work was supported by GAČR under the grants
ICC/08/E010, and 201/09/1891, and by MŠMT ČR under grants 1M0572 and
2C06019.

## References

1. Cano, A., Moral, S.: Heuristic algorithms for the triangulation of graphs. In:
   Bouchon-Meunier, B., Yager, R.R., Zadeh, L.A. (eds.) IPMU 1994. LNCS,
   vol. 945, Springer, Heidelberg (1995)
2. Dempster, A.: Upper and lower probabilities induced by a multi-valued mapping. Ann. of Math. Stat. 38, 325–339 (1967)
3. Dempster, A.P., Kong, A.: Uncertain evidence and artificial analysis. J. Stat.
   Planning and Inference 20, 355–368 (1988); These are very powerful models for
   representing and manipulating belief functions in multidimensional spaces
4. Jensen, F.V.: Bayesian Networks and Decision Graphs. IEEE Computer Society
   Press, New York (2001)
5. Jiroušek, R.: Composition of probability measures on finite spaces. In: Geiger,
   D., Shenoy, P.P. (eds.) Proc. of the 13th Conf. Uncertainty in Artificial Intelligence UAI 1997. Morgan Kaufmann Publ., San Francisco (1997)
6. Jiroušek, R.: A note on local computations in Dempster-Shafer theory of evidence. Accepted to 7th Symposium on Imprecise Probabilities and Their Applications, Innsbruck (2011)
7. Jiroušek, R., Vejnarová, J.: Compositional models and conditional independence in evidence theory. Int. J. Approx. Reason. 52, 316–334 (2011)
8. Jiroušek, R., Vejnarová, J., Daniel, M.: Compositional models of belief functions. In: de Cooman, G., Vejnarová, J., Zaffalon, M. (eds.) Proc. of the 5th
   Symposium on Imprecise Probabilities and Their Applications, Praha (2007)

9. Lauritzen, S.L.: Graphical Models. Oxford University Press, Oxford (1996)
10. Lauritzen, S.L., Spiegelhalter, D.J.: Local computation with probabilities on graphical structures and their application to expert systems. J. of Royal Stat. Soc. series B 50, 157–224 (1988)
11. Shafer, G.: A Mathematical Theory of Evidence. Princeton University Press, Princeton (1976)
12. Shenoy, P.P., Shafer, G.: Axioms for Probability and Belief-Function Propagation. In: Shachter, R.D., Levitt, T., Lemmer, J.F., Kanal, L.N. (eds.) Uncertainty in Artificial Intelligence, vol. 4, North-Holland, Amsterdam (1990)

# Some New Entropies on the Interval-Valued Fuzzy Set

Wenyi Zeng, Hongxing Li, and Shuang Feng

**Abstract.** In this paper, we review some existing entropies of interval-valued fuzzy set, and propose some new formulas to calculate the entropy of interval-valued fuzzy set. Finally, we give one comparison with some existing entropies to illustrate our proposed entropies reasonable.

**Keywords:** Fuzzy set, Interval-valued fuzzy set, Normalized distance, Similarity measure, Entropy.

## 1 Introduction

Since the fuzzy set was introduced by Zadeh [13], fuzzy set theory has become an important approach to treat imprecision and uncertainty. Another well-known generalization of an ordinary fuzzy set is the interval-valued fuzzy set, which was first introduced by Zadeh [14, 15, 16]. Since then, many researchers have investigated this topic and have established some meaningful conclusions. For example, Wang et al. [11] investigated the combination and normalization of the interval-valued belief structures, Deschrijver [5] investigated the arithmetic operators of the interval-valued fuzzy set theory. Moreover, some researchers have pointed out that there is a strong connection

Wenyi Zeng
College of Information Science and Technology, Beijing Normal University, Beijing, 100875, P.R. China
e-mail: zengwy@bnu.edu.cn

Hongxing Li
School of Control Science and Engineering, Dalian University of Technology, Dalian, 116024, P.R. China
e-mail: lihx@dlut.edu.cn

Shuang Feng
School of Applied Mathematics, Beijing Normal University, Zhuhai, Guangdong, 519087, P.R. China

S. Li (Eds.): Nonlinear Maths for Uncertainty and its Appli., AISC 100, pp. 189–196.
springerlink.com      © Springer-Verlag Berlin Heidelberg 2011

between Atanassov's intuitionistic fuzzy sets and the interval-valued fuzzy sets. For more details, readers can refer to [4, 6, 9].

The entropy of fuzzy set is an important topic in the fuzzy set theory. The entropy of a fuzzy set describes the fuzziness degree of a fuzzy set and was first mentioned by Zadeh [13] in 1965. Several scholars have studied it from different points of view. For example, in 1972, De Luca and Termini [3] first introduced some axioms which captured people's intuitive comprehension to describe the fuzziness degree of a fuzzy set. Kaufmann [8] proposed a method for measuring the fuzziness degree of a fuzzy set by a metric distance between its membership function and the membership function of its nearest crisp set. Another method presented by Yager [12] was to view the fuzziness degree of a fuzzy set in terms of a lack of distinction between the fuzzy set and its complement. Based on these concepts and their axiomatic definitions, Zeng and Li [18] investigated the relationship among the inclusion measure, the similarity measure, and the fuzziness of fuzzy sets.

Aimed at the concept of entropy of fuzzy set, some researchers extended these concept to the interval-valued fuzzy set theory and investigated its related topic from different points of view. For example, Burillo and Bustince [2] introduced the concept of entropy of Atanassov's intuitionistic fuzzy set and the interval-valued fuzzy set in 1996. Zeng and Li [17] introduced the entropy of the interval-valued fuzzy set by using a different method and investigated the relationship between the similarity measure and the entropy of the interval-valued fuzzy sets. Wang and Li [10] studied the integral representation of the interval-valued fuzzy degree and the interval-valued similarity measure. Grzegorzewski [7] proposed a definition of the distance of interval-valued fuzzy sets based on the Hausdroff metric. In this paper, we propose some new formulas to calculate the entropy of interval-valued fuzzy set and compare with some existing entropies to illustrate our proposed entropies reasonable.

The organization of our work is as follows. In section 2, some basic notions of interval-valued fuzzy set are reviewed. In section 3, we propose some new entropies of interval-valued fuzzy set based on the distance and the similarity measure between interval-valued fuzzy sets, and do some comparisons between these entropies. The conclusion is given in the last section.

## 2   Some Notions

Throughout this paper, we use $X = \{x_1, x_2, \cdots, x_n\}$ to denote the discourse set, and IVFSs stand for the set of all interval-valued fuzzy subsets in $X$. $A$ expresses an interval-valued fuzzy set, and the operation "$c$" stands for the complement operation.

Let $L = [0, 1]$ and $[L]$ be the set of all closed subintervals of the interval $[0, 1]$. Especially for an arbitrary element $a \in [0, 1]$, we assume that $a$ is the same as $[a, a]$, namely, $a = [a, a]$. Then, according to Zadeh's extension principle [13], for any $\bar{a} = [a^-, a^+], \bar{b} = [b^-, b^+] \in [L]$, we can popularize some operators such as $\bigvee, \bigwedge$, and $c$ to $[L]$ and have $\bar{a} \bigvee \bar{b} = [a^- \bigvee b^-, a^+ \bigvee b^+], \bar{a} \bigwedge \bar{b} = [a^- \bigwedge b^-, a^+ \bigwedge b^+], \bar{a}^c = [1 - a^+, 1 - a^-], \bigvee_{t \in W} \overline{a_t} = [\bigvee_{t \in W} a_t^-, \bigvee_{t \in W} a_t^+]$ and $\bigwedge_{t \in W} \overline{a_t} = [\bigwedge_{t \in W} a_t^-, \bigwedge_{t \in W} a_t^+]$, where $W$ denotes an arbitrary index set. Furthermore, we have $\bar{a} = \bar{b} \iff a^- = b^-, a^+ = b^+, \bar{a} \leqslant \bar{b} \iff a^- \leqslant b^-, a^+ \leqslant b^+$, and $\bar{a} < \bar{b} \iff \bar{a} \leqslant \bar{b}$ and $\bar{a} \neq \bar{b}$; then there exists a minimal element $\bar{0} = [0, 0]$ and a maximal element $\bar{1} = [1, 1]$ in $[L]$.

We call a mapping, $A : X \longrightarrow [L]$ an interval-valued fuzzy set in $X$. For every $A \in$ IVFSs and $x \in X$, then $A(x) = [A^-(x), A^+(x)]$ is the degree of membership of an element $x$ to the interval-valued fuzzy set $A$. Thus, fuzzy sets $A^- : X \to [0, 1]$ and $A^+ : X \to [0, 1]$ are called low and upper fuzzy sets of the interval-valued fuzzy set $A$, respectively. For simplicity, we denote $A = [A^-, A^+]$, $\mathcal{F}(X)$ and $\mathcal{P}(X)$ stand for the set of all fuzzy sets and crisp sets in $X$, respectively.

If $A, B \in$ IVFSs, then the following operations can be found in Zeng and Li [17].

$A \subseteq B$ iff $\forall x \in X, A^-(x) \leq B^-(x)$ and $A^+(x) \leq B^+(x)$,
$A = B$ iff $\forall x \in X, A^-(x) = B^-(x)$ and $A^+(x) = B^+(x)$,
$(A)^c(x) = [(A^+(x))^c, (A^-(x))^c], \forall x \in X$,
$(A \cap B)(x) = [A^-(x) \wedge B^-(x), A^+(x) \wedge B^+(x)], \forall x \in X$,
$(A \cup B)(x) = [A^-(x) \vee B^-(x), A^+(x) \vee B^+(x)], \forall x \in X$.

**Definition 1.** For any positive real number $n$, $A \in$ IVFSs, we order $A^n \in$ IVFSs, and its membership function is defined as follows.

$$A^n(x) = [(A^-(x))^n, (A^+(x))^n], \text{ for every } x \in X$$

For some linguistic hedges such that "very", "more or less" and "slightly", we frequently use the mathematical models $A^n$ to represent the modifiers of linguistic variables. For example, we define the concentration and dilation of the interval-valued fuzzy set $A$ as follows.

$$\text{Concentration}: C(A) = A^2, \quad \text{Dilation}: D(A) = A^{\frac{1}{2}}$$

Therefore, we have some mathematical models in the following.

$$\text{Very } A = C(A) = A^2, \text{ more or less } A = D(A) = A^{\frac{1}{2}}, \text{ Very very } A = A^4$$

For example, let $X = \{6, 7, 8, 9, 10\}$, $A = \{(6, [0.1, 0.2]), (7, [0.3, 0.5]), (8, [0.6, 0.8]), (9, [0.9, 1]), (10, [1, 1])\}$, according to our mathematical models, then we have:

$$A^{\frac{1}{2}} = \{(6, [0.316, 0.448]), (7, [0.548, 0.707]), (8, [0.775, 0.894]), (9, [0.949, 1]), (10, [1, 1])\}$$

$$A^2 = \{(6, [0.01, 0.04]), (7, [0.09, 0.25]), (8, [0.36, 0.64]),$$
$$(9, [0.81, 1]), (10, [1, 1])\}$$

$$A^4 = \{(6, [0.0, 0.0016]), (7, [0.0081, 0.065]), (8, [0.1296, 0.4096]),$$
$$(9, [0.6561, 1]), (10, [1, 1])\}$$

## 3   Main Results

In 1996, Burillo and Bustince [2] introduced the concept of the entropy of the interval-valued fuzzy set. Later, Zeng and Li [17] used a different approach from Burillo and Bustince's [2] and extended De Luca and Termini [3] axioms for the fuzzy set to the interval-valued fuzzy set.

**Definition 2.** [17]   A real function $E : \text{IVFSs} \longrightarrow [0, 1]$ is called an entropy of an interval-valued fuzzy set, if $E$ satisfies the following properties:
   (E1) $E(A) = 0$ if $A$ is a crisp set;
   (E2) $E(A) = 1$   iff   $A^-(x_i) + A^+(x_i) = 1, \forall i = 1, 2, \cdots, n$;
   (E3) $E(A) \leq E(B)$ if $A$ is less fuzzy than $B$, i.e., $A^-(x_i) \leq B^-(x_i)$ and $A^+(x_i) \leq B^+(x_i)$ for $B^-(x_i) + B^+(x_i) \leq 1, \forall i = 1, 2, \cdots, n$ or $A^-(x_i) \geq B^-(x_i)$ and $A^+(x_i) \geq B^+(x_i)$ for $B^-(x_i) + B^+(x_i) \geq 1, \forall i = 1, 2, \cdots, n$;
   (E4) $E(A) = E(A^c)$.

In the following, we will give two formulas to calculate the entropy of the interval-valued fuzzy set $A$.

$$E_1(A) = 1 - \frac{1}{n} \sum_{i=1}^{n} |A^-(x_i) + A^+(x_i) - 1| \tag{1}$$

$$E_2(A) = 1 - \sqrt{\frac{1}{n} \sum_{i=1}^{n} \left( A^-(x_i) + A^+(x_i) - 1 \right)^2} \tag{2}$$

**Definition 3.** [17]   A real function $N : \text{IVFSs} \times \text{IVFSs} \longrightarrow [0, 1]$ is called the similarity measure of the interval-valued fuzzy sets, if $N$ satisfies the following properties:
   (N1) $N(A, A^c) = 0$ if $A$ is a crisp set;
   (N2) $N(A, B) = 1$   iff   $A = B$;
   (N3) $N(A, B) = N(B, A)$;
   (N4) For all $A, B, C \in \text{IVFSs}$, if $A \subseteq B \subseteq C$, then $N(A, C) \leq N(A, B)$, $N(A, C) \leq N(B, C)$.

$$N_1(A, B) = 1 - \frac{1}{2n} \sum_{i=1}^{n} \left( |A^-(x_i) - B^-(x_i)| + |A^+(x_i) - B^+(x_i)| \right) \tag{3}$$

In many theoretical and practical problems, people want to numerically express the differences of two objects (notions) by means of the distance of the corresponding fuzzy sets. For example, Grzegorzewski [7] proposed a distance for the interval-valued fuzzy sets based on the Hausdroff metric. For the interval-valued fuzzy sets $A$ and $B$, Atanassov [1] proposed the following formulas to calculate the normalized Hamming distance and the normalized Euclidean distance between the interval-valued fuzzy sets $A$ and $B$.

Normalized Hamming distance $d_1(A, B)$

$$d_1(A, B) = \frac{1}{2n} \sum_{i=1}^{n} \left( |A^-(x_i) - B^-(x_i)| + |A^+(x_i) - B^+(x_i)| \right) \qquad (4)$$

Normalized Euclidean distance $d_2(A, B)$

$$d_2(A, B) = \sqrt{\frac{1}{2n} \sum_{i=1}^{n} \left( (A^-(x_i) - B^-(x_i))^2 + (A^+(x_i) - B^+(x_i))^2 \right)} \qquad (5)$$

Obviously, they are straightforward generalizations of distances used in the classical set, which are obtained by replacing the characteristic functions of the classical sets with the membership functions of the interval-valued fuzzy sets.

**Theorem 1.** *Given a real function $f : [0, 1] \to [0, 1]$, if $f$ is a strictly monotone decreasing function, and $d$ is the normalized distance of the interval-valued fuzzy sets, for $A, B \in IVFSs$, then*

$$N(A, B) = \frac{f(d(A, B)) - f(1)}{f(0) - f(1)}$$

*is the similarity measure of the interval-valued fuzzy sets $A$ and $B$.*

*Proof.* (N1) If $A$ is a crisp set, then known by the definition of the normalized distance of the interval-valued fuzzy sets, we have $d(A, A^c) = 1$, therefore, $N(A, A^c) = 0$.

(N2) Known by the expression of $N(A, B)$,

$$\begin{aligned} N(A, B) = 1 \quad &\text{iff} \quad f(d(A, B)) - f(1) = f(0) - f(1) \\ &\text{iff} \quad d(A, B) = 0 \\ &\text{iff} \quad A = B \end{aligned}$$

(N3) Known by the definition of the normalized distance $d(A, B)$, we have $d(A, B) = d(B, A)$, therefore, $N(A, B) = N(B, A)$.

(N4) Since $A \subseteq B \subseteq C$, then we have $d(A, B) \le d(A, C)$, $d(B, C) \le d(A, C)$, and $f$ is a strictly monotone decreasing function. Thus, we have $f(d(A, B)) \ge f(d(A, C))$ and $f(d(B, C)) \ge f(d(A, C))$, and therefore, we obtain $N(A, C) \le N(A, B)$, $N(A, C) \le N(B, C)$.

Hence, we complete the proof of Theorem 1.                               $\square$

Now, our problem is how to select a useful and reasonable function $f$. In general, the simplest function $f$ is chosen as $f(x) = 1 - x$. Thus, the corresponding similarity measure of the interval-valued fuzzy sets $A$ and $B$ is $N(A, B) = 1 - d(A, B)$.

Considering that an exponential operation is highly useful in dealing with a similarity relation and cluster analysis in the fuzzy set theory, therefore, we choose $f(x) = e^{-x}$. Then, the corresponding similarity measure of the interval-valued fuzzy sets $A$ and $B$ is

$$N_e(A, B) = \frac{e^{-d(A,B)} - e^{-1}}{1 - e^{-1}}. \tag{6}$$

On the other hand, we may choose $f(x) = \dfrac{1}{1+x}$, then the corresponding similarity measure of the interval-valued fuzzy sets $A$ and $B$ is

$$N_c(A, B) = \frac{1 - d(A, B)}{1 + d(A, B)}. \tag{7}$$

And if we choose $f(x) = 1 - x^2$, then we can obtain the corresponding similarity measure of the interval-valued fuzzy sets $A$ and $B$,

$$N_d(A, B) = 1 - d^2(A, B). \tag{8}$$

**Theorem 2.** *Suppose that $d$ and $N$ are the normalized distance and the similarity measure of the interval-valued fuzzy sets, respectively, for $A \in IVFSs$, then $E(A) = N(A, A^c)$ is the entropy of the interval-valued fuzzy set $A$.*

We choose Eq.(4) being the distance of interval-valued fuzzy sets, and Eq. (3), (6), (7) and (8) being the similarity measure of interval-valued fuzzy sets, respectively, then we have the following formulas to calculate the entropy of interval-valued fuzzy set $A$.

$$E_1(A) = 1 - \frac{1}{n} \sum_{i=1}^{n} |A^-(x_i) + A^+(x_i) - 1| \tag{9}$$

$$E_e(A, B) = \frac{exp(-\frac{1}{n} \sum_{i=1}^{n} |A^-(x_i) + A^+(x_i) - 1|) - exp(-1)}{1 - exp(-1)} \tag{10}$$

$$E_c(A, B) = \frac{1 - \frac{1}{n} \sum_{i=1}^{n} |A^-(x_i) + A^+(x_i) - 1|}{1 + \frac{1}{n} \sum_{i=1}^{n} |A^-(x_i) + A^+(x_i) - 1|} \tag{11}$$

$$E_d(A, B) = 1 - (\frac{1}{n} \sum_{i=1}^{n} |A^-(x_i) + A^+(x_i) - 1|)^2 \tag{12}$$

Thus, we have the following comparison result.

**Example 1.** Let $X = \{6, 7, 8, 9, 10\}$, $A = \{(6, [0.1, 0.2]), (7, [0.3, 0.5]), (8, [0.6, 0.8]), (9, [0.9, 1]), (10, [1, 1])\}$, then we have:

$$A^{\frac{1}{2}} = \{(6, [0.316, 0.448]), (7, [0.548, 0.707]), (8, [0.775, 0.894]), (9, [0.949, 1]), (10, [1, 1])\}$$

$$A^2 = \{(6, [0.01, 0.04]), (7, [0.09, 0.25]), (8, [0.36, 0.64]), (9, [0.81, 1]), (10, [1, 1])\}$$

$$A^4 = \{(6, [0.0, 0.0016]), (7, [0.0081, 0.065]), (8, [0.1296, 0.4096]), (9, [0.6561, 1]), (10, [1, 1])\}$$

**Table 1** Results of the entropy of interval-valued fuzzy set with different entropy formulas

| IVFSs | d | $E_1$ | $E_e$ | $E_c$ | $E_d$ |
|---|---|---|---|---|---|
| $A^{\frac{1}{2}}$ | 0.622 | 0.378 | 0.267 | 0.233 | 0.613 |
| $A$ | 0.64 | 0.36 | 0.252 | 0.22 | 0.59 |
| $A^2$ | 0.684 | 0.316 | 0.216 | 0.188 | 0.532 |
| $A^4$ | 0.808 | 0.192 | 0.123 | 0.106 | 0.347 |

From the viewpoint of concentration and dilation operators, the entropies of these interval-valued fuzzy sets have the following requirement:

$$E(A^{\frac{1}{2}}) \geq E(A) \geq E(A^2) \geq E(A^4)$$

Obviously, known by the results of Table 1, we find that our calculating conclusion is very accordance with our knowledge.

Correspondingly, the readers can use the similar method and extend these conclusions as above to the continuous set $X = [a, b]$.

# 4  Conclusions

Considering the importance of the entropy of interval-valued fuzzy set, in this paper, we propose some new formulas to calculate the entropy of interval-valued fuzzy set. We believe that some different normalized distances and similarity measures of interval-valued fuzzy sets will induce more formulas to calculate the entropy of interval-valued fuzzy set. on the other hand, known by Zeng and Li [17], some different entropies of interval-valued fuzzy set will also induce more formulas to calculate the similarity measure of interval-valued fuzzy sets based on the transformation between the similarity measure and the entropy of interval-valued fuzzy sets. These conclusions will rich information measure of interval-valued fuzzy sets. Therefore, our conclusions can be extensively applied in many fields such as pattern recognition, image processing, approximate reasoning, fuzzy control, and so on.

**Acknowledgements.** This work is supported by grants from the National Natural Science Foundation of China (No. 10971243). And corresponding author is Wenyi Zeng(zengwy@bnu.edu.cn).

# References

1. Atanassov, K.: Intuitionistic Fuzzy Sets: Theory and Applications. Physica-Verlag, Heidelberg (1999)
2. Bustince, H., Burillo, P.: Entropy on intuitionistic fuzzy sets and on interval-valued fuzzy sets. Fuzzy Sets and Systems 78, 305–316 (1996)
3. De Luca, A., Termini, S.: A definition of non-probabilistic entropy in the setting of fuzzy sets theory. Inform. and Control 20, 301–312 (1972)
4. Deschrijver, G., Kerre, E.E.: On the relationship between some extensions of fuzzy set theory. Fuzzy Sets and Systems 133, 227–235 (2003)
5. Deschrijver, G.: Arithmetic operators in interval-valued fuzzy set theory. Information Sciences 177, 2906–2924 (2007)
6. Deschrijver, G., Kerre, E.E.: On the position of intuitionistic fuzzy set theory in the framework of theories modelling imprecision. Information Sciences 177, 1860–1866 (2007)
7. Grzegorzewski, P.: Distances between intuitionistic fuzzy sets and/or interval-valued fuzzy sets based on the Hausdorff metric. Fuzzy Sets and Systems 148, 319–328 (2004)
8. Kaufmann, A.: Introduction to the Theory of Fuzzy Subsets-Fundamental Theoretical Elements, vol. 1. Academic Press, New York (1975)
9. Wang, G.J., He, Y.Y.: Intuitionistic fuzzy sets and L-fuzzy sets. Fuzzy Sets and Systems 110, 271–274 (2000)
10. Wang, G.J., Li, X.P.: On the IV-fuzzy degree and the IV-similarity degree of IVFS and their integral representation. J. Engineering Mathematics 21, 195–201 (2004)
11. Wang, Y.M., Yang, J.B., Xu, D.L., Chin, K.S.: On the combination and normalization of interval-valued belief structures. Information Sciences 177, 1230–1247 (2007)
12. Yager, R.R.: On the measure of fuzziness and negation, Part I: Membership in the unit interval. Internat. J. General Systems 5, 189–200 (1979)
13. Zadeh, L.A.: Fuzzy sets. Inform. Control 8, 338–353 (1965)
14. Zadeh, L.A.: The concept of a linguistic variable and its application to approximate reasoning (I). Inform. Sci. 8, 199–249 (1975)
15. Zadeh, L.A.: The concept of a linguistic variable and its application to approximate reasoning (II). Inform. Sci. 8, 301–357 (1975)
16. Zadeh, L.A.: The concept of a linguistic variable and its application to approximate reasoning (III). Inform. Sci. 9, 43–80 (1975)
17. Zeng, W.Y., Li, H.X.: Relationship between similarity measure and entropy of interval-valued fuzzy sets. Fuzzy Sets and Systems 157, 1477–1484 (2006)
18. Zeng, W.Y., Li, H.X.: Inclusion measure, similarity measure and the fuzziness of fuzzy sets and their relations. International J. Intelli. Systems 21, 639–653 (2006)

# Bottleneck Combinatorial Optimization Problems with Fuzzy Scenarios

Adam Kasperski and Paweł Zieliński

**Abstract.** In this paper a class of bottleneck combinatorial optimization problems with unknown costs is discussed. A scenario set containing all the costs realizations is specified and a possibility distribution in this scenario set is provided. Several models of uncertainty with solution algorithms for each uncertainty representation are presented.

**Keywords:** Bottleneck optimization, Uncertainty, Possibility theory.

## 1 Introduction

A *combinatorial optimization problem* consists of a finite set of elements $E = \{e_1, \ldots, e_n\}$ and a set $\Phi \subseteq 2^E$ of subsets of $E$, called the set of *feasible solutions*. Each element $e \in E$ has a cost $c_e$ and we seek a solution $X \in \Phi$ whose *bottleneck cost* $f(X) = \max_{e \in X} c_e$ is minimal. Such formulation encompasses a large variety of classical combinatorial optimization problems, for instance: the bottleneck path, assignment, spanning tree, *etc.* (see, e.g. [1]).

In this paper we study the case in which the element costs are uncertain. In general, this uncertainty can be modeled by specifying a *scenario set* $\Gamma$, which contains all possible cost realizations, called *scenarios*. Several ways of defining the scenario set $\Gamma$ have been proposed in literature. Among the

Adam Kasperski
Institute of Industrial Engineering and Management, Wrocław University of Technology, Wybrzeże Wyspiańskiego 27, 50-370, Wrocław, Poland
e-mail: adam.kasperski@pwr.wroc.pl

Paweł Zieliński
Institute of Mathematics and Computer Science, Wrocław University of Technology, Wybrzeże Wyspiańskiego 27, 50-370, Wrocław, Poland
e-mail: pawel.zielinski@pwr.wroc.pl

S. Li (Eds.): Nonlinear Maths for Uncertainty and its Appli., AISC 100, pp. 197–204.
springerlink.com © Springer-Verlag Berlin Heidelberg 2011

most popular are the *discrete* and *interval* uncertainty representations (see,
e.g. [7]). In the former, $\Gamma$ contains a finite number of distinct cost scenarios.
In the latter one, a closed interval is specified for each element cost and
$\Gamma$ is the Cartesian product of all these intervals. The discrete and interval
scenario representations model different types of uncertainty. The discrete
representation allows us to express a *global uncertainty*, connected with some
events having a global influence on the element costs. On the other hand, the
interval scenario representation is appropriate for modeling a *local uncertainty*
resulting from an imprecise nature of a single cost. In this case the value of
a cost may vary independently on the values of the remaining costs.

The method of choosing a solution depends on additional information provided with the scenario set $\Gamma$. If a probability distribution in $\Gamma$ is given, then
*stochastic optimization* is appropriate. If a possibility distribution in $\Gamma$ is
specified, then *fuzzy optimization* techniques can be applied [5]. Finally, if no
distribution in the scenario set is given, then *robust optimization* approach
can be used [2,7]. In this paper we assume that a possibility distribution in the
scenario set $\Gamma$ is provided. So, in order to choose a solution, we use the fuzzy
possibilistic optimization techniques (see, e.g., [6]). Our approach also generalizes the robust *minmax* and *minmax regret* models (see, e.g., [2,3,7]). We
propose several descriptions of the scenario set in which both global and local
uncertainty are taken into account. An optimal solution to all the problems
considered can be computed in polynomial time if only their deterministic
counterparts are polynomially solvable.

## 2   Robust Bottleneck Problems

In this section we recall some known facts on the robust bottleneck combinatorial optimization problems. Let $\Gamma$ be a scenario set and let $c_e(S)$ be the
cost of element $e \in E$ under scenario $S \in \Gamma$. We use $f(X,S) = \max_{e \in X} c_e(S)$
to denote the bottleneck cost of solution $X$ under scenario $S$. We also denote
by $f^*(S)$ the bottleneck cost of an optimal solution under $S$. The quantity
$\delta(X,S) = f(X,S) - f^*(S)$ is called a *deviation* of solution $X$ under $S$. In the
robust approach two criteria of choosing a solution are widely applied. The
first, called a *minmax criterion*, leads to the following problem:

$$\text{ROB1:}\quad \min_{X \in \Phi} \max_{S \in \Gamma} f(X,S).$$

The second, called a *minmax regret criterion*, leads to the following problem:

$$\text{ROB2:}\quad \min_{X \in \Phi} \max_{S \in \Gamma} \delta(X,S).$$

A deeper discussion on both robust criteria can be found in [7].

There are two popular models of uncertainty representation. In the *discrete uncertainty representation* the set of scenarios $\Gamma = \{S_1, \ldots, S_K\}$ contains $K \geq 1$ distinct cost scenarios. Such a description of the scenario set is appropriate if we wish to model a global uncertainty, where each scenario $S_j \in \Gamma$ corresponds to some event having a global influence on the costs. Solving ROB1 (resp. ROB2) can be then reduced to computing an optimal solution to the deterministic problem with the element costs $c_e^* = \max_{1 \leq j \leq K} c_e(S_j)$ (resp. $c_e^* = \max_{1 \leq j \leq K}(c_e(S_j) - f^*(S_j)))$, $e \in E$, (see [2]).

In the *interval uncertainty representation*, for each element $e \in E$ a closed interval $[\underline{c}_e, \overline{c}_e]$ is specified, which contains all possible values of the cost of $e$. Then $\Gamma = \times_{e \in E}[\underline{c}_e, \overline{c}_e]$ is the Cartesian product of all the uncertainty intervals. This representation allows us to model a local uncertainty, which represents a varying nature of the element costs. We have to assume, however, that the element costs are unrelated, which means that the value of an element cost does not depend on the values of the remaining element costs. In this case, ROB1 is trivial, since $\max_{S \in \Gamma} f(X, S) = \max_{e \in X} \overline{c}_e$, and it is enough to solve the deterministic problem with the element costs $\overline{c}_e$, $e \in E$. The ROB2 problem is more complex. Let $S^e \in \Gamma$ be the scenario in which the cost of $e$ is $\overline{c}_e$ and the costs of all $f \neq e$ are $\underline{c}_f$. It was shown in [3], then solving ROB2 reduces to computing an optimal solution of its deterministic counterpart with the element costs $c_e^* = \max\{0, \overline{c}_e - f^*(S^e)\}$, $e \in E$.

## 3   Possibilistic Bottleneck Problems

We first recall some basic notions on *possibility theory* (a more detailed description of this theory can be found in [4]). Let $\tilde{Z}$ be an unknown real valued quantity. We specify for $\tilde{Z}$ a *possibility distribution* $\pi_{\tilde{Z}}$, where $\pi_{\tilde{Z}}(z) = \Pi(\tilde{Z} = z)$ is a possibility of the event that $\tilde{Z}$ will take the value of $z$. Let $\tilde{A}$ be a fuzzy set in $\mathbb{R}$ with membership function $\mu_{\tilde{A}}$. Then "$\tilde{Z} \in \tilde{A}$" is a *fuzzy event* and the *possibility* that it will occur can be computed as follows:

$$\Pi(\tilde{Z} \in \tilde{A}) = \sup_{z \in \mathbb{R}} \min\{\pi_{\tilde{Z}}(z), \mu_{\tilde{A}}(z)\}. \tag{1}$$

The *necessity* of the event "$\tilde{Z} \in \tilde{A}$" can be computed in the following way:

$$N(\tilde{Z} \in \tilde{A}) = 1 - \Pi(\tilde{Z} \in \tilde{A}'), \tag{2}$$

where $\tilde{A}'$ is the complement of $\tilde{A}$ with the membership function $1 - \mu_{\tilde{A}}(z)$. We obtain a particular case of the possibility distribution if we assume that $\tilde{Z}$ is a *fuzzy interval*. Then $\pi_{\tilde{Z}}$ is quasi concave, upper semicontinuous and has a bounded support. These assumptions imply that each $\lambda$-cut of $\tilde{Z}$, i.e. the set $\tilde{Z}^\lambda = \{z : \pi_{\tilde{Z}}(z) \geq \lambda\} = [\underline{z}(\lambda), \overline{z}(\lambda)]$, $\lambda \in (0, 1]$, is a closed interval (see, e.g., [4]). We will also assume that $\tilde{Z}^0$ is the smallest closed set containing the support of $\tilde{Z}$.

We now describe scenario set $\Gamma$ by a possibility distribution $\pi_\Gamma$ in $\mathbb{R}^n$. Namely, the value of $\pi_\Gamma(S) \in [0,1]$, $S \in \mathbb{R}^n$, is a possibility of the event that the cost scenario $S = (c_e(S))_{e \in E}$ will occur. The possibility distribution $\pi_\Gamma$ coincides with the membership function $\mu_\Gamma$ of the fuzzy subset $\Gamma$ of $\mathbb{R}^n$, $\pi_\Gamma = \mu_\Gamma$. We assume that $\mu_\Gamma(S) = 1$ for at least one $S \in \mathbb{R}^n$, $\Gamma$ is a bounded fuzzy set, i.e. $\forall \lambda > 0$, $\exists c \in \mathbb{R}$ $\{S \in \mathbb{R}^n : \mu_\Gamma(S) \geq \lambda\} \subset \{S \in \mathbb{R}^n : \|S\| \leq c\}$, and $\mu_\Gamma$ is an upper semicontinuous function. Now the possibility distributions $\pi_{\tilde{f}(X)}$ and $\pi_{\tilde{\delta}(X)}$ for the uncertain cost $\tilde{f}(X)$ and deviation $\tilde{\delta}(X)$ of a given solution $X$ are defined as follows:

$$
\begin{aligned}
\pi_{\tilde{f}(X)}(z) = \Pi(\tilde{f}(X) = z) = \sup_{\{S \in \Gamma : f(X,S) = z\}} \pi_\Gamma(S), \\
\pi_{\tilde{\delta}(X)}(z) = \Pi(\tilde{\delta}(X) = z) = \sup_{\{S \in \Gamma : \delta(X,S) = z\}} \pi_\Gamma(S).
\end{aligned}
\tag{3}
$$

In order to choose a solution we need an additional information from a decision maker. Assume that he/she expresses his/her preferences about the solution cost (deviation) by means of a *fuzzy goal* $\tilde{G}$, which is a fuzzy interval with a bounded support and a nonincreasing upper semicontinuous membership function $\mu_{\tilde{G}} : \mathbb{R} \to [0,1]$. The value of $\mu_{\tilde{G}}(g)$ is the degree of acceptance of the solution cost (deviation) equal to $g$. We will denote by $\tilde{G}'$ the complement of $\tilde{G}$ with the membership function $1 - \mu_{\tilde{G}}(z)$. Now $\tilde{f}(X) \in \tilde{G}$ and $\tilde{\delta}(X) \in \tilde{G}$ are fuzzy events and it is natural to compute a solution $X$ which maximizes the necessity of these events (see also [6]). We thus need to solve the optimization problems $\max_{X \in \Phi} N(\tilde{f}(X) \in \tilde{G})$ and $\max_{X \in \Phi} N(\tilde{\delta}(X) \in \tilde{G})$. By applying (2) we get the equivalent formulations of both problems:

FUZZY ROB1 : $\min_{X \in \Phi} \Pi(\tilde{f}(X) \in \tilde{G}')$ and FUZZY ROB2 : $\min_{X \in \Phi} \Pi(\tilde{\delta}(X) \in \tilde{G}')$.

The following theorem is crucial for solving both fuzzy problems:

**Theorem 1.** *An optimal solution of* FUZZY ROB1 *(resp.* FUZZY ROB2*) can be obtained by computing an optimal solution of its deterministic counterpart with the costs* $c_e^* = \sup_S \min\{\pi_\Gamma(S), \mu_{\tilde{G}'}(c_e(S))\}$, $e \in E$ *(resp.* $c_e^* = \sup_S \min\{\pi_\Gamma(S), \mu_{\tilde{G}'}(c_e(S) - f^*(S))\}$, $e \in E$*).*

*Proof.* The theorem follows from (1), (3) and the properties of $\mu_{\tilde{G}'}$.  □

Theorem 1 allows us to transform the fuzzy problems into the equivalent deterministic ones. However, computing the costs $c_e^*$ for $e \in E$ requires of solving an optimization problem, which may be a difficult task. We now show some additional results on both problems which can help to simplify the computations. Let $\Gamma^\lambda$, $\lambda \in (0,1]$ be the set of all the scenarios whose possibility of occurrence is not less than $\lambda$, i.e. $\Gamma^\lambda = \{S : \pi_\Gamma(S) \geq \lambda\}$. Note that $\Gamma^\lambda$ is a crisp subset of $\mathbb{R}^n$. Let us define the bounds $\overline{f}^\lambda(X) = \sup_{S \in \Gamma^\lambda} f(X,S)$ and $\overline{\delta}^\lambda(X) = \sup_{S \in \Gamma^\lambda} \delta(X,S)$ for a given $X \in \Phi$.

**Theorem 2.** *The following equalities hold:*

$$\Pi(\tilde{f}(X) \in \tilde{G}') = \inf\{\lambda : \lambda \geq \mu_{\tilde{G}'}(\overline{f}^{\lambda}(X))\},$$
$$\Pi(\tilde{\delta}(X) \in \tilde{G}') = \inf\{\lambda : \lambda \geq \mu_{\tilde{G}'}(\overline{\delta}^{\lambda}(X))\}.$$

*Proof.* The theorem follows from (1) and the properties of $\mu_{\tilde{G}'}$.  □

Theorem 3, the boundedness of $\Gamma$, the upper semicontinuity of $\mu_{\Gamma}$ and the continuity of $f(X, S)$ with respect to $S$ show that FUZZY ROB1 and FUZZY ROB2 can be represented as the following optimization problems:

$$
\begin{array}{ll}
\min \lambda & \min \lambda \\
\lambda \geq \mu_{\tilde{G}'}(\overline{f}^{\lambda}(X)), & \lambda \geq \mu_{\tilde{G}'}(\overline{\delta}^{\lambda}(X)), \\
X \in \Phi, & X \in \Phi, \\
\lambda \in [0, 1], & \lambda \in [0, 1],
\end{array}
\tag{4}
$$

Consider now the following problems for a fixed $\lambda \in (0, 1]$:

$$\min_{X \in \Phi} \overline{f}^{\lambda}(X) = \min_{X \in \Phi} \sup_{S \in \Gamma^{\lambda}} f(X, S), \tag{5}$$

$$\min_{X \in \Phi} \overline{\delta}^{\lambda}(X) = \min_{X \in \Phi} \sup_{S \in \Gamma^{\lambda}} \delta(X, S). \tag{6}$$

By combining (4) and (5) we get that FUZZY ROB1 can be solved by applying a binary search on $\lambda \in [0, 1]$ (see Algorithm 4). If problem (5) can be solved in $O(f(n))$ time, then FUZZY ROB1 can be solved in $O(f(n) \log \epsilon^{-1})$ time with a given precision $\epsilon > 0$. A similar algorithm can be used to solve FUZZY ROB2. It is enough to modify lines 3 and 4 in Algorithm 4.

**Algorithm 4.** The algorithm for solving FUZZY ROB1
1  $\lambda := 0.5$, $\lambda_1 := 0$, $k := 2$, $X := \emptyset$
2  **while** $|\lambda - \lambda_1| > \epsilon$ **do**
3  | Find an optimal solution $Y$ to problem (5)
4  | **if** $\lambda \geq \mu_{\tilde{G}'}(\overline{f}^{\lambda}(Y))$ **then** $\lambda := \lambda - \frac{1}{2^k}$, $X := Y$ **else** $\lambda := \lambda + \frac{1}{2^k}$
5  | $k := k + 1$, $\lambda_1 := \lambda$
6  **return** $X$

**Model I.** Suppose that one of the pairwise distinct scenarios $S_1, \ldots, S_K$ can occur and the possibility that it will be $S_j$ equals $\pi_j$. Hence the possibility distribution $\pi_{\Gamma}(S) = \pi_j$ if $S = S_j$ for some $j = 1, \ldots, K$ and $\pi_{\Gamma}(S) = 0$ otherwise. A sample problem of this type is shown in Fig. 1a. This is the shortest path problem, where $E = \{e_1, \ldots, e_5\}$ is the set of arcs of the sample graph $G$ and $\Phi = \{\{e_1, e_4\}, \{e_1, e_3, e_5\}, \{e_2, e_5\}\}$ contains all the paths between nodes $s$ and $t$ in $G$. There are three possible cost scenarios, denoted by $S_1$, $S_2$ and $S_3$, each of them can occur with a given possibility (see Fig. 1b). So, in this model only the global uncertainty is taken into account. Theorem 1 now implies:

**Theorem 3.** *An optimal solution to* FUZZY ROB1 *(resp.* FUZZY ROB2*) can be obtained by computing an optimal solution of its deterministic counterpart with the costs* $\hat{c}_e = \max_{1 \le j \le K} \min\{\pi_j, \mu_{\tilde{G}'}(c_e(S_j))\}$, $e \in E$ *(resp.* $\hat{c}_e = \max_{1 \le j \le K} \min\{\pi_j, \mu_{\tilde{G}'}(c_e(S_j) - f^*(S_j))\}$, $e \in E$).

Hence both fuzzy problems are polynomially solvable if only their deterministic counterparts are polynomially solvable.

**Model II.** Consider again the sample problem shown in Fig. 1a. Three cost scenarios shown in this example correspond to three events which globally influence the arc costs. However, it is quite possible that the costs under each event are still unknown. For example, the costs may represent the arc traveling times and it is reasonable to assume that we only know their lower and upper bounds. This situation is shown in Fig. 1c. The three columns represent three possible events each of which can occur with a given possibility $\pi_j$. Furthermore, the $j$-th event results in an interval scenario set $\Gamma_j$, which is the Cartesian product of interval costs. Consequently, each scenario $S \in \Gamma_j$ has the possibility of occurrence equal to $\pi_j$ and one can compute $\pi_\Gamma(S)$ by taking the maximum $\pi_j$ over all $j$ such that $S \in \Gamma_j$. For instance, the scenario $S = (3, 3, 4, 1, 4)$ can occur if the first or the second event will happen. Thus $\pi_\Gamma(S) = \max\{\pi_1, \pi_2\} = 1$.

Let us now formalize the model. Let $\Gamma = \Gamma_1 \cup \Gamma_2 \cup \cdots \cup \Gamma_K$, where $\Gamma_j$ is the Cartesian product of the intervals $[\underline{c}_e^j, \overline{c}_e^j]$ for $e \in E$. Each scenario $S \in \Gamma_j$ has a possibility of occurrence equal to $\pi_j$. Then

$$\pi_\Gamma(S) = \max_{\{1 \le j \le K : S \in \Gamma_j\}} \pi_j \qquad (7)$$

and $\pi_\Gamma(S) = 0$ if $S \notin \Gamma_j$ for all $j = 1, \ldots, K$.

**Theorem 4.** *An optimal solution of* FUZZY ROB1 *(resp.* FUZZY ROB2*) can be obtained by computing an optimal solution of its deterministic counterpart with the costs* $\hat{c}_e = \max_{1 \le j \le K} \min\{\pi_j, \mu_{\tilde{G}'}(\overline{c}_e^j)\}$, $e \in E$ *(resp.* $\hat{c}_e = \max_{1 \le j \le K} \min\{\pi_j, \mu_{\tilde{G}'}(\overline{c}_e^j - f^*(S_j^e))\}$, $e \in E$, *where* $S_j^e$ *is the scenario such that* $c_e(S_j^e) = \overline{c}_e^j$ *and* $c_f(S_j^e) = \underline{c}_f^j$ *for all* $f \ne e$).

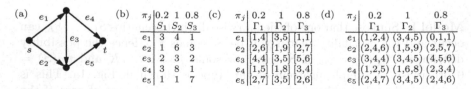

**Fig. 1** A sample shortest path problem with three models of uncertainty.

*Proof.* We present the proof for FUZZY ROB2 (the proof for FUZZY ROB1 is similar). We show that $\hat{c}_e = c_e^*$ for all $e \in E$, where $c_e^*$ are the element costs from Theorem 1. Let $j$ maximize the right hand side of the formula for $\hat{c}_e$. Then $S_j^e \in \Gamma_j$, $\pi_\Gamma(S_j^e) \geq \pi_j$ (see (7)) and $\mu_{\tilde{G}'}(c_e(S_j^e) - f^*(S_j^e)) = \mu_{\tilde{G}'}(\bar{c}_e^j - f^*(S_j^e))$. Therefore, $\hat{c}_e = \min\{\pi_j, \mu_{\tilde{G}'}(\bar{c}_e^j - f^*(S_j^e))\} \leq \min\{\pi_\Gamma(S_j^e), \mu_{\tilde{G}'}(c_e(S_j^e) - f^*(S_j^e))\} \leq c_e^*$. On the other hand, let $S$ be a scenario that maximizes the right hand side of the formula for $c_e^*$ (see Theorem 1). Suppose that $\pi_\Gamma(S) = \pi_j$ and $S \in \Gamma_j$ (see (7)). Since $\mu_{\tilde{G}'}$ is nondecreasing, $\mu_{\tilde{G}'}(c_e(S_j^e) - f^*(S_j^e)) \geq \mu_{\tilde{G}'}(c_e(S) - f^*(S))$. Therefore, $c_e^* = \min\{\pi_\Gamma(S), \mu_{\tilde{G}'}(c_e(S) - f^*(S))\} \leq \min\{\pi_j, \mu_{\tilde{G}'}(c_e(S_j^e) - f^*(S_j^e))\} \leq \hat{c}_e$. $\square$

Notice that the presented model is a generalization of Model I. Moreover, if $K = 1$, then FUZZY ROB1 (resp. FUZZY ROB2) is equivalent to ROB1 (resp. ROB2) with the interval uncertainty representation.

**Model III.** Now we generalize Model II by allowing the costs to be modeled as fuzzy intervals (recall that a closed interval is a special case of a fuzzy one). Let us consider again the shortest path problem shown in Fig. 1a. There are three events that can globally influence the arc costs and each event has some possibility of occurrence. Additionally, the arc costs under each event are unknown and they are modeled by fuzzy intervals (see Fig. 1d). For example, if the first event will happen, then the cost of arc $e_1$ has the possibility distribution in the form of the triangular fuzzy interval $(1, 2, 4)$.

This model can be formalized as follows. Assume that under the $j$-th event, $1 \leq j \leq K$, the cost of element $e \in E$ has a possibility distribution $\mu_e^j$. Consider a cost scenario $S = (s_e)_{e \in E}$. If the $j$-th even happens, then the possibility that $S$ will occur is $\pi_\Gamma(S|j) = \Pi(\bigwedge_{e \in E}(c_e = s_e)|j) = \min_{e \in E} \Pi(c_e = s_e|j) = \min_{e \in E} \mu_e^j(s_e)$. Since there are $K$ possible events and each event $j$ has a possibility of occurrence $\pi_j$, the possibility that $S$ will occur can be computed as follows:

$$\pi_\Gamma(S) = \max_{1 \leq j \leq K} \min\{\pi_j, \pi_\Gamma(S|j)\}. \tag{8}$$

In order to solve the fuzzy problems we apply Algorithm 4. So we must only provide the methods for solving problems (5) and (6). Let $S = (s_e)_{e \in E}$. By (8), $\pi_\Gamma(S) \geq \lambda$ if and only if there exists $j \in \{1, \dots, K\}$ such that $\pi_j \geq \lambda$ and $\pi_\Gamma(S|j) \geq \lambda$. Consequently, $\pi_j \geq \lambda$ and $\mu_e^j(s_e) \geq \lambda$ for all $e \in E$ or, equivalently, $\pi_j \geq \lambda$ and $s_e \in [\underline{c}_e^j(\lambda), \overline{c}_e^j(\lambda)]$ for all $e \in E$. Hence $\Gamma^\lambda$ can be computed in the following way. First, determine the set $U^\lambda = \{j : 1 \leq j \leq K, \pi_j \geq \lambda\}$. Then, for each $j \in U^\lambda$ define $\Gamma_j$ as the Cartesian product $\times_{e \in E} [\underline{c}_e^j(\lambda), \overline{c}_e^j(\lambda)]$ and, finally, $\Gamma^\lambda = \bigcup_{j \in U^\lambda} \Gamma_j$. Now

$$\overline{f}^\lambda(X) = \max_{S \in \Gamma^\lambda} f(X, S) = \max_{j \in U^\lambda} \max_{S \in \Gamma_j} \max_{e \in X} c_e(S) = \max_{e \in X} \max_{j \in U^\lambda} \overline{c}_e^j(\lambda).$$

Consequently, problem (5) can be solved by computing an optimal solution for the deterministic problem with costs $c_e^* = \max_{j \in U^\lambda} \overline{c}_e^j(\lambda)$, $e \in E$. A similar reasoning shows that solving problem (6) can be reduced to computing an optimal solution for its deterministic counterpart with the costs $c_e^* = \max\{0, \overline{c}_e^j(\lambda) - f^*(S_e^j(\lambda))\}$ for $e \in E$, where $S_e^j(\lambda)$ is the scenario in which $c_e(S_e^j(\lambda)) = \overline{c}_e^j(\lambda)$ and $c_f(S_e^j(\lambda)) = \underline{c}_f^j(\lambda)$ for all the elements $f \neq e$.

## 4 Conclusions

In this paper we have investigated the bottleneck combinatorial optimization problems with uncertain costs. This uncertainty has been modeled by a possibility distribution over a scenario set. We have proposed several models, with different descriptions of the scenario set and different criteria of choosing a solution, in the possibilistic setting. We have generalized the known robust models. An optimal solution to all the problems considered can be computed in polynomial time if only the deterministic version of the problem is polynomially solvable.

**Acknowledgements.** The second author was partially supported by Polish Committee for Scientific Research, grant N N206 492938.

## References

1. Ahuja, R.K., Magnanti, T.L., Orlin, J.B.: Network Flows: theory, algorithms, and applications. Prentice Hall, Englewood Cliffs (1993)
2. Aissi, H., Bazgan, C., Vanderpooten, D.: Min–max and min–max regret versions of combinatorial optimization problems: A survey. European Journal of Operational Research 197, 427–438 (2009)
3. Averbakh, I.: Minmax regret solutions for minimax optimization problems with uncertainty. Operations Research Letters 27, 57–65 (2000)
4. Dubois, D., Prade, H.: Possibility Theory: an approach to computerized processing of uncertainty. Plenum Press, New York (1988)
5. Inuiguchi, M., Ramik, J.: Possibilistic linear programming: a brief review of fuzzy mathematical programming and a comparison with stochastic programming in portfolio selection problem. Fuzzy Sets and Systems 111, 3–28 (2000)
6. Inuiguchi, M., Sakawa, M.: Robust optimization under softness in a fuzzy linear programming problem. International Journal of Approximate Reasoning 18, 21–34 (1998)
7. Kouvelis, P., Yu, G.: Robust Discrete Optimization and its applications. Kluwer Academic Publishers, Dordrecht (1997)

# About the Probability-Field-Intersections of Weichselberger and a Simple Conclusion from Least Favorable Pairs

Martin Gümbel

**Abstract.** In the frame of probability theory of Weichselberger there are probability fields and operations on probability fields. We look at the probability-field-intersection and present a simple conclusion for this operation, if there exists a least favorable pair of probabilities.

**Keywords:** Interval probability, Probability-field-intersection, Least favorable pairs.

## 1 Some Brief Extracts from the Theory of Interval Probability of Weichselberger

In the first section we will present some definitions following Weichselberger in his theory of interval valued probabilities. Basic for his concept of probability are probability fields. These fields have restricting inequalities (the intervals) and probabilities in the classical sense which are contained in them. On the probability-fields operations can be defined. We will introduce the `probability-field-intersection` operation. In view of the `probability-field-intersection` are there sufficient conditions for the cut of the intervals and the cut of the set of probabilities to be empty or nonempty?

The derivation by the axioms is lent from [8]. The interval-field-intersection example is from [5].

Martin Gümbel
Hagelmühlweg 15, 86316 Friedberg Germany
e-mail: martin.guembel@t-online.de

Let $\Omega = \{\omega_i \,|\, i \in I\}$ be a (discrete countable) measure space[1] with a countable indexset $I$. The subsets $\Omega$ containing just one element are noted $E_i$. With $\mathcal{P}(\Omega)$ we have a $\sigma$-algebra on $(\Omega, \mathcal{P}(\Omega))$.

**Definition 1.1** *A function* $p : \mathcal{P}(\Omega) \to \mathbb{R}$ *is called* **K-function**, *if*

  (K.I)   $\forall_{A \in \mathcal{P}(\Omega)} \, p(A) \geq 0$,
  (K.II)  $p(\Omega) = 1$,
  (K.III) *for all sequences* $(A_i)_{i \in \mathbb{N}}$ *of pairwise disjoint events (of* $\mathcal{P}(\Omega)$*):*
            $p\left(\bigcup_{i=1}^{\infty} A_i\right) = \sum_{i=1}^{\infty} p(A_i)$

*hold.*[2]

A K-function is a (classical) simple valued probability in form of a real number greater equal 0 and less equal 1.

We introduce now interval valued probabilities $P$: It seems `reasonable` to work with intervals such that a K-function is contained in the defining intervals:

**Definition 1.2** *Choosing* $(\Omega, \mathcal{P}(\Omega))$ *as before we define* $P$: $\mathcal{P}(\Omega) \to \mathbb{R}^2$. *Such a* $P(A) = [L(A), U(A)]$, *for all* $A \in \mathcal{P}(\Omega)$, *is called* **R-field**, *if the following axioms hold:*

  (T.IV)   $0 \leq L(A) \leq U(A) \leq 1$.
  (T.V)    *The set* $\mathcal{M}$ *of K-functions* $p(\cdot)$ *over* $\mathcal{P}(\Omega)$ *with*
            $\forall_{A \in \mathcal{P}(\Omega)} L(A) \leq p(A) \leq U(A)$ *is not empty.*

*The set* $\mathcal{M}$ *is called* **structure** *of the R-field.*

**Definition 1.3** *An R-field with structure* $\mathcal{M}$ *is called* **F-field** *('F' for feasible), if additionally the following axiom holds:*

  (T.VI)   *For all* $A \in \mathcal{P}(\Omega)$:
            $L(A) = \inf_{p \in \mathcal{M}} p(A)$   *and*   $U(A) = \sup_{p \in \mathcal{M}} p(A)$.

The F-fields have some well known specializations:

**Definition 1.4** *A F-field is called*
**C-field**, *if the two-monotone inequality*

$$\forall_{A,B \in \mathcal{P}(\Omega)} L(A) + L(B) \leq L(A \cap B) + L(A \cup B) \tag{1}$$

*holds.*

The C is an abbreviation for `Choquet`, see [3].

The totally monotone fields derived by evidences (with the belief-functions taken as $L(\cdot)$ and the plausibility-functions as $U(\cdot)$ introduced by Dempster-Shafer, see [4]) are two-monotone C-fields and therefore F-fields. (The

---

[1] Many of the results in the theory of interval valued theory of Weichselberger hold on more general spaces, see [1].

[2] These are the Kolmogorov axioms.

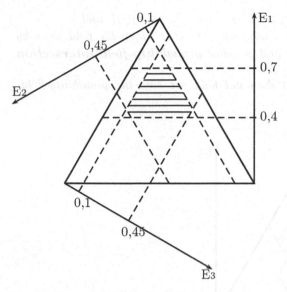

**Fig. 1 Graphic example F-field**
$I = \{1, 2, 3\}$, the structure (represented by the striped region in the diagram) is a convex set between the extrempoints. One probability in the F-field is $\mathbf{p} = (0,4; 0,15; 0,45)$. For example $P(E_1) = [0,4; 0,7]$.

Dempster-Shafer-theory has an eminent meaning because of the combination rule. To cite one application of the combination rule we refer f.e. to [9]). The feasible probability intervals (probability fields, where all the limiting intervals can be deduced from intervals imposed only on the sets $E_i \subseteq \{\Omega\}$, $i \in I$, abbreviated F-PRI, see [2]) are C-Fields and therefore F-fields. The restrictions for R-fields and F-fields by the intervals are less general than the restrictions by linear previsions proposed by Walley [7].

Let $\mathcal{R}_1$ be an R-field with the intervals $[L_1(\cdot); U_1(\cdot)]$ and $\mathcal{R}_2$ an R-Field with the intervals $[L_2(\cdot); U_2(\cdot)]$. The following condition is of interest:

**Definition 1.5** *The **Interval-cuts-condition** contains the following inequalities:*

$$\forall_{A \in \mathcal{P}(\Omega)} \, [L_1(A), U_1(A)] \cap [L_2(A), U_2(A)] \neq \emptyset. \qquad (2)$$

Now we have the

**Definition 1.6 *Probability-field-intersection***
*Let $\mathcal{R}_1$ be an R-field with the intervals $[L_1(\cdot); U_1(\cdot)]$ and $\mathcal{R}_2$ an R-Field with the intervals $[L_2(\cdot); U_2(\cdot)]$. If (2) (the interval-cuts-condition) holds and if the field*

$\widetilde{\mathcal{R}} = \mathcal{R}_1 \cap \mathcal{R}_2$ *with the intervals* $\widetilde{L}(\cdot) = \max\{L_1(\cdot), L_2(\cdot)\}$ *and*
$\widetilde{U}(\cdot) = \min\{U_1(\cdot), U_2(\cdot)\}$ *has a nonempty structure, then the field given by*
*the intervals* $\widetilde{\mathcal{R}}$ *is an R-field and is called* **probability-field-intersection**
*of the fields* $\mathcal{R}_1$ *and* $\mathcal{R}_2$.
*If* $\widetilde{\mathcal{R}}$ *is not an R-field or (2) does not hold, we say: the probability-field-*
*intersection is empty.*

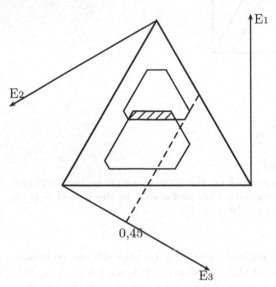

**Fig. 2 visualization-example of a probability-field-intersection**
In the graphical example the probability-field-intersection of two F-fields is an R-
field. $\widetilde{U(E_3)}$ is not reached by a probability and is the dashed line. The structure
of the probability-field-intersection is the striped area.

The interval-cuts-condition is a necessary condition to have an nonempty
probability-field-intersection. But it is not sufficient as the following example
demonstrates:
$I = \{1, 2, 3, 4\}$, the field $\mathcal{F}_0$ with the intervals $[L_0\,;\,U_0]$ and $\mathcal{F}_1$ with the
intervals $[L_1\,;\,U_1]$:

| intervals: | $[L_0\,;\,U_0]$ | $[L_1\,;\,U_1]$ | $\left[\widetilde{L}\,;\,\widetilde{U}\right]$ |
|---|---|---|---|
| $P(\emptyset)$ = | $[0\,;\,0]$ | $[0\,;\,0]$ | $[0\,;\,0]$ |
| $P(E_1)$ = | $[0,1\,;\,0,35]$ | $[0,35\,;\,0,5]$ | $[0,35\,;\,0,35]$ |
| $P(E_2)$ = | $[0,15\,;\,0,4]$ | $[0,15\,;\,0,3]$ | $[0,15\,;\,0,3]$ |
| $P(E_3)$ = | $[0,1\,;\,0,3]$ | $[0,15\,;\,0,3]$ | $[0,15\,;\,0,3]$ |
| $P(E_4)$ = | $[0,2\,;\,0,4]$ | $[0\,;\,0,2]$ | $[0,2\,;\,0,2]$ |
| $P(E_1 \cup E_2)$ = | $[0,5\,;\,0,5]$ | $[0,5\,;\,0,7]$ | $[0,5\,;\,0,5]$ |
| $P(E_1 \cup E_3)$ = | $[0,4\,;\,0,55]$ | $[0,5\,;\,0,8]$ | $[0,5\,;\,0,55]$ |

$$
\begin{array}{llll}
 & & [L_0\,;U_0] & [L_1\,;U_1] & \left[\tilde{L}\,;\tilde{U}\right] \\
P(E_1 \cup E_4) & = & [0,3\,;0,7] & [0,5\,;0,55] & [0,5\,;0,55] \\
P(E_2 \cup E_3) & = & [0,3\,;0,7] & [0,45\,;0,5] & [0,45\,;0,5] \\
P(E_2 \cup E_4) & = & [0,45\,;0,6] & [0,2\,;0,5] & [0,45\,;0,5] \\
P(E_3 \cup E_4) & = & [0,5\,;0,5] & [0,3\,;0,5] & [0,5\,;0,5] \\
P(E_1 \cup E_2 \cup E_3) & = & [0,6\,;0,8] & [0,8\,;1] & [0,8\,;0,8] \\
P(E_1 \cup E_2 \cup E_4) & = & [0,7\,;0,9] & [0,7\,;0,85] & [0,7\,;0,85] \\
P(E_1 \cup E_3 \cup E_4) & = & [0,6\,;0,85] & [0,7\,;0,85] & [0,7\,;0,85] \\
P(E_2 \cup E_3 \cup E_4) & = & [0,65\,;0,9] & [0,5\,;0,65] & [0,65\,;0,65] \\
P(\Omega) & = & [1\,;1] & [1\,;1] & [1\,;1]\;.
\end{array}
$$

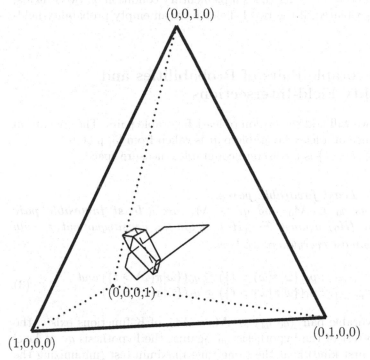

Fig. 3 $\mathcal{F}_0$ and $\mathcal{F}_1$ of the example

We regard the following K-functions:

$$
\begin{array}{cc}
\mathcal{F}_0: & \mathcal{F}_1: \\
(\,0,35\,;0,15\,;0,2\,;0,3\,) & (\,0,35\,;0,3\,;0,15\,;0,2\,) \\
(\,0,3\,;0,2\,;0,1\,;0,4\,) & (\,0,5\,;0,2\,;0,3\,;0\,) \\
(\,0,3\,;0,2\,;0,2\,;0,3\,) & (\,0,35\,;0,15\,;0,3\,;0,2\,) \\
(\,0,1\,;0,4\,;0,3\,;0,2\,)
\end{array}
$$

Every lower and upper intervalbound can be reached in $\mathcal{F}_0$ and $\mathcal{F}_1$ by some probability. So $\mathcal{F}_0$ and $\mathcal{F}_1$ are F-fields.

The F-fields satisfies the interval-cuts-condition (2). In the intervals $\left[\tilde{L}\,;\,\tilde{U}\right]$ there cannot be a probability otherwise we would get:

$$
\begin{aligned}
2 &= p(E_1 \cup E_2) + p(E_1 \cup E_3) + p(E_2 \cup E_3) + 2p(E_4) \\
&\leq \tilde{U}(E_1 \cup E_2) + \tilde{U}(E_1 \cup E_3) + \tilde{U}(E_2 \cup E_3) + 2\tilde{U}(E_4) \\
&= 0,5 + 0,55 + 0,5 + 0,4 \\
&= 1,95,
\end{aligned}
$$

which is a contradiction and this means that two F-fields can satisfy (2) having an empty probability-field-intersection.

In the next section we will give a supplementary condition on two F-fields, such that (2) is violated, if the two F-fields have an empty probability-field-intersection.

## 2 Least Favorable Pairs of Probabilities and Probability-Field-Intersections

In this section we will add the notion of least favorable pairs. The conclusion with the existence of a least favorable pair is taken from [5] p.66-68.
Again $\Omega = \{\omega_i \,|\, i \in I\}$ is a (discrete countable) measure space.

**Definition 2.1 Least favorable pairs:**
*Two K-functions $q_0 \in \mathcal{M}_0$ and $q_1 \in \mathcal{M}_1$ are a **least favorable pair** for the test $\mathcal{F}_0$ ($H_0$) against $\mathcal{F}_1$ ($H_1$), if for the densityquotient $\pi$ with $\pi(A) = \frac{q_1(A)}{q_0(A)}$ almost everywhere we have*

$$
\begin{aligned}
&\forall_{t \geq 0} \forall_{p_0 \in \mathcal{M}_0} : p_0(\{\omega | \pi(\omega) > t\}) \leq q_0(\{\omega | \pi(\omega) > t\}) \text{ and} \\
&\forall_{t \geq 0} \forall_{p_1 \in \mathcal{M}_1} : p_1(\{\omega | \pi(\omega) > t\}) \geq q_1(\{\omega | \pi(\omega) > t\}) \,.
\end{aligned}
\tag{3}
$$

If a least favorable pair $(q_0, q_1) \in \mathcal{M}_0 \times \mathcal{M}_1$ of K-functions exists, the likelihood-ratio-test of the hypothesis $q_0$ against the hypothesis $q_1$ at level $\alpha$ (error of the first kind) is at the same time maximin-test (minimizing the maximal $\beta$, error of the second kind) of $\mathcal{F}_0$ against $\mathcal{F}_1$ (F-fields) at level $\alpha$. Compare [1], p. 98, Proposition 3.5 .

If $\mathcal{F}_0$ and $\mathcal{F}_1$ have a K-function $p$ in common ($\mathcal{M}_0 \cap \mathcal{M}_1 \neq \emptyset$), then the two F-fields represent no separable pair of hypothesis.

With the existence of a least favorable pair there is a consequence for the intervalcuts of two F-fields with disjoint structures $\mathcal{M}_0$ and $\mathcal{M}_1$:

**Theorem 2.2 *Existence of an empty intervalcut for general F-Fields:***
*If $\mathcal{F}_0$ and $\mathcal{F}_1$ are two F-Fields with structures $\mathcal{M}_0$ and $\mathcal{M}_1$ and $\mathcal{M}_0 \cap \mathcal{M}_1 = \emptyset$ and a least favorable pair $(q_0, q_1) \in \mathcal{M}_0 \times \mathcal{M}_1$ exists, then there is a set $B$ with*

$$U_0(B) < L_1(B) \,, \tag{4}$$

*and consequently for this set $B$ $[L_1(B), U_1(B)] \cap [L_0(B), U_0(B)] = \emptyset$, the interval-cuts-condition is violated.*

*Proof.* Since $\mathcal{M}_0 \cap \mathcal{M}_1 = \emptyset$, the K-functions $q_0$ and $q_1$ (the least favorable pair) are not identical and with the likelihood-ratio-test $q_0$ against $q_1$ at level $q_0(A_{t_0})$ ($\alpha$, error of the first kind) has $\beta$-value (error of the second kind) $q_1(\neg A_{t_0})$ for some critical value $t_0$ and this likelihood-ratio-test is unbiased or

$$q_0(A_{t_0}) < q_1(A_{t_0})$$

holds, since the probability of the set $A_{t_0}$ to reject $q_0$ must be less than the acceptance of $q_1$. [3] But refering to the definition of least favorable pairs and (3) with $(T.VI)$ this yields

$$q_0(A_{t_0}) = U_0(A_{t_0}) < L_1(A_{t_0}) = q_1(A_{t_0}).$$

Choose $B = A_{t_0}$    □

Since for $\mathcal{F}_0$ and $\mathcal{F}_1$ of the example in section 1 the interval-cuts-condition holds, in consequence of theorem 2.2 there cannot be a least favorable pair for these two fields! [4] For C-fields there is the

**Theorem 2.3 *Huber- and Strassen-Theorem:***
*If $\mathcal{F}_0$ and $\mathcal{F}_1$ are C-fields and $\mathcal{M}_0 \cap \mathcal{M}_1 = \emptyset$ , a least favorable pair $(q_0, q_1) \in \mathcal{M}_0 \times \mathcal{M}_1$ exists.*

See [6] (also for more general measure spaces under continuity conditions). As a consequence of theorem 2.2 and of theorem 2.3 we have

**Theorem 2.4 *Condition for the probability-field-intersection of C-fields:***
*The probability-field-intersection of two C-fields $\mathcal{F}_0$ and $\mathcal{F}_1$ is nonempty if and only if the interval-cuts-condition (2) holds.*

This is also valid for the totally monotone fields and the feasible probability intervals (F-PRI).

---

[3] Compare also p. 59 [5].
[4] Another example, where two F-fields have no least favorable pair, is shown with linear optimization in [1].

# References

1. Augustin, T.: Optimale Tests bei Intervallwahrscheinlichkeit. Vandenhoeck und Ruprecht, Göttingen (1998)
2. de Campos, L.M., Huete, J.F., Moral, S.: Probability intervals, a tool for uncertain reasoning. International Journal of Uncertainty and Knowledge-based Systems 2, 167–185 (1993)
3. Choquet, G.: Theory of capacities. Annales de l'Institut Fourier 5, 131–195 (1954)
4. Dempster, A.P.: Upper and lower probabilities induced by a multivalued mapping. The Annals of Mathematical Statistics 37, 325–339 (1967)
5. Gümbel, M.: Über die effiziente Anwendung von F-PRI - Ein Beitrag zur Statistik im Rahmen eines allgemeineren Wahrscheinlichkeitsbegriffs. Pinusdruck, Christiane u. Karl Jürgen Mühlberger, Augsburg (2009)
6. Huber, P.J., Strassen, V.: Minimax tests and the Neyman-Pearson lemma for capacities. Annals of Statistics 1, 251–263 (1973)
7. Walley, P.: Statistical reasoning with imprecise probabilities. Chapmann & Hall, London (1991)
8. Weichselberger, K.: Elementare Grundbegriffe einer allgemeineren Wahrscheinlich-keitsrechnung I. Physica, Heidelberg (2001)
9. Zouhal, L.M., Denoeux, T.: An evidence-theoretic k-NN rule with parameter optimization. Technical Report Heudiasyc 97/46. IEEE Transactions on Systems, Man and Cybernetics (1998)

# Using Imprecise and Uncertain Information to Enhance the Diagnosis of a Railway Device

Zohra L. Cherfi, Latifa Oukhellou, Etienne Côme, Thierry Denœux, and Patrice Aknin

**Abstract.** This paper investigates the use of partially reliable information elicited from multiple experts to improve the diagnosis of a railway infrastructure device. The general statistical model used to perform the diagnosis task is based on a noiseless Independent Factor Analysis handled in a soft-supervised learning framework.

**Keywords:** Belief function theory, Soft-supervised learning, Independent Factor Analysis, EM algorithm, Fault diagnosis.

## 1 Introduction

When a pattern recognition approach is adopted to solve diagnosis problems, it involves using machine learning techniques to assign the measured signals to one of several predefined classes of defects. In most real world applications, a large amount of data is available but their labeling is generally a time-consuming and expensive task. However, it can be taken advantage of

Zohra L. Cherfi, Etienne Côme, and Patrice Aknin
UPE, IFSTTAR, GRETTIA, 2 rue de la Butte Verte, 93166 Noisy-le-Grand, France
e-mail: zohra.cherfi@ifsttar.fr, etienne.come@ifsttar.fr,
patrice.aknin@ifsttar.fr

Latifa Oukhellou
LISSI, UPE Créteil, 61 av du Gal de Gaulle, 94100 Créteil, France
e-mail: latifa.oukhellou@ifsttar.fr

Thierry Denœux
HEUDIASYC, UTC, UMR CNRS 6599, B.P 20529, 60205 Compiègne, France
e-mail: thierry.denoeux@hds.utc.fr

S. Li (Eds.): Nonlinear Maths for Uncertainty and its Appli., AISC 100, pp. 213–220.
springerlink.com © Springer-Verlag Berlin Heidelberg 2011

*expert knowledge* to label the data. In this case, the class labels can be subject to imprecision and uncertainty. A solution to deal with imprecise and uncertain class labels have been proposed in [3] [6]. In this framework, this paper presents a fault diagnosis application using partially labeled data to learn a statistical model based on Independent Factor Analysis (IFA) [2]. Learning of this statistical model is usually performed in an unsupervised way. The idea investigated in this paper is to incorporate additional information on the class membership of some samples to estimate the parameters of the IFA model, using an extension of the EM algorithm [3] [6].

This paper is organized as follows. Background material on belief functions is first recalled in Sect. 2. Learning the IFA model from data with soft labels is then addressed in Sect. 3. Sect. 4 describes the application under study and introduces the diagnosis problem in greater detail. Experimental results are finally reported in Sect. 5, and Sect. 6 concludes the paper.

## 2 Background on Belief Functions

This section provides a brief account of the fundamental notions of the Dempster-Shafer theory of belief functions, also referred to as Evidence Theory [4, 9]. A particular interpretation of this theory has been proposed by Smets [11], under the name of the Transferable Belief Model (TBM).

### 2.1 Belief Representation

Let $\Omega = \{\omega_1, ..., \omega_n\}$ be a finite *frame of discernment*, defined as a set of exclusive and exhaustive hypotheses about some question $Q$ of interest. Partial information about the answer to question $Q$ can be represented by a *mass function* $m : 2^{\Omega} \rightarrow [0,1]$ such that $\sum_{A \subseteq \Omega} m(A) = 1$. The quantity $m(A)$ represents a measure of the belief that is assigned to subset $A \subseteq \Omega$ given the available evidence and that cannot be committed to any strict subset of $A$. Every $A \subseteq \Omega$ such that $m(A) > 0$ is called a *focal set* of $m$. A mass function or a bba (for *basic belief assignment*) is said to be:

- *normalized* if $\emptyset$ is not a focal set (condition not imposed in the TBM);
- *dogmatic* if $\Omega$ is not a focal set;
- *vacuous* if $\Omega$ is the only focal set (it then represents total ignorance);
- *simple* if it has at most two focal sets and, if it has two, $\Omega$ is one of those;
- *categorical* if it is both simple and dogmatic.

A simple bba such that $m(A) = 1 - w$ for some $A \neq \Omega$ and $m(\Omega) = w$ can be noted $A^w$. Thus, the vacuous bba can be noted $A^1$ for any $A \subset \Omega$, and a categorical bba can be noted $A^0$ for some $A \neq \Omega$.

The information contained in a bba $m$ can be equivalently represented thanks to the *plausibility* function $pl(A) = \sum_{B \cap A \neq \emptyset} m(B), \forall A \subseteq \Omega$. The quantity $pl(A)$ is an upper bound on the degree of support that could be assigned to $A$ if more specific information became available.

## 2.2 Information Combination

Let $m_1$ and $m_2$ be two bbas defined over a common frame of discernment $\Omega$, they may be combined using a suitable operator. The most common ones are the conjunctive and disjunctive rules of combination defined, respectively, as:

$$(m_1 \textcircled{\cap} m_2)(A) = \sum_{B \cap C = A} m_1(B) m_2(C), \quad \forall A \subseteq \Omega. \tag{1}$$

$$(m_1 \textcircled{\cup} m_2)(A) = \sum_{B \cup C = A} m_1(B) m_2(C), \quad \forall A \subseteq \Omega. \tag{2}$$

The mass assigned to the empty set may be interpreted as a *degree of conflict* between the two sources. An extension of the conjunctive rule proposed by Yager assumes that, in case of conflict, the result is not reliable but the solution must be in $\Omega$. The mass on $\emptyset$ is thus redistributed to $\Omega$ which leads to a normalized bba [12].

The conjunctive and disjunctive rules of combination assume the independence of the data sources. In [5], Denœux introduced the cautious rule of combination $\textcircled{\wedge}$ to combine bbas provided by non independent sources. Although the cautious rule can be applied to any non dogmatic bba, it will be recalled here only in the case of *separable* bba, i.e., bbas that can be decomposed as the conjunctive combination of simple bbas [9] [10]. Let $m_1$ and $m_2$ be two such bbas. They can be written as $m_1 = \textcircled{\cap}_{A \subset \Omega} A^{w_1(A)}$ and $m_2 = \textcircled{\cap}_{A \subset \Omega} A^{w_2(A)}$, where $A^{w_1(A)}$ and $A^{w_2(A)}$ are simple bbas, $w_1(A) \in (0,1]$ and $w_2(A) \in (0,1]$ for all $A \subset \Omega$. Their combination using the cautious rule is defined as:

$$(m_1 \textcircled{\wedge} m_2)(A) = \textcircled{\cap}_{A \subset \Omega} A^{w_1(A) \wedge w_2(A)}, \tag{3}$$

where $\wedge$ denotes the minimum operator. This rule avoids to double-count common evidence when combining non distinct bbas (idempotence property) i.e., it verifies $m \textcircled{\wedge} m = m$ for all $m$.

# 3   Statistical Model and Learning Method

## 3.1   Independent Factor Analysis

IFA is based on a generative model that makes it possible to recover independent latent components (sources) from their observed linear mixtures [2]. In its noiseless formulation (used throughout this paper), the IFA model can be expressed as $\mathbf{y} = H\mathbf{z}$, where $H$ is a nonsingular square matrix of size $S$, $\mathbf{y}$ is the observed random vector whose elements are the $S$ mixtures and $\mathbf{z}$ the random vector whose elements are the $S$ latent components. Each source density is a mixture of Gaussians (MOG), so that a wide class of densities can be approximated. The pdf of $\mathbf{z}$ is thus given by $f^{\mathcal{Z}}(\mathbf{z}) = \prod_{j=1}^{S} \sum_{k=1}^{K^j} \pi_k^j \varphi(z^j; \mu_k^j, \nu_k^j)$, where $z^j$ denotes the $j$-th component of vector $\mathbf{z}$, $\varphi(.; \mu, \nu)$ denotes the pdf of a Gaussian random variable of mean $\mu$ and variance $\nu$; $\pi_k^j$, $\mu_k^j$ and $\nu_k^j$ are the proportion, mean and variance of component $k$ for source $j$, and $K^j$ is the number of components for source $j$. In the classical unsupervised setting used in IFA, the problem is to estimate both the mixing matrix $H$ and the MOG parameters from the observed variables $\mathbf{y}$ alone. Maximum likelihood estimation of the model parameters can be achieved by an alternating optimization strategy [1].

## 3.2   Soft-Supervised Learning in IFA

This section considers the learning of the IFA model in a soft-supervised learning context where partial knowledge of the cluster membership of some samples is available in the form of belief functions. In the general case, we will consider a learning set $\mathbf{M} = \{(\mathbf{y}_1, m_1^1, \ldots, m_1^S), \ldots, (\mathbf{y}_N, m_N^1, \ldots, m_N^S)\}$, where $m_i^1, \ldots, m_i^S$ is a set of bbas encoding uncertain knowledge on the cluster membership of sample $i = 1 \ldots N$ for each one of the $S$ sources. Each bba $m_i^j$ is defined on the frame of discernment $\mathcal{U}^j = \{c_1, \ldots, c_{K^j}\}$ composed of all possible clusters for source $j$. Let us denote by $\mathbf{x}_i = (\mathbf{y}_i, u_i^1, \ldots, u_i^S)$ the completed data where $\mathbf{y}_i \in \mathbb{R}^S$ are the observed variables and $u_i^j \in \mathcal{U}^j$, $\forall j \in \{1, \ldots, S\}$ are the cluster membership variables which are ill-known. In this model two independence assumptions are made. The random generation process induces the stochastic independence assumption between realizations:

$$f(\mathbf{X}; \boldsymbol{\Psi}) = \prod_{i=1}^{N} f(\mathbf{x}_i; \boldsymbol{\Psi}), \tag{4}$$

where $\boldsymbol{\Psi}$ is the IFA parameter vector, $\mathbf{X} = (\mathbf{x}_1, \ldots, \mathbf{x}_N)$ is the complete sample vector and $f(\mathbf{x}_i)$ the pdf of a complete observation according to the IFA model. Additionally, the imperfect perception of cluster memberships induces the following cognitive independence assumption (see [9, page 149]):

$$pl(\mathbf{X}) = \prod_{i=1}^{N} pl_i(\mathbf{x}_i) = \prod_{i=1}^{N}\prod_{j=1}^{S} pl_i^j(u_i^j), \tag{5}$$

where $pl(\mathbf{X})$ is the plausibility that the complete sample vector is equal to $\mathbf{X}$, $pl_i(\mathbf{x}_i)$ is the plausibility that the complete data for instance $i$ is $\mathbf{x}_i$ and $pl_i^j(u_i^j)$ is the plausibility that the source $j$ for example $i$ was generated from component $u_i^j$. Under the assumptions (4) and (5) and following [6], the observed data log likelihood can be written as:

$$l(\boldsymbol{\Psi}; \mathbf{M}) = -N\log(|\det(H)|) + \sum_{i=1}^{N}\sum_{j=1}^{S}\log\left(\sum_{k=1}^{K^j} pl_{ik}^j \pi_k^j \varphi((H^{-1}\mathbf{y}_i)^j; \mu_k^j, \nu_k^j)\right) \tag{6}$$

where $pl_{ik}^j = pl_i^j(c_k)$ is the plausibility that sample $i$ belongs to cluster $k$ of latent variable $j$. This criterion must be maximized with respect to $\boldsymbol{\Psi}$ to compute parameter estimates. An extension of the EM algorithm called E$^2$M for Evidential EM can be used to perform this task [6].

# 4 Diagnosis Approach

## 4.1 Problem Description

The track circuit is an essential component of the automatic train control system [8]. Its main function is to detect the presence or absence of vehicle traffic on a given section of the railway track. For this purpose, the railway track is divided into different sections (Fig. 1); each section is equipped with a specific track circuit consisting of: a transmitter connected to one of the two section ends, the two rails that can be considered as a transmission line, a receiver at the other end of the track section and trimming capacitors connected between the two rails at constant spacing to compensate the inductive behavior of the track. A train is detected when the wheels and axles short-circuit the track. It induces the loss of the track circuit signal and the drop of the received signal below a threshold indicates that the section is occupied. The different parts of the track circuit can be subject to malfunctions that must be detected as soon as possible to maintain the system at the required safety and availability levels. In the most extreme cases, an unfortunate attenuation of the transmitted signal may induce important signaling problems (a section can be considered as occupied even if it is not). The objective of diagnosis is to avoid such inconvenience on the basis of inspection signals analysis [8]. For this purpose, an inspection vehicle is able to deliver a measurement signal (denoted as Icc) linked to electrical characteristics of the system (Fig. 1). This paper describes the approach adopted for the diagnosis of track circuit from real inspection signals, it will focus on trimming capacitor faults.

**Fig. 1** Track circuit representation and examples of inspection signals (Icc) simulated along a 1500 m track circuit: one of them corresponds to a fault-free system, while the others correspond to a signal with one defective capacitor, and respectively a signal with two defective capacitors.

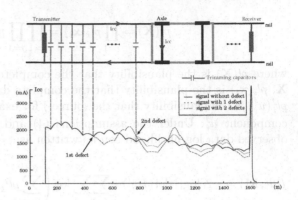

## 4.2 Diagnosis Methodology

A track circuit can be considered as a complex system made up of a series of $S$ spatially related subsystems, where each subsystem correspond to a trimming capacitor. A defect on one subsystem can be represented either by its capacitance or by a discrete value if considering a finite number of operating modes. The approach adopted here consists in extracting features from the measurement signal, and building a generative model as shown in Fig. 2. In this model, the variables $y_i^j$ are observed variables extracted by approximating each arch of the inspection signal (Icc) by a quadratic polynomial. The variables $z_i^j$ are continuous latent variables corresponding to continuous values describing the subsystem defects (capacitances), while the discrete latent variables $u_i^j$ correspond to the membership of the subsystem operating mode to one of the following three states: fault-free, medium defect, major defect. Assuming that a linear relationship exists between observed and latent variables and that each latent variable can be modeled semi-parametrically by a MOG, the involved generative model can be considered as an IFA model [2].

**Fig. 2** Generative model for the diagnosis of track circuits represented by a graphical model

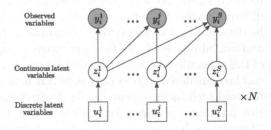

# 5 Results and Discussion

The diagnosis system was assessed using a real data set consisting of 422 inspection signals provided by the French National Railway Company (SNCF). A labeling campaign was organized with the aim of presenting separately the signals to four experts for labeling. Three classes were considered for the labeling operation, corresponding to the three operating modes of the subsystems. Experts were allowed to specify imprecise and uncertain labels that were represented as belief functions. The labels from individual experts were combined using each of the rules described in Sect. 2.2, and the IFA model was fit using the combined labels thanks to the $E^2M$ algorithm [6].

Because of the lack of the ground truth about the real state of the components under diagnosis, we chose to take as reference another labeling of the same database obtained thanks to a third-party expertise. This particular labeling was achieved simultaneously by three observers and in favorable conditions to provide a reference tool. This reference labeling will be referred to as REF in the following. The classification results reported in Table 1 reveal a good classification performance despite some confusions between contiguous classes. The confusion matrices corresponding to individual experts provide some information on expert skills. Indeed, experts 1 and 4 seem to better detect major defects, while experts 2 and 3 are more accurate for the detection of medium defects. The combination of expert opinions makes it possible to improve the detection of both types of defects. The best results were achieved by the cautious rule, which suggests that the expert opinions cannot be regarded as independent. The confusion between contiguous classes can be explained by the fact that identification of medium defects is a difficult exercise due to the continuous nature of the real states.

**Table 1** Confusion matrices between true classes $\omega_0$, $\omega_1$ and $\omega_2$ (defined by the REF labeling) and their estimates $d_0$, $d_1$ and $d_2$ (computed on ten cross validation test sets)

| | $\omega_0$ | $\omega_1$ | $\omega_2$ | | $\omega_0$ | $\omega_1$ | $\omega_2$ | | $\omega_0$ | $\omega_1$ | $\omega_2$ | | $\omega_0$ | $\omega_1$ | $\omega_2$ |
|---|---|---|---|---|---|---|---|---|---|---|---|---|---|---|---|
| $d_0$ | 98.8 | 33.1 | 2.1 | $d_0$ | 98.9 | 34.7 | 3.0 | $d_0$ | 98.7 | 22.1 | 2.1 | $d_0$ | 98.8 | 34.6 | 3.3 |
| $d_1$ | 0.9 | 51.1 | 6.9 | $d_1$ | 0.8 | 58.8 | 12.2 | $d_1$ | 1.1 | 63.6 | 13.8 | $d_1$ | 1.0 | 49.6 | 5.8 |
| $d_2$ | 0.2 | 15.8 | 90.9 | $d_2$ | 0.3 | 6.5 | 84.7 | $d_2$ | 0.2 | 14.3 | 84.1 | $d_2$ | 0.2 | 15.8 | 90.9 |
| | *(Expert 1)* | | | | *(Expert 2)* | | | | *(Expert 3)* | | | | *(Expert 4)* | | |

| | $\omega_0$ | $\omega_1$ | $\omega_2$ | | $\omega_0$ | $\omega_1$ | $\omega_2$ | | $\omega_0$ | $\omega_1$ | $\omega_2$ |
|---|---|---|---|---|---|---|---|---|---|---|---|
| $d_0$ | 98.9 | 30.7 | 2.9 | $d_0$ | 98.9 | 20.2 | 2.9 | $d_0$ | 98.9 | 20.4 | 2.6 |
| $d_1$ | 0.9 | 58.0 | 7.7 | $d_1$ | 1.0 | 64.2 | 6.5 | $d_1$ | 1.0 | 65.3 | 4.9 |
| $d_2$ | 0.2 | 11.3 | 89.4 | $d_2$ | 0.1 | 15.6 | 90.6 | $d_2$ | 0.1 | 14.2 | 92.4 |
| | *(Disjonctive rule)* | | | | *(Conjonctive rule)* | | | | *(Cautious rule)* | | |

# 6   Conclusions

The particular application that was considered concerns the diagnosis of railway track circuits. Experiments were carried out with real signals labeled by four different human experts. Experts' uncertain knowledge about the state of each subsystem was encoded as belief functions, which were pooled using different combination rules. These combined opinions were shown to yield better classification results than those obtained from each individual expert, especially with the cautious rule of combination [5], which can be explained by the existence of common knowledge shared among the experts.

This work can be extended in several directions. The approach relies on expert knowledge elicitation in the belief function framework, an important problem that has not received much attention until now. More sophisticated combination schemes could also be considered: for instance, discount rates could be learned from the data to take into account the competence of each individual expert [7].

# References

1. Amari, S., Cichocki, A., Yang, H.: A New Learning Algorithm for Blind Signal Separation. In: Proceedings of NIPS. MIT Press, Cambridge (1996)
2. Attias, H.: Independent factor analysis. Neural Computation 11(4), 803–851 (1999)
3. Côme, E., Oukhellou, L., Denoeux, T., Aknin, P.: Learning from partially supervised data using mixture models and belief functions. Pattern Recognit. 42(3), 334–348 (2009)
4. Dempster, A.: Upper and lower probabilities induced by a multivalued mapping. Ann. Math. Stat. 38, 325–339 (1967)
5. Denoeux, T.: Conjunctive and disjunctive combination of belief functions induced by non distinct bodies of evidence. Artificial Intelligence 172, 234–264 (2008)
6. Denœux, T.: Maximum likelihood from evidential data: An extension of the EM algorithm. In: Borgelt, C., González-Rodríguez, G., Trutschnig, W., Lubiano, M.A., Gil, M.Á., Grzegorzewski, P., Hryniewicz, O., et al. (eds.) Combining Soft Computing and Statistical Methods in Data Analysis. AISC, vol. 77, pp. 181–188. Springer, Heidelberg (2010)
7. Elouedi, Z., Mellouli, K., Smets, P.: Assessing sensor reliability for multisensor data fusion within the Transferable Belief Model. IEEE Trans. on Syst. Man and Cybern. B. 34, 782–787 (2004)
8. Oukhellou, L., Debiolles, A., Denoeux, T., Aknin, P.: Fault diagnosis in railway track circuits using Dempster-Shafer classifier fusion. Eng. Appl. of Artif. Intell. 23(1), 117–128 (2010)
9. Shafer, G.: A mathematical theory of evidence. Princeton Univ. Press, Princeton (1976)
10. Smets, P.: The canonical decomposition of a weighted belief. In: Int. Joint Conf. on Artif. Intell. Morgan Kaufmann, San Francisco (1995)
11. Smets, P., Kennes, R.: The Transferable Belief Model. Artif. Intell. 66, 191–234 (1994)
12. Yager, R.: On the Dempster-Shafer framework and new combination rules. Information Sciences 41(2), 93–137 (1987)

# Pricing CDS with Jump-Diffusion Risk in the Intensity-Based Model

Ruili Hao and Zhongxing Ye

**Abstract.** In this paper, we mainly discuss the pricing of credit default swap (CDS) in intensity-based models with counterparty risk. The default intensity of firm depends on the stochastic interest rate driven by the jump-diffusion process and the default states of counterparty firms. Moreover, we apply the hyperbolic function to illustrate the attenuation effect of correlated defaults between counterparties. Our models are extensions of the models in Jarrow and Yu (2001) and Bai, Hu and Ye (2007). In the model, we make use of the techniques in Park (2008) to obtain some important results and derive the explicit prices of bond and CDS in the primary-secondary and looping default frameworks respectively.

**Keywords:** Jump-diffusion process, Credit default swap, Bond, Counterparty risk, Hyperbolic attenuation function.

## 1 Introduction

Recently, credit securities are actively traded and the valuation of credit securities has called for more effective models according to the real market. Until now, there have been mainly two basic models: the structural model and the reduced-form model.

The structural approach considers that the firm's default is governed by the value of its assets and debts. It was pioneered by Merton (1974), then extended by Black and Cox (1976), Longstaff and Schwartz (1995). In their models, the asset process was all driven by the Brownian motion. For the valuation of credit derivatives involving jump-diffusion process, it is still

Ruili Hao and Zhongxing Ye
Department of Mathematics, Shanghai Jiaotong University, No.800 Dongchuan Road, Shanghai, 200240, China
e-mail: haoruili13@163.com,zxye@sjtu.edu.cn

S. Li (Eds.): Nonlinear Maths for Uncertainty and its Appli., AISC 100, pp. 221–229.
springerlink.com                                © Springer-Verlag Berlin Heidelberg 2011

difficult to get explicit results in the event of defaulting before the maturity date, even if using the above approaches.

Comparing with the structural approach, the reduced-form approach is flexible and tractable. It was pioneered by Jarrow, Lando and Turnbull (1994) and Duffie and Singleton (1995). They considered the default as a random event which is controlled by an exogenous intensity process.

Davis and Lo (2001) firstly proposed the model of credit contagion to account for concentration risk in large portfolios of defaultable securities. Later, motivated by a series of events such as the South Korean banking crisis and so on, Jarrow and Yu (2001) thought that the traditionally structural and reduced-form models all ignored the firm's specific source of credit risk. They applied the Davis's contagious model and introduced the concept of counterparty risk which is from the defaults of firm's counterparties. Their models paid more attention to the primary-secondary framework in which the defaults of firms were correlated with the interest rate. Besides, there are also other similar applications such as Leung and Kwork (2005), Bai, Hu and Ye (2007) and so on. In recent years, some authors applied this approach into portfolio credit securities such as Yu (2007) and Leung and Kwok (2009). Nevertheless, the stochastic interest rate in the above models still was driven by diffusion processes.

This paper mainly studies the pricing of CDS in primary-secondary and looping default frameworks, extending the models in Jarrow and Yu (2001). Our models consider the risk-free interest rate as the macroeconomic variable which presents the interaction between credit risk and market risk. However, the stochastic interest rate follows a jump-diffusion process rather than the continuous diffusion process in Jarrow and Yu (2001). Thus, our models not only reflect the real market much better, but more precisely to identify the impact of counterparty risk on the valuation of credit securities.

## 2   Model

Let $(\Omega, \mathscr{F}, \{\mathscr{F}_t\}_{t=0}^{T^*}, P)$ be the filtered probability space satisfying the usual conditions, where $\mathscr{F} = \mathscr{F}_{T^*}$, $T^*$ is large enough but finite and $P$ is an equivalent martingale measure.

On $(\Omega, \mathscr{F}, \{\mathscr{F}_t\}_{t=0}^{T^*}, P)$, $X = \{X_t\}_{t=0}^{T^*}$ represent economy-wide state variables. There are two firms with two point processes $N^i, i = A, B$ ($N_0^i = 0$) which represent the default processes of two firms respectively. When $N^i$ first jumps from 0 to 1, we call that the firm $i$ defaults and denote $\tau^i$ be the default time of firm $i$. Thus, $N_t^i = 1_{\{\tau^i \leq t\}}$ where $1_{\{.\}}$ is the indicator function. The filtration $\mathscr{F}_t = \mathscr{F}_t^X \vee \mathscr{F}_t^A \vee \mathscr{F}_t^B$ is generated by the state variables and the default processes of two firms where

$$\mathscr{F}_t^X = \sigma(X_s, 0 \leq s \leq t) \quad and \quad \mathscr{F}_t^i = \sigma(N_s^i, 0 \leq s \leq t), \quad i = A, B.$$

Let

$$\mathcal{H}_t^A = \mathcal{F}_t^A \vee \mathcal{F}_{T^*}^X \vee \mathcal{F}_{T^*}^B, \mathcal{H}_t^B = \mathcal{F}_t^B \vee \mathcal{F}_{T^*}^X \vee \mathcal{F}_{T^*}^A.$$

We assume that the default time $\tau^i (i = A, B)$ possesses a strictly positive $\mathcal{H}_0^i$-adapted intensity process $\lambda_t^i$ satisfying $\int_0^t \lambda_s^i ds < \infty, P - a.s.$ for all $t \in [0, T^*]$. The conditional survival probability distribution of primary firm $i$ is given by

$$P(\tau^i > T^* | \mathcal{H}_0^i) = \exp\left( -\int_t^{T^*} \lambda_s^i ds \right), \quad t \in [0, T^*].$$

We suppose that the state variable $X_t$ only contains the risk-free spot rate $r_t$ following the jump-diffusion process

$$dr_t = \alpha(K - r_t)dt + \sigma dW_t + q_t dY_t, \tag{1}$$

where $W_t$ is a standard Brownian motion on the probability space $(\Omega, \mathcal{F}, P)$ and $Y_t$ is a Possion process under $P$ with intensity $\mu$. $q_t$ is a deterministic function and $\alpha, \sigma, K$ are constants. $W_t$ and $Y_t$ are mutually independent. In fact, from Park (2008), we know that for $u \geq t$,

$$r_t = r_0 e^{-\alpha t} + \alpha K \int_0^t e^{-\alpha(t-s)} ds + \sigma \int_0^t e^{-\alpha(t-s)} dW_s + \int_0^t q_s e^{-\alpha(t-s)} dY_s$$

$$= f(t, u) + \int_t^u \alpha K e^{\alpha(v-u)} dv + \int_t^u \sigma e^{\alpha(v-u)} dW_v + \int_t^u q_v e^{\alpha(v-u)} dY_v,$$

where $f(0, u) = r_0 e^{-\alpha u}$ and

$$f(t, u) = f(0, u) + \int_0^t \alpha K e^{\alpha(v-u)} dv + \int_0^t \sigma e^{\alpha(v-u)} dW_v + \int_0^t q_v e^{\alpha(v-u)} dY_v \tag{2}$$

We now give two important lemmas which is important to the pricing of CDS.

**Lemma 1.** *Suppose $R_{t,T} = \int_t^T r_s ds$ be the cumulative interest from time $t$ to $T$. Let $E_t[\cdot]$ denotes the expectation conditional on $\mathcal{F}_t$ and $E_t[e^{-aR_{t,T}}] = g(a, t, T)$ for all $a \in \mathcal{R}$, then we obtain*

$$g(a, t, T) = e^{\int_t^T [-af(t,u) + \frac{1}{2}\sigma^2 a^2 c_T^2(u) + \mu(e^{-aq_u c_T(u)} - 1)]du - aK(T-t) + aK c_T(t)},$$

*where*

$$c_v(u) = -\frac{1}{\alpha}(e^{\alpha(u-v)} - 1), \quad 0 \leq v, u \leq T. \tag{3}$$

*In particular, $g(1, t, T)$ is the time-t price of zero-coupon bond.*

*Proof.* Omitted. The detail proof is given in Ref [5].

**Lemma 2.**  *Assume $R_{t,T}$ be the cumulative interest from time $t$ to $T$. For $0 \leq t_0 \leq t_1 \leq t_2 \leq t_3 \leq T$, let $E_{t_0}[e^{-m_1 R_{t_0,t_1}-m_2 R_{t_1,t_2}}] = H(t_0,t_1,t_2;m_1,m_2)$ and $E_{t_0}[e^{-m_1 R_{t_0,t_1}-m_2 R_{t_1,t_2}-m_3 R_{t_2,t_3}}] = G(t_0,t_1,t_2,t_3;m_1,m_2,m_3)$ where $m_1, m_2$ and $m_3$ are real numbers. Then*

$$H(t_0,t_1,t_2;m_1,m_2)$$

$$= \exp\left(-m_1 \int_{t_0}^{t_1} f(t_0,u)du - F(m_2,t_0,t_1,t_2) - \sum_{i=1}^{2} m_i K(t_i - t_{i-1})\right)$$

$$\cdot \exp(m_1 K c_{t_1}(t_0) + m_2(K - r_0)d(t_1,t_2,0))$$

$$\cdot \exp\left[\mu \sum_{i=1}^{2} \int_{t_{i-1}}^{t_i} [e^{-q_u(m_i c_{t_i}(u)+m_{i+1}d(t_i,t_{i+1},u)1_{\{i+1\leq 2\}})} - 1]du\right]$$

$$\cdot \exp\left[\frac{1}{2}\sigma^2 \sum_{i=1}^{2} \int_{t_{i-1}}^{t_i} (m_i c_{t_i}(u) + m_{i+1}d(t_i,t_{i+1},u)1_{\{i+1\leq 2\}})^2 du\right].$$

*and*

$$G(t_0,t_1,t_2,t_3;m_1,m_2,m_3)$$

$$= \exp\left(-m_1 \int_{t_0}^{t_1} f(t_0,u)du - \sum_{i=2}^{3} F(m_i,t_{i-2},t_{i-1},t_i)\right)$$

$$\cdot \exp\left(\sum_{i=2}^{3} m_i(K - r_0)d(t_{i-1},t_i,0) - \sum_{i=1}^{3} m_i K(t_i - t_{i-1}) + m_1 K c_{t_1}(t_0)\right)$$

$$\cdot \exp\left[\mu \sum_{i=1}^{3} \int_{t_{i-1}}^{t_i} [e^{-q_u(m_i c_{t_i}(u)+m_{i+1}d(t_i,t_{i+1},u)1_{\{i+1\leq 3\}})} - 1]du\right]$$

$$\cdot \exp\left[\frac{1}{2}\sigma^2 \sum_{i=1}^{3} \int_{t_{i-1}}^{t_i} (m_i c_{t_i}(u) + m_{i+1}d(t_i,t_{i+1},u)1_{\{i+1\leq 3\}})^2 du\right],$$

*where $f(t_i,u), c_{t_i}(u)$ are given by (2), (3) and*

$$d(t_i,t_{i+1},u) = -\frac{1}{\alpha}e^{\alpha u}(e^{-\alpha t_{i+1}} - e^{-\alpha t_i})$$

$$F(m_i,t_{i-2},t_{i-1},t_i) = \int_0^{t_{i-2}} \sigma m_i d(t_{i-1},t_i,u)dW_u + \int_0^{t_{i-2}} m_i q_u d(t_{i-1},t_i,u)dY_u.$$

*Proof.*  Omitted. The detail proof is presented in Ref [6].

# 3   Main Results

In this section, we begin to price CDS. Firm $C$ holds a bond issued by the reference firm $A$ with the maturity date $T$. To decrease the possible loss, firm $C$ buys protection with the maturity date $T_1$ ($T_1 \leq T$) from firm $B$ on condition that firm $C$ gives the payments to firm $B$ at a fixed swap rate in time while firm $B$ promises to make up firm $C$ for the loss caused by the default of firm $A$ at a certain rate. Each party has the obligation to make payments until its own default. The source of credit risk may be from three parties: the issuer of bond, the buyer of CDS and the seller of CDS.

In the following, we consider a simple situation which only contains the risk from reference firm $A$ and firm $B$. To make the calculation convenient, we suppose that the face value of the bond issued by firm $A$ is 1 dollar. In the event of firm $A$'s default, firm $B$ compensates firm $C$ for 1 dollar if it doesn't default, otherwise 0 dollar. Thus, the swap rate could be obtained by the principle that the time-0 market value of firm $C$'s fixed rate payment equals to the time-0 market value of firm $B$'s promised payment in the event of $A$'s default. Denoted the swap rate by a constant $c$ and interest rate by $r_t$. There are four cases for the defaults of firm $A$ and firm $B$:

1. The defaults of firm $A$ and firm $B$ are mutually independent conditional on the risk-free interest rate.
2. Firm $A$ is the primary party whose default only depends on the risk-free interest rate and the firm $B$ is the secondary party whose default depends on the risk free interest rate and the default state of firm $A$.
3. Firm $B$ is the primary party and the firm $A$ is the secondary party.
4. The defaults of firm $A$ and firm $B$ are mutually contagious (looping default).

## 3.1   Pricing CDS in the Primary-Secondary Framework

The primary-secondary model was proposed by Jarrow and Yu (2001). In their model, the interest rate $r_t$ was driven by the diffusion process and the pricing formula of bond was obtained. We generalize their model and allow the interest rate to follow a jump-diffusion process. Let the default times of firm $A$ and $B$ be $\tau^A$ with the intensity $\lambda^A$ and $\tau^B$ with the intensity $\lambda^B$ respectively.

Case.1 The defaults of firm $A$ and firm $B$ are mutually independent conditioning on the common factor which is interest rate in this paper. The intensities are given by

$$\lambda_t^A = b_0^A + b_1^A r_t,$$
$$\lambda_t^B = b_0^B + b_1^B r_t.$$

Since this model in case 1 could be considered as a simple special case of case 2,3 in primary-secondary framework or case 4 and the price of CDS could be derived by the similar method. Therefore, this case is not discussed in details.

Case 2. Firm $A$ is the primary firm and firm $B$ is secondary firm. We assume that their intensity processes respectively satisfy linear relations below:

$$\lambda_t^A = b_0^A + b_1^A r_t, \tag{4}$$

$$\lambda_t^B = b_0^B + b_1^B r_t + b^B 1_{\{\tau^A \le t\}}, \tag{5}$$

where $b_0^A, b_1^A, b_0^B, b_1^B$ and $b^B$ are positive constants.

When $b^B = 0$, it reduces to the model in case 1. First, we give the pricing formulas of defaultable bonds issued by firm $A$ and $B$.

**Lemma 3.** *Suppose that the bonds issued by firm $A$ and $B$ have the same maturity date $T$ and recovery rates both equal to zero. If the intensity processes $\lambda_t^A$ and $\lambda_t^B$ satisfy (4) and (5) and no defaults occur up to time $t$, then the time-$t$ price of bond issued by primary firm $A$ is*

$$V^A(t,T) = g(1 + b_1^A, t, T)e^{-b_0^A(T-t)};$$

*and the time-$t$ price of bond issued by secondary firm $B$ is*

$$V^B(t,T) = g(1 + b_1^B, t, T)e^{-(b_0^B + b^B)(T-t)} + b^B e^{-(K + K b_1^B + b_0^B + b^B)T}$$

$$\cdot e^{(1 + b_0^B + b_0^A + b_1^B + b_1^A)t} \int_t^T e^{-(1 + b_1^B + b_1^A) \int_t^s f(t,u)du + (b^B - b_1^A K - b_0^A)s}$$

$$\cdot e^{(1 + b_1^B + b_1^A)K c_s(t) + (1 + b_1^B)(K - r_0)d(s,T,0) - f_1(t,s) + M(s)} ds,$$

*where for $\forall k, v, u \in [0,T]$, $c_v(u), d(k,v,u)$ is given by (3) and Lemma 2,*

$$f_1(t,s) = \int_0^t \sigma(1 + b_1^B)d(s,T,u)dW_u + \int_0^t (1 + b_1^B)q_u d(s,T,u)dY_u,$$

$$M(s) = \int_s^T [\frac{1}{2}\sigma^2(1 + b_1^B)^2 c_T^2(u) + \mu(e^{-(1+b_1^B)q_u c_T(u)} - 1)]du$$

$$+ \int_t^s \frac{1}{2}\sigma^2[(1 + b_1^B + b_1^A)c_s(u) + (1 + b_1^B)d(s,T,u)]^2 du$$

$$+ \int_t^s \mu[e^{-q_u((1+b_1^B+b_1^A)c_s(u) + (1+b_1^B)d(s,T,u))} - 1]du.$$

*Proof.* Omitted. The details are given in Ref [5].

Next, we give the pricing formula of CDS.

**Theorem 1.** *Suppose that the risk-free interest rate $r_t$ satisfies (1) and the intensities $\lambda^A$ and $\lambda^B$ satisfy (4) and (5) respectively. Then, the swap rate $c$ has the following expression*

$$c = \frac{V^B(0,T_1) - e^{-(b_0^B + b_0^A)T_1} g(1 + b_1^B + b_1^A, 0, T_1)}{\int_0^{T_1} g(1,0,s)ds}, \tag{6}$$

*where $g(\cdot,\cdot,\cdot)$ and $V^B(0,T_1)$ are given in Lemma 1 and Lemma 3 respectively.*

*Proof.* The process is omitted and the details are given in Ref [5].

Case 3. Firm $B$ is the primary firm and firm $A$ is secondary firm. Suppose that the default intensities satisfy

$$\lambda_t^A = b_0^A + b_1^A r_t + b^A 1_{\{\tau^B \le t\}}, \tag{7}$$

$$\lambda_t^B = b_0^B + b_1^B r_t, \tag{8}$$

where $b_0^A, b_1^A, b_0^B, b_1^B$ and $b^A$ are positive constants. By the similar approach in Theorem 1, we could obtain the swap rate

$$c = \frac{g(1 + b_1^B, 0, T_1)e^{-b_0^B T_1} - e^{-(b_0^B + b_0^A)T_1} g(1 + b_1^B + b_1^A, 0, T_1)}{\int_0^{T_1} g(1,0,s)ds},$$

where $g(\cdot,\cdot,\cdot)$ are given by Lemma 1. We omit the details.

## 3.2 Pricing CDS in the Looping Default Framework

Case 4. The defaults of firm $A$ and firm $B$ are mutually contagious. The defaults of firm $A$ and $B$ have direct linkage. Namely, default risk may occur when one firm holds large amounts of debt issued by the other firm. The default intensities satisfy the relations

$$\lambda_t^A = b_0^A + b_1^A r_t + b^A 1_{\{\tau^B \le t\}},$$

$$\lambda_t^B = b_0^B + b_1^B r_t + b^B 1_{\{\tau^A \le t\}},$$

where $b_0^A, b_1^A, b^A, b_0^B, b_1^B$ and $b^B$ are positive constants. Since this model is a special case of the following more general model with attenuate effect, therefore this model will not be discussed in details.

Now, we turn to the more general model with attenuate effect. Suppose that their intensity processes satisfy

$$\lambda_t^A = b_0^A + b_1^A r_t + 1_{\{\tau^B \le t\}} \frac{b_2^A}{b_3^A(t - \tau^B) + 1}, \tag{9}$$

$$\lambda_t^B = b_0^B + b_1^B r_t + 1_{\{\tau^A \le t\}} \frac{b_2^B}{b_3^B(t - \tau^A) + 1}, \tag{10}$$

where $b_0^A, b_1^A, b_3^A, b_0^B, b_1^B$ and $b_3^B$ are nonnegative real numbers, $b_2^A, b_2^B$ are real numbers satisfying $b_0^A + b_1^A + b_2^A > 0, b_0^B + b_1^B + b_2^B > 0$. If firm $A$ is a competitor(or copartner) of firm $B$, $b_2^A < 0$(or $b_2^A > 0$). In the event of firm $B$' default, the intensity $\lambda_t^A$ jumps abruptly and the effect will attenuate until disappear with hyperbolic speed as time goes on. Reversely, the impact of firm $A$ to $B$ is similar to above situation. Parameters $b_2^A$ and $b_2^B$ reflect the impact of counterparty's default to intensities. Parameters $b_3^A$ and $b_3^B$ reflect the attenuation speed. When $b_1^A = 0$ and $b_1^B = 0$, the model becomes the one in Bai, Hu and Ye (2007). When $b_3^A = 0$ and $b_3^B = 0$, it becomes the model of case 4.

For simplicity, we assume $-b_2^A = b_3^A = c^A > 0$ and $-b_2^B = b_3^B = c^B > 0$ in the following discussion.

**Lemma 4.** *We suppose that the bond issued by firm $B$ have the maturity date $T$, recovery rate equals zero and no default occur up to time $t$. If the intensity process $\lambda_t^B$ satisfy (10), then the time-$t$ price of bond issued by firm $B$ is given by*

$$V^B(t,T) = (c^B(T-t) + 1)e^{-b_0^B(T-t)}g(1 + b_1^B, t, T) - c^B e^{-b_0^B T + (b_0^B + b_0^A)t}$$

$$\cdot \int_t^T (c^A(s-t) + 1)e^{-b_0^A s}H(t,s,T; 1 + b_1^A + b_1^B, 1 + b_1^B)ds + c^A c^B e^{-b_0^B T + 2b_0^B t}$$

$$\cdot e^{b_0^A t} \int_t^T \int_t^s e^{-b_0^A s - b_0^B u}G(t,u,s,T; 1 + b_1^A + 2b_1^B, 1 + b_1^B + b_1^A, 1 + b_1^B)du\,ds,$$

*where $R_{\cdot,\cdot}, g(\cdot, t, T), H(t, s, T; \cdot, \cdot)$ and $G(t, u, s, T; \cdot, \cdot, \cdot)$ are given in Lemma 1 and Lemma 2.*

*Proof.* The detail proof is given in Ref [6]. We omit it.

If no defaults occur up to time $t$, then the time-$t$ price of bond issued by firm $A$ can be obtained similarly by Lemma 4.

**Theorem 2.** *Suppose that the intensities $\lambda^A$ and $\lambda^B$ satisfy (9) and (10). If no defaults occur up to time $t$, then the swap rate*

$$c = \frac{V^B(0, T_1) - e^{-(b_0^B + b_0^A)T_1}g(1 + b_1^B + b_1^A, 0, T_1)}{\int_0^{T_1} g(1, 0, s)ds},$$

*where $g(\cdot, \cdot, \cdot)$ and $V^B(0, T_2)$ are given in Lemma 1 and Lemma 4.*

*Proof.* Omitted. The proof is referred to [6].

**Remark.** The interest rate $r_t$ in our model is an extension of Vasicek model. It may cause negative intensity. We could use the similar method in Jarrow and Yu (2001) to avoid this case. For example, we could assume $\lambda_t^A = \max\{b_0^A + b_1^A r_t, 0\}, \lambda_t^B = \max\{b_0^B + b_1^B r_t + b^B 1_{\{\tau^A \le t\}}, 0\}$. The other cases are similar. We shall discuss it elsewhere.

**Acknowledgements.** The research was supported Supported by the National Basic Research Program of China (973 Program No. 2007CB814903).

# References

1. Black, F., Cox, J.C.: Valuing corporate securities: some effects of bond indenture provisions. Journal of Finance 31, 351–367 (1976)
2. Bai, Y.F., Hu, X.H., Ye, Z.X.: A model for dependent default with hyperbolic attenuation effect and valuation of credit default swap. Applied Mathematics and Mechanics (English Edition) 28(12), 1643–1649 (2007)
3. Duffie, D., Singleton, K.J.: Modeling term structures of defaultable bonds. Working paper, Stanford University Business School (1995)
4. Davis, M., Lo, V.: Infectious defaults. Quantitative Finance, 382-387 (2001)
5. Hao, R.L., Ye, Z.X.: The intensity model for pricing credit securities with jump-diffusion and counterparty risk. Mathematical Problems in Engineering, Article ID 412565, 16 pages (2011), doi:10.1155/2011/412565
6. Hao, R.L., Ye, Z.X.: Pricing credit securities in intensity model with jump-diffusion risk and hyperbolic attenuation effect. Working paper, Shanghai Jiao Tong University (2010)
7. Jarrow, R.A., Yu, F.: Counterparty risk and the pricing of defaultable securities. Journal of Finance 56(5), 1765–1799 (2001)
8. Jarrow, R.A., Lando, D., Turnbull, S.: A Markov model for the term structure of credit risk spreads. Working paper, Conell University (1994)
9. Jarrow, R.A., Lando, D., Yu, F.: Default risk and diversification: theory and emperical implications. Mathematical Finance 15(1), 1–26 (2005)
10. Longstaff, F.A., Schwartz, E.S.: A simple approach to valuing risky fixed and floating rate debt. Journal of Finance 50, 789–819 (1995)
11. Leung, S.Y., Kwork, Y.K.: Credit default swap valuation with counterparty risk. Kyoto Economic Review 74(1), 25–45 (2005)
12. Leung, K.S., Kwok, Y.K.: Counterparty risk for credit default swaps: Markov chain interacting intensities model with stochastic intensity. Asia-Pacific Finan Markets 16, 169–181 (2009)
13. Merton, R.C.: On the pricing of corporate debt: the risk structure of interest rates. Journal of Finance 29, 449–470 (1974)
14. Park, H.S.: The survival probability of mortality intensity with jump-diffusion. Journal of the Korean Statistical Society 37, 355–363 (2008)
15. Yu, F.: Correlated defaults in intensity-based models. Mathematical Finance 17(2), 155–173 (2007)

# Pricing Convertible Bonds Using the CGMY Model

Coenraad C.A. Labuschagne and Theresa M. Offwood

**Abstract.** This paper looks at using the CGMY stock price process to price European convertible bonds. We compare the prices given by the CGMY model to prices given by the popular geometric Brownian motion model.

**Keywords:** Convertible bonds, Brownian motion, Levy process, CGMY process, CGMY model, Component model, Monte Carlo method.

## 1 Introduction

Convertible bonds are complex financial instruments, which despite their name, have more in common with derivatives than with conventional bonds. As the pricing techniques are moving away from geometric Brownian motion (GBM) models, it is important to look at the valuation of convertible bonds under different densities. This paper has chosen to compare the CGMY model prices of convertible bonds to the GBM model prices. It will have a look at two methods: the component model and the Monte Carlo technique, to price European convertible bonds.

The paper is set up as follows. In Section 2, we define convertible bonds and how they are structured. Section 3 looks at the CGMY process. The two pricing methods are described in Section 4. Our results are then given in Section 5 and Section 6 concludes.

Coenraad C.A. Labuschagne and Theresa M. Offwood
School of Computational and Applied Mathematics,
University of the Witwatersrand, South Africa
e-mail: coenraad.labuschagne@wits.ac.za, tmoffwood@gmail.com

S. Li (Eds.): Nonlinear Maths for Uncertainty and its Appli., AISC 100, pp. 231–238.
springerlink.com                    © Springer-Verlag Berlin Heidelberg 2011

## 2 Convertible Bonds

A convertible bond is a corporate debt security that gives the holder the right to exchange future coupon payments and the principal repayment for a prescribed number of shares of equity. Thus it can be seen as a hybrid security with elements of both debt and equity. Due to its hybrid nature, convertibles appeal to both issuers and investors with different preferences of risk.

From the issuer's perspective, a convertible bond reduces the cost of debt funding compared to straight debt alone. From the investor's perspective, a convertible bond offers advantages particular to both stocks and bonds. On the one hand they offer a greater stability of income than regular stock, while on the other hand if the company does well, they can convert to equity and receive the benefits of holding stock.

The paper written by Goldman, Sachs & Co [4] summarises nicely what quantities and features are typically included in convertible bond contracts:

- **Principal** (N): The face value of the convertible bond, i.e. the redemption value.
- **Coupon** (c): The annual interest rate as a percentage of the principal.
- **Conversion Ratio** ($\gamma_t$): The number of shares of the underlying stock that the convertible bond can be exchanged into. This ratio is usually determined at issue and is only changed to keep the total equity value constant, eg. when dividends or stock splits occur.
- **Conversion Price:** The price of each underlying share paid on conversion, assuming the bond principal is used to pay for the shares received.

$$\text{Conversion Price} = \frac{\text{Principal}}{\text{Conversion Ratio}} \qquad (1)$$

- **Conversion Value** ($\gamma_t S_t$): The conversion value is generally determined on a daily basis as the closing price of the stock multiplied by the conversion ratio.
- **Call Provisions:** A call provision gives the issuer the right to buy the bond back at the call price, which is specified in the call schedule. Generally, convertible bonds are call protected for a certain number of years and only become callable after a certain date.
- **Put provisions:** A put provision gives the investor the right to sell the bond back at the put price on certain dates prior to maturity. This provides the investor with extra downside protection.

In this paper, however, we will only look at European convertible bonds with no call or put provisions. We leave this for future work. The convertible bond will pay constant coupons at regular times and will be exchangeable into a certain number of shares at the discretion of the investor.

Some further notation:

- $r_{t,T}$ = continuously compounded risk-free interest rate from t to T
- $V_t$ = fair value of the convertible bond
- $T$ = maturity of the convertible bond
- $S_t$ = price of the underlying equity at time t
- $d_S$ = continuously compounded dividend yield of the underlying equity
- $\kappa$ = final redemption ratio at time T in percentage points of the face value
- $t_1, t_2, ..., t_n$ = coupon payment dates

## 3  CGMY Model

Let $(\Omega, \mathcal{F}, (\mathcal{F}_t)_{t\geq 0}, P)$ be a filtered probability space satisfying the usual conditions. Let $T \in [0, \infty]$ denote the time horizon. We consider a Lévy process $X$. The following theorem, the formula in which is known as the *Lévy-Khintchine formula*, plays an important role in our discussion of the CGMY proccess.

**Theorem 1.** *The distribution $f_X$ of a random variable $X$ is infinitely divisible if and only if there exists a triplet $(b, c, \nu)$, with $b \in \mathbb{R}, c \in \mathbb{R}^+$ and a measure satisfying $\nu(0) = 0$ and $\int_{\mathbb{R}}(1 \wedge |x|^2)\nu(dx) < \infty$ such that*

$$\mathcal{E}[e^{iu.X}] = exp\Big[ibu - \frac{u^2c}{2} + \int_{\mathbb{R}}(e^{iux-1-iux\mathbf{1}_{\{|x|<1\}}})\nu(dx)\Big]. \qquad (2)$$

The triplet $(b, c, \nu)$ is called the Lévy or characteristic triplet and the exponent in $(2).\kappa(u) = ibu - \frac{u^2c}{2} + \int_{\mathbb{R}}(e^{iux-1-iuh(x)})\nu(dx)$ is called the *Lévy* or *characteristic exponent*. Moreover, $b \in \mathbb{R}$ is called the *drift term*, $c \in \mathbb{R}^+$ the *diffusion coefficient* and $\nu$ the Lévy measure.

In the literature, several choices for the Lévy process in the stock returns process have been considered. Madan and Seneta [6] proposed the variance gamma (VG) Lévy process, Eberlein and Keller [5] used a hyperbolic model and the Normal inverse Gaussian (NIG) model was introduced by Barndorff-Nielsen [1]. Carr et al. [3] in 2002 introduced the CGMY model, of which the VG model is a special case. See [7] for more details.

### 3.1  CGMY Process

The Lévy density of the CGMY process is given by

$$\kappa_{CGMY}(x) = \begin{cases} \frac{Ce^{-G|x|}}{|x|^{1-Y}} & \text{if } x < 0 \\ \frac{Ce^{-M|x|}}{|x|^{1+Y}} & \text{if } x > 0, \end{cases}$$

where $C > 0$, $G$, $M \geq 0$ and $Y < 2$. We denote by $X_{CGMY}(t, C, G, M, Y)$ the infinitely divisible process of independent increments with the above Lévy density.

The parameters play an important role in capturing the various aspects of the stochastic process. The parameter $C$ can be seen as the measure of overall level of activity in the process. The parameters $G$ and $M$ control the rate of exponential decay on the right and left of the Lévy density, leading to skewed distributions when they are unequal. The parameter $Y$ is useful in characterising the monotonicity of the process including whether the process has finite or infinite activity and finite or infinity variation.

The characteristic function of the CGMY process is given by

$$\phi_{CGMY}(u, t, C, G, M, Y) = e^{tC\Gamma(-Y)[(M-iu)^Y - M^Y + (G+iu)^Y - G^Y]}.$$

The CGMY stock price process is given by

$$S_t(\omega) = S_0 e^{(\mu+\omega)t + X_{CGMY}(t, C, G, M, Y)},$$

where $\mu$ is the mean rate of returns on the stock and $\omega$ is a 'convexity correction'.

## 3.2  Pricing Options Given the Characteristic Function

If the density of our stock price process is known, then pricing options is easy as you just need to calculate the expected value. However, if only the characteristic function is known, then Carr and Madan [2] showed that the price of a European call option $C(T, K)$ with maturity $T$ and strike $K$ is given by

$$C(T, K) = \frac{e^{-\alpha log(K)}}{\pi} \int_0^\infty e^{-iv log(K)} \rho(v) dv,$$

where $\rho(v) = \frac{e^{-rT}\phi(v-(\alpha+1)i)}{\alpha^2 + \alpha - v^2 + i(2\alpha+1)v}$ and $\alpha$ is a positive constant such that the $\alpha$th moment of the stock price exists (typically a value of $\alpha = 0.75$ will do). Using fast Fourier transforms, it is possible to compute within seconds the complete option surface.

## 4  Pricing Convertible Bonds

### 4.1  The Component Model

In practice, a very popular method for pricing convertible bonds is via the component model, also called the synthetic model. The convertible bond is divided into a straight bond component denoted by $B_t$ and a call option $K_t$

on the conversion value $\gamma_t S_t$ with strike $X_t = B_t$. The fair value of the bond is calculated using the standard formulae, i.e. the sum of the present value of its cashflows:

$$B_t = Ne^{-(r_{t,T}+\xi_t)(T-t)} + Nc\sum_{i=1}^{n} e^{-(r_{t_i,T}+\xi_{t_i})(t_i-t)}. \qquad (3)$$

The call can be priced either using the standard Black-Scholes formula or via the fast Fourier transform mentioned in section 3.2. Then the value of the convertible bond is just the sum of the two: $V_t = B_t + K_t$.

## 4.2 Monte Carlo Methods

As with most complex derivatives, it is possible to approximate the price of a convertible bond using Monte Carlo simulations. Since convertible bonds are American in style, a technique to find the optimal stopping time needs to be added to the usual Monte Carlo method. For example in the case of a vanilla convertible bond, at every conversion time, the investor compares the payoff from immediate conversion to the expected present value of future payoffs from the bond to decide whether he should convert or not.

As in this paper we are only considering European convertible bonds, this condition only needs to be checked at maturity. These values are then averaged and discounted to today to result in an approximate price.

## 5 Results

There are various assumptions that were made to simplify the implementation of these models for the purpose of this paper. First of all, interest rates are assumed to be constant. Similarly, constant volatilities are also assumed. The stock volatilities used should be based on historical volatilities.

Since the convertible bond pays regular coupons, a 360 day year is assumed. This allows for the assumption that, for example, semi-annual coupon payments happen every 180 days. To make it more accurate one would need to work with the actual dates.

All models work with a conversion ratio instead of a conversion price. In practice, it seems that conversion prices are more common then ratios. It is not difficult to change the models to work with prices instead of ratios.

For the purpose of this paper, the following parameters will be used throughout the next few sections to illustrate the advantages and challenges of implementing the models described in this paper:

- $S_0 = 50$
- $\sigma_S = 30\%$
- $d_S = 3\%$
- $r = 8\%$

- $N = 100$
- $\kappa = 1$
- $\gamma = 2$
- $c = 10\%$ NACS (Nominal annual compounded semi-annually)

In Figures 1 and 2 we used $C = 4$, $G = 40$, $M = 70$ and $Y = 0.75$ as the values for the CGMY parameters. For more accuracy, one would need to calibrate the CGMY model to the stock returns.

The first aspect we inspected was the convergence rate of both the Monte Carlo using the GBM paths and the Monte Carlo using the CGMY paths. This is shown in Figure 1.

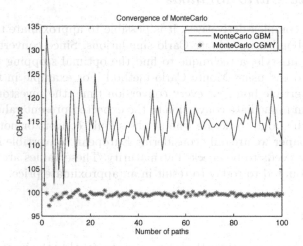

**Fig. 1** Conversion rate of Monte Carlo methods.

It is immediately obvious that the CGMY Monte Carlo converges instantaneously, while the Monte Carlo based on GBM does not seem to converge at all. Note as well, that they both give significantly different prices. The price calculated using the CGMY paths is lower than that given by the GBM paths.

In Figure 2 and 3 we have plotted the price of the convertible bond as a function of time to maturity. The prices given by the GBM-component model, CGMY-component model, GBM Monte Carlo model and CGMY Monte Carlo model are shown.

Looking only at the component model prices, the CGMY component model price is slightly lower than the GBM price, however, they have a similar shape. The two Monte Carlo methods differ greatly. The Monte Carlo price based on the CGMY model is lower and steeper than the Monte Carlo price based on GBM. The CGMY Monte Carlo price starts close to where the component prices start, while the GBM Monte Carlo price starts higher.

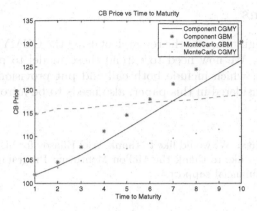

**Fig. 2** Component and Monte Carlo pricing techniques with coupons using a negatively skewed CGMY process.

Next we decided to adjust the CGMY parameters to see how the prices would change. Remember that the GBM model is a symmetric model, while the CGMY model with the above mentioned values is negatively skewed. This would explain why the CGMY prices are in general lower than the GBM prices.

In Figure 3, we interchanged the values for $G$ and $M$, thereby making it positively skewed.

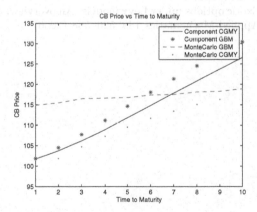

**Fig. 3** Component and Monte Carlo pricing techniques with coupons using a positively skewed CGMY process.

As can be seen, the CGMY Monte Carlo price changes significantly. At $T = 1$, the positively skewed CGMY price is slightly lower, and at $T = 10$, the price is quite a bit higher, as one would expect.

# 6  Conclusions

This is only the beginning into the research of using the CGMY model to price convertible bonds. We now need to extend these model to price American convertible bonds, which include both call and put provisions. Credit risk, which was not considered in this paper, also needs to be introduced into the pricing.

**Acknowledgements.** We would like to thank Ariel Eliasov for his CGMY Matlab code. We would also like to thank the Mellon Mentoring Programme, DAAD and the NRF for their financial support.

# References

1. Barndorff-Nielsen, O.E.: Normal inverse Gaussian distribution and stochastic volatility models. Scandinavian Journal of Statistics 24, 1–13 (1997)
2. Carr, P., Madan, D.H.: Option valuation using the fast Fourier transform. Journal of Computational Finance 2, 61–73 (1998)
3. Carr, P., Geman, H., Madan, D.H., Yor, M.: The fine structure of asset returns: an empirical investigation. Journal of Business 75, 305–332 (2002)
4. Goldman, Sachs & Co: Valuing convertible bonds as derivatives. Quantitative Strategies Research Notes (1994)
5. Eberlein, E., Keller, U.: Hyperbolic distributions in finance. Bernoulli 1, 281–299 (1995)
6. Madan, D.B., Seneta, E.: The V.G. model for share market returns. Journal of Business 63, 511–524 (1990)
7. Schoutens, W.: Exotic options under Lévy models: An overview. Journal of Computational and Applied Mathematics 189, 526–538 (2006)

# The Dividend Problems for Compound Binomial Model with Stochastic Return on Investments

Jiyang Tan, Xiangqun Yang, Youcai Zhang, and Shaoyue Liu

**Abstract.** We consider a discrete time risk process with stochastic return on investments based on the compound binomial model, and we are interested in the expected present value of all dividends paid out until ruin occurs when the insurer uses a simple barrier strategy.

**Keywords:** Compound binomial process, Return on investment, Dividend barrier.

## 1 Introduction

In the compound binomial model, the number of insurance claims is governed by a binomial process $N(n), n = 0, 1, 2, \cdots$. In any time period, the probability of a claim is $p, 0 < p < 1$, and the probability of no claim is $q = 1 - p$. We denote by $\xi_n = 1$ the event where *a claim occurs in the time period (n-1,n]* and denote by $\xi_n = 0$ the event where *no claim occurs in the time period (n-1,n]*. Then $N(n) = \sum_{k=1}^{n} \xi_k$ for $n \geq 1$ and $N(0) = 0$. The occurrences of claims in different time periods are independent events. The claim amounts $X_1, X_2, X_3, \cdots$ are mutually independent, identically distributed, positive and integer-valued random variables; they are independent

Jiyang Tan and Shaoyue Liu
Department of mathematics, Xiangtan University, Xiangtan 411105, China
e-mail: tanjiyang15@163.com, liusy@xtu.edu.cn

Xiangqun Yang
Department of mathematics, Hunan Normal University, Changsha 410081, China
e-mail: xqyang@hunnu.edu.cn

Youcai Zhang
Hengyang Finance Economics and Industry Vocational College, Hengyang, 421003, China
e-mail: zyc7012@163.com

S. Li (Eds.): Nonlinear Maths for Uncertainty and its Appli., AISC 100, pp. 239–246.
springerlink.com © Springer-Verlag Berlin Heidelberg 2011

of the binomial process $\{N(n)\}$. Let the initial surplus be $u$, which is a non-negative integer. Assume that the premium received in each time period is $c$, a positive integer. For $n = 1, 2, \cdots$, the surplus at time $n$ is

$$U(n) = u + cn - S_n, \tag{1}$$

where $S_n = X_1\xi_1 + X_2\xi_2 + \cdots + X_n\xi_n$, and $U(0) = u$.

In this paper, we consider the compound binomial model modified by the inclusion of stochastic return on the investments by an insurer. We mainly investigate the expected value of all dividends paid out until ruin occurs for a barrier strategy. The dividend policy is that when the surplus is bigger than a positive integer $b$, the insurer pays out all of the surplus over $b$ as dividends. For more risk models in the presence of dividend payments, see [5] [2] [9] [3] [4] [1] [7] [8]. Some discrete time risk models with dividend has been seen in literature. For example, Tan and Yang [11] considered a compound binomial model with randomized decisions on paying dividends; Kim et al [6] proposed a discrete time model with dividend payments affected by an Markov environment process. It is worth mentioning that Paulsen and Gjessing [10] considered a continuous time risk model with stochastic return on investments and dividend payments. They were interested mainly in expected present value of all dividends paid until ruin occurs, and this expected value was found by solving a boundary value problem for an integro-differential equation. Xiong and Yang [12] considered the Cramér-Lundberg model with investments in a risky asset, and proved that $\psi(u) = 1$. We consider the compound binomial model with investments and dividend payments, and obtain the expected present value of all dividends paid until ruin occurs.

Our paper is organized as follows. We introduce the model in Section 2. In Section 3, we find that the expected present value of the first dividend prior to ruin satisfies a set of linear equations, and obtain its solution. Similarly, we find some results about all dividends up to the ruin time.

## 2   The Model and Preliminaries

Consider a discrete time risk process based on the compound binomial model (1). The surplus process is described as follows.

Let $Y_{in}$ $(n = 1, 2, \cdots)$ denote the return on the investment by an insurer in the time period $(n - 1, n]$ when the surplus is $i$ at time $n - 1$. And assume that for all $i$ $(i = 0, 1, 2, \cdots)$, $\{Y_{in}, n = 1, 2, \cdots\}$ is a sequence of mutually independent, identically distributed, integer-valued random variables, and independent of $\{S_n, n = 1, 2, \cdots\}$. Besides, for $i \neq j$ and $n \neq m$, $Y_{in}$ is independent of $Y_{jm}$. Let $U(n)$ denote the surplus at time $n(n = 0, 1, \cdots)$ when no dividend is considered. Then

$$U(n) = U(n-1) + c + Y_{U(n-1),n} - X_n\xi_n, \tag{2}$$

where $U(0) = u$.

We introduce a constant dividend barrier into the model (2). Assume that any surplus of the insurer above the level $b$ (a positive integer) is immediately paid out to the shareholders so that the surplus is brought back to the level $b$. When the surplus is below, nothing is done. Once the surplus is negative, the insurer is ruined and the process stops. Let $V(n)$ denote the surplus at time $n$. Then

$$V(n) = \min\{V(n-1) + c + Y_{V(n-1),n} - X_n\xi_n, \ b\}, \tag{3}$$

where $V(0) = u$. We are interested in the expected present value of the accumulated dividends up to the time of ruin in the risk model (3).

It is reasonable that the incomes from investments are brought in. The income $Y_{in}(i \geq 0, n \geq 1)$ is certainly related to the capital $i$, and when $i = 0$ the income should be 0, i.e. $Y_{0n} = 0$ with probability 1. In general, $\Pr(Y_{in} \geq -i) = 1$. We should point out the assumption that $Y_{in}$ is restricted to be integer-valued is not completely identical with reality, but as the unit of money decreases it is closer and closer to reality.

Put $X = X_1$ and let

$$f(k) = \Pr(X = k), \quad k = 1, 2, 3, \cdots \tag{4}$$

be the common probability function of the claim amounts. (The value of $f(k)$ is zero if $k$ is not a positive integer.) Let

$$F(n) = \sum_{k=-\infty}^{n} f(k); \quad \bar{F}(n) = 1 - F(n). \tag{5}$$

Let

$$g_i(k) = \Pr(Y_{in} = k), \quad k = 0, \pm 1, \pm 2, \cdots; \quad i = 0, 1, 2, \cdots, \tag{6}$$

$$G_i(n) = \sum_{k=-\infty}^{n} g_i(k); \quad \bar{G}_i(n) = 1 - G_i(n); \tag{7}$$

where $g_0(0) = \Pr(Y_{0n} = 0) = 1$.

Define

$$T = \inf\{t \geq 1 : V(t) < 0\} \quad (\inf \emptyset = \infty) \tag{8}$$

as the time of ruin and

$$\tau = \inf\{t \geq 1 : U(t) > b\} \quad (\inf \emptyset = \infty). \tag{9}$$

Assume

$$\Pr(X_n > c) > 0. \tag{10}$$

The assumption is reasonable because an insurer should face a real risk process. Because of the assumption (10) and $b < \infty$, $\Pr(T < \infty) = 1$. We assume throughout this paper that $0 \le u \le b$.

## 3    The Expected Present Value of Dividends

Let $Z_{in} = Y_{in} - X_n$, and $h_i(z)$ denote $Z_{in}$'s probability function. Then

$$h_i(z) = \Pr(Z_{in} = z) = \sum_{x=1}^{\infty} f(x) g_i(z + x), \quad z = 0, \pm 1, \pm 2, \cdots. \qquad (11)$$

Let

$$H_i(n) = \sum_{k=-\infty}^{n} h_i(k); \quad \bar{H}_i(n) = 1 - H_i(n).$$

Let $D_1(u)$ denote the expected present value of the first dividend, and $D(u)$ denote the expected present value of all dividends up to the ruin time with a discounted factor $v$ $(0 < v \le 1)$.

By conditioning on the time 1, we see that

$$D_1(u) = vq \left[ \sum_{y=-u}^{b-(u+c)} D_1(u+c+y) g_u(y) + \sum_{y=b+1-u-c}^{\infty} (u+c+y-b) g_u(y) \right]$$

$$+ vp \left[ \sum_{z=-(u+c)}^{b-(u+c)} D_1(u+c+z) h_u(z) + \sum_{z=b+1-u-c}^{\infty} (u+c+z-b) h_u(z) \right],$$

$$\qquad (12)$$

and

$$D(u) = vq \left[ \sum_{y=-u}^{b-(u+c)} D(u+c+y) g_u(y) + \sum_{y=b+1-u-c}^{\infty} (u+c+y-b) g_u(y) \right.$$

$$\left. + D(b) \bar{G}_u(b-u-c) \right] + vp \left[ \sum_{z=-(u+c)}^{b-(u+c)} D(u+c+z) h_u(z) \right.$$

$$\left. + \sum_{z=b+1-u-c}^{\infty} (u+c+z-b) h_u(z) + D(b) \bar{H}_u(b-u-c) \right]. \qquad (13)$$

Note that $g_u(y) = 0$ for $y < -u$. We change equivalently Eq. (12) into

$$D_1(u) = \sum_{z=0}^{b} v D_1(z) [q g_u(z-u-c) + p h_u(z-u-c)]$$

$$+ \sum_{y=b+1-(u+c)}^{\infty} v(u+c+y-b)[qg_u(y)+ph_u(y)]. \tag{14}$$

Eq. (14) can be rewritten as

$$(\mathbf{I} - vq\mathbf{Q_1} - vp\mathbf{P_1}) \begin{pmatrix} D_1(0) \\ D_1(1) \\ \vdots \\ D_1(b) \end{pmatrix} = \mathbf{v}, \tag{15}$$

where $\mathbf{I}$ is a $(b+1) \times (b+1)$ unit matrix; $\mathbf{Q_1}$ is

$$\begin{pmatrix} g_0(-c) & g_0(1-c) & \cdots & g_0(b-c) \\ g_1(-1-c) & g_1(-c) & \cdots & g_1(b-1-c) \\ g_2(-2-c) & g_2(-1-c) & \cdots & g_2(b-2-c) \\ \cdots & \cdots & \cdots\cdots \\ g_b(-b-c) & g_b(1-b-c) & \cdots & g_b(-c) \end{pmatrix};$$

$\mathbf{P_1}$ is equal to

$$\begin{pmatrix} f(c) & f(c-1) & \cdots & f(c-b) \\ h_1(-1-c) & h_1(-c) & \cdots & h_1(b-1-c) \\ h_2(-2-c) & h_2(-1-c) & \cdots & h_2(b-2-c) \\ \cdots & \cdots & \cdots\cdots \\ h_b(-b-c) & h_b(1-b-c) & \cdots & h_b(-c) \end{pmatrix};$$

and $\mathbf{v}$ is the column vector

$$\begin{pmatrix} v\sum_{y=b+1-c}^{\infty}(c+y-b)[qg_0(y)+ph_0(y)] \\ v\sum_{y=b-c}^{\infty}(1+c+y-b)[qg_1(y)+ph_1(y)] \\ \vdots \\ v\sum_{y=1-c}^{\infty}(c+y)[qg_b(y)+ph_b(y)] \end{pmatrix}.$$

**Theorem 1.** *Under the assumption that* $\Pr(Y_{in} \leq 0) > 0$ *($i = 0, 1, \cdots, b$) or $v < 1$, the set of linear equations (15) has a solution and the solution is unique, i.e.,*

$$(D_1(0), D_1(1), \cdots, D_1(b))^T = (\mathbf{I} - vq\mathbf{Q_1} - vp\mathbf{P_1})^{-1}\mathbf{v}. \tag{16}$$

*Proof.* First, we consider the case $\Pr(Y_{in} \leq 0) > 0$. Let

$$\mathbf{R} = (r_{kj}) = \mathbf{I} - vq\mathbf{Q_1} - vp\mathbf{P_1}.$$

Owing to the assumption (10). we have

$$\exists x_0 > c, \quad s.t. \quad f(x_0) > 0,$$

which leads to that

$$f(c) + f(c-1) + \cdots + f(c-b) < 1.$$

Therefore,

$$vq \sum_{k=0}^{b} g_0(k-c) + vp \sum_{k=0}^{b} f(c-k)] < 1,$$

which leads to that

$$|r_{11}| > \sum_{j \neq 1} |r_{1j}|.$$

Hence, there exists $\lambda_1 \in (0,1)$ such that $\lambda_1 |r_{11}| > \sum_{j \neq 1} |r_{1j}|$. Let $\mathbf{S_1} = diag(\lambda_1, 1, \cdots, 1)_{b+1}$ and $\mathbf{R_1} = (r_{kj}^1) = \mathbf{R S_1}$. Because of (10), we have for every $i$ $(0 \leq i \leq b)$

$$\exists x < -c \ s.t. \ h_i(x) > 0. \tag{17}$$

It is easily seen that

$$|r_{kk}^1| > \sum_{j \neq k} |r_{kj}^1| \quad (k = 1, 2)$$

because of (17).

Owing to $|r_{22}^1| > \sum_{j \neq 2} |r_{2j}^1|$, there exists $\lambda_2 \in (0,1)$ such that $\lambda_2 |r_{22}^1| > \sum_{j \neq 2} |r_{2j}^1|$. Let $\mathbf{S_2} = diag(1, \lambda_2, 1, \cdots, 1)_{b+1}$ and $\mathbf{R_2} = (r_{kj}^2) = \mathbf{R_1 S_2}$. We have that

$$|r_{kk}^2| > \sum_{j \neq k} |r_{kj}^2| \quad (k = 1, 2, 3).$$

We continue with the above program, and get $\lambda_3, \lambda_4, \cdots, \lambda_b$ in turn and $\mathbf{S_k} = diag(1, \cdots, 1, \lambda_k, 1, \cdots, 1)_{b+1}$ $(k = 3, 4, \cdots)$ with the $k$-th element being $\lambda_k$. Finally, we obtain

$$\mathbf{R_b} = (r_{kj}^b) = \mathbf{R S}, \tag{18}$$

where $\mathbf{S} = \mathbf{S_1 S_2} \cdots \mathbf{S_b} = diag(\lambda_1, \lambda_2, \cdots, \lambda_b, 1)$. Note that

$$|r_{kk}^b| > \sum_{j \neq k} |r_{kj}^b| \quad (k = 1, 2, \cdots, b+1).$$

Thus, $\mathbf{R_b}$ is a (row) strictly diagonally dominant matrix, which is nonsingular. Hence, the coefficient matrix $\mathbf{R} = \mathbf{I} - vq\mathbf{Q_1} - vp\mathbf{P_1}$ (called as H-matrix) in the linear equations (15) is also a nonsingular matrix, which leads to the result.

For the case $v < 1$, $\mathbf{R}$ is a strictly diagonally dominant matrix. Owing to such a fact, the result also holds.                                                                              □

**Remark.** *The condition* $\Pr(Y_{in} \leq 0) > 0$ *implies that the insurer selects risky investments.*

Similarly, from Eq. (13), we obtain

$$(\mathbf{I} - vq\mathbf{Q} - vp\mathbf{P})\,(D(0), D(1), \cdots, D(b))^T = \mathbf{v}, \qquad (19)$$

where $\mathbf{Q}$ is

$$\begin{pmatrix} g_0(-c) & g_0(1-c) & \cdots & g_0(b-1-c) & \bar{G}_0(b-1-c) \\ g_1(-1-c) & g_1(-c) & \cdots & g_1(b-2-c) & \bar{G}_1(b-2-c) \\ g_2(-2-c) & g_2(-1-c) & \cdots & g_2(b-3-c) & \bar{G}_2(b-3-c) \\ \cdots & \cdots & \cdots\cdots & \cdots & \\ g_b(-b-c) & g_b(1-b-c) & \cdots & g_b(-1-c) & \bar{G}_b(-1-c) \end{pmatrix};$$

$\mathbf{P}$ is

$$\begin{pmatrix} f(c) & f(c-1) & \cdots & f(c-b-1) & F(c-b) \\ h_1(-1-c) & h_1(-c) & \cdots & h_1(b-2-c) & \bar{H}_1(b-2-c) \\ h_2(-2-c) & h_2(-1-c) & \cdots & h_2(b-3-c) & \bar{H}_2(b-3-c) \\ \cdots\cdots & & \cdots\cdots & \cdots & \\ h_b(-b-c) & h_b(1-b-c) & \cdots & h_b(-1-c) & \bar{H}_b(-1-c) \end{pmatrix}.$$

Thus, we have the following theorem.

**Theorem 2.** *Under the assumption that* $\Pr(Y_{in} \leq 0) > 0$ *($i = 0, 1, \cdots, b$) or $v < 1$, the set of linear equations (19) has a solution and the solution is unique, i.e.,*

$$(D(0), D(1), \cdots, D(b))^T = (\mathbf{I} - vq\mathbf{Q} - vp\mathbf{P})^{-1}\mathbf{v}. \qquad (20)$$

*Proof.* Similar to Theorem 1. □

**Acknowledgements.** This work was supported by the Natural Sciences Foundation of China (grant no. 10871064), the Open Fund Project of Key Laboratory in Hunan Universities (09K026) and the Research Project of Department of Science and Technology of Hunan Province (2009FJ3141).

# References

1. Albrecher, H., Clarmol, M.: On the distribution of dividend payments in Sparre Andersen model with generalized Erlang(n) interclaim times. Insurance: Mathematics and Economics 37, 324–334 (2005)
2. Albrecher, H., Kainhofer, R.: Risk theory with a nonlinear dividend barrier. Computing 68, 289–311 (2002)
3. Hojgaard, B., Taksar, M.: Optimal dynamic portfolio selection for a corporation with controllable risk and dividend distribution policy. Quantitative Finance 4, 315–327 (2004)

4. Frostig, E.: The expected time to ruin in a risk process with constant barrier via martingales. Insurance: Mathematics and Economics 37, 216–228 (2005)
5. Gerber, H.U., Cheng, S., Yan, Y.: An Introduction to Mathematical Risk Theory. WPC, Beijing (1997) (in Chinese)
6. Kim, B., Kim, H.S., Kim, J.: A risk model with paying dividends and random environment. Insurance: Mathematics and Economics 42, 717–726 (2008)
7. Li, S., Dickson, D.C.M.: The maximum surplus before ruin in an Erlang(n) risk process and related problems. Insurance: Mathematics and Economics 38, 529–539 (2006)
8. Lin, X.S., Pavlova, K.P.: The compound Poisson risk model with a threshold dividend strategy. Insurance: Mathematics and Economics 38, 57–80 (2006)
9. Lin, X.S., Willmot, G.E., Drekic, S.: The classical risk model with a constant dividend barrier: Analysis of the Gerber-Shiu discounted penalty function. Insurance: Mathematics and Economics 33, 551–566 (2003)
10. Paulsen, J., Gjessing, H.K.: Optimal choice of dividend barriers for a risk process with stochastic return on investments. Insurance: Mathematics and Economics 20, 215–223 (1997)
11. Tan, J., Yang, X.: The compound binomial model with randomized decisions on paying dividends. Insurance: Mathematics and Economics 39, 1–18 (2006)
12. Xiong, S., Yang, W.S.: Ruin probability in the Cramér-Lundberg model with risky investments. Stochastic Processes and their Applications 121, 1125–1137 (2011)

# Pricing Formulas of Compound Options under the Fractional Brownian Motion

Chao Zhang, Jizhou Zhang, and Dongya Tao

**Abstract.** In this paper, the pricing formulas of the compound options under the fractional Brownian motion are given by the method of partial differential equation.

**Keywords:** Fractional Brownian motion, Compound option, Black-Scholes formula.

## 1 Introduction

It is well known that the Brownian motion was first introduced to finance by Bachelier [8]. Black and Scholes [3] and, independently, Merton [10] proposed the model of the prices of stock options on the basis of the geometric Brownian motion. Nevertheless, Black and Scholes models are far from perfection. Two apparent problems exist in the Black-Scholes formulation, namely financial processes are not Gaussian and Markovian in distribution. Thus the fractional Brownian motion (for short FBM) is a generalization of the more well-known process of Brownian motion. The fractional Brownian motion was originally introduced by Kolmogorov [7]. But Kolmogorov did not use the name "fractional Brownian motion" and he called the process "Wiener spiral". The name "fractional Brownian motion" came from the influential paper by Mandelbrot and Van Ness [9].

Chao Zhang and Jizhou Zhang
Department of Mathematics, Shanghai Normal University, Shanghai, 200234
e-mail: zcxz1977@163.com, zhangjz@shnu.edu.cn

Dongya Tao
School of Mathematical Sciences, Xuzhou Normal University
e-mail: dytao770929@163.com

S. Li (Eds.): Nonlinear Maths for Uncertainty and its Appli., AISC 100, pp. 247–254.
springerlink.com　　　　　　　　　　　　　　© Springer-Verlag Berlin Heidelberg 2011

The FBM is a continuous zero mean Gaussian process with stationary increments. It has long-range dependency property which makes the FBM a plausible model in mathematical finance. Cheridito [2], Hu and Oksendal [5] and Guasoni [4] have proved that there is no arbitrage in the market under FBM. Necula [11] deduced the option pricing formula under FBM by using the technology of fractal geometry. In addition, Hu and Oksendal [5] have also obtained the pricing formula by the method of measure transformation.

A compound option is an option on an option. There are four main types of compound option, namely, a call on a call, a call on a put, a put on a call, a put on a put. In this paper, we will derive the pricing formula of a call on a call under FBM by the method of partial differential equation(for short PDE). We can deal with other compound options in the same manner.

This paper is organized as followed. The next section is for basic assumptions. In section 3, we study the formulas of compound options. The final section is for concluding remarks.

## 2  Basic Assumptions

We first give the following assumptions.

(1) The governing stochastic differential equation for the price of underlying asset $S_t$ is given by

$$dS_t = \mu S_t dt + \sigma S_t dW_t^H,$$

where $\mu$ is the instantaneous expected return rate of $S_t$, $\sigma$ is its volatility, $W_t^H$ is FBM, $E(dW_t^H) = 0$ and $Var(dW_t^H) = t^{2H}$. In particular, if $H = \dfrac{1}{2}$, then $W_t^H$ is a standard Brownian motion.

(2) Riskless interest rate $r$ is constant.

(3) Dividend rate is $q$.

(4) No transaction cost and tax.

(5) The market is complete.

## 3  Main Results

Let $V_{c,c}^{co}(S, t)$ denotes the price of the compound option at time $t \in [0, T_1]$ under FBM and $\hat{K}$ is its strike price. We set the following portfolio $\Pi$ and its value at time $t$ is given by

$$\Pi = V_{c,c}^{co} - \Delta C,$$

where $C = C(S,t)$ denotes a call value at time $t \in [T_1, T_2](T_2 > T_1)$ under FBM whose strike price is $K$. By choosing suitable $\Delta$, we can ensure that $\Pi$ is riskless such that

$$d\Pi = r\Pi dt = r(V_{c,c}^{co} - \Delta C)dt.$$

According to the Itô formula of FBM, we can obtain that

$$
\begin{aligned}
d\Pi &= dV_{c,c}^{co} - \Delta dC \\
&= (\frac{\partial V_{c,c}^{co}}{\partial t} + Ht^{2H-1}\sigma^2 S^2 \frac{\partial^2 V_{c,c}^{co}}{\partial S^2} + \mu S \frac{\partial V_{c,c}^{co}}{\partial S})dt + \sigma S \frac{\partial V_{c,c}^{co}}{\partial S} dS \\
&\quad - \Delta((\frac{\partial C}{\partial t} + Ht^{2H-1}\sigma^2 S^2 \frac{\partial^2 C}{\partial S^2} + \mu S \frac{\partial C}{\partial S})dt + \sigma S \frac{\partial C}{\partial S} dS).
\end{aligned}
\tag{1}
$$

Let $\Delta = \dfrac{\partial V_{c,c}^{co}}{\partial S} / \dfrac{\partial C}{\partial S}$. In addition, $C$ satisfies (see [9])

$$\frac{\partial C}{\partial t} + Ht^{2H-1}\sigma^2 S^2 \frac{\partial^2 C}{\partial S^2} + (r-q)S\frac{\partial C}{\partial S} - rC = 0. \tag{2}$$

Thus, we have from (1) and (2) that

$$
\begin{cases}
\dfrac{\partial V_{c,c}^{co}}{\partial t} + Ht^{2H-1}\sigma^2 S^2 \dfrac{\partial^2 V_{c,c}^{co}}{\partial S^2} + (r-q)S\dfrac{\partial V_{c,c}^{co}}{\partial S} - rV_{c,c}^{co} = 0, \quad 0 < t < T_1, \\
V^{co}|_{t=T_1} = (C(S,T_1) - K)^+,
\end{cases}
\tag{3}
$$

where (see [9])

$$C(S,T_1) = Se^{-q(T_2-T_1)}N(\hat{d}_1) - Ke^{-r(T_2-T_1)}N(\hat{d}_2),$$

$$\hat{d}_1 = \frac{\ln \frac{S}{K} + (r-q)(T_2-T_1) + \frac{\sigma^2}{2}(T_2^{2H} - T_1^{2H})}{\sigma\sqrt{T_2^{2H} - T_1^{2H}}}, \quad \hat{d}_2 = \hat{d}_1 - \sigma\sqrt{T_2^{2H} - T_1^{2H}}.$$

In the following, our main result on the problem (3) is given.

**Theorem 3.1.** The value of the compound option (a call on a call) at time $t$ is

$$
\begin{aligned}
V_{c,c}^{co}(S,t) &= e^{-q(T_2-T_1)} \exp\{-\frac{B_2}{2\sigma^2(T_2^{2H} - T_1^{2H})\tau}\}M(a_2, b_2; \rho) \\
&\quad - Ke^{-r(T_2-T_1)}\exp\{-\frac{B_1}{2\sigma^2(T_2^{2H} - T_1^{2H})\tau}\}M(a_1, b_1; \rho) \\
&\quad - \hat{K}N\left(\frac{\ln S - x^*}{\sigma\sqrt{T_1^{2H} - t^{2H}}}\right),
\end{aligned}
\tag{4}
$$

where $M(a, b; \rho), a_1, a_2, b_1, b_2, \rho, B_1, B_2, \tau$ and $x^*$ are defined below.

**Proof.** Let $x = \ln S$, $\tau = \rho(t)$, $\eta(t) = x + \alpha(t)$ and $W(\eta, \tau) = V^{co}_{c,c}(x, t)e^{\beta(t)}$. Then the equation (3) becomes

$$\rho'(t)\frac{\partial W}{\partial \tau} + H\sigma^2 t^{2H-1}\frac{\partial^2 W}{\partial \eta^2} + (r - q - H\sigma^2 t^{2H-1} + \alpha'(t))\frac{\partial W}{\partial \eta} - (r + \beta'(t))W = 0.$$

$$(5)$$

Let $\rho(t) = T_1^{2H} - t^{2H}$, $\beta(t) = r(T_1 - t)$ and $\alpha(t) = (r - q)(T_1 - t) - \frac{\sigma^2}{2}(T_1^{2H} - t^{2H})$. We know from (5) that

$$\begin{cases} \dfrac{\partial W}{\partial \tau} - \dfrac{\sigma^2}{2}\dfrac{\partial^2 W}{\partial \eta^2} = 0, & -\infty < \eta < +\infty, \\ W(\eta, 0) = (C(\eta, T_1) - \hat{K})^+, \end{cases}$$

where

$$C(\eta, T_1) = e^{\eta(T_1) - q(T_2 - T_1)}N(\tilde{d}_1) - Ke^{-r(T_2 - T_1)}N(\tilde{d}_2),$$

$$\tilde{d}_1 = \frac{\eta(T_1) - \ln K + (r - q)(T_2 - T_1) + \dfrac{\sigma^2}{2}(T_2^{2H} - t^{2H})}{\sigma\sqrt{T_2^{2H} - T_1^{2H}}},$$

$$\tilde{d}_2 = \tilde{d}_1 - \sigma\sqrt{T_2^{2H} - T_1^{2H}}.$$

Denote by $x^*$ the root of the equation $e^{\eta(T_1) - q(T_2 - T_1)}N(\tilde{d}_1) - Ke^{-r(T_2 - T_1)}N(\tilde{d}_2) - \hat{K} = 0$. By using the Poisson formula, we obtain

$$W(\eta, \tau)$$
$$= \frac{1}{\sigma\sqrt{2\pi\tau}}\int_{-\infty}^{+\infty} e^{-\frac{(x-\xi)^2}{2\sigma^2\tau}}(e^{\eta(T_1) - q(T_2 - T_1)}N(\tilde{d}_1) - Ke^{-r(T_2 - T_1)}N(\tilde{d}_2) - \hat{K})^+ d\xi$$
$$= \frac{1}{\sigma\sqrt{2\pi\tau}}\int_{x^*}^{+\infty} e^{-\frac{(x-\xi)^2}{2\sigma^2\tau}}(e^{\eta(T_1) - q(T_2 - T_1)}N(\tilde{d}_1) - Ke^{-r(T_2 - T_1)}N(\tilde{d}_2) - \hat{K})^+ d\xi$$
$$= I_1 + I_2 + I_3,$$

$$(6)$$

where

$$I_1 = e^{-q(T_2 - T_1)}\int_{x^*}^{+\infty}\frac{1}{\sigma\sqrt{2\pi\tau}}e^{-\frac{(x-\xi)^2}{2\sigma^2\tau}}e^\xi N(\tilde{d}_1)d\xi,$$

$$I_2 = -Ke^{-r(T_2 - T_1)}\int_{x^*}^{+\infty}\frac{1}{\sigma\sqrt{2\pi\tau}}e^{-\frac{(x-\xi)^2}{2\sigma^2\tau}}N(\tilde{d}_2)d\xi,$$

$$I_3 = -\hat{K}\int_{x^*}^{+\infty}\frac{1}{\sigma\sqrt{2\pi\tau}}e^{-\frac{(x-\xi)^2}{2\sigma^2\tau}}d\xi.$$

In the following, we will derive $I_1$, $I_2$ and $I_3$, respectively.

First, we verify $I_3$ by

$$I_3 = -\hat{K} \int_{x^*}^{+\infty} \frac{1}{\sigma\sqrt{2\pi\tau}} e^{-\frac{(x-\xi)^2}{2\sigma^2\tau}} d\xi = -\hat{K} \int_{-\infty}^{\frac{x-x*}{\sigma\sqrt{\tau}}} \frac{1}{\sqrt{2\pi}} e^{-\frac{y^2}{2}} dy \quad (7)$$

$$= -\hat{K}N(\frac{x-x*}{\sigma\sqrt{\tau}}),$$

where $N(x)$ is the standard normal distribution function. Next,

$$I_2 = -Ke^{-r(T_2-T_1)} \int_{x^*}^{+\infty} \frac{1}{\sigma\sqrt{2\pi\tau}} e^{-\frac{(x-\xi)^2}{2\sigma^2\tau}} N(\tilde{d}_2) d\xi$$

$$= -Ke^{-r(T_2-T_1)} \int_{x^*}^{+\infty} \frac{1}{\sigma\sqrt{2\pi\tau}} e^{-\frac{(x-\xi)^2}{2\sigma^2\tau}} \left( \int_{-\infty}^{\tilde{d}_2} \frac{1}{\sqrt{2\pi}} e^{-\frac{z^2}{2}} dz \right) d\xi$$

$$= -Ke^{-r(T_2-T_1)} \int_{x^*}^{+\infty} \frac{1}{\sigma\sqrt{2\pi\tau}} e^{-\frac{(x-\xi)^2}{2\sigma^2\tau}}$$

$$\left( \int_{-\infty}^{\frac{\xi-\ln K+(r-q)(T_2-T_1)-\frac{\sigma^2}{2}(T_2^{2H}-T_1^{2H})}{\sigma\sqrt{T_2^{2H}-T_1^{2H}}}} \frac{1}{\sqrt{2\pi}} e^{-\frac{z^2}{2}} dz \right) d\xi$$

$$= -Ke^{-r(T_2-T_1)} \int_{x^*}^{+\infty} \int_{-\infty}^{\xi} \frac{1}{2\pi\sigma^2\sqrt{(T_2^{2H}-T_1^{2H})\tau}}$$

$$\exp\left( -\frac{(x-\xi)^2}{2\sigma^2\tau} - \frac{(z-\ln K+(r-q)(T_2-T_1)-\frac{\sigma^2}{2}(T_2^{2H}-T_1^{2H}))^2}{2\sigma^2(T_2^{2H}-T_1^{2H})} \right) dzd\xi.$$

We make the variable transformation as follows, $\xi = -u$, $z = v - u$. Then we have

$$I_2 = -Ke^{-r(T_2-T_1)} \frac{1}{2\pi\sigma^2\sqrt{(T_2^{2H}-T_1^{2H})\tau}} \int_{-\infty}^{-x^*} \int_{-\infty}^{0}$$

$$\exp\left( -\frac{(x+u)^2}{2\sigma^2\tau} - \frac{(v-u-\ln K+(r-q)(T_2-T_1)-\frac{\sigma^2}{2}(T_2^{2H}-T_1^{2H}))^2}{2\sigma^2(T_2^{2H}-T_1^{2H})} \right) dudv$$

$$(8)$$

Denote by $P$ the exponential part of (8), that is,

$$P = -\frac{(x+u)^2}{2\sigma^2\tau} - \frac{(v-u-\ln K+(r-q)(T_2-T_1)-\frac{\sigma^2}{2}(T_2^{2H}-T_1^{2H}))^2}{2\sigma^2(T_2^{2H}-T_1^{2H})}$$

$$= -\frac{1}{2\sigma^2(T_2^{2H}-T_1^{2H})\tau}\Big\{(T_2^{2H}-T_1^{2H}+\tau)u^2 + 2u\{x(T_2^{2H}-T_1^{2H})$$

$$+\tau((r-q)(T_2-T_1) - \frac{\sigma^2}{2}(T_2^{2H}-T_1^{2H}) - \ln K)\} + \tau v^2$$

$$+2\tau v((r-q)(T_2-T_1) - \frac{\sigma^2}{2}(T_2^{2H}-T_1^{2H}) - \ln K) - 2\tau uv$$

$$+x^2(T_2^{2H}-T_1^{2H}) + \tau((r-q)(T_2-T_1) - \frac{\sigma^2}{2}(T_2^{2H}-T_1^{2H}) - \ln K)^2\Big\}.$$

$$(9)$$

For convenience, We assume

$$P = -\frac{1}{2\sigma^2(T_2^{2H} - T_1^{2H})\tau}\left((au+c_1)^2 - 2\rho(au+c_1)(bv+d_1) + (bv+d_1)^2 + B_1\right).$$

Due to (9), we have

$$a = \sqrt{T_2^{2H} - T_1^{2H} + \tau} = \sqrt{T_2^{2H} - t^{2H}},$$

$$b = \sqrt{\tau} = \sqrt{T_1^{2H} - t^{2H}},$$

$$\rho = \frac{\sqrt{\tau}}{\sqrt{T_2^{2H} - T_1^{2H} + \tau}} = \sqrt{\frac{T_1^{2H} - t^{2H}}{T_2^{2H} - t^{2H}}},$$

$$c_1 = x\sqrt{T_2^{2H} - t^{2H}} +$$

$$\frac{2\tau\sqrt{T_2^{2H} - t^{2H}}\left((r-q)(T_2 - T_1) - \frac{\sigma^2}{2}(T_2^{2H} - T_1^{2H}) - \ln K\right)}{T_2^{2H} - T_1^{2H}},$$

$$d_1 = \tau((r-q)(T_2 - T_1) - \frac{\sigma^2}{2}(T_2^{2H} - T_1^{2H}) - \ln K)$$

$$+\sqrt{\tau}(x + \frac{2\tau\left((r-q)(T_2 - T_1) - \frac{\sigma^2}{2}(T_2^{2H} - T_1^{2H}) - \ln K\right)}{T_2^{2H} - T_1^{2H}}),$$

$$B_1 = x^2(T_2^{2H} - T_1^{2H}) +$$

$$\tau((r-q)(T_2 - T_1) - \frac{\sigma^2}{2}(T_2^{2H} - T_1^{2H}) - \ln K)^2 - c_1^2 + 2\rho c_1 d_1 - d_1^2.$$

Therefore, we have from (8) that

$$I_2 = -Ke^{-r(T_2 - T_1)}\frac{1}{2\pi\sigma^2\sqrt{(T_2^{2H} - T_1^{2H})\tau}} \times$$

$$e^{-\frac{B_1}{2\sigma^2(T_2^{2H} - T_1^{2H})\tau}}\int_{-\infty}^{-x^*}\int_{-\infty}^{0} e^P \, du\, dv. \tag{10}$$

Let

$$u^* = \frac{\sqrt{T_2^{2H} - t^{2H}}\,u + c_1}{\sigma\sqrt{(T_2^{2H} - T_1^{2H})(T_1^{2H} - t^{2H})}}, \qquad v^* = \frac{\sqrt{T_1^{2H} - t^{2H}}\,v + d_1}{\sigma\sqrt{(T_2^{2H} - T_1^{2H})(T_1^{2H} - t^{2H})}}.$$

From (10), we obtain that

$$I_2 = -Ke^{-r(T_2-T_1)}e^{-\dfrac{B_1}{2\sigma^2(T_2^{2H}-T_1^{2H})\tau}}\int_{-\infty}^{a_1}\int_{-\infty}^{b_1}f(u^*,v^*)dudv$$

$$= -Ke^{-r(T_2-T_1)}e^{-\dfrac{B_1}{2\sigma^2(T_2^{2H}-T_1^{2H})\tau}}M(a_1,b_1;\rho),$$

(11)

where $M(a,b;\rho)$ is standard bivariate normal distribution function with correlation coefficient $\rho$, $f(x,y)$ is its probability density function, and $a_1$, $b_1$ are given by

$$a_1 = \frac{-\sqrt{T_2^{2H}-T_1^{2H}+\tau}x^*+c_1}{\sigma\sqrt{(T_2^{2H}-T_1^{2H})\tau}}, \qquad b_1 = \frac{d_1}{\sigma\sqrt{(T_2^{2H}-T_1^{2H})\tau}}.$$

Finally, we can derive by the above similar way that

$$I_1 = e^{-q(T_2-T_1)}e^{-\dfrac{B_2}{2\sigma^2(T_2^{2H}-T_1^{2H})\tau}}M(a_2,b_2;\rho),$$

(12)

where

$$a_2 = \frac{-\sqrt{T_2^{2H}-T_1^{2H}+\tau}x^*+c_2}{\sigma\sqrt{(T_2^{2H}-T_1^{2H})\tau}}, \qquad b_2 = \frac{d_2}{\sigma\sqrt{(T_2^{2H}-T_1^{2H})\tau}},$$

$$c_2 = (x+2\sigma^2\tau)\sqrt{T_2^{2H}-T_1^{2H}+\tau}$$
$$+ \frac{2\tau\sqrt{T_2^{2H}-T_1^{2H}+\tau}\left((r-q)(T_2-T_1)+\frac{\sigma^2}{2}(T_2^{2H}-T_1^{2H})-\ln K\right)}{T_2^{2H}-T_1^{2H}},$$

$$d_2 = \sqrt{\tau}((r-q)(T_2-T_1)+\frac{\sigma^2}{2}(T_2^{2H}-T_1^{2H})-\ln K$$
$$+\sqrt{\tau}(x+2\sigma^2\tau+\frac{2\tau\left((r-q)(T_2-T_1)+\frac{\sigma^2}{2}(T_2^{2H}-T_1^{2H})-\ln K\right)}{T_2^{2H}-T_1^{2H}}),$$

$$B_2 = x^2(T_2^{2H}-T_1^{2H})+\tau((r-q)(T_2-T_1)+\frac{\sigma^2}{2}(T_2^{2H}-T_1^{2H})-\ln K)^2$$
$$-c_2^2+2\rho c_2 d_2-d_2^2.$$

Therefore, substituting (7), (11), (12) into (6) and backing to the original variables, we obtain the conclusion (4).

By a similar argument, we may obtain the results on the other compound options (a call on a put). Here, we omit the proof process and the results on the a put on a call and a put on a put.

**Theorem 3.2.** The value formulaes for other European compound options (a call on a pu, a put on a call and a put on a put, respectively) at time $t$ are

$$V_{c,p}^{co}(S,t) = e^{-q(T_2-T_1)}\exp\{-\frac{B_2}{2\sigma^2(T_2^{2H}-T_1^{2H})\tau}\}M(a_2,-b_2;-\rho)$$
$$-Ke^{-r(T_2-T_1)}\exp\{-\frac{B_1}{2\sigma^2(T_2^{2H}-T_1^{2H})\tau}\}M(a_1,-b_1;-\rho)$$
$$+\hat{K}N\left(\frac{\ln S-x^*}{\sigma\sqrt{T_1^{2H}-t^{2H}}}\right),$$

**Remark.** In the Theorems above, if we take $H = \dfrac{1}{2}$, then we obtain the results in [10] for the compound options.

## 4 Concluding Remarks

In this paper, we used the method of PDE to obtain the formulas of four types of compound options under FBM. They could be regarded by the generalization of the existing results under the standard fractional Brownian motion. The approach is also suitable for other types of options.

**Acknowledgments.** This project is supported by the National Basic Research Program of China(2007CB814903), originality and perspectiveness advanced research of Shanghai Normal University, Shanghai Leading Academic Discipline Project (No.S30405) and Special Funds for Major Specialties of Shanghai Education Committee.

## References

1. Bender, C., Sottinen, T., Valkeila, E.: Arbitrage with fractional Brownian Motion. Theory of Stochastics Process 12, 3–4 (2006)
2. Cheridito, P.: Arbitrage in fractional Brownian Motion models. Finance and Stochastics 7, 533–553 (2003)
3. Fischer, B., Myron, S.: The pricing of option and corporate liabilities. J. Political Economy 81, 637–654 (1973)
4. Guasoni, P.: No arbitrage under transaction costs with fractional Brownian motion and beyond. Mathematical Finance 16, 569–582 (2006)
5. Hu, Y.Z., Oksendal, B.: Fractional white noise calculus and application to finance. infinite dimensional analysis. Quantum Probability and Related Topics 6, 1–32 (2003)
6. Jiang, L.S.: Mathematical Modeling and Methods of Option Pricing. World Scientific Publishing Company, Singapore (2005)
7. Kolmogorov, A.: Wiensersche Spiralen und einige andere interessante Kurven in Hilbertschen Raum. Comptes Rendus (Doklady) de l'Academie des Sciences de l'URSS 26, 115-118 (1940)
8. Louis, B.: Theorie de la speculation. Annales Scientifiques de l'Ecole Normale Superieure 17, 21–86 (1900)
9. Mandelbrot, B.B., Van Ness, J.W.: Fractional Brownian motions.Fractional Noises and Applications. Siam Review 10, 422–437 (1968)
10. Merton, R.C.: Theory of rational option pricing, Bell. J. Econ. & Manag. Sci. 4, 141–183 (1973)
11. Necula, C.: Option pricing in a fractional Brownian Motion environment. Draft. Academy of Economic Studies 12, 1–18 (2002)
12. Nualart, D.: Stochastic Integration with Respect to Fractional Brownian Motion and Applications (2004) (preprint)
13. Nualart, D., Rascanu, A.: Differential equations driven by fractional Brownian Motion. Collectanea Mathematica 53(1), 55–81 (2002)

# Influence of Risk Incentive by Limited Dividend Provision

Hiroshi Inoue, Masatoshi Miyake, and Li Guan

**Abstract.** Moral hazard problem, which has been broadly studied in economics, financial engineering and other areas, is understood as one of inefficiency to distribute the resources. It is interpreted that after some contract was concluded, one person who has more information than the other may change his behaviour and attitude toward his investment planning, causing some trouble to the other person. In this paper, we study a risk incentive problem between a creditor and shareholders, whose right to make a claim is different from each other, for corporate profits. In particular, we refer to and discuss limited provision on dividend, which may play a role to be able to solve the incentive problem. Some numerical examples are examined to illustrate the problem mentioned.

**Keywords:** Barrier option, Limited dividend provision, Shareholders and creditor, Option pricing, Risk-shifting incentive.

## 1 Introduction

The moral hazard can be applied to financial contract since there exists information asymmetry between a principal as a lender and an agent as a borrower. The agent having an information predominance over the principal,

Hiroshi Inoue and Masatoshi Miyake
School of Management, Tokyo University of Science, Kuki-shi, Saitama 346-8512,
Japan
e-mail: inoue@ms.kuki.tus.ac.jp, toshimiya1@yahoo.co.jp

Li Guan
Department of Applied Mathematics, Beiging University of Technology,
100 Pingleyuan Chaoyang District, Beiging 100124, China
e-mail: guanli@bjut.edu.cn

S. Li (Eds.): Nonlinear Maths for Uncertainty and its Appli., AISC 100, pp. 255–262.
springerlink.com                           © Springer-Verlag Berlin Heidelberg 2011

after the contract was concluded, may have incentive to pursue his own profit by taking some behavior which is not perceived by the principal, concealing the information and telling lies. Jensen and Meckling [4] study incentive problems that, when an agent's behavior cannot be perceived by a principal, the agent breaks their contract and invests in a risky project with higher return after the agent borrowed money from the principal. As a mean to mitigate risk incentive problem occurred in firms a creditor may set covenant or collateral and give a limitation to the behavior of firms or utilize a convertible bond and a warrant bond. Also, limited dividend provision to protect creditor's right taken up in this study is one of possible methods. For the study of risk incentive problem and limited dividend provision Isagawa and Yamashita [3] is referred, in which the limited dividend provision attached on a bond can give some possibility to remove agency cost incurred by the debt of the shareholders' risk incentive problem.

In this paper, we consider the risk-shifting problem between shareholders and a creditor in a framework of option pricing theory, in which the shareholders start investing activities after he raised investing funds from the creditor. In particular, we refer to and discuss limited provision on dividend, which may play a role to be able to solve the incentive problem. Our analysis is based on option pricing theory, so that the methodology is different from the model developed by Isagawa and Yamashita. We note that for the option pricing formula of Black-Scholes model some barriers need to be attached to the formula since in the standard B-S equity valuation model shareholders always select infinite-volatility projects.

## 2  Structure of Contract

We focus on all-equity firm which consists of shareholders' equity.

- Presently, shareholders plan to commit their funds to a venture, but don't have surplus fund. Therefore, they raise investing funds $I$ by issuing discount bonds, whose maturity is the end of the term, in exchange for a promise that the shareholders repay redemption (face value) $D$ $(=I)$ of the discount bonds to the creditor.
- Next, the shareholders start investing activities after they raised investing funds from the creditor. The shareholders can choose and invest all funds in any one of a series of projects $n$. In this case, the value of the project is equal to the investing fund $(S_0 = I)$ and the expected profit rate is constant with non-risk rate $r$. Also, assume the projects take different risk $\sigma$. Further, the value of asset $S_i$ follows the next geometric Brownian motion.

$$dS_i = rS_idt + \sigma_iS_idW \quad (i = 1...n) \tag{1}$$

- The project brings profit twice, at the mid and the end of a period. The profit $x(> 0)$ through the period is the same for all projects. Denote the asset value at maturity time by $S(\sigma)$. On the other hand, the profit at the end of the period is fluctuated since the asset follows equation (1) so that the profits are different among projects. At maturity time, the profit of the shareholders become $x(\sigma, D)$ and that of the creditor is expressed as the difference that the asset value minus shareholders' profit, $S(\sigma) - x(\sigma, D)$. The shareholders has limited liability.

- Also, when the value of asset drops at default boundary $L$ the default of obligation occurs and by exercising a security right the asset $L$ is collected. Then, the shareholders transfer the asset to the creditor and the return of the shareholders becomes 0.

At this time, applying option pricing theory to the pay-off structure, as the volatility in Black-Scholes gets high the value of call option which the shareholders possess may increase. On the contrary, the creditor has loss when the project failed while the creditor can not receive the gain form upside when the project with high risk is successful. This is because the creditor who has short position of call option make their value high by controlling the volatility.

Therefore, in this paper, some consideration for the asset dynamics of project is taken. When the asset value reaches the default boundary before maturity time, the creditor becomes aware of the situation and the project is immediately suspended. After that, the creditor collects the asset by performing a security right and the shareholders transfer the asset to the creditor, so that the shareholders' gain becomes 0.

## 3 Gain of Shareholders and Creditor and Optimal Risk Without Limited Dividend Provision

In a case that there is no limited dividend provision, when left profit during the period to shareholders' discretion each shareholder pays himself a dividend. Hence, letting redemption fund be $D$ and the profit at maturity $x(\sigma, D)$, the gain of the shareholders at $T$-th term $V_{shareholder}(\sigma, D, x^+)$ is obtained as below.

$$V_{shareholder}(\sigma, D, x^+) = x + x(\sigma, D) \tag{2}$$

Assuming that profit at the end of the period follows geometric Brownian motion of (1) we have

$$x(\sigma, D) = \begin{cases} \max[S_T - D, 0] & \min_{0 \leq \tau \leq T}(S_\tau) > L \\ 0 & otherwise \end{cases}$$

$$= e^{rT} \left\{ S_0 N(d_1) - De^{-rT} N(d_1 - \sigma\sqrt{T}) - S_0 \left(\frac{S_0}{L}\right)^{-\frac{2r}{\sigma^2}-1} N(d_2) \right.$$

$$\left. + \left(\frac{S_0}{L}\right)^{-\frac{2r}{\sigma^2}+1} De^{-rT} N(d_2 - \sigma\sqrt{T}) \right\}, \tag{3}$$

where $d_1 = \frac{\ln(S_0/D)+(r+\sigma^2/2)T}{\sigma\sqrt{T}}, d_2 = \frac{\ln(L^2/S_0D)+(r+\sigma^2/2)T}{\sigma\sqrt{T}}$.

The rate of change of shareholders' gain for risk $\partial V_{shareholder}(\sigma, D, x^+)/\partial\sigma$ can be found, and the optimal risk which maximizes the shareholders' gain $\sigma^*_{shareholder}$ is found as follows.

$$\left.\frac{\partial V_{shareholder}(\sigma, D, x^+)}{\partial\sigma}\right|_{\sigma=\sigma^*_{shareholder}} = 0 \tag{4}$$

The optimal value of $\sigma^*_{shareholder}$ can't be analytically obtained and need to use numerical computation which is showed later.

The gain of the creditor at $T$-th term $V_{creditor}(\sigma, D, x^+)$ is expressed as a difference that the whole asset $x + S(\sigma)$ minus the gain of the shareholders of (2) $V_{shareholder}(\sigma, D, x^+)$, hence the gain of the creditor is obtained as below,

$$V_{creditor}(\sigma, D, x^+) = x + S(\sigma) - V_{shareholder}(\sigma, D, x^+) \tag{5}$$

By introducing knock-out condition, the value of the asset at the end of the period $S(\sigma)$ becomes, denoting its barrier by $L$,

$$S(\sigma) = \begin{cases} \min[S_T, D] & \min_{0 \leq \tau \leq T}(S_\tau) > L \\ L & otherwise \end{cases}$$

$$= S_0 e^{rT} N(d_3) - \left(\frac{L}{S_0}\right)^{\frac{2r}{\sigma^2}-1} \frac{L^2}{S_0} e^{rT} N(d_4) + L$$

$$- \left(LN(d_3 - \sigma\sqrt{T}) - \left(\frac{L}{S_0}\right)^{\frac{2r}{\sigma^2}-1} LN(d_4 - \sigma\sqrt{T})\right), \tag{6}$$

where $d_3 = \frac{\ln(S_0/L)+(r+\sigma^2/2)T}{\sigma\sqrt{T}}, d_4 = \frac{\ln(L/S_0)+(r+\sigma^2/2)T}{\sigma\sqrt{T}}$.

The optimal risk which maximizes the gain of the creditor $\sigma^*_{creditor}$ is

$$V_{creditor}(\sigma, D, x^+)\big|_{\max \sigma \geq \sigma^*_{creditor}} = I. \tag{7}$$

It means to find the largest value of $\sigma$ for which the creditor's gain is equivalent to $I$.

Assume two different projects, one is project $R$ which has the optimal risk $\sigma^*_{shareholder}$ of shareholders and the other one is project $S$ which has the optimal risk $\sigma^*_{creditor}$ of creditor. Here, project $S$ is referred as a safety project while $R$ is referred a risky project. In other word, $\sigma_R > \sigma_S$. On this supposition, if the shareholders implement project $S$ the debt for the creditor is cleared off. Hence, it is the best way for the creditor to implement project $S$ with $\sigma_S = \sigma^*_{creditor}$

On the other hand, to the shareholders the relation between the value of the shareholders $V_{shareholder}(\sigma_S, D, x^+)$ and the value of the shareholders $V_{shareholder}(\sigma_R, D, x^+)$, when implementing project $S$ or when implementing project $R$, is described, $V_{shareholder}(\sigma_R, D, x^+) \geq V_{shareholder}(\sigma_S, D, x^+)$. Also, the shareholders have risk incentive to do project $R$ with high risk against the creditor's will, in order to maximize his own profit, as a result the profit of the creditor is hampered, causing risk incentive. But, if the creditor is sensible he may predict in advance risk incentive that the shareholders do project $R$ and claims redemption fund $D^*(\geq D)$ which satisfies the following relation.

$$V_{creditor}(\sigma_R, D, x^+)\big|_{D=D^*} = I \qquad (8)$$

Replacing

$$V_{creditor}(\sigma_R, D^*, x^+) = I$$

by $x + S(\sigma_R) - V_{shareholder}(\sigma_R, D^*, x^+) = I$, the gain of the shareholders is

$$V_{shareholder}(\sigma_R, D^*, x^+) = x + S(\sigma_R) - I. \qquad (9)$$

The relation of the gains of the shareholders between when choosing project $R$, under redemption fund followed to creditor's claim, and when choosing project $S$ is expressed as follows. Since $S(\sigma)$ is a decreasing function with respect to $\sigma$,

$$V_{shareholder}(\sigma_S, D, x^+) - V_{shareholder}(\sigma_R, D^*, x^+) = S(\sigma_S) - S(\sigma_R) > 0 \quad (10)$$

holds. The difference in (10) is agency cost which the shareholders bear. Thus, as the difference of the optimal risk between the shareholders and the creditor gets greater the agency cost which the shareholders bear increases.

Let the investing cost for both projects be $I=100$ and the redemption fund $D=100$. Let the profit during the period be $x=20$. Also, let the asset value of the project at the initial term be $S_0=100$ and the expected profit rate $r=0.2$. Further, assume project period is $T=2$, the default boundary is $L=70$. Then, each risk for project $S$ and $R$ is obtained from (4) and (7), $\sigma_S=0.036887$, $\sigma_R=0.474947$, respectively. As showed in table 1, if choosing project $S$ the creditor can collect the promised redemption money in full while if choosing project with high risk $R$ against the creditor's will, the

gain of the shareholders goes up from 69.1825 to 73.8781 but the gain of the
creditor goes down from 100 to 84.1224, resulting in situation of default. If the
creditor is sensible he may in advance predict shareholders' risk incentive and
claim $D^*$=130.125 for investing fund $I = 100$. In other words, the creditor set
required profit rate $100 \times \exp(2 \times r_d) = 130.125 \rightarrow r_d$=0.131663. In this case,
the creditor's gain becomes 100 while the shareholders' gain will be 58.0006
by (9). Hence, when project $S$ was chosen the difference from both gains of
the shareholders is 69.1825-58.0006=11.1819. Note the difference is agency
cost which the shareholders bear.

**Table 1** Gains of shareholders and creditor without limited dividend provision

|  | Project $S$ | Project $R$ (Cum div. Redm. money $D$) | Project $R$ (Non div. Redm. money $D^*$) |
|---|---|---|---|
| Gain of shareholders | 69.1825 | 73.8781 | 58.0006 |
| Gain of creditor | 100 | 84.1224 | 100 |

## 4  Gains of Shareholders and Creditor and Optimal Risk with Limited Dividend Provision

When dividend provision is limited the gains of the shareholders and creditor
is found with optimal risks. Limited dividend provision is applied and the
profit during the period is reserved till the end, then the gain of the shareholders at $T$-th term will be below. For the case of $\min_{0 \leq \tau \leq T} (S_\tau) > L$, the value
that calculated as the sum of the profit during the period and the asset value
minus redemption funds becomes the gain of the shareholders. Otherwise,
because the default of obligation occurs the total sums of collected asset $L$
and the profit during the period become the underlying asset, so that the
redemption funds become strike price.

$$V_{shareholder}(\sigma, D, x^-) = \begin{cases} \max[x + S_T - D, 0] & \min_{0 \leq \tau \leq T}(S_\tau) > L \\ \max[x + L - D, 0] & otherwise \end{cases} \tag{11}$$

On the other hand, the creditor's gain at $T$-th term $V_{creditor}(\sigma, D, x^-)$ is obtained as the difference for which the shareholders' gain $V_{shareholder}(\sigma, D, x^-)$
is subtracted from $x + S(\sigma)$. Hence, the creditor's gain is

$$V_{creditor}(\sigma, D, x^-) = x + S(\sigma) - V_{shareholder}(\sigma, D, x^-) \tag{12}$$

Then, the gains of the shareholders and the creditor is expressed as in Table 2.
Note that $x(\sigma, D - x)$ is obtained by replacing $D$ in (3) by $D - x$. For $D \leq x$
and $x < D \leq x + L$, the profit during the period is reserved till the end of the
period, then the creditor can collect the promised redemption money in full.

**Table 2** Gains for shareholders and creditor

| | $D \leq x$ | | $x < D$ |
|---|---|---|---|
| | | $D \leq x + L$ | $x + L < D$ |
| Gain of shareholders | $x + S(\sigma) - D$ | $x + S(\sigma) - D$ | $x(\sigma, D - x)$ |
| Gain of creditor | $D$ | $D$ | $x + S(\sigma) - x(\sigma, D - x)$ |
| total | $x + S(\sigma)$ | $x + S(\sigma)$ | $x + S(\sigma)$ |

To do so, limited dividend provision may be applied and hence there does not occur any risk incentive problem between the shareholders and the creditor. On the other hand, for $x + L < D$ even though the gain during the period is reserved there exists risk incentive problem between the shareholders and the creditor. The optimal risk which maximizes the shareholders' gain in $x + L < D$ $\sigma^{**}_{shareholder}$ is

$$\frac{\partial V_{shareholder}(\sigma, D, x^-)}{\partial \sigma}\bigg|_{\sigma = \sigma^{**}_{shareholder}} = 0. \tag{13}$$

The optimal risk $\sigma^{**}_{creditor}$ which maximizes the creditor's gain is

$$V_{creditor}(\sigma, D, x^-)\big|_{\max \sigma \geq \sigma^{**}_{creditor}} = I. \tag{14}$$

Similarly as before, assume two different projects $S$, $R$ which have the optimal risk of the shareholders and the creditor $\sigma^{**}_{shareholder}$ and $\sigma^{**}_{creditor}$. On this supposition, it is the best choice for the creditor to choose project $S$ with $\sigma_S = \sigma^{**}_{creditor}$. On the other hand, the relation between the shareholders' gain when choosing project $S$ and the shareholders' gain when choosing project $R$ is expressed as $V_{shareholder}(\sigma_R, D, x^-) \geq V_{shareholder}(\sigma_S, D, x^-)$. For the shareholders, to choose project $R$ is the best choice with $\sigma_R = \sigma^{**}_{shareholder}$. As a consequence, the gain of the creditor is hampered and a incentive problem appears. A sensible creditor predicts in advance risk incentive of the shareholders and claims the redemption funds $D^{**}$ which satisfy $V_{creditor}(\sigma_R, D, x^-)\big|_{D=D^{**}} = I$,

Next, let examine with numerical example. Using the same numerical values as before. Respective risk for project $S$ and $R$ are obtained from (13), (14), respectively, $\sigma_S = 0.0581002$, $\sigma_R = 0.147411$. If project $S$ was selected the creditor can collect the promised redemption funds while project $R$ was selected against the creditor's will, the gain of the shareholders increases from 69.1825 to 69.1845 but the creditor's gain decreases 100 to 99.9834, so that default of obligation appears. If the creditor is sensible he may predict in advance risk incentive of the shareholders and claim $D^{**} = 120.287$ for investing funds $I = 100$. In other words, the creditor set required profit rate $r_d = 0.0923552$. In this case, the gain of creditor becomes 100 while that of the shareholders becomes 69.168. Hence, the difference from the shareholders' gain for project $S$ is 69.1825 − 69.168 = 0.0145. When we compare the agency cost without limited divided provision with that of limited dividend provision, the shareholders' agency cost is decreased by 11.1819-0.0145=11.1674.

**Table 3** Gains of shareholders and creditor with limited dividend provision

|  | Project $S$ | Project $R$ (Cum div. Redm. money $D$) | Project $R$ (Non div. Redm. money $D^{**}$) |
|---|---|---|---|
| Gain of shareholders | 69.1825 | 69.1845 | 69.168 |
| Gain of creditor | 100 | 99.9834 | 100 |

## 5 Concluding Remarks

We studied a risk incentive problem between shareholders and creditor with/without limitation of dividend provision. The gains of the shareholders and the creditor is derived by using option pricing evaluation method with knock out condition numerical examples. It is found that without dividend constraint, there is a difference of the gains of the both sides between project $S$ and project $R$, in particular, the difference is noticeable when the redemption amount is optimal. However, when the dividend is preserved the circumstance is different from that of unlimited dividend. The gains of both sides for safe project and risky project are compared under dividend constraints, so that the creditor do not lose much gain even risky project was selected.

## References

1. Black, F., Cox, J.C.: Valuing corporate securities: Some effects of bond indenture provisions. Journal of Finance 31, 351–367 (1976)
2. Chesney, M., Gibson-Asner, R.: Reducing asset substitution with warrant and convertible debt issues. Journal of Derivatives 9(1), 39–52 (2001)
3. Isagawa, N., Yamashita, T.: Lender's risk incentives and borrower's risk incentives. Keiei Zaimu Kenkyu 23(1), 77–87 (2003) (in Japanese)
4. Jensen, M., Meckling, W.: Theory of the firm: managerial behavior, agency cost and capital structure. Journal of Financial Economics 3, 3065–3360 (1976)

# An Application of the Forward Integral to an Insider's Optimal Portfolio with the Dividend

Yong Liang, Weiyin Fei, Hongjian Liu, and Dengfeng Xia

**Abstract.** This paper discusses an insider's optimal portfolio with dividend, in the case that the performance is measured in terms of the logarithm of the terminal wealth minus a term measuring the roughness and the growth of the portfolio. The stochastic calculus of the forward integral is employed. The explicit solution in a special case is obtained.

**Keywords:** Forward integrals, Optimal portfolio, Insider trading, Dividend, Filtration.

## 1 Introduction

In a financial market, an insider possesses more information than the information generated by the financial market. In recent years, there has been an increasing interest in the insider trading (see, e.g., Biagani and Øksendal [2], Fei and Wu [4, 5]). Some of them might trade, directly or indirectly, on the asset and make profit from the privileged information.

Karatzas and Shreve [7] introduce the dividend into their model of portfolio. Fei [3] studies optimal consumption and portfolio choice with ambiguity and anticipation. In Back [1] and Kyle [8], the impact of trading strategies on prices is explained by the presence of an insider. Hu and Øksendal [6] study optimal smooth portfolio selection for an insider. In probabilistic terminology, information is generally represented by a filtration. Usually an investor can only use the filtration generated by the market to make a decision. We call

Yong Liang, Weiyin Fei, Hongjian Liu, and Dengfeng Xia
School of Mathematics and Physics, Anhui Polytechnic University,
Anhui Wuhu 241000, P.R. China
e-mail: liangyong@ahpu.edu.cn, wyfei@ahpu.edu.cn, hjliu2006@ahpu.edu.cn,
dengfengxia@ahpu.edu.cn

S. Li (Eds.): Nonlinear Maths for Uncertainty and its Appli., AISC 100, pp. 263–270.
springerlink.com                              © Springer-Verlag Berlin Heidelberg 2011

such investors *honest*. An insider has a larger filtration (more information) available to her and can use this larger filtration to maximize her portfolio.

To simplify our presentation we assume that the market consists of following two assets over the time period $[0, T]$. The first one is a bond whose price follows $dS_0(t) = r(t)S_0(t)dt, 0 \leq t \leq T$. Another asset is the stock whose price follows the following geometric Brownian motion $dS(t) = S(t)[\mu(t)dt + \sigma(t)dB(t)], 0 \leq t \leq T$, where $T > 0$ is constant and $r(t), \mu(t)$ and $\sigma(t)$ are given $\mathcal{F}_t$-adapted processes, $B(t) = B_t(\omega), 0 \leq t \leq T$, is a Brownian motion and $dB(t)$ denotes the Itô type stochastic differential. Denote $\mathcal{F}_t = \sigma(B_s, 0 \leq s \leq t)$, which is the information generated by the market. Assume for example that at the beginning $(t = 0)$ the insider knows in addition the future value of the underlying Brownian motion at time $T_0$, where $T_0 > T$. Then her information filtration is given by $\mathcal{G}_t = \sigma(B_s, 0 \leq s \leq t) \vee \sigma(B_{T_0})$, the filtration generated by the Brownian motion up to time $t$ and $B_{T_0}$. The insider may use the filtration $\mathcal{G}_t$ (rather than as usual use only the filtration $\mathcal{F}_t$) to optimize her portfolio.

In what follows, just as Karatzas and Shreve [7] introduce the dividend into their model, we shall consider the financial market with the dividend rate $\delta(t)$ which is an $\mathcal{F}_t$-adapted process. So our model extends the one of Hu and Øksendal [6] where the dividend is not considered.

Let us express the portfolio in terms of the fraction $\pi(t)$ of the total wealth invested in the stock at time $t$. Let $X^\pi(t)$ denote the corresponding wealth at time $t$. Similar to Pikovsky and Karatzas [10], we consider the problem of maximizing the expectation of the logarithmic utility of terminal wealth

$$\Phi_{\mathcal{G}} \stackrel{\triangle}{=} \sup_{\pi}\{\mathbf{E}\left[\log(X^\pi(T))\right]\}, \tag{1}$$

where the supremum is taken over all $\mathcal{G}_t$-adapted portfolios $\pi(\cdot)$. It is easy to obtain that in this case the optimal insider portfolio is

$$\pi^*(t) = \frac{\mu(t) + \delta(t) - r(t)}{\sigma^2(t)} + \frac{B(T_0) - B(t)}{\sigma(t)(T_0 - t)}. \tag{2}$$

Moreover, the corresponding maximal expected utility $\Phi_{\mathcal{G}}$ is given by

$$\Phi_{\mathcal{G}} = \mathbf{E}\left[\int_0^T \left\{r(s) + \frac{1}{2}\frac{(\mu(s) + \delta(s) - r(s))^2}{\sigma^2(s)} + \frac{1}{2(T_0 - s)}\right\}ds\right], \quad T_0 \geq T.$$

In particular, if $T_0 = T$ we get $\Phi_{\mathcal{G}} = \infty$, which is clearly an unrealistic result. If $T_0 = T$ we see by (2) that the optimal portfolio $\pi^*$ needed to achieve $\Phi_{\mathcal{G}} = \infty$ will converge towards the derivative of $B(t)$ at $t = T_0^-$. Thus $\pi^*(t)$ will consist of more and more wild fluctuations as $t \to T_0^-$. This is both practically impossible and also undesirable from the point of view of the insider: She does not want to explore a too conspicuous portfolio, compared to that of the honest trader, which in the optimal case is just

$$\pi^*_{\text{honest}}(t) = \frac{\mu(t) + \delta(t) - r(t)}{\sigma^2(t)}.$$

To model this constraint we propose to modify the problem (1) to the following:

**Problem 1.** Find $\pi^* \in \mathcal{A_G}$ and $\Phi$ such that

$$\Phi = \sup_{\pi \in \mathcal{A_G}} \mathbf{E}\left[\log(X^{(\pi)}(T)) - \tfrac{1}{2}\int_0^T |\mathbf{Q}\pi(s)|^2 ds\right]$$
$$= \mathbf{E}\left[\log(X^{(\pi^*)}(T)) - \tfrac{1}{2}\int_0^T |\mathbf{Q}\pi^*(s)|^2 ds\right],$$

where $\mathcal{A_G}$ is a suitable family of admissible $\mathcal{G}_t$-adapted portfolios $\pi$. Here $\mathbf{Q} : \mathcal{A_G} \to \mathcal{A_G}$ is some linear operator measuring the size and/or the fluctuations of the portfolio.

For example we could have $\mathbf{Q}\pi(s) = \lambda_1(s)\pi(s)$, where $\lambda_1(s) \geq 0$ is some given weight function. This models the situation where the insider is penalized for large volumes of trade.

An alternative choice of $\mathbf{Q}$ would be $\mathbf{Q}\pi(s) = \lambda_2(s)\frac{d}{ds}\pi(s)$, for some weight function $\lambda_2(s) \geq 0$. In this case the insider is penalized for large trade fluctuations. Other choices of $\mathbf{Q}$ are also possible.

Now this paper is arranged as follows. Section 2 studies the smooth portfolio of an insider investor in a financial market with the dividend. Finally, Section 3 concludes.

## 2    An Insider's Optimal Smooth Portfolio with Dividend

In this section, we discuss Problem 1 in the introduction. Assume that an investor selects one riskless bond and one risk stock whose prices dynamics follows as those in the introduction. The related concepts and properties of the forward integral can refer to Hu and Øksendal [6].

Let $\mathcal{G}_t \supset \mathcal{F}_t$ be the information filtration available to the insider and let $\pi(t)$ be the portfolio chosen by the insider, measured in terms of the fraction of the total wealth $X(t) = X^{(\pi)}(t)$ invested in the stock at time $t \in [0,T]$. Then the corresponding wealth $X(t) = X^{(\pi)}(t)$ at time $t$ is modeled by the forward stochastic differential equation

$$\begin{aligned}
dX(t) &= (1 - \pi(t))X(t)r(t)dt \\
&\quad + \pi(t)X(t)\left[(\mu(t) + \delta(t))dt + \sigma(t)d^-B(t)\right] \\
&= X(t)\{[r(t) + (\mu(t) + \delta(t) - r(t))\pi(t)]dt \\
&\quad + \sigma(t)\pi(t)d^-B(t)\}.
\end{aligned} \qquad (3)$$

For simplicity we assume $X(0) = 1$. This forward integral model for the anticipating stochastic differential equation (3) shows the forward integral as a limit of Riemann sums of the Itô type, i.e. where the $i$-th term has the form $\phi(t_i)(B(t_{i+1}) - B(t_i))$ with $\phi$ evaluated at the left end point $t_i$ of the interval $[t_i, t_{i+1}]$. Moreover, if $B(t)$ happens to be a semimartingale with respect to $\mathcal{G}_t$, then indeed the forward integral coincides with the semimartingale integral.

We now specify the set $\mathcal{A} = \mathcal{A}_{\mathcal{G}}$ of the admissible portfolios $\pi$ as follows.

**Definition 1.** Let $\mathcal{A} = \mathcal{A}_{\mathcal{G}}$ denote the space of all stochastic processes $\pi(t)$ such that (4)-(7) hold, where

$$\pi(t) \text{ is } \mathcal{G}_t - \text{adapted and the } \sigma - \text{algebra generated by}$$

$$\{\pi(t); \pi \in \mathcal{A}\} \text{ is equal to } \mathcal{G}_t, \text{ for all } t \in [0, T], \tag{4}$$

$$\pi \in L^{1,2} \text{ and } \pi \text{ belongs to the domain of } \mathbf{Q}, \tag{5}$$

$$\sigma(t)\pi(t) \text{ is forward integrable}, \tag{6}$$

$$\mathbf{E}\left[\int_0^T |\mathbf{Q}\pi(t)|^2 dt\right] < \infty. \tag{7}$$

With these definitions we can now specify Problem 1 as follows.

**Problem 2.** Find $\Phi$ and $\pi^* \in \mathcal{A}$ such that

$$\Phi = \sup_{\pi \in \mathcal{A}} J(\pi) = J(\pi^*),$$

where

$$J(\pi) = \mathbf{E}\left[\log(X^{(\pi)}(T)) - \frac{1}{2}\int_0^T |\mathbf{Q}\pi(t)|^2 dt\right],$$

$\mathbf{Q}: \mathcal{A} \to \mathcal{A}$ being a given linear operator ($\mathbf{E}$ denotes the expectation with respect to P). We call $\Phi$ the value of the insider and $\pi^* \in \mathcal{A}$ an optimal portfolio (if it exists).

We now proceed to solve Problem 2: We get that the solution of (3) is

$$X(t) = \exp\left(\int_0^t \{r(s) + (\mu(s) + \delta(s) - r(s))\pi(s)\right.$$
$$\left. - \tfrac{1}{2}\sigma^2(s)\pi^2(s)\} ds + \int_0^t \sigma(s)\pi(s)d^- B(s)\right).$$

Therefore we get

$$J(\pi) = \mathbf{E}\left[\int_0^T \{r(t) + (\mu(t) + \delta(t) - r(t))\pi(t) - \tfrac{1}{2}\sigma^2(t)\pi^2(t)\} dt \right.$$
$$\left. + \int_0^T \sigma(t)\pi(t)d^- B(t) - \tfrac{1}{2}\int_0^T |\mathbf{Q}\pi(t)|^2 dt\right]. \tag{8}$$

To maximize $J(\pi)$ we use a calculus of variation technique, as follows: Suppose an optimal insider portfolio $\pi = \pi^*$ exists (in the following we omit the $*$). Let $\theta \in \mathcal{A}$ be another portfolio. Then the function

$$f(y) \overset{\triangle}{=} J(\pi + y\theta), \; y \in \mathbf{R}$$

is maximal for $y = 0$ and hence

$$0 = f'(0) = \tfrac{d}{dy}[J(\pi + y\theta)]_{y=0}$$

$$= \mathbf{E}\left[\int_0^T \left\{(\mu(t) + \delta(t) - r(t))\theta(t) - \sigma^2(t)\pi(t)\theta(t)\right\} dt \right. \tag{9}$$

$$\left. + \int_0^T \sigma(t)\theta(t)d^- B(t) - \int_0^T \mathbf{Q}\pi(t)\mathbf{Q}\theta(t)dt\right].$$

Let $\mathbf{Q}^*$ denote the adjoint of $\mathbf{Q}$ in the Hilbert space $L^2([0,T] \times \Omega)$, namely,

$$\mathbf{E}\left[\int_0^T \alpha(t)(\mathbf{Q}\beta)(t)dt\right] = \mathbf{E}\left[\int_0^T (\mathbf{Q}^*\alpha)(t)\beta(t)dt\right]$$

for all $\alpha$ and $\beta$ in $\mathcal{A}$. Then we can rewrite (9) as

$$\mathbf{E}[\int_0^T \left\{\mu(t) + \delta(t) - r(t) - \sigma^2(t)\pi(t) \right. \\ \left. -\mathbf{Q}^*\mathbf{Q}\pi(t)\right\}\theta(t)dt + \int_0^T \sigma(t)\theta(t)d^- B(t)] = 0. \tag{10}$$

Let

$$M(t) \overset{\triangle}{=} \int_0^t \left\{\mu(s) + \delta(s) - r(s) - \sigma^2(s)\pi(s) \right. \\ \left. -\mathbf{E}[\mathbf{Q}^*\mathbf{Q}\pi(s)|\mathcal{G}_s]\right\} ds + \int_0^t \sigma(s)dB(s). \tag{11}$$

Now we give the following theorem.

**Theorem 1.** *Suppose an insider's optimal portfolio $\pi \in \mathcal{A}$ for Problem 2 exists. Then*

$$dB(t) = d\hat{B}(t) - \tfrac{1}{\sigma(t)}\left\{\mu(t) + \delta(t) - r(t) - \sigma^2(t)\pi(t) - \mathbf{E}[\mathbf{Q}^*\mathbf{Q}\pi(t)|\mathcal{G}_t]\right\} dt$$

*where $\hat{B}(t) \overset{\triangle}{=} \int_0^t \sigma^{-1}(s)dM(s)$ is a $\mathcal{G}_t$-Brownian motion. In particular, $B(t)$ is a semimartingale with respect to $\mathcal{G}_t$.*

Proof. Apply (10) to a special choice of $\theta$: Fix $t \in [0,T]$ and $h > 0$ such that $t + h < T$ and choose $\theta(s) = \theta_0(t)\chi_{[t,t+h]}(s); s \in [0,T]$, where $\theta_0(t)$ is $\mathcal{G}_t$-measurable. Since $\sigma(t)$ is $\mathcal{F}_t$-adapted, we have

$$\mathbf{E}\left[\int_0^T \sigma(s)\theta(s)d^- B(s)\right] = \mathbf{E}\left[\int_t^{t+h} \sigma(s)\theta_0(t)d^- B(s)\right] \\ = \mathbf{E}\left[\theta_0(t)\int_t^{t+h} \sigma(s)dB(s)\right].$$

We know that $r(t), \mu(t), \delta(t)$ and $\sigma(t)$ are given $\mathcal{F}_t$-adapted processes, so using the property of conditional expectation in conjunction with (10) we get

$$\mathbf{E}\left[\left(\int_t^{t+h} \{\mu(s) + \delta(s) - r(s) - \sigma^2(s)\pi(s) \right.\right.$$
$$\left.\left. -\mathbf{Q}^*\mathbf{Q}\pi(s)\} \, ds + \int_t^{t+h} \sigma(s)dB(s)\right) \theta(t)\right] = 0.$$

Since this holds for all such $\theta(t)$ we conclude that $\mathbf{E}[M(t+h) - M(t)|\mathcal{G}_t] = 0$. Due to $\sigma(t) \neq 0$, the proof of Theorem is complete.                    □

From Theorem 1, we now find an equation for an optimal portfolio $\pi$: Assume that there exists a function $\gamma_t(s, \omega)$ such that $\gamma_t(s)$ is $\mathcal{G}_t$-measurable for all $s \leq t$ and $t \to \int_0^t \gamma_t(s)ds$ is of finite variation a.s. and

$$N(t) \stackrel{\triangle}{=} B(t) - \int_0^t \gamma_t(s)ds \text{ is a martingale with repect to } \mathcal{G}_t. \quad (12)$$

Suppose that $\pi \in \mathcal{A}$ satisfies the following equation

$$\sigma^2(t)\pi(t) + \mathbf{E}[\mathbf{Q}^*\mathbf{Q}\pi(t)|\mathcal{G}_t]$$
$$= \mu(t) + \delta(t) - r(t) + \sigma(t)\frac{d}{dt}\left(\int_0^t \gamma_t(s)ds\right). \quad (13)$$

The following theorem can be obtained.

**Theorem 2.** *Suppose (12) holds and that $\pi \in \mathcal{A}$ is optimal for Problem 2. Then $\pi$ solves the equation (13).*

*Proof.* By comparing (11) and (12) we get that $\sigma(t)dN(t) = dM(t)$, which deduces

$$-\sigma(t)\frac{d}{dt}\left(\int_0^t \gamma_t(s)ds\right) = \mu(t) + \delta(t) - r(t) - \sigma^2(t)\pi(t) - \mathbf{E}[\mathbf{Q}^*\mathbf{Q}\pi(t)|\mathcal{G}_t].$$

Thus we can get (13). Therefore, the claim is proved.                    □

Next we turn to a partial converse of Theorem 2.

**Theorem 3.** *Suppose (12) holds. Let $\pi(t)$ be a process solving the equation (13). Assume $\pi \in \mathcal{A}$. Then $\pi$ is optimal for Problem 2.*

*Proof.* Note that $\pi_t \in \mathcal{A}$ is $\mathcal{G}_t$-adapted. Substituting

$$dB(t) = dN(t) + \frac{d}{dt}\left(\int_0^t \gamma_t(s)ds\right)dt$$

and $\sigma(t)\pi(t)d^- B(t) = \sigma(t)\pi(t)dN(t) + \sigma(t)\pi(t)\frac{d}{dt}\left(\int_0^t \gamma_t(s)ds\right) dt$ into (8), we get

$$J(\pi) = \mathbf{E}\left[\int_0^T \{r(t) + (\mu(t) + \delta(t) - r(t))\pi(t)\right.$$
$$\left. - \tfrac{1}{2}\sigma^2(t)\pi^2(t) + \sigma(t)\pi(t)\tfrac{d}{dt}\left(\int_0^t \gamma_t(s)ds\right) - \tfrac{1}{2}|\mathbf{Q}\pi(t)|^2 dt\right],$$

(14)

which is a concave functional of $\pi$, so if we can find $\pi = \pi^* \in \mathcal{A}$ such that

$$\frac{d}{dy}[J(\pi^* + y\theta)]_{y=0} = 0, \quad \text{for all } \theta \in \mathcal{A},$$

then $\pi^*$ is optimal. By a computation similar to the one leading to (10) we get

$$\frac{d}{dy}[J(\pi^* + y\theta)]_{y=0}$$
$$= \mathbf{E}\left[\int_0^T \{\mu(t) + \delta(t) - r(t) - \sigma^2(t)\pi^*(t)\right.$$
$$\left. + \sigma(t)\tfrac{d}{dt}\int_0^t \gamma_t(s)ds - \mathbf{Q}^*\mathbf{Q}\pi^*(t)\}\theta(t)dt\right],$$

which is 0 if $\pi = \pi^*$ solves equation (13). Thus we complete the proof.  □

Now we consider a special case. Let

$$\mathbf{Q}\pi(t) = \lambda_1(t)\sigma(t)\pi(t),$$

(15)

where $\lambda_1(t) \geq 0$ is deterministic and

$$\pi(t) = \pi^*(t) = \frac{\mu(t) + \delta(t) - r(t) + \sigma(t)\frac{d}{dt}\int_0^t \gamma_t(s)ds}{\sigma^2(t)[1 + \lambda_1^2(t)]}.$$

(16)

Then we have the following theorem.

**Theorem 4.** *Suppose (12) and (15) holds. Let $\pi^*(t)$ be given by (16). If $\pi^* \in \mathcal{A}$ then $\pi^*$ is optimal for Problem 2. Moreover, the insider value is*

$$\Phi = J(\pi^*)$$
$$= \mathbf{E}\left[\int_0^T \left\{r(t) + \tfrac{1}{2}(1 + \lambda_1^2(t))^{-1}\left(\tfrac{\mu(t)+\delta(t)-r(t)}{\sigma(t)} + \tfrac{d}{dt}\int_0^t \gamma_t(s)ds\right)^2\right\} dt\right].$$

*Proof.* From (13) we gets the form

$$\sigma^2(t)\pi(t) + \lambda_1^2(t)\sigma^2(t)\pi(t) = \mu(t) + \delta(t) - r(t) + \sigma(t)\frac{d}{dt}\int_0^t \gamma_t(s)ds,$$

which deduces (16). Substituting (16) into the formula (14) for $J(\pi)$ we obtain the claim.  □

## 3 Conclusions

In this paper, the optimal portfolio of an insider investor in a financial market with the dividend is studied. In the case that the performance is measured in terms of the logarithm of the terminal wealth minus a term measuring the roughness and the growth of the portfolio, we characterize the optimal smooth portfolio of the insider investor. The stochastic calculus of the forward integral is employed. The explicit solution in a special case is obtained which shows that the optimal policy depends on the dividend of a stock.

**Acknowledgements.** This paper is supported by National Natural Science Foundation of China (10826098), Anhui Natural Science Foundation (090416225), Anhui Key Natural Science Foundation of Universities (KJ2010A037) and Youth Foundation of Anhui Polytechnic University (2009YQ034).

## References

1. Back, K.: Insider trading in continuous time. Rev. Financ Stud. 5, 387–409 (1992)
2. Biagani, F., Øksendal, B.: A general stochastic calculus approach to insider trading. Appl. Math. Optim. 52, 167–181 (2005)
3. Fei, W.Y.: Optimal consumption and portfolio choice with ambiguity and anticipation. Information Sciences 177, 5178–5190 (2007)
4. Fei, W.Y., Wu, R.Q.: Anticipative potfolio optimization under constraints and a higher interest rate for borrowing. Stochastic Analysis and Applications 20, 311–345 (2002)
5. Fei, W.Y., Wu, R.Q.: Optimization of utility for 'large investor' with anticipation. Stochastic Analysis and Applications 21, 329–358 (2003)
6. Hu, Y.Z., Øksendal, B.: Optimal smooth portfolio selection for an insider. Journal of Applied Probability 44, 742–752 (2007)
7. Karatzas, I., Shreve, S.E.: Methods of Mathematical Finance. Springer, New York (1998)
8. Kyle, A.: Continuous auctions and insider trading. Econometrica 53, 1315–1335 (1985)
9. Nualart, D., Pardoux, É.: Stochastic calculus with anticipating integrands. Probab. Theory Related Fields 78, 535–581 (1998)
10. Pikovsky, I., Karatzas, I.: Anticipative portfolio optimization. Adv. Appl. Probab. 28, 1095–1122 (1996)
11. Russo, F., Vallois, P.: Forward, backward and symmetric stochastic integration. Probab. Theory Related Fields 97, 403–421 (1993)
12. Russo, F., Vallois, P.: Stochastic calculus with respect to continuous finite quadratic variation processes. Stochastics and Stochastics Reports 70, 1–40 (2000)

# The Nonlinear Terminal-Boundary Problems for Barrier Options

Yulian Fan

**Abstract.** In the framework of stochastic optimal control theory, we get the nonlinear terminal-boundary value problems satisfied, in the sense of viscosity solutions, by the worst case values of the barrier options with uncertain volatilities. We also prove that the out-in parity does not hold in uncertain volatility model.

**Keywords:** Volatility uncertainty, Barrier option, Nonlinear terminal-boundary value problem.

## 1 Introduction

[2] and [6] introduced the uncertainty in volatility (uncertain volatility model, abbreviated UVM): instead of choosing a pricing model that incorporates a complete view of the forward volatility as a single number, or a predetermined function of time and price, or even a stochastic process with given statistics, they assume the market operates under the less stringent assumption that the volatility of future prices is not known, but is assumed to lie in a fixed interval $[\underline{\sigma}, \bar{\sigma}]$. The authors obtained a generalization of the duality formula by stochastic control techniques in the case of European options with payoffs that depend only on the terminal values of the underlying asset $S_T$. The discrete-time case has been studied recently in [4].

After the seminal works by [2] and [6], [1] and [3] have considered the pricing in the UVM of a basket of options in which includes barrier and American options written on a single asset. Still in the single-asset case: [8] has also studied the pricing of American options.

Yulian Fan
School of Science, North China University of Technology,
No.5 Jinyuanzhuang Road, Shijingshan District, Beijing, 100144, P.R. China
e-mail: fanyl@ncut.edu.cn

S. Li (Eds.): Nonlinear Maths for Uncertainty and its Appli., AISC 100, pp. 271–277.
springerlink.com      © Springer-Verlag Berlin Heidelberg 2011

In the literature of the pricing of barrier options without volatility uncertainty, the probabilistic approach is widely used. But in the uncertain volatility model, there is no specific pricing measure. In this paper, we'll use the stochastic optimal control theory to get the nonlinear terminal-boundary value problems satisfied by the values of the barrier options, and prove that the out-in parity does not hold in uncertain volatility model.

The paper is organized as follows. In section 2, we briefly introduce the uncertain volatility model. In section 3, we deduce the nonlinear terminal-boundary value problems satisfied by the values of the barrier options.

## 2  The Uncertain Volatility Model

Let $(\Omega, \mathcal{F}, P)$ be a probability space and $(W_t)_{t \geq 0}$ be a Brownian motion. The filtration generated by $W_t$ is denoted by $\mathcal{F}_t := \sigma\{W_\nu, 0 \leq \nu \leq t\} \vee \mathcal{N}$, where $\mathcal{N}$ is the collection of P-null subsets.

Assume a risky asset follows a controlled diffusion process

$$dS_\nu^{t,s} = \mu_\nu S_\nu d\nu + \sigma_\nu S_\nu dW_\nu, S_t = s, \tag{1}$$

where $\mu$ and $\sigma$ are adapted functions such that $\underline{\sigma} \leq \sigma \leq \bar{\sigma}$. The constant $\underline{\sigma}$ and $\bar{\sigma}$ represent the upper and lower bounds on the volatility that should be input in the model according to the user's expectation about future price fluctuations.

If there is no arbitrage, the forward stock price dynamics under any pricing measure should satisfy the modified risk-neutral Itô equation

$$dS_\nu^{t,s} = rS_\nu d\nu + \sigma_\nu S_\nu dW_\nu, S_t = s, \tag{2}$$

where $r$ is the riskless interest rate

In the next section, we'll consider the pricing PDEs of the barrier options based on this asset.

## 3  The PDEs for the Barrier Options

### 3.1  Single Barrier Options

Let's first consider the down-out call option with maturity T and strike price K. Let $L \leq s$ be the predetermined barrier level. If $S_\nu^{t,s} \leq L$ for some $\nu \in [t, T]$, the option is called "knock out" and expires worthless; otherwise, the option has the same payoff as a vanilla call option. The terminal payoff of the option can thus be written as $(S_T^{t,s} - K)^+ \mathbf{1}_{\{S_\nu^{t,s} > L, \nu \in [t,T]\}}$, where $\mathbf{1}$ is the indicator function.

Define a stopping time

$$\tau = \tau(t, x) := \inf\{\nu : S_\nu^{t,s} = L\} \wedge T,$$

and a function $\Psi_{do}(t, s, K)$

$$\Psi_{do}(t, s, K) = \begin{cases} (s - K)^+, & (t,s) \in \{T\} \times (L, +\infty), \\ 0, & (t,s) \in [0,T] \times \{L\}. \end{cases}$$

Then if $S_\tau^{t,s} \in (L, \infty)$, we have $\tau = T$. Hence

$$\Psi_{do}(\tau, S_\tau^{t,s}, K) = (S_\tau^{t,s} - K)^+ \mathbf{1}_{\{S_\nu^{t,s} > L, \nu \in [t,T]\}}.$$

Denote $\Sigma[t, T] = \{\sigma = (\sigma_\nu)_{t \le \nu \le T} \mid \sigma \text{ is } \mathcal{F}_\nu\text{-adapted with } \underline{\sigma} \le \sigma_\nu \le \bar{\sigma}, t \le \nu \le T\}$. Then at time t, the worst case price of the option for the seller is

$$\begin{aligned} C_{do}(t, s, K) &= \inf_{\sigma \in \Sigma[t,T]} E[e^{-rT}(S_T^{t,s} - K)^+ \mathbf{1}_{\{S_\nu^{t,s} > L, \nu \in [t,T]\}}] \\ &= \inf_{\sigma \in \Sigma[t,T]} E[e^{-r\tau}\Psi_{do}(\tau, S_\tau^{t,s}, K)]. \end{aligned} \tag{3}$$

Consider the following nonlinear terminal-boundary value problem:

$$\begin{cases} \frac{\partial C_{do}}{\partial t} + \inf_{\sigma \in [\underline{\sigma}, \bar{\sigma}]}\{\frac{1}{2}\sigma^2 s^2 \frac{\partial^2 C_{do}}{\partial s^2}\} + rs\frac{\partial C_{do}}{\partial s} - rC_{do} = 0, s > L, 0 \le t < T \\ C_{do}(t, L, K) = 0, 0 \le t < T, \\ C_{do}(T, s, K) = (s - K)^+, s > L \end{cases} \tag{4}$$

Using the usual argument of viscosity solution, we can prove $C_{do}(t, s, K)$ is the unique viscosity solution of (4) (refer to [10] theorem 4.2-4.4).

For the up-out call option with maturity T, strike price K and upper barrier H, if $S_\nu^{t,s} \ge H$ for some $\nu \in [t, T]$, the option is called "knock out" and expires worthless; otherwise, the option has the same payoff as a vanilla call option. The terminal payoff of the option can thus be written as

$$(S_T^{t,s} - K)^+ \mathbf{1}_{\{S_\nu^{t,s} < H, \nu \in [t,T]\}},$$

where $\mathbf{1}$ is the indicator function.

For the case of $H \le K$, before the option is in-the-money it has to hit the barrier H and be knock out, so in this case the price of the option is zero. We just consider the case of $H > K$. In this case, by the same argument, we get the similar nonlinear PDE

$$\begin{cases} \frac{\partial C_{uo}}{\partial t} + \inf_{\sigma \in [\underline{\sigma}, \bar{\sigma}]}\{\frac{1}{2}\sigma^2 s^2 \frac{\partial^2 C_{uo}}{\partial s^2}\} + rs\frac{\partial C_{uo}}{\partial s} - rC_{uo} = 0, s < H, 0 \le t < T \\ C_{uo}(t, H, K) = 0, 0 \le t < T, \\ C_{uo}(T, s, K) = (s - K)^+, s < H. \end{cases} \tag{5}$$

Now let's see the knock in barrier options. Take the down-in option as an example. If $S_\nu^{t,s} \le L$ for some $\nu \in [t, T]$, the option is called "knock in" and has the same payoff as a vanilla call option; otherwise, expires worthless. The terminal payoff of the option can thus be written as

$$C(\tau, S_\tau^{t,s}, K)\mathbf{1}_{\{S_\tau^{t,s} = L\}},$$

where $C(t, s, K)$ is the price of the vanilla European call option. Define function $\Psi_{di}(t, s, K)$

$$\Psi_{di}(t, s, K) = \begin{cases} 0, & (t, s) \in \{T\} \times (L, +\infty), \\ C(t, s, K), & (t, s) \in [0, T] \times \{L\}. \end{cases}$$

Then

$$\Psi_{di}(\tau, S_\tau^{t,s}, K) = C(\tau, S_\tau^{t,s}, K)\mathbf{1}_{\{S_\tau^{t,s}=L\}}.$$

The worst case price of this option is

$$C_{di}(t, s, K) = \inf_{\sigma \in \Sigma[t,T]} E[e^{-r\tau}\Psi_{di}(\tau, S_\tau^{t,s}, K)]. \tag{6}$$

Similarly, we can prove that the price of the down-in call option is the unique viscosity solution of the following nonlinear PDE:

$$\begin{cases} \frac{\partial C_{di}}{\partial t} + \inf_{\sigma \in [\underline{\sigma}, \bar{\sigma}]}\{\frac{1}{2}\sigma^2 s^2 \frac{\partial^2 C_{di}}{\partial s^2}\} + rs\frac{\partial C_{di}}{\partial s} - rC_{di} = 0, s > L, 0 \le t < T, \\ C_{di}(t, L, K) = C(t, L, K), 0 \le t < T, \\ C_{di}(T, s, K) = 0, s > L. \end{cases} \tag{7}$$

In the model without uncertain volatility, the out-in parity for European style option is a well known result, i.e.,

$$\bar{C}(t, s, K) = \bar{C}_{do}(t, s, K) + \bar{C}_{di}(t, s, K), \tag{8}$$

where $\bar{C}, \bar{C}_{do}, \bar{C}_{di}$ are the prices of the vanilla European call option, the down-out barrier call option and the down-in barrier call option respectively.

Does the out-in parity still hold in uncertain volatility model? Let

$$\Psi(t, s, K) = \begin{cases} (s - K)^+, & t = T, s \ge 0, \\ C(t, s, K), & t < T, s \ge 0. \end{cases}$$

Then $\Psi$ is the payoff function (when $t = T$) or the price (when $t < T$) of the vanilla European call option in uncertain volatility model. It is easily to see

$$\Psi(t, s, K) = \Psi_{do}(t, s, K) + \Psi_{di}(t, s, K), \forall t \in [0, T], s \ge L. \tag{9}$$

That is the payoff of the vanilla call option equals to the payoff of the down-out barrier call option plus that of the down-in barrier call option. But the worst case pricing means that

$$C(t, s, K) = \inf_{\sigma. \in \Sigma[t,T]} \Psi(t, s, K) = \inf_{\sigma. \in \Sigma[t,T]} [\Psi_{do}(t, s, K) + \Psi_{di}(t, s, K)]$$

$$\ge C_{do}(t, s, K) + C_{di}(t, s, K)], \forall t \in [0, T], s \ge 0. \tag{10}$$

Hence, in general, the out-in parity does not hold in uncertain volatility model.

The price of the vanilla European call option $C(t, s, K)$ is the solution of the following nonlinear PDE

$$\begin{cases} \frac{\partial C}{\partial t} + \inf_{\sigma \in [\underline{\sigma}, \bar{\sigma}]} \{\frac{1}{2}\sigma^2 s^2 \frac{\partial^2 C}{\partial s^2}\} + rs\frac{\partial C}{\partial s} - rC = 0, \\ C(T, s, K) = (s - K)^+. \end{cases}$$

Since $C(t, s, K)$ is a convex function of the stock price, we have $\frac{\partial^2 C}{\partial s^2} \geq 0$. Hence the above PDE is

$$\begin{cases} \frac{\partial C}{\partial t} + \frac{1}{2}\underline{\sigma}^2 s^2 \frac{\partial^2 C}{\partial s^2} + rs\frac{\partial C}{\partial s} - rC = 0, \\ C(T, s, K) = (s - K)^+, \end{cases}$$

which is the Black-Scholes equation with volatility $\underline{\sigma}$.

Similarly, the up-in call option satisfies the following PDE

$$\begin{cases} \frac{\partial C_{ui}}{\partial t} + \inf_{\sigma \in [\underline{\sigma}, \bar{\sigma}]} \{\frac{1}{2}\sigma^2 s^2 \frac{\partial^2 C_{ui}}{\partial s^2}\} + rs\frac{\partial C_{ui}}{\partial s} - rC_{ui} = 0, s < H, 0 \leq t < T, \\ C_{ui}(t, L, K) = C(t, L, K), 0 \leq t < T, \\ C_{ui}(T, s, K) = 0, s < H. \end{cases} \quad (11)$$

For the put barrier options, using similar argument, we can get the corresponding nonlinear PDEs. One type of put options should be paid more attention to. For the down-out put options, in the case that the lower barrier is higher than or equals to the strike price, i.e., $L \geq K$, the option has to be knocked out before it is in the money, so in this case the price of the option is zero. For the down-out put option, we assume $L < K$. We get the PDE of the prices of the down-out(up-out) put option $P_{do}(t, s, K)(P_{uo}(t, s, K))$ with the terminal condition $(s - K)^+$ replaced by $(K - s)^+$ in (4)((5)), and the PDE of the prices of the down-in(up-in) put option $P_{di}(t, s, K)(P_{ui}(t, s, K))$ with the boundary condition $C(t, s, K)$ replaced by $P(t, s, K)$ in (7)((11)), where $P(t, s, K)$ is the price of the vanilla European put option.

Also, we have that the worst case value of a vanilla put option is larger than or equals to that of a down-out put option plus that of a down-in put option, i.e.,

$$P(t, s, K) \geq P_{do}(t, s, K) + P_{di}(t, s, K). \quad (12)$$

## 3.2 Double Barrier Options

A double barrier option is characterized by two barriers, L (lower barrier) and H (upper barrier). The double barrier knock out call option is knocked out if either barrier is touched. Otherwise, the option gives at maturity T the standard Black-Scholes payoff: $\max(0, S_T^{t,s} - K)$.

For the case that the upper barrier $H$ is lower than or equals to the strike price $K$, to be in-the-money, the call option has to be knocked out first, so in this case the price of the barrier option is zero. We just consider the case of $H > K$.

The terminal payoff of the option can thus be written as

$$(S_T^{t,s} - K)^+ 1_{\{L < S_\nu^{t,s} < H, \nu \in [t,T]\}},$$

where $1$ is the indicator function.

Let $G = (L, H)$. Define a stopping time

$$\tau = \tau(t, x) := \inf\{\nu : S_\nu^{t,s} \notin G\} \wedge T,$$

and a function $\Psi_{dbo}(t, s, K)$

$$\Psi_{dbo}(t, s, K) = \begin{cases} (s - K)^+, & (t, s) \in \{T\} \times int(G), \\ 0, & (t, s) \in [0, T] \times \partial G. \end{cases}$$

Then at time t, the worst case price of the option is

$$C_{dbo}(t, s, K) = \inf_{\sigma \in \Sigma[t,T]} E[e^{-r\tau} \Psi_{dbo}(\tau, S_\tau^{t,s}, K)] \tag{13}$$

By the theorem of [10] Theorem 4.2-4.4, $C_{dbo}(t, s, K)$ is the unique viscosity solution of the following nonlinear PDE:

$$\begin{cases} \frac{\partial C_{dbo}}{\partial t} + \inf_{\sigma \in [\underline{\sigma}, \bar{\sigma}]} \{\frac{1}{2}\sigma^2 s^2 \frac{\partial^2 C_{dbo}}{\partial s^2}\} + rs\frac{\partial C_{dbo}}{\partial s} - rC_{dbo} = 0, H > K \\ C_{dbo}|_{\partial G} = 0 \\ C_{dbo}|_{t=T} = (s - K)^+. \end{cases} \tag{14}$$

Similarly, the price of the double barrier knock in call option $C_{dbi}(t, s, K)$ should satisfy the following PDE

$$\begin{cases} \frac{\partial C_{dbi}}{\partial t} + \inf_{\sigma \in [\underline{\sigma}, \bar{\sigma}]} \{\frac{1}{2}\sigma^2 s^2 \frac{\partial^2 C_{dbi}}{\partial s^2}\} + rs\frac{\partial C_{dbi}}{\partial s} - rC_{dbi} = 0, \\ C_{dbi}|_{\partial G} = C(t, s, K) \\ C_{dbi}|_{t=T} = 0. \end{cases} \tag{15}$$

We also have that the worst case value of a vanilla call option is larger than or equals to that of a double out call option plus that of a double in call option, i.e.,

$$C(t, s, K) \geq C_{dbo}(t, s, K) + C_{dbi}(t, s, K). \tag{16}$$

For the corresponding put options, we still can get the PDE of the prices of the double out put option $P_{dbo}(t, s, K)$ with the terminal condition $(s - K)^+$ replaced by $(K - s)^+$ in (14), and the PDE of the prices of the double in put option $P_{dbi}(t, s, K)$ with the boundary condition $C(t, s, K)$ replaced by

$P(t, s, K)$ in (15), where $P(t, s, K)$ is the price of the vanilla European put option. And we also have

$$P(t, s, K) \geq P_{dbo}(t, s, K) + P_{dbi}(t, s, K). \tag{17}$$

**Acknowledgements.** This work is supported by Beijing Natural Science Foundation(Granted No. 1112009).

# References

1. Avellaneda, M., Buff, R.: Combinatorial implications of nonlinear uncertain-volatility models: the case of barrier options. Applied Mathematical Finance 6, 1–18 (1999)
2. Avellaneda, M., Levy, A., Paras, A.: Pricing and hedging derivative securities in markets with uncertain volatilities. Applied Mathematical Finance 2, 73–88 (1995)
3. Buff, R.: Uncertain Volatility Models-Theory and Application. Springer Finance Lecture Notes. Springer, Heidelberg (2002)
4. Delbaen, F.: Coherent measures of risk on general probability space. In: Sandmann, K., Schobucher, P.J. (eds.) Advances in Finance and Stochastics, Essays in Honor of Dieter Sondermann, pp. 1–37. Springer, Berlin (2002)
5. Forsyth, A., Labahn, G.: Numerical Methods for Controlled Hamilton-Jacobi-Bellman PDEs in Finance. The Journal of Computational Finance 11(2), 1–44 (2007)
6. Lyons, T.J.: Uncertain volatility and the risk-free synthesis of derivatives. Journal of Applied Finance 2, 117–133 (1995)
7. Meyer, G.: The Black-Scholes Barenblatt Equation for Options with Uncertain Volatility and its Application to Static Hadging. International Journal of Theoretical and Applied Finance 9(5), 673–703 (2006)
8. Smith, A.T.: American options under uncertain volatility. Applied Mathematical Finance 9(2), 123–141 (2002)
9. Zhang, K., Wang, S.: A computational scheme for uncertain volatility in option pricing. Applied Numerical Mathematics 59, 1754–1767 (2009)
10. Yong, J., Zhou, X.: Stochastic Controls: Hamiltonian Systems and HJB Equations. Springer, Heidelberg (1999)

# European Option Pricing with Ambiguous Return Rate and Volatility

Junfei Zhang and Shoumei Li

**Abstract.** In this paper, we consider the problem of option pricing when return rate and volatility are ambiguous. Firstly we illustrate how to describe this ambiguous option pricing model by using set-valued differential inclusion and how to change the discussion of pricing bound problems of options into that of maximal and minimal conditional expectations. Secondly we discuss the properties of maximal and minimal conditional expectations, especially the representation theorem of maximal and minimal expectations. Finally we give the bounds of the European option pricing by using above theorems.

**Keywords:** Set-valued stochastic differential inclusion, Martingale measures, Maximal and minimal conditional expectations, Bounds of option prices.

## 1 Introduction

In 1973, Black and Sholes [3] provided with the famous pricing formula for European options under the assumption that the price of risky underlying asset $\{S_t : t \in [0, T]\}$ is described as $\frac{dS_t}{S_t} = \mu dt + \sigma dW_t$, where $\mu$ is a constant and called the expected return rate, $\sigma$ is also a constant and called volatility, $\{W_t : t \in [0, T]\}$ is a standard Brownian motion, $T$ is maturity ($0 < T < \infty$). Following their work, many authors discussed various option pricing problems under more general model

$$\frac{dS_t}{S_t} = \mu_t dt + \sigma_t dW_t, \tag{1}$$

Junfei Zhang and Shoumei Li
Department of Applied Mathematics, Beijing University of Technology
e-mail: junfeizhang@emails.bjut.edu.cn, lisma@bjut.edu.cn

S. Li (Eds.): Nonlinear Maths for Uncertainty and its Appli., AISC 100, pp. 279–286.
springerlink.com                                    © Springer-Verlag Berlin Heidelberg 2011

where $\{\mu_t : t \in [0,T]\}$ and $\{\sigma_t : t \in [0,T]\}$ are two given predict processes with respect to the filtration $\{\mathcal{F}_t : t \in [0,T]\}$, induced by the Brownian motion $\{W_t : t \in [0,T]\}$, and the risk-free bond is described as $dB_t = r_t B_t dt$, where $r_t$ is determined interest rate function on [0,T] with $B_0 = 1$ (e.g. [11]). The pricing theory of options shows that, if the market is complete and arbitrage-free, there exists a unique risk neutral martingale measure Q defined by

$$\frac{dQ}{dP} = \exp\{-\frac{1}{2}\int_0^T \left(\frac{\mu_s - r_s}{\sigma_s}\right)^2 ds + \int_0^T \left(\frac{\mu_s - r_s}{\sigma_s}\right) dW_s\}$$

such that for any contingent claim $\xi_T$ at time $T$, the value of $\xi_T$ at any time $t \in [0,T)$ is given by

$$V_t(\xi_T) = B_t E_Q[\xi_T B_T^{-1}|\mathcal{F}_t] = E_Q\left[e^{-\int_t^T r_u du}\xi_T|\mathcal{F}_t\right],$$

and, in particular, the current price is $V_0(\xi_T) = E_Q\left[e^{-\int_0^T r_u du}\xi_T\right]$. Note that we use the notation $E_Q$ to denote expectation with respect to the probability measure Q.

In the real world, however, one is difficult to observe exactly $\{\mu_t : t \in [0,T]\}$ and $\{\sigma_t : t \in [0,T]\}$, since the factors of affecting market are too complex. But it is possible to estimate the lower bound and upper bound of the expected return rate, i.e. the expected return rate is within some interval, for example, $\mu_t \in [a,b] := U$ with $b > a$. Similarly it happens for the volatility $\sigma_t \in [c,d] := V$ with $d > c > 0$. In this case, the model is with ambiguity since $U, V$ are subsets of $\mathbb{R}$, the set of all real numbers. It can be rewritten as the following set-valued differential inclusion

$$\frac{dS_t}{S_t} \in U dt + V dW_t. \tag{2}$$

Our problem is how to estimate the prices of the options under this set-valued model? At least, we should estimate the lower bound and upper bound of option prices.

Before we go to the bound estimation, let us look at the right hand of (2). It relates two kind integrals: Itô integral of a set-valued stochastic process with respect to a real-valued Brownian motion $W_t$, and the Lebesgue integral of a set-valued process with respect to time $t$. For the first type, Kisielewicz introduced the definition in [8]. More related works may refer to [6], [9], [12], [14] and so on. There are also some works of the Lebesgue integral of a set-valued stochastic process with respect to time $t$, readers may refer to [10], [13] and their related references. Concerning the definitions, existence of the strong solutions and weak solutions of set-valued stochastic differential inclusions, readers may refer to [1], [2], [9] and [14].

The general set-valued stochastic differential inclusion is given by

$$dx_t \in F(t, x_t)dt + G(t, x_t)dW_t, \tag{3}$$

where $F(t, x_t), G(t, x_t)$ are set-valued functions of $\mathbb{R}$, $\{x_t : t \in [0, T]\}$ is an adapted stochastic process. A continuous stochastic process $x = \{x_t : t \in [0, T]\}$ is called a strong solution of (3), with an initial value $x_0 = \eta$, a.e., if there exist square integrable selections $f = \{f(t, x_t) : t \in [0, T]\}, g = \{g(t, x_t) : t \in [0, T]\}$ of $\{F(t, x_t) : t \in [0, T]\}$ and $\{G(t, x_t) : t \in [0, T]\}$ respectively, such that for any $t \in [0, T]$,

$$x_t = x_0 + \int_0^t f(s, x_s)ds + \int_0^t g(s, x_s)dW_s, \quad a.e..$$

In general, the strong solutions of (3) are a set of stochastic processes.

Now let us back to the model (2). To make the model simple, we assume that the risk-free bond has constant interest rate $r$ in $[0, T]$. If continuous stochastic processes $\{\mu_t\}, \{\sigma_t\}$ satisfy $a \le \mu_t \le b$, $c \le \sigma_t \le d$ and

$$\frac{dS_t}{S_t} = \mu_t dt + \sigma_t dW_t,$$

with $S_0 = s > 0$, then we may obtain its related strong solutions. Thus, we may get a set of strong solutions of (2), denoted by $\mathcal{S}$.

Let $\nu_t := \frac{\mu_t - r}{\sigma_t}$. If $a \ge r$, then $\frac{a-r}{d} \le \nu_t \le \frac{b-r}{c}$. If $a < r < b$, then $\frac{a-r}{c} \le \nu_t \le \frac{b-r}{c}$. If $r \ge b$, then $\frac{a-r}{c} \le \nu_t \le \frac{b-r}{d}$. Without any loss in generality, we suppose $\nu_t \in [k_1, k_2]$. In this case, the risk neutral martingale measures are no longer unique, they belong to the following set of probabilities

$$\mathcal{P} = \{Q^\nu : \frac{dQ^\nu}{d\mathbb{P}} = \exp\{-\frac{1}{2}\int_0^T \nu_s^2 ds + \int_0^T \nu_s dW_s, \ k_1 \le \nu_s \le k_2, \ 0 \le s \le T\}\},$$

which is the set of the risk neutral martingale measures with respect to the strong solutions set $\mathcal{S}$ of the model (2). Hence, at any time $t \in [0, T)$, we may define the minimal value and maximal value of a contingent claim $\xi_T$ as

$$\overline{V}_t[\xi_T] = \sup_{Q \in \mathcal{P}} e^{-r(T-t)} E_Q[\xi_T | \mathcal{F}_t], \quad \underline{V}_t[\xi_T] = \inf_{Q \in \mathcal{P}} e^{-r(T-t)} E_Q[\xi_T | \mathcal{F}_t]. \quad (4)$$

In particular, the current maximal price and minimal price of $\xi_T$ are

$$\overline{V}_0[\xi_T] = \sup_{Q \in \mathcal{P}} E_Q[e^{-rT}\xi_T], \quad \underline{V}_0[\xi_T] = \inf_{Q \in \mathcal{P}} E_Q[e^{-rT}\xi_T]. \quad (5)$$

Note that, $k_1, k_2$ are determined by $a, b, c, d$, so the pricing bounds about model (2) can be obtained from (4) and (5).

If $\xi_T$ is an European option, can we give the formula of its current maximal price and minimal price? Furthermore, it is nature to ask how to calculus $\overline{V}_t[\xi_T], \underline{V}_t[\xi_T]$ for any $t \in [0, T]$? To do it, we have to investigate the maximal and minimal conditional expectations.

We especially would like to thank Chen and Kulperger's excellent work [5]. In their paper, they gave the model starting from multiple prior probability measures, introduced the concept of the maximal and minimal conditional expectations, discussed the martingale representation theorem under a symmetrical assumption $|v_t| \leq k$ ($k$ is a given constant). However, they didn't give any explain why they should have this symmetrical assumption under the ambiguous model. As we have known from above introduction by using the view of set-valued stochastic differential inclusion, $|v_t|$ usually is asymmetric.

In our paper, we shall discuss some properties of above maximal and minimal conditional expectations related to (4), especially martingale representation theorem. Our proofs are enlightened by [5] but we treat asymmetrical case. We shall also calculus the lower bound price and upper bound price of European options.

We organize our paper as follows. In Section 2, we shall discuss the properties of maximal and minimal conditional expectations, especially their martingale representation theorem. Section 3 will present the formula of the maximal and minimal values of options whose payoff are monotone functions of terminal risky asset price $S_T$. In Section 4, we shall also calculate the lower bound prices and upper of European options. Since the page limitation, we have to omit the proofs of theorems in this paper.

## 2 The Properties of Maximal and Minimal Conditional Expectations

Assume that $(\Omega, \mathcal{F}, \mathbb{P})$ is a completed probability space, $W =: \{W_t : t \in [0, T]\}$ is a real-valued Brownain motion defined on $(\Omega, \mathcal{F}, \mathbb{P})$ and $\{\mathcal{F}_t : 0 \leq t \leq T\}$ is natural filtration generated by the Brownian motion $W$. Furthermore, we take $\mathcal{F}$ to be $\mathcal{F}_T$ for simplicity. Let $L^2(\Omega, \mathcal{F}, \mathbb{P}) = \{\xi : \xi$ is a $\mathcal{F}$-measurable random variable s.t. $E[|\xi|^2] < \infty\}$. By using the notation in Section 1, we have the following definition.

**Definition 1.** (cf. [5])For any $\xi \in L^2(\Omega, \mathcal{F}, \mathbb{P})$, we define the maximal and minimal conditional expectations of $\xi$ with respect to $\mathcal{F}_t$, $\overline{\mathcal{E}}[\xi|\mathcal{F}_t]$, denoted by $\underline{\mathcal{E}}[\xi|\mathcal{F}_t]$ respectively, as

$$\overline{\mathcal{E}}[\xi|\mathcal{F}_t] = ess \sup_{Q \in \mathcal{P}} E_Q[\xi|\mathcal{F}_t], \quad \underline{\mathcal{E}}[\xi|\mathcal{F}_t] = ess \inf_{Q \in \mathcal{P}} E_Q[\xi|\mathcal{F}_t].$$

where ess means essential. And the maximal and minimal expectations of $\xi$ is

$$\overline{\mathcal{E}}[\xi] = \sup_{Q \in \mathcal{P}} E_Q[\xi], \quad \underline{\mathcal{E}}[\xi] = \inf_{Q \in \mathcal{P}} E_Q[\xi].$$

**Lemma 1.** *For any* $\xi \in L^2(\Omega, \mathcal{F}, \mathbb{P})$, $\overline{\mathcal{E}}[\xi|\mathcal{F}_0] = \overline{\mathcal{E}}[\xi]$ *and* $\underline{\mathcal{E}}[\xi|\mathcal{F}_0] = \underline{\mathcal{E}}[\xi]$.

**Theorem 1.** *For any* $\xi \in L^2(\Omega, \mathcal{F}, \mathbb{P})$, *then* $\overline{\mathcal{E}}[\xi|\mathcal{F}_t] \in L^2(\Omega, \mathcal{F}_t, \mathbb{P})$ *and* $\underline{\mathcal{E}}[\xi|\mathcal{F}_t] \in L^2(\Omega, \mathcal{F}_t, \mathbb{P})$, *in specially,* $\overline{\mathcal{E}}[|\xi|] < \infty$, $\underline{\mathcal{E}}[|\xi|] < \infty$.

Next, we state the representation theorem of minimal and maximal conditional expectations by using the solution of backward stochastic differential equation (BSDE for short). For more general case, readers may refer to [7].

**Theorem 2.** *Let* $\xi \in L^2(\Omega, \mathcal{F}, \mathbb{P})$.

(i) *If* $Y_t := \overline{\mathcal{E}}[\xi|\mathcal{F}_t]$, *then there exists an adapted process* $\{z_t : t \in [0, T]\}$ *such that* $(Y_t, z_t)$ *is the solution of BSDE*

$$Y_t = \xi + \int_t^T (k_1 z_s I_{\{z_s < 0\}} + k_2 z_s I_{\{z_s \geq 0\}}) ds - \int_t^T z_s dW_s, \quad 0 \leq t \leq T.$$

(ii) *If* $y_t := \underline{\mathcal{E}}[\xi|\mathcal{F}_t]$, *then there exists an adapted process* $\{x_t : t \in [0, T]\}$ *such that* $(y_t, x_t)$ *is the solution of BSDE*

$$y_t = \xi - \int_t^T (k_1 x_s I_{\{x_s < 0\}} + k_2 x_s I_{\{x_s \geq 0\}}) ds - \int_t^T x_s dW_s, \quad 0 \leq t \leq T.$$

## 3   Calculations of Maximal and Minimal Conditional Expectations

For simplicity, we firstly study the properties of $\overline{\mathcal{E}}[\cdot|\mathcal{F}_t]$, the results of $\underline{\mathcal{E}}[\cdot|\mathcal{F}_t]$ can be obtained in the same way.

The general stochastic differential equation (SDE) is given by

$$\begin{cases} dX_t = b(t, X_t)dt + \sigma(t, X_t)dW_t, \\ X_0 = x, \end{cases} \tag{6}$$

where $b$ and $\sigma$ : $[0, T] \times R \to R$ are continuous in $(t, x)$ and Lipshictz continuous in $x$. It is well-known that the SDE (6) has a unique solution $X = \{X_t : t \in [0, T]\}$ with $X_T \in L^2(\Omega, \mathcal{F}, \mathbb{P})$.

**Lemma 2.** *(cf. [4]) Let* $X$ *be the solution of the SDE (6). Assume* $\Phi$ *is a function such that* $\Phi(X_T) \in L^2(\Omega, \mathcal{F}, \mathbb{P})$. *Consider the BSDE*

$$y_t = \Phi(X_T) + \int_t^T a_s z_s ds - \int_t^T z_s dW_s, 0 \leq t \leq T$$

*and it has a unique pair solution, denoted by* $(y_t, z_t)$. *Since this solution depends on the adapted process* $\{a_s\}$, *we write it as* $(y_t^a, z_t^a)$.

(i) *If* $\Phi$ *is an increasing function, then* $z_t \sigma(t, X_t) \geq 0$, *a.e.* $t \in [0, T)$.

(ii)*If* $\Phi$ *is a decreasing function, then* $z_t \sigma(t, X_t) \leq 0$, *a.e.* $t \in [0, T)$.

**Theorem 3.** *Assume the conditions of Lemma 2 hold, and $\sigma(t,x) > 0$ for all $0 \le t \le T$ and $x \in R$, then there are equivalent martingale measures $Q_2$ and $Q_1$, such that*

*(i) If $\Phi$ is an increasing function, then*

$$\overline{\mathcal{E}}[\Phi(X_T)|\mathcal{F}_t] = E_{Q_2}[\Phi(X_T)|\mathcal{F}_t], \quad \underline{\mathcal{E}}[\Phi(X_T)|\mathcal{F}_t] = E_{Q_1}[\Phi(X_T)|\mathcal{F}_t], \quad t \in [0,T].$$

*In particular,*

$$\overline{\mathcal{E}}[\Phi(X_T)] = E_{Q_2}[\Phi(X_T)] \quad and \quad \underline{\mathcal{E}}[\Phi(X_T)] = E_{Q_1}[\Phi(X_T)].$$

*(ii) If $\Phi$ is a decreasing function, then*

$$\overline{\mathcal{E}}[\Phi(X_T)|\mathcal{F}_t] = E_{Q_1}[\Phi(X_T)|\mathcal{F}_t], \quad \underline{\mathcal{E}}[\Phi(X_T)|\mathcal{F}_t] = E_{Q_2}[\Phi(X_T)|\mathcal{F}_t], \quad t \in [0,T].$$

*In particular,*

$$\overline{\mathcal{E}}[\Phi(X_T)] = E_{Q_1}[\Phi(X_T)] \quad and \quad \underline{\mathcal{E}}[\Phi(X_T)] = E_{Q_2}[\Phi(X_T)].$$

*Where the equivalent martingale measures $Q_2$ and $Q_1$ are defined by*

$$\begin{cases} \frac{dQ_2}{dP} = \exp\{-\frac{1}{2}k_2^2 T + k_2 W_T\}, \\ \frac{dQ_1}{dP} = \exp\{-\frac{1}{2}k_1^2 T + k_1 W_T\}. \end{cases} \tag{7}$$

By Theorem 3, we have the following corollary because of the monotone property of $\Phi(x) = e^{-rT}(x - K)^+$ and $\Phi(x) = e^{-rT}(K - x)^+$.

**Corollary 1.** *Assume that $\{X_t\} = \{S_t\}$ in SDE (6) is the price of a stock and $b(s, X_s) = \mu_s S_s$ with $a \le \mu_s \le b$ and $\sigma(s, X_s) = \sigma_s S_s$ with $c \le \sigma_s \le d$ and $X_0 = x > 0$, for all $s \in [0, T]$. Let $(S_T - K)^+$ and $(K - S_T)^+$ be the payoff of the European call option and put option respectively. Then*

*(i) the upper bound and the lower bound of the European call option price are given respectively by*

$$\overline{V}_0[(S_T - K)^+] = \overline{\mathcal{E}}[e^{-rT}(S_T - K)^+] = E_{Q_2}[e^{-rT}(S_T - K)^+]$$

*and*

$$\underline{V}_0[(S_T - K)^+] = \underline{\mathcal{E}}[e^{-rT}(S_T - K)^+] = E_{Q_1}[e^{-rT}(S_T - K)^+];$$

*(ii) the upper bound and the lower bound of the European put option price are given respectively by*

$$\overline{V}_0[(K - S_T)^+] = \overline{\mathcal{E}}[e^{-rT}(K - S_T)^+] = E_{Q_1}[e^{-rT}(K - S_T)^+]$$

*and*

$$\underline{V}_0[(S_T - K)^+] = \underline{\mathcal{E}}[e^{-rT}(S_T - K)^+] = E_{Q_2}[e^{-rT}(S_T - K)^+].$$

*where $K$ is strike price, and $Q_1, Q_2$ are defined by Eq. (7).*

# 4  European Option Pricing Under Ambiguous Model

In this section, we will obtain the formula of upper and lower bounds of the European call and put option by the results in Section 3.

From Theorem 3 and Eq. (7), we have $F_P(W_T) = N(0,T)$, $F_{Q_2}(W_T - k_2T) = N(0,T)$, and $F_{Q_1}(W_T - k_1T) = N(0,T)$. Therefor, $F_{Q_2}(W_T) = N(k_2T,T)$ and $F_{Q_1}(W_T) = N(k_1T,T)$, where $F_P(\xi)$ means the distribution of random variable $\xi$ with respect to probability measure $P$. We also use the notation $N(\mu, \sigma^2)$ to denote the normal distribution with mean $\mu$ and variance $\sigma^2$, and let $F_0$ be the standard normal distribution function.

In the following, we consider the European call option pricing bounds formula. Assume that the stock price $S_t$ satisfies geometric Brownian motion

$$\frac{dS_t}{S_t} = \mu dt + \sigma dW_t, \quad S_0 = x > 0.$$

For simplicity, we assume interest rate $b > r > a$ (that is $k_1 = \frac{a-r}{c}$ and $k_2 = \frac{b-r}{c}$) is constant. However, our conclusion can be easily extended to the case where $r$ is not constant. By Corollary 1, we have the formula of the bounds of European call option pricing.

$$\begin{aligned}
\overline{V}_0[(S_T - K)^+] &= \overline{\mathcal{E}}[B_T^{-1}(S_T - K)^+] \\
&= e^{-rT} E_{Q_2}[(xe^{(\mu - \frac{1}{2}\sigma^2)T + \sigma W_T} - K)^+] \\
&= e^{-rT} E_P[(xe^{(k_2\sigma + \mu - \frac{1}{2}\sigma^2)T + \sigma W_T} - K)^+] \\
&= e^{-rT} \int_{d_2}^{\infty} (xe^{(k_2\sigma + \mu - \frac{1}{2}\sigma^2)T + \sigma y} - K)\frac{1}{\sqrt{2\pi T}}e^{-\frac{y^2}{2T}}dy \\
&= e^{-rT}\left\{ xe^{(k_2\sigma + \mu)T} F_0(-\frac{d_2 - \sigma T}{\sqrt{T}}) - KF_0(-\frac{d_2}{\sqrt{T}}) \right\}
\end{aligned}$$

Because equivalent probability measure is $Q_2$, where $\mu = b, \sigma = c$. And

$$\begin{aligned}
\underline{V}_0[(S_T - K)^+] &= \underline{\mathcal{E}}[(B_T^{-1}(S_T - K)^+] \\
&= e^{-rT} E_{Q_1}[(xe^{(\mu - \frac{1}{2}\sigma^2)T + \sigma W_T} - K)^+] \\
&= e^{-rT} E_P[(xe^{(k_1\sigma + \mu - \frac{1}{2}\sigma^2)T + \sigma W_T} - K)^+] \\
&= e^{-rT} \int_{d_1}^{\infty} (xe^{(k_1\sigma + \mu - \frac{1}{2}\sigma^2)T + \sigma y} - K)\frac{1}{\sqrt{2\pi T}}e^{-\frac{y^2}{2T}}dy \\
&= e^{-rT}\left\{ xe^{(k_1\sigma + \mu)T} F_0(-\frac{d_1 - \sigma T}{\sqrt{T}}) - KF_0(-\frac{d_1}{\sqrt{T}}) \right\}
\end{aligned}$$

Because equivalent probability measure is $Q_1$, where $\mu = a, \sigma = c$. And where

$$d_2 = \frac{\log\frac{K}{x} - (k_2 c + b - \frac{1}{2}c^2)T}{c\sqrt{T}}, d_1 = \frac{\log\frac{K}{x} - (k_1 c + a - \frac{1}{2}c^2)T}{c\sqrt{T}}.$$

With the same way, we have the formula of the bounds of European put option pricing

$$\overline{P}_0[(K - X_T)^+] = e^{-rT}\{KF_0(\frac{d_1}{\sqrt{T}}) - xe^{(k_1\sigma+\mu)T}F_0(\frac{d_1 - \sigma T}{\sqrt{T}})\}$$

and

$$\underline{P}_0[(K - X_T)^+] = e^{-rT}\{KF_0(\frac{d_2}{\sqrt{T}}) - xe^{(k_1\sigma+\mu)T}F_0(\frac{d_2 - \sigma T}{\sqrt{T}})\}$$

where $d_1, d_2$ are same as above.

**Acknowledgements.** This research is supported by PHR (No. 201006102) and Beijing Natural Science Foundation (Stochastic Analysis with uncertainty and applications finance).

# References

1. Ahmed, N.U.: Nonlinear stochastic differential inclusions on Banach space. Stoch. Anal. Appl. 12, 1–10 (1994)
2. Aubin, J.P., Prato, G.D.: The viability theorem for stochastic differenrial inclusions. Stoch. Anal. Appl. 16, 1–15 (1998)
3. Black, F., Scholes, M.: The pricing of options and corporate liabilities. J. Polit. Econ. 81, 637–654 (1973)
4. Chen, Z., Kulperger, R., Wei, G.: A comononic theorem of BSDEs and its applications. Stoch. Proc. Appl. 115, 41–54 (2005)
5. Chen, Z., Kulperger, R.: Minmax pricing and Choquet pricing. Insur. Math. Econ. 38, 518–528 (2006)
6. Jung, E.J., Kim, J.H.: On set-valued stochastic integrals. Stoch. Anal. Appl. 21, 401–418 (2003)
7. Karoui, N., El, P.S., Quenez, M.C.: Backward stochatic differentail equation in finance. Math. Fina. 7, 1–71 (1997)
8. Kisielewicz, M.: Set-valued stochastic integrals and stochastic inclusions. Discuss. Math. 13, 119–126 (1993)
9. Kisielewicz, M.: Weak compactness of solution sets to stochastic differential inclusions with non-convex right-hand sides. Stoch. Anal. Appl. 23, 871–901 (2005)
10. Li, J., Li, S., Ogura, Y.: Strong solution of Itô type set-valued stochastic differential equation. Math. Sin. English Series 26, 1739–1748 (2010)
11. Shreve, S.E.: Stochastic Calculus for Finance. Springer, London (2004)
12. Zhang, J., Li, S., Mitoma, I., Okazaki, Y.: On set-valued stochastic integrals in an M-type 2 Banach space. J. Math. Anal. Appl. 350, 216–233 (2009)
13. Zhang, J., Li, S., Mitoma, I., Okazaki, Y.: On the solution of set-valued stochastic differential equation in M-type 2 Banach space. Tohoku Math. J. 61, 417–440 (2009)
14. Zhang, W., Li, S., Wang, Z., Gao, Y.: An Introduction of Set-Valued Stochastic Processes. Science Press, Beijing (2007)

# Knightian Uncertainty Based Option Pricing with Jump Volatility

Min Pan and Liyan Han

**Abstract.** In the viewpoint of Knightian uncertainty, this paper deals with option pricing with jump volatility. First, we prove that the jump volatility model is a Knightian uncertainty problem; then we identify the factors which reflect the Knighitan uncertainty based on $k$-Ignorance. We find that the option price under Knightian uncertainty is not unique but an interval. Through theoretical analysis and simulation, we conclude that the intensity of Poisson, the jump size, and the maturity date determine the price interval.

**Keywords:** Jump, Knightian Uncertainty, $k$-Ignorance, Option Pricing.

## 1 Introduction

Knightian uncertainty is different from risk. Unlike under risk, we can not get the precise probability distribution under Knightian uncertainty. There are two methods to deal with Knightian uncertainty. The first one is proposed by Bewely[1] which adopts the "inertia" assumption instead of the completeness assumption. Similarly, Epstein & Wang[4] propose a family of probability distributions based on individual "belief". The second one, Gilboa[6], Schmeidler[10], and Gilboa & Schmeidler[7] relax the "certainty rule" and use nonadditive measure based on Choquet integral. Additionally, Chen & Epstein[3] use a $k$-Ignorance to contain the probability distributions. Inspired by their theories, we establish a model with jump volatility and identify the factors affecting the $k$-Ignorance. We also get a price interval to reflect the Knightian uncertainty.

Min Pan and Liyan Han
School of Economics and Management, Beihang University, Beijing, 100191, China
e-mail: xjpanmin@163.com, hanly1@163.com

S. Li (Eds.): Nonlinear Maths for Uncertainty and its Appli., AISC 100, pp. 287–294.
springerlink.com                    © Springer-Verlag Berlin Heidelberg 2011

On option pricing, the focus of research shifts from risk to Kngihtian uncertainty. In Black and Scholes'[2] model, they deal with an "ideal conditions" without Knightian uncertainty. Merton[8] establishes a model with time-varying and non-stochastic volatility. Although the probability distribution is unique, we can find that the parameters of it are more complex. Merton[9] establishes a model where the underlying stock has a Poisson jump process; the probability distribution of the stock is not unique but depending on the jump process. There is a family of probability distributions of the underlying asset. In the viewpoint of this paper, there is Knightian uncertainty. Hull & White[5] develop a stochastic volatility model. In their model, the volatility of the underlying stock follows an independent process. Every possible path of the volatility decides a probability distribution of the stock; in fact, the probability distribution of the underlying stock is unknown; the stochastic volatility causes the Knightian uncertainty. In this view, we can say these models are treating the Knightian uncertainty.

This paper deals with the European call option pricing with jump volatility based on Knightian uncertainty. We have two contributions. (1) We prove that the jump volatility model can be transferred into a model of Knightian uncertainty. We first show the ambiguity of the model. After choosing a reference model, we create a family of models. Then, we get the option price interval based on $k$-Ignorance. (2) We indicate the factors which affect the $k$-Ignorance. And we prove that the factors affecting the $k$-Ignorance also affect the option price interval. This conclusion captures the feature of the Knightian uncertainty.

The following is the structure of this paper. Section 1 is the introduction. In section 2, we establish our model and get the solution. Section 3, after investigating the price bias through Monte Carlo simulation, we illustrate that our option price interval captures the feature of the Knightian uncertainty well. Section 4 is the conclusion.

## 2  Model

In a risk-neutral world, consider a European call option $C$ with maturity date $T$, strike price $K$, and depending on the security $S$, which price obeys the following stochastic processes:

$$dS/S = rdt + vdw \tag{1}$$

$$dv = v\,(Y - 1)\,dq \tag{2}$$

In equation (1), $r$ is the risk-free return of the bond, and $v$ is the instantaneous volatility of the return of the stock; $dw$ is a standard Wiener process. Equation (2) is a jump process describing the change of volatility $v$; $q\,(t)$ is a Poisson process with intensity $\lambda$; $dw$ and $dq$ are assumed to be independent of each other. We can treat $\lambda$ as the average number of events occurred in unit time; $(Y - 1)$ is the random variable percentage change in the instantaneous volatility when the event occurred. $Y \sim Unif\left[1 - \theta, 1 + \theta\right], \theta > 0$. If the Poisson event does not occur, (2) can be written as $dv = 0$; if the Poisson event does occur, $dv = (Y - 1)v$.

With jump volatility process $v(t)$, given a path of volatility, we can divide the interval $[0, T]$ by jumps. Suppose there are $n$ jumps and we get a partition $\tau$: $0 = t_0, t_1, \cdots, t_j, t_{j+1}, \cdots, t_n, t_{n+1} = T$. $t_j\,(1 \leqslant j \leqslant n)$ means the time when the $j$th jump happens. In every interval $[t_j, t_{j+1}]$, the volatility keeps constant and the distribution of $\log\{S(t_{j+1})/S(t_j)\}$ is normal. Because $\log\{S(T)/S(0)\} = \sum_{0}^{n} \log\{S(t_{j+1})/S(t_j)\}$, the distribution of $\log\{S(T)/S(0)\}$ is normal, too. But there are infinite paths of $v(t)$, so we can not know the distribution of $\log\{S(T)/S(0)\}$ precisely. The jump process of volatility causes the shift of the distribution. There is Knightian uncertainty.

The process $v(t)$ makes the probability distribution unknown. In the light of Bewley's theory, we follow three steps to deal with this Knightian uncertainty: (1) The individual selects a reference probability measure $P$ according to his "belief", "inertia", or even "animal spirit"; (2) he knows the probability measure $P$ is not precise; by disturbing $P$, he gets a family of probability measures P; (3) because of the non-unique of probability distribution, he can not get a precise option price; he can only get an interval. The option's price should be in this interval.

It is a natural choice to select the average of $v(t)$ as a reference. Averagely, we know that there are $\lambda t$ jumps in $[0, t]$ due to the character of Poisson process. Because $EY = 1$, we have $E(v) = v(0)$. Under the probability measure $P$, the reference model should be:

$$dS/S = rdt + v(0)\,dw \tag{3}$$

The individual knows that the model is inaccurate, there may be other probability measure $\tilde{P}$. We define $z(t) \triangleq d\tilde{P}/dP$, so $z(t)$ is the Radon-Nikodym derivative of $\tilde{P}$ and $P$. Due to Girsanov Theorem we can note $z(t)$ as $z(t) = \exp\left[\int_0^t h(\tau)\,dw(\tau) - \frac{1}{2}\int_0^t h^2(\tau)\,d\tau\right]$, with Novikov condition $E\left[\exp\left[\int_0^t h^2(\tau)\,d\tau\right]\right] < \infty$. To make the problem easier, we treat the $h(t)$ as some unknown and deterministic process in this paper.

According to Girsanov Theorem, we know $dw = h(t)\,dt + d\tilde{w}$, where $d\tilde{w}$ is a standard Brownian motion under another possible probability measure $\tilde{P}$. Substituting this transform into (3), we get a process under another possible risk-neutral probability measure $\tilde{P}$:

$$dS/S = (r + v(0)h(t))\,dt + v(0)\,d\tilde{w} \tag{4}$$

Equation (4) does not mean one process, but a family of possible processes with the same form. Different $h(t)$ means different process. To get the value of European call $C(T, S(t))$ which underlying stock price follows equation (4), we have the proposition:

**Proposition 1.** [1] *Suppose that, in a risk-neutral world, a stock price $S(t)$ with time-varying, nonrandom interest rate follows the stochastic process $dS(t) = \tilde{r}(t) S(t) dt + v(0) S(t) d\tilde{w}$. The value of European call with the maturity date $T$ is $C(T, S(t)) = BSM\left(T - t, S(t); K, \frac{1}{T-t} \int_t^T \tilde{r}(\tau) d\tau, v^2(0)\right)$. Here the $BSM(t^*, S; K, r^*, v^*)$ is the value of Black-Scholes-Merton European call with expiration time $t^*$, initial stock price $S$, exercise price $K$, constant interest rate $r^*$ and constant volatility $v^*$.*

According to this proposition, if the price of underlying stock follows the form of (4), the price of European call option is:

$$C(T, S(t)) = BSM(T - t, S(t); K, r^*, v(0)) \tag{5}$$

where

$$r^* = r + \frac{v(0)}{(T - t)} \int_t^T h(\tau) d\tau \tag{6}$$

Because (4) means a family of processes, (5) means a family of prices. We know that $h(t)$ determines the interval of price by (5) and (6). We deal with the Knightian uncertainty via a control-set $H = \{h(t) \mid |h(t)| \leq k\}$,[2] which is called as $k$-Ignorance by Chen and Epstein. $k$-Ignorance reflects the Knightian uncertainty, and gives the probability measure family a constraint. $k$ reflects the range of the Knightian uncertainty. But they did not identify the factors upon which $k$ depends. Here we know that the Poisson jump causes the Knightian uncertainty. So the Poisson process and the size of jumps decide the $k$-Ignorance. $k$ should reflect the characters of $q(t)$ and $Y$.

We define $k \triangleq \lambda^a \theta^b t^\gamma$, $a \geqslant 0, b \geqslant 0, |\gamma| < 1$ to show this relationship. $\lambda$ reflects the character of $q(t)$; and $\theta$ reflects the size of jumps. $k \triangleq \lambda^a \theta^b t^\gamma$ is a reasonable definition. A larger value of $\lambda$ means that the jump will happen with a larger probability. Similarly, a larger $\theta$ may contribute to a larger size for every jump. What's more, $a$ and $b$ reflect the individual sensitivity to $q(t)$ and $Y$. And $t^\gamma$ reflects the possible change of $k$-Ignorance; if $\gamma = 0$, it means the individual has a constant $k$-Ignorance. Then, we have $|h(t)| \leqslant \lambda^a \theta^b t^\gamma$. And we get the range for $h(t)$: $-\lambda^a \theta^b t^\gamma \leqslant h(t) \leqslant \lambda^a \theta^b t^\gamma$.

According to the features of $BSM(t^*, S; K, r^*, v^*)$, we know that the larger the value of $r^*$ is, the larger the price of option will be. When $h(t) = \lambda^a \theta^b t^\gamma$, we get the upper bound of the price, we note it as $\overline{C(T, S(t))}$; when $h(t) = -\lambda^a \theta^b t^\gamma$, we get the lower bound of the price $\underline{C(T, S(t))}$. They are:

---

[1] It is a simplification version of Shreve[11], exercise 5.4.

[2] If we consider equation (6), $h(t)$ should be noted as $h(\tau)$, and we have $t \leq \tau \leq T$.

$$\overline{C\left(T,S\left(t\right)\right)} = BSM\left(T-t,S\left(t\right);K,\overline{r},v\left(0\right)\right) \tag{7}$$

where,

$$\overline{r} = r + \frac{v\left(0\right)\lambda^a\theta^b}{\left(T-t\right)\left(\gamma+1\right)}\left(T^{\gamma+1} - t^{\gamma+1}\right) \tag{8}$$

$$\underline{C\left(T,S\left(t\right)\right)} = BSM\left(T-t,S\left(t\right);K,\underline{r},v\left(0\right)\right) \tag{9}$$

where,

$$\underline{r} = r + \frac{v\left(0\right)\lambda^a\theta^b}{\left(T-t\right)\left(\gamma+1\right)}\left(t^{\gamma+1} - T^{\gamma+1}\right) \tag{10}$$

The option price will be in the interval $\left[\underline{C\left(T,S\left(t\right)\right)},\overline{C\left(T,S\left(t\right)\right)}\right]$. In (7), (8), (9) and (10), we can see that if $\lambda = 0$ or $\theta = 0$, the interval will degenerate to a point; it means that there is no Knightian uncertainty in that situation. Also, we can find that the larger $\lambda$ and $\theta$ become, or the smaller $t$ (larger $T-t$) turns, the wider the price interval will be. This shows that it will be more difficult for the individual to get a precise price if the Poisson process is more intensive, the jump is more severe, or the maturity date is farther. A more violent situation or a longer time-latitude always brings larger Knightian uncertainty, and it is more difficult for the individual to get a precise price. In the next section, we will use Monte Carlo simulation to identify this character.

## 3   Simulation

Black-Scholes-Merton option pricing formula is widely used in reality (B-S price for short), but in fact, the volatility can not be observed. Commonly, people always assume that the volatility is constant and use some "average" value. In our viewpoint, if the Knightian uncertainty caused by jump volatility exists, there should be clearly bias between the B-S price and the true price. So we can identify some characters of pricing option with Knightian uncertainty by analyzing this price bias. Also, we can see that our price interval captures some important features of Kinghtian uncertainty.

We use Monte Carlo simulation to calculate the option's true price. As we assumed in section 2, the equations (1), (2) can be written as:

$$dS = rSdt + vSdw$$

$$v\left(t\right) = v\left(0\right)\prod_{n=0}^{q(t)} Y_n$$

where $Y_n \sim Unif\left[1-\theta, 1+\theta\right]$; $dw$, $q\left(t\right)$, and $Y_n$ are independent of each other.

We simulate the jump times $\tau_1, \tau_2, \cdots$ explicitly by generating the time interval for next jump, from the exponential distribution with mean $1/\lambda$. In the interval between two close jumps $(\tau_{i-1}, \tau_i]$, $S(t)$ evolves like a geometric Brownian motion with constant volatility $v(\tau_{i-1})$, so we have the following equations for the paths:

$$v(\tau_{i-1}) = v(\tau_{i-2}) Y_{i-1}$$

$$S(\tau_i) = S(\tau_{i-1}) \exp\left\{ \left(r - \tfrac{1}{2} v(\tau_{i-1})^2\right)(\tau_i - \tau_{i-1}) + v(\tau_{i-1})\left[w(\tau_i) - w(\tau_{i-1})\right] \right\}$$

We can get one "sample value" of the option price:

$$e^{-rT} \max[S(T) - K, 0]$$

Let $r = 0.2$, $v(0) = 0.2$, $S(0) = 100$ be constant for our simulation. And calculate every combination with different $\lambda$, $\theta$, $K$ and $T$ as Table 1. One combination means one option. After doing 10,000 simulations for every combination, we get a group of prices for different options.

**Table 1** Parameters of Monte Carlo simulation

| Parameter | Value |
|---|---|
| $\lambda$ | 60, 120, 180, 360 |
| $\theta$ | 0.02, 0.1 |
| $K$ | 80, 85, 90, 95, 100, 105, 110, 115, 120 |
| $T$ | 15, 30, 60, 90, 120, 150, 180, 210, 240, 270, 300, 330, 360 |
| Other Parameters | $r = 0.2$, $v(0) = 0.2$, $S(0) = 100$ |

When calculating the B-S price, we use the average volatility value: $E(v(t)) = v(0)$. The price bias has the following characters. First, the value of $\lambda$ dramatically affects the price bias. From Fig. 1, we can see the price bias clearly. With different $\lambda$ (See Fig.1), we can see that the larger the value of $\lambda$ is, the larger the bias will be. Because $\lambda$ reflects the impact of the Poisson process, a higher frequency of the jump will come along with a larger value of $\lambda$. There will be more Knightian Uncertainty and it is more difficult to know the precise probability distribution. Formulas (7), (8), (9), and (10) illustrate that the individual will use a wider price interval to accommodate the possible larger bias.

Second, the maturity date is important for the price bias. Fig.2 shows the price bias when expiration time is 15 days. We can see that there is almost no price bias. Comparing Fig.1 with Fig.2, we find that the price bias will disappear when the maturity date is closing. In our formulas, it means a narrower price interval.

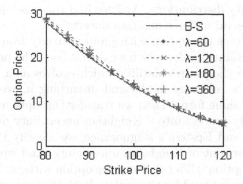

**Fig. 1** Price bias when $T=360$ days, $\theta=0.1$

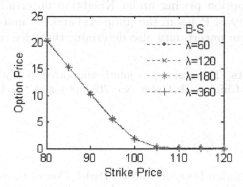

**Fig. 2** Price bias when $T=15$ days, $\theta=0.1$

Third, the jump size will affect the price bias. In the extreme case, if $\theta = 0$, and according to our formulas, the price interval will degenerate to a point. We can expect that there will be no price bias and vise versa. As Fig.3 shows, when we choice a lager $\theta$, the price bias will be clearly. Our simulation proves this character.

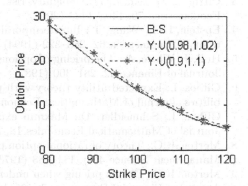

**Fig. 3** Price bias when $T=360$ days, $\lambda=360$

## 4 Conclusion

Frank Knight distinguished the risk and the Knightian uncertainty. Under Knightian uncertainty, the individual can not get a precise probability distribution because of information missing, ambiguity, etc. So he has to face a family of probability distributions. And we find that the option price is not unique, but an interval under Knightian uncertainty.

This paper analyzes the option with jump volatility from the perspective of Knightian uncertainty, under which we do not know the precise probability distribution due to the jump volatility which evolves as a Poisson process. Through selecting a reference model and disturbing it, we get a family of processes with the same form. Then we transfer this option pricing problem with jump volatility process into a Knightian uncertainty model.

Based on Chen and Epstein's $k$-Ignorance, we identify the factors which determine the constraint of Knightian uncertainty. And we get the formula for European call option. Differing from the option without Knightian uncertainty, we get a price interval for the option. Both the theoretical analysis and the Monte Carlo simulation show that our price interval captures the main characters of the option pricing under Knightian uncertainty. The factors, that is, the intensity of Poisson, the jump size and the maturity date which reflect the Knightian uncertainty also determine the price interval.

**Acknowledgements.** This research is jointly supported by National Natural Science Foundation of China with Grant No. 70831001 and No. 70671005 and No. 70821061.

# References

1. Bewley, T.: Knightian Decision Theory: Part I. Cowles Foundation Discussion Paper No. 807. Yale University, Massachusetts (1986)
2. Black, F., Scholes, M.: The valuation of option and corporate liabilities. Journal of Political Economy 81, 637–654 (1973)
3. Chen, Z., Epstein, L.G.: Ambiguity, risk, and asset returns in continuous time. Econometrica 70, 1403–1443 (2002)
4. Epstein, L.G., Wang, T.: Intertermporal asset pricing under knightian uncertainty. Econometrica 62, 283–322 (1994)
5. Hull, J., White, A.: The pricing of options on assets with stochastic volatilities. Journal of Finance 42, 281–300 (1987)
6. Gilboa, I.: Expected utility theory with purely subjective non-additive probabilities. Journal of Mathematical Economics 16, 65–88 (1987)
7. Gilboa, I., Schmeidler, D.: Maxmin expected utility with non-unique prior. Journal of Mathematical Economics 18, 141–153 (1989)
8. Merton, R.C.: Theory of rational option pricing. Bell Journal of Economics and Management Science 4(1), 141–183 (1973)
9. Merton, R.C.: Option pricing when underlying stock returns are discontinuous. Journal of Financial Economics 3, 125–144 (1976)
10. Schmeidler, D.: Subjective probability and expected utility without additivity. Econometrica 57, 571–587 (1989)
11. Shreve, S.E.: Stochastic Calculus for Finance II: Continuous-time Models. Springer, New York (2004)

# Conditional Ruin Probability with a Markov Regime Switching Model

Xuanhui Liu, Li-Ai Cui, and Fangguo Ren

**Abstract.** Ruin probabilities have been of a major interest in mathematical insurance. The diffusion process is used to the model of risk reserve of an insurance company usually. In this paper, we introduce a Markov chain and extend the Reserve processes to a jump-diffusion model and research the ruin probabilities. By using stochastic calculus techniques and the Martingale method a partial differential equation satisfied by the finite time horizon conditional ruin probability is obtained.

**Keywords:** Ruin probability, Regime-switching, Martingale, Jump-diffusion.

## 1 Introduction

For a long period of time, ruin probabilities have been of a major interest in mathematical insurance and have been investigated by many authors. The earliest work on this problem can be, at least, tracked back to Lundberg [4]. When we consider the ruin problems, the quantity of interest is the amount of surplus (by surplus, we mean the excess of some initial fund plus premiums collected over claims paid). We say ruin happens when the surplus becomes negative In order to track surplus, we need to model

Xuanhui Liu
School of Science, Xi'an Polytechnic University, Xian 710048, China
e-mail: lxhl112011@163.com

Li-Ai Cui
School of Science, Xi'an Polytechnic University, Xian 710048, China

Fangguo Ren
College of Mathematics and Information Science, Shaanxi Normal University,
Xi'an 710062, China
e-mail: rfangguo@snnu.edu.cn

S. Li (Eds.): Nonlinear Maths for Uncertainty and its Appli., AISC 100, pp. 295–300.
springerlink.com

the claim payments, premiums collected, investment incomes, and expenses, along with any other items that affect the cash flow. The classic risk metric model is a compound Poisson process. Iglehart [2] have showed that under the proper limit, the compound Poisson model could be approximated by a diffusion model. Recently, many papers have been using the diffusion model to model the surplus process of an insurance company. See for example, Asmussen and Taksar [1], Paulsen [5] , [6].Yang Hailiang [9] considered a diffusion model for the risk reserve and obtained a partial differential equation. Because insurance company's investment and operation is operated in market economy environment.Recently,regime-switching,orMarkov-modulated,models have received much attention among both researchers and market practitioners.shuxiangxie [8],Robert,J. [7],Ka-fai [3].The optimal portfolio selection is investigated in Markovian regime-switching financial market.the Markovian regime-switching model Provides a natural way of describe the impact of structural changes in (macro)-economic condition on asset price dynamics and risk. we consider risk model jointly driven by Brownian motion and Poisson process and Markovian regime-switching model is more realistic and practicable. We use Brownian motion describing the influence affected by the Market continuous information to the risk level, and use the Poisson process describing the influence affected by emergency information. Applicaton Markovian regime-switching the impact of structural changes in(macro)-economic condition on risk. This has more important theoretical significance to ruin probability's analysis and research.

## 2 The Dynamic Formulation

**Lemma 2.1.** Let $x(t)$ satisfy

$$dx(t) = b(t, x(t), r(t)dt + \sigma(t, x(t), r(t))dW(t)$$

and $\psi(t, x(t), i) \in C^2([0, \infty) \times R^n), i = 1, \ldots, l$ be given.Then,

$$E\{\varphi(T, x(T), r(T)) - \varphi(s, x(s), r(s))|r(s) = i\}$$
$$= E\{\int_s^T [\varphi_t(t, x(t), r(t)) + \Gamma\varphi(t, x(t), r(t))]dt|r(s) = i\}, \tag{1}$$

where

$$\Gamma\varphi(t, x, i) = \frac{1}{2}tr[\sigma(t, x, i)^T \varphi_{xx}(t, x, i)] + b(t, x, i)^T \varphi_x(t, x, i) + \sum_{j=1}^{l} q_{ij}\varphi(t, x, j).$$

Let the risk reserve level at time $t \geq 0$ be denoted by $R(t), \pi(t, \alpha(t), R(t))$, being the aggregate rate at which premiums are cashed at that time. Let $\mu(t)$ be the average aggregate claim rate at time $t$ and let $\beta(t)$ be the force of interest at time $t$ .Let $\alpha(t)$ be a continuous-time stationary Markov chain. $\alpha(t)$ takes value in a finite state space $M = \{1, 2, \ldots, l\}$.

$$\begin{cases} dR(t) = [\pi(t, \alpha(t), R(t)) + \beta(t)R(t) - \mu(t)]dt + \sigma(t, \alpha(t), R(t))dW(t) \\ \qquad\qquad\qquad + \varphi(t, \alpha(t), R(t))dN(t) \\ R(0) = r_0, \alpha(0) = i, (i = 1, 2, \ldots, l) \end{cases} \tag{2}$$

Where $W(t)$ is a standard Brownian motion and $W(t)$ is a Poisson process. Let $F_t^W$ be the completion of $\sigma\{W_s : 0 \leq s \leq t\}$ , the $\sigma$ fields generated by $W(t)$ . Let $F_t^N$ be the completion of $\sigma\{N_s : 0 \leq s \leq t\}$ , the $\sigma$ fields generated by $N(t)$. $W(t)$ , $N(t)$ and $\alpha(t)$ are independent of each other. The Markov chain $\alpha(t)$ has a generator $Q(q_{ij})_{l \times l}$ and stationary transition probabilities: $P_{ij}(t) = P\{\alpha(t) = j | \alpha(0) = i\}, t \geq 0, i, j = 1, 2, \ldots, l$ And generator is defined as:

$$q_{ij} = \begin{cases} \lim_{t \to 0^+} \frac{P_{ij}}{t}, i \neq j \\ \lim_{t \to 0^+} \frac{P_{ij}-1}{t}, i = j \end{cases} i, j = 1, 2, \ldots, l \tag{3}$$

$\sigma(t, \alpha(t), R(t))$ depends on $R(t)$. $\varphi(t, \alpha(t), R(t))$ is relatively jumping height under emergency information. Notice that the coefficients of the above stochastic differential equation depend on history only through $R(t)$ . Therefore, the reserve process $R(t)$ is a Markov process.

# 3  Partial Differential Equation Satisfied by the Conditional Ruin Probability

In this section, we will drive the partial differential equation satisfied by the conditional ruin probability. First, let us give some definitions.

**Definition 3.1.** The probability of ruin that occurs between time $t$ and $T$, given the reserve at time $t$ is $r$ and the Markov chain at time $t$ is $k$, is denoted by

$$\psi(t, k, r) = P(\inf_{t \leq s \leq T} R(s) < 0 | \alpha(t) = k, R(t) = r). \tag{4}$$

**Definition 3.2.** The time of the first ruin occurring is a stopping time, and is defined by $\tau = \inf\{t \geq 0 | R(t) < 0\}$. $\tag{5}$

In this paper, we will consider both the probability of ruin over a finite time horizon and over an infinity time horizon. For $0 < T \leq T \leq \infty, T < \infty$, corresponding to the finite time horizon, and $T = \infty$ corresponding to the infinity time horizon case.

**Lemma 3.1.** $\psi(t, \alpha(t), R(t))$ is a $F_{t \wedge \tau}$ martingale.

**Proof**

$$
\begin{aligned}
\psi(t \wedge \tau, \alpha(t \wedge \tau), R(t \wedge \tau)) &= P(\inf_{t \wedge \tau \leq s \leq T} R(s) < 0 | \alpha(t \wedge \tau), R(t \wedge \tau)) \\
&= E[I\{\inf_{t \wedge \tau \leq s \leq T} R(s) < 0 | \alpha(t \wedge \tau), R(t \wedge \tau)\}] \\
&= E[I\{\tau < T | \alpha(t \wedge \tau), R(t \wedge \tau)\}] \\
&= E[I\{\tau < T | F_{t \wedge \tau}\}], \qquad\qquad (6)
\end{aligned}
$$

Which is a $F_{t \wedge \tau}$ martingale and the last equalitdy holds because of the diffusion process $R(t)$ has strong Markovian property. The following theorem gives the main result of this section.

**Theorem 3.1.** The conditional ruin probability $\psi(t, k, r)$ satisfies the following parabolic partial differential equation:

$$
\begin{aligned}
&\tfrac{\partial \psi}{\partial t}(t, k, r) + [\pi(t, k, r) + \beta(t)r - \mu(t)]\tfrac{\partial \psi}{\partial r}(t, k, r) + \tfrac{1}{2}[\sigma(t, k, r) \times \\
&\tfrac{\partial^2 \psi}{\partial^2 r}(t, k, r)\sigma(t, k, r)] + \sum_{j=1}^{l} q_{ij} \psi(t, j, r) + \lambda E[\psi(t, k, r \\
&+ \phi(t, k, r)) - \psi(t, k, r)] = 0, \qquad\qquad (7)
\end{aligned}
$$

and the following condition: $\psi(t, k, r) = 1$ if $r < 0$. $\qquad\qquad (8)$

**Proof:** From It'o formula and Lemma 2.1, we have that

$$
\begin{aligned}
d\psi(t, k, r) &= [\tfrac{\partial \psi}{\partial s} + [\pi(s, k, r) + \beta(s)r - \mu(s)]\tfrac{\partial \psi}{\partial r} + \tfrac{1}{2}tr[\sigma(s, k, r)\tfrac{\partial^2 \psi}{\partial^2 r}\sigma(s, k, r)] \\
&\quad + \sum_{j=1}^{l} q_{ij} \psi(t, j, r)]ds + \sigma(s, k, r)\tfrac{\partial \psi}{\partial r}dW(s) \\
&\quad + [\psi(s, k, r + \phi(s, k, r)) - \psi(s, k, r)]dN(s) \\
&= [\tfrac{\partial \psi}{\partial s} + [\pi(s, k, r) + \beta(s)r - \mu(s)]\tfrac{\partial \psi}{\partial r} + \tfrac{1}{2}[\sigma(s, k, r)\tfrac{\partial^2 \psi}{\partial^2 r}\sigma(s, k, r)] \\
&\quad + \sum_{j=1}^{l} q_{ij} \psi(t, j, r)]ds + \sigma(s, k, r)\tfrac{\partial \psi}{\partial r}dW(s) \\
&\quad + \lambda E[\psi(s, k, r + \phi(s, k, r)) - \psi(s, k, r)]ds
\end{aligned}
$$

Where $\lambda E[\psi(s, k, r + \phi(s, k, r)) - \psi(s, k, r)]ds$ is Ruin probality's variation produced by Poisson jumping. Therefore,

$$\psi(t \wedge \tau, \alpha(t \wedge \tau), R(t \wedge \tau)) - \psi(0, \alpha(0), R(0))$$

$$= \int_0^{t \wedge \tau} [\frac{\partial \psi}{\partial s} + [\pi(s, k, r) + \beta(s)r - \mu(s)]\frac{\partial \psi}{\partial r} + \frac{1}{2}[\sigma(s, k, r)\frac{\partial^2 \psi}{\partial^2 r}\sigma(s, k, r)]$$

$$+ \sum_{j=1}^{l} q_{ij}\psi(s, j, r)ds + \int_0^{t \wedge \tau} \sigma(s, k, r)\frac{\partial \psi}{\partial r}dW(s)$$

$$+ \int_0^{t \wedge \tau} \lambda E[\psi(s, k, r + \phi(s, k, r)) - \psi(s, k, r)]ds$$

$$= \int_0^{t \wedge \tau} [\frac{\partial \psi}{\partial s} + [\pi(s, k, r) + \beta(s)r - \mu(s)]\frac{\partial \psi}{\partial r} + \frac{1}{2}[\sigma(s, k, r)\frac{\partial^2 \psi}{\partial^2 r}\sigma(s, k, r)]$$

$$+ \sum_{j=1}^{l} q_{ij}\psi(s, j, r) + \lambda E[\psi(s, k, r + \phi(s, k, r)) - \psi(s, k, r)]]ds$$

$$+ \int_0^{t \wedge \tau} \sigma(s, k, r)\frac{\partial \psi}{\partial r}dW(s).$$

By Lemma 3.1, we know that $\psi(t \wedge \tau, \alpha(t \wedge \tau))$ is a $F_{t \wedge \tau}$ martingale, so $\psi(t \wedge \tau, \alpha(t \wedge \tau), R(t \wedge \tau)) - \psi(0, i, r_0))$ is a zero initial valued martingale (obviously it is square integrable). Therefore,

$$\int_0^{t \wedge \tau} [\frac{\partial \psi}{\partial s} + [\pi(s, k, r) + \beta(s)r - \mu(s)]\frac{\partial \psi}{\partial r} + \frac{1}{2}[\sigma(s, k, r)\frac{\partial^2 \psi}{\partial^2 r}\sigma(s, k, r)]$$

$$+ \sum_{j=1}^{l} q_{ij}\psi(s, j, r)ds + \lambda E[\psi(s, k, r + \phi(s, k, r)) - \psi(s, k, r)]]ds = 0. \tag{9}$$

Since $t$ can be any nonnegative real number, we have

$$\frac{\partial \psi}{\partial s} + [\pi(s, k, r) + \beta(s)r - \mu(s)]\frac{\partial \psi}{\partial r} + \frac{1}{2}[\sigma(s, k, r)\frac{\partial^2 \psi}{\partial^2 r}\sigma(s, k, r)]$$

$$+ \sum_{j=1}^{l} q_{ij}\psi(s, j, r)ds + \lambda E[\psi(s, k, r + \phi(s, k, r)) - \psi(s, k, r)] = 0 \tag{10}$$

The condition (2.8) is obvious from the definition of $\psi(t, k, r)$.

# 4  Concluding Remarks

In this paper, we take the reserve level of a insurance company as the risk measurement size and study the problem of ruin probability driven by Brownian motion and Poisson process. By using generalized formula and Martingale method, a partial differential equations satisfied by the ruin probability

influenced by the Market continuous information and emergency information
is obtained. This has important theoretical significance on the finance.

**Acknowledgements.** This subject was supported by the National Natural Science Foundation of Shaanxi ( No. 2010JQ1013) and Natural Science Foundation of Shaanxi Province of China( No. JK0499).

# References

1. Asmussen, S.: ControlledDiffusionModelsfor Optimal Dividend Pay-Out. Insurance. Math Econ. 20, 1–15 (1997)
2. Iglehart, D.L.: DiffusionApproximations inCollective Risk Theory. J. App. Probability. 6, 258–262 (1969)
3. Ka-Fai: Optimal portfolios with regime-switching and Value-at-risk constrain. Automatica 46, 979–989 (2010)
4. Lundberg, F.: Approximerad Framstallning av Sannolikhetsfunktionen. Almqvist and Wiksell. Uppsala (1903)
5. Paulsen, J.: Ruin probability with stochastic Return on Investment. Adv. Appl. Prob. 29, 965–985 (1997)
6. Paulsen, J.: Ruin Theory with Compounding Assets -A Survey Insurance. Math. Econ. 22, 3–16 (1998)
7. Robert, J.: Portfolio risk minimization and differential games. Nonlinear Analysis 71, 2127–2135 (2009)
8. Xie, S.: Continuous-time mean-variance portfolio selection with liability and regime-switching, Insurance. Math. Econ. 45, 148–155 (2009)
9. Yang, H.: Conditional ruin probability with stochastic interest rate. Stochastic Analysis and Applications 19(2), 207–214 (2001)

# A Property of Two-Parameter Generalized Transition Function

Yuquan Xie

**Abstract.** We find that a generalized transition function $P_{s,t}(x, y, z, A)$ with two parameters $s, t > 0$ on a measurable space $(E, \mathscr{E})$ has a very interesting and important property: its total transition probability $P_{s,t}(x, y, z, E)$ is only related to the product $st$ of the two parameters $s, t > 0$ and is unrelated to the states $x, y, z \in E$. To be more exact, there is a constant $0 \leq \lambda \leq +\infty$ such that

$$P_{s,t}(x, y, z, E) \equiv \exp(-\lambda st), \ \forall s, t > 0, x, y, z \in E.$$

**Keywords:** Markov processes, Two parameters, Transition function.

## 1  Introduction

M.R.Cairoli [2] put forward a kind of transition function with two parameters which is often referred to as *three points transition function* or *tri-point transition function* [11]. It plays very important roles in Markov processes with two parameters[3 ~ 4], such as Brown sheet, Poisson sheet, Lévy sheet, Ornstein-Uhlenbeck processes, Bessel processes and so on. There are some investigations about the two-parameter transition functions (see, for examples, [5 ~ 11]). In this paper, we use mathematical analysis to find a very interesting and important property that a two-parameter generalized transition function has and an one-parameter one does not have.

**Definition 1.** [2] *Let* $(E, \mathscr{E})$ *be a general measurable space. A function family*

$$P_{s,t}(x, y, z, A), \ \forall \ s, t > 0, x, y, z \in E, A \in \mathscr{E} \tag{1}$$

Yuquan Xie
School of Mathematics and Computational Science in Xiangtan University,
Xiangtan, P.R. China, 411105
e-mail: xyqxyq@xtu.edu.cn

S. Li (Eds.): Nonlinear Maths for Uncertainty and its Appli., AISC 100, pp. 301–308.
springerlink.com                                    © Springer-Verlag Berlin Heidelberg 2011

is called a **two-parameter transition function** or **three points transition function** on $(E, \mathscr{E})$, if the following four conditions $(A) \sim (D)$ hold true

(A)  $\forall s, t > 0, x, y, z \in E$, the function $A \to P_{s,t}(x, y, z, A)$ is a probability measure on $(E, \mathscr{E})$.

(B)  $\forall s, t > 0, A \in \mathscr{E}$, the function $(x, y, z) \to P_{s,t}(x, y, z, A)$ is $\mathscr{E}^3$-measurable.

(C)  **Horizontal Chapman-Kolmogorov Equation:** $\forall s, h, t > 0, x, y, z, \in E, A \in \mathscr{E}$, we have

$$P_{s+h,t}(x, y, z, A) = \int_{\eta \in E} P_{s,t}(x, y, \xi, d\eta) P_{h,t}(\xi, \eta, z, A), \ \forall \xi \in E. \qquad (2)$$

(D)  **Vertical Chapman-Kolmogorov Equation:** $\forall s, t, h > 0, x, y, z, \in E, A \in \mathscr{E}$, we have

$$P_{s,t+h}(x, y, z, A) = \int_{\eta \in E} P_{s,t}(x, \xi, z, d\eta) P_{s,h}(\xi, y, \eta, A), \ \forall \xi \in E. \qquad (3)$$

If the condition $(A)$ is substituted by $(A')$ as follows:

$(A')$  $\forall s, t > 0, x, y, z \in E$, the function $A \to P_{s,t}(x, y, z, A)$ is a nonnegative measure on $(E, \mathscr{E})$ with

$$0 \le P_{s,t}(x, y, z, E) \le 1. \qquad (4)$$

Then $P_{s,t}(x, y, z, A)$ is called a *two-parameter generalized transition function*

If the condition $(A)$ is substituted by $(A'')$ as follows:

$(A'')$  $\forall s, t > 0, x, y, z \in E$, the function $A \to P_{s,t}(x, y, z, A)$ is a measure on $(E, \mathscr{E})$ with

$$1 \le P_{s,t}(x, y, z, E) \le +\infty. \qquad (5)$$

Then $P_{s,t}(x, y, z, A)$ is called a *two-parameter second generalized transition function*

## 2  Main Results

**Theorem 2.** Let $(E, \mathscr{E})$ be a general measurable space, $P_{s,t}(x, y, z, A)$ be a two-parameter generalized transition function on $(E, \mathscr{E})$. Then there exists a constant number $0 \le \lambda \le +\infty$ such that

$$P_{s,t}(x, y, z, E) \equiv \exp(-\lambda st), \ \forall s, t > 0, x, y, z \in E.$$

**Theorem 3.** Let $(E, \mathscr{E})$ be a general measurable space, $P_{s,t}(x, y, z, A)$ be a two-parameter second generalized transition function on $(E, \mathscr{E})$. Then there exists a constant number $0 \le \lambda \le +\infty$ such that

$$P_{s,t}(x, y, z, E) \equiv \exp(\lambda st), \ \forall s, t > 0, x, y, z \in E.$$

It is especially worth noting that, the one-parameter generalized transition functions does not have similar results as Theorem 2 and 3.

## 3  Several Lemmas

*In order to prove Theorem 2 and 3 above, we firstly give several lemmas as follows:*

**Lemma 4.** Let $f(s)$ be a monotonic function in $(0, +\infty)$. Then the discontinuous points of the function $f(s)$ in $(0, +\infty)$ form at most a countable set and both the left limit $f(s-)$, $\forall s > 0$ and the right limit $f(s+)$, $\forall s \geq 0$ are existent.

**Lemma 5.** Let $f(s)$ be a monotonic decreasing function in $(0, +\infty)$. If $f(s + h) = f(s)f(h)$, $\forall s, h > 0$, then there exists a constant number $0 \leq \lambda \leq +\infty$ such that $f(s) = \exp(-\lambda s)$, $\forall s > 0$.

**Lemma 6.** Let $f(s)$ be a monotonic decreasing function in $(0, +\infty)$, $g(s)$ be a continuous everywhere function in $(0, +\infty)$, and $C$ be a dense subset in $(0, +\infty)$. If $f(s+) = g(s)$, $\forall s \in C$, then the function $f(s) = g(s)$, $\forall s > 0$.

**Lemma 7** Let $f(s), g(s)$ be two monotonic decreasing functions in $(0, +\infty)$. If $f(s + h) = f(s)g(h)$, $\forall s, h > 0$, then the function $f(s)$ is continuous everywhere in $(0, +\infty)$.

The above Lemma 4 and 5 are well-known, their proofs can be found in general textbooks, but Lemma 6 and 7 may be new, despite their proofs are not very difficult.

**Proof of Lemma 6.** *For arbitrary $s > 0$, because $C$ is dense in $(0, +\infty)$, we always can take two series $\{s_n\}, \{t_n\} \subseteq C$ such that $s_n < s < t_n$, $\forall n \geq 1$ and $s_n \uparrow s$, $t_n \downarrow s$ as $n \uparrow \infty$. Because the function $f(s)$ is monotonic decreasing, we have*

$$g(s_n) = f(s_n+) \geq f(s-) \geq f(s) \geq f(s+) \geq f(t_n+) = g(t_n), \ \forall \, n \geq 1.$$

*Letting $n \uparrow +\infty$, we have $g(s-) \geq f(s) \geq g(s+)$. By $g(s)$ is continuous in $(0, +\infty)$, we have $g(s-) = g(s) = g(s+)$. Thus $f(s) = g(s)$, $\forall s > 0$.*

**Proof of Lemma 7.** *Because $f(s)$ is a monotonic decreasing function in $(0, +\infty)$, by Lemma 4, we know that the right limit $f(s+)$, $\forall s \geq 0$ is existent. By the conditions $f(s+h) = f(s)g(h)$, $\forall s, h > 0$, we have $f(s+) = f(0+)g(s)$ and $f((s + h)+) = f(s+)g(h)$, $\forall \, s, h > 0$. Thus*

$$f(0+)g(s + h) = f(0+)g(s)g(h), \ \forall \, s, h > 0.$$

*If $f(0+) = 0$, then $f(s+) \equiv 0$, $\forall \, s > 0$.*
*If $f(0+) \neq 0$, then $g(s + h) = g(s)g(h)$, $\forall \, s, h > 0$. By Lemma 5, there exists a constant number $0 \leq \lambda \leq +\infty$ such that $g(s) = \exp(-\lambda s)$, $\forall s > 0$.*

*Thus*
$$f(s+) = f(0+)\exp(-\lambda s), \ \forall s > 0.$$

*Because the function $f(s)$ is monotonic decreasing, by Lemma 6, we have*

$$f(s) \equiv 0, \ \forall s > 0 \quad or \quad f(s) = f(0+)\exp(-\lambda s), \ \forall s > 0.$$

*It is shown the function $f(s)$ is continuous everywhere in $(0, +\infty)$.*

## 4 Proofs of Theorem

**Proof of Theorem 2.** *$\forall s, t, h > 0, x, y, z \in E$, by (3) and (4), we have*

$$P_{s,t+h}(x, y, z, E) \leq P_{s,t}(x, \xi, z, E), \ \forall \xi \in E. \tag{6}$$

*Specially, when $\xi = y$, we have*

$$P_{s,t+h}(x, y, z, E) \leq P_{s,t}(x, y, z, E).$$

*It is shown that the function $t \to P_{s,t}(x, y, z, E)$ is monotonic decreasing. By Lemma 4, the limit*

$$P_{s,t+}(x, y, z, E) \triangleq \lim_{h \to 0+} P_{s,t+h}(x, y, z, E), \tag{7}$$

*is existent. By (6), we easy check that*

$$P_{s,t+h}(x, y, z, E) \leq P_{s,t+}(x, \xi, z, E), \quad \forall \xi \in E, \tag{8}$$
$$P_{s,t+}(x, y, z, E) \leq P_{s,t}(x, \xi, z, E), \quad \forall \xi \in E, \tag{9}$$
$$P_{s,t+}(x, y, z, E) \leq P_{s,t+}(x, \xi, z, E), \quad \forall y, \xi \in E. \tag{10}$$

*By (10) and the symmetries of $y, \xi \in E$, we have*

$$P_{s,t+}(x, y, z, E) = P_{s,t+}(x, \xi, z, E), \quad \forall y, \xi \in E. \tag{11}$$

*It is shown that $P_{s,t+}(x, y, z, E)$ is unrelated to $y \in E$.*
*For arbitrary $s, t, h, r > 0, x, y, z, \xi \in E$, by (2),(8) and (11), we have*

$$P_{s+h,t+r}(x, y, z, E)$$
$$= \int_{\eta \in E} P_{s,t+r}(x, y, \xi, d\eta) P_{h,t+r}(\xi, \eta, z, E)$$

$$\leq \int_{\eta \in E} P_{s,t+r}(x, y, \xi, d\eta) P_{h,t+}(\xi, \eta, z, E)$$
$$= P_{s,t+r}(x, y, \xi, E) P_{h,t+}(\xi, \eta, z, E), \ \forall \eta \in E,$$
$$\leq P_{s,t+}(x, y, \xi, E) P_{h,t+}(\xi, \eta, z, E), \ \forall \eta \in E,$$

*Letting $r \downarrow 0$, for arbitrary $\xi, \eta \in E$, we have*

$$P_{s+h,t+}(x,y,z,E) \leq P_{s,t+}(x,y,\xi,E)P_{h,t+}(\xi,\eta,z,E). \qquad (12)$$

*Again by (2),(9) and (11), we have*

$$P_{s+h,t}(x,y,z,E)$$
$$= \int_{\eta \in E} P_{s,t}(x,y,\xi,d\eta)P_{h,t}(\xi,\eta,z,E)$$
$$\geq \int_{\eta \in E} P_{s,t}(x,y,\xi,d\eta)P_{h,t+}(\xi,\eta,z,E)$$
$$= P_{s,t}(x,y,\xi,E)P_{h,t+}(\xi,\eta,z,E), \quad \forall \eta \in E,$$
$$\geq P_{s,t+}(x,y,\xi,E)P_{h,t+}(\xi,\eta,z,E), \quad \forall \eta \in E,$$

*that is,*

$$P_{s+h,t}(x,y,z,E) \geq P_{s,t+}(x,y,\xi,E)P_{h,t+}(\xi,\eta,z,E). \qquad (13)$$

*By (12) and (13), we have*

$$P_{s+h,t+}(x,y,z,E) \leq P_{s,t+}(x,y,\xi,E)P_{h,t+}(\xi,\eta,z,E) \leq P_{s+h,t}(x,y,z,E). \quad (14)$$

*For arbitrary $t > 0$, we always can take a series $\{t_n\} \subseteq (t,+\infty)$ such that $t_n \downarrow t$ as $n \uparrow \infty$. By (14), we have*

$$P_{s+h,t_n+}(x,y,z,E) \leq P_{s,t_n+}(x,y,\xi,E)P_{h,t_n+}(\xi,\eta,z,E) \leq P_{s+h,t_n}(x,y,z,E),$$

$$P_{s+h,t+}(x,y,z,E) \leq P_{s,t+}(x,y,\xi,E)P_{h,t+}(\xi,\eta,z,E) \leq P_{s+h,t+}(x,y,z,E).$$

*Thus*

$$P_{s+h,t+}(x,y,z,E) = P_{s,t+}(x,y,\xi,E)P_{h,t+}(\xi,\eta,z,E). \qquad (15)$$

*Specially, when $\xi = z$, we have*

$$P_{s+h,t+}(x,y,z,E) = P_{s,t+}(x,y,z,E)P_{h,t+}(z,\eta,z,E). \qquad (16)$$

*By (4) and (16), we have*

$$P_{s+h,t+}(x,y,z,E) \leq P_{s,t+}(x,y,z,E).$$

*It is shown that the function $s \to P_{s,t+}(x,y,z,E)$ also is monotonic decreasing. By Lemma 4, the limit*

$$P_{s+,t+}(x,y,z,E) \hat{=} \lim_{h \to 0+} P_{s+h,t+}(x,y,z,E), \qquad (17)$$

*is existent. By (4) and (15), we have*

$$P_{s+,t+}(x,y,z,E) \leq P_{s+,t+}(x,y,\xi,E), \ \forall \ z,\xi \in E.$$

*By the symmetries of $z,\xi \in E$, we have*

$$P_{s+,t+}(x,y,z,E) = P_{s+,t+}(x,y,\xi,E), \ \forall \ z,\xi \in E.$$

*It is shown that $P_{s+,t+}(x,y,z,E)$ is not only unrelated to $y \in E$, but also unrelated to $z \in E$.*

*By (16) and Lemma 7, we know that the function $s \to P_{s,t+}(x,y,z,E)$ is continuous everywhere in $(0,+\infty)$ and is unrelated to $y, z \in E$. For simplicity, we will denote $P_{s,t+}(x,y,z,E)$ by $\varphi_{s,t}(x)$, that is,*

$$\varphi_{s,t}(x) \hat{=} P_{s,t+}(x,y,z,E), \ \forall \ x,y,z \in E, s,t > 0.$$

*By (15), we have*

$$\varphi_{s+h,t}(x) = \varphi_{s,t}(x)\varphi_{h,t}(\xi), \ \forall \ x,\xi \in E.$$

*By Lemma 7, we know that the function $s \to \varphi_{s,t}(x)$ is continuous in $(0,+\infty)$. By (4), we have*

$$\varphi_{h,t}(x) = \varphi_{h+,t}(x) = \lim_{s \downarrow 0} \varphi_{s+h,t}(x) \leq \varphi_{h,t}(\xi), \ \forall \ x,\xi \in E.$$

*By the symmetries of $x,\xi \in E$, we know that $\varphi_{s,t}(x)$ also is unrelated to $x \in E$. Let $\varphi_{s,t} \hat{=} \varphi_{s,t}(x), \ \forall \ x \in E$. Then*

$$\varphi_{s+h,t} = \varphi_{s,t}\varphi_{h,t}, \ \forall \ s,h > 0.$$

*By Lemma 5, there exists a constant number $0 \leq \mu(t) \leq +\infty$ such that*

$$\varphi_{s,t} = \exp(-\mu(t)s), \ \forall \ s,t > 0,$$

*that is,*

$$P_{s,t+}(x,y,z,E) = \exp(-\mu(t)s), \ \forall \ s,t > 0. \tag{18}$$

*Similarly, for arbitrary $s > 0$, there also exists a constant number $0 \leq \lambda(s) \leq +\infty$ such that*

$$P_{s+,t}(x,y,z,E) = \exp(-\lambda(s)t), \ \forall \ s,t > 0. \tag{19}$$

*Let $C$ be a subset of right continuous point of the function $t \to P_{1,t}(x,y,z,E)$. By (18), we know that the function $s \to P_{s,t+}(x,y,z,E)$ is continuous everywhere in $(0,+\infty)$. Thus*

$$P_{1,t}(x,y,z,E) = P_{1,t+}(x,y,z,E) = P_{1+,t+}(x,y,z,E)$$
$$\leq P_{1+,t}(x,y,z,E) \leq P_{1,t}(x,y,z,E), \ \forall \ t \in C.$$

*So*

$$P_{1,t+}(x,y,z,E) = P_{1+,t}(x,y,z,E) = P_{1,t}(x,y,z,E), \ \forall\, t \in C.$$

*By (18) and (19), we have*

$$\exp(-\mu(t)) = \exp(-\lambda(1)t) = \exp(-\lambda t), \ \forall\, t \in C,$$

*where $\lambda \hat{=} \lambda(1)$. Thus*

$$P_{s,t+}(x,y,z,E) = \exp(-\lambda st), \ \forall\, s > 0, \ t \in C. \tag{20}$$

*By Lemma 4, we know that the subset $C$ is dense in $(0,+\infty)$. Thus, by (20) and Lemma 6, we have*

$$P_{s,t}(x,y,z,E) = \exp(-\lambda st), \ \forall\, s,t > 0. \qquad \Box$$

**Proof of Theorem 3.** *It is completely similar to the proofs of Theorem 2.*

## 5 Application of Theorem

Theorem 2 and 3 are very interesting and important results, we give some application examples as follows:

**Example 8.** If $P_{s,t}(x,y,z,A) \neq 0$ is a horizontal constant type (or vertical constant type) two-parameter generalized transition function on $(E,\mathscr{E})$, that is, $P_{s,t}(x,y,z,A)$ is unrelated to $s > 0$ (or $t > 0$), then $P_{s,t}(x,y,z,A)$ is a two-parameter transition function on $(E,\mathscr{E})$.

**Proof.** *By Theorem 2, there exists a constant $0 \leq \lambda < +\infty$ such that $P_{s,t}(x,y,z,E) = exp(-\lambda st)$. Because $P_{s,t}(x,y,z,E)$ is unrelated to $s > 0$ (or $t > 0$), thus $\lambda = 0$ or $+\infty$. However $P_{s,t}(x,y,z,E) \neq 0$, thus $\lambda = 0$ and $P_{s,t}(x,y,z,E) \equiv 1$.* $\qquad \Box$

**Remark 9.** *If $P_{s,t}(x,y,z,A)$ is a two-parameter transition function on $(E,\mathscr{E})$, then under some conditions, we can prove that both limits*

$$\lim_{s\downarrow 0} P_{s,t}(x,y,z,A) \quad and \quad \lim_{s\uparrow\infty} P_{s,t}(x,y,z,A)$$

*are two horizontal constant type two-parameter generalized transition functions on $(E,\mathscr{E})$. Theorem 2 more further shows that two limits above are two horizontal constant type two-parameter transition functions.*

**Example 10.** *As all know that an one-parameter generalized transition function $P(s,x,A)$ on a measurable space $(E,\mathscr{E})$ can be turn into an one-parameter transition function $\tilde{P}(s,x,A)$ on another measurable space $(\tilde{E},\tilde{\mathscr{E}})$ by appending a new state $\delta$ into $E$ as follows:*

$$\tilde{P}(s,x,A) \hat{=} \begin{cases} P(s,x,A), & \text{when } x \in E, A \in \mathscr{E}, \\ 0, & \text{when } x = \delta, A \in \mathscr{E}, \\ 1 - P(s,x,E), & \text{when } x \in E, A = \Delta, \\ 1, & \text{when } x = \delta, A = \Delta, \end{cases} \quad (21)$$

where $\delta \notin E, \Delta = \{\delta\}, \tilde{E} = E \bigcup \Delta, \tilde{\mathscr{E}} = \sigma(\mathscr{E} \bigcup \Delta)$.

However, for a two-parameter generalized transition function $P_{s,t}(x,y,z,A)$ on $(E,\mathscr{E})$, according to Theorem 2, only need to take

$$\bar{P}_{s,t}(x,y,z,A) = exp(\lambda st) P_{s,t}(x,y,z,A),$$

we can turn $P_{s,t}(x,y,z,A)$ into a two-parameter transition function $\bar{P}_{s,t}(x,y, z,A)$ on the same measurable space $(E,\mathscr{E})$.

# References

1. Bodnariu, M.: Three-point transition functions and Markov properties. Math.Rep. Bucur. 5(55), 219–232 (2003)
2. Cairoli, M.R.: Une classe de processus de Markov. C.R. Acad. Sc, Paris, Ser A. 273, 1071–1074 (1971)
3. Guo, J.Y., Yang, Y.Q.: Three-point transition function for two-parameter Markov chains and their four systems of partial differential equations. Science in China, Series A 35(7), 806–818 (1992)
4. Khoshnevisan, D.: Multiparameter Processes. An Introduction to Random Fields. Springer, New York (2002)
5. Liu, S.Y.: Density functions of three points transition functions of two-parameter Markov chains. Natural science journal of Xiangtan University 15(3), 26–27 (1993) (in Chinese)
6. Xie, Y.Q.: Standard tri-point transition function. Science in China, Series A 48(7), 904–914 (2005)
7. Xie, Y.Q.: Tri-point transition function with three states. Journal of Engineering mathematics 23(4), 733–737 (2006) (in Chinese)
8. Xie, Y.Q.: A family of tri-point transition function with two states. Scientific research monthly 5(5), 45–46 (2007)
9. Yang, Y.Q.: Wide-past homogeneous Markov processes with two parameters and two states three-point transition function. Chinese Science Bulletin 41(2), 192 (1996) (in Chinese)
10. Yang, X.Q., Li, Y.Q.: Markov Processes with Two Parameters. Nunan Science and Technology Press (1996) (in chinese)
11. Wang, Z.K.: Ornstein-Uhlenbeck processes with two parameters. Acta Mathematica Scientia 4(1), 1–12 (1984)

# The Extinction of a Branching Process in a Varying or Random Environment

Yangli Hu, Wei Hu, and Yue Yin

**Abstract.** A sufficient condition and a necessary condition for extinction of BPVE are given. Then a splitting statistical regularity of individual in every generation of BPRE is given. In the last example, the speed of extinction of BPRE is contrasted with that of BPVE.

**Keywords:** Random environment, Varying environment, Branching process, Extinction.

## 1 Introduction

As an extension of classical branching process, the research on a branching process in a random environment (BPRE) and a branching process in a varying environment (BPVE) dated from 1960s. It is one of the most fruitful fields about the research on branching processes. A series of profound results were acquired (details sees in [2] [3] [7] [5] [6] [1] [4] and their bibliographies). The mathematical formulation of BPRE was given in [3]. In [3], the measurability related to the definition was proved, an equivalent theorem was given and the existence of this model was proved. This laid a solid foundation of the further research on it. In [2], bounds of the extinction probability of BPRE and BPVE were obtained, then a sufficient and necessary condition for extinction of BPVE and a sufficient condition and a necessary condition for certain extinction of BPRE were formed. But they were not proved and the proofs are not trivial. In this paper, we give a series of corresponding proofs and then obtain a sufficient condition and a necessary condition (Theorem 1

Hu Yangli, Hu Wei, and Yin Yue
College of Mathematics and Computing Science, Changsha University of Science and Technology, Changsha 410004, China
e-mail: `huyangli76@sohu.com`, `huwei001168@sohu.com`,
`yinyuejisuan0501@126.com`

S. Li (Eds.): Nonlinear Maths for Uncertainty and its Appli., AISC 100, pp. 309–315.
springerlink.com &copy; Springer-Verlag Berlin Heidelberg 2011

and 2). Thus the sufficient and necessary condition given by Agresti in 1975 becomes a corollary. And we give a splitting statistical regularity of individual in every generation of BPRE. The last example shows that how to use the extinction probability of BPVE to describe the extinction probability of BPRE.

## 2  BPVE

Let $\{Z_n : n \geq 0\}$ be BPVE, satisfying that

$$Z_0 = 1, Z_{n+1} = \sum_{i=1}^{Z_n} X_{ni}, \tag{1}$$

where for any fixed $n \geq 0$, $\{X_{ni} : i \geq 1\}$ is a sequence of independent and identically distributed random variables taking values in the set of nonnegative integers, but $\{X_{ni} : n \geq 0, i \geq 1\}$ is just a sequence of independent random variables. For any fixed $n \geq 0, i \geq 1$, $X_{ni}$ denotes the number of the offspring produced by the $i^{\text{th}}$ individual in the $n^{\text{th}}$ generation with he probability generating function(pgf) $f_n(s) = E(s^{X_{ni}})$. A necessary condition for certain extinction of BPVE was given in the next. Let

$$T = \min\{n : Z_n = 0\}, \quad q = \lim_{n \to \infty} P(Z_n = 0), \quad P_n = \prod_{j=0}^{n-1} f_j'(1),$$

$$M = \sup_{n \geq 0} \frac{f_n''(1)}{f_n'(1)}, N = \sum_{j=0}^{\infty} \frac{1}{P_{j+1}}, \delta(n) = \inf_{j \geq n} \frac{f_j''(0)}{f_j'(1)}.$$

**Theorem 1.** If $M < \infty, f_j''(1) < \infty, j \geq 0, q = 1$, then $N = \infty$.

*Proof.* Assuming that $N < \infty$, then

$$P_n \to \infty, n \to \infty. \tag{2}$$

According to (2.4) in [2], we have

$$P(T \leq n) \leq 1 - [P_n^{-1} + \sum_{j=0}^{n-1} \frac{f_j''(1)}{f_j'(1)P_{j+1}}]^{-1}$$

$$\leq 1 - [P_n^{-1} + M \sum_{j=0}^{n-1} P_{j+1}^{-1}]^{-1} \leq 1 - [P_n^{-1} + NM]^{-1}. \tag{3}$$

Taking limit in both sides of (3) as $n \to \infty$, we have

$$q = \lim_{n \to \infty} P(T \leq n) \leq 1 - \frac{1}{MN} < 1.$$

It contradict the condition $q = 1$. □

Conversely, we can also obtain a sufficient condition for certain extinction.

**Theorem 2.** *If* $f_j''(1) < \infty, j \geq 0, N = \infty$ , *and there exists a finite nonnegative integer* $n_0$ *such that* $\delta(n_0) > 0$, *then* $q = 1$.

*Proof.* According to (2.4) in [2], we have

$$P(T \leq n) \geq 1 - [P_n^{-1} + \frac{1}{2} \sum_{j=0}^{n-1} \frac{f_j''(0)}{f_j'(1)P_{j+1}}]^{-1}$$

$$\geq 1 - [P_n^{-1} + \frac{1}{2} \sum_{j=0}^{n_0-1} \frac{f_j''(0)}{f_j'(1)P_{j+1}} + \frac{1}{2}\delta(n_0) \sum_{j=n_0}^{n-1} P_{j+1}^{-1}]^{-1}$$

$$\geq 1 - [\frac{1}{2}\delta(n_0) \sum_{j=n_0}^{n-1} P_{j+1}^{-1}]^{-1}, n > n_0. \tag{4}$$

Because $\delta(n_0) > 0, N = \infty$, taking limit in both sides of (4) as $n \to \infty$, we have

$$q - \lim_{n \to \infty} P(T \leq n) \geq 1,$$

so $q = 1$. □

The following corollary can be obtained from Theorem 1 and Theorem 2. This is just the conclusion of Theorem 1 in [2].

**Corollary 1.** *If* $f_j''(1) < \infty, j \geq 0, M < \infty$ *and exist a finite nonnegative integer* $n_0$ *such that* $\delta(n_0) > 0$, *then* $q = 1 \Leftrightarrow N = \infty$.

# 3 BPRE

Let $\xi = \{\xi_n : n \geq 0\}$ be a sequence of independent random variables defined on a probability space $(\Omega, \mathfrak{S}, P)$ and taking values in a measurable space $(\Theta, \Sigma)$. Let $P(\cdot|\xi) = P_\xi(\cdot)$ and $E(\cdot|\xi) = E_\xi(\cdot)$.

Let $\{Z_n : n \geq 0\}$ be BPRE, satisfying (1). For a given random environment $\xi$, $\{X_{ni} : n \geq 0, i \geq 1\}$ is a sequence of independent random variables taking value in the set of non-negative integers and satisfying

$$P_\xi(X_{nj} = r_{nj}, 1 \leq j \leq l, 0 \leq n \leq m) = \prod_{n=0}^{m} \prod_{j=1}^{l} P_{\xi_n}(X_{nj} = r_{nj}).$$

The conditional pgf of the number of offspring created by an individual in the $n^{\text{th}}$ generation is denoted by

$$f_n(s) \equiv f_{\xi_n}(s) = E_\xi(s^{X_{ni}}).$$

For a given random environment $\xi$, the conditional pgf of the number of all the individuals in the $n^{\text{th}}$ generation of the process $\{Z_n, n \geq 0\}$ is denoted by $\varphi_n(\xi; s) = E_\xi(s^{Z_n})$, then by Theorem 2.1 in [3], we have $\varphi_n(\xi; s) = f_{\xi_0}(f_{\xi_1}(\cdots f_{\xi_{n-1}}(s)\cdots))$. Set $\varphi_n(s) = E(s^{Z_n})$. Let

$$U = \sup_{j \geq 0} E\left(\frac{f_j''(1)}{(f_j'(1))^2}\right), V = \sum_{j=1}^{\infty} \prod_{i=0}^{j-1} E\left(\frac{1}{f_i'(1)}\right),$$

$$\gamma(n) = \inf_{j \geq n} \frac{Ef_j''(0)}{Ef_j'(1)}, W = \sum_{j=0}^{\infty} \prod_{i=0}^{j} \frac{1}{Ef_i'(1)}, A = \sup_{j \geq 1} E\left(\frac{1}{\prod_{i=0}^{j-1} f_i'(1)}\right)^2,$$

we can also obtain some conditions for certain extinction.

**Theorem 3.** *(i) If $U < \infty, q = 1$, then $V = \infty$.*

*(ii)Let $\xi$ be a sequence of independent and identically distributed random variables, if $q = 1, E\left(\frac{f_0''(1)}{(f_0'(1))^2}\right) < \infty$, then $E\left(\frac{1}{f_0'(1)}\right) \geq 1$.*

*Proof.* (i) Assuming that $V < \infty$, then we have

$$\prod_{i=0}^{\infty} E\left(\frac{1}{f_i'(1)}\right) = 0. \tag{5}$$

According to (3.4) in [2], we have

$$P(T \leq n) \leq 1 - \left[\prod_{j=0}^{n-1} E\left(\frac{1}{f_j'(1)}\right) + \sum_{j=1}^{n-1} E\left(\frac{f_j''(1)}{(f_j'(1))^2}\right) \prod_{i=0}^{j-1} E\left(\frac{1}{f_i'(1)}\right)\right]^{-1}$$

$$\leq 1 - \left[\prod_{j=0}^{n-1} E\left(\frac{1}{f_j'(1)}\right) + U \sum_{j=1}^{n-1} \prod_{i=0}^{j-1} E\left(\frac{1}{f_i'(1)}\right)\right]^{-1}. \tag{6}$$

Taking limit in both sides of (6) as $n \to \infty$, we have

$$q = \lim_{n \to \infty} P(T \leq n) \leq 1 - (UV)^{-1} < 1,$$

It contradict the condition $q = 1$.

(ii) Because we have the fact that

$$U = E\left(\frac{f_0''(1)}{(f_0'(1))^2}\right), \tag{7}$$

$$V = \sum_{j=1}^{\infty} \prod_{i=0}^{j-1} E\left(\frac{1}{f_i'(1)}\right) = \sum_{j=1}^{\infty} \left[E\left(\frac{1}{f_0'(1)}\right)\right]^j. \tag{8}$$

We can get the conclusion from (7), (8) and (i). □

**Theorem 4.** (i) If $W = \infty$, and there exists a finite nonnegative integer $n_0$ such that $\gamma(n_0) > 0$, then $q = 1$.

(ii) Let $\xi$ be a sequence of independent and identically distributed random variables, if

$$\frac{E f_0''(0)}{E f_0'(1)} > 0, E f_0'(1) \le 1,$$

then $q = 1$.

*Proof.* (i) According to (3.2) in [2], for all $n \ge n_0$, we have

$$P(T \le n) \ge 1 - \left[\left(\prod_{j=0}^{n-1} E f_j'(1)\right)^{-1} + \frac{1}{2} \sum_{j=0}^{n-1} \frac{E f_j''(0)}{E f_j'(1) \cdot \prod_{i=0}^{j} E f_i'(1)}\right]^{-1}$$

$$\ge 1 - \left[\left(\prod_{j=0}^{n-1} E f_j'(1)\right)^{-1} + \frac{1}{2} \sum_{j=0}^{n_0-1} \frac{E f_j''(0)}{E f_j'(1) \cdot \prod_{i=0}^{j} E f_i'(1)}\right.$$

$$\left.+ \frac{\gamma(n_0)}{2} \sum_{j=n_0}^{n-1} \prod_{i=0}^{j} \frac{1}{E f_i'(1)}\right]^{-1}$$

$$\ge 1 - \left[\frac{\gamma(n_0)}{2} \sum_{j=n_0}^{n-1} \prod_{i=0}^{j} \frac{1}{E f_i'(1)}\right]^{-1}. \tag{9}$$

Taking limit in both sides of (9) as $n \to \infty$, since $\gamma(n_0) > 0, W = \infty$, we have

$$q = \lim_{n \to \infty} P(T \le n) \ge 1,$$

so $q = 1$.

(ii) Because we have the fact that

$$\gamma(n) \equiv \frac{E f_0''(0)}{E f_0'(1)}, \tag{10}$$

$$W = \sum_{j=0}^{\infty} \left[\frac{1}{E f_0'(1)}\right]^{j+1}, \tag{11}$$

so we can get the conclusion from (10), (11) and (i). □

**Theorem 5.** Let $\xi$ be a sequence of independet random variables, if $\inf_{j \ge 0} P(f_j'(1) \le 1) > 0$, $A < \infty$ and $V = \infty$, then as $j \to \infty$, there exist an integer $k \ge j$ such that $P(f_k'(1) \le 1) = 1$.

*Proof.* By the assumption, we have

$$\sum_{j=1}^{\infty} E\left(\frac{1}{\prod_{i=0}^{j-1} f_i'(1)}\right) = V = \infty,$$

and

$$\infty = \sum_{j=1}^{\infty} E\left(\frac{1}{\prod_{i=0}^{j-1} f_i'(1)}\right) P\left(\frac{1}{f_j'(1)} \geq 1\right)$$

$$\leq \sum_{j=1}^{\infty} \left[E\left(\frac{1}{\prod_{i=0}^{j-1} f_i'(1)}\right)^2\right]^{\frac{1}{2}} P\left(\frac{1}{f_j'(1)} \geq 1\right)$$

$$\leq A^{\frac{1}{2}} \sum_{j=1}^{\infty} P\left(\frac{1}{f_j'(1)} \geq 1\right). \tag{12}$$

But $A < \infty$, so we have

$$\sum_{j=1}^{\infty} P\left(\frac{1}{f_j'(1)} \geq 1\right) = \infty.$$

Since $\{\frac{1}{f_j'(1)} \geq 1 : j \geq 1\}$ is a sequence of independent random variables, using Borel $0-1$ law, we have $P(f_j'(1) \leq 1, i.o.) = 1$, $j = 1, 2, \cdots$. So we can obtain the conclusion directly. $\qquad\square$

*Example 1.* A model : For a given random environment $\xi$, each individual in the $n^{\text{th}}$ generation produces offspring according to a Poisson distribution with the mean $\lambda(\xi_n) \sim \Gamma(\alpha_n, \beta_n)$.

*Proof.* We set $\lambda(\xi_i) = \lambda_i$, then $p_k(\xi_i) = \frac{(\lambda_i)^k}{k!} e^{-\lambda_i}, i \geq 0$. Since $f_i'(1) = \lambda_i$, $\lambda_i \sim \Gamma(\alpha_i, \beta_i)$, so

$$E\left(\frac{1}{\lambda}\right) = \int_0^{\infty} \frac{1}{x} \cdot \frac{\alpha^\beta x^{\beta-1}}{\Gamma(\beta) e^{\alpha x}} dx = \int_0^{\infty} \frac{\alpha^\beta x^{\beta-2}}{\Gamma(\beta) e^{\alpha x}} dx = \frac{\alpha^\beta}{\Gamma(\beta)} \int_0^{\infty} \frac{x^{\beta-2}}{e^{\alpha x}} dx$$

$$= \frac{\alpha^\beta}{\Gamma(\beta)} \cdot \left(\frac{1}{\alpha}\right)^{\beta-2} \cdot \frac{1}{\alpha} \int_0^{\infty} \frac{\mu^{\beta-2}}{e^\mu} d\mu = \frac{\alpha^\beta}{\Gamma(\beta)} \cdot \left(\frac{1}{\alpha}\right)^{\beta-1} \cdot \Gamma(\beta-1) = \frac{\alpha}{\beta-1},$$

where $\mu = \alpha \cdot x$. So $(E(\frac{1}{f_i'(1)}))^{-1} = \frac{\beta_i-1}{\alpha_i}$. By Jensen's inequality, this process becomes extinct slowerly than the one in a varying environment with means $\{(\beta_i - 1)/\alpha_i, i \geq 0\}$.

For a fixed $\xi_j$, $f_j(s)$ is a increasing convex function. By Jensen's inequality and the smoothing property of conditional expectation,

$$\begin{aligned}
\varphi_n(s) = E\varphi_n(\xi; s) &= E(f_0(f_1(\cdots(f_{n-1}(s)\cdots))) \\
&= E\{E[f_0(f_1(\cdots(f_{n-1}(s)\cdots))|\xi]\} \\
&\geq E\{f_0(E[f_1(\cdots f_{n-1}(s)\cdots)|\xi])\} \\
&\geq \cdots \geq Ef_0(Ef_1(\cdots Ef_{n-1}(s)\cdots)).
\end{aligned} \tag{13}$$

Since $f_i(s) = e^{\lambda_i(s-1)}, \lambda_i \sim \Gamma(\alpha_i, \beta_i)$, so we have

$$\begin{aligned}
Ee^{\lambda(s-1)} &= \int_0^\infty e^{x(s-1)} \cdot \frac{\alpha^\beta x^{\beta-1}}{\Gamma(\beta)e^{\alpha x}} dx \\
&= \frac{\alpha^\beta}{\Gamma(\beta)}(\frac{1}{\alpha+1-s})^\beta \int_0^\infty e^{(s-1-\alpha)x} x^{\beta-1} dx \\
&= \frac{\alpha^\beta}{\Gamma(\beta)} \int_0^\infty \mu^{\beta-1} e^{-\mu} du = (\frac{\alpha}{\alpha+1-s})^\beta,
\end{aligned}$$

where $\mu = (\alpha + 1 - s)x$. Obviously $(\alpha/(\alpha + 1 - s))^\beta$ is a pgf, so by (13) and (3.1) in [1], this process becomes extinct fasterly than the one in a varying environment with pgf's $\{(\frac{\alpha_i}{\alpha_i+1-s})^{\beta_i}, i \geq 0\}$. □

**Acknowledgements.** This paper is supported by NNSF of China (Grant No. 10771021), Research Fund for the Doctoral Program of Higher Education of China(Grant No.20104306110001), the Planned Science and Technology Project of Hunan Province(Grant No.2010fj6036, 2009fi3098) and the Scientific Research Fund of Hunan Provincial Education Department(Grant No.08C120, 09C113, 09C059).

# References

1. Agresti, A.: Bounds on the extinction time distribution of a branching process. Adv. Appl. Prob. 6, 322–325 (1974)
2. Agresti, A.: On the extinction time of varying and random environment branching processes. J. Appl. Prob. Sci. 12(1), 39–46 (1975)
3. Hu, Y.L., Yang, X.Q., Li, Y.Q.: Theorem of Equivalence on the Branching Process in Random Environments. Acta Mathematicae Applicatae Sinica 30(3), 411–421 (2007)
4. Hu, Y.L., Wu, Q.P., Li, Y.Q.: Explosiveness of Age-Dependent Branching Processes in Random Environments. Acta Mathematica Sinica 53(5), 1027–1034 (2010)
5. Japers, P.: Galton-Watson processes in varying environments. J. Appl. Prob. 11, 174–178 (1974)
6. Li, Y.Q., Li, X., Liu, Q.S.: A random walk with a branching system in random environments. Science in China (Series A) 50(5), 698–704 (2007)
7. Smith, W.L., Wilkinson, W.: On branching processes in random environments. Ann. Math. Statis. 40(3), 814–827 (1969)

# Metric Adjusted Skew Information and Metric Adjusted Correlation Measure

Kenjiro Yanagi and Shigeru Furuichi

**Abstract.** We show that a Heisenberg type or a Schrödinger type uncertainty relation for Wigner-Yanase-Dyson skew information proved by Yanagi can hold for an arbitrary quantum Fisher information under some conditions. One of them is a refinement of the result of Gibilisco and Isola.

**Keywords:** Heisenberg uncertainty relation, Schrödinger uncertainty relation, Wigner-Yanase-Dyson skew information, Operator monotone fnction, Quantum Fisher information.

## 1 Introduction

Wigner-Yanase skew information

$$I_\rho(H) = \frac{1}{2}Tr\left[\left(i\left[\rho^{1/2}, H\right]\right)^2\right] = Tr[\rho H^2] - Tr[\rho^{1/2}H\rho^{1/2}H]$$

was defined in [9]. This quantity can be considered as a kind of the degree for non-commutativity between a quantum state $\rho$ and an observable $H$. Here we denote the commutator by $[X, Y] = XY - YX$. This quantity was generalized by Dyson

Kenjiro Yanagi
Division of Applied Mathematical Science, Graduate School of Science and Engineering, Yamaguchi University, 2-16-1, Tokiwadai, Ube, 755-8611, Japan
e-mail: yanagi@yamaguchi-u.ac.jp

Shigeru Furuichi
Department of Computer Science and System Analysis,
College of Humanities and Sciences, Nihon University, 3-25-40, Sakurajyousui, Setagaya-ku, Tokyo, 156-8550, Japan
e-mail: furuichi@chs.nihon-u.ac.jp

S. Li (Eds.): Nonlinear Maths for Uncertainty and its Appli., AISC 100, pp. 317–324.
springerlink.com                                    © Springer-Verlag Berlin Heidelberg 2011

$$I_{\rho,\alpha}(H) = \frac{1}{2}Tr[(i[\rho^\alpha, H])(i[\rho^{1-\alpha}, H])] = Tr[\rho H^2] - Tr[\rho^\alpha H \rho^{1-\alpha} H], \alpha \in [0,1]$$

which is known as the Wigner-Yanase-Dyson skew information. Recently it is shown that these skew informations are connected to special choices of quantum Fisher information in [2]. The family of all quantum Fisher informations is parametrized by a certain class of operator monotone functions $\mathcal{F}_{op}$ which were justified in [7]. The Wigner-Yanase skew information and Wigner-Yanase-Dyson skew information are given by the following operator monotone functions

$$f_{WY}(x) = \left(\frac{\sqrt{x}+1}{2}\right)^2, \quad f_{WYD}(x) = \alpha(1-\alpha)\frac{(x-1)^2}{(x^\alpha - 1)(x^{1-\alpha} - 1)}, \quad \alpha \in (0,1),$$

respectively. In particular the operator monotonicity of the function $f_{WYD}$ was proved in [8]. On the other hand the uncertainty relation related to Wigner-Yanase skew information was given by Luo [6] and the uncertainty relation related to Wigner-Yanase-Dyson skew information was given by Yanagi [10], respectively. In this paper we generalize these uncertainty relations to the uncertainty relations related to quantum Fisher informations.

## 2  Operator Monotone Functions

Let $M_n(\mathbb{C})$(resp. $M_{n,sa}(\mathbb{C})$) be the set of all $n \times n$ complex matrices (resp. all $n \times n$ self-adjoint matrices), endowed with the Hilbert-Schmidt scalar product $\langle A, B \rangle = Tr(A^*B)$. Let $M_{n,+}(\mathbb{C})$ be the set of strictly positive elements of $M_n(\mathbb{C})$ and $M_{n,+,1}(\mathbb{C})$ be the set of stricly positive density matrices, that is $M_{n,+,1}(\mathbb{C}) = \{\rho \in M_n(\mathbb{C}) | Tr\rho = 1, \rho > 0\}$. If it is not otherwise specified, from now on we shall treat the case of faithful states, that is $\rho > 0$.

A function $f : (0, +\infty) \to \mathbb{R}$ is said operator monotone if, for any $n \in \mathbb{N}$, and $A, B \in M_n$ such that $0 \le A \le B$, the inequalities $0 \le f(A) \le f(B)$ hold. An operator monotone function is said symmetric if $f(x) = xf(x^{-1})$ and normalized if $f(1) = 1$.

**Definition 1.** $\mathcal{F}_{op}$ is the class of functions $f : (0, +\infty) \to (0, +\infty)$ such that

(1)  $f(1) = 1$,
(2)  $tf(t^{-1}) = f(t)$,
(3)  $f$ is operator monotone.

*Example 1.* Examples of elements of $\mathcal{F}_{op}$ are given by the following list

$$f_{RLD}(x) = \frac{2x}{x+1}, \quad f_{WY}(x) = \left(\frac{\sqrt{x}+1}{2}\right)^2, \quad f_{BKM}(x) = \frac{x-1}{\log x},$$

$$f_{SLD}(x) = \frac{x+1}{2}, \quad f_{WYD}(x) = \alpha(1-\alpha)\frac{(x-1)^2}{(x^\alpha - 1)(x^{1-\alpha} - 1)}, \quad \alpha \in (0,1).$$

**Remark 1.** Any $f \in \mathcal{F}_{op}$ satisfies

$$\frac{2x}{x+1} \leq f(x) \leq \frac{x+1}{2}, \quad x > 0.$$

For $f \in \mathcal{F}_{op}$ define $f(0) = \lim_{x \to 0} f(x)$. We introduce the sets of regular and non-regular functions

$$\mathcal{F}_{op}^r = \{f \in \mathcal{F}_{op} | f(0) \neq 0\}, \quad \mathcal{F}_{op}^n \{f \in \mathcal{F}_{op} | f(0) = 0\}$$

and notice that trivially $\mathcal{F}_{op} = \mathcal{F}_{op}^r \cup \mathcal{F}_{op}^n$.

**Definition 2.** For $f \in \mathcal{F}_{op}^r$ we set

$$\tilde{f}(x) = \frac{1}{2}\left[(x+1) - (x-1)^2 \frac{f(0)}{f(x)}\right], \quad x > 0.$$

**Theorem 1** ( [1], [2], [5]). *The correspondence* $f \to \tilde{f}$ *is a bijection between* $\mathcal{F}_{op}^r$ *and* $\mathcal{F}_{op}^n$.

## 3 Metric Adjusted Skew Information and Metric Adjusted Correlation Measure

In Kubo-Ando theory of matrix means one associates a mean to each operator monotone function $f \in \mathcal{F}_{op}$ by the formula

$$m_f(A, B) = A^{1/2} f(A^{-1/2} B A^{-1/2}) A^{1/2},$$

where $A, B \in M_{n,sa}(\mathbb{C})$. Using the notion of matrix means one may define the class of monotone metrics (also said quantum Fisher informtions) by the following formula

$$\langle A, B \rangle_{\rho,f} = Tr(A \cdot m_f(L_\rho, R_\rho)^{-1}(B)),$$

where $L_\rho(A) = \rho A, R_\rho(A) = A\rho$. In this case one has to think of $A, B$ as tangent vectors to the manifold $M_{n,+,1}(\mathbb{C})$ at the point $\rho$ (see [7], [2]).

**Definition 3.** For $A, B \in M_{n,sa}$ and $\rho \in M_{n,+,1}(\mathbb{C})$, we define the following quantities:

$$Corr_\rho^f(A, B) = \frac{f(0)}{2} \langle i[\rho, A], i[\rho, B]\rangle_{\rho,f}, \quad I_\rho^f(A) = Corr_{\rho,f}(A, A),$$

$$C_{\rho,f}(A, B) = Tr[m_f(L_\rho, R_\rho)(A)B], \quad C_\rho^f(A) = C_{\rho,f}(A, A),$$

$$U_\rho^f(A) = \sqrt{V_\rho(A)^2 - (V_\rho(A) - I_\rho^f(A))^2},$$

The quantity $I_\rho^f(A)$ is known as metric adjusted skew information [4] and the metric adjusted correlation measure $Corr_\rho^f(A, B)$ was also previously defined in [4].

Then we have the following proposition.

**Proposition 1.** ( [1], [3]) *For $A, B \in M_{n,sa}(\mathbb{C})$ and $\rho \in M_{n,+,1}(\mathbb{C})$, we have the following relations, where we put $A_0 = A - Tr[\rho A]I$ and $B_0 = B - Tr[\rho B]$.*

(1) $\quad I_\rho^f(A) = I_\rho^f(A_0) = Tr(\rho A_0^2) - Tr(m_{\tilde{f}}(L_\rho, R_\rho)(A_0) \cdot A_0)$
$\quad = V_\rho(A) - C_\rho^{\tilde{f}}(A_0),$

(2) $\quad J_\rho^f(A) = Tr(\rho A_0^2) + Tr(m_{\tilde{f}}(L_\rho, R_\rho)(A_0) \cdot A_0) = V_\rho(A) + C_\rho^{\tilde{f}}(A_0),$

(3) $\quad 0 \leq I_\rho^f(A) \leq U_\rho^f(A) \leq V_\rho(A),$

(4) $\quad U_\rho^f(A) = \sqrt{I_\rho^f(A) \cdot J_\rho^f(A)}.$

(5) $\quad Corr_\rho^f(A, B) = Corr_\rho^f(A_0, B_0)$
$\quad = \frac{1}{2}Tr[\rho A_0 B_0] + \frac{1}{2}Tr[\rho B_0 A_0] - Tr[m_{\tilde{f}}(L_\rho, R_\rho)(A_0)B_0]$
$\quad = \frac{1}{2}Tr[\rho A_0 B_0] + \frac{1}{2}Tr[\rho B_0 A_0] - C_\rho^{\tilde{f}}(A_0, B_0).$

## 4   The Main Result

**Theorem 2.** *For $f \in \mathcal{F}_{op}^r$, if*

$$\frac{x+1}{2} + \tilde{f}(x) \geq 2f(x), \tag{1}$$

*then it holds*

$$U_\rho^f(A) \cdot U_\rho^f(B) \geq f(0)|Tr(\rho[A, B])|^2, \tag{2}$$

$$U_\rho^f(A) \cdot U_\rho^f(B) \geq 4f(0)|Corr_\rho^f(A, B)|^2, \tag{3}$$

*where $A, B \in M_{n,sa}(\mathbb{C})$ and $\rho \in M_{n,+,1}(\mathbb{C})$.*

In order to prove Theorem 2, we use several lemmas.

**Lemma 1.** *If (1) holds, then the following inequality is satisfied;*

$$\left(\frac{x+y}{2}\right)^2 - m_{\tilde{f}}(x, y)^2 \geq f(0)(x - y)^2.$$

**Proof.** By (1) we have

$$\frac{x+y}{2} + m_{\tilde{f}}(x, y) \geq 2m_f(x, y).$$

Then

$$\left(\frac{x+y}{2}\right)^2 - m_{\tilde{f}}(x,y)^2 = \left\{\frac{x+y}{2} - m_{\tilde{f}}(x,y)\right\}\left\{\frac{x+y}{2} + m_{\tilde{f}}(x,y)\right\}$$

$$\geq \frac{f(0)(x-y)^2}{2m_f(x,y)}2m_f(x,y) = f(0)(x-y)^2.$$

□

**Lemma 2.** *Let* $\{|\phi_1\rangle, |\phi_2\rangle, \cdots, |\phi_n\rangle\}$ *be a basis of eigenvectors of* $\rho$, *corresponding to the eigenvalues* $\{\lambda_1, \lambda_2, \cdots, \lambda_n\}$. *We put* $a_{jk} = \langle \phi_j | A_0 | \phi_k \rangle, b_{jk} = \langle \phi_j | B_0 | \phi_k \rangle$. *By Corollary 6.1 in [1],*

$$I_\rho^f(A) = \frac{1}{2}\sum_{j,k}(\lambda_j + \lambda_k)a_{jk}a_{kj} - \sum_{j,k}m_{\tilde{f}}(\lambda_j,\lambda_k)a_{jk}a_{kj}$$

$$= 2\sum_{j<k}\left\{\frac{\lambda_j+\lambda_k}{2} - m_{\tilde{f}}(\lambda_j,\lambda_k)\right\}|a_{jk}|^2.$$

$$J_\rho^f(A) = \frac{1}{2}\sum_{j,k}(\lambda_j + \lambda_k)a_{jk}a_{kj} + \sum_{j,k}m_{\tilde{f}}(\lambda_j,\lambda_k)a_{jk}a_{kj}$$

$$\geq 2\sum_{j<k}\left\{\frac{\lambda_j+\lambda_k}{2} + m_{\tilde{f}}(\lambda_j,\lambda_k)\right\}|a_{jk}|^2.$$

$$(U_\rho^f(A))^2 = \frac{1}{4}\left(\sum_{j,k}(\lambda_j+\lambda_k)|a_{jk}|^2\right)^2 - \left(\sum_{j,k}m_{\tilde{f}}(\lambda_j,\lambda_k)|a_{jk}|^2\right)^2$$

*and*

$$Corr_\rho^f(A,B)$$

$$= \frac{1}{2}\sum_{j,k}\lambda_j a_{jk}b_{kj} + \frac{1}{2}\sum_{j,k}\lambda_k a_{jk}b_{kj} - \sum_{j,k}m_{\tilde{f}}(\lambda_j,\lambda_k)a_{jk}b_{kj}$$

$$= \sum_{j<k}\left(\frac{\lambda_j+\lambda_k}{2} - m_{\tilde{f}}(\lambda_j,\lambda_k)\right)a_{jk}b_{kj} + \sum_{j<k}\left(\frac{\lambda_j+\lambda_k}{2} - m_{\tilde{f}}(\lambda_k,\lambda_j)\right) \quad (4)$$

We are now in a position to prove Theorem 2.

**Proof of Theorem 2.** Since

$$Tr(\rho[A,B]) = Tr(\rho[A_0, B_0]) = \sum_{j,k}(\lambda_j - \lambda_k)a_{jk}b_{kj},$$

we have

$$f(0)|Tr(\rho[A,B])|^2$$

$$\leq \left( \sum_{j,k} f(0)^{1/2} |\lambda_j - \lambda_k| |a_{jk}| |b_{kj}| \right)^2$$

$$\leq \left( \sum_{j,k} \left\{ \left( \frac{\lambda_j + \lambda_k}{2} \right)^2 - m_{\tilde{f}}(\lambda_j, \lambda_k)^2 \right\}^{1/2} |a_{jk}| |b_{kj}| \right)^2$$

$$\leq \left( \sum_{j,k} \left\{ \frac{\lambda_j + \lambda_k}{2} - m_{\tilde{f}}(\lambda_j, \lambda_k) \right\} |a_{jk}|^2 \right) \times$$

$$\left( \sum_{j,k} \left\{ \frac{\lambda_j + \lambda_k}{2} + m_{\tilde{f}}(\lambda_j, \lambda_k) \right\} |b_{kj}|^2 \right)$$

$$= I_\rho^f(A) J_\rho^f(B).$$

Hence we have the Heisenberg type inequality (2). On the other hand, by (4), we have

$$|Corr_\rho^f(A,B)|$$

$$\leq \sum_{j<k} \left| \left( \frac{\lambda_j + \lambda_k}{2} - m_{\tilde{f}}(\lambda_j, \lambda_k) \right) a_{jk} b_{kj} \right| + \sum_{j<k} \left| \left( \frac{\lambda_j + \lambda_k}{2} - m_{\tilde{f}}(\lambda_k, \lambda_j) \right) a_{kj} b_{jk} \right|$$

$$\leq \sum_{j<k} \left| \frac{\lambda_j + \lambda_k}{2} - m_{\tilde{f}}(\lambda_j, \lambda_k) \right| |a_{jk}| |b_{kj}| + \sum_{j<k} \left| \frac{\lambda_j + \lambda_k}{2} - m_{\tilde{f}}(\lambda_k, \lambda_j) \right| |a_{kj}| |b_{jk}|$$

$$= 2 \sum_{j<k} \left| \frac{\lambda_j + \lambda_k}{2} - m_{\tilde{f}}(\lambda_j, \lambda_k) \right| |a_{jk}| |b_{kj}|$$

$$\leq \sum_{j<k} |\lambda_j - \lambda_k| |a_{jk}| |b_{kj}|.$$

Then we have

$$f(0)|Corr_\rho^f(A,B)|^2$$

$$\leq \left( \sum_{j<k} f(0)^{1/2} |\lambda_j - \lambda_k| |a_{jk}| |b_{kj}| \right)^2$$

$$\leq \left( \sum_{j<k} \left\{ \left( \frac{\lambda_j + \lambda_k}{2} \right)^2 - m_{\tilde{f}}(\lambda_j, \lambda_k)^2 \right\}^{1/2} |a_{jk}||b_{kj}| \right)^2$$

$$\leq \left( \sum_{j<k} \left\{ \frac{\lambda_j + \lambda_k}{2} - m_{\tilde{f}}(\lambda_j, \lambda_k) \right\} |a_{jk}|^2 \right) \times$$

$$\left( \sum_{j<k} \left\{ \frac{\lambda_j + \lambda_k}{2} + m_{\tilde{f}}(\lambda_j, \lambda_k) \right\} |b_{kj}|^2 \right)$$

$$= \frac{1}{4} I_\rho^f(A) J_\rho^f(B).$$

By the similar way we also have the Schrödinger type inequality (3). □

By putting

$$f_{WYD}(x) = \alpha(1-\alpha) \frac{(x-1)^2}{(x^\alpha - 1)(x^{1-\alpha} - 1)}, \quad \alpha \in (0,1),$$

we obtain the following uncertainty relation;

**Corollary 1 ( [10]).** *For $A, B \in M_{n,sa}(\mathbb{C})$ and $\rho \in M_{n,+,1}(\mathbb{C})$,*

$$U_\rho^{f_{WYD}}(A) U_\rho^{f_{WYD}}(B) \geq \alpha(1-\alpha)|Tr(\rho[A,B])|^2,$$

$$U_\rho^{f_{WYD}}(A) U_\rho^{f_{WYD}}(B) \geq 4\alpha(1-\alpha)|Corr_{\rho,\alpha}(A,B)|^2,$$

*where*

$$Corr_{\rho,\alpha}(A,B) =$$
$$\frac{1}{2} Tr[\rho A_0 B_0] + \frac{1}{2} Tr[\rho B_0 A_0] - \frac{1}{2} Tr[\rho^\alpha A_0 \rho^{1-\alpha} B_0] - \frac{1}{2} Tr[\rho^\alpha B_0 \rho^{1-\alpha} A_0].$$

**Proof.** Since

$$f_{WYD}(x) = \alpha(1-\alpha) \frac{(x-1)^2}{(x^\alpha - 1)(x^{1-\alpha} - 1)},$$

it is clear that

$$\tilde{f}_{WYD}(x) = \frac{1}{2} \{ x + 1 - (x^\alpha - 1)(x^{1-\alpha} - 1) \}.$$

By Lemma 3.3 in [10] we have for $0 \leq \alpha \leq 1$ and $x > 0$,

$$(1 - 2\alpha)^2 (x-1)^2 - (x^\alpha - x^{1-\alpha})^2 \geq 0.$$

Then we can rewrite as follows;

$$(x^{2\alpha} - 1)(x^{2(1-\alpha)} - 1) \geq 4\alpha(1 - \alpha)(x - 1)^2.$$

Thus

$$\frac{x+1}{2} + \tilde{f}_{WYD}(x) = x + 1 - \frac{1}{2}(x^\alpha - 1)(x^{1-\alpha} - 1) = \frac{1}{2}(x^\alpha + 1)(x^{1-\alpha} + 1)$$

$$\geq 2\alpha(1 - \alpha)\frac{(x - 1)^2}{(x^\alpha - 1)(x^{1-\alpha} - 1)} = 2f_{WYD}(x).$$

It follows from Theorem 2 that we can give the aimed result.                    □

**Remark 2.** In [3], the following result was given. Even if (1) does not necessarily hold, then

$$U_\rho^f(A)U_\rho^f(B) \geq f(0)^2|Tr[(\rho[A, B])]|^2, \tag{5}$$

$$U_\rho^f(A)U_\rho^f(B) \geq 4f(0)^2|Corr_\rho^f(A, B)|^2, \tag{6}$$

where $A, B \in M_{n,sa}(\mathbb{C})$ and $\rho \in M_{n,+,1}(\mathbb{C})$. Since $f(0) < 1$, it is easy to show (5), (6) are weaker than (2), (3), respectively.

# References

1. Gibilisco, P., Imparato, D., Isola, T.: Uncertainty principle and quantum Fisher information, II. J. Math. Phys. 48, 72109 (2007)
2. Gibilisco, P., Hansen, F., Isola, T.: On a correspondence between regular and non-regular operator monotone functions. Linear Algebra and its Applications 430, 2225–2232 (2009)
3. Gibilisco, P., Isola, T.: On a refinement of Heisenberg uncertainty relation by means of quantum Fisher information. J. Math. Anal. Appl. 375, 270–275 (2011)
4. Hansen, F.: Metric adjusted skew information. Proc. Nat Acad. Sci. 105, 9909–9916 (2008)
5. Kubo, F., Ando, T.: Means of positive linear operators. Math. Ann. 246, 205–224 (1980)
6. Luo, S.: Heisenberg uncertainty relation for mixed states. Phys. Rev. A 72, 42110 (2005)
7. Petz, D.: Monotone metrics on matrix spaces. Linear Algebra and its Applications 244, 81–96 (1996)
8. Petz, D., Hasegawa, H.: On the Riemannian metric of $\alpha$-entropies of density matrices. Lett. Math. Phys. 38, 221–225 (1996)
9. Wigner, E.P., Yanase, M.M.: Information content of distribution. Proc. Nat. Acad. Sci. 49, 910–918 (1963)
10. Yanagi, K.: Uncertainty relation on Wigner-Yanase-Dyson skew information. J. Math. Anal. Appl. 365, 12–18 (2010)

# Integral-Based Modifications of OWA-Operators

Erich Peter Klement and Radko Mesiar

**Abstract.** An OWA-operator (ordered weighted averaging aggregation operator) can be seen as a discrete Choquet integral with respect to a symmetric monotone measure. Based on this representation and using universal integrals, several modifications of OWA-operators are introduced and discussed.

**Keywords:** Choquet integral, Sugeno integral, Universal integral, OWA operator, Monotone measure, Symmetric monotone measure.

## 1 Introduction

The *ordered weighted averaging operator* (*OWA-operator* for short, see [12]) $\mathrm{OWA_w} \colon [0,1]^n \to [0,1]$ based on a weight vector $\mathbf{w} = (w_1, w_2, \ldots, w_n)$ with $w_i \geq 0$ and $\sum_{i=1}^n w_i = 1$ is given by

$$\mathrm{OWA_w}(\mathbf{x}) = \sum_{i=1}^n w_i \cdot x_{(i)}. \tag{1}$$

Here $(\cdot)$ is a permutation of $(1, \ldots, n)$ making the input vector $\mathbf{x} = (x_1, \ldots, x_n)$ non-increasing, i.e., $x_{(1)} \geq x_{(2)} \geq \cdots \geq x_{(n)}$. OWA operators have attracted

Erich Peter Klement
Department of Knowledge-Based Mathematical Systems,
Johannes Kepler University, 4040 Linz (Austria)
e-mail: ep.klement@jku.at

Radko Mesiar
Department of Mathematics and Descriptive Geometry,
Faculty of Civil Engineering, Slovak University of Technology, 81 368 Bratislava
(Slovakia), and Institute for Research and Applications of Fuzzy Modeling,
University of Ostrava, 70103 Ostrava (Czech Republic)
e-mail: mesiar@math.sk

S. Li (Eds.): Nonlinear Maths for Uncertainty and its Appli., AISC 100, pp. 325–331.
springerlink.com                                    © Springer-Verlag Berlin Heidelberg 2011

a lot of attention in several applications — we only recall two edited volumes on this topic [13, 14].

In [4] an integral representation of OWA-operators was given. Indeed, the $n$-dimensional OWA-operator $\text{OWA}_\mathbf{w}$ can be seen as the Choquet integral [2] with respect to the symmetric monotone measure $m\colon 2^{\{1,\ldots,n\}} \to [0,1]$ given by $m(\emptyset) = 0$ and, for $E \neq \emptyset$, by

$$m(E) = \sum_{i=1}^{|(|E)} w_i.$$

Note also that, as a consequence of the axiomatic characterization of the discrete Choquet integral given in [8, 9], OWA-operators are exactly symmetric comonotone additive aggregation functions (on $[0, 1]$).

Finally note that applying the Sugeno integral [11] based on the symmetric monotone measure $m$ (as given above) to an $n$-dimensional vextor $\mathbf{x} \in [0, 1]^n$ one gets

$$\text{Su}_m(\mathbf{x}) = \bigvee_{i=1}^{n} (x_{(i)} \wedge v_i), \tag{2}$$

where, for each $i \in \{1, \ldots, n\}$, $v_i = \sum_{j=1}^{i} w_j$, i.e., $v_i = m(E)$ for each $E \subseteq \{1, \ldots, n\}$ with $|(|E) = i$. Hence $\text{Su}_m$ can be called an *ordered weighted maximum operator* (*OWMax-operator* for short).

The aim of this paper is a further modification of OWA-operators, where also other kinds of fuzzy integrals will be considered. To specify, we will take into account four classes of discrete universal integrals recently introduced in [5]. The paper is organized as follows. In the following section, discrete universal integrals are recalled, in particular the copula-based, Benvenuti, the smallest and the greatest universal integrals. The modified OWA-operators obtained in this way are characterized, and some examples are given.

## 2   Discrete Universal Integrals

For functions with values in the nonnegative real numbers, universal integrals which can be defined on arbitrary measurable spaces and for arbitrary monotone measures were introduced and investigated in [5]. We restrict our considerations to discrete universal integrals, i.e., $X = \{1, \ldots, n\}$ is a finite space (equipped with the $\sigma$-algebra $\mathcal{A} = 2^X$). Moreover, we will require that each discrete universal integral acts on $[0, 1]^n$ as an idempotent aggregation function, i.e., it assigns the output $c$ to the constant input $\mathbf{c} = (c, \ldots, c)$ for each $c \in [0, 1]$.

**Definition 1.** Let $\odot\colon [0, 1]^2 \to [0, 1]$ be a non-decreasing function with neutral element 1 (i.e., $\odot$ is a semicopula [3]). Let $m\colon 2^{\{1,\ldots,n\}} \to [0, 1]$ be a monotone measure, i.e., a non-decreasing set function such that $m(\emptyset) = 0$ and $m(\{1, \ldots, n\}) = 1$. Each idempotent aggregation function $I_{\odot,m}\colon [0, 1]^n \to$

$[0, 1]$ satisfying $I_{\odot,m}(c \cdot 1_E) = c \odot m(E)$ for each $c \in [0,1]$ and $E \subseteq \{1, \ldots, n\}$ is called a *discrete universal integral* (*based on* $\odot$).

The monotonicity of the universal integral implies that the smallest $\odot$-based discrete universal integral $I^s_{\odot,m} \colon [0,1]^n \to [0,1]$ is given by

$$I^s_{\odot,m}(\mathbf{x}) = \bigvee_{t \in [0,1]} (t \odot m(\{i \in \{1, \ldots, n\} \mid x_i \geq t\}))$$

$$= \bigvee_{t \in [0,1]} x_{(i)} \odot m(\{(1), \ldots, (i)\}),$$

where the same notation $x_{(i)}$ is used as in (1) and (2) for the OWA- and OWMax-operators.

Recall that $I^s_{\wedge,m}$ is the Sugeno integral, while $I^s_{\Pi,m}$ is the Shilkret integral [10] with respect to the monotone measure $m$. Similarly, the greatest $\odot$-based universal integral $I^g_{\odot,m} \colon [0,1]^n \to [0,1]$ can be introduced. Here the *support*, $\mathrm{supp}(\mathbf{x}) = \{i \in \{1, \ldots, n\} \mid x_i > 0\}$, and the *essential supremum* $\mathbf{x}^{(m)}$ of $\mathbf{x}$ with respect to a monotone measure $m$ given by

$$\mathbf{x}^{(m)} = \sup\{t \in [0,1] \mid m(\{i \in \{1, \ldots, n\} \mid x_i \geq t\}) > 0\}$$

play a crucial role. Observe that $\mathbf{x}^{(m)} = x_{(1)} = \max(\mathbf{x})$ if $m(E) = 0$ implies $E = \emptyset$. Using these notations we obtain

$$I^g_{\odot,m}(\mathbf{x}) = \mathbf{x}^{(m)} \odot m(\mathrm{supp}(\mathbf{x})),$$

and for each $\odot$-based discrete universal integral $I_{\odot,m}$ we have

$$I^s_{\odot,m} \leq I_{\odot,m} \leq I^g_{\odot,m}.$$

A copula-based discrete universal integral is based on a (two-dimensional) copula $C \colon [0,1]^2 \to [0,1]$, i.e., a semicopula $C$ satisfying the property of 2-monotonicity, i.e., for all $x, x^*, y, y^* \in [0,1]$ with $x \leq x^*$ and $y \leq y^*$

$$C(x^*, y^*) - C(x^*, y) \geq C(x, y^*) - C(x, y).$$

Note that the copulas form a convex compact subclass of the class of pseudo-multiplications, and that for each copula $C$, the $C$-based discrete universal integral $K_{C,m} \colon [0,1]^n \to [0,1]$ is given by

$$K_{C,m}(\mathbf{x}) = \sum_{i=1}^{n} (C(x_{(i)}, m(\{(1), \ldots, (i)\})) - C(x_{(i)}, m(\{(1), \ldots, (i-1)\}))),$$

using the convention $\{(1), (0)\} = \emptyset$.

Recall that for the greatest copula $\text{Min}(= \wedge)$, $K_{\text{Min},m}$ is the discrete Sugeno integral, while the product copula $\Pi$ yields $K_{\Pi,m}$, the discrete Choquet integral with respect to the monotone measure $m$.

The last class of discrete universal integrals we will introduce here is based on ideas in [1]. Let $\oplus\colon [0,a]^2 \to [0,a]$ be a pseudo-addition on $[0,a]$ with $a \in [1,\infty]$, i.e., a continuous non-decreasing associative function with neutral element 0 (then the symmetry of $\oplus$ follows, see [7]), and let $\otimes\colon [0,a]^2 \to [0,a]$ be a non-decreasing mapping with annihilator 0, i.e., $0 \otimes u = u \otimes 0 = 0$ for each $u \in [0,a]$, which is left-distributive over $\oplus$, i.e., $(u \oplus v) \otimes w = (u \otimes w) \oplus (v \otimes w)$ for all $u,v,w \in [0,a]$, and for which $u \otimes 1 = 1 \otimes u$ for all $u \in [0,1]$. Then the *Benvenuti integral* $B_{\oplus,\otimes,m}\colon [0,1]^n \to [0,1]$ with respect to a monotone measure $m\colon 2^{\{1,\dots,n\}} \to [0,1]$ is given by

$$B_{\oplus,\otimes,m}(\mathbf{x}) = \bigoplus_{i=1}^{n} \left( x_{(i)} \ominus x_{(i+1)} \right) \otimes m(\{(1),\dots,(i)\}),$$

where $x_{(i)} \ominus x_{(i+1)} = \sup\{z \in [0,a] \mid x_{(i+1)} \oplus z = x_{(i)}\}$, and $x_{(n+1)} = 0$ by convention.

Observe that $B_{+,\cdot,m}$ (i.e., $+$ and $\cdot$ are the standard arithmetic operations) is just the discrete Choquet integral, while $B_{\vee,\wedge,m}$ is the discrete Sugeno integral with respect to the monotone measure $m$. Moreover, $B_{\vee,\otimes,m} = I^s_{\odot,m}$ whenever $\odot = \otimes|_{[0,1]^2}$.

## 3 Modified OWA-Operators

For each universal integral $I_{\odot,m}$ with respect to a symmetric monotone measure $m$, the function $I_{\odot,m}\colon [0,1]^n \to [0,1]$ can be understood as a modification of an OWA-operator. Recall once more that a symmetric monotone measure $m\colon 2^{\{1,\dots,n\}} \to [0,1]$ is determined by a weight vector $\mathbf{w} = (w_1,\dots,w_n)$ via $m(E) = \sum_{i=1}^{|(|E)} w_i = v_{|(|E)}$. OWA-operators contain as a special case three basic aggregation functions:

(i)     Min is related to $\mathbf{w}_{\text{Min}} = (0,\dots,0,1)$,
(ii)    Max is related to $\mathbf{w}_{\text{Max}} = (1,0,\dots,0)$,
(iii)   the arithmetic mean AM is related to $\mathbf{w}_{\text{AM}} = (\frac{1}{n},\frac{1}{n}\dots,\frac{1}{n})$.

The corresponding symmetric monotone measures are then given by

(i)     $m_{\text{Min}}(E) = \begin{cases} 1 & \text{if } E = \{1,\dots,n\}, \\ 0 & \text{otherwise,} \end{cases}$

(ii)    $m_{\text{Max}}(E) = \begin{cases} 0 & \text{if } E = \emptyset, \\ 1 & \text{otherwise,} \end{cases}$

(iii)   $m_{\text{AM}}(E) = \frac{1}{n}|(|E)$.

It is not difficult to check that $I^s_{\odot,m_{\text{Min}}} = \text{Min} = I^g_{\odot,m_{\text{Min}}}$, and thus always $I_{\odot,m_{\text{Min}}} = \text{Min}$. Similarly one can show that $I_{\odot,m_{\text{Max}}} = \text{Max}$. On the other

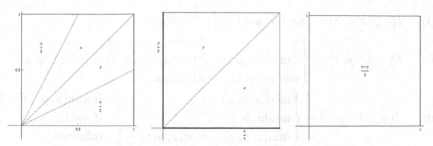

**Fig. 1** $\Pi$-based modifications of the arithmetic mean: $I^s_{\Pi,m_{AM}}$ (left), $I^g_{\Pi,m_{AM}}$ (center), and $K_{\Pi,m_{AM}}$

hand, $I^s_{\odot,m_{AM}}(\mathbf{x}) = \bigvee^n_{i=1} x_{(i)} \odot \frac{i}{n}$ and $I^g_{\odot,m_{AM}}(\mathbf{x}) = \mathrm{Max}(\mathbf{x}) \odot \frac{1}{n}|(|\mathrm{supp}(\mathbf{x}))$ are different, in general.

For a general symmetric monotone measure $m$ characterized by a weight vector $\mathbf{w}$ we obtain the following modifications of OWA-operators:

(i)   $I^s_{\odot,m}(\mathbf{x}) = \bigvee^n_{i=1} x_{(i)} \odot v_i,$

(ii)  $I^g_{\odot,m}(\mathbf{x}) = x^{(m)} \odot v_{|(|\mathrm{supp}(\mathbf{x}))},$

(iii) $K_{C,m}(\mathbf{x}) = \sum^n_{i=1}(C(x_{(i)}, v_i) - C(x_{(i)}, v_{i-1})),$ where $v_0 = 0$ by convention,

(iv)  $B_{\oplus,\otimes,m}(\mathbf{x}) = \bigoplus^n_{i=1}(x_{(i)} \ominus x_{(i+1)}) \otimes v_i.$

*Example 1.* Fix $n = 2$ and consider the standard product $\Pi$ on $[0,1]$ (or $[0,\infty]$). Then we have the following modifications of the arithmetic mean (see Figure 1):

(i)   $I^s_{\Pi,m_{AM}}(x,y) = \min(x,y) \vee \frac{\max(x,y)}{2},$

(ii)  $I^g_{\Pi,m_{AM}}(x,y) = \begin{cases} \frac{\max(x,y)}{2} & \text{if } \min(x,y) = 0, \\ \max(x,y) & \text{otherwise,} \end{cases}$

(iii) $K_{\Pi,m_{AM}}(x,y) = B_{+,\cdot,m_{AM}}(x,y) = \frac{x+y}{2} = \mathrm{AM}(x,y),$

However, if $\oplus_2 \colon [0,\infty]^2 \to [0,\infty]$ denotes the pseudo-addition given by

$$u \oplus_2 v = \sqrt{u^2 + v^2},$$

then we get $B_{\oplus_2,\cdot,m_{AM}}(x,y) = \sqrt{\frac{x^2+y^2}{2}}.$

*Example 2.* Fix $n = 3$, $\mathbf{w} = (0, \frac{1}{3}, \frac{2}{3})$, let $m$ be the symmetric monotone measure determined by $\mathbf{w}$, and consider the smallest copula $T_{\mathbf{L}}$ given by $T_{\mathbf{L}}(x,y) = \max(x+y-1,0)$. Then

(i)   $I^s_{T_L,m}(x,y,z) = (\mathrm{med}(x,y,z) - \tfrac{2}{3}) \vee \min(x,y,z),$

(ii)  $I^g_{T_L,m}(x,y,z) = \begin{cases} 0 & \text{if } \mathrm{med}(x,y,z) = 0, \\ \tfrac{1}{3}\mathrm{med}(x,y,z) & \text{if } 0 = \min(x,y,z) < \mathrm{med}(x,y,z), \\ \mathrm{med}(x,y,z) & \text{otherwise,} \end{cases}$

(iii) $K_{T_L,m}(x,y,z) = \begin{cases} \mathrm{med}(x,y,z) & \text{if } \min(x,y,z) \geq \tfrac{2}{3}, \\ \min(x,y,z) & \text{if } \mathrm{med}(x,y,z) \leq \tfrac{2}{3}, \\ \min(x,y,z) + \mathrm{med}(x,y,z) - \tfrac{2}{3} & \text{otherwise.} \end{cases}$

Note that there is no Benvenuti integral $B_{\oplus,\otimes}$ such that $\otimes|_{[0,1]^2} = T_L$.

As already mentioned, OWA-operators can be characterized as symmetric comonotone additive aggregation functions [4]. Each of the classes of modified OWA-operators introduced here are symmetric idempotent aggregation functions linked to some discrete universal integral and based on a monotone measure $m$ which can be identified with the aggregation of the characteristic function vector.

We have the following axiomatic characterization of the OWA modifications introduced here. For more details see [1, 5, 6].

**Theorem 1.** *A symmetric idempotent aggregation function* $A: [0,1]^n \to [0,1]$ *is*

(i)   *the smallest discrete universal integral if and only if it is comonotone maxitive and if* $A(1_E) = A(1_F)$ *implies* $A(t \cdot 1_E) = A(t \cdot 1_F)$ *for all* $t \in [0,1]$;

(ii)  *a copula-based discrete universal integral if and only if it is comonotone modular;*

(iii) *a* $(\oplus, \odot)$*-based discrete Benvenuti integral if and only if it is comonotone* $\oplus$*-additive and if* $A(t \cdot 1_E) = t \odot A(1_E)$ *for all* $E \subseteq \{1, \ldots, n\}$ *and* $t \in [0,1]$.

## 4   Concluding Remarks

We have introduced several modifications of OWA-operators which have the form of a discrete universal integral with respect to a symmetric monotone measure. Note that some classes of the aggregation functions discussed here can be already found in the literature, although sometimes under different names. For example, the smallest universal integrals can be seen as (N)-fuzzy integrals as introduced in [15]. Copula-based discrete universal integrals were recently shown to coincide with OMA-operators (ordered modular averages) [6]. We are convinced that all the modified OWA-operators considered here will offer a wider choice of models for decision procedures in any domain where OWA-operators have been applied succesfully (information science, engineering, social choice, economics, image processing, etc.)

**Acknowledgements.** The research summarized in this paper was supported by the Grants VEGA 1/0080/10, VZ MSM 6198898701, APVV-0073-10, and TUCI 0601-66.

# References

1. Benvenuti, P., Mesiar, R., Vivona, D.: Monotone set functions-based integrals. In: Pap, E. (ed.) Handbook of Measure Theory., vol. II, ch. 33, pp. 1329–1379. Elsevier Science, Amsterdam (2002)
2. Choquet, G.: Theory of capacities. Ann. Inst. Fourier (Grenoble) 5, 131–292 (1953/1954)
3. Durante, F., Sempi, C.: Semicopulæ. Kybernetika (Prague) 41, 315–328 (2005)
4. Grabisch, M.: Fuzzy integral in multicriteria decision making. Fuzzy Sets and Systems 69, 279–298 (1995)
5. Klement, E.P., Mesiar, R., Pap, E.: A universal integral as common frame for Choquet and Sugeno integral. IEEE Trans. Fuzzy Systems 18, 178–187 (2010)
6. Mesiar, R., Mesiarová-Zemánková, A.: The ordered modular averages. IEEE Trans. Fuzzy Systems 19, 42–50 (2011)
7. Mostert, P.S., Shields, A.L.: On the structure of semi-groups on a compact manifold with boundary. Ann. of Math., II. Ser. 65, 117–143 (1957)
8. Schmeidler, D.: Integral representation without additivity. Proc. Amer. Math. Soc. 97, 255–261 (1986)
9. Schmeidler, D.: Subjective probability and expected utility without additivity. Econometrica 57, 571–587 (1989)
10. Shilkret, N.: Maxitive measure and integration. Indag. Math. 33, 109–116 (1971)
11. Sugeno, M.: Theory of Fuzzy Integrals and its Applications. PhD Thesis, Tokyo Institute of Technology (1974)
12. Yager, R.R.: On ordered weighted averaging aggregation operators in multicriteria decisionmaking. IEEE Trans. Systems Man Cybernet. 18, 183–190 (1988)
13. Yager, R.R., Kacprzyk, J. (eds.): The Ordered Weighted Averaging Operators. Theory and Applications. Springer, Berlin (1997)
14. Yager, R.R., Kacprzyk, J., Beliakov, G. (eds.): Recent Developments in the Ordered Weighted Averaging Operators: Theory and Practice. Springer, Berlin (2011)
15. Zhao, R. (N)-Fuzzy integral. J. Math. Res. Expo. 2, 55–72 (1981)

# Fuzzy Similarity Measure Model for Trees with Duplicated Attributes

Dianshuang Wu and Guangquan Zhang

**Abstract.** In many business situations, complex user profiles are described by tree structures, and evaluating the similarity between these trees is essential in many applications, such as recommender systems. This paper proposes a fuzzy similarity measure model for trees with duplicated attributes. In this model, the conceptual similarity between attributes and the weights of nodes are expressed by linguistic terms. To deal with duplicated attributes in the trees, nodes with the same concept are clustered. The most conceptual corresponding cluster pairs among two trees are identified. Based on the corresponding cluster pairs, the conceptual similarity and the value similarity between two trees are evaluated, and the final similarity measure is assessed as a weighted sum of their conceptual and value similarities.

**Keywords:** Tree similarity measure, Fuzzy similarity measure, Trees with duplicated attributes.

## 1 Introduction

Due to a huge amount of products available on e-commerce websites, recommender systems are essential in e-business environment nowadays [1]. The basic idea of recommender systems is to recommend a customer the items which are preferred by the customer's similar users. Therefore, effective evaluation of the similarity between users is vital to the success of recommender systems. In many business situations, users' profiles are so complex that they

Dianshuang Wu and Guangquan Zhang
Decision Systems & e-Service Intelligence (DeSI) Lab, Centre for Quantum
Computation & Intelligent Systems (QCIS), Faculty of Engineering and Information
Technology, University of Technology, Sydney, P.O. Box 123, Broadway, NSW 2007,
Australia
e-mail: Dianshuang.Wu@uts.edu.au, zhangg@it.uts.edu.au

S. Li (Eds.): Nonlinear Maths for Uncertainty and its Appli., AISC 100, pp. 333–340.
springerlink.com      © Springer-Verlag Berlin Heidelberg 2011

can only be described by hierarchical tree structures [7]. For example, two business user profiles in telecom industry are shown in Fig.1. Taking $T_1$ in the figure as an example, the business user has two accounts: one mobile account and one landline account. Each account has many services. The mobile account has four services, three $49 plans and one $59 plan. The landline account has two services, one $45 plan and one $55 plan. The average spending per month of each service is also listed in the bracket under the service. Different accounts or services also have different importance degrees. To compare this kind of complex user profiles comprehensively, a tree similarity measure method is proposed in this study.

Tree structures are used for information representation in various areas, such as e-business [7,2], XML Schema matching [3] and case-based reasoning [6]. The similarity measure between trees is essential in these applications. In these researches, trees are modeled as node labeled [3], edge labeled, weighted trees [2,6]. These tree models are not sufficient enough to represent the abundant information in our situation. In [7], a hierarchical item tree is defined, in which each node is associated with an attribute, a weight and a value. When evaluating the similarity between two trees, the tree structures, nodes' weights, concepts and values are all considered. A maximum correspondence tree mapping is constructed to identify the concept corresponding node pairs of two trees, and the conceptual similarity between two trees is evaluated. The value similarity between two trees is evaluated based on the mapping. The final similarity measure between two trees is assessed as a weighted sum of their conceptual and value similarities.

There are two problems applying the method in [7] to our problem. First, the maximum correspondence tree mapping in the method is a one to one mapping, which is useful to the trees that every node has a distinct attribute. However, there are duplicated nodes in our trees. For example, nodes $v_4$, $v_5$, $v_6$ in $T_1$ cannot be distinguished by their concepts. These duplicated attribute nodes can only be mapped randomly when constructing the maximum correspondence tree mapping, which cannot fully express the correspondence between two trees. Thus, a similarity measure method for trees with duplicated attributes is needed. Second, in practical situations, the conceptual

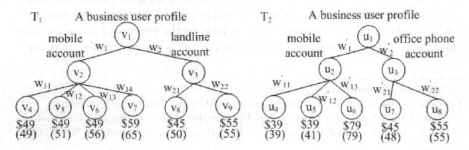

**Fig. 1** Two hierarchical user profiles

similarity measures between any two attributes are usually evaluated by domain experts. The similarity degrees between different attributes are hard to rate by exact numbers. Linguistic terms, such as 'very similar', 'absolutely different', are suitable to describe them. The weights of nodes are also assigned by users or experts. Linguistic variables, such as 'import', 'more important', are used to evaluate them. In such a case, precise mathematical approaches are not enough to tackle such linguistic variables, but fuzzy set theory can be applied to deal with the situation [5, 10]. Therefore, a fuzzy similarity measure of trees with duplicated attributes is proposed.

The rest of this paper is organized as follows. Section 2 describes the features of the hierarchical tree structured user profiles formally. A comprehensive fuzzy similarity measure of trees with duplicated attributes is proposed in Section 3. Finally, conclusions and future studies are discussed in Section 4.

## 2  Hierarchical Profile Tree (HP-Tree)

To describe the features of the tree structured user profile more formally, a hierarchical profile tree (HP-tree) is defined in this section.

**Definition 1.** HP-tree. A hierarchical profile tree is a structure $T = (V, E, A, W, R)$, in which $V$ is a finite set of nodes; $E$ is a binary relation on $V$ where each pair $(u, v) \in E$ represents the parent-child relationship between two nodes $u, v \in V$; $A$ is a set of attributes assigned to each node in $V$; $W$ are the weights assigned to each node to represent their importance degrees to their siblings, which are expressed by linguistic terms; and $R$ are the values assigned to every leaf node to describe the relevant attribute.

The two user profiles in Fig. 1 are two examples of HP-trees. As seen in Fig. 1, 'landline' in $T_1$ and 'office phone' in $T_2$ almost represent the same thing but with different terms. To identify these conceptual similar terms in different trees, a conceptual similarity measure between attributes is introduced as in [8]. The conceptual similarity is expressed by linguistic terms. Let $A_1$ and $A_2$ be two attribute sets, $a_1 \in A_1$, and $a_2 \in A_2$. The conceptual similarity measure between $a_1$ and $a_2$ is defined, denoted as $sc_{A_1,A_2}(a_1, a_2)$. For convenience, the subscript $A_1$, $A_2$ is omitted if there is no confusion.

The HP-tree is an extension of the HI-tree defined in [7]. First, there are conceptual duplicated nodes in HP-trees. Second, the nodes' weights and the conceptual similarities between attributes are represented by linguistic terms. In this paper, the linguistic terms in set $Weight$ are used to describe the weights.

$Weight$={$Very\ low\ (VL)$, $Low\ (L)$, $Medium\ low\ (ML)$, $Medium\ (M)$, $Medium\ high\ (MH)$, $High\ (H)$, $Very\ high\ (VH)$}.

The linguistic terms in set $S$ are used to describe the similarity measures between attribute terms.

$S$={$Absolutely\ different\ (AD)$, $Very\ different\ (VD)$, $Different\ (D)$, $Medium\ (M)$, $Similar\ (S)$, $Very\ similar\ (VS)$, $Absolutely\ similar\ (AS)$}.

Fuzzy numbers are applied to deal with these linguistic terms. Based on research results in [10, 9], we can use any forms of fuzzy numbers, called general fuzzy numbers, to describe these linguistic terms. This study defines fuzzy numbers $a_1, a_2, ..., a_7$ to describe the terms in $Weight$ respectively, where $a_1 < a_2 < ... < a_7$ , and defines normalized fuzzy numbers $b_1, b_2, ..., b_7$ to describe the terms in $S$ respectively, where $b_1 < b_2 < ... < b_7$.

## 3 Fuzzy Similarity Measure Model for HP-Trees

A fuzzy similarity measure model for HP-trees is presented in this section. In this model, the nodes with the same concept are clustered, and the most conceptual corresponding cluster pairs among two trees are identified. Then, the conceptual similarity and the value similarity between two trees are evaluated, and the final similarity measure is assessed as a weighted sum of their conceptual and value similarities.

### 3.1 Weights Normalization

The weights of nodes in HP-trees are expressed by general fuzzy numbers and need to be normalized. For each internal node $v$, let $C(v)$ be $v$'s children. For any $u \in C(v)$, let $w_u$ be $u$'s weight, its normalization $w_u^*$ is computed as:

$$w_u^* = \frac{w_u}{\sum_{t \in C(v)} w_{t0}^R}. \tag{1}$$

### 3.2 Conceptual Similarity

As the conceptual hierarchy of the HP-tree, the root node can represent the whole tree. Therefore, the conceptual similarity between two trees, $sct(T_1, T_2)$, can be defined by the concept correspondence degree between their root nodes, $cord(root(T_1), root(T_2))$:

$$sct(T_1, T_2) = cord(root(T_1), root(T_2)). \tag{2}$$

The concept correspondence degree, $cord()$, needs to be defined. Given two nodes $v$ and $u$ in two trees, there are three situations based on the nodes' structures [7]: 1) both $v$ and $u$ are leaves; 2) $v$ is a leaf and $u$ is an internal node; 3) both $v$ and $u$ are internal nodes.

In the first two cases, $cord()$ is defined as [7]. In case 1), $cord(v, u) = sc(a_v, a_u)$, where $a_v$ represents the attribute of node $v$. In case 2), let $u_1, u_2, ..., u_q$ be $u$'s children, $cord(v, u) = \alpha \cdot sc(a_v, a_u) + (1 - \alpha) \cdot \sum_{i=1}^{q} w_{u_i}^* \cdot cord(v, u_i)$, where $\alpha$ is the influence factor of the parent node and $w_{u_i}^*$ is the weight of $u_i$.

In case 3), duplicated attributes should be treated. Let $C(v) = \{v_1, v_2, ..., v_p\}$ and $C(u) = \{u_1, u_2, ..., u_q\}$ be the children sets of $v$ and $u$ respectively. The duplicated attributes in the trees should be identified first. Clusters of duplicated nodes in $C(v)$ and $C(u)$ are constructed as $CC(v)$ and $CC(u)$ respectively, where $CC(v) = \{vc_1, vc_2, ..., vc_m\}$ and $CC(u) = \{uc_1, uc_2, ..., uc_n\}$. $vc_i, i = 1, 2, ..., m$, and $uc_j, j = 1, 2, ..., n$, are node set, such that

$$\forall x, y \in vc_i, \forall t \in C(u), cord(x,t) = cord(y,t), \tag{3}$$

$$\forall x, y \in uc_j, \forall t \in C(v), cord(x,t) = cord(y,t). \tag{4}$$

Let $x \in vc_i$ and $y \in uc_j$, the concept correspondence degree between $vc_i$ and $uc_j$, $cord_c(vc_i, uc_j)$, is defined as $cord(x,y)$.

A bipartite graph $G_{v,u} = (V, E)$, induced by $CC(v)$ and $CC(u)$, is constructed as follows: $V = CC(v) \bigcup CC(u)$, $E = \{(s,t) : s \in CC(v), t \in CC(u)\}$. The weights of edges are defined based on the concept correspondence degree between relevant node clusters. A fuzzy positive-ideal value $r^*$ and a fuzzy negative-ideal value $r^-$ are defined as: $r^*=1$ and $r^-=0$. For edge $(s,t)$, the distances between $cord_c(s,t)$ and $r^*$, $cord_c(s,t)$ and $r^-$ are calculated as $d^*_{s,t} = d(cord_c(s,t), r^*)$ and $d^-_{s,t} = d(cord_c(s,t), r^-)$, where $d(\cdot)$ is the distance between two fuzzy numbers. The weight of edge $(s,t)$ is defined as $w_{s,t} = 1/2(d^-_{s,t} + (1 - d^*_{s,t}))$. To find most corresponding node cluster pairs between $CC(v)$ and $CC(u)$, a maximum weighted bipartite matching (MWBM) problem [4] of $G_{v,u}$ is resolved. A MWBM of $CC(v)$ and $CC(u)$, $M_{v,u}$ is constructed.

The concept correspondence degree between two internal nodes $v$ and $u$, $cord(v,u)$, is defined as:

$$cord(v,u) = \alpha \cdot sc(a_v, a_u) + (1-\alpha) \cdot \sum_{(vc_i, uc_j) \in M_{v,u}} \tfrac{1}{2}(w^*_{vc_i} + w^*_{uc_j}) \cdot cord_c(vc_i, uc_j)$$

$$\tag{5}$$

where $w^*_{vc_i} = \sum_{x \in vc_i} w^*_x$, $w^*_{uc_j} = \sum_{y \in uc_j} w^*_y$.

According to the above formulas, the conceptual similarity between the roots of two trees can be calculated, and the conceptual similarity between two trees is obtained.

During the computation process of the conceptual similarity between two trees, the children clusters of each internal node are recorded. For two corresponding internal nodes in two trees, the maximum concept correspondence cluster mapping of their children is also recorded. Based on the records, the most corresponding nodes among two trees can be identified. The roots of two trees are corresponding node pairs. Then the corresponding nodes in the children of two roots are identified based on two roots' children's maximum concept correspondence cluster mapping. Other corresponding nodes can be identified in the same way.

## 3.3 Value Similarity between Two Trees

In this section, values of nodes are taken into account, and the value similarity between two HP-trees is evaluated. In the HP-tree, only leaf nodes are assigned values initially, and internal nodes' values are computed by aggregating their children's. As the root is the representation of the whole tree, the value similarity between two trees, $svt(T_1, T_2)$, can be defined by the value similarity between two root nodes, $sv(root(T_1), root(T_2))$:

$$svt(T_1, T_2) = sv(root(T_1), root(T_2)). \tag{6}$$

The value similarity between nodes within two trees should be defined first. For any corresponding node pair $(v, u)$ in two trees, there are two cases: 1) $v$ or $u$ is a leaf node; 2) both $v$ and $u$ are internal nodes.

In case 1), $sv(v, u) = s(value(v), value(u))$, where $value(v)$ denotes $v$'s value and $s(\cdot)$ denotes a value similarity measure. If $v$ is a leaf node, $value(v)$ is assigned initially. Otherwise, it is computed by aggregating its children's values. $s(\cdot)$ can be defined according to the specific applications. In this study, the similarity between two values $a$ and $b$ is defined as $s(a, b) = 1 - |a - b|/max(a, b)$. For example, the value similarity between $v_4$ and $u_4$ in Fig. 1 is 0.796.

In case 2), the value similarity between $v$ and $u$ is evaluated based on their children's value similarities. According to $v$ and $u$'s children cluster mapping $M_{v,u}$ recorded in last section, the most corresponding clusters are identified. The value similarity between $v$ and $u$ is evaluated by aggregating the value similarity between these corresponding clusters of their children. Let $CC(v) = \{vc_1, vc_2, ..., vc_m\}$ and $CC(u) = \{uc_1, uc_2, ..., uc_n\}$ be node clusters of $v$ and $u$'s children respectively. Within each cluster, different nodes have different values. To fully reflect the value similarity between two trees, we should compare the values node by node, rather than cluster by cluster. Let $vc_i$ and $uc_j$ be a corresponding cluster pair. A bipartite graph is constructed by $vc_i$ and $uc_j$ as follows: $V = vc_i \bigcup uc_j, E = \{(x, y) : x \in vc_i, y \in uc_j\}$. The weights of edges are derived from the value similarity between the endpoints. As the value similarity is also a fuzzy concept, the fuzzy positive-ideal value $r^*$ and the fuzzy negative-ideal value $r^-$ are introduced. For edge $(x, y)$, the distances between $sv(x, y)$ and $r^*$, $sv(x, y)$ and $r^-$ are calculated as $d^*_{x,y} = d(sv(x, y), r^*)$ and $d^-_{x,y} = d(sv(x, y), r^-)$, where $d(\cdot)$ is the distance between two fuzzy numbers. The weight of edge $(x, y)$ is defined as $w_{x,y} = 1/2(d^-_{x,y} + (1 - d^*_{x,y}))$. Then, a maximum weighted bipartite matching (MWBM) $M_{vc_i, uc_j}$ is constructed. The value similarity between clusters $vc_i$ and $uc_j$ is calculated by:

$$sv_c(vc_i, uc_j) = \sum_{(x,y) \in M_{vc_i, uc_j}} \tfrac{1}{2}(w^*_x + w^*_y) \cdot sv(x, y). \tag{7}$$

The value similarity between $v$ and $u$ is computed by:

$$sv(v, u) = \sum_{(vc_i, uc_j) \in M_{v,u}} sv_c(vc_i, uc_j). \tag{8}$$

Based on the above formulas, the value similarity between two roots can be evaluated, and the value similarity between two trees is obtained.

## 3.4   Similarity Measure between Two Trees

The final comprehensive similarity measure of two trees $T_1$ and $T_2$ is defined as follows:

$$sim(T_1, T_2) = \alpha_1 \cdot sct(T_1, T_2) + \alpha_2 \cdot svt(T_1, T_2). \tag{9}$$

where $\alpha_1 + \alpha_2 = 1$.

The similarity measures are normalized fuzzy numbers, and their ranges belong to closed interval $[0, 1]$. A closeness coefficient is defined to rank the similarity measures. Let the similarity between $T$ and $T_i, i = 1, ..., n$ be $sim(T, T_i)$. A fuzzy positive-ideal similarity value $s^*$ and a fuzzy negative-ideal similarity value $s^-$ are defined respectively: $s^* = 1$ and $s^- = 0$. The distances between $sim(T, T_i)$ and $s^*$, $sim(T, T_i)$ and $s^-$ are calculated as $d_i^* = d(sim(T, T_i), s^*)$ and $d_i^- = d(sim(T, T_i), s^-)$, where $d(\cdot)$ is the distance between two fuzzy numbers. The closeness coefficient of $T_i$ is defined as:

$$c_i = \tfrac{1}{2}(d_i^- + (1 - d_i^*)). \tag{10}$$

The user profile $T_i$ that corresponds to larger $c_i$ is more similar to $T$.

## 4   Conclusions and Future Work

This study proposes a fuzzy similarity measure model for trees with duplicated attributes. In this model, the conceptual similarity between attributes and the weights of nodes are expressed by linguistic terms. To deal with duplicated attributes in the trees, nodes with the same concept are clustered. The most conceptual corresponding cluster pairs among two trees are identified during the conceptual similarity computation process. Based on the corresponding cluster pairs, the value similarity between two trees is evaluated, and the final similarity measure is assessed as a weighted sum of their conceptual and value similarities. Further study includes developing a recommender system based on the proposed model and experimentally evaluating the model.

**Acknowledgements.** The work presented in this paper was supported by the Australian Research Council (ARC) under Discovery Project DP110103733.

# References

1. Adomavicius, G., Tuzhilin, A.: Toward the next generation of recommender systems: a survey of the state-of-the-art and possible extensions. IEEE Transactions on Knowledge and Data Engineering 17, 734–749 (2005)
2. Bhavsar, V.C., Boley, H., Yang, L.: A weighted-tree similarity algorithm for multi-agent systems in e-business environments. Computational Intelligence 20, 584–602 (2004)
3. Jeong, B., Lee, D., Cho, H., Lee, J.: A novel method for measuring semantic similarity for XML schema matching. Expert Systems with Applications 34, 1651–1658 (2008)
4. Jungnickel, D.: Graphs, networks, and algorithms. Springer, Heidelberg (2007)
5. Marimin, U.M., Hatono, I., Tamura, H.: Linguistic labels for expressing fuzzy preference relations in fuzzy group decision making. IEEE Transactions on Systems, Man, and Cybernetics, Part B: Cybernetics 28, 205–218 (1998)
6. Ricci, F., Senter, L.: Structured cases, trees and efficient retrieval. Advances in Case-Based Reasoning 1488, 88–99 (1998)
7. Wu, D., Lu, J., Zhang, G.: A hybrid recommendation approach for hierarchical items. In: International Conference on Intelligent Systems and Knowledge Engineering (ISKE), pp. 492–497 (2010)
8. Xue, Y., Wang, C., Ghenniwa, H., Shen, W.: A tree similarity measuring method and its application to ontology comparison. Journal of Universal Computer Science 15, 1766–1781 (2009)
9. Zadeh, L.A.: The concept of a linguistic variable and its application to approximate reasoning–I 1. Information Sciences 8, 199–249 (1975)
10. Zhang, G., Lu, J.: Using general fuzzy number to handle uncertainty and imprecision in group decision-making. Intelligent Sensory Evaluation: Methodologies and Applications, 51-70 (2004)

# Linking Developmental Propensity Score to Fuzzy Sets: A New Perspective, Applications and Generalizations

Xuecheng Liu, Richard E. Tremblay, Sylvana Cote, and Rene Carbonneau

**Abstract.** First, we outline the group-based trajectory models for longitudinal data; second, we briefly describe the concept of propensity scores based on these models; third, we give a new perspective of propensity scores in fuzzy sets; fourth, we apply operations of fuzzy sets to propensity scores; fifth, we generalize propensity scores to trajectories based on fuzzy and possibilistic clusterings.

**Keywords:** Group-based trajectory modeling, Group membership posterior probability, Propensity score, Fuzzy set, Operation of fuzzy sets, Fuzzy clustering, Possibilistic clustering

## 1 Introduction

Liu et al. [7] proposes the concept of individual developmental propensity scores based on group-based trajectory models (special mixture models) for

Xuecheng Liu, Richard E. Tremblay, Sylvana Cote, and Rene Carbonneau
Research Unit on Children's Psychosocial Maladjustment,
University of Montreal, Canada
e-mail: xuecheng.liu@umontreal.ca

Richard E. Tremblay
School of Public Health and Population Sciences, University College Dublin, Ireland
e-mail: richard.ernest.tremblay@umontreal.ca

Sylvana Cote
International Laboratory for Child and Adolescent Mental Health,
University of Montreal, Canada and INSERM U669, France
e-mail: sylvana.cote@umontreal.ca

Rene Carbonneau
Department of Pediatrics, Faculty of Medicine, University of Montreal, Canada
e-mail: rene.carbonneau@umontreal.ca

S. Li (Eds.): Nonlinear Maths for Uncertainty and its Appli., AISC 100, pp. 341–348.
springerlink.com      © Springer-Verlag Berlin Heidelberg 2011

longitudinal data. The individual developmental propensity score (or propensity score for short) measures the degree of an individual to follow a given group trajectory identified by the group-based trajectory model. As a continuous variable, the propensity score has more variation/information than the categorical variable of classified groups based on group membership posterior probabilities, further it is robust to the choice of the number of groups.

This paper links the concept of propensity scores to fuzzy sets and is organized as follows. In Section 2, we outline the group-based trajectory models for longitudinal data. In Section 3, we briefly describe the concept of propensity scores based on the group-based trajectory models. In section 4, we give a new perspective of propensity scores in fuzzy sets. In Section 5, based on the new perspective of propensity scores, we apply operations of fuzzy sets to propensity scores. In Section 6, we generalize propensity scores to trajectories based on fuzzy and possibilistic clusterings.

## 2 Outline of Group-Based Trajectory Models for Longitudinal Data

We outline group-based trajectory statistical models for longitudinal data (e.g. Nagin [8,9]) as follows. The data are collected at several times $t_1, ..., t_K$ over a period $[t_1, t_K]$. For each individual (or, in a more general term, subject) indexed by $i = 1, 2, ..., N$, the data are denoted by $y_{i1}, ..., y_{iK}$. We assume that there are $G$ latent groups corresponding to $G$ trajectories indexed by $g = 1, ..., G$. Each individual follows the group trajectory $g$ with prior probability $\pi_g$. (So, $\pi_g \geq 0$ and $\pi_1 + \cdots + \pi_G = 1$.) Each group trajectory $g$ is usually specified by a polynomial as

$$c_g(t) \stackrel{\text{def}}{=} \alpha_g + \beta_g t + \gamma_g t^2 \tag{1}$$

for quadratic form, for example, where $\alpha_g, \beta_g$ and $\gamma_g$ are parameters.

Let $Y_{ik}$ denote the random variable corresponding to $y_{ik}$ for all $i$ and $k$. Fix $G$, conditional on a group trajectory $g$ and time $t_k$, we assume that $Y_{ik}$ has distribution $P_{gk}(Y_{ik} = y_{ik})$ with the mean $c_j(t_k)$ (or, sometimes, a function of $c_j(t_k)$, depending on assumption of data distributions) and other parameters. Let $\theta$ denote the set of all parameters. Under assumption that, conditional on a group trajectory, for each individual, observations in different times are independent, the likelihood of the models is

$$L(\theta|\text{Data}) \stackrel{\text{def}}{=} \prod_{i=1}^{N} \sum_{g=1}^{G} \left[ \pi_g \prod_{k=1}^{K} P_{gk}(Y_{ik} = y_{ik}) \right] \tag{2}$$

Note that the choice of the distributions of $Y_{ik}$ largely depend on the data types. For example, with continuous data, we could assume that the distribution is normal or censored normal; with count data, Poisson distribution; and with dichotomized data, Bernoulli distribution; etc.

Denoted by $\hat{\theta}$ the maximal likelihood estimate (MLE) of $\theta$, and (possible) missing data are assumed to be missing at random (MAR, Rubin [10]), and the model is usually estimated with full information maximum likelihood (FIML) method. For each individual $i$, the group membership posterior probabilities are

$$p_{ig} \overset{\text{def}}{=} \frac{\hat{\pi}_g \prod_{k=1}^{K} P_{gk}(Y_{ik} = y_{ik})}{\sum_{g=1}^{G} \hat{\pi}_g \prod_{k=1}^{K} P_{gk}(Y_{ik} = y_{ik})} \tag{3}$$

where $g = 1, ..., G$.

There is a controversial/challenging issue in group-based trajectory modeling, or more general mixture modeling: how to decide "the" number of groups $G$. Since $G$ is latent, we need to fix it when we estimate a model. With different choices of $G$, we could choose one model as the final model according to some criteria, like model BICs.

## 3 Brief Description of Propensity Scores

The motivation of proposing propensity scores is as follows. Usually, we are more interested in comparison between the highest level group trajectory group vs other groups in two approaches: One is to create a categorical variable by classifying individuals based on group membership posterior probabilities; the other is to use continuous variable of membership posterior probability directly. For the classifying approach, sometimes it is hard to classify certain individuals, and very often to classify very similar individuals into different groups, in addition we lose information in the group membership posterior probabilities. For the posterior probability approach, the posterior probabilities are not stable: they depend on the number of groups, closeness between the interested group trajectories, and so on.

In order to overcome above disadvantages of the group-based trajectory models (or more general mixture models), Liu et al. [7] propose serval scores (continuous variables) which measure propensities for an individual to follow any given trajectory among the group trajectories. They use all information in estimates of group trajectories and individual's group membership posterior probabilities. They improve the group membership posterior probabilities in many aspects, and has no disadvantages as discussed above. In addition, it is robust to the number of groups.

**Assumptions and notations.** We always assume that the variables $Y_{ik}$ are non-negative in discussion below. (It is true for almost situations for real data. If not, we may transform data in order to have such property.) To simplify our discussion, we also assume that the group trajectories are "completely ordered", i.e., there exist no two intersecting group trajectories. (If not, the discussions are similar.) In addition, for each pair of group trajectories $g, g'$, we use $A(g, g')$ to denote the area of the region bounded by the group trajectories $g$, $g'$ and the vertical lines $t = T_1$, $t = T_K$. We denote $A(g, g')$ by $A(g)$ when $g'$ is identified or defined by $y = 0$ (time-axis). The highest and the lowest group trajectories will be denoted by $g_{\text{high}}$ and $g_{\text{low}}$ respectively.

**The propensity score: general form.** We define the propensity score in two steps. First, for each group trajectory $g$, define

$$K_g \overset{\text{def}}{=} A(g)/A(g_{\text{high}}). \tag{4}$$

We have

$$0 \le K_g \le 1 \quad \text{and} \quad K_{g_{\text{high}}} = 1, \ (g = 1, ..., G).$$

Second, we define the propensity score for the individual $i$ as

$$\text{PS}(i) \overset{\text{def}}{=} K_1 p_{i1} + K_2 p_{i2} + \cdots + K_G p_{iG}, \tag{5}$$

where $p_{ig}$'s are the group membership posterior probabilities defined in (3).

The propensity score is a continuous variable valued in $[\min\{K_1, ..., K_G\}, 1]$, and usually, $\min\{K_1, ..., K_G\} > 0$.

**An alternative propensity score.** We can define an alternative propensity score (APS for short) as follows. First, for $g = 1, ..., G$, define

$$K_g^{(a)} \overset{\text{def}}{=} A(g, g_{\min})/A(g_{\text{high}}, g_{\min}). \tag{6}$$

Then, in the same way as (5),

$$\text{APS}(i) \overset{\text{def}}{=} K_1^{(a)} p_{i1} + K_2^{(a)} p_{i2} + \cdots + K_G^{(a)} p_{iG}. \tag{7}$$

From (6), $K_{g_{\text{low}}}^{(a)} = 0$ and $K_{g_{\text{high}}}^{(a)} = 1$. So, APS takes values in $[0, 1]$.

**Propensity score to medium level group trajectories.** Up to now, we focus on our discussion of propensity scores to the highest level group trajectory. We can apply the "dual" approach to the lowest level group trajectory in a dual way. Now we propose a propensity score to any of medium level group trajectories, denoted by $g_m$, in the following way. First, For each $g$, define

$$K_g^{(g_m)} \overset{\text{def}}{=} 1 - A(g, g_m)/M_m; \tag{8}$$

where

$$M_m = \max\{A(g_m, g_{\text{high}}), \ A(g_m, g_{\text{low}})\}. \tag{9}$$

Then, for each individual $i$, define the propensity score to the group trajectory $g_m$ as

$$\text{PS}^{(g_m)}(i) \overset{\text{def}}{=} K_1^{(g_m)} p_{i1} + K_2^{(g_m)} p_{i2} + \cdots + K_G^{(g_m)} p_{iG}. \tag{10}$$

We can show that APS is identical to $PS^{(g_m)}$ when $g_m$ is selected as $g_{high}$ in (10).

# 4 A New Perspective of Propensity Scores of Fuzzy Sets

In this section, we only consider the general version of propensity scores defined by (5). Discussion for the other propensity scores is similar.

To the general version of propensity score, $K_g$ can be interpreted as a measure of *closeness* between the group trajectory $g$ and $g_{high}$. If an individual follows exactly the group trajectory $g$, the propensity score is just $K_g$. For any individual, the propensity score is the weighted sum of $K_g$s with the weights of individual's group membership posterior probabilities.

We can have a new perspective of the general version propensity score in the frame of fuzzy set (Zadeh [11]) in the following way. First, define the universal set (often called $G$−simplex) as

$$\Delta_G \overset{\text{def}}{=} \{[p_1, ..., p_G]'; \; p_i \geq 0, p_1 + \cdots + p_G = 1\}, \tag{11}$$

Second, we identify each individual $i$ by the group membership posterior probability vector $[p_{i1}, ..., p_{iG}]' \in \Delta_G$.

Third, let $H$ be the fuzzy set over $\Delta_G$ of individuals whose individual-trajectories are close to the highest group trajectory $g_{high}$.

Fourth, the propensity score defined in (5) is used to define the membership degrees of an individual to the fuzzy set $H$. It is clear such assignment of membership degree is appropriate through the definitions of propensity score.

**Note.** The definition of $K_g$'s is based on the same idea used in Liu [6] to define similarity between fuzzy sets, which is the complement of distance between fuzzy sets.

# 5 Applying Operations of Fuzzy Sets to Propensity Scores

In the new perspective of propensity scores in the frame of fuzzy sets, in this section, we illustrate with examples how to apply operations of fuzzy sets to form new general propensity scores.

**Applying Operations of fuzzy sets to propensity scores.** Suppose that we identified 4 group trajectories indexed as 1, 2, 3 and 4, labeled respectively as "low", "medium-low", "medium-high" and "high" respectively. Denote by $A_1$, $A_2$, $A_3$ and $A_4$ the fuzzy sets to the propensity scores to these 4 group trajectories. With operations of union ($\cup$), intersection ($\cap$) and complement

($^c$) of fuzzy sets, we can define the following *new* propensity scores of an individual to follow

- either the "medium-high" or the "high" level group trajectory by the fuzzy set $A_3 \cup A_4$;
- the "medium-high" and the "high" level group trajectories by the fuzzy set $A_3 \cap A_4$;
- the "medium" level group trajectory by the fuzzy set $A_2 \cup A_3$;
- not the "low" level group trajectory by the fuzzy set $[A_1]^c$;
- not any of these 4 group trajectories by the fuzzy set $[A_1 \cup A_2 \cup A_3 \cup A_4]^c$,

just to mention a few.

**Applying linguistic hedge to propensity scores.** Linguistic hedges could apply to fuzzy sets to create new fuzzy sets (Zadeh [12]). For example, applying the hedges "very" and "somewhat" to fuzzy set $A_4$ mentioned above, we can define propensity scores of an individual $i$ to follow a "very high" and "somewhat" high group trajectories as

$$\mathrm{PS}_{\mathrm{veryhigh}}(i) \stackrel{\mathrm{def}}{=} [A_4(i)]^2 \tag{12}$$

and

$$\mathrm{PS}_{\mathrm{somewhat}}(i) \stackrel{\mathrm{def}}{=} [A_4(i)]^{0.5} \tag{13}$$

respectively.

**Note.** By combining the operations of fuzzy sets, union, intersection, complement, linguistic hedges, we can define more new "operations" of fuzzy sets and these new operations can also be applying in defining new propensity scores as needed.

## 6  Generalization of Propensity Cores to Fuzzy and Possibilistic Clustering-Based Group Trajectories

In this section, we generalize the propensity scores based on the statistical modeling group trajectories described in Section 3 to ones on two non-statistical, fuzzy and possibilistic, clustering group trajectories. We also briefly compare these statistical and non-statistical clustering approaches.

**Propensity scores to fuzzy clustering-based group trajectories.** Assume that the data $y_{ik}$ for all $i$ and $k$ are as the same as in Section 2. Each individual $i$ corresponds to a $K$-dimensional point, i.e., $\mathbf{y}_i \stackrel{\mathrm{def}}{=} [y_{i1}, ..., y_{iK}]' \in \mathbb{R}^K$. Fuzzy clustering (also called soft clustering) is used to assign each individual $i$ into each of groups (or, clusters, classes, etc) $g$ ($g = 1, ..., G$) with membership degree $u_{ig}$ in $[0, 1]$ such that

$$\sum_{g=1}^{G} u_{ig} = 1. \tag{14}$$

There are many fuzzy clustering algorithms, but they share similar ideas. Here we briefly describe Fuzzy C-Means (FCM) Algorithm (Bezdek [1]). Let $d$ be the Euclidean distance in $\mathbb{R}^K$, and $G$ be the number of groups, starting from an initial clusters of individuals, iteratively, we search the optimal solution to minimize the objective function below:

$$\sum_{g=1}^{G} \sum_{i=1}^{N} u_{ig}^m [d(\mathbf{y}_i, \mathbf{c}_g)]^2, \tag{15}$$

where $\mathbf{c}_g \in \mathbb{R}^K$ is the center of the points in the group $g$, $g = 1, ..., G$, and $m \geq 1$ is the fuzziness parameter, usually selected as 2.

Since $u_{ig}$ plays a similar role as the posterior probability $p_{ig}$ in the statistical group-based trajectory modeling in Section 3, with the fuzzy clustering groups (clusters), we can define the group trajectory to the group $g$ as the broken line joining the points $(t_1, \bar{y}_1^{(g)}), ..., (t_K, \bar{y}_K^{(g)})$, where, for $k = 1, ..., K$,

$$\bar{y}_k^{(g)} \overset{\text{def}}{=} \sum_{i=1}^{N} u_{ig} y_{ik}. \tag{16}$$

Taking the general version of a statistical modeling propensity score in Section 3 as an example, it can be generalized to fuzzy clustering propensity score as

$$\text{PS}_{\text{fc}}(i) \overset{\text{def}}{=} K_1 u_{i1} + K_2 u_{i2} + \cdots + K_G u_{iG}, \tag{17}$$

where $K_g$ $(g = 1, ..., G)$ is defined in the same way as in Section 3.

**Propensity scores to possibilistic clustering-based group trajectories: an immature solution.** Possibilistic clustering (Krishnapuram et al. [4, 5]) is as the same as fuzzy clustering except that we drop the constrain that, for each individual, the sum of the membership degrees is 1. To avoid the trivial and unrealistic solution with the objective function in (14), The objective function is modified by adding an adjusting term as follows:

$$\sum_{g=1}^{G} \sum_{i=1}^{N} u_{ig}^m [d(\mathbf{y}_i, \mathbf{c}_g)]^2 + \sum_{g=1}^{G} \beta_g \sum_{i=1}^{N} (1 - u_{ig})^m, \tag{18}$$

where the parameters $\beta_g > 0$, $(g = 1, ..., G)$.

Since $u_{ig}$ $(g = 1, 2, ..., G)$ have nothing to do with probabilities in the sense that $\sum_{g=1}^{G} u_{ig}$ is not necessary 1, to define propensity scores with possibilistic clustering, one approach is to find the best probabilities transformed from the possibilistic distribution $u_{ig}$ ($g = 1, 2, ..., G$) for each individual $i$. There are many literatures in discussing probability-possibility transformation, for example, see Dubois et al. [2].

**Comparison of the statistical and non-statistical modeling clustering approaches.** The advantage of statistical modeling approach is that when we know the distributions of data, it is more effective. Further, we need not to impute missing data, since under the assumption of missing at random, the model will take care of the missing data automatically. (Its disadvantage is that when we "*incorrectly*" specify the distributions of the data, it is less effective.)

The advantages of fuzzy and possibilistic clustering approaches are that we do not need to know the distribution of the data, that is to say, it is robust to the distributions of the data. Its disadvantage is that we need to treat missing data either repairing them before clustering or carrying out incrementally in each iteration (see more recent review paper by Garcia-Laencina et al. [3] and references therein).

# References

1. Bezdek, J.C.: Pattern Recognition with Fuzzy Objective Function Algoritms. Plenum Press, New York (1981)
2. Dubois, D., et al.: Probability-Possibility Transformations, Triangular Fuzzy Sets, and Probabilistic Inequalities. Reliable Computing 10, 273–297 (2004)
3. Garcia-Laencina, P.J., et al.: Pattern classification with missing data: a review. Neural Computing and Applications 19, 263–282 (2010)
4. Krishnapuram, R., et al.: A possibilistic approach to clustering. IEEE Trans. on Fuzzy Systems 1, 98–110 (1993)
5. Krishnapuram, R., et al.: The possibilistic C-Means algorithm: insights and recommendations. IEEE Trans. on Fuzzy Systems 4, 385–393 (1996)
6. Liu, X.: Entropy, distance measure and similarity measure of fuzzy sets and their relations. Fuzzy Sets and Systems 52, 305–318 (1992)
7. Liu, X. et al.: Measuring individual propensity to follow a developmental trajectory. University of Montreal, Technical report (2011)
8. Nagin, D.S.: Analyzing Developmental Trajectories: Semi-Parametric, Group-Based Approach. Psychological Methods 4, 139–177 (1999)
9. Nagin, D.S.: Group-based Modeling of Development. Harvard University Press, Cambridge (2005)
10. Rubin, D.B.: Inference and missing data (with discussion). Biometrika 63, 581–592 (1976)
11. Zadeh, L.A.: Fuzzy Sets. Inf. Control 8, 338–353 (1965)
12. Zadeh, L.A.: Fuzzy Set Theoretic Interpretation Linguistic Hedges. J. of Cybernetics 2, 4–34 (1972)

# Chaos in a Fractional-Order Dynamical Model of Love and Its Control

Rencai Gu and Yong Xu

**Abstract.** This paper aims at investigating the dynamics of fractional-order model of love in the fact that fractional-order derivatives could possess memories by which romantic relationships are naturally impacted. Based on the discussions of properties including the stability of equilibrium points, chaotic behaviors and typical bifurcations, we found rich dynamics exhibited by the fractional-order love system with proper fractional order and model parameters. Besides, the control problems were studied theoretically and the simulation results illustrated the effectiveness of the proposed methods.

**Keywords:** Fractional-order model of love, Chaotic behaviors, Control.

## 1 Introduction

The theory of fractional-order derivatives and the applications of fractional calculus in physics and engineering are just a recent focus of interests [5]. Fractional differential equations could be used to describe the operations of a variety psychological and life sciences processes [2] [6] [7]. Recently, many efforts have been devoted to the studies of chaotic dynamics and control of fractional-order differential system [4] [16] [1] [11] [10]. Indeed, many investigations have obtained chaos control in fractional-order chaotic systems [8].

Love affairs, one kind of typical psychology activities can be described as a series of ordinary differential equations [15] [13]. Fractional-order love models are more realistic with fractional-order to denote memories. In this paper we are to examine the dynamics of the fractional-order love model with the its control problems.

Rencai Gu and Yong Xu
Northwestern Polytechnical University, Xi'an, China
e-mail: `rencai2050@163.com, hsux3@nwpu.edu.cn`

S. Li (Eds.): Nonlinear Maths for Uncertainty and its Appli., AISC 100, pp. 349–356.
springerlink.com                    © Springer-Verlag Berlin Heidelberg 2011

## 2  Dynamical Model of Love

The integer-order dynamical love model [14] is given as

$$\frac{dR_J}{dt} = aR_J + bf_1(J - G), \frac{dJ}{dt} = cf_1(R_J) + dJ,$$

$$\frac{dR_G}{dt} = aR_G + bf_1(G - J), \frac{dG}{dt} = ef_1(R_G) + fG, \tag{1}$$

where $f_1(x) = x(1 - |x|)$. This model describes a 'Love-triangle', in which Romeo ($R$) is involved in romantic relationships with Juliet ($J$) and Guinevere ($G$), where $R_J$ is Romeo's feeling for Juliet and $R_G$ is Romeo's feeling for Guinevere.

Here we can take the nonlinearity in (1) as $f_1(x) = x(1 - x^2)$, and then corresponding fractional-order love system can be rewritten as:

$$\frac{d^{q_1}R_J}{dt^{q_1}} = aR_J + b(J - G)(1 - (J - G)^2),$$

$$\frac{d^{q_2}J}{dt^{q_2}} = cR_J(1 - R_J{}^2) + dJ,$$

$$\frac{d^{q_3}R_G}{dt^{q_3}} = aR_G + b(G - J)(1 - (G - J)^2), \tag{2}$$

$$\frac{d^{q_4}G}{dt^{q_4}} = eR_G(1 - R_G{}^2) + fG,$$

where $d^{q_i}/dt^{q_i} = D^{q_i}, q_i \in (0,1], (i = 1,2,3,4)$, the operator $D^{\theta}$ is generally called "$\theta$-order Caputo differential operator", which can be defined as:

$$D^{\theta}x(t) = \frac{1}{\Gamma(m-\theta)} \int_0^t (t - \tau)^{m-\theta-1} x^{(m)}(\tau)d\tau, \tag{3}$$

where $0 \leq m - 1 < q < m$, $\Gamma$ stands for Gamma function.

On the basis of [12], the order $q_i$ has some practical physical meanings which represents the impact factor of memory (IFM) of an individual. It is noteworthy that the conception of IFM is proposed here to denote a measurement of how profound an individual is influenced by his/her past experiences. When one's IFM is low, his/her past experiences may have little influence on his/her present and future life; while when the IFM is high, it might be difficult for him/her to escape from the past experiences, in despite of nightmares or sweet memories.

## 3  Dynamics of the Fractional-Order Love System

It is clear that the integer-order model can be viewed as a special case from the more general fractional-order model of $q_i = 1.0, i = 1,2,3,4$. The numerical algorithm of the initial value problem for fractional-order systems is shown in [9]. Using Benettin method to calculate the top Lyapunov exponent $\lambda_1$ of the system (2) when $q_i = 1.0, i = 1, 2, 3, 4$, we get $\lambda_1$ is 0.44 for $a = -3, b = 10, c = -6, d = e = 2, f = -1$, it indicates that the integer-order love system is chaotic. With the change of the order of the equations of system (2), we can analyze the chaotic dynamics of the fractional order love system.

## 3.1  The Stability of Equilibrium Points

One can find a necessary and sufficient condition in [3] to judge the stability of the fractional order systems, which is as follows:

**Theorem 1.** *System (2) is locally asymptotically stable if all the eigenvalues of the Jacobian matrix $J = \partial f(x)/\partial x$ evaluated at the equilibrium points satisfy:*

$$|\arg(eig(J))| > \theta\pi/2 \,, \theta = \max(q_1, q_2, q_3, q_4), \tag{4}$$

*where $eig(J)$ represents the eigenvalues of matrix $J$.*

Obviously, $s_0(0, 0, 0, 0)$ is a equilibrium point of the system (2), then the Jacobian matrix of Eq. (2) at $s_0$ is

$$A = \begin{pmatrix} a & b & 0 & -b \\ c & d & 0 & 0 \\ 0 & -b & a & b \\ 0 & 0 & e & f \end{pmatrix}, \tag{5}$$

the characteristic equation of the Jacobian matrix $A$ is

$$\lambda^4 + \alpha_1\lambda^3 + \alpha_2\lambda^2 + \alpha_3\lambda + \beta = 0, \tag{6}$$

where

$$\begin{aligned}
\alpha_1 &= -(2a + f + d), \\
\alpha_2 &= df + 2af + 2ad + a^2 - be - bc, \\
\alpha_3 &= -2adf - a^2f - a^2d + ade + bde + adc + bcf, \\
\beta &= a^2df - abed - abcf,
\end{aligned} \tag{7}$$

when the parameters $a = -3, b = 10, c = -6, d = e = 2, f = -1$, the eigenvalues of the Eq.(6) are $\lambda_{1,2} = 0.2882 \pm 6.0334i$, $\lambda_3 = -3.0$, $\lambda_4 = -2.576$. If $\theta \in [0.97, 1.0]$, the Eq.(4)cannot be satisfied, thus $s_0$ is unstable.

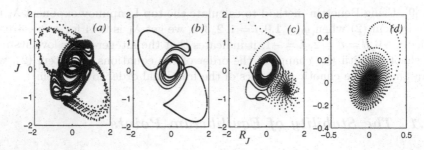

**Fig. 1** Attractors of system (2) for $q_1 = q_2 = q_3 = q_4 = \theta$ with the same parameters and initial conditions as in Fig.1: (a)$\theta = 0.99$;(b)$\theta = 0.98$;(c)$\theta = 0.97$;(d)$\theta = 0.96$.

## 3.2 Analysis of Chaotic Dynamics of the System

### Let system parameters fixed, $q_i, i = 1, 2, 3, 4$ varied

Fix $a = -3, b = 10, c = -6, d = e = 2, f = -1$, Fig.1 shows several typical attractors' projection in the $R_J - J$ plane for $q_1 = q_2 = q_3 = q_4 = \theta$. When $\theta = 0.98$, the trajectory of system (2) takes on periodicity, as Fig. 1(b) shows; If $\theta$ increased to 0.99, the system behaves chaotically, see Fig. 1(a); The system (2) stabilizes to a fixed point s for $\theta = 0.97$, as Fig. 1(c) shows; Fig.1(d) indicates the system will tend to the origin $s_0(0, 0, 0, 0)$ when $\theta = 0.96$, which corresponds to the conclusion mentioned above. Corresponding to psychology, we may explain that in real world, people with low IFM has simpler psychological activity than someone who with high IFM. If the IFM is high enough, one becomes more sensitive and his/her feeling turns to be more complex and unpredictable.

Take $a = -3, b = 10, c = -6, d = e = 2, f = -1, q_2 = q_3 = q_4 = 1.0$, phase diagrams of the $R_J - J$ plan of variable $R_J$ for $q_1 = 0.99, 0.985, 0.98$ are shown in Fig. 2(a-c), respectively. From which we can observe three different dynamics. When $\theta = 0.985$, the trajectory of system (2) takes on periodicity, as reflected in Fig. 2(b); If $\theta$ increased to 0.99, the system behaves chaotically, see Fig. 2(a); The system (2) stabilizes to a fixed point $s(1.01, -0.68, -1.01, 0.454)$ for $\theta = 0.98$, as Fig. 2(c) shows. For the cases of Fig.1(c) and Fig.2(c), Romeo eventually loves Juliet and hates Guinevere, while Juliet hates him and Guinevere loves him.

### Let Fractional-Order $q_i$ Fixed, System Parameters Varied.

Here, the fractional-orders $q_1, q_3$ are equal and fixed at 0.99, while $q_2, q_4$ are equal and fixed at 0.98. The parameters $a = -3, c = -6, d = e = 2, f = -1$ and $b$ is varied from 11.2 to 11.7. The initial states of the fractional-order love system are $R_J(0) = 0.1, J(0) = 0.3, R_G(0) = 0.2, G(0) = 0.4$, and the maximum value of $R_J$ in conditions of $R_J < 1.0, J < 0.5$ as ordinate,

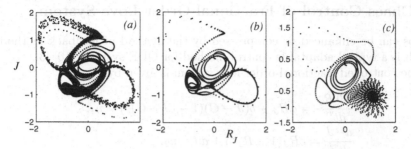

**Fig. 2** Attractors of system (2) for $q_2 = q_3 = q_4 = 1.0, q_1 = \theta$ with the same parameters and initial conditions as in Fig.1: (a)$\theta = 0.99$;(b)$\theta = 0.985$;(c)$\theta = 0.98$.

**Fig. 3** Bifurcation diagram of the fractional-order love system with $b$.

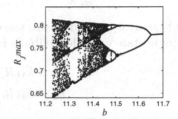

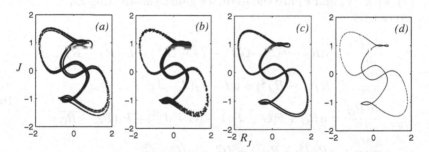

**Fig. 4** Attractors of the fractional-order love system for $q_1 = q_3 = 0.99, q_2 = q_4 = 0.98$ with $a = -3, c = 6, d = e = 3, f = -1$ and the same initial conditions as in Fig.1: (a) $b = 11.33$; (b) $b = 11.4$;(c) $b = 11.55$;(d) $b = 11.7$.

bifurcation diagram in Fig.3 was obtained. It shows that the fractional-order love system is chaotic when $b = 11.4$, see Fig.4 (b). There exists a tangent bifurcation when $b \approx 11.36$, and a flip bifurcation when $b \approx 11.51$. Fig.4 (a) shows a period 3 attractor of the fractional-order love system with $b = 11.33$, which implies there exist chaos. If $b = 11.55$ and $11.7$, the system appear period 2 and period 1 attractor, respectively, as given in Fig.4(c) and Fig.4(d).

# 4 Chaos Control of Fractional-Order Love System

Chaos can be enhanced or compressed by the method of feedback method, which is a typical adaptable control method for ODE systems [10] [3].

The controlled fractional-order love system is given by

$$
\begin{aligned}
\frac{d^{q_1} R_J}{dt^{q_1}} &= aR_J + b(J - G)(1 - (J - G)^2) - u_1, \\
\frac{d^{q_2} J}{dt^{q_2}} &= cR_J(1 - R_J{}^2) + dJ - u_2, \\
\frac{d^{q_3} R_G}{dt^{q_3}} &= aR_G + b(G - J)(1 - (G - J)^2) - u_3, \\
\frac{d^{q_4} G}{dt^{q_4}} &= eR_G(1 - R_G{}^2) + fG - u_4,
\end{aligned}
\tag{8}
$$

where $u_1, u_2, u_3$ and $u_4$ are the external control terms, and the control law of single state variable feedback has the following form

$$
\begin{aligned}
u_1 &= k_1(R_J - \overline{R_J}), u_2 = k_2(J - \overline{J}), \\
u_3 &= k_3(R_G - \overline{R_G}), u_4 = k_4(G - \overline{G}),
\end{aligned}
\tag{9}
$$

where $(\overline{R_J}, \overline{J}, \overline{R_G}, \overline{G})$ is the desired unstable equilibrium point of the chaotic Eq. (2), $k_1, k_2, k_3$ and $k_4$ are the feedback gains, substituting Eq. (9) into Eq. (8), we can get

$$
\begin{aligned}
\frac{d^{q_1} R_J}{dt^{q_1}} &= aR_J + b(J - G)(1 - (J - G)^2) - k_1(R_J - \overline{R_J}), \\
\frac{d^{q_2} J}{dt^{q_2}} &= cR_J(1 - R_J{}^2) + dJ - k_2(J - \overline{J}), \\
\frac{d^{q_3} R_G}{dt^{q_3}} &= aR_G + b(G - J)(1 - (G - J)^2) - k_3(R_G - \overline{R_G}), \\
\frac{d^{q_4} G}{dt^{q_4}} &= eR_G(1 - R_G{}^2) + fG - k_4(G - \overline{G}).
\end{aligned}
\tag{10}
$$

It is clear that Eq. (10) has one equilibrium point $(\overline{R_J}, \overline{J}, \overline{R_G}, \overline{G})$. Substituting the coordinate of $s_0$ into (10), we get the Jacobean matrix as follows

$$
J(s_0) = \begin{pmatrix}
a - k_1 & b & 0 & -b \\
c & d - k_2 & 0 & 0 \\
0 & -b & a - k_3 & b \\
0 & 0 & e & f - k_4
\end{pmatrix}.
\tag{11}
$$

The characteristic equation of the Jacobian matrix $J(s_0)$ is given by

$$
p(\lambda) = \lambda^4 - \overline{\alpha_1}\lambda^3 + \overline{\alpha_2}\lambda^2 - \overline{\alpha_3}\lambda + \overline{\beta} = 0,
\tag{12}
$$

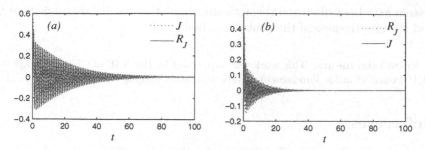

**Fig. 5** Time responses for the states $R_J, J$ of the controlled Eq. (2) stabilizing the equilibrium point $s_0$ (a) $\theta = 0.99, \kappa = 0.6$; (b) $\theta = 0.98, k_1 = -0.6$.

where

$$\overline{\alpha_1} = 2a + f + d - (k_1 + k_2 + k_3 + k_4), \qquad \overline{\alpha_2} = (a - k_3)(f - k_4) +$$
$$(a + d - k_1 - k_2)(a + f - k_3 - k_4) + (a - k_1)(d - k_2) - be - bc$$
$$\overline{\alpha_3} = (a - k_3)(f - k_4)(a + d - k_1 - k_2) + (a - k_1)(d - k_2)(a + f - k_3 - k_4)$$
$$- bc(a + d - k_1 - k_2) - bc(a + f - k_3 - k_4),$$
$$\overline{\beta} = (a - k_1)(d - k_2)(a - k_3)(f - k_4) - be(a - k_1)(d - k_2) - bc(a - k_3)(f - k_4).$$
$$(13)$$

Let $a = -3, b = 10, c = -6, d = e - 2, f = 1, k_2 = k_4 = 0$ and $k_1 = k_3 = \kappa$, when $0.7 \leq \kappa \leq 34.3$, the real part of all the eigenvalues are negative, so system (2) is stable at $s_0 \forall \theta \in (0,1)$. If we select $\theta = 0.99$ and $0.472 \leq \kappa \leq 35.3$, then the eigenvalues satisfy Eq.(4), thus the system is also stable at $s_0$ in this case. Similarly, let $k_2 = k_4 = 1.0, k_1 = 1 - k_3$, when $-0.4 \leq k_1 \leq 3.44$, the real part of all the eigenvalues are negative, then system (2) is stable at $s_0, \forall \theta \in (0,1)$. The eigenvalues satisfy Eq.(4) if $\theta = 0.98$ and $-0.82 \leq k_1 \leq 3.52$. Simulation results are presented in Fig. 5 for (a) $\theta = 0.99$ with $\kappa = 0.6$, (b) $\theta = 0.98$ with $k_1 = -0.6$, the initial states are taken as $R_J(0) = 0.1, J(0) = 0.3, R_G(0) = 0.2, G(0) = 0.4$. It is evident that the designed controller can effectively control the fractional-order love system to asymptotically stable at equilibrium points $s_0$.

## 5   Concluding Remarks

The aim of this paper is to create interests and spark research efforts in the field of "psychology and life sciences", where fractional-order modelling might offer more insights towards the understanding of the dynamical behaviors of these systems. We considered a more realistic dynamical model with fractional-order, and control problems of the fractional-order love system have been discussed theoretically.Simulation results are carried out to find the effectiveness of the control method.

**Acknowledgements.** This work was supported by the NSF of China (10972181), NPU Foundation for Fundamental Research, and Aoxiang Star Plan of NPU.

# References

1. Arena, P., Caponetto, R., Fortuna, L., Porto, D.: Bifurcation and chaos in noninteger order cellular neural networks. Int. J. Bifurcat. Chaos 7, 1527–1539 (1998)
2. Bagley, R., Calico, R.A.: Fractional order state equations for the control of viscoelastically damped structures. J. Guid. Control dynam. 14, 304–311 (1991)
3. Chen, A.M., Lu, J.A., Lü, J.H., Yu, S.M.: Generating hyperchaotic Lü attractor via state feedback control. Physica A 364, 103–110 (2006)
4. Hartley, T.T., Lorenzo, C.F., Qammer, H.: Chaos in a fractional order Chua's system. IEEE Trans. CAS-I 42, 485–490 (1995)
5. Hilfer, R.: Applications of Fractional Calculus in Physics. World Scientific, New Jersey (2000)
6. Koeller, R.C.: Application of fractional calculus to the theory of viscoelasticity. J. Appl. Mech. 51, 294–298 (1984)
7. Koeller, R.C.: Polynomial operators, Stieltjes convolution, and fractional calculus in hereditary mechanics. Acta Mech. 58, 251–264 (1986)
8. Li, C.P., Chen, G.: Chaos in the fractional order Chen system and its control. Chaos Soliton Fract 22, 549–554 (2004)
9. Li, C.P., Peng, G.J.: Chaos in Chen's system with a fractional order. Chaos Soliton Fract 22, 443–450 (2004)
10. Matouk, A.E.: Dynamical analysis feedback control and synchronization of Liu dynamical system. Nonlinear Anal. 69, 3213–3224 (2008)
11. Pecora, L.M., Carroll, T.L.: Synchronization in chaotic systems. Phys. Rev. Lett. 64, 821–824 (1990)
12. Song, L., Xu, S.Y., Yang, M.J.: Dynamical models of happiness with fractional order. Commun. Nonlinear Sci. Numer. Simulat. 15, 616–628 (2010)
13. Sprott, J.C.: Dynamical models of love. Nonlinear Dyn. Psychol. Life Sci. 3, 303–314 (2004)
14. Sprott, J.C.: Dynamical models of happiness. Nonlinear Dyn. Psychol. Life Sci. 9, 23–36 (2005)
15. Strogatz, S.: Love affairs and differential equations. Math. Mag. 61, 35 (1988)
16. Yu, Y., Li, H., Wang, S., Yu, J.: Dynamic analysis of a fractional-order Lorenz chaotic system. Chaos Soliton Fract 42, 1181–1189 (2009)

# The Semantics of *wlp* and *slp* of Fuzzy Imperative Programming Languages

Hengyang Wu and Yixiang Chen

**Abstract.** In this paper, we focus on the weakest liberal precondition semantics (*wlp*, for short) and the strongest liberal postcondition semantics (*slp*, for short) of fuzzy imperative programming languages and discuss their some basic properties.

**Keywords:** Fuzzy imperative language, The weakest liberal precondition, The strongest liberal postcondition, Fuzzy logic.

## 1 Introduction

Fuzzy programming languages can be used to describe fuzzy algorithms and fuzzy control rules. In 1980, J. M. Adamo in [1, 2] gave a fuzzy programming language **L.P.l** and introduced its syntax and semantics in detail; In 1991, D. F. Clark and A. Kandle in [6] gave another fuzzy programming language **HALO**; In 1993 and 1997, R. M. Bueno et al in [4, 5] gave fuzzy programming languages **L** and **XL**, respectively. **XL** language allows an indefinite loop statement (**while** statement); In 2004, D. S. Alvarez et al in [3] gave an imperative fuzzy programming language and studied its denotational and operational semantics; Recently, T. Vetterlein et al in [18] gave a fuzzy

Hengyang Wu
Information Engineer College, Hangzhou Dianzi University,
Hangzhou 310018, P.R. China
Shanghai Key Laboratory of Trustworthy Computing,
EastChina Normal University, Shanghai 200062, P.R. China
e-mail: wuhengy_1974@yahoo.com.cn

Yixiang Chen
Shanghai Key Laboratory of Trustworthy Computing,
EastChina Normal University, Shanghai 200062, P.R China
e-mail: yxchen@sei.ecnu.edu.cn

S. Li (Eds.): Nonlinear Maths for Uncertainty and its Appli., AISC 100, pp. 357–364.
springerlink.com      © Springer-Verlag Berlin Heidelberg 2011

programming language **FAS**, which can be used in medicine. We also in [7] gave an approach to investigate the semantics of fuzzy programming languages. However, these researches about the semantics haven't reached a very high abstract level like the ones of the classical programming languages [9,13,14,15,16] and the probabilistic programming languages [10,11,12,17,19].

Main goal of this paper is setting up the theory of predicate calculus of fuzzy programming languages. We focus on the weakest liberal preconditions semantics and the strongest liberal postconditions semantics of fuzzy imperative programming languages, which support the stepwise refinement of programs, and get the desired results. In our discussion, fuzzy logic is a foundation, where the standard negation $\neg$, t-norm $\otimes$, the corresponding t-conorm $\oplus$ and the $S-$implication $\rightarrow$ satisfying the adjoint condition are used to model fuzzy **not**, **conjunctivity**, **disjunctivity** and **implication**, respectively.

## 2   Preliminaries

In this section, we introduce fuzzy logic and recall the notions of fuzzy predicates and fuzzy relations.

### 2.1   Fuzzy Logic

In this paper, fuzzy logic is a kind of logic consisting of those formulae in the syntax as follows: $\varphi := \neg\varphi \mid \varphi \vee \varphi \mid \varphi \wedge \varphi \mid \varphi \rightarrow \varphi \mid (\forall x)\varphi \mid (\exists x)\varphi$. Each formula $\varphi$ is assigned a real number in the unit interval $[0,1]$, representing the possibility truth value degree. We use the $\otimes$, $\oplus$ and $\rightarrow$ as the logical connectives of the truth value for conjunction, disjunction and implication, respectively, where $\otimes$, $\oplus$ and $\rightarrow$ are a t-norm, the corresponding t-conorm and a $S-$implication satisfying the adjoint condition (i.e., for any $a, b, c \in [0, 1]$, $a \rightarrow b = 1 - a \otimes (1 - b)$ and $a \otimes b \leq c$ iff $a \leq b \rightarrow c$), respectively. Moreover, in this paper, we need the $\otimes$ to be continuous. Łukasiewicz logical system satisfies the previous conditions: for any $a, b, c \in [0, 1]$, $\neg a = 1 - a$, $a \rightarrow b = \min(1 - a + b, 1)$, $a \otimes b = \max(a + b - 1, 0)$ and $a \oplus b = \min(a + b, 1)$.

### 2.2   Fuzzy Predicates and Fuzzy Relations

A *fuzzy predicate* on a state space $X$ is just a fuzzy set [20] on $X$. That is, a mapping from $X$ to the unit interval. We denote the set of fuzzy predicates on $X$ by $\mathcal{F}(X)$. The partial order $\sqsubseteq$ on $\mathcal{F}(X)$ is defined pointwise: for $A, B \in \mathcal{F}(X)$, $A \sqsubseteq B$ iff for any $x \in X$, $A(x) \leq B(x)$.

The join $\sqcup_{i \in I} A_i$ and meet $\sqcap_{i \in I} A_i$ of a family $\{A_i \mid i \in I\}$ of fuzzy predicates are given by $(\sqcup_{i \in I} A_i)(x) = \sup_{i \in I} A_i(x)$ and $(\sqcap_{i \in I} A_i)(x) = \inf_{i \in I} A_i(x)$. The bottom element of $\mathcal{F}(X)$ is $\mathbf{0}$ such that $\mathbf{0}(x) = 0$ for any $x \in X$, and its top element is $\mathbf{1}$ such that $\mathbf{1}(x) = 1$ for any $x \in X$.

For any $r \in [0,1]$, we define the constant fuzzy predicate $\mathbf{r}$ as the constant function $\mathbf{r}$ from $X$ to $[0,1]$(i.e., $\mathbf{r}(x) = r$ for all $x \in X$), and a point fuzzy predicate $\eta_x$ as $\eta_x(x') = 1$ if $x' = x$ and $0$ otherwise for all $x \in X$.

In addition to the greatest lower bound and the least upper bound, we introduce the logical connectives: negation $\neg$, t-norm $\otimes$, t-conorm $\oplus$ and the corresponding implication $\rightarrow$ which are pointwise defined as follows: for all $A, B \in \mathcal{F}(X)$ and $x \in X$, $(\neg A)(x) = 1 - A(x)$, $(A \otimes B)(x) = A(x) \otimes B(x)$, $(A \oplus B)(x) = A(x) \oplus B(x)$ and $(A \rightarrow B)(x) = A(x) \rightarrow B(x)$. In particular, $(\mathbf{r} \otimes A)(x) = r \otimes A(x)$, where $r \in [0,1]$. Clearly, for any fuzzy predicate $A$, $A = \sqcup_{x \in X} A(x) \otimes \eta_x$ or $A = \sqcap_{x \in X} \eta_x \rightarrow A(x)$.

Fuzzy relations can be used to describe the semantics of fuzzy programming statements. Given two state spaces $X$ and $Y$, a *fuzzy relation* $P$ from $X$ to $Y$ is a fuzzy set on the product space $X \times Y$. The order between fuzzy relations is defined pointwise. The bottom element is $\bot$ such that $\bot(x,y) = 0$ for any $(x,y) \in X \times Y$, and its top element is $\top$ such that $\top(x,y) = 1$ for any $(x,y) \in X \times Y$.

The join $\sqcup_{i \in I} P_i$ and meet $\sqcap_{i \in I} P_i$ of a family $\{P_i \mid i \in I\}$ of fuzzy relations on $X \times Y$ are given by $(\sqcup_{i \in I} P_i)(x,y) = \sup_{i \in I} P_i(x,y)$ and $(\sqcap_{i \in I} P_i)(x,y) = \inf_{i \in I} P_i(x,y)$. The identity relation $Id$ in $X \times X$ is defined as: $Id(x,y) = 1$ if $y = x$ and $0$ otherwise.

The composition of fuzzy relations is defined as $\sup - \min$ ($\wedge$) composition by Zadeh in [20]. Latter, some people replace the operation $\wedge$ by $\otimes$. For fuzzy relations $P$ in $X \times Y$ and $Q$ in $Y \times Z$, sup-$\otimes$ composition $P \bullet Q$ in $X \times Z$ is defined by $(P \bullet Q)(x,z) = \sup_{y \in Y} P(x,y) \otimes Q(y,z)$ for all $x \in X$ and $z \in Z$.

## 3 Denotational Semantics

The syntax of fuzzy language fragment is defined as follows:

$$S ::= \text{skip}|X := FUZZ(X)|S_1; S_2|S_1 \;[\!]\; S_2|\text{if } b \text{ then } S_1 \text{ else } S_2|\text{while } b \text{ do } S$$

where $X := FUZZ(X)$ is a fuzzy assignment statement associating with a fuzzy relation $P$ and $b$ is a boolean condition, i.e., a mapping from a state space to $\{0,1\}$. The program $S_1 \;[\!]\; S_2$ nondeterministically executes either $S_1$ or $S_2$.

*Remark 1.* Fuzzy assignment statement can be seen as an extension of the classical assignment statement. For example [5], $X := FUZZ(X)$ associates with a fuzzy relation $P$, which is defined as follows:

$$P(x,y) = \begin{cases} 1, & \text{if } y = x - 1 \\ 0.5, & \text{if } y = x - 2 \\ 0, & \text{otherwise.} \end{cases}$$

If the state space is the positive integer set, then when the input $x = 3$, the output is a fuzzy set $A = 1/2 + 0.5/1$. This fuzzy set can be understood as "the states that 3 can reach", $A(1) = 0.5$ means the possibility is 0.5 that 3 reaches 1.

The denotation of a fuzzy imperative program $[\![S]\!]$ is a fuzzy relation on the state space, i.e., $[\![S]\!] : X \times X \rightarrow [0,1]$, which is defined inductively by the following semantic clauses.

- $[\![skip]\!] = Id$
- $[\![X := FUZZ(X)]\!] = P$, where $P$ is the associated fuzzy relation of $X := FUZZ(X)$.
- $[\![S_1; S_2]\!] = [\![S_1]\!] \bullet [\![S_2]\!]$
- $[\![S_1 [\!] S_2]\!] = [\![S_1]\!] \sqcup [\![S_2]\!]$
- $[\![\text{if } b \text{ then } S_1 \text{ else } S_2]\!](x, y) = b(x) \otimes [\![S_1]\!](x, y) \oplus \neg b(x) \otimes [\![S_2]\!](x, y)$
- $[\![\text{while } b \text{ do } S]\!] = \mu F$

where $\mu F$ is the least fixed point of the functional $F$ and $F : (X \times X \rightarrow [0,1]) \rightarrow (X \times X \rightarrow [0,1])$ defined by $F(R)(x,y) = b(x) \otimes (S \bullet R)(x,y) \oplus (1 - b(x)) \otimes Id(x,y)$ for any $R \in (X \times X \rightarrow [0,1])$ and $(x,y) \in X \times X$.

Since $(X \times X \rightarrow [0,1])$ is a complete lattice and the functional $F$ is an order-preserving mapping, we have the following proposition.

**Proposition 1.** *The functional $F$ has the following properties:*
*(1) $F$ preserves sup;*
*(2) $F$ has the least fixed point $\mu F$ and $\mu F = \sqcup_{n \in \omega} F^n(\bot)$, where $F^n(R) = F(F^{n-1}(R))$ and $F^0(R) = R$ for any $R \in (X \times X \rightarrow [0,1])$.*

## 4 The *wlp* and *slp* Semantics

In the classical case, in Dijkstra's opinion [8], a partial logic of Hoare triple $P\{S\}Q$ is valid means: for an input $x$, if $x$ makes predicate $P$ true, then the program $S$ either terminates a state, say $y \in X$ which makes predicate $Q$ true, or nontermination, where $P$ and $Q$ are subsets of state space. This can be formalized as follows: $P(x) = 1 \Longrightarrow \forall y, S(x,y) = 1$ **implies** $Q(y) = 1$. Further, this can be translated as follows: $P(x) \leq \inf\{S(x,y) \rightarrow Q(y) \mid y \in X\}$.

Now, we generalize the notion "true" to the fuzzy case and consider fuzzy Hoare triple $A\{S\}B$ where $A$ and $B$ are fuzzy sets and $S$ a fuzzy program. $A\{S\}B$ is valid means: $A(x) \leq \inf\{S(x,y) \rightarrow B(y) \mid y \in X\}$. Thus, for any postcondition $B$ and the command $S$, we can get the weakest liberal precondition (i.e., the largest $A$ such that $A\{S\}B$ is valid) as follows: $wlp(S,B)(x) = \inf_{y \in X} S(x,y) \rightarrow B(y)$.

**Theorem 1.** *$wlp(S,B)$ satisfies the following equations*

- $wlp(skip, B) = B$.
- $wlp(X := FUZZ(X), B)(x) = \inf_{y \in X} P(x,y) \rightarrow B(y)$.

- $wlp(S_1; S_2, B) = wlp(S_1, wlp(S_2, B))$.
- $wlp(S_1 \parallel S_2, B) = wlp(S_1, B) \sqcap wlp(S, B)$.
- $wlp(\text{if } b \text{ then } S_1 \text{ else } S_2, B) = b \otimes wlp(S_1, B) \oplus \neg b \otimes wlp(S_2, B)$.
- $wlp(\text{while } b \text{ do } S, B) = \nu G$ where $\nu G$ is the greatest fixed point of $G$ and $G(A)(x) = b(x) \otimes wlp(S, A)(x) \oplus \neg b(x) \otimes B(x)$.

*Proof.* We choose to prove $wlp(\text{while } b \text{ do } S, B) = \nu G$.

Firstly

$$wlp(\text{while } b \text{ do } S, B)(x)$$
$$= \inf_{y \in X}(\sqcup_{n \in \omega} F^n(\bot))(x, y) \to B(y)$$
$$= \inf_{y \in X} \inf_{n \in \omega} F^n(\bot)(x, y) \to B(y)$$
$$= \inf_{y \in X} \inf_{n \in \omega} F^{n+1}(\bot)(x, y) \to B(y)$$
$$= \inf_{n \in \omega} \inf_{y \in X} F^{n+1}(\bot)(x, y) \to B(y)$$
$$= \inf_{n \in \omega} \inf_{y \in X} F(F^n(\bot))(x, y) \to B(y)$$
$$= \inf_{n \in \omega} \inf_{y \in X}[b(x) \otimes (S \bullet F^n(\bot))(x, y) \oplus (1 - b(x)) \otimes Id(x, y)] \to B(y)$$
$$= \inf_{n \in \omega}[\inf_{y \in X}[b(x) \otimes [(S \bullet F^n(\bot))(x, y) \to B(y)]] \oplus [(1 - b(x)) \otimes B(x)]]$$
$$= \inf_{n \in \omega}[\inf_{y \in X}[b(x) \otimes [\sup_{z \in X} S(x, z) \otimes F^n(\bot)(z, y) \to B(y)]]$$
$$\oplus [(1 - b(x)) \otimes B(x)]]$$
$$= \inf_{n \in \omega}[\inf_{z \in X}[b(x) \otimes \inf_{y \in X}[S(x, z) \to (F^n(\bot)(z, y) \to B(y))]]$$
$$\oplus [(1 - b(x)) \otimes B(x)]]$$
$$= \inf_{n \in \omega}[\inf_{z \in X}[b(x) \otimes [S(x, z) \to \inf_{y \in X}[F^n(\bot)(z, y) \to B(y)]]$$
$$\oplus [(1 - b(x)) \otimes B(x)]]$$
$$= \inf_{n \in \omega}[b(x) \otimes \inf_{z \in X}(S(x, z) \to A_n(z)) \oplus [(1 - b(x)) \otimes B(x)]]$$
$$= \inf_{n \in \omega}[b(x) \otimes wlp(S, A_n)(x) \oplus (1 - b(x)) \otimes B(x)]$$

where $A_n(z) = \inf_{y \in X} F^n(\bot)(z, y) \to B(y)$. Then, one can verify that $A_0(x) = 1$ for any $x$ and $A_{n+1}(x) = b(x) \otimes wlp(S, A_n)(x) \oplus (1 - b(x)) \otimes B(x)$. That is,

$$wlp(\text{while } b \text{ do } S, B) = \sqcap_{n \in \omega} A_n. \tag{1}$$

Secondly, $\sqcap_{n \in \omega} A_n$ is the fixed point of $G$.

$$G(\sqcap_{n \in \omega} A_n)(x) = \sqcap_{n \in \omega} G(A_n)(x)$$
$$= \sqcap_{n \in \omega} b(x) \otimes wp(S, A_n)(x) \oplus (1 - b(x)) \otimes B(x)(\sqcap_{n \in \omega} A_{n+1})(x)$$
$$= (\sqcap_{n \in \omega} A_n)(x) \quad (\text{since } A_0 = 1)$$

Finally, suppose that $C$ is any fixed point of $G$, now we prove $\sqcap_{n \in \omega} A_n \sqsupseteq C$. We use the mathematical introduction to prove. (1)$n = 0$, clearly $A_0 \sqsupseteq C$. (2)Suppose $n = k$, $A_k \sqsupseteq C$. Then

$$A_{k+1}(x)$$
$$= b(x) \otimes wlp(S, A_k)(x) \oplus (1 - b(x)) \otimes B(x)$$
$$\geq b(x) \otimes wp(S, C)(x) \oplus (1 - b(x)) \otimes B(x)$$
$$= G(C)(x)$$
$$= C(x) \quad (\text{since } C \text{ is the fixed point of } G)$$

Hence, for any $n$, $A_n \sqsupseteq C$. Then, $\sqcap_{n\in\omega} A_n \sqsupseteq C$. Thus, $\sqcap_{n\in\omega} A_n$ is the greatest fixed point of $G$.

**Theorem 2.** $wlp(S, B)$ *has the following properties*

- **Termination.** $wlp(S, \mathbf{1}) = \mathbf{1}$.
- **Monotonicity.** $wlp(S, B_1) \sqsubseteq wlp(S, B_2)$ *if* $B_1 \sqsubseteq B_2$.
- **Implication.** $wlp(S, \mathbf{r} \to B) = \mathbf{r} \to wlp(S, B)$ *where* $r \in [0, 1]$.
- **Conjunctivity.** $wlp(S, \sqcap_{i\in I} B_i) = \sqcap_{i\in I} wlp(S, B_i)$ *if* $\{B_i : i \in I\} \subseteq \mathcal{F}(X)$.

Now, we consider the strongest liberal postcondition $slp(A, S)$ for a given command $S$ and a precondition $A$. That is, finding the strongest liberal post-condition $sp(A, S)$ which makes the Hoare triple $A\{S\}sp(A, S)$ valid. We know, if $sp(A, S) \sqsubseteq B$, then $A\{S\}B$ is valid. Hence, the strongest postcondition means the smallest $B$ such that $A\{S\}B$ is valid.

**Theorem 3.** *Let* $A \in \mathcal{F}(X)$ *and* $S$ *a command. Then* $slp(A, S)(y) = \sup_{x\in X} S(x, y) \otimes A(x)$.

**Theorem 4.** $slp(A, S)$ *satisfies the following equations*

- $slp(A, skip) = A$.
- $slp(A, X := FUZZ(X))(y) = \sup_{x\in X} P(x, y) \otimes A(x)$.
- $slp(A, S_1; S_2) = slp(slp(A, S_1), S_2)$.
- $slp(A, S_1 \;[\!]\; S_2) = slp(A, S_1) \sqcup slp(A, S_2)$.
- $slp(A, if\ b\ then\ S_1\ else\ S_2) = b \otimes slp(A, S_1) \oplus \neg b \otimes slp(A, S_2)$.
- $slp(A, while\ b\ do\ S) = \mu T$ *where* $\mu T$ *is the least fixed point of* $T$ *and* $T(C)(x) = b(x) \otimes slp(C, S)(x) \oplus \neg b(x) \otimes A(x)$.

**Theorem 5.** $slp(A, S)$ *has the following properties*

- **Strictness.** $slp(\mathbf{0}, S) = \mathbf{0}$.
- **Monotonicity.** $slp(A_1, S) \sqsubseteq slp(A_2, S)$ *if* $A_1 \sqsubseteq A_2$.
- **Homogeneity.** $slp(\mathbf{r} \otimes A, S) = \mathbf{r} \otimes slp(A, S)$ *where* $r \in [0, 1]$.
- **Disjunctivity.** $slp(\sqcup_{i\in I} A_i, S) = \sqcup_{i\in I} slp(A_i, S)$ *if* $\{A_i : i \in I\} \subseteq \mathcal{F}(X)$.

## 5   Duality and Logic

Given the functional $T : \mathcal{F}(X) \to \mathcal{F}(X)$, which has implication property and is conjunctivity. That is, $T(\mathbf{r} \to A) = \mathbf{r} \to T(A)$ and $T(\sqcap_{i\in I} A_i) = \sqcap_{i\in I} T(A_i)$ for any $A \in \mathcal{F}(X)$, $r \in [0, 1]$ and $\{A_i : i \in I\} \subseteq \mathcal{F}(X)$. Define $T^* : X \times X \to [0, 1]$ such that

$$T^*(x, y) = 1 - T(\neg \eta_y)(x) \tag{2}$$

for any $(x, y) \in X \times X$.

Given a fuzzy relation $R : X \times X \to [0,1]$, define $R^\circ : \mathcal{F}(X) \to \mathcal{F}(X)$ such that

$$R^\circ(A)(x) = \inf_{y \in X} R(x,y) \to A(y) \qquad (3)$$

for any $A \in \mathcal{F}(X)$ and $x \in X$. Clearly, $R^\circ$ has implication property and is conjunctivity.

**Theorem 6.** $(R^\circ)^\star = R$ and $(T^\star)^\circ = T$.

*Proof.* Firstly, for any $(x,y) \in X \times X$,

$$
\begin{aligned}
&(R^\circ)^\star(x,y) \\
&= 1 - R^\circ(\neg \eta_y)(x) \\
&= 1 - \inf_{z \in X} R(x,z) \to (1 - \eta_y(z)) \\
&= 1 - (R(x,y) \to 0) \\
&= R(x,y).
\end{aligned}
$$

Hence, $(R^\circ)^\star = R$.

Secondly, for any $A \in \mathcal{F}(X)$ and $x \in X$,

$$
\begin{aligned}
&(T^\star)^\circ(A)(x) \\
&= \inf_{y \in X} T^\star(x,y) \to A(y) \\
&= \inf_{y \in X}(1 - T(\neg \eta_y)(x)) \to A(y) \\
&= \inf_{y \in X}[\neg A(y) \to T(\neg \eta_y)(x)] \\
&= \inf_{y \in X} T(\neg A(y) \to \neg \eta_y)(x) \\
&= \inf_{y \in X} T(\eta_y \to A(y))(x) \\
&= T(\inf_{y \in X} \eta_y \to A(y))(x) \\
&= T(A)(x).
\end{aligned}
$$

So, $(T^\star)^\circ = T$.

In this case, we call $R$ and $R^\circ$ to be dual, similarity for $T$ and $T^\star$. We now consider how we can use the duality related the logic and semantics. The semantics of a fuzzy program $\mathbf{S}$ is defined as a fuzzy relation $X \times X \to [0,1]$, the weakest liberal precondition semantics is a function $\mathcal{F}(X) \to \mathcal{F}(X)$, which has implication property and is conjunctivity. The semantics, $[\![\mathbf{S}]\!]$, is dual to the functional given by the weakest liberal preconditions semantics, namely $\lambda A.wlp(\mathbf{S}, A)$. Symbolically this is just

$$[\![\mathbf{S}]\!]^\circ = \lambda A.wlp(\mathbf{S}, A). \qquad (4)$$

Moreover from the previous duality we know that $(\lambda A.wlp(\mathbf{S}, A))^\star = [\![\mathbf{S}]\!]$.

**Acknowledgements.** The first author acknowledges support from the Shanghai Postdoctoral Scientific Program under Grant No.09R21412400, the Shanghai Key Laboratory of Trustworthy Computing under Grant No.07BZ22304, the Project of the Excellent Youth Fund of the University of Zhejiang Province and the Hangzhou

Dianzi University fund under Grant No.KYS181507093. The second author acknowledges support from 973 Project Grant No.2011CB302802 and the National Nature Science Foundation of China under Grant No.61021004.

# References

1. Adamo, J.M.: L.p.l. A fuzzy programming language: 1. syntactic aspects. Fuzzy Sets and Systems 3, 151–179 (1980)
2. Adamo, J.M.: L.p.l. A fuzzy programming language: 2. semantic aspects. Fuzzy Sets and Systems 3, 261–289 (1980)
3. Alvarez, D.S., Antonio, F., Gomez, S.: A fuzzy language. Fuzzy Sets and Systems 141, 335–390 (2004)
4. Bueno, R.M., Clares, B., Conejo, R., Perez de la Cruz, J.L.: An elementary fuzzy programming language. Fuzzy Sets and Systems 58, 55–73 (1993)
5. Bueno, R.M., Perez de la Cruz, J.L., Conejo, R., Clares, B.: A family of fuzzy programming languages. Fuzzy Sets and Systems 87, 167–179 (1997)
6. Clark, D.F., Kandel, A.: HALO-a fuzzy programming language. Fuzzy Sets and Systems 44, 199–208 (1991)
7. Chen, Y.X., Wu, H.Y.: Domain semantics of possibility computations. Information Sciences 178, 2661–2679 (2008)
8. Dijkstra, E.W.: A Discipline of Programming. Prentice Hall International, Englewood Cliffs (1976)
9. Hoare, C.A.R.: Some properties of predicate transformers. Journal of the Association for Computing Machinery 25, 461–480 (1978)
10. He, J.F., Seidel, K., McIver, A.K.: Probabilistic models for the guarded command language. Science of Computer Programming 28, 171–192 (1997)
11. Jones, C.: Probabilistic non-determinism. PhD thesis, University of Edinburgh, Edinburgh (1990)
12. Morgan, C., McIver, A.K., Seidel, K.: Probabilistic predicate transformers. ACM Trans. Programming Languages and Systems 8, 325–353 (1996)
13. Plotkin, G.D.: A powerdomains construction. SIAM Journal on Computing 5, 452–487 (1976)
14. Plotkin, G.D.: An structural approach to operational semantics. Technical Report DAIMI FN-19. Computer Science Department, Aarhus University (1981)
15. Smyth, M.B.: Powerdomain. Journal of Computer and Systems Sciences 16, 23–36 (1978)
16. Stoy, J.E.: Denotational semantics: The Scott-Strachey approach to programming language theory. MIT Press, Cambridge (1977)
17. Tix, R., Keimel, K., Plotkin, G.D.: Semantics domains for combining probability and non-determinism. Electronic Notes in Theoretical Computer Science 129, 1–104 (2005)
18. Vetterlein, T., Mandl, H., Adlassnig, K.P.: Fuzzy Arden Syntax: A fuzzy programming language for medicine. Artificial Intelligence in Medicine 49, 1–10 (2010)
19. Ying, M.S.: Reasoning about probabilistic sequential programs in a probabilistic logic. Acta Informatica 39, 315–389 (2003)
20. Zadeh, L.A.: Fuzzy sets. Information and Control 8, 338–353 (1965)

# Public-Key Cryptosystem Based on $n$-th Residuosity of $n = p^2q$

Tomoko Adachi

**Abstract.** We set $n = pq$ where $p$ and $q$ are odd primes. A number $z$ is said to be an $n$-th residue modulo $n^2$ if there exists a nonnegative integer $y$ such that $z = y^n \bmod n^2$, where $y$ is less than $n^2$ and is coprime to $n^2$. In this paper, we investigate $n$-th residues in the case of $n = p^2q$ where $p$ and $q$ are odd primes. We will give $n$-th residues in the case of $n = 3^2 \cdot 5$, in particular.

**Keywords:** $n$-th residues, Public key cryptosystem.

## 1 Introduction

Suppose $p$ is an odd prime and $a$ is an integer. $a$ is defined to be a quadratic residue modulo $p$ if $a \not\equiv 0 \pmod{p}$ and the congruence $y^2 \equiv a \pmod{p}$ has a solution $y$ where nonnegative $y$ is less than $n$. It is well-known that a quadratic residue is adopted to public key cryptosystems. For example, we show Rabin Cryptosystem [5]. Let $n = pq$, where $p$ and $q$ are primes, and $p, q \equiv 3 \pmod 4$. The value $n$ is the public key, while $p$ and $q$ are the private key. For a plaintext $m < n$, we define the cipertext $c = m^2 \pmod n$. Quadratic residuosity is adopted in a trapdoor mechanism of this public key cryptosystem. As well, the public key cryptosystem by Kurosawa et. al. [2] also utilized quadratic residuosity. Moreover, the public key cryptosystem by Naccache and Stern [3] utilized higher residuosity. In this paper, we adopt $n$-th residuosity.

We set $n = pq$ where $p$ and $q$ are odd primes. A number $z$ is said to be an $n$-th residue modulo $n^2$ if there exists a nonnegative integer $y$ such that $z = y^n \bmod n^2$, where $y$ is less than $n^2$ and is coprime to $n^2$. The problem

Tomoko Adachi
Toho University, Department of Information Sciences, 2-2-1 Miyama, Funabashi, Chiba, 274-8510, Japan
e-mail: adachi@is.sci.toho-u.ac.jp

S. Li (Eds.): Nonlinear Maths for Uncertainty and its Appli., AISC 100, pp. 365–370.
springerlink.com                          © Springer-Verlag Berlin Heidelberg 2011

of deciding $n$-th residuosity, that is, distinguishing $n$-th residues from non $n$-th residues will be denoted by CR$[n]$. Paillier [4] introduced public key cryptosystems based on CR$[n]$. Here, we investigate more general cases. In this paper, we investigate $n$-th residues in the case of $n = p^2 q$ where $p$ and $q$ are odd primes. We will give $n$-th residues in the case of $n = 3^2 \cdot 5$, in particular.

## 2   Review: $n$-th Residuosity of $n = pq$

In this section, we describe $n$-th residuosity of $n = pq$ and its cryptosystem. All the results in this section are due to [4].

We set $n = pq$ where $p$ and $q$ are large primes. In this case, we denote by $\phi(n) = (p-1)(q-1)$ Euler's function and by $\lambda(n) = \mathrm{lcm}(p-1, q-1)$. We adopt $\lambda$ instead of $\lambda(n)$ for visual comfort. We denote by $Z_{n^2}$ a residue class ring modulo $n^2$ and by $Z_{n^2}^*$ its invertible element set. The set $Z_{n^2}^*$ is a multiplicative subgroup of $Z_{n^2}$ of order $\phi(n^2) = n\phi(n) = pq(p-1)(q-1)$. For any $w \in Z_{n^2}^*$, $w^\lambda = 1 \pmod{n}$ and $w^{n\lambda} = 1 \pmod{n^2}$ hold.

**Definition 1.** A number $z$ is said to be an $n$-th residue modulo $n^2$ if there exists a number $y \in Z_{n^2}^*$, such that $z = y^n \pmod{n^2}$.

The set of $n$-th residues is a multiplicative subgroup of $Z_{n^2}^*$ of order $\phi(n)$. The problem of deciding $n$-th residuosity, that is, distinguishing $n$-th residues from non $n$-th residues will be denoted by CR$[n]$. As for prime residuosity, deciding $n$-th residuosity, is believed to be computationally hard.

Let $g$ be some element of $Z_{n^2}^*$ and denote by $\varepsilon_g$ the integer-valued function defined by

$$Z_n \times Z_n^* \ \rightarrow \ Z_{n^2}^*$$
$$(x, y) \longmapsto g^x y^n \pmod{n^2}.$$

Here, depending on $g$, $\varepsilon_g$ may feature an interesting property such as the following lemma.

**Lemma 1.** *If the order of $g$ is a nonzero multiple of $n$ then $\varepsilon_g$ is bijection.*

We denote by $\mathcal{B}_\alpha \subset Z_{n^2}^*$ the set of elements of order $n\alpha$ and by $\mathcal{B}$ their disjoint union for $\alpha = 1, \cdots, \lambda$.

**Definition 2.** Assume that $g \in \mathcal{B}$. For $w \in Z_{n^2}^*$, we call $n$-th residuosity class of $w$ with respect to $g$ the unique integer $x \in Z_n$ for which there exists $y \in Z_n^*$, such that

$$\varepsilon_g(x, y) = w.$$

Adopting Benaloh's notations [1], the class of $w$ is denoted $[[w]]_g$. It is worthwhile noticing the following property.

**Lemma 2.** $[[w]]_g = 0$ *if and only if $w$ is an $n$-th residue modulo $n^2$. Furthermore,*

$$\forall w_1, w_2 \in Z_{n^2}^* \qquad [[w_1 w_2]]_g = [[w_1]]_g + [[w_2]]_g \pmod{n}$$

*that is, the class function $w \longmapsto [[w]]_g$ is a homomorphism from $(Z_{n^2}^*, \times)$ to $(Z_n, +)$ for any $g \in \mathcal{B}$.*

By Lemma 2, it can easily be shown that, for any $w \in Z_{n^2}^*$ and $g_1, g_2 \in \mathcal{B}$, we have

$$[[w]]_{g_1} = [[w]]_{g_2} [[g_2]]_{g_1} \pmod{n}, \tag{1}$$

which yields $[[g_1]]_{g_2} = [[g_2]]_{g_1}^{-1} \bmod n$ and thus $[[g_2]]_{g_1}$ is invertible modulo $n$.

The set

$$S_n = \{u < n^2 \mid u = 1 \pmod{n}\}$$

is a multiplicative subgroup of integers modulo $n^2$ over which the function $L$ such that

$$\forall u \in S_n \qquad L(u) = \frac{u-1}{n}$$

is clearly well-defined.

**Lemma 3.** *For any $w \in Z_{n^2}^*$, $L(w^\lambda \pmod{n^2}) = \lambda[[w]]_{1+n} \pmod{n}$.*

By Lemma 3, for any $g \in \mathcal{B}$ and $w \in Z_{n^2}^*$, we can compute

$$\frac{L(w^\lambda \pmod{n^2})}{L(g^\lambda \pmod{n^2})} = \frac{\lambda[[w]]_{1+n}}{\lambda[[g]]_{1+n}} = \frac{[[w]]_{1+n}}{[[g]]_{1+n}} = [[w]]_g \pmod{n}, \tag{2}$$

by virtue of Equation 1.

Now, we describe the public key cryptosystem based on the $n$-th residuosity class problem.

Set $n = pq$ and randomly select a base $g \in \mathcal{B}$: as shown before, this can be done efficiently by checking whether $\gcd(L(g^\lambda \pmod{n^2}), n) = 1$. Now, consider $(n, g)$ as public parameters whilst the pair $(p, q)$ remains private. The cryptosystem is depicted below. For a plaintext $m < n$, we select a random $r < n$, and compute the ciphertext $c = g^m r^n \pmod{n^2}$. That is to say, we employ $\varepsilon_g$ as an encryption function. For a ciphertext $c < n^2$, we compute the plaintext $m = \frac{L(c^\lambda \pmod{n^2})}{L(g^\lambda \pmod{n^2})} \bmod n$, by Equation 2.

## 3   $n$-th Residuosity of $n = p^2q$

In this section, we will investigate $n$-th residues in the case of $n = p^2q$ where $p$ and $q$ are odd primes, and $p \neq q$. Especially, we will give some $n(= p^2q)$-th residues in the case of $p = 3$ and $q = 5$.

We set $n = p^2q$ where $p$ and $q$ are large primes. In this case, we denote by $\phi(n) = p(p-1)(q-1)$ Euler's function and by $\lambda(n) = \mathrm{lcm}(p, p-1, q-1)$. We adopt $\lambda$ instead of $\lambda(n)$ for visual comfort. We denote by $Z_{n^2}$ a residue class ring modulo $n^2$ and by $Z_{n^2}^*$ its invertible element set. The set $Z_{n^2}^*$ is a multiplicative subgroup of $Z_{n^2}$ of order $\phi(n^2) = n\phi(n) = p^3q(p-1)(q-1)$. For any $w \in Z_{n^2}^*$, the following equations hold:

$$w^\lambda = 1 \quad (\mathrm{mod}\ n),\ w^{n\lambda} = 1 \quad (\mathrm{mod}\ n^2).$$

For $n = p^2q$, we define $n$-th residue modulo $n^2$ in the same way as Definition 1 for $n = pq$. Let $g$ be some element of $Z_{n^2}^*$ and $\varepsilon_g$ be the function defined by

$$\begin{aligned} Z_n \times Z_n^* &\rightarrow Z_{n^2}^* \\ (x, y) &\longmapsto \varepsilon_g(x, y) = g^x y^n \quad (\mathrm{mod}\ n^2). \end{aligned} \tag{3}$$

If the order of $g$ is a nonzero multiple of $n$ then $\varepsilon_g$ is bijection. We denote by $\mathcal{B}_\alpha \subset Z_{n^2}^*$ the set of elements of order $n\alpha$ and by $\mathcal{B}$ their disjoint union for $\alpha = 1, \cdots, \lambda$. For $n = p^2q$, we define $n$-th residuosity class in the same way as Definition 2 for $n = pq$.

For $n = p^2q$, we define the function $L$ as follows:

$$\begin{aligned} S_n = \{u < n^2 \mid u = 1 \quad (\mathrm{mod}\ n)\} &\rightarrow Z_n \\ u &\longmapsto L(u) = \tfrac{u-1}{n}. \end{aligned} \tag{4}$$

In the case of $n = pq$, we have public-key cryptosystem, that is, for any $g \in \mathcal{B}$, the function $\varepsilon_g$ is encryption and $\frac{L(c^\lambda \ (\mathrm{mod}\ n)^2)}{L(g^\lambda \ (\mathrm{mod}\ n)^2)}$ mod $n$ is decryption. However, in the case of $n = p^2q$, for any $g \in \mathcal{B}$, the function $\varepsilon_g$ is NOT encryption. Because, in the case of $n = 3^2 \cdot 5$, when we choose $g = 19 \in \mathcal{B}$, $L(g^\lambda \ (\mathrm{mod}\ n)^2) = 3$ is not invertible element modulo $n^2$ and we cannot decrypt. Hence, we investigate public-key cryptosystem based on $n(= p^2q)$-th residuosity in order to be well-defined.

Here, we give $n(= p^2q)$-th residuosity class in the case of $p = 3$ and $q = 5$. We obtain $\lambda(n) = \mathrm{lcm}(p, p-1, q-1) = 12$. For $g \in Z_{n^2}^*$, the value $\alpha$ such that an order of $g$ is equal to $n\alpha$ is 1, 2, 3, 4, 6 or 12. Each $\mathcal{B}_\alpha$ is as follows, and $\mathcal{B} = \mathcal{B}_1 \cup \mathcal{B}_2 \cup \mathcal{B}_3 \cup \mathcal{B}_4 \cup \mathcal{B}_6 \cup \mathcal{B}_{12}$.

$\mathcal{B}_1 = \{$ 46, 91, 181, 316, 361, 496, 586, 631, 721, 766, 856, 991, 1036, 1171, 1261, 1306, 1396, 1441, 1531, 1666, 1711, 1846, 1981 $\}$

$\mathcal{B}_2 = \{$ 19, 38, 44, 64, 71, 89, 116, 154, 179, 206, 289, 314, 334, 341, 359, 386, 469, 494, 521, 559, 584, 604, 611, 629, 656, 694, 719, 739, 746, 764, 791, 809, 854, 881, 964, 989, 1009, 1016, 1034, 1061, 1144, 1169, 1196, 1234, 1259, 1279, 1286, 1304, 1331, 1369, 1394, 1414, 1421, 1439, 1466, 1504, 1529, 1556, 1639, 1664, 1684, 1691, 1709, 1736, 1819, 1844, 1871, 1909, 1934, 1954, 1961, 1979, 2006 $\}$

$\mathcal{B}_3 = \{$ 16, 31, 61, 106, 121, 166, 196, 211, 241, 256, 286, 331, 346, 391, 421, 436, 466, 481, 511, 556, 571, 616, 646, 661, 691, 706, 736, 781, 796, 841, 871,

886, 916, 931, 961, 1006, 1021, 1066, 1096, 1111, 1141, 1156, 1186, 1231, 1246, 1291, 1321, 1336, 1366, 1381, 1411, 1456, 1471, 1516, 1546, 1561, 1591, 1606, 1636, 1981, 1696, 1741, 1771, 1786, 1816, 1831, 1861, 1906, 1921, 1966, 1996, 2011 }

$\mathcal{B}_4$ = { 8, 17, 37, 62, 73, 98, 127, 152, 172, 197, 208, 233, 253, 262, 287, 388, 397, 413, 422, 442, 467, 478, 503, 548, 577, 602, 613, 658, 667, 683, 692, 737, 748, 773, 802, 827, 847, 872, 883, 908, 928, 937, 953, 962, 1063, 1072, 1088, 1097, 1117, 1142, 1153, 1178, 1198, 1223, 1252, 1277, 1288, 1313, 1342, 1358, 1367, 1387, 1412, 1423, 1448, 1477, 1502, 1547, 1558, 1583, 1603, 1612, 1628, 1637, 1738, 1747, 1763, 1772, 1792, 1817, 1828, 1853, 1873, 1898, 1927, 1952, 1963, 1988, 2008, 2017 }

$\mathcal{B}_6$ = { 4, 11, 14, 29, 34, 41, 56, 59, 79, 86, 94, 104, 119, 131, 139, 146, 164, 169, 184, 191, 194, 209, 214, 221, 229, 236, 239, 254, 259, 266, 281, 284, 304, 311, 319, 329, 344, 356, 364, 371, 389, 394, 409, 416, 419, 434, 439, 446, 454, 461, 464, 479, 484, 491, 506, 509, 529, 536, 544, 554, 569, 581, 583, 596, 614, 619, 634, 641, 644, 659, 664, 671, 679, 686, 689, 704, 709, 716, 731, 734, 754, 761, 769, 779, 794, 806, 814, 821, 839, 844, 859, 866, 869, 884, 889, 896, 904, 911, 914, 929, 934, 941, 953, 979, 994, 1004, 1019, 1031, 1039, 1046, 1064, 1069, 1084, 1091, 1094, 1109, 1114, 1121, 1129, 1136, 1139, 1154, 1159, 1166, 1181, 1184, 1204, 1211, 1219, 1229, 1244, 1256, 1264, 1271, 1289, 1294, 1309, 1316, 1319, 1334, 1339, 1346, 1354, 1361, 1364, 1379, 1384, 1391, 1406, 1409, 1429, 1436, 1444, 1454, 1469, 1481, 1489, 1496, 1514, 1519, 1534, 1541, 1544, 1559, 1564, 1571, 1579, 1586, 1589, 1604, 1609, 1616, 1631, 1634, 1654, 1661, 1669, 1679, 1694, 1706, 1714, 1721, 1739, 1744, 1759, 1766, 1769, 1784, 1789, 1796, 1804, 1811, 1814, 1829, 1834, 1841, 1856, 1859, 1879, 1886, 1894, 1904, 1939, 1946, 1964, 1909, 1984, 1991, 2009. 2014, 2021 }

$\mathcal{B}_{12}$ = { 2, 22, 23, 38, 47, 52, 58, 67, 77, 83, 88, 92, 97, 103, 112, 113, 122, 128, 133, 137, 142, 148, 158, 167, 173, 178, 187, 202, 203, 212, 223, 227, 238, 247, 248, 263, 272, 277, 283, 292, 302, 308, 313, 317, 322, 328, 337, 338, 347, 353, 358, 362, 367, 373, 383, 392, 398, 403, 412, 427, 428, 437, 448, 452, 463, 472, 473, 488, 497, 502, 508, 517, 527, 533, 538, 542, 547, 553, 562, 563, 572, 578, 583, 587, 592, 598, 608, 617, 623, 628, 637, 652, 653, 662, 673, 677, 688, 697, 698, 713, 722, 727, 733, 742, 752, 758, 763, 767, 772, 778, 787, 788, 797, 803, 808, 812, 817, 823, 833, 842, 848, 853, 862, 877, 878, 887, 898, 902, 913, 922, 923, 938, 947, 952, 958, 967, 977, 983, 988, 992, 997, 1003, 1012, 1013, 1022, 1028, 1033, 1037, 1042, 1048, 1058, 1067, 1073, 1078, 1087, 1102, 1103, 1112, 1123, 1127, 1138, 1147, 1148, 1163, 1172, 1177, 1183, 1192, 1202, 1208, 1213, 1217, 1222, 1228, 1237, 1238, 1247, 1253, 1258, 1262, 1267, 1273, 1283, 1292, 1298, 1303, 1312, 1327, 1328, 1337, 1348, 1352, 1363, 1372, 1373, 1388, 1397, 1402, 1408, 1417, 1427, 1433, 1438, 1442, 1447, 1453, 1462, 1463, 1472, 1478, 1483, 1487, 1492, 1498, 1508, 1517, 1523, 1528, 1537, 1552, 1553, 1562, 1573, 1577, 1588, 1597, 1598, 1613, 1622, 1627, 1633, 1642, 1652, 1658, 1663, 1667, 1672, 1678, 1687, 1688, 1697, 1703, 1708, 1712, 1717, 1723, 1733, 1742, 1748, 1753, 1762, 1777, 1778, 1787, 1798, 1802, 1813, 1822, 1823, 1838, 1847, 1852, 1858, 1867, 1877, 1883, 1888, 1892, 1897, 1903, 1912, 1913, 1922, 1928,

1933, 1937, 1942, 1948, 1958, 1967, 1973, 1978, 1987, 2002, 2003, 2012, 2023
}

For all $g \in \mathcal{B}$, we check whether $L(g^\lambda \pmod{n^2})$ is invertible element modulo $n^2$ or not. Therefore, we obtain the following result: If $g$ is an element of the set $\mathcal{B}_3$, $\mathcal{B}_6$ or $\mathcal{B}_{12}$, then $L(g^\lambda \pmod{n^2})$ is invertible element modulo $n^2$. If $g$ is an element of the set $\mathcal{B}_1$, $\mathcal{B}_2$ or $\mathcal{B}_4$, then $L(g^\lambda \pmod{n^2})$ is not invertible element modulo $n^2$.

We set $\mathcal{B}' = \mathcal{B}_3 \cup \mathcal{B}_6 \cup \mathcal{B}_{12}$. For any $w \in Z_{n^2}^*$, we have $L(w^\lambda \pmod{n}) = \lambda[[w]]_{1+n} \pmod{n}$. Therefore, for any $g \in \mathcal{B}'$ and $w \in Z_{n^2}^*$, we can compute

$$\frac{L(w^\lambda \pmod{n^2})}{L(g^\lambda \pmod{n^2})} = \frac{\lambda[[w]]_{1+n}}{\lambda[[g]]_{1+n}} = \frac{[[w]]_{1+n}}{[[g]]_{1+n}} = [[w]]_g \pmod{n}.$$

Hence, we get the following theorem.

**Theorem 1.** *We set $n = p^2 q$ and $\lambda = lcm(p, p-1, q-1)$. In the case of $p = 3$ and $q = 5$, for any $g \in \mathcal{B}' = \mathcal{B}_3 \cup \mathcal{B}_6 \cup \mathcal{B}_{12}$, we obtain public-key cryptosystem as public keys $(n, g)$ and private keys $(p, q)$. For a plaintext $m < n$, we select a random $r < n$, and compute the cipertext $c$ by Equation 5. For a cipertext $c < n^2$, we compute the plaintext $m$ by Equation 6.*

$$c = g^m r^n \pmod{n^2}, \tag{5}$$

$$m = \frac{L(c^\lambda \pmod{n^2})}{L(g^\lambda \pmod{n^2})} \pmod{n}. \tag{6}$$

For $n = p^2 q$, we obtain the public key cryptosystem based on the $n$-th residuosity class problem.

# References

1. Benaloh, J.C.: Veryfiable Secret-Ballot Ellections, PhD Thesis, Yale University (1988)
2. Kurosawa, K., Itoh, T., Takeuchi, M.: Public key cryptosystem using a reciprocal number with the same intractability as factoring a large number. Electronics Letters 23(15), 809–810 (1987)
3. Naccache, D., Stern, J.: A new public-key Cryptosystem Based on Higher Residues. In: 5th ACM Conference on Computer and Communications Security, pp. 59–66 (1998)
4. Paillier, P.: Public-key cryptosystems based on composite degree residuosity classes. In: Stern, J. (ed.) EUROCRYPT 1999. LNCS, vol. 1592, pp. 223–238. Springer, Heidelberg (1999)
5. Rabin, M.O.: Digitized signatures and public-key functions as intractable as factorization. MIT Laboratory for Computer Science Technical Report, LCS/TR-212 (1979)

# Shrinking Projection Method for a Family of Quasinonexpansive Mappings with a Sequence of Subsets of an Index Set

Yasunori Kimura

**Abstract.** Using a sequence of subsets of an index set for a family of quasi-nonexpansive mappings, we propose an iterative scheme generated by the shrinking projection method for finding their common fixed point. We prove strong convergence of this scheme under appropriate conditions.

**Keywords:** Common fixed point, Iterative scheme, Quasinonexpansive, Shrinking projection method, Mosco convergnence.

## 1 Introduction

Finding a common fixed point of various types of nonlinear operators is one of the most developed topics in nonlinear analysis. A large number of iterative schemes have been proposed by many researchers. In particular, we will focus on the following result proved by Takahashi, Takeuchi, and Kubota.

**Theorem 1 (Takahashi-Takeuchi-Kubota [8]).** *Let $H$ be a Hilbert space and $C$ a nonempty closed convex subset of $H$. Let $\{T_\lambda : \lambda \in \Lambda\}$ be a family of nonexpansive mappings of $C$ into itself and $\{S_n\}$ a sequence of nonexpansive mappings of $C$ into itself satisfying*

$$\emptyset \neq \bigcap_{\lambda \in \Lambda} F(T_\lambda) \subset \bigcap_{n=1}^{\infty} F(S_n),$$

*where $F(T)$ is the set of fixed points of a mapping $T$. Suppose that $\{S_n\}$ satisfies the NST condition (I) with $\{T_\lambda\}$, that is, for each bounded sequence $\{w_n\} \subset C$, it holds that $\lim_{n\to\infty}\|w_n - T_\lambda w_n\| = 0$ for all $\lambda \in \Lambda$ whenever*

Yasunori Kimura
Department of Information Science, Toho University, Miyama, Funabashi,
Chiba 274-8510, Japan
e-mail: yasunori@is.sci.toho-u.ac.jp

S. Li (Eds.): Nonlinear Maths for Uncertainty and its Appli., AISC 100, pp. 371–378.
springerlink.com &copy; Springer-Verlag Berlin Heidelberg 2011

$\lim_{n\to\infty}\|w_n - S_n w_n\| = 0$. Let $\{\alpha_n\}$ be a sequence in $[0, a]$, where $0 < a < 1$. For an arbitrary point $x \in H$, generate a sequence $\{x_n\}$ by the following iterative scheme: $x_1 \in C$, $C_1 = C$, and

$$y_n = \alpha_n x_n + (1 - \alpha_n) S_n x_n,$$
$$C_{n+1} = \{z \in C_n : \|z - y_n\| \leq \|z - x_n\|\},$$
$$x_{n+1} = P_{C_{n+1}} x$$

for $n \in \mathbb{N}$. Then, $\{x_n\}$ converges strongly to $P_F x \in C$, where $F = \bigcap_{\lambda \in \Lambda} F(T_\lambda)$ and $P_K$ is the metric projection of $H$ onto a nonempty closed convex subset $K$ of $H$.

This method is called the shrinking projection method and has been modified and generalized to Banach spaces and others; see [3, 6, 9, 7, 4, 11, 2].

Kimura and Tahahashi [4] proved a strong convergence theorem for a family of quasinonexpansive mappings in a reflexive Banach spece with certain conditions by using the technique of convergence of sets. In this paper, we use a sequence of subsets of an index set and obtain a strong convergence theorem, which generalizes the result in [4] in the setting where the underlying space is a Hilbert space. We also show several convergence theorems deduced from the main result.

## 2  Preliminaries

Throughout the present paper, a Hilbert space is over the real scalar field. Let $C$ be a nonempty closed convex subset of a Hilbert space $H$. A mapping $T : C \to H$ is said to be quasinonexpansive if $F(T)$ is nonempty and $\|Tx - z\| \leq \|x - z\|$ for all $x \in C$ and $z \in F(T)$, where $F(T)$ denotes the set of fixed points of $T$; $F(T) = \{z \in C : z = Tz\}$. We know that $F(T)$ is closed and convex if $T$ is quasinonexpansive.

For a nonempty closed convex subset of $K$ in $H$, We denote by $P_K$ the metric projection of $H$ onto $K$. Namely, for $x \in H$, a point $P_K x \in K$ satisfies that $\|x - P_K x\| \leq \|x - y\|$ for every $y \in K$, which is always uniquely determined in a Hilbert space.

Let $\{K_n\}$ be a sequence of nonempty closed convex subsets of $H$. We define subsets s-Li$_n$ $K_n$ and w-Ls$_n$ $K_n$ as follows: $x \in$ s-Li$_n$ $K_n$ if and only if there exists $\{x_n\} \subset H$ such that $\{x_n\}$ converges strongly to $x$ and that $x_n \in K_n$ for all $n \in \mathbb{N}$. On the other hand, $y \in$ w-Ls$_n$ $K_n$ if and only if there exist a subsequence $\{K_{n_i}\}$ of $\{K_n\}$ and a sequence $\{y_i\} \subset H$ such that $\{y_i\}$ converges weakly to $y$ and that $y_i \in K_{n_i}$ for all $i \in \mathbb{N}$. We define that $\{K_n\}$ converges to $K_0$ in the sense of Mosco [5] if $K_0$ satisfies that $K_0 =$ s-Li$_n$ $K_n =$ w-Ls$_n$ $K_n$. In this case, we write $K_0 =$ M-$\lim_{n\to\infty} K_n$. It is easy to see that if a sequence of nonempty closed convex subsets $\{K_n\}$ is decreasing with respect to inclusion and $\bigcap_{n=1}^{\infty} K_n$ is nonempty, then $\{K_n\}$ converges to $\bigcap_{n=1}^{\infty} K_n$ in the sense of Mosco. For more details, see [1].

The following theorem proved by Tsukada [10] plays an important role in our results. Obviously, this theorem can be applied to the case where the underlying space is a Hilbert space.

**Theorem 2 (Tsukada [10]).** *Let $E$ be a smooth, reflexive, and strictly convex Banach space. Suppose that $E$ has the Kadec-Klee property, that is, every weakly convergent sequence $\{x_n\}$ in $E$ with the limit $x_0 \in E$ converges strongly to $x_0$ whenever $\{\|x_n\|\}$ converges to $\|x_0\|$. Let $\{K_n\}$ be a sequence of nonempty closed convex subsets of $E$. If $K_0 = \text{M-lim}_{n \to \infty} K_n$ exists and is nonempty, then $\{P_{K_n}x\}$ converges strongly to $P_{K_0}x$ for every $x \in C$.*

## 3  Strong Convergence of an Iterative Sequence

Let $\Lambda$ be an index set. In the main result, we use a sequence of subsets $\{\Lambda_n\}$ of $\Lambda$ and assume that $\Lambda_0 = \bigcap_{n=1}^{\infty} \bigcup_{k=n}^{\infty} \Lambda_k$ is nonempty. We will show some examples of $\{\Lambda_n\}$ and $\Lambda_0$.

*Example 1.* If $\{\Lambda_n\}$ is an increasing sequence of subsets of an index set $\Lambda$, then $\Lambda_0 = \bigcup_{n=1}^{\infty} \Lambda_n$.

*Example 2.* Let $\Lambda$ be an arbitrary index set and $\{I_k : k = 0, 1, \ldots, N-1\}$ a finite number of subsets of $\Lambda$. For $n \in \mathbb{N}$, let $\Lambda_n = I_{(n \bmod N)}$. Then it follows that $\Lambda_0 = \bigcup_{k=0}^{N-1} I_k$. As a special case of the result above, if $\Lambda = \{0, 1, \ldots, N-1\}$ and $I_k = \{k\}$ for $k = 0, 1, \ldots, N-1$, then $\Lambda = \Lambda_0$.

*Example 3.* Let $\{I_k\}$ be a sequence of subsets of an arbitrary index set $\Lambda$. Let $i : \mathbb{N} \to \mathbb{N}$ be such that

$$
\begin{aligned}
&i(1) = 1, \\
&i(2) = 1, \quad i(3) = 2, \\
&i(4) = 1, \quad i(5) = 2, \quad i(6) = 3, \\
&i(7) = 1, \quad i(8) = 2, \quad i(9) = 3, \quad i(10) = 4, \ldots.
\end{aligned}
$$

More presicely, for each $k \in \mathbb{N}$, let $l(k)$ be a unique natural number satisfying that

$$
\sum_{j=0}^{l(k)-1} j < k \le \sum_{j=0}^{l(k)} j
$$

and define $i(k) = k - \sum_{j=0}^{l(k)-1} j$. Using this function, we let $\Lambda_n = I_{i(n)}$ for $n \in \mathbb{N}$. Then, we have that $\Lambda_0 = \bigcup_{n=1}^{\infty} I_n$. In particular, if $\Lambda = \mathbb{N}$ and $\Lambda_n = \{i(n)\}$ for every $n \in \mathbb{N}$, then $\Lambda_0 = \Lambda = \mathbb{N}$.

Now we prove a strong convergence theorem for an iterative scheme generated by a family of quasinonexpansive mappings to their common fixed point. This result generalizes various types of iterative schemes by the shrinking projection method.

**Theorem 3.** *Let $\Lambda$ be an index set and $\{\Lambda_n\}$ a sequece of nonempty subsets of $\Lambda$ such that $\Lambda_0 = \bigcap_{n=1}^{\infty} \bigcup_{k=n}^{\infty} \Lambda_k$ is nonempty. Let $H$ be a Hilbert space and $C$ a nonempty closed convex subset of $H$. For $\lambda \in \Lambda$, let $T_\lambda : C \to H$ be a quasinonexpansive mapping and suppose that, if both $\{v_n\}$ and $\{T_\lambda v_n\}$ have the same strong limit $v_0 \in C$ for a sequence $\{v_n\}$ of $C$, then $v_0 \in F(T_\lambda)$. For each $\lambda \in \Lambda$, let $\{\alpha_{n,\lambda} : n \in N_\lambda\}$ be a family of real numbers such that*

$$0 \leq \sup_{\substack{n \in N_\lambda \\ n \geq k}} \alpha_{n,\lambda} < 1$$

*for some $k \in \mathbb{N}$ depending on $\lambda$, where $N_\lambda = \{n \in \mathbb{N} : \lambda \in \Lambda_n\}$. Let $u \in H$ and generate a sequence $\{x_n\}$ in $C$ and a sequence $\{C_n\}$ of subsets in $C$ as follows: $x_1 \in C$, $C_1 = C$, and*

$$y_{n,\lambda} = \alpha_{n,\lambda} x_n + (1 - \alpha_{n,\lambda}) T_\lambda x_n \text{ for each } \lambda \in \Lambda_n,$$

$$C_{n+1} = \left\{ z \in H : \sup_{\lambda \in \Lambda_n} \|y_{n,\lambda} - z\| \leq \|x_n - z\| \right\} \cap C_n,$$

$$x_{n+1} = P_{C_{n+1}} u$$

*for every $n \in \mathbb{N}$. If $F = \bigcap_{\lambda \in \Lambda} F(T_\lambda)$ is nonempty, then $\{x_n\}$ is well defined and it converges strongly to $x_0 \in F_0$, where $F_0 = \bigcap_{\lambda \in \Lambda_0} F(T_\lambda)$. In particular, if $\Lambda = \Lambda_0$, then $x_0 = P_F u$.*

We remark that, in the theorem above, if there is no $n$ satisfying $n \in N_\lambda$ and $n \geq k$, then the condition $0 \leq \sup_{n \in N_\lambda, n \geq k} \alpha_{n,\lambda} < 1$ is supposed to be true. In other words, we assume that $\sup_{(n,\lambda) \in \emptyset} \alpha_{n,\lambda} = 0$.

*Proof.* Firstly, we prove that $\{x_n\}$ and $\{C_n\}$ are well defined by induction. Suppose that $\{x_1, \ldots, x_m\}$ and $\{C_1, \ldots, C_m\}$ are defined and $F = \bigcap_{\lambda \in \Lambda} F(T_\lambda) \subset C_m$ for $m \in \mathbb{N}$. Then since it holds that

$$\left\{ z \in H : \sup_{\lambda \in \Lambda_m} \|y_{m,\lambda} - z\| \leq \|x_m - z\| \right\}$$

$$= \bigcap_{\lambda \in \Lambda_m} \left\{ z \in H : \|y_{m,\lambda} - z\|^2 \leq \|x_m - z\|^2 \right\}$$

$$= \bigcap_{\lambda \in \Lambda_m} \left\{ z \in H : \langle 2(x_m - y_{m,\lambda}), z \rangle \leq \|x_m\|^2 - \|y_{m,\lambda}\|^2 \right\},$$

we have that it is closed and convex. Since $C_m$ is closed and convex, so is $C_{m+1}$. Let $z \in F$. Then we have that

$$\|y_{m,\lambda} - z\| = \|\alpha_{m,\lambda} x_m + (1 - \alpha_{m,\lambda}) T_\lambda x_m - z\|$$

$$\leq \alpha_{m,\lambda} \|x_m - z\| + (1 - \alpha_{m,\lambda}) \|T_\lambda x_m - z\|$$

$$\leq \alpha_{m,\lambda} \|x_m - z\| + (1 - \alpha_{m,\lambda}) \|x_m - z\|$$

$$= \|x_m - z\|$$

for all $\lambda \in \Lambda_m$ and thus $z \in C_{m+1}$. Since $F$ is nonempty, we obtain that $C_{m+1}$ is also nonempty. Therefore there exists a metric projection $P_{C_{m+1}}$ of $H$ onto $C_{m+1}$ and hence $x_{m+1}$ is also well defined. Hence we have that $\{x_n\}$ and $\{C_n\}$ are both well defined.

By definition, $\{C_n\}$ is a decreasing sequence with respect to inclusion. It follows that M-$\lim_{n \to \infty} C_n = C_0 = \bigcap_{n=1}^{\infty} C_n$. Since $C_0$ includes $F$ and thus it is nonempty, Theorem 2 implies that $\{x_n\} = \{P_{C_n} u\}$ converges strongly to $x_0 = P_{C_0} u$.

Fix $\lambda \in \Lambda_0 = \bigcap_{n=1}^{\infty} \bigcup_{k=n}^{\infty} \Lambda_k$ arbitrarily and we will show that $x_0 \in F(T_\lambda)$. From the definition of $\Lambda_0$, there exists a subsequence $\{n_j\}$ of $\mathbb{N}$ such that $\lambda \in \Lambda_{n_j}$ for every $j \in \mathbb{N}$. Since $x_0 \in C_0 = \bigcap_{n=1}^{\infty} C_n$, we have that

$$
\begin{aligned}
\|x_{n_j} - x_0\|^2 &\geq \|y_{n_j, \lambda} - x_0\|^2 \\
&= \|\alpha_{n_j, \lambda} x_{n_j} + (1 - \alpha_{n_j, \lambda}) T_\lambda x_{n_j} - x_0\|^2 \\
&= \alpha_{n_j, \lambda} \|x_{n_j} - x_0\|^2 + (1 - \alpha_{n_j, \lambda}) \|T_\lambda x_{n_j} - x_0\|^2 \\
&\quad - \alpha_{n_j, \lambda}(1 - \alpha_{n_j, \lambda}) \|T_\lambda x_{n_j} - x_{n_j}\|^2
\end{aligned}
$$

and thus

$$
\|T_\lambda x_{n_j} - x_0\|^2 - \alpha_{n_j, \lambda} \|T_\lambda x_{n_j} - x_{n_j}\|^2 \leq \|x_{n_j} - x_0\|^2
$$

for sufficiently large $j \in \mathbb{N}$. Tending $j \to \infty$, we get that

$$
\limsup_{j \to \infty} (1 - \alpha_{n_j, \lambda}) \|T_\lambda x_{n_j} - x_0\|^2 \leq 0
$$

and hence $\{T_\lambda x_{n_j}\}$ converges strongly to $x_0$. From the assumption of $T_\lambda$, we have that $x_0 \in F(T_\lambda)$, which implies that $x_0 \in F_0 = \bigcap_{\lambda \in \Lambda_0} F(T_\lambda)$.

For the case where $\Lambda = \Lambda_0$, it follows that

$$
x_0 \in F_0 = \bigcap_{\lambda \in \Lambda_0} F(T_\lambda) = \bigcap_{\lambda \in \Lambda} F(T_\lambda) = F \subset \bigcap_{n=1}^{\infty} C_n = C_0.
$$

Since $x_0 = P_{C_0} u$, we obtain that $x = P_F u$, which is the desired result.

## 4 Deduced Results

In this section, we prove several results deduced from the main result proved in the previous section.

Suppose that $\Lambda_n = \Lambda$ for all $n \in \mathbb{N}$. In this case, we may change the assumption for a coefficients $\{\alpha_{n, \lambda}\}$ into milder one. Namely, we obtain the following result, which was essentially proved by Kimura and Takahashi [4] in the setting where the underlying space is a Banach space with certain properties.

**Corollary 1 (Kimura-Takahashi [4]).** *Let $H$, $C$, $\Lambda$, and $\{T_\lambda\}$ be the same as in Theorem 3. For each $\lambda \in \Lambda$, let $\{\alpha_{n,\lambda} : n \in \mathbb{N}\}$ be a sequence in $[0,1]$ such that $\liminf_{n\to\infty} \alpha_{n,\lambda} < 1$. Let $u \in H$ and generate a sequence $\{x_n\}$ in $C$ and a sequnce $\{C_n\}$ of subsets in $C$ as follows: $x_1 \in C$, $C_1 = C$, and*

$$y_{n,\lambda} = \alpha_{n,\lambda} x_n + (1 - \alpha_{n,\lambda}) T_\lambda x_n \text{ for each } \lambda \in \Lambda,$$

$$C_{n+1} = \left\{ z \in H : \sup_{\lambda \in \Lambda} \|y_{n,\lambda} - z\| \le \|x_n - z\| \right\} \cap C_n,$$

$$x_{n+1} = P_{C_{n+1}} u$$

*for every $n \in \mathbb{N}$. If $F = \bigcap_{\lambda \in \Lambda} F(T_\lambda)$ is nonempty, then $\{x_n\}$ is well defined and it converges strongly to $x_0 = P_F u$.*

*Proof.* Fix $\lambda \in \Lambda$. Then, since $\liminf_{n\to\infty} \alpha_{n,\lambda} < 1$, there exists an infinite subset $S_\lambda$ of $\mathbb{N}$ such that $\sup_{n\in S_\lambda} \alpha_{n,\lambda} < 1$. Let $\Lambda' = (\mathbb{N} \cup \{0\}) \times \Lambda$ and define a subset $\Lambda'_n$ of $\Lambda$ as

$$\Lambda'_n = \bigcup_{\lambda \in \Lambda} (\{(0, \lambda) : n \in S_\lambda\} \cup \{(n, \lambda) : n \notin S_\lambda\}).$$

For each $n \in \mathbb{N}$ and $(k, \lambda) \in \Lambda'_n$, let $T_{(k,\lambda)} = T_\lambda$ and $\alpha_{n,(k,\lambda)} = \alpha_{n,\lambda}$ for $k \in \mathbb{N} \cup \{0\}$. Then, it is easy to see that the iterative scheme $\{x_n\}$ coincides with that defined in Theorem 3 with the index set $\Lambda'$. Hence we obtain the desired result.

Next, we will assume that the index set $\Lambda$ is a countable set. For the case where $\Lambda$ is finite, we may apply Example 2 with the main result.

**Corollary 2.** *Let $H$ and $C$ be the same as in Theorem 3, and $\Lambda = \{0, 1, \ldots, N-1\}$. For $k \in \Lambda$, let $T_k : C \to H$ be a quasinonexpansive mapping and suppose that, if both $\{v_n\}$ and $\{T_k v_n\}$ have the same strong limit $x_0 \in C$ for a sequence $\{v_n\}$ of $C$, then $v_0 \in F(T_k)$. Let $\{\alpha_n : n \in \mathbb{N}\}$ be a sequence of nonnegative real numbers such that $\sup_{n\in\mathbb{N}} \alpha_n < 1$. Let $u \in H$ and generate a sequence $\{x_n\}$ in $C$ and a sequence $\{C_n\}$ of subsets in $C$ as follows: $x_1 \in C$, $C_1 = C$, and*

$$y_n = \alpha_n x_n + (1 - \alpha_n) T_{(n \bmod N)} x_n,$$

$$C_{n+1} = \{z \in H : \|y_n - z\| \le \|x_n - z\|\} \cap C_n,$$

$$x_{n+1} = P_{C_{n+1}} u$$

*for every $n \in \mathbb{N}$. If $F = \bigcap_{k=0}^{N-1} F(T_k)$ is nonempty, then $\{x_n\}$ is well defined and it converges strongly to $P_F u$.*

For the case where $\Lambda$ is countably infinite, we have the following result by using Example 3.

**Corollary 3.** *Let $H$ and $C$ be the same as in Theorem 3. For $k \in \mathbb{N}$, let $T_k : C \to H$ be a quasinonexpansive mapping and suppose that, if both $\{v_n\}$ and $\{T_k v_n\}$ have the same strong limit $x_0 \in C$ for a sequence $\{v_n\}$ of $C$, then $v_0 \in F(T_k)$. Let $\{\alpha_n : n \in \mathbb{N}\}$ be a sequence of nonnegative real numbers such that $\sup_{n \in \mathbb{N}} \alpha_n < 1$. Let $u \in H$ and generate a sequence $\{x_n\}$ in $C$ and a sequence $\{C_n\}$ of subsets in $C$ as follows: $x_1 \in C$, $C_1 = C$, and*

$$y_n = \alpha_n x_n + (1 - \alpha_n) T_{i(n)} x_n,$$
$$C_{n+1} = \{z \in H : \|y_n - z\| \leq \|x_n - z\|\} \cap C_n,$$
$$x_{n+1} = P_{C_{n+1}} u$$

*for every $n \in \mathbb{N}$, where $i : \mathbb{N} \to \mathbb{N}$ is the mapping defined in Example 3. If $F = \bigcap_{k=0}^{\infty} F(T_k)$ is nonempty, then $\{x_n\}$ is well defined and it converges strongly to $P_F u$.*

We may apply the index set shown in Example 1 with the case where $\Lambda = \mathbb{N}$ and consequently we get the following corollary.

**Corollary 4.** *Let $H$, $C$, and $\{T_k\}$ be the same as in Theorem 3. For each $k \in \mathbb{N}$, let $\{\alpha_{n,k} : n \geq k\}$ be a family of real numbers such that $\limsup_{n \to \infty} \alpha_{n,k} < 1$. Let $u \in H$ and generate a sequence $\{x_n\}$ in $C$ and a sequence $\{C_n\}$ of subsets in $C$ as follows: $x_1 \in C$, $C_1 = C$, and*

$$y_{n,k} = \alpha_{n,k} x_n + (1 - \alpha_{n,k}) T_k x_n \text{ for each } 1 \leq k \leq n,$$
$$C_{n+1} = \left\{ z \in H : \sup_{1 \leq k \leq n} \|y_{n,k} - z\| \leq \|x_n - z\| \right\} \cap C_n,$$
$$x_{n+1} = P_{C_{n+1}} u$$

*for every $n \in \mathbb{N}$. If $F = \bigcap_{k=1}^{\infty} F(T_k)$ is nonempty, then $\{x_n\}$ is well defined and it converges strongly to $P_F u$.*

*Remark 1.* Using the technique in Corollary 1, we may weaken the assumption for the coefficients $\{\alpha_{n,\lambda}\}$ and $\{\alpha_n\}$ for Corollaries 2, 3, and 4. For example, the condition that $\sup_{n \in \mathbb{N}} \alpha_n < 1$ appearing in Corollary 4 can be replaced with that $\{\alpha_{n,k}\} \subset [0, 1]$ and $\liminf_{n \to \infty} \alpha_{n,k} < 1$ for $k \in \mathbb{N}$.

**Acknowledgements.** The author is supported by Grant-in-Aid for Scientific Research No. 22540175 from Japan Society for the Promotion of Science.

# References

1. Beer, G.: Topologies on Closed and Closed Convex Sets. Kluwer Academic Publishers Group, Dordrecht (1993)
2. Kimura, Y.: Convergence of a sequence of sets in a Hadamard space and the shrinking projection method for a real Hilbert ball. Abstr. Appl. Anal., Art. ID 582,475, 11 (2010)

3. Kimura, Y., Nakajo, K., Takahashi, W.: Strongly convergent iterative schemes for a sequence of nonlinear mappings. J. Nonlinear Convex Anal. 9(3), 407–416 (2008)
4. Kimura, Y., Takahashi, W.: On a hybrid method for a family of relatively nonexpansive mappings in a Banach space. J. Math. Anal. Appl. 357, 356–363 (2009)
5. Mosco, U.: Convergence of convex sets and of solutions of variational inequalities. Adv. in Math. 3, 510–585 (1969)
6. Plubtieng, S., Ungchittrakool, K.: Hybrid iterative methods for convex feasibility problems and fixed point problems of relatively nonexpansive mappings in Banach spaces. Fixed Point Theory Appl., Art. ID 583,082, 19 (2008)
7. Qin, X., Cho, Y.J., Kang, S.M.: Convergence theorems of common elements for equilibrium problems and fixed point problems in Banach spaces. J. Comput. Appl. Math. 225(1), 20–30 (2009)
8. Takahashi, W., Takeuchi, Y., Kubota, R.: Strong convergence theorems by hybrid methods for families of nonexpansive mappings in Hilbert spaces. J. Math. Anal. Appl. 341, 276–286 (2008)
9. Takahashi, W., Zembayashi, K.: Strong convergence theorem by a new hybrid method for equilibrium problems and relatively nonexpansive mappings. Fixed Point Theory Appl., Art. ID 528,476, 11 (2008)
10. Tsukada, M.: Convergence of best approximations in a smooth Banach space. J. Approx. Theory 40(4), 301–309 (1984)
11. Wattanawitoon, K., Kumam, P.: Strong convergence theorems by a new hybrid projection algorithm for fixed point problems and equilibrium problems of two relatively quasi-nonexpansive mappings Nonlinear. Anal. Hybrid Syst. 3(1), 11–20 (2009)

# Note on Generalized Convex Spaces

Xiaodong Fan and Yue Cheng

**Abstract.** Some examples are given to show that some known generalized convex spaces are so abstract that some basic properties related to the convexity are lost. In order to improve the convexity structure for applications, the concepts of path-convex space, path-convex set and path-convex function are introduced. And their properties are discussed.

**Keywords:** Generalized convex spaces, Path-convex space, Path-convex function.

## 1 Introduction

Convexity plays an important role in many aspects of mathematics. There have been many generalizations of the concept of linear convex space under the circumstances without linear structure, such as, H-spaces introduced by Horvath [5] [4], L-spaces due to Ben-El-Mechaiekh et al. [1], spaces having property [H] due to Huang [6], FC-spaces due to Ding [2] [3]. The most general ones seem to be the G-convex space introduced by Park and Kim [12] and the interval spaces due to Stachó [13]. Recently, Park proposed a new generalized convex space, called KKM space [8] [10] [11], which is a generalization of G-convex spaces. All the generalized convex spaces mentioned above contain H-space as a special case. Observing these spaces, it is not difficult to find that many important properties related to the convexity are lost.

In Section 2, we give some examples to show that interval spaces (H-spaces, or Wu spaces, respectively) don't has some basic properties for practical

Xiaodong Fan
Department of Mathematics, Beijing University of Technology,
Beijing 100124, P.R. China
e-mail: fxd@emails.bjut.edu.cn

Yue Cheng
College of Elect & Control Engineering, Beijing University of Technology,
Beijing 100124, P.R. China
e-mail: chengyue0721@emails.bjut.edu.cn

S. Li (Eds.): Nonlinear Maths for Uncertainty and its Appli., AISC 100, pp. 379–386.
springerlink.com                                    © Springer-Verlag Berlin Heidelberg 2011

applications. In order to improve the convexity structure for applications, we introduce a new generalized convex space, called path-convex space. And some properties of path-convex set are discussed in Section 3. We also give the concepts of path-convex functions and path-quasiconvex functions, and investigate their properties.

## 2   Examples of Generalized Convex Spaces

A topological space $E$ is *contractible* if there exists a continuous map $h :$ $E \times [0,1] \to E$, such that $h(x,0)$ is a constant map and $h(x,1)$ is the identity map. A contractible space is path-connected, and so connected.

**Definition 2.1.** [5] [4]Let $E$ be a topological space, $A$ and $A'$ be finite subsets of $E$. An ordered pair $(E, \{\Gamma_A\})$ is said to be *H-space* if $\Gamma_A$ is a given family of nonempty contractible subsets indexed by all the finite subset of $E$ such that $A \subset A'$ implies $\Gamma_A \subset \Gamma_{A'}$. A subset $D \subset E$ is said to be H-convex if for any finite subset $A \subset D$ it follows $\Gamma_A \subset D$.

Denote $\mathscr{C}(E)$ as the family of all connected subsets of $E$.

**Definition 2.2.** [7] [13] An *interval space* is a pair $(E, \langle \cdot, \cdot \rangle)$ where $E$ is a topological space and $\langle \cdot, \cdot \rangle : E \times E \to \mathscr{C}(E)$ is a map such that $x, y \in \langle x, y \rangle$ for all $(x, y) \in E \times E$. If furthermore $\langle x, y \rangle = \langle y, x \rangle$ for all $x, y \in E$, then $(E, \langle \cdot, \cdot \rangle)$ is called symmetric. Symmetric interval spaces were introduced by Stachó [13]. Subsets $C \subset E$ are called *interval-convex* if $x, y \in C$ implies $\langle x, y \rangle \subset C$.

**Definition 2.3.** [7]A pair $(E, P)$ is called a *Wu space*, if $E$ is a topological space and $P : [0,1] \times E \times E \to E$ is a map such that for any pairs $x, y \in E$ we have that $P(\cdot, x, y)$ is continuous, $P(0, x, y) = x$ and $P(1, x, y) = y$. The map $P$ is called *Wu map*.

Every Wu space $(E, P)$ gives rise to an interval space $(E, \langle \cdot, \cdot \rangle)$ with $\langle x, y \rangle = \{P(t, x, y) : t \in [0, 1]\}$. Subsets $C \subset E$ are called Wu interval-convex if $\{P(t, x, y) : t \in [0, 1]\} \subset C$ for any $x, y \in C$.

Now, we give some examples to show that interval spaces (H-spaces, or Wu spaces, respectively) don't has some basic properties for practical applications.

**Example 2.1**

1) For any $x, y \in \mathbb{R}^2$, define $\langle \cdot, \cdot \rangle : \mathbb{R}^2 \to \mathscr{C}(\mathbb{R})$ as in Definition 2.2 by

$$\langle x, y \rangle = B(\frac{x+y}{2}, \frac{\| x - y \|}{2}) = \{z \in \mathbb{R}^2 : \| z - \frac{x+y}{2} \| \le \frac{\| x - y \|}{2}\}.$$

It is obvious that $\langle x, y \rangle$ is contractible, and so connected. We can assert the following two consequences:

a) $\langle x, y \rangle$ is not interval-convex for any $x \neq y$.

b) The minimal interval-convex set which contains $x$ and $y$ is the space $\mathbb{R}^2$ itself.

In fact, take $u, v \in \langle x, y \rangle$ such that $u, v, x, y$ are distinct from each other and $\| v - \frac{x+y}{2} \| = \| u - \frac{x+y}{2} \| = \frac{\|x-y\|}{2}$. By the definition of $\langle u, v \rangle$, it follows that $\langle u, v \rangle \not\subset \langle x, y \rangle$. Thus $\langle x, y \rangle$ is not interval-convex set. See Fig.2.1.

By the symmetries, it is suffice to consider the vertical direct as in Fig.2.1. Suppose that $x = (a, 0), y = (-a, 0)$, the maximal extending attains when $s = (-\frac{\sqrt{2}}{2}a, \frac{\sqrt{2}}{2}a)$ and $t = (\frac{\sqrt{2}}{2}a, \frac{\sqrt{2}}{2}a)$.

Hence we obtain a bigger disc with the center at $\frac{x+y}{2}$ and the radius being $\sqrt{2}a$. It is not interval-convex for the same reason as a). By repeating the above process, we get a sequence of discs with radius converging to $+\infty$.

Therefore, there are only three kinds of interval-convex sets: the empty set, $\mathbb{R}^2$ and all singletons.

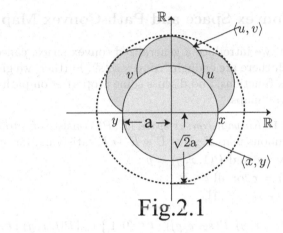

$$\text{Fig.2.1}$$

2) For any $A = \{x^1, x^2, \cdots x^n\} \subset \mathbb{R}^2$, define $\Gamma_A$ as in Definition 2.1 by

$$\Gamma_A = \bigcap \{C \subset \mathbb{R}^2 : C \text{ is contractible and } \langle x^i, x^j \rangle \subset C \text{ for all } 1 \le i, j \le n\}$$

where $\langle x^i, x^j \rangle = B(\frac{x^i+x^j}{2}, \frac{\|x^i-x^j\|}{2})$. Thus $(\mathbb{R}^2, \{\Gamma_A\})$ is a H-space. The conclusion that only empty set, $\mathbb{R}^2$ and singletons are H-convex sets follows from the similar discusses as in 1).

## Example 2.2

1). For any $x = (x_1, x_2), y = (y_1, y_2) \in \mathbb{R}^2$ and $t \in [0, 1]$, let

$$P(t, x, y) = \begin{cases} \frac{x+y}{2} + \frac{\|x-y\|}{2} e^{((1-t)\pi + \arctan(\frac{x_2-y_2}{x_1-y_1}))i} & x_1 < y_1 \\[2mm] \frac{x+y}{2} + \frac{\|x-y\|}{2} e^{(t\pi + \arctan(\frac{x_2-y_2}{x_1-y_1}))i} & x_1 > y_1 \\[2mm] \frac{x+y}{2} + \frac{\|x-y\|}{2} e^{(t\pi - \frac{\pi}{2})i} & x_1 = y_1, x_2 \le y_2 \\[2mm] \frac{x+y}{2} + \frac{\|x-y\|}{2} e^{((1-t)\pi - \frac{\pi}{2})i} & x_1 = y_1, x_2 > y_2 \end{cases}$$

where $e^{\theta i} = (\cos\theta, \sin\theta)$. That is, $\{P(t,x,y) : t \in [0,1]\}$ is the semi-circle of the disc in Example 2.1. One can deduce that the set $\{P(t,x,y) : t \in [0,1]\}$ and all bounded sets are not Wu interval-convex.

2) For any $x, y \in \mathbb{R}^2$ and $t \in [0,1]$, let

$$P(t,x,y) = \begin{cases} (1-2t)x, \ 0 \le t < \frac{1}{2} \\ \\ (2t-1)y, \ \frac{1}{2} \le t \le 1. \end{cases}$$

Then $\{P(t,x,x) : t \in [0,1]\} = \{\lambda x : \lambda \in [0,1]\}$ which implies that singletons are not Wu interval-convex for any $x \ne (0,0)$.

# 3 Path-Convex Space and Path-Convex Map

In this section, we introduce a generalized convex space, called path-convex space, in which there are enough "convex sets". Further, we give the concept of path-convex functions, and discuss some properties on path-convex spaces and path-convex maps.

**Definition 3.1.** An *path-convex space* $(E; P)$ consists of a topological space $E$ and a continuous map $P : \mathbb{R} \times E \times E \to E$ satisfying, for any $x, y \in E$,

i) $P(0,x,y) = x$ and $P(1,x,y) = y$ ;

ii) $P(t,x,x) = x$ for all $t \in \mathbb{R}$;

iii) for any $s_1, s_2 \in [0,1]$,

$$\{P(t, P(s_1,x,y), P(s_2,x,y)) : t \in [0,1]\} \subset \{P(t,x,y) : t \in [0,1]\}.$$

$P$ is said to be a *path-convex structure* on $E$.

A subset $C$ of $E$ is said to be *path-convex* if and only if $P(t,x,y) \in C$ for all $x, y \in C$ and $0 < t < 1$.

For any subset $D$ of $E$, the *path-convex hull* of $D$, which is denoted as $co_P(D)$, is given by

$$co_P(D) = \bigcap \{C : D \subset C \ and \ C \ is \ path-convex\}.$$

**Remark 3.1.** Example 2.2 implies that conditions ii) and iii) are necessary to ensure that all singletons and $\{P(t,x,y) : t \in [0,1]\}$ are path-convex, respectively.

**Example 3.1**

1) Every convex set in topological vector space $E$ is path-convex by taking a path-convex structure $P(t,x,y) = tx + (1-t)y$.

2) We give an example of the path-convex structure in $\mathbb{R}^2$ as follows.

For any $x = (x_1, x_2), y = (y_1, y_2) \in \mathbb{R}^2$ and $t \in \mathbb{R}$, let

$$P(t, x, y) := \begin{cases} m(x, y) + (2t - 1)(y - m(x, y)), \, t > \frac{1}{2} \\ \\ m(x, y) + (1 - 2t)(x - m(x, y)), \, t \leq \frac{1}{2} \end{cases}$$

where $m(x, y) = (\min\{x_1, y_1\}, \min\{x_2, y_2\})$.

It is obvious that $co_P(\{x, y\}) = \{P(t, x, y) : 0 \leq t \leq 1\}$ is path-convex in $(\mathbb{R}^2, P)$. However it is not convex.

**Proposition 3.1**

(a) The empty set, E and all singletons are path-convex.

(b) If $\{L_\lambda : \lambda \in \Lambda\}$ is an arbitrary family of path-convex sets, then $\bigcap_\lambda L_\lambda$ is path-convex.

(c) If $\{L_\lambda : \lambda \in \Lambda\}$ is a family of path-convex sets such that for any $\lambda_1, \lambda_2 \in \Lambda$, there is $\lambda_3 \in \Lambda$ with $L_{\lambda_1} \cup L_{\lambda_2} \subset L_{\lambda_3}$, then $\bigcup_\lambda L_\lambda$ is path-convex.

**Proposition 3.2.** The following properties hold:

(a) $co_P(\emptyset) = \emptyset$, $co_P(E) = E$, $co_P(\{x\}) = \{x\}$ for any $x \in E$.

(b) For all $D \subset E$, $D \subset co_P(D)$ and $co_P(co_P(D)) = co_P(D)$.

(c) For all $D_1, D_2 \in E$, if $D_1 \subset D_2$, then $co_P(D_1) \subset co_P(D_2)$.

Propositions 3.1 and 3.2 are rather standard in the context of generalized convexities. The proof is obvious and is omitted.

**Theorem 3.1.** The closure of a path-convex subset of $E$ is path-convex.

Proof. Let $C$ be a path-convex subset of $E$ and $\overline{C}$ be its closure. If $x = \lim\limits_{n \to \infty} x_n$ and $y = \lim\limits_{n \to \infty} y_n$ with $x_n, y_n \in C$ for all $n$, from the continuity of $P$, one can deduce that $P(t, x, y) = \lim\limits_{n \to \infty} P(t, x_n, y_n)$ for all $t \in [0, 1]$. Furthermore, by the path-convexity hypothesis of $C$, it follows that $P(t, x_n, y_n) \in C$ for all $n$. Hence $P(t, x, y) \in \overline{C}$. $\qquad\square$

**Theorem 3.2.** Let $D$ be a compact subset of $E$. Then the path-convex hull $co_P(D)$ is compact.

Proof. Let $\{U_\lambda : \lambda \in \Lambda\}$ be any open covering of $co_P(D)$, then $\{P^{-1}(U_\lambda) : \lambda \in \Lambda\}$ is an open covering of $[0, 1] \times D \times D$ in $\mathbb{R} \times E \times E$ by the continuity of $P$. Since $[0, 1] \times D \times D$ is compact, there is a finite subcovering $\{P^{-1}(U_{\lambda_i}) : 1 \leq i \leq n\}$. Note $co_P(\{x, y\}) = \{P(t, x, y) : 0 \leq t \leq 1\}$ for all $x, y \in D$. Hence $\{U_{\lambda_i} : 1 \leq i \leq n\}$ is a finite subcovering of $co_P(D)$. $\qquad\square$

**Remark 3.2.** Example 2.1 and 2.2 imply that Theorem 3.2 do not hold in interval spaces, H-spaces and Wu spaces, respectively.

**Definition 3.2.** Let $(E_1; P_1)$, $(E_2; P_2)$ be two path-convex spaces, a map $T : E_1 \to E_2$ is said to be *path-affine map* if

$$T(P_1(t, x, y)) = P_2(t, T(x), T(y)),$$

for each $x, y \in E$ and $t \in \mathbb{R}$.

**Theorem 3.3.** Let $(E_1; P_1)$, $(E_2; P_2)$ be two path-convex spaces, and $T$ : $E_1 \to E_2$ be a path-affine map. Then for all path-convex set $C$ of $E_2$, the inverse image $T^{-1}(C)$ is a path-convex set in $E_1$. And for all path-convex set $D$ of $E_1$, the image $T(C)$ is a path-convex set in $E_2$.

Proof. For any $x_1, x_2 \in T^{-1}(C)$, $T(x_1), T(x_2) \in C$. $C$ is path-convex in $(E_2; P_2)$. Then $P_2(t, T(x_1), T(x_2)) \in C$ for all $t \in (0,1)$. Since

$$T(P_1(t, x_1, x_2)) = P_2(t, T(x_1), T(x_2)), \quad \forall t \in (0,1),$$

$P_1(t, x_1, x_2) \in T^{-1}(C)$. Hence $T^{-1}(C)$ is a path-convex set of $(E_1, P_1)$.

   For any $y_1, y_2 \in T(D)$, there are $x_1, x_2 \in D$ such that $y_1 = T(x_1)$, $y_2 = T(x_2)$. $D$ is path-convex in $(E_1; P_1)$. Then $P_1(t, x_1, x_2) \in D$. Since

$$T(P_1(t, x_1, x_2)) = P_2(t, T(x_1), T(x_2)) = P_2(t, y_1, y_2),$$

$P_2(t, y_1, y_2) \in T(D)$. Hence $T(D)$ is a path-convex set of $(E_2, P_2)$. $\qquad\square$

**Definition 3.3.** A function $f : E \to \mathbb{R}$ is said to be *path-convex* on a path-convex set $D \subset E$ if and only if

$$f(P(t, x, y)) \leq tf(x) + (1 - t)f(y)$$

for each $x, y \in D$ and $t \in [0,1]$. If $-f$ is path-convex, then $f$ is called *path-concave* on $D$. If the inequality is strict for any two distinct points $x$ and $y$, $f$ is called *strictly path-convex* and *strictly path-concave*, respectively.

   Let $(E; P)$, $(E'; P')$ be two path-convex spaces and $E \times E'$ be equipped with the product topology. It is easy to prove that $(E \times E', P \times P')$ is also a path-convex space. Therefore, a subset $S \subset E \times E'$ is path-convex in $(E \times E', P \times P')$ if we have

$$(P(t, x, y), P'(t, x', y')) \in S, \quad t \in (0,1)$$

for any $(x, x'), (y, y') \in S$. Especially, a subset $S \subset E \times \mathbb{R}$ is path-convex if $(x, a), (y, b) \in S$ imply

$$(P(t, x, y), ta + (1 - t)b) \in S, \quad t \in (0,1).$$

   Let $f$ be a function from $E$ to $\mathbb{R}$, we call the subset

$$epi(f) := \{(x, a) : x \in E, a \in \mathbb{R}, f(x) \leq a\}.$$

the epigraph of $f$. The sets

$$S(f, \lambda) := \{x \in E : f(x) \leq \lambda\}, \quad \lambda \in \mathbb{R}$$

are said to be sections of $f$.

   Next we give characterizations of a path-convex function $f$ in terms of its epigraph and sections.

**Theorem 3.4.** A function $f : E \to \mathbb{R}$ is path-convex if and only if its epigraph $epi(f)$ is path-convex in $E \times \mathbb{R}$.

Proof. Suppose that $f$ is path-convex. We show that the set $epi(f)$ is path-convex. For any $(x, a), (y, b) \in epi(f)$, one has $f(x) \leq a, f(y) \leq b$. $f$ is path-convex. Then $f(P(t, x, y)) \leq tf(x) + (1 - t)f(y) \leq ta + (1 - t)b$. Hence $(f(P(t, x, y)), ta + (1 - t)b) \in epi(f)$, $epi(f)$ is path-convex.

Conversely, assume that $epi(f)$ is path-convex. For any $x, y \in E$, $(x, f(x))$, $(y, f(y)) \in epi(f)$. By the path-convexity of $epi(f)$,

$$(P(t, x, y), tf(x) + (1 - t)f(y)) \in epi(f).$$

Therefore $f(P(t, x, y)) \leq tf(x) + (1 - t)f(y)$ and $f$ is path-convex.    □

The following properties are obvious.

**Propositions 3.3.** Suppose that $(E; P)$, $(E'; P')$ are path-convex spaces, and the functions $f, g$ from $E$ to $\mathbb{R}$ are path-convex. The following results hold.

 a) $f + g$ is path-convex.
 b) If $a > 0$, then $af$ is path-convex.
 c) If $T$ is a path-affine map from $E'$ to $E$, then $f \circ T$ is path-convex.

**Theorem 3.5.** Assume that $\{f_i\}_{i \in I}$ is a family of path-convex functions which are bounded above about $i \in I$ on a path-convex set $C \subset E$. Then the function $f(x) = \sup_{i \in I} f_i(x)$ is path-convex on $C$.

Proof. Since each $f_i$ is a path-convex function on $C$, the epigraph of $f_i$ is a path-convex set in $E \times \mathbb{R}$. From the definitions of $f$ and $epi(f)$, one has $epi(f) = \cap_{i \in I} epi(f_i)$. Hence, by Proposition 3.1(b) and Theorem 3.4, $f$ is a path-convex function on $C$.    □

**Theorem 3.6.** Let $g$ be a path-convex function from $(E \times E', P \times P')$ to $\mathbb{R}$ that is bounded below about $y \in E'$. Then the function $f$ from $E$ to $\mathbb{R}$ defined by

$$f(x) := \inf_{y \in E'} g(x, y)$$

is path-convex.

Proof. Suppose that $x_1, x_2 \in E$. Then for any $\epsilon > 0$, there exist $y_1, y_2 \in E'$ such that

$$g(x_i, y_i) < f(x_i) + \epsilon, \quad i = 1, 2.$$

It follows from the path-convexity hypothesis of $g$ that

$$f(P_1(t, x_1, x_2)) \leq g(P_1(t, x_1, x_2), P_2(t, y_1, y_2)) < tf(x_1) + (1 - t)f(x_2) + \epsilon$$

for each $t \in [0, 1]$. Let $\epsilon$ tend to 0 and the proof is completed.    □

**Definition 3.4.** A function $f : E \to \mathbb{R}$ is said to be *path-quasiconvex* if

$$f(P(t, x, y)) \leq \max\{f(x), f(y)\}$$

for all $x, y \in E$ and $t \in [0,1]$. If $f(P(t,x,y)) \geq \min\{f(x), f(y)\}$ for all $x, y \in E$ and $t \in [0,1]$, $f$ is called *path-quasiconcave*. If the strict inequalities holds for any two distinct points $x$ and $y$, $f$ is called *strictly path-quasiconvex* or *strictly path-quasiconcave*, respectively.

**Theorem 3.7.** A function $f : E \to \mathbb{R}$ is path-quasiconvex if and only if its section $S(f, \lambda) := \{x \in E : f(x) \leq \lambda\}$ is path-convex for each $\lambda \in \mathbb{R}$.

Proof. Suppose that $f$ is path-quasiconvex. Then, for any $x, y \in S(f, \lambda)$ and $t \in [0,1]$, one has $f(P(t,x,y)) \leq \max\{f(x), f(y)\} \leq \lambda$. It follows that $P(t,x,y) \in S(f, \lambda)$ and the set $S(f, \lambda)$ is path-convex.

Conversely, assume that $S(f, \lambda)$ is a path-convex set for each $\lambda \in \mathbb{R}$. For any $x, y \in E$, let $\lambda = \max\{f(x), f(y)\}$. Then $x, y \in S(f, \lambda)$. Since $S(f, \lambda)$ is path-convex, $P(t,x,y) \in S(f, \lambda)$ for all $t \in [0,1]$ which implies that $f(P(t,x,y)) \leq \lambda = \max\{f(x), f(y)\}$ for all $t \in [0,1]$.                    □

**Corollary 3.1.** If $f$ is a path-convex function from $E$ to $\mathbb{R}$, then its sections $S(f, \lambda)$ are path-convex.

# References

1. Ben-El-Mechaiekh, H., Chebbi, S.: Abstract convexity and fixed points. Journal of Mathematical Analysis and Applications 222, 138–150 (1998)
2. Ding, X.P.: Maximal element theorems in product FC-spaces and generalized games. Journal of Mathematical Analysis and Applications 305, 29–42 (2005)
3. Ding, X.P.: Generalized KKM type theorems in FC-spaces with applications. Journal of Global Optimization 36, 581–596 (2006)
4. Horvath, C.D.: Contractibility and generalized convexity. Journal of Mathematical Analysis and Applications 156, 341–357 (1991)
5. Horvath, C.D.: Extension and selection theorems in topological spaces with a generalized convexity structure. Annales de Faculté des Sciences de Toulouse 2, 253–269 (1993)
6. Huang, J.: The matching theorems and coincidence theorems for generalized R-KKM mapping in topological spaces. Journal of Mathematical Analysis and Applications 312, 1374–3821 (2005)
7. Kindler, J., Trost, R.: Minimax theorems for interval spaces. Acta Mathematica Hungarica 54, 38–49 (1989)
8. Park, S.: On generalizations of the KKM principle on abstract convex spaces. Nonlinear Analysis Forum 11, 67–77 (2006)
9. Park, S.: Various subclasses of abstract convex spaces for the KKM theory. Proceedings of the National Institute of Mathematical Science 2, 35–47 (2007)
10. Park, S.: Elements of the KKM theory on abstract convex spaces. Journal of the Korean Mathematical Society 45, 1–27 (2008)
11. Park, S.: Equilibrium existence theorems in KKM spaces. Nonlinear Analysis, Theory, Methods and Applications 69, 4352–4364 (2008)
12. Park, S., Kim, H.: Admissible classes of multifunctions on generalized convex spaces. Proceedings College of Natural Science Seoul National University 18, 1–21 (1993)
13. Stachó, L.L.: Minimax theorems beyond topological vector spaces. Acta Scientiarum Mathematicarum 42, 157–164 (1980)

# Halpern's Iteration for a Sequence of Quasinonexpansive Type Mappings

Koji Aoyama

**Abstract.** In this paper, we prove strong convergence of a Halpern's iteration generated by a sequence of quasinonexpansive type mappings in a Hilbert space. Then using the result, we establish convergence theorems for a $\lambda$-hybrid mapping and a maximal monotone operator.

**Keywords:** Quasinonexpansive mapping, $\lambda$-hybrid mapping, Fixed point.

## 1 Introduction

Let $H$ be a Hilbert space and $C$ a closed convex subset of $H$. This paper is devoted to the study of convergence of an iterative sequence $\{x_n\}$ defined by an arbitrary point $x_1 \in C$ and

$$x_{n+1} = \alpha_n u + (1 - \alpha_n) S_n x_n \tag{1}$$

for $n \in \mathbb{N}$, where $u$ is a point in $C$, $\alpha_n$ is a real number in $[0, 1]$, and $S_n$ is a self-mapping of $C$ for $n \in \mathbb{N}$. In particular, we focus on the case where each $S_n$ is given by the following form:

$$S_n = \frac{1}{n} \sum_{k=1}^{n} T^{k-1},$$

where $T$ is some quasinonexpansive self-mapping of $C$. Then we prove that, under some assumptions, $\{x_n\}$ converges strongly to a fixed point of $T$.

Strong convergence of the iteration defined by (1) was established by Shimizu and Takahashi [11] when $S_n$ is generated by two commutative

Koji Aoyama
Department of Economics, Chiba University, Yayoi-cho, Inage-ku, Chiba-shi, Chiba 263-8522, Japan
e-mail: aoyama@le.chiba-u.ac.jp

S. Li (Eds.): Nonlinear Maths for Uncertainty and its Appli., AISC 100, pp. 387–394.
springerlink.com © Springer-Verlag Berlin Heidelberg 2011

nonexpansive mappings; by Kamimura and Takahashi [6] when $S_n$ is a resolvent of a maximal monotone operator; by Kurokawa and Takahashi [9] when $T$ is a nonspreading mapping introduced by Kohsaka and Takahashi [8]; by Osilike and Isiogugu [10] when $T$ is a strictly pseudo nonspreading mapping; see also [5], [13], and [14].

This paper is organized as follows: In §2, we establish some preliminaries that we need. In §3, we first prove a strong convergence theorem for a sequence of quasinonexpansive type mappings (Theorem 1). Then using Theorem 1, we obtain a convergence theorem for a $\lambda$-hybrid mapping (Theorem 2), which is closely related to the results in [9, 10]. Furthermore, Theorem 1 can also be applicable to the proof of the convergence theorem in [6]; see Corollary 2.

## 2  Preliminaries

Throughout the present paper, $H$ denotes a real Hilbert space, $\langle \cdot , \cdot \rangle$ the inner product of $H$, $\| \cdot \|$ the norm of $H$, $C$ a nonempty closed convex subset of $H$, $I$ the identity mapping on $H$, and $\mathbb{N}$ the set of positive integers. Strong convergence of a sequence $\{x_n\}$ in $H$ to $x$ is denoted by $x_n \to x$ and weak convergence by $x_n \rightharpoonup x$. It is clear that the following inequality holds for all $x, y \in H$:

$$\|x + y\|^2 \leq \|x\|^2 + 2 \langle y, x + y \rangle . \tag{2}$$

The metric projection of $H$ onto $C$ is denoted by $P_C$, that is, $P_C(x) \in C$ and $\|P_C(x) - x\| \leq \|y - x\|$ for all $x \in H$ and $y \in C$. It is known that

$$\langle x - P_C(x), y - P_C(x) \rangle \leq 0 \tag{3}$$

for all $x \in H$ and $y \in C$; see [12].

The set of fixed points of a mapping $T \colon C \to C$ is denoted by $F(T)$. A mapping $T \colon C \to C$ is said to be nonexpansive if $\|Tx - Ty\| \leq \|x - y\|$ for all $x, y \in H$. A mapping $T \colon C \to C$ is said to be quasinonexpansive if $F(T)$ is nonempty and $\|Tx - z\| \leq \|x - z\|$ for all $x \in C$ and $z \in F(T)$. It is known that $F(T)$ is closed and convex if $T$ is a quasinonexpansive mapping. Let $\lambda$ be a real number. A mapping $T \colon C \to C$ is said to be $\lambda$-hybrid [2, 3] if

$$\|Tx - Ty\|^2 \leq \|x - y\|^2 + 2(1 - \lambda) \langle x - Tx, y - Ty \rangle \tag{4}$$

for all $x, y \in C$. It is clear that $T$ is quasinonexpansive if $T$ is a $\lambda$-hybrid mapping with a fixed point.

Let $\kappa$ be a real number with $0 \leq \kappa < 1$. Then a mapping $T \colon C \to C$ is said to be $\kappa$-strictly pseudononspreading [10] if

$$\|Tx - Ty\|^2 \leq \|x - y\|^2 + 2 \langle x - Tx, y - Ty \rangle + \kappa \|x - Tx - (y - Ty)\|^2$$

for all $x, y \in C$. We know the following result.

**Lemma 1 ( [3, Lemma 2.5]).** *Let $H$ be a Hilbert space, $C$ a nonempty closed convex subset of $H$, $\kappa$ and $\beta$ real numbers with $0 \leq \kappa \leq \beta < 1$, and $T: C \to C$ a $\kappa$-strict pseudononspreading mapping. Then $T_\beta = \beta I + (1-\beta)T$ is $-\beta/(1-\beta)$-hybrid.*

The following lemma is a variant of [1, Lemma 3.1] and [3, Lemma 2.7].

**Lemma 2.** *Let $H$ be a Hilbert space, $C$ a nonempty closed convex subset of $H$, $T: C \to C$ a mapping, $\{y_{m,n}\}$ a double sequence in $C$, $\{\xi_{m,n}\}$ a double sequence of real numbers, $\{z_n\}$ a sequence in $C$ defined by $z_n = (1/n)\sum_{k=1}^{n} y_{k,n}$ for $n \in \mathbb{N}$, and $z$ a weak cluster point of $\{z_n\}$. Suppose that $\{y_{1,n}\}$ is bounded,*

$$\xi_{m,n} \leq \|y_{m,n} - z\|^2 - \|y_{m+1,n} - Tz\|^2$$

*for all $m, n \in \mathbb{N}$, and $(1/n)\sum_{k=1}^{n}\xi_{k,n} \to 0$ as $n \to \infty$. Then $z$ is a fixed point of $T$.*

*Proof.* By assumption, it is clear that

$$\begin{aligned}
\xi_{k,l} &\leq \|y_{k,l} - z\|^2 - \|y_{k+1,l} - Tz\|^2 \\
&= \|y_{k,l} - Tz + Tz - z\|^2 - \|y_{k+1,l} - Tz\|^2 \\
&= \|y_{k,l} - Tz\|^2 - \|y_{k+1,l} - Tz\|^2 + 2\langle y_{k,l} - Tz, Tz - z \rangle + \|Tz - z\|^2
\end{aligned}$$

for every $k, l \in \mathbb{N}$. Summing these inequalities from $k = 1$ to $n$ and dividing by $n$, we have

$$\frac{1}{n}\sum_{k=1}^{n}\xi_{k,l} \leq \frac{1}{n}\left(\|y_{1,l} - Tz\|^2 - \|y_{n+1,l} - Tz\|^2\right)$$

$$+ 2\left\langle \frac{1}{n}\sum_{k=1}^{n}y_{k,l} - Tz, Tz - z \right\rangle + \|Tz - z\|^2$$

$$\leq \frac{1}{n}\|y_{1,l} - Tz\|^2 + 2\left\langle \frac{1}{n}\sum_{k=1}^{n}y_{k,l} - Tz, Tz - z \right\rangle + \|Tz - z\|^2$$

for every $n, l \in \mathbb{N}$. Since $z$ is a weak cluster point of $\{z_n\}$, there is a subsequence $\{z_{n_i}\}$ of $\{z_n\}$ such that $z_{n_i} \rightharpoonup z$. Replacing both $n$ and $l$ by $n_i$ in the above inequality, we obtain

$$\frac{1}{n_i}\sum_{k=1}^{n_i}\xi_{k,n_i} \leq \frac{1}{n_i}\|y_{1,n_i} - Tz\|^2 + 2\langle z_{n_i} - Tz, Tz - z \rangle + \|Tz - z\|^2.$$

Since $(1/n_i)\sum_{k=1}^{n_i}\xi_{k,n_i} \to 0$, $\{y_{1,n_i}\}$ is bounded, and $z_{n_i} \rightharpoonup z$, we conclude that

$$0 \leq 2\langle z - Tz, Tz - z \rangle + \|Tz - z\|^2 = -\|Tz - z\|^2$$

and hence $Tz = z$. $\qquad\square$

Let $\{T_n\}$ be a sequence of self-mappings of $C$ and $F$ a nonempty closed convex subset of $H$. Then we say that $\{T_n\}$ satisfies the *condition (S) with respect to $F$* if every weak cluster point of $\{T_n x_n\}$ belongs to $F$ whenever $\{x_n\}$ is a bounded sequence in $C$.

**Lemma 3.** *Let $H$ be a Hilbert space, $C$ a nonempty closed convex subset of $H$, $\lambda$ a real number, $T: C \to C$ a $\lambda$-hybrid mapping with a fixed point, and $S_n: C \to C$ a mapping defined by $S_n = (1/n) \sum_{k=1}^{n} T^{k-1}$ for $n \in \mathbb{N}$, where $T^0 = I$. Then $\{S_n\}$ satisfies the condition (S) with respect to $F(T)$.*

*Proof.* We know that $T$ is quasinonexpansive, so $F(T)$ is closed and convex. Let $\{x_n\}$ be a bounded sequence in $C$ and set $z_n = S_n(x_n)$ for $n \in \mathbb{N}$. Let $\{z_{n_i}\}$ be a subsequence of $\{z_n\}$ such that $z_{n_i} \rightharpoonup z$. Note that $z \in C$. Since $T$ is $\lambda$-hybrid, it follows from (4) that

$$\xi_{m,n} \leq \left\| T^{m-1} x_n - z \right\|^2 - \left\| T^m x_n - Tz \right\|^2$$

for every $m, n \in \mathbb{N}$, where $\xi_{m,n} = -2(1-\lambda) \langle T^{m-1} x_n - T^m x_n, z - Tz \rangle$. It is clear that $z_n = (1/n) \sum_{k=1}^{n} T^{k-1} x_n$ for every $n \in \mathbb{N}$ and $\{T^0 x_n\} = \{x_n\}$ is bounded. Since $T$ is quasinonexpansive, $\{T^n x_n\}$ is bounded and thus

$$\frac{1}{n} \sum_{k=1}^{n} \xi_{k,n} = \frac{-2(1-\lambda)}{n} \sum_{k=1}^{n} \langle T^{k-1} x_n - T^k x_n, z - Tz \rangle$$

$$= -2(1-\lambda) \left\langle \frac{x_n - T^n x_n}{n}, z - Tz \right\rangle \to 0$$

as $\to \infty$. Therefore, Lemma 2 implies that $z$ is a fixed point of $T$. This means that $\{S_n\}$ satisfies the condition (S) with respect to $F(T)$.                          $\square$

A mapping $A: H \to 2^H$ is said to be an operator and we can identify an operator $A: H \to 2^H$ with a subset of $H \times H$. An operator $A \subset H \times H$ is said to be monotone if $\langle x - x', y - y' \rangle \geq 0$ for all $(x,y), (x',y') \in A$; a monotone operator $A \subset H \times H$ is said to be maximal if $A = B$ whenever $B \subset H \times H$ is a monotone operator with $A \subset B$. Let $A \subset H \times H$ be a maximal monotone operator, $\rho$ a positive real number, and $A^{-1}0$ the set of zero points of $A$, that is, $A^{-1}0 = \{z \in H : (z,0) \in A\}$. Then it is known that $(I + \rho A)^{-1}$ is a single-valued nonexpansive self-mapping of $H$ and $F\left((I + \rho A)^{-1}\right) = A^{-1}0$. Such a mapping $(I + \rho A)^{-1}$ is called the resolvent of $A$; see [12] for more details.

The following lemma is a special case of [4, Lemma 3.6].

**Lemma 4.** *Let $H$ be a Hilbert space, $A \subset H \times H$ a maximal monotone operator such that $A^{-1}0$ is nonempty, and $\{\rho_n\}$ a sequence of positive real numbers such that $\rho_n \to \infty$. Then $\left\{(I + \rho_n A)^{-1}\right\}$ satisfies the condition (S) with respect to $A^{-1}0$.*

To prove our result, we also need the following; see, for example, [14].

**Lemma 5.** *Let $\{\xi_n\}$ be a sequence of nonnegative real numbers, $\{\alpha_n\}$ a sequence in $[0,1]$, and $\{\gamma_n\}$ a sequence of real numbers. Suppose that $\xi_{n+1} \leq (1 - \alpha_n)\xi_n + \alpha_n\gamma_n$ for all $n \in \mathbb{N}$, $\limsup_{n \to \infty} \gamma_n \leq 0$, and $\sum_{n=1}^{\infty} \alpha_n = \infty$. Then $\xi_n \to 0$.*

## 3   Strong Convergence Theorems

In this section, we first prove strong convergence of a Halpern's iteration generated by a sequence of quasinonexpansive type mappings.

**Theorem 1.** *Let $H$ be a Hilbert space, $C$ a nonempty closed convex subset of $H$, $u$ a point in $C$, and $\{\alpha_n\}$ a sequence in $[0,1]$ such that $\alpha_n \to 0$ and $\sum_{n=1}^{\infty} \alpha_n = \infty$. Let $\{S_n\}$ be a sequence of self-mappings of $C$ and $F$ a nonempty closed convex subset of $C$. Suppose that*

$$\|S_n y - z\| \leq \|y - z\| \tag{5}$$

*for all $n \in \mathbb{N}$, $y \in C$, and $z \in F$, and that $\{S_n\}$ satisfies the condition (S) with respect to $F$. Let $\{x_n\}$ be a sequence in $C$ defined by $x_1 \in C$ and*

$$x_{n+1} = \alpha_n u + (1 - \alpha_n)S_n x_n \tag{6}$$

*for $n \in \mathbb{N}$. Then $\{x_n\}$ converges strongly to $P_F(u)$.*

*Proof.* Set $w = P_F(u)$. We first show that $\{x_n\}$ and $\{S_n x_n\}$ are bounded. Since $w \in F$, it follows from (6) and (5) that

$$
\begin{aligned}
\|x_{n+1} - w\| &\leq \alpha_n \|u - w\| + (1 - \alpha_n)\|S_n x_n - w\| \\
&\leq \alpha_n \|u - w\| + (1 - \alpha_n)\|x_n - w\|,
\end{aligned}
$$

so, by induction on $n$,

$$\|S_n x_n - w\| \leq \|x_n - w\| \leq \max\{\|x_1 - w\|, \|u - w\|\}$$

for every $n \in \mathbb{N}$. Therefore, $\{x_n\}$ and $\{S_n x_n\}$ are bounded.

We next show $\limsup_{n \to \infty} \langle u - w, x_{n+1} - w \rangle \leq 0$. By the boundedness of $\{S_n x_n\}$, it follows from (6) and $\alpha_n \to 0$ that

$$x_{n+1} - S_n x_n = \alpha_n(u - S_n x_n) \to 0. \tag{7}$$

The boundedness of $\{S_n x_n\}$ also implies that there exists a strictly increasing sequence $\{n_i\}$ in $\mathbb{N}$ such that $S_{n_i} x_{n_i} \rightharpoonup v$ and

$$\limsup_{n \to \infty} \langle u - w, x_{n+1} - w \rangle = \lim_{i \to \infty} \langle u - w, x_{n_i+1} - w \rangle.$$

Since $\{x_n\}$ is bounded and $\{S_n\}$ satisfies the condition (S) with respect to $F$, it follows from (7) that $x_{n_i+1} \rightharpoonup v \in F$. Therefore, by (3), we conclude that

$$\limsup_{n \to \infty} \langle u - w, x_{n+1} - w \rangle = \langle u - w, v - w \rangle \le 0. \tag{8}$$

Lastly, we show that $\{x_n\}$ converges strongly to $w$. Since $\alpha_n \in [0, 1]$, it follows from (2) and (5) that

$$
\begin{aligned}
\|x_{n+1} - w\|^2 &= \|(1 - \alpha_n)(S_n x_n - w) + \alpha_n(u - w)\|^2 \\
&\le \|(1 - \alpha_n)(S_n x_n - w)\|^2 + 2\langle \alpha_n(u - w), x_{n+1} - w \rangle \\
&= (1 - \alpha_n)^2 \|S_n x_n - w\|^2 + 2\alpha_n \langle u - w, x_{n+1} - w \rangle \\
&\le (1 - \alpha_n)\|x_n - w\|^2 + 2\alpha_n \langle u - w, x_{n+1} - w \rangle
\end{aligned}
$$

for every $n \in \mathbb{N}$. Thus Lemma 5 and (8) imply that $x_n \to w$.          □

*Remark 1.* In Theorem 1, if we assume that each $S_n$ is quasinonexpansive and $F = \bigcap_{n=1}^{\infty} F(S_n)$ is nonempty, then it is clear that (5) holds for all $y \in C$ and $z \in F$.

Using Theorem 1, we obtain the following strong convergence theorem for a $\lambda$-hybrid mapping.

**Theorem 2.** *Let $H$, $C$, $u$, and $\{\alpha_n\}$ be the same as in Theorem 1. Let $\lambda$ be a real number, $T: C \to C$ a $\lambda$-hybrid mapping with a fixed point, and $S_n: C \to C$ a mapping defined by*

$$S_n = \frac{1}{n} \sum_{k=1}^{n} T^{k-1}$$

*for $n \in \mathbb{N}$, where $T^0 = I$. Let $\{x_n\}$ be a sequence in $C$ defined by $x_1 \in C$ and (6) for $n \in \mathbb{N}$. Then $\{x_n\}$ converges strongly to $P_{F(T)}(u)$.*

*Proof.* Since $T$ is quasinonexpansive, it is clear that

$$
\begin{aligned}
\|S_n y - z\| &= \left\| \frac{1}{n} \sum_{k=1}^{n} T^{k-1} y - z \right\| \\
&\le \frac{1}{n} \sum_{k=1}^{n} \|T^{k-1} y - z\| \\
&\le \frac{1}{n} \sum_{k=1}^{n} \|y - z\| = \|y - z\|
\end{aligned}
$$

for all $y \in C$ and $z \in F(T)$. Thus (5) holds. Moreover, Lemma 3 shows that $\{S_n\}$ satisfies the condition (S) with respect to $F(T)$. Therefore, Theorem 1 implies the conclusion.          □

A direct consequence of Theorem 2 is the following result.

**Corollary 1 ( [10, Theorem 3.2]).** *Let $H$, $C$, $u$, and $\{\alpha_n\}$ be the same as in Theorem 1. Let $\kappa$ and $\beta$ be real numbers with $0 \le \kappa \le \beta < 1$, $T: C \to C$ a $\kappa$-strictly pseudononspreading mapping with a fixed point, and $S_n: C \to C$ a mapping defined by*

$$S_n = \frac{1}{n} \sum_{k=1}^{n} (T_\beta)^{k-1}$$

*for $n \in \mathbb{N}$, where $T_\beta = \beta I + (1 - \beta)T$ and $(T_\beta)^0 = I$. Let $\{x_n\}$ be a sequence in $C$ defined by $x_1 \in C$ and (6) for $n \in \mathbb{N}$. Then $\{x_n\}$ converges strongly to $P_{F(T)}(u)$.*

*Proof.* By Lemma 1, we know that $T_\beta$ is a $-\beta/(1-\beta)$-hybrid self-mapping of $C$. It is obvious that $F(T_\beta) = F(T) \neq \emptyset$. Therefore, Theorem 2 implies the conclusion. □

*Remark 2.* In [10, Theorem 3.2], $\{\alpha_n\}$ is assumed to be a sequence in $[0, 1)$. If $\kappa = \beta = 0$ in Corollary 1, then we obtain [9]*Theorem 4.1.

Theorem 1 also implies the following result proved in [6]; see also [7].

**Corollary 2 ( [6, Theorem 1]).** *Let $H$ and $\{\alpha_n\}$ be the same as in Theorem 1. Let $u$ be a point in $H$, $A \subset H \times H$ a maximal monotone operator with a zero point, and $\{\rho_n\}$ a sequence of positive real numbers such that $\rho_n \to \infty$. Let $\{x_n\}$ be a sequence in $C$ defined by $x_1 \in C$ and*

$$x_{n+1} = \alpha_n u + (1 - \alpha_n)(I + \rho_n A)^{-1} x_n$$

*for $n \in \mathbb{N}$. Then $\{x_n\}$ converges strongly to $P_{A^{-1}0}(u)$.*

*Proof.* Set $S_n = (I + \rho_n A)^{-1}$ for $n \in \mathbb{N}$. Since $S_n$ is nonexpansive and $F(S_n) = A^{-1}0$, it follows that $\|S_n y - z\| \le \|y - z\|$ for all $y \in H$ and $z \in A^{-1}0$. Thus (5) holds. Moreover, Lemma 4 shows that $\{S_n\}$ satisfies the condition (S) with respect to $A^{-1}0$. Therefore, Theorem 1 implies the conclusion.

*Remark 3.* In [6, Theorem 1], the initial point $x_1$ is assumed to be $u$.

# References

1. Akatsuka, M., Aoyama, K., Takahashi, W.: Mean ergodic theorems for a sequence of nonexpansive mappings in Hilbert spaces. Sci. Math. Jpn. 68, 233–239 (2008)
2. Aoyama, K., Iemoto, S., Kohsaka, F., Takahashi, W.: Fixed point and ergodic theorems for λ-hybrid mappings in Hilbert spaces. J. Nonlinear Convex Anal. 11, 335–343 (2010)

3. Aoyama, K., Kohsaka, F.: Fixed point and mean convergence theorems for a family of λ-hybrid mappings. Journal of Nonlinear Analysis and Optimization: Theory & Applications 2, 85–92 (2011)
4. Aoyama, K., Kohsaka, F., Takahashi, W.: Proximal point methods for monotone operators in Banach spaces. Taiwanese Journal of Mathematics 15, 259–281 (2011)
5. Halpern, B.: Fixed points of nonexpanding maps. Bull. Amer. Math. Soc. 73, 957–961 (1967)
6. Kamimura, S., Takahashi, W.: Approximating solutions of maximal monotone operators in Hilbert spaces. J. Approx. Theory 106, 226–240 (2000)
7. Kohsaka, F., Takahashi, W.: Strong convergence of an iterative sequence for maximal monotone operators in a Banach space. Abstr. Appl. Anal., 239–249 (2004)
8. Kohsaka, F., Takahashi, W.: Fixed point theorems for a class of nonlinear mappings related to maximal monotone operators in Banach spaces. Arch. Math (Basel) 91, 166–177 (2008)
9. Kurokawa, Y., Takahashi, W.: Weak and strong convergence theorems for nonspreading mappings in Hilbert spaces. Nonlinear Anal. 73, 1562–1568 (2010)
10. Osilike, M.O., Isiogugu, F.O.: Weak and strong convergence theorems for nonspreading-type mappings in Hilbert spaces. Nonlinear Anal. 74, 1814–1822 (2011)
11. Shimizu, T., Takahashi, W.: Strong convergence to common fixed points of families of nonexpansive mappings. J. Math. Anal. Appl. 211, 71–83 (1997)
12. Takahashi, W.: Introduction to nonlinear and convex analysis. Yokohama Publishers, Yokohama (2009)
13. Wittmann, R.: Approximation of fixed points of nonexpansive mappings. Arch. Math (Basel) 58, 486–491 (1992)
14. Xu, H.K.: Iterative algorithms for nonlinear operators. J. London Math. Soc. 66(2), 240–256 (2002)

# Convergence of Iterative Methods for an Infinite Family of Pseudo-contractions

Yuanheng Wang and Jiashuai Dong

**Abstract.** In this paper, we establish some strong convergence theorems for an infinitely countable family of Lipschitzian pseudo-contractions in Hilbert spaces by proposing some kinds of new iterative methods. The results here extend and improve the corresponding results of other authors', such as Haiyun Zhou [Convergence theorems of fixed points for Lipschitz pseudo-contractions in Hilbert spaces, J Math Anal Appl 343: 546-556 ], Marino G and Xu H K [Weak and strong convergence theorems for strict pseudo-contractions in Hilbert spaces, J Math Anal Appl 329(1): 336-346 ], Rhoades B E [Fixed point iterations using infinite matrices, Trans Amer Math Soc 196: 162-176].

**Keywords:** Infinite family of pseudo-contractions, Lipschitz mapping, Iterative algorithm, Strong convergence theorem.

## 1 Introduction

Mann's iteration algorithm which was introduced by Mann [2] generates a sequence $\{x_n\}$ by the following manner:

$$\forall x_0 \in C, x_{n+1} = \alpha_n x_n + (1 - \alpha_n)Tx_n, n \geq 0,$$

where $\{\alpha_n\}$ is a real sequence in (0,1) which satisfies certain control conditions.

In 1967, Browder and Petryshyn [1] established the first convergence result for $\kappa$-strict pseudo-contractions in real Hilbert spaces. They proved weak and strong convergence theorems by using Mann's algorithm with a constant control sequence $\alpha_n = \alpha$ for all n. Afterward, Rhoades [5] generalized in part the corresponding results in [1] in the sense that a variable control sequence

Yuanheng Wang and Jiashuai Dong
Department of Mathematics, Zhejiang Normal University, Jinhua 321004, China
e-mail: Wangyuanheng@yahoo.com.cn, yhwang@zjnu.cn

S. Li (Eds.): Nonlinear Maths for Uncertainty and its Appli., AISC 100, pp. 395–402.
springerlink.com                                    © Springer-Verlag Berlin Heidelberg 2011

was taken into consideration. Under assumption that the domain of mapping
T is compact convex, he established a strong convergence theorem by using Mann's algorithm with a control sequence $\{\alpha_n\}$ satisfying the conditions
$\alpha_0 = 1$, $0 < \alpha_n < 1$, $\Sigma_{n=0}^{\infty}(1 - \alpha_n) = \infty$ and $\limsup_{n\to\infty} \alpha_n = \alpha > \kappa$.
However, without compact assumption on the domain of mapping T, in gen-
eral, one cannot expect to infer any weak convergence results from Rhoades's
convergence theorem. In 2007, Marino and Xu [3] have proved a weak convergence theorem by using Mann's algorithm with a control sequence $\{\alpha_n\}$
satisfying the conditions $\kappa < \alpha_n < 1$ and $\Sigma_{n=0}^{\infty}(\alpha_n - \kappa)(1 - \alpha_n) = \infty$.
Their convergence theorem extends and improves the corresponding results
in [2,5]. In 2009, Zhou [7] improves and extends Marino and Xu's convergence
theorems (Theorems 3.1 and 4.1) by virtue of new analysis techniques. The
purpose of this paper is to extend the results of Zhou [6] to an an infinitely
countable family of Lipschitzian pseudo-contractions in Hilbert spaces by
proposing some kinds of new iterative methods. The results here improve the
corresponding results of other authors'( [1] [5] [3] [4][6]).

## 2   Preliminaries

Throughout this paper, we assume that H is a real Hilbert space with inner
product $\langle \cdot, \cdot \rangle$ and C is a nonempty closed convex subset of H. Let $\omega_\omega(x_n) =$
$\{x : \exists x_{n_k} \rightharpoonup x\}$ denote the weak $\omega$-limit set of $\{x_n\}$ and let N denote the set
$\{1, 2, \cdots, n, \cdots\}$. Recall that T: $C \to C$ is called a $\kappa$-strict pseudo-contraction
if there exists a constant $\kappa \in [0, 1)$ such that

$$\|Tx - Ty\|^2 \leq \|x - y\|^2 + \kappa\|(I - T)x - (I - T)y\|^2$$

for all x, y $\in$ C.

When $\kappa = 0$, T is said to be nonexpansive; when $\kappa = 1$, T is said to be
pseudo-contractive. Clearly, the class of $\kappa$-strict pseudo-contractions falls into
the one between classes of nonexpansive mappings and pseudo-contractions.
T is said to be strongly pseudo-contractive, if there exists a positive constant
$\lambda \in (0, 1)$ such that $T + \lambda I$ is pseudo-contractive. We remark that the class of
strongly pseudo-contractive mappings is independent of the class of $\kappa$-strict
pseudo-contractions. We also remark in passing that if T is a $\kappa$-strict pseudo-
contraction, then it is Lipschitz continuous, and a pseudo- contraction may
be not continuous.

**Lemma 1.** ([6]) (Demi-closedness principle) Let C be a nonempty closed
convex subset of a real Hilbert space H and $T : C \to C$ be a demicontinuous
pseudo-contractive self-mapping from C into itself. Then F(T) is a closed
convex subset of C and I-T is demiclosed at zero.

**Lemma 2.** ([3]) Let H be a real Hilbert space. $\forall x, y \in H$, there hold the
following identities

(a) $\|x \pm y\|^2 = \|x\|^2 \pm 2\langle x, y\rangle + \|y\|^2$;
(b) $\|tx + (1 - t)y\|^2 = t\|x\|^2 + (1 - t)\|y\|^2 - t(1 - t)\|x - y\|^2$;
(c) $\|x + y\|^2 + \|x - y\|^2 = 2(\|x\|^2 + \|y\|^2)$.

**Lemma 3.** ([6]) Let C be a nonempty closed convex subset of a real Hilbert space H. For every $x \in H$, there exists a unique point $z \in C$ such that $\|x - z\| \leq \|x - y\|$, for all $y \in C$. Define $P_C : H \to C$ by $z = P_C x$. Then $z = P_C x$ if and only if for all $y \in C$, the following inequality holds: $\langle x - z, y - z\rangle \leq 0$.

**Lemma 4.** ([8]) Let C be a nonempty closed convex subset of a real Hilbert space H and $P_C : H \to C$ be the metric projection from H onto C. Then for all $x \in H$, $y \in C$, the following inequality holds

$$\|y - P_C x\|^2 + \|x - P_C x\|^2 \leq \|x - y\|^2.$$

## 3   Main Results

In this section, we propose several iterative algorithms for an infinitely countable family of Lipschitzian pseudo-contractions in Hilbert spaces.

**Theorem 1.** Let C be a nonempty closed convex subset of a real Hilbert space H and let $\{T_i\}_{i=1}^{\infty} : C \to C$ be an infinitely countable family of Lipschitzian pseudo-contractions such that $F = \bigcap_{i=1}^{\infty} F(T_i) \neq \phi$. Assume the control sequences $\{\alpha_{n,i}\}$, $\{\beta_{n,i}\}$ are chosen in (0,1) satisfying the conditions:

(a) $\beta_{n,i} \leq \alpha_{n,i}, \forall n \geq 0, i \in N$.
(b) $\liminf_{n\to\infty} \alpha_{n,i} > 0, i \in N$.
(c) $\limsup_{n\to\infty} \alpha_{n,i} \leq \alpha_i \leq \frac{1}{\sqrt{1+L_i^2}+1}, i \in N$.

where $L_i$ is the Lipschitzian constant of $T_i$. The sequence $\{x_n\}$ is given by the following manner:

$$
\begin{cases}
x_0 = x \in C, \\
y_{n,i} = (1 - \alpha_{n,i})x_n + \alpha_{n,i}T_ix_n, \\
z_{n,i} = (1 - \beta_{n,i})x_n + \beta_{n,i}T_iy_{n,i}, \\
C_{n,i} = \{z \in C : \|z_{n,i} - z\|^2 \leq \|x_n - z\|^2 - \alpha_{n,i}\beta_{n,i}(1 - 2\alpha_{n,i} \\
\qquad\qquad - L_i^2\alpha_{n,i}^2)\|x_n - T_ix_n\|^2\}, C_n = \bigcap_{i=1}^{\infty} C_{n,i}, \\
Q_0 = C, Q_n = \{z \in Q_{n-1} : \langle z - x_n, x - x_n\rangle \leq 0\}, n \geq 1, \\
x_{n+1} = P_{C_n \cap Q_n}x, n = 0, 1, 2, \cdots, i \in N.
\end{cases}
\tag{1}
$$

Then the sequence $\{x_n\}$ converges strongly to $P_F x$, where $P_F$ is the metric projection from H onto F.

**Proof:** We split the proof into seven steps.

**Step1.** Show that $P_F$ is well defined for every x $\in$ C. By Lemma 1, we know $F(T_i)$ is closed and convex for every i $\in$ N. $F = \bigcap_{i=1}^{\infty} F(T_i)$ is nonempty by our assumption. Therefore, F is a nonempty closed convex subset of C. Consequently, $P_F$ is well defined for every x $\in$ C.

**Step2.** Show that $C_n$ and $Q_n$ are closed and convex for all n $\geq$ 0. This follows from the constructions of $C_n$ and $Q_n$. We omit the details.

**Step3.** Show that $F \subset C_n \bigcap Q_n$ for all n $\geq$ 0. We first prove $F \subset C_n$. Let p $\in$ F and let n $\geq$ 0. Then, using the definition of pseudo-contraction, (1) and Lemma 2, we have

$$
\begin{aligned}
\|z_{n,i} - p\|^2 &= \|(1 - \beta_{n,i})(x_n - p) + \beta_{n,i}(T_i y_{n,i} - p)\|^2 \\
&= (1 - \beta_{n,i})\|x_n - p\|^2 + \beta_{n,i}\|T_i y_{n,i} - p\|^2 - \beta_{n,i}(1 - \beta_{n,i})\|x_n - T_i y_{n,i}\|^2 \\
&\leq (1 - \beta_{n,i})\|x_n - p\|^2 + \beta_{n,i}(\|y_{n,i} - p\|^2 + \|T_i y_{n,i} - y_{n,i}\|^2) \\
&\quad - \beta_{n,i}(1 - \beta_{n,i})\|x_n - T_i y_{n,i}\|^2.
\end{aligned}
\tag{2}
$$

we also have

$$
\begin{aligned}
\|T_i y_{n,i} - y_{n,i}\|^2 &= \|(1 - \alpha_{n,i})(x_n - T_i y_{n,i}) + \alpha_{n,i}(T_i x_n - T_i y_{n,i})\|^2 \\
&= (1 - \alpha_{n,i})\|x_n - T_i y_{n,i}\|^2 + \alpha_{n,i}\|T_i x_n - T_i y_{n,i}\|^2 - \alpha_{n,i}(1 - \alpha_{n,i})\|x_n - T_i x_n\|^2 \\
&\leq (1 - \alpha_{n,i})\|x_n - T_i y_{n,i}\|^2 + \alpha_{n,i} L_i^2 \|x_n - y_{n,i}\|^2 - \alpha_{n,i}(1 - \alpha_{n,i})\|x_n - T_i x_n\|^2 \\
&= (1 - \alpha_{n,i})\|x_n - T_i y_{n,i}\|^2 + \alpha_{n,i}^3 L_i^2 \|x_n - T_i x_n\|^2 - \alpha_{n,i}(1 - \alpha_{n,i})\|x_n - T_i x_n\|^2 \\
&= (1 - \alpha_{n,i})\|x_n - T_i y_{n,i}\|^2 + \alpha_{n,i}(\alpha_{n,i}^2 L_i^2 + \alpha_{n,i} - 1)\|x_n - T_i x_n\|^2,
\end{aligned}
\tag{3}
$$

and

$$
\begin{aligned}
\|y_{n,i} - p\|^2 &= \|(1 - \alpha_{n,i})(x_n - p) + \alpha_{n,i}(T_i x_n - p)\|^2 \\
&= (1 - \alpha_{n,i})\|x_n - p\|^2 + \alpha_{n,i}\|T_i x_n - p\|^2 - \alpha_{n,i}(1 - \alpha_{n,i})\|x_n - T_i x_n\|^2 \\
&\leq (1 - \alpha_{n,i})\|x_n - p\|^2 + \alpha_{n,i}\|x_n - p\|^2 + \alpha_{n,i}\|x_n - T_i x_n\|^2 - \\
&\quad \alpha_{n,i}(1 - \alpha_{n,i})\|x_n - T_i x_n\|^2 = \|x_n - p\|^2 + \alpha_{n,i}^2\|x_n - T_i x_n\|^2.
\end{aligned}
\tag{4}
$$

Substituting (3) and (4) into (2) yields

$$
\begin{aligned}
\|z_{n,i} - p\|^2 &\leq (1 - \beta_{n,i})\|x_n - p\|^2 + \beta_{n,i}\|x_n - p\|^2 \\
&\quad + \beta_{n,i}\alpha_{n,i}^2\|x_n - T_i x_n\|^2 + \beta_{n,i}(1 - \alpha_{n,i})\|x_n - T_i y_{n,i}\|^2 \\
&\quad + \alpha_{n,i}\beta_{n,i}(\alpha_{n,i}^2 L_i^2 + \alpha_{n,i} - 1)\|x_n - T_i x_n\|^2 - \beta_{n,i}(1 - \beta_{n,i})\|x_n - T_i y_{n,i}\|^2 \\
&= \|x_n - p\|^2 + \beta_{n,i}(\beta_{n,i} - \alpha_{n,i})\|x_n - T_i y_{n,i}\|^2 \\
&\quad + \alpha_{n,i}\beta_{n,i}(\alpha_{n,i}^2 L_i^2 + 2\alpha_{n,i} - 1)\|x_n - T_i x_n\|^2.
\end{aligned}
\tag{5}
$$

From condition (a), we have $\beta_{n,i}(\beta_{n,i} - \alpha_{n,i}) \leq 0$, and hence

$$
\begin{aligned}
|z_{n,i} - p\|^2 &= \|x_n - p\|^2 + \alpha_{n,i}\beta_{n,i}(\alpha_{n,i}^2 L_i^2 + 2\alpha_{n,i} - 1)\|x_n - T_i x_n\|^2 \\
&= \|x_n - p\|^2 - \alpha_{n,i}\beta_{n,i}(1 - 2\alpha_{n,i} - \alpha_{n,i}^2 L_i^2)\|x_n - T_i x_n\|^2,
\end{aligned}
\tag{6}
$$

which shows that $p \in C_{n,i}$ for all n $\geq 0$, i $\in$ N. This proves that $F \subset C_{n,i}$ for all n $\geq 0$, and hence $F \subset C_n$ for all n $\geq 0$.

Next we prove $F(T) \subset Q_n$ for all n $\geq 0$. We prove this by induction. For n=0, we have $F(T) \subset C = Q_0$. Assume that $F(T) \subset Q_n$. Since $x_{n+1}$ is the projection of x onto $C_n \bigcap Q_n$, by Lemma 3, we have $\langle x_{n+1} - z, x - x_{n+1} \rangle \geq 0$, for any $z \in C_n \bigcap Q_n$. As $F \subset C_n \bigcap Q_n$ by the induction assumption, the last inequality holds, in particular, for all $p \in F$. This together with the definition of $Q_{n+1}$ implies that $F \subset Q_{n+1}$. Hence, $F \subset C_n \bigcap Q_n$ for all n $\geq 0$.

**Step4.** Show that $\lim_{n \to \infty} \|x_n - x\|$ exists. In view of (2) and Lemma 3, we have $x_n = P_{Q_n} x$, which means that for any $z \in Q_n$, $\|x_n - x\| \leq \|z - x\|$. Since $x_{n+1} \in Q_n$ and $p \in F(T) \subset Q_n$, we obtain $\|x_n - x\| \leq \|x_{n+1} - x\|$, and $\|x_n - x\| \leq \|p - x\|$, for all n $\geq 0$. Consequently, $\lim_{n \to \infty} \|x_n - x\|$ exists.

**Step5.** Show that $x_n \to p \in C$, as $n \to \infty$. For m > n, by the definition of $Q_n$, we see that $Q_m \subset Q_n$. Noting that $x_m = P_{Q_m} x$ and $x_n = P_{Q_n} x$, by Lemma 4, we conclude that $\|x_m - x_n\|^2 \leq \|x_m - x\|^2 - \|x_n - x\|^2$. In view of step 4, we deduce that $x_m - x_n \to 0$ as $m, n \to \infty$, which means that $\{x_n\}$ is cauchy. Since H is complete and C is closed, we can assume that $x_n \to p \in C$, $n \to \infty$. In particular, we obtain that $\|x_{n+1} - x_n\| \to 0$ as $n \to \infty$.

**Step6.** Show that $\|x_n - T_i x_n\| \to 0$, as $n \to \infty$, i $\in$ N. By the fact $x_{n+1} \in C_{n,i}$, we have

$$
\begin{aligned}
\|x_{n+1} - z_{n,i}\|^2 &\leq \|x_{n+1} - x_n\|^2 \\
&\quad - \alpha_{n,i}\beta_{n,i}(1 - 2\alpha_{n,i} - \alpha_{n,i}^2 L_i^2)\|x_n - T_i x_n\|^2,
\end{aligned}
\tag{7}
$$

$$
\begin{aligned}
\|x_{n+1} - z_{n,i}\|^2 &= \|x_{n+1} - x_n\|^2 \\
&\quad + 2\langle x_{n+1} - x_n, x_n - z_{n,i} \rangle + \|x_n - z_{n,i}\|^2.
\end{aligned}
\tag{8}
$$

Combining (7) and (8), and noting that $z_{n,i} = (1 - \beta_{n,i})x_n + \beta_{n,i} T_i y_{n,i}$, we get

$$
\begin{aligned}
\beta_{n,i}^2 \|x_n - T_i y_{n,i}\|^2 &+ 2\beta_{n,i}\langle x_{n+1} - x_n, x_n - T_i y_{n,i} \rangle \\
&\leq -\alpha_{n,i}\beta_{n,i}(1 - 2\alpha_{n,i} - \alpha_{n,i}^2 L_i^2)\|x_n - T_i x_n\|^2.
\end{aligned}
\tag{9}
$$

Since $\beta_{n,i} > 0$ for all n $\geq 0$, canceling $\beta_{n,i}$ in (9), we get

$$
\begin{aligned}
\beta_{n,i}\|x_n - T_i y_{n,i}\|^2 &+ 2\langle x_{n+1} - x_n, x_n - T_i y_{n,i} \rangle \\
&\leq -\alpha_{n,i}(1 - 2\alpha_{n,i} - \alpha_{n,i}^2 L_i^2)\|x_n - T_i x_n\|^2.
\end{aligned}
\tag{10}
$$

From the assumption on $\{\alpha_n\}$ and noting that $\{T_i y_{n,i}\}$ is bounded, we see that there exist positive constants $a_i$, $b_i$ and $M_i$ such that

$$b_i(1 - 2a_i - a_i^2 L_i^2)\|x_n - T_i x_n\|^2 \leq M_i\|x_{n+1} - x_n\|. \tag{11}$$

for n large enough, where $M_i = 2\sup\{\|x_n - T_i y_n\| : n \geq 0\}$. Indeed, we can choose $a_i \in (\alpha_i, \frac{1}{\sqrt{1+L_i^2}+1})$. For such $a_i$, $\alpha_{n,i} < a_i$ for all n $\geq 0$. Noting that the function $f_i(t) = 1 - 2t - L_i t^2$ is strictly decreasing in (0,1), we infer that

$$(1 - 2\alpha_{n,i} - \alpha_{n,i}^2 L_i^2) > (1 - 2a_i - a_i^2 L_i^2) > 0. \tag{12}$$

Choosing $b_i \in (0, c_i)$, where $c_i = \liminf_{n\to\infty} \alpha_{n,i}$, we have that $\alpha_{n,i} > b_i$ for all n $\geq 0$. This together with (10) and (12), deduces to (11). It follows from step 5 that $\|x_n - T_i x_n\| \to 0$, as $n \to \infty$, $i \in N$.

**Step7.** Show that $p = P_F x$. Since $x_n \to p$ and $\forall i \in N$, $\|x_n - T_i x_n\| \to 0$, as $n \to \infty$, we have that $p = T_i p$, and hence $p \in F$. From step 3, we know that $F \subset Q_n$ for all n $\geq 0$, hence, by using Lemma 3, for arbitrary $z \in F$, we have $\langle z - x_n, x - x_n\rangle \leq 0$. This leads $\langle z - p, x - p\rangle \leq 0$ for all $z \in F$. By Lemma 3, we conclude that $p = P_F x$.

**Remark 1.** From the proof process of Theorem 1, we can see that the construction of $Q_n$ is a strong condition which ensures that $\{x_n\}$ is cauchy, and hence $\{x_n\}$ strongly converges a point in C. In the next theorem, we modify the algorithm used in Theorem 1 so that the strong convergence is still obtained but the assumption on $Q_n$ is weakened.

**Theorem 2.** Let C be a nonempty closed convex subset of a real Hilbert space H and let $\{T_i\}_{i=1}^\infty : C \to C$ be an infinitely countable family of Lipschitzian pseudo-contractions such that $F = \bigcap_{i=1}^\infty F(T_i) \neq \phi$. Assume the control sequences $\{\alpha_{n,i}\}$, $\{\beta_{n,i}\}$ are chosen in (0,1) satisfying the conditions

  (a) $\beta_{n,i} \leq \alpha_{n,i}$, $\forall n \geq 0$, $i \in N$;
  (b) $\liminf_{n\to\infty} \alpha_{n,i} > 0$, $i \in N$;
  (c) $\limsup_{n\to\infty} \alpha_{n,i} \leq \alpha_i \leq \frac{1}{\sqrt{1+L_i^2}+1}$, $i \in N$,

where $L_i$ is the Lipschitzian constant of $T_i$. The sequence $\{x_n\}$ is given by the following manner:

$$\begin{cases} x_0 = x \in C, \\ y_{n,i} = (1 - \alpha_{n,i})x_n + \alpha_{n,i}T_i x_n, \\ z_{n,i} = (1 - \beta_{n,i})x_n + \beta_{n,i}T_i y_{n,i}, \\ C_{n,i} = \{z \in C : \|z_{n,i} - z\|^2 \leq \|x_n - z\|^2 - \alpha_{n,i}\beta_{n,i}(1 - 2\alpha_{n,i} \\ \qquad - L_i^2\alpha_{n,i}^2)\|x_n - T_i x_n\|^2\}, C_n = \bigcap_{i=1}^\infty C_{n,i}, \\ Q_n = \{z \in C : \langle z - x_n, x - x_n\rangle \leq 0\}, \\ x_{n+1} = P_{C_n \cap Q_n}x, n = 0, 1, 2, \cdots, i \in N. \end{cases} \tag{13}$$

Then the sequence $\{x_n\}$ converges strongly to $P_F x$, where $P_F$ is the metric projection from H onto F.

**Proof:** We split the proof into nine steps. Following the proof lines of Theorem 1, we can also show which are right from **Step1** to **Step6**.

**Step7.** $\omega_\omega(x_n) \subset F$. In fact, since $\{x_n\}$ is a bounded set in H, we see that $\omega_\omega(x_n) \neq \phi$, consequently, there exists a subsequence $\{x_{n_k}\}$ of $\{x_n\}$ converging weakly to p for each $p \in \omega_\omega(x_n)$. By Step 6, we have $\|x_{n_k} - T_i x_{n_k}\| \to 0$, as $k \to \infty$. By Lemma 1, we get $p = T_i p$, and hence $p \in F$.

**Step8.** $\{x_n\}$ converges weakly to $v = P_F x$. By Lemma 1, we see that F is a nonempty closed convex subset of H, hence $v = P_F x$ is determined uniquely. Suppose that any subsequence $\{x_{n_k}\}$ of $\{x_n\}$ such that $x_{n_k} \rightharpoonup p \in F$. In view of the weak lower semi-continuity of the norm, we get

$$\|x - v\| \leq \|x - p\| \leq \liminf_{k \to \infty} \|x - x_{n_k}\| \leq \limsup_{k \to \infty} \|x - x_{n_k}\| \leq \|x - v\|.$$

Thus, we obtain $\|x - p\| = \|x - v\|$, this implies $p = v$. Therefore, $\{x_n\}$ converges weakly to $v = P_F x$.

**Step9.** $\{x_n\}$ converges strongly to $v = P_F x$. By Lemma 2, we have

$$\begin{aligned}
\|x_n - x_m\|^2 &= \|x_n - x + x - x_m\|^2 \\
&= 2\|x_n - x\|^2 + 2\|x - x_m\|^2 - \|x_n + x_m - 2x\|^2 \\
&= 2\|x_n - x\|^2 + 2\|x - x_m\|^2 - 4\|\frac{x_n + x_m}{2} - x\|^2.
\end{aligned}$$

By step 8, we have known that $\frac{x_n + x_m}{2}$ converges weakly to $v$ as m, n $\to \infty$. By the weak lower semi-continuity of the norm, we obtain

$$\liminf_{m,n \to \infty} \|\frac{x_n + x_m}{2} - x\|^2 \geq \|x - v\|^2.$$

Taking the superior limit on the both sides, we get

$$\limsup_{m,n \to \infty} \|x_n - x_m\|^2 \leq 2 \lim_{n \to \infty} \|x_n - x\|^2 + 2 \lim_{n \to \infty} \|x - x_m\|^2$$

$$- 4 \liminf_{m,n \to \infty} \|\frac{x_n + x_m}{2} - x\|^2 \leq 2d^2 + 2d^2 - 4\|v - x\|^2$$

$$\leq 4(d^2 - \|v - x\|^2) \leq 0.$$

Hence, $\{x_n\}$ is a cauchy sequence. It follows from step 7 in Theorem 1 that $x_n \to v$ as n $\to \infty$. This completes the proof.

**Remark 2.** Theorem 2 extends Theorem 3.6 in Zhou[6] to an infinitely countable family of Lipschitzian pseudo-contractions.

**Acknowledgements.** This work was supported by the National Science Foundation of China(11071169) and Zhejiang Province(Y6100696).

# References

1. Browder, F.E., Petryshyn, W.V.: Construction of fixed points of nonlinear mappings in Hilbert spaces. J. Math. Anal. Appl. 20, 197–228 (1967)
2. Mann, W.R.: Mean value methods in iterations. Proc. Amer. Math. Soc. 4, 506–510 (1953)
3. Marino, G., Xu, H.K.: Weak and strong convergence theorems for strict pseudo-contractions in Hilbert spaces. J. Math. Anal. Appl. 329(1), 336–346 (2007)
4. Reich, S.: Asymptotic behavior of contractions in Banach spaces. J. Math. Anal. Appl. 44, 57–70 (1973)
5. Rhoades, B.E.: Fixed point iterations using infinite matrices. Trans. Amer. Math. Soc. 196, 162–176 (1974)
6. Wang, Y.H., Zeng, L.C.: Convergence of generalized projective modified iteration methods in Banach spaces. Chinese Ann. Math. Ser. A 30(1), 55–62 (2009) (in Chinese)
7. Zhou, H.Y.: Convergence theorems of fixed points for Lipschitz pseudo-contractions in Hilbert spaces. J. Math. Anal. Appl. 343, 546–556 (2008)
8. Zhou, H.Y., Su, Y.F.: Strong convergence theorems for a family of quasi-asymptotic pseudo-contractions in Hilbert spaces. Nonlinear Anal. 70(11), 4047–4052 (2009)

# Existence of Fixed Points of Nonspreading Mappings with Bregman Distances

Fumiaki Kohsaka

**Abstract.** We propose a notion of nonspreading mapping with respect to a Bregman distance in a Banach space. We investigate the existence of a fixed point of such a mapping. The proposed class of mappings is related to zero point problems for monotone operators in Banach spaces.

**Keywords:** Banach space, Bregman distance, $D_g$-nonspreading mapping, firmly nonexpansive mapping, fixed point theorem.

## 1 Introduction

Many nonlinear problems in optimization and nonlinear analysis can be formulated as the problem of solving the inclusion $0 \in Au$ for a maximal monotone operator $A \colon X \to 2^{X^*}$ defined in a real Banach space $X$. Such a point $u$ is called a zero point of $A$ and the set of all zero points of $A$ is denoted by $A^{-1}0$.

The resolvent $T_A = (\nabla g + A)^{-1}\nabla g$ of the operator $A$ plays a central role in the study of approximation of a zero point of $A$, where $g \colon X \to (-\infty, +\infty]$ is a suitably chosen differentiable convex function. The mapping $T_A$ is $D_g$-firm in the sense of Bauschke, Borwein, and Combettes [1], where $D_g$ denotes the Bregman distance [3] with respect to $g$, and the fixed point set $F(T_A)$ of $T_A$ coincides with $A^{-1}0$.

In this paper, we propose a notion of $D_g$-nonspreading mapping; see formula (5). As we see in Section 2, every $D_g$-firm mapping is $D_g$-nonspreading. In the particular case when $g = \| \cdot \|^2/2$ in a smooth Banach space $X$, the notions of $D_g$-nonspreading mapping and $D_g$-firm mapping coincide

Fumiaki Kohsaka
Department of Computer Science and Intelligent Systems, Oita University,
Dannoharu, Oita-shi, Oita 870-1192, Japan
e-mail: f-kohsaka@oita-u.ac.jp

S. Li (Eds.): Nonlinear Maths for Uncertainty and its Appli., AISC 100, pp. 403–410.
springerlink.com © Springer-Verlag Berlin Heidelberg 2011

with the notions of nonspreading mapping [7] and firmly nonexpansive-type mapping [6], respectively. We study the existence of fixed points of $D_g$-nonspreading mappings in reflexive Banach spaces. The obtained results generalize the corresponding results obtained in [6,7,10].

## 2   Preliminaries

In what follows, unless otherwise specified, $X$ is a reflexive real Banach space and $C$ is a nonempty, closed, and convex subset of $X$. We denote the conjugate space of $X$ by $X^*$. For $x \in X$ and $x^* \in X^*$, $x^*(x)$ is also denoted by $\langle x, x^* \rangle$. The norms of $X$ and $X^*$ are denoted by $\| \cdot \|$. The closed ball with radius $r > 0$ centered at $0 \in X$ is denoted by $rB_X$. The interior and the closure of a subset $E$ of $X$ (with respect to the norm topology of $X$) are denoted by $\operatorname{Int} E$ and $\overline{E}$, respectively. The set of all real numbers and the set of all positive integers are denoted by $\mathbb{R}$ and $\mathbb{N}$, respectively. We also denote the set $\mathbb{R} \cup \{+\infty\}$ by $(-\infty, +\infty]$. The set of all proper, lower semicontinuous, and convex functions of $X$ into $(-\infty, +\infty]$ is denoted by $\Gamma(X)$. The fixed point set of a mapping $T \colon C \to X$ is denoted by $F(T)$.

Let $g \in \Gamma(X)$ be a function. The effective domain of $g$ is the set $D(g) = \{x \in X : g(x) \in \mathbb{R}\}$. The function $g$ is said to be strictly convex on a nonempty subset $E$ of $D(g)$ if $g(\alpha x + (1 - \alpha)y) < \alpha g(x) + (1 - \alpha)g(y)$ whenever $x, y \in E$, $x \neq y$, and $0 < \alpha < 1$. The function $g$ is said to be coercive (resp. strongly coercive) if $g(x_n) \to +\infty$ (resp. $g(x_n)/\|x_n\| \to +\infty$) whenever $\{x_n\}$ is a sequence of $X$ such that $\|x_n\| \to +\infty$. The function $g$ is said to be bounded on bounded sets if $g(rB_X)$ is bounded for all $r > 0$. In particular, this condition implies that $D(g) = X$. For $g \in \Gamma(X)$, we denote the subdifferential of $g$ and the conjugate function of $g$ by $\partial g$ and $g^*$, respectively. It is well-known that $g^* \in \Gamma(X^*)$ and $\partial g^* = (\partial g)^{-1}$, where the latter follows from the reflexivity of $X$.

Let $g \in \Gamma(X)$ be a function and let $x \in \operatorname{Int} D(g)$ be a point. The directional derivative of $g$ at $x$ in the direction $h \in X$ is defined by $d^+g(x)(h) = \lim_{t \downarrow 0}\big(g(x + th) - g(x)\big)/t$. The mapping $d^+g(x) \colon X \to \mathbb{R}$ is continuous and sublinear. The function $g$ is also said to be Gâteaux differentiable at $x$ if $d^+g(x)(-h) = -d^+g(x)(h)$ for all $h \in X$. In this case, $d^+g(x)$ is also denoted by $\nabla g(x)$. Under the assumptions on $g$ and $x$, this is equivalent to any one of the following assertions holds: $d^+g(x) \in X^*$; $\partial g(x)$ consists of one point. Thus $\partial g(x) = \nabla g(x)$ holds. The function $g$ is said to be Gâteaux differentiable on $\operatorname{Int} D(g)$ if it is Gâteaux differentiable at any $x \in \operatorname{Int} D(g)$. See [2,4,11] for more information on convex analysis. We need the following propositions:

**Proposition 1 (see [9]).** *Every coercive function in $\Gamma(X)$ has a minimizer over $X$.*

**Proposition 2 (see [2,11]).** *Let $g \in \Gamma(X)$ be a function. Then $g$ is strongly coercive if and only if $g^*$ is bounded on bounded sets.*

**Proposition 3.** *Let $g \in \Gamma(X)$ be a real valued and strongly coercive function which is Gâteaux differentiable and strictly convex on $X$. Then $\nabla g \colon X \to X^*$ is a bijection, $g^*$ is Gâteaux differentiable on $X^*$, and $\nabla g^* = (\nabla g)^{-1}$.*

*Proof.* By assumption, we have $\partial g(x) = \nabla g(x)$ for all $x \in X$. We first show that $\nabla g$ is onto. Let $x^* \in X^*$ be given. Since $X$ is reflexive and the function $f = g - x^* \in \Gamma(X)$ is strongly coercive, Proposition 1 implies that there exists $u \in X$ such that $f(u) = \inf f(X)$, and consequently, $\nabla g(u) = x^*$. Thus $\nabla g$ is surjective. The strict convexity of $g$ on $X$ implies that $\nabla g$ is injective; see [4]. By Proposition 2, we also know that $D(g^*) = X^*$. Since $X$ is reflexive, we have $\partial g^* = (\partial g)^{-1} = (\nabla g)^{-1}$. Since $\nabla g$ is a bijection, we know that $\partial g^*(x^*)$ consists of one point for all $x^* \in X^*$. Consequently, $g^*$ is Gâteaux differentiable on $X^*$ and $\nabla g^* = (\nabla g)^{-1}$. $\qquad\Box$

Let $g \in \Gamma(X)$ be a function such that $g$ is Gâteaux differentiable on $\operatorname{Int} D(g)$. The Bregman distance [3] $D_g \colon X \times \operatorname{Int} D(g) \to [0, +\infty]$ with respect to $g$ is defined by $D_g(y, x) = g(y) - \langle y - x, \nabla g(x) \rangle - g(x)$ for all $y \in X$ and $x \in \operatorname{Int} D(g)$. It is obvious that $D_g(y, x) \geq 0$ and $D_g(x, x) = 0$ for all $y \in X$ and $x \in \operatorname{Int} D(g)$. The following follows from the definition of $D_g$:

$$\langle x - y, \nabla g(z) - \nabla g(w) \rangle = D_g(x, w) + D_g(y, z) - D_g(x, z) - D_g(y, w) \quad (1)$$

for all $x, y \in D(g)$ and $z, w \in \operatorname{Int} D(g)$. In particular,

$$\langle x - z, \nabla g(z) - \nabla g(y) \rangle = D_g(x, y) - D_g(x, z) - D_g(z, y) \quad (2)$$

holds for all $x \in D(g)$ and $y, z \in \operatorname{Int} D(g)$. If $g$ is also strictly convex on $\operatorname{Int} D(g)$, then $y = x$ whenever $y, x \in \operatorname{Int} D(g)$ and $D_g(y, x) = 0$. See [2, 4] for more information on Bregman distance.

Let $X$ be a smooth, strictly convex, and reflexive Banach space, let $p$ be a real number such that $1 < p < +\infty$, and let $g = \| \cdot \|^p / p$. Then it obviously holds that $g \in \Gamma(X)$, $D(g) = X$, $g$ is strongly coercive, and bounded on bounded sets. We also know that $g$ is Gâteaux differentiable and strictly convex on $X$; see [2, 11]. In this case, $D_g \colon X \times X \to [0, +\infty)$ is given by

$$D_g(y, x) = \frac{1}{p} \left\{ \|y\|^p - p\langle y, J_p x \rangle + (p - 1)\|x\|^p \right\} \quad (3)$$

for all $y, x \in X$, where $J_p$ denotes the duality mapping of $X$ into $X^*$ with respect to the weight function $\omega(t) = t^{p-1}$ given by

$$J_p x = \left\{ x^* \in X^* : \langle x, x^* \rangle = \|x\| \|x^*\|, \ \|x^*\| = \|x\|^{p-1} \right\} \quad (4)$$

for all $x \in X$.

Let $E$ be a nonempty subset of $X$ and let $g \in \Gamma(X)$ be a function such that $g$ is Gâteaux differentiable on $\operatorname{Int} D(g)$ and $E \subset \operatorname{Int} D(g)$. Then we say that a mapping $T \colon E \to \operatorname{Int} D(g)$ is $D_g$-nonspreading (or nonspreading with respect to $D_g$) if

$$D_g(Tx, Ty) + D_g(Ty, Tx) \leq D_g(Tx, y) + D_g(Ty, x) \qquad (5)$$

for all $x, y \in E$. A mapping $T: E \to \operatorname{Int} D(g)$ is also said to be $D_g$-firm (or firm with respect to $D_g$) [1] if

$$\langle Tx - Ty, \nabla g(Tx) - \nabla g(Ty) \rangle \leq \langle Tx - Ty, \nabla g(x) - \nabla g(y) \rangle$$

for all $x, y \in E$. It follows from (1) that every $D_g$-firm mapping is $D_g$-nonspreading. In fact, $T: E \to \operatorname{Int} D(g)$ is $D_g$-firm if and only if

$$D_g(Tx, Ty) + D_g(Ty, Tx) \leq D_g(Tx, y) + D_g(Ty, x)$$
$$- D_g(Tx, x) - D_g(Ty, y)$$

for all $x, y \in E$. If $g = \| \cdot \|^2/2$ and $X$ is a Hilbert space, then $T$ is $D_g$-firm if and only if it is firmly nonexpansive in the classical sense, i.e., $\|Tx - Ty\|^2 \leq \langle Tx - Ty, x - y \rangle$ for all $x, y \in E$.

Let $g \in \Gamma(X)$ be a strongly coercive function such that $g$ is Gâteaux differentiable and strictly convex on $\operatorname{Int} D(g)$ and $C \subset \operatorname{Int} D(g)$. Fix $x \in \operatorname{Int} D(g)$. By Proposition 1 and the strict convexity of $g$, there exists a unique $y_x \in C$ such that $D_g(y_x, x) = \inf_{y \in C} D_g(y, x)$. The mapping $\Pi_C^g: \operatorname{Int} D(g) \to C$ defined by $\Pi_C^g x = y_x$ for all $x \in \operatorname{Int} D(g)$ is called the Bregman projection of $\operatorname{Int} D(g)$ onto $C$. It is known that if $x \in \operatorname{Int} D(g)$ and $z \in C$, then

$$z = \Pi_C^g x \iff \sup_{y \in C} \langle y - z, \nabla g(x) - \nabla g(z) \rangle \leq 0. \qquad (6)$$

It is also known [1] that $\Pi_C^g$ is $D_g$-firm and $F(\Pi_C^g) = C$.

An operator $A: X \to 2^{X^*}$ is said to be monotone if $\langle x - y, x^* - y^* \rangle \geq 0$ whenever $x^* \in Ax$ and $y^* \in Ay$. The domain and the range of $A$ are defined by $D(A) = \{x \in X : Ax \neq \emptyset\}$ and $R(A) = \bigcup_{x \in X} Ax$, respectively. Let $g \in \Gamma(X)$ be a function such that $g$ is Gâteaux differentiable and strictly convex on $\operatorname{Int} D(g)$ and $C \subset \operatorname{Int} D(g)$. Suppose that $A: X \to 2^{X^*}$ is a monotone operator such that $D(A) \subset C \subset (\nabla g)^{-1} R(\nabla g + A)$. Then for each $x \in C$, there exists a unique $z_x \in D(A)$ such that $\nabla g(x) \in \nabla g(z_x) + Az_x$. The mapping $T_A: C \to C$ defined by $T_A x = z_x$ for all $x \in C$ is called the resolvent of $A$ with respect to $g$. In other words, $T_A x = (\nabla g + A)^{-1} \nabla g(x)$ for all $x \in C$. It is known [1] that $T_A$ is $D_g$-firm and $F(T_A) = A^{-1} 0$. If we additionally assume that $D(g) = X = C$, $g$ is strongly coercive, bounded on bounded sets, and $A$ is maximal monotone, then $R(\nabla g + A) = X^*$; see [5].

The following proposition shows that there exists a discontinuous $D_g$-nonspreading mapping:

**Proposition 4.** *Let $X$ be a smooth, strictly convex, and reflexive Banach space, let $p$ be a real number such that $1 < p < +\infty$, and let $g$ be the function defined by $g = \| \cdot \|^p/p$. Let $S_1, S_2: X \to X$ be $D_g$-firm mappings such that $S_1(X)$ and $S_2(X)$ are contained by $rB_X$ for some $r > 0$, let $\delta$ be a positive real number with $\delta^{p-1}\{(p-1)\delta - pr\}/p \geq 4r^p$, and let $T: X \to X$ be the mapping* *defined by $Tx = S_1x$ if $x \in \delta B_X$; $S_2x$ otherwise. Then $T$ is $D_g$-nonspreading. Further, if $S_1(X)$ and $\overline{S_2(X)}$ are disjoint, then $T$ is discontinuous at any point $z \in X$ with $\|z\| = \delta$ with respect to the norm topology of $X$*

*Proof.* Let $x, y \in X$ be given. If either $x, y \in \delta B_X$ or $x, y \notin \delta B_X$, then (5) obviously holds. Suppose that $x \in \delta B_X$ and $y \notin \delta B_X$. Then we have from (3) and (4) that

$$D_g(Tx, Ty) + D_g(Ty, Tx) \leq \frac{2}{p}\{r^p + pr^p + (p-1)r^p\} = 4r^p.$$

On the other hand, we have

$$D_g(Tx, y) + D_g(Ty, x) \geq D_g(Tx, y)$$
$$\geq \frac{1}{p}\{\|S_1x\|^p - p\|S_1x\|\|y\|^{p-1} + (p-1)\|y\|^p\}$$
$$\geq \frac{\|y\|^{p-1}}{p}\{(p-1)\|y\| - p\|S_1x\|\} \geq \frac{\delta^{p-1}}{p}\{(p-1)\delta - pr\}.$$

Thus (5) holds and hence $T$ is $D_g$-nonspreading.

Suppose that $S_1(X) \cap \overline{S_2(X)} = \emptyset$ and fix $z \in X$ with $\|z\| = \delta$. Set $z_n = (1 + 1/n)z$ for all $n \in \mathbb{N}$. Note that $Tz = S_1z \in S_1(X)$ and $Tz_n = S_2z_n \in S_2(X)$ for all $n \in \mathbb{N}$. By assumption, $\{Tz_n\}$ does not converge strongly to $Tz$. Therefore $T$ is not norm-to-norm continuous at $z$. □

Proposition 4 immediately implies the following corollary:

**Corollary 1.** *Let $C_1$ and $C_2$ be nonempty, closed, and convex subsets of a smooth, strictly convex, and reflexive Banach space $X$ such that $C_1$ and $C_2$ are contained by $rB_X$ for some $r > 0$, let $S_i = \Pi_{C_i}^g$ for $i \in \{1, 2\}$, and let $p$, $g$, $\delta$, and $T$ be the same as in Proposition 4. Then $T$ is $D_g$-nonspreading. Further, if $C_1$ and $C_2$ are disjoint, then $T$ is discontinuous at any point $z \in X$ with $\|z\| = \delta$ with respect to the norm topology of $X$*


## 3 Existence of a Fixed Point

In this section, we obtain a fixed point theorem for a nonspreading mapping with respect to a Bregman distance. The following proposition directly follows from (2) and (5):

**Proposition 5.** *Let $E$ be a nonempty subset of $X$, let $g \in \Gamma(X)$ be a function such that $g$ is Gâteaux differentiable on $\text{Int } D(g)$ and $E \subset \text{Int } D(g)$, and let $T: E \to \text{Int } D(g)$ be a mapping. Then $T$ is $D_g$-nonspreading if and only if*

$$0 \leq D_g(Ty, y) + D_g(Ty, x) - D_g(Ty, Tx) + \langle Tx - Ty, \nabla g(Ty) - \nabla g(y)\rangle$$

*for all $x, y \in E$.*

Motivated by the techniques in [6, 7, 8], we show the following fixed point theorem:

**Theorem 1.** *Let $g \in \Gamma(X)$ be a function such that $g$ is Gâteaux differentiable and strictly convex on $\text{Int}\, D(g)$ and $C \subset \text{Int}\, D(g)$, let $T \colon C \to C$ be a $D_g$-nonspreading mapping, and let $\{U_n^T\}$ be the sequence of mappings defined by $U_n^T = \sum_{k=1}^n T^k/n$ for each $n \in \mathbb{N}$. Then the following assertions are equivalent:*

(a)   *$T$ has a fixed point;*
(b)   *$\{T^n x\}$ is bounded for some $x \in C$;*
(c)   *$\{U_n^T x\}$ has a bounded subsequence for some $x \in C$.*

*Proof.* If $T$ has a fixed point, then $\{T^n p\}$ is bounded for all $p \in F(T)$. Thus (a) implies (b). It is also clear that (b) implies (c).

Thus it remains to be seen if (c) implies (a). To see this, we suppose that there exists $x \in C$ such that $\{U_n^T x\}$ has a bounded subsequence $\{U_{n_i}^T x\}$. Set $z_n = U_n^T x$ for all $n \in \mathbb{N}$. Since $\{z_{n_i}\}$ is bounded and $X$ is reflexive, there exists a subsequence $\{z_{n_{i_j}}\}$ of $\{z_{n_i}\}$ which converges weakly to some $u \in X$. Since $C$ is closed and convex, we have $u \in C$. According to Proposition 5, the following inequality

$$
0 \leq D_g(Tu, u) + D_g(Tu, T^k x) - D_g(Tu, T^{k+1} x) \\
+ \langle T^{k+1} x - Tu, \nabla g(Tu) - \nabla g(u) \rangle \tag{7}
$$

holds whenever $k \in \mathbb{N} \cup \{0\}$, where $T^0$ denotes the identity mapping on $C$. Multiplying $1/n$ by the inequality obtained after summing up (7) with respect to $k \in \{0, 1, \ldots, n-1\}$, we have

$$
0 \leq D_g(Tu, u) + \frac{1}{n}\{D_g(Tu, x) - D_g(Tu, T^n x)\} \\
+ \langle z_n - Tu, \nabla g(Tu) - \nabla g(u) \rangle
$$

for all $n \in \mathbb{N}$. This gives us that

$$
0 \leq D_g(Tu, u) + \frac{1}{n_{i_j}} D_g(Tu, x) + \left\langle z_{n_{i_j}} - Tu, \nabla g(Tu) - \nabla g(u) \right\rangle \tag{8}
$$

for all $j \in \mathbb{N}$. Letting $j \to +\infty$ in (8), we obtain

$$
0 \leq D_g(Tu, u) + \langle u - Tu, \nabla g(Tu) - \nabla g(u) \rangle = -D_g(u, Tu),
$$

where the last equality follows from (1). Accordingly, we have $D_g(u, Tu) = 0$. The strict convexity of $g$ on $C$ implies that $u$ is a fixed point of $T$. $\qquad\square$

Theorem 1 immediately implies the following corollaries:

**Corollary 2.** *Let $g$, $T$, and $\{U_n^T\}$ be the same as in Theorem 1. Then the following assertions hold:*

(a)   *If $C$ is bounded, then $T$ has a fixed point;*
(b)   *$T$ has no fixed point if and only if $\|U_n^T x\| \to +\infty$ for all $x \in C$.*

**Corollary 3.** *Let $C$ be a nonempty, closed, and convex subset of a smooth, strictly convex, and reflexive Banach space $X$, let $p$ be a real number such that $1 < p < +\infty$, and let $g$ be the function defined by $g = \| \cdot \|^p/p$. Let $T: C \to C$ be a $D_g$-nonspreading mapping and let $\{U_n^T\}$ be the sequence of mappings defined by $U_n^T = \sum_{k=1}^{n} T^k/n$ for each $n \in \mathbb{N}$. Then the following assertions are equivalent: $T$ has a fixed point; $\{T^n x\}$ is bounded for some $x \in C$; $\{U_n^T x\}$ has a bounded subsequence for some $x \in C$.*

*Remark 1.* Corollary 3 is a generalization of the corresponding result in [7], where the equivalence between the first two assertions was shown for $p = 2$.

# 4   Characterization of the Existence of a Fixed Point

Inspired by the discussion in [10], we show the following theorem:

**Theorem 2.** *Let $g \in \Gamma(X)$ be a real valued and strongly coercive function such that $g$ is Gâteaux differentiable and strictly convex on $X$. Then the following assertions are equivalent:*

(a)   *$C$ is bounded;*
(b)   *every $D_g$-nonspreading self mapping on $C$ has a fixed point;*
(c)   *every $D_g$-firm self mapping on $C$ has a fixed point.*

*Proof.* According to Corollary 2, we know that (a) implies (b). Since every $D_g$-firm mapping on $C$ is also $D_g$-nonspreading, (b) clearly implies (c).

We show that (c) implies (a). Suppose that $C$ is not bounded. Then the uniform boundedness theorem ensures the existence of $x^* \in X^*$ such that $\inf x^*(C) = -\infty$. Proposition 3 ensures that $\nabla g: X \to X^*$ is a bijection, $g^*$ is Gâteaux differentiable on $X^*$, and $\nabla g^* = (\nabla g)^{-1}$. Let $S: C \to X$ and $T: C \to C$ be the mappings defined by $Sx = \nabla g^*(\nabla g(x) - x^*)$ and $Tx = \Pi_C^g Sx$ for all $x \in C$. Using the fact that $\Pi_C^g: X \to X$ is $D_g$-firm, we can show that $T$ is $D_g$-firm. We next show that $T$ has no fixed point. If $u \in C$, then the choice of $x^*$ implies the existence of $y \in C$ such that $x^*(y) < x^*(u)$. Consequently, we obtain $\langle y - u, \nabla g(Su) - \nabla g(u) \rangle = \langle u - y, x^* \rangle > 0$. Thus (6) yields that $\Pi_C^g(Su) \neq u$. Consequently, $T$ has no fixed point. □

**Corollary 4.** *Let $C$ be a nonempty, closed, and convex subset of a smooth, strictly convex, and reflexive Banach space $X$, let $p$ be a real number such that $1 < p < +\infty$, and let $g$ be the function defined by $g = \| \cdot \|^p/p$. Then the following assertions are equivalent: $C$ is bounded; every $D_g$-nonspreading self mapping on $C$ has a fixed point; every $D_g$-firm self mapping on $C$ has a fixed point.*

*Remark 2.* In the case when $p = 2$, Corollary 4 is reduced to the corresponding result in [10].

# References

1. Bauschke, H.H., Borwein, J.M., Combettes, P.L.: Bregman monotone optimization algorithms. SIAM J. Control Optim. 42, 596–636 (2003)
2. Borwein, J.M., Vanderwerff, J.D.: Convex Functions: Constructions, Characterizations and Counterexamples. Cambridge University Press, Cambridge (2010)
3. Bregman, L.M.: The relaxation method of finding the common point of convex sets and its application to the solution of problems in convex programming. USSR Comput. Math. Math. Phys. 7, 200–217 (1967)
4. Butnariu, D., Iusem, A.N.: Totally Convex Functions for Fixed Points Computation and Infinite Dimensional Optimization. Kluwer Academic Publishers, Dordrecht (2000)
5. Kohsaka, F., Takahashi, W.: Proximal point algorithms with Bregman functions in Banach spaces. J. Nonlinear Convex Anal. 6, 505–523 (2005)
6. Kohsaka, F., Takahashi, W.: Existence and approximation of fixed points of firmly nonexpansive-type mappings in Banach spaces. SIAM J. Optim. 19, 824–835 (2008)
7. Kohsaka, F., Takahashi, W.: Fixed point theorems for a class of nonlinear mappings related to maximal monotone operators in Banach spaces. Arch. Math (Basel) 91, 166–177 (2008)
8. Takahashi, W.: A nonlinear ergodic theorem for an amenable semigroup of nonexpansive mappings in a Hilbert space. Proc. Amer. Math. Soc. 81, 253–256 (1981)
9. Takahashi, W.: Nonlinear Functional Analysis. Yokohama Publishers, Yokohama (2000)
10. Takahashi, W., Yao, J.C., Kohsaka, F.: The fixed point property and unbounded sets in Banach spaces. Taiwanese J. Math. 14, 733–742 (2010)
11. Zălinescu, C.: Convex Analysis in General Vector Spaces. World Scientific Publishing Co. Inc., River Edge (2002)

# The $(h, \varphi)-$Generalized Second-Order Directional Derivative in Banach Space

Wenjuan Chen, Caozong Cheng, and Caiyun Jin

**Abstract.** In this paper, we introduce the conception of $(h, \varphi)-$generalized second-order directional derivatives of real function in Banach space and discuss some of their properties and their relationships. In order to define above conceptions, we generalize the Ben-Tal's generalized algebraic operations in Euclidean space to Banach space.

**Keywords:** $(h, \varphi)-$generalized second-order directional derivatives, Ben-Tal's generalized algebraic operations, Banach space.

## 1 Introduction

In 1977, Ben-Tal[1] introduced the generalized algebraic operations, and they have been applied in optimization widely. In 2001, Q. X. Zhang[2] introduced $(h, \varphi)-$generalized directional derivatives and $(h, \varphi)-$generalized gradient of function in $R^n$ which involve generalized algebraic operations. In 2006, Y. H. Xu[3] introduced $(h, \varphi)-$Lipschitz function in $R^n$, and then defined $(h, \varphi)-$directional derivative of $(h, \varphi)-$Lipschitz function in $R^n$ and $(h, \varphi)-$generalized gradient at $x$. However, Ben-Tal's generalized algebraic operations define on Euclidean space $R^n$, that limits us in Euclidean space $R^n$ on the research of $(h, \varphi)-$generalized directional derivative and $(h, \varphi)-$generalized gradient. For break the limitation, we generalize generalized algebraic

Wenjuan Chen
College of Applied Sciences, Beijing University of Technology
No.6 Middle School Linyi
e-mail: wenjuan0829@sina.com

Caozong Cheng and Caiyun Jin
College of Applied Sciences, Beijing University of Technology
e-mail: czcheng@bjut.edu.cn, jincaiyun@bjut.edu.cn

S. Li (Eds.): Nonlinear Maths for Uncertainty and its Appli., AISC 100, pp. 411–418.
springerlink.com          © Springer-Verlag Berlin Heidelberg 2011

operations to Banach space. On the basis of above work, we introduce the $(h, \varphi)$−generalized second-order directional derivatives of real function in Banach space and discuss some properties. At last, we discuss the relations between $(h, \varphi)$− generalized Hessian and $(h, \varphi)$−generalized second-order directional derivatives in Hilbert space.

## 2  Generalized Algebraic Operations in Banach

According to the Ben-Tal generalized addition and scalar multiplication in [1], we define Ben-Tal generalized addition and scalar multiplication in Banach space $X$ as follows:

1) Let $H$ be a subset of $X$ and $K$ be a complex domain. Let $h : H \to R^n$ be a continuous function and its inverse function $h^{-1}$ exists. $h$−vector addition and $h$−scalar multiplication are definited, respectively

$$x \oplus y = h^{-1}(h(x) + h(y)), \ \forall \ x, \ y \in H \ ;$$
$$\lambda \odot x = h^{-1}(\lambda h(x)), \ \forall \ x \in H, \lambda \in K.$$

Especially, for convenient, we write generalized addition and scalar multiplication in $R$ as follows:

2) Let $\varphi$ be a continuous real valued function of $\Phi \subset R$, and has inverse function $\varphi^{-1}$, $\varphi$−addition is definited as $\alpha[+]\beta = \varphi^{-1}(\varphi(\alpha) + \varphi(\beta))$, $\forall \alpha, \beta \in \Phi$; and $\varphi$− scalar multiplication is definite as $\lambda[\cdot]\alpha = \varphi^{-1}(\lambda\varphi(\alpha))$, $\forall \ \alpha \in \Phi, \ \lambda \in R$.

3) Let $X$ be an inner product space with inner $(\cdot, \cdot)$, $(h, \varphi)$−inner product is definited as $(x, y)_{h,\varphi} = \varphi^{-1}(h(x), h(y))$, $\forall \ x, y \in H$.

Following from the above definitions, generalized subtraction in Banach space and $R$ are definite as $x \ominus y = x \oplus ((-1) \odot y) = h^{-1}(h(x) - h(y))$; $\alpha[-]\beta = \alpha[+]((-1)[\cdot]\beta) = \varphi^{-1}(\varphi(\alpha) - \varphi(\beta))$, respectively.

In the following sections, we always suppose that $h : X \to X$ is 1-1 onto continuous function; $\varphi : R \to R$ is 1-1 onto monotone increasing function. If $f$ is real function on $X$, we denote $\hat{f}(t) = \varphi(f(h^{-1}(t))) = \varphi f h^{-1}(t)$.

## 3  $(h, \varphi)$−Generalized Second-Order Directional Derivatives

In this section, we will introduce some $(h, \varphi)$−generalized second-order directional derivatives in real Banach space and discuss their properties. For convenience, we recall the definition of generalized first-order directional derivatives and generalized second-order directional derivatives.

Let $X$ be a real Banach space and $f : X \to R$ . Generalized directional derivative[4] of function $f$ at $x$ in direction $u \in X$ and generalized second-order directional derivative[5] of function $f$ at $x$ in direction $(u, v) \in X \times X$ are, respectively

$$f^0(x; u) = \limsup_{\substack{y \to x \\ t \downarrow 0}} \frac{f(y + tu) - f(y)}{t},$$

$$f^{00}(x; u, v) = \limsup_{\substack{y \to x \\ s, t \downarrow 0}} \frac{1}{st} \Big\{ f(y + su + tv) - f(y + su) - f(y + tv) + f(y) \Big\}.$$

Next, we will use generalized algebraic operations on Banach space generalize the above definitions.

**Definition 1.** Let $X$ be a real Banach space and $f : X \to R$ . $(h, \varphi)$–generalized directional derivative of function $f$ at $x$ in direction $u \in X$ and $(h, \varphi)$–generalized second-order directional derivatives in direction $(u, v) \in X \times X$ are, respectively

$$f^*(x; u) = \limsup_{\substack{y \to x \\ t \downarrow 0}} \frac{1}{t} [\cdot] \{ f(y \oplus t \odot u)[-]f(y) \} ,$$

$f^{**}(x; u, v)$

$$= \limsup_{\substack{y \to x \\ s, t \downarrow 0}} \frac{1}{st} [\cdot] \{ f(y \oplus s \odot u \oplus t \odot v)[-]f(y \oplus s \odot u)[-]f(y \oplus t \odot v)[+]f(y) \} .$$

It is easy to verify that lemma 3.1 in [3] also hold in Banach space, and according to the monotonicity of $\varphi^{-1}$, we can rewritten it as follow.

**Lemma 1.** Let $g$ be a real valued function of Banach space $X$ and $x_0 \in X$, then

$$\limsup_{x \to x_0} \varphi^{-1} \Big( g(x) \Big) = \varphi^{-1} \Big( \limsup_{x \to x_0} g(x) \Big).$$

On the lines of proposition 3.1 of [3], we can prove that $(h, \varphi)$–generalized directional derivative has the following relation with original generalized directional derivative.

**Proposition 1.** Let $X$ be a real Banach space and $f : X \to R$, then $\forall x, u \in X$,

$$f^*(x; u) = \varphi^{-1} \Big( \hat{f}^0 \Big( h(x); h(u) \Big) \Big).$$

For the relationship of $(h, \varphi)$–generalized second-order directional derivative and original generalized directional derivative, we have:

**Proposition 2.** *Let* $X$ *be a real Banach space and* $f : X \to R$, *then* $\forall (x, u, v) \in X \times X \times X, \forall x, u \in X,$

$$f^{**}(x; u, v) = \varphi^{-1}\left( \hat{f}^{00}\left(h(x); h(u), h(v)\right)\right).$$

*Proof.* From $\lambda[\cdot]\alpha = \varphi^{-1}(\lambda\varphi(\alpha))$, we have that $\forall (x, u, v) \in X \times X \times X,$

$$f^{**}(x; u, v)$$

$$= \limsup_{\substack{y \to x \\ s,t\downarrow 0}} \varphi^{-1}\left\{ \frac{1}{st}\varphi\left(f(y\oplus s\odot u\oplus t\odot v)[-]f(y\oplus s\odot u)[-]f(y\oplus t\odot v)[+]f(y)\right)\right\}.$$

Furthermore, following from the definition of Ben-Tal's generalized algebraic operations that

$$f^{**}(x; u, v) = \limsup_{\substack{y \to x \\ s,t\downarrow 0}} \varphi^{-1}\left\{ \frac{1}{st}\varphi\left(\varphi^{-1}\left(\varphi f(y \oplus s \odot u \oplus t \odot v) - \varphi f(y \oplus s \odot u)\right.\right.\right.$$

$$\left.\left.\left. -\varphi f(y \oplus t \odot v) + \varphi f(y))\right)\right)\right\}$$

$$= \limsup_{\substack{y \to x \\ s,t\downarrow 0}} \varphi^{-1}\left\{ \frac{1}{st}\left(\varphi f(y \oplus s \odot u \oplus t \odot v) - \varphi f(y \oplus s \odot u)\right.\right.$$

$$\left.\left. -\varphi f(y \oplus t \odot v) + \varphi f(y))\right)\right\}$$

$$= \limsup_{\substack{y \to x \\ s,t\downarrow 0}} \varphi^{-1}\left\{ \frac{1}{st}\left(\varphi f h^{-1}(h(y) + sh(u) + th(v))\right.\right.$$

$$\left. -\varphi f h^{-1}(h(y) + sh(u)) - \varphi f h^{-1}(h(y) + th(v)) + \varphi f h^{-1}(h(y)))\right\}.$$

Let $z = h(y), \hat{f} = \varphi f h^{-1}$, then following from Lemma 3.1, that

$$f^{**}(x; u, v) = \varphi^{-1}\left( \hat{f}^{00}\left(h(x); h(u), h(v)\right)\right).$$

Which completes the proof.

## 4   Propositions of $(h, \varphi)$−Generalized Second-Order Directional Derivative

**Definition 2.** *Let* $X$ *be a real Banach space and* $Y$ *be a subset of* $X$, $f : Y \to R$ *is said to be*

1) $(h, \varphi)-$ positive homogeneous $\Leftrightarrow f(t \odot x) = t[\cdot]f(x), \forall t > 0, \forall x \in Y;$

2) $(h, \varphi)-$ subadditivity $\Leftrightarrow f(x \oplus y) \leq f(x)[+]f(y), \forall x, y \in Y;$

3) $(h, \varphi)-$ sublinear $\Leftrightarrow f$ is both $(h, \varphi)-$ positive homogeneous and $(h, \varphi)-$sub-additivity.

**Theorem 1.** *Let$X$ be a real Banach space and $f : X \to R$ . Then
1) $\forall$ $(x, u, v) \in X \times X \times X$, mapping $(u, v) \to f^{**}(x; u, v)$ is symmetrical and $(h, \varphi)-$bisublinear, $((h, \varphi)-$ sublinear in $u, v$, respectively);
2) mapping $x \to f^{**}(x; u, v)$ is upper semicontinuous, $\forall$ $(u, v) \in X \times X$;
3) $f^{**}(x; u, (-1) \odot v) = f^{**}(x; (-1) \odot u, v) = ([-]f)^{**}(x; u, v)$, $\forall$ $(x, u, v) \in X \times X \times X$.*

**Proof:** 1) Since $\hat{f} : X \to R$ , $h(u) \in X$, $h(v) \in X$, $h(x) \in X$, from [5], we know that $\hat{f}^{00}(h(x); h(u), h(v))$ is symmetrical in $(h(u), h(v))$, then

$$f^{**}(x; u, v) = \varphi^{-1}\left( \hat{f}^{00}\left( h(x); h(u), h(v) \right) \right)$$

$$= \varphi^{-1}\left( \hat{f}^{00}\left( h(x); h(v), h(u) \right) \right) = f^{**}(x; v, u)$$

That is, $f^{**}(x; u, v)$ is symmetrical in $(u, v) \in X \times X$.

Next, we only proof that $f^{**}(x; \cdot, v)$ is $(h, \varphi)-$positive homogeneous and $(h, \varphi)-$subadditivity,

$$f^{**}(x; \lambda \odot u, v) = \varphi^{-1}(\hat{f}^{00}(h(x); h(\lambda \odot u), h(v))) = \varphi^{-1}(f^{00}(h(x); \lambda h(u), h(v))) ,$$

furthermore, by proposition 1.2 of [5], we know that

$$f^{**}(x; \lambda \odot u, v) = \varphi^{-1}\left( \lambda \hat{f}^{00}\left( h(x); h(u), h(v) \right) \right)$$

$$= \varphi^{-1}\left( \lambda \varphi f^{**}(x; u, v) \right) = \lambda[\cdot] f^{**}(x; u, v).$$

Which implies that $f^{**}(x; \cdot, v)$ is $(h, \varphi)-$positive homogeneous.

Next, we prove that $f^{**}(x; \cdot, v)$ is $(h, \varphi)-$subadditivity. In fact, $\forall$ $u, w \in X$,

$$f^{**}(x; u \oplus w, v) = \varphi^{-1}\left( \hat{f}^{00}\left( h(x); h(u \oplus w), h(v) \right) \right)$$

$$= \varphi^{-1}\left( \hat{f}^{00}\left( h(x); h(u) + h(w), h(v) \right) \right).$$

Following from proposition 1.2 of [5], we have that

$$\hat{f}^{00}\left( h(x); h(u) + h(w), h(v) \right) \leq \hat{f}^{00}\left( h(x); h(u), h(v) \right) + \hat{f}^{00}\left( h(x); h(w), h(v) \right).$$

Since $\varphi^{-1}$ is monotonically increasing,

$$f^{**}(x; u \oplus w, v) \leq \varphi^{-1}\left( \hat{f}^{00}\left( h(x); h(u), h(v) \right) + \hat{f}^{00}\left( h(x); h(w), h(v) \right) \right)$$

$$= \varphi^{-1}\Big(\varphi f^{**}(x;u,v) + \varphi f^{**}(x;w,v)\Big) = f^{**}(x;u,v)[+]f^{**}(x;w,v).$$

2) Let $\{x_i\}_{i=1}^{\infty} \subset X$ be a sequence which converges to $x$. Then by $h$ is continuous, we have that $h(x_i) \to h(x)$. From proposition 1.2 of [5], we have that $\forall\, u,v \in X$, $h(x) \to \hat{f}^{00}(h(x);h(u),h(v))$ is upper semicontinuous, that is

$$\limsup_{i\to\infty} \hat{f}^{00}\Big(h(x_i);h(u),h(v)\Big) \le \hat{f}^{00}\Big(h(x);h(u),h(v)\Big).$$

Furthermore by $\varphi^{-1}$ is monotonically increasing, we have that,

$$\varphi^{-1}(\limsup_{i\to\infty} \hat{f}^{00}\Big(h(x_i);h(u),h(v)\Big) \le \varphi^{-1}\Big(\hat{f}^{00}\Big(h(x);h(u),h(v)\Big)\Big).$$

From Lemma 3.1, we can get that

$$\limsup_{i\to\infty} \varphi^{-1}\Big(\hat{f}^{00}\Big(h(x_i);h(u),h(v)\Big)\Big) \le \varphi^{-1}\Big(\hat{f}^{00}\Big(h(x);h(u),h(v)\Big)\Big),$$

that is,

$$\limsup_{i\to\infty} f^{**}(x_i;u,v) \le f^{**}(x;u,v).$$

Then $x \to f^{**}(x;u,v)$ is upper semicontinuous.

3) Following from Proposition 3.2 that $\forall\, (x,u,v) \in X \times X \times X$,

$$f^{**}(x;u,(-1)\odot v) = \varphi^{-1}\Big(\hat{f}^{00}\Big(h(x);h(u),h((-1)\odot v)\Big)\Big)$$

$$= \varphi^{-1}\Big(\hat{f}^{00}\Big(h(x);h(u),-h(v)\Big)\Big).$$

Furthermore by proposition 1.2 of [5], we have

$$f^{**}(x;u,(-1)\odot v) = \varphi^{-1}\Big((-\hat{f})^{00}\Big(h(x);h(u),h(v)\Big)\Big).$$

On the other hand

$$([-]f)^{**}(x;u,v) = \varphi^{-1}\Big((-\hat{f})^{00}\Big(h(x);h(u),h(v)\Big)\Big).$$

Which implies that

$$f^{**}(x;u,(-1)\odot v) = ([-]f)^{**}(x;u,v).$$

In the same way, we can prove that $f^{**}(x;(-1)\odot u,v) = ([-]f)^{**}(x;u,v)$. Then 3)holds.

**Definition 3.** Function $f : X \to R$ is $(h,\varphi)-$twice $C-$differentiable at $x$, if $f^{**}(x;u,\cdot)$ is lower semicontinuous , $\forall\, u \in X$. If $\forall\, x \in X$, $f^{**}(x;u,\cdot)$

is $(h, \varphi)$—twice $C$—differentiable at $x$, then we called $f$ is $(h, \varphi)$—twice $C$—differentiable on $X$.

**Theorem 2.** *Function $f : X \to R$ is $(h, \varphi)$—convex, if $\forall\ x_1, x_2 \in X$, $\forall\ \lambda \in [0, 1]$, we have*

$$f((1 - \lambda) \odot x_1 \oplus \lambda \odot x_2) \leq (1 - \lambda)[\cdot]f(x_1)[+]\lambda[\cdot]f(x_2).$$

*Remark 1.* If $f$ twice $C$—differentiable at $x$ ($\forall\ u \in X$, $f^{00}(x; u, \cdot)$ is lower semicontinuous), then $f$ is $(h, \varphi)$—twice $C$—differentiable at $x$. Since we just take $h(x) = x, x \in X, \varphi(t) = t, t \in R$. But the converse of this proposition is not holds. For example, $\forall\ x \in R$, let $h(x) = x, f(x) = x^{\frac{4}{3}}, \varphi(x) = x^{\frac{3}{4}}$, by [5], we know that $f$ isn't twice $C$—differentiable at $x = 0$, and it is easy to verify that $f$ is $(h, \varphi)$—twice $C$—differentiable at$x = 0$.

**Theorem 3.** *Let $f : X \to R$ be $(h, \varphi)$—convex, $(h, \varphi)$—twice $C$—differentiable on $X$, and $\varphi^{-1}(0) \geq 0$. Then $f^{**}(x; u, u) \geq 0$, $\forall\ x, u \in X$.*

*Proof.* From Corollary 4.3 of [6], we can get $\hat{f} : X \to R$ is convex. By Proposition 3.2, we can get $\hat{f} : X \to R$ is $(h, \varphi)$—twice $C$—differentiable at $h(x)$, then following from Proposition 4.3 of [5], that $\hat{f}^{00}(h(x); h(u), h(u)) \geq 0, \forall\ h(u) \in X$. Furthermore, since $\varphi^{-1}$ is monotonically increasing function, we have $f^{**}(x; u, u) = \varphi^{-1}(\hat{f}^{00}(h(x); h(u), h(u))) \geq 0, \forall\ x, u \in X$.

**Proposition 3.** *Let $f, g : X \to R$. Then $\forall x, u, v \in X$, we have*

$$(f[+]g)^{**}(x; u, v) \leq f^{**}(x; u, v)[+]g^{**}(x; u, v).$$

*Proof.* Let $f[+]g = M$, by Proposition 3.2, we have

$$M^{**}(x; u, v) = \varphi^{-1}\left(\hat{M}^{00}\Big(h(x); h(u), h(v)\Big)\right).$$

Since $M = \varphi^{-1}(\varphi f + \varphi g)$, we have $\hat{M} = \varphi(\varphi^{-1}(\varphi f + \varphi g))h^{-1} = \varphi f h^{-1} + \varphi g h^{-1}$. Furthermore by Proposition 3.2, that

$$M^{**}(x; u, v) = \varphi^{-1}\left((\varphi f h^{-1} + \varphi g h^{-1})^{00}\Big(h(x); h(u), h(v)\Big)\right)$$

$$\leq \varphi^{-1}\left((\varphi f h^{-1})^{00}\Big(h(x); h(u), h(v)\Big) + (\varphi g h^{-1})^{00}\Big(h(x); h(u), h(v)\Big)\right)$$

$$= \varphi^{-1}\left(\varphi f^{**}(x; u, v) + \varphi g^{**}(x; u, v)\right) = f^{**}(x; u, v)[+]g^{**}(x; u, v).$$

**Lemma 2.** *Let $X$ be a real Banach space, $f : X \to R$ be a continuous function, $x, u, v \in X$ and $f^0(\cdot; u)$ be finite near $x$. Then*

$$\left(f^0(\cdot; u)\right)^0(x; v) = \limsup_{\substack{y \to x \\ s \downarrow 0}} \frac{1}{s}\Big(f^0(y + sv; u) - f^0(y; u)\Big) \leq f^{00}(x; u, v).$$

**Theorem 4.** *Let $X$ be a real Banach space, $f : X \to R$ be a continuous function, $x, u, v \in X$ and $\hat{f}^0(\cdot; u)$ be finite near $x$. Then*

$$\left(f^*(\cdot; u)\right)^*(x; v) = \limsup_{\substack{y \to x \\ s \downarrow 0}} \frac{1}{s}[\cdot]\left(f^*\left(y \oplus s \odot v; u\right)[-]f^*(y; u)\right) \leq f^{**}(x; u, v).$$

*Proof.* Since $f : X \to R$ is continuous, then $\hat{f} : X \to R$ is continuous too. By Lemma 4.1, we can get

$$\limsup_{\substack{h(y) \to h(x) \\ s \downarrow 0}} \frac{1}{s}\left\{\hat{f}^0\left(h(y)+sh(v); h(u)\right)-\hat{f}^0\left(h(y); h(u)\right)\right\} \leq f^{00}\left(h(x); h(u), h(v)\right).$$

By Lemma 3.1, we can get

$$\limsup_{\substack{h(y) \to h(x) \\ s \downarrow 0}} \frac{1}{s}[\cdot]\varphi^{-1}\left(\hat{f}^0\left(h(y) + sh(v); h(u)\right)\right)[-]\varphi^{-1}\left(\hat{f}^0\left(h(y); h(u)\right)\right)$$

$$= \limsup_{\substack{h(y) \to h(x) \\ s \downarrow 0}} \varphi^{-1}\left\{\frac{1}{s}\left(\hat{f}^0\left(h(y) + sh(v); h(u)\right) - \hat{f}^0\left(h(y); h(u)\right)\right)\right\}$$

$$\leq \varphi^{-1}\left(f^{00}\left(h(x); h(u), h(v)\right)\right).$$

From Proposition 3.1 and Proposition 3.2, we have

$$\left(f^*(\cdot; u)\right)^*(x; v) = \limsup_{\substack{y \to x \\ s \downarrow 0}} \frac{1}{s}[\cdot]\left(f^*(y \oplus s \odot v; u)[-]f^*(y; u)\right) \leq f^{**}(x; u, v).$$

# References

1. Avriel, M.: Nonlinear Programming: Analysis and Method. Prentice-Hall, Englewood Cliffs (1976)
2. Zhang, Q.X.: On sufficiency and duality of solutions for nonsmooth $(h, \varphi)$−semi-infinite programming. Acta Math. Appl. Sinica (Chinese Series) 24(1), 129–138 (2001)
3. Xu, Y.H., Liu, S.Y.: The $(h, \varphi)$−lipschitz function, its generalized directional derivative and generalized gradient (chinese). Acta Math. Sci. 26A(2), 212–222 (2006)
4. Clarke, F.H.: Optimization and Nonsmooth Analysis. Wiley Interscience, New York (1983)
5. Cominetti, R., Correa, R.: A generalized second-order derivative in nonsmooth optimization. SIAM J. Control Optim. 28(4), 789–809 (1990)
6. Ben-tal, A.: On generalized means and generalized convex functions. J. Optim. Theory Appl. 21, 1–13 (1977)

# Estimation of Bessel Operator Inversion by Shearlet

Lin Hu and Youming Liu

**Abstract.** Curvelets are used to deal with the inverse problem of recovering a function $f$ from noisy Bessel data $B_\alpha f$ by Candès and Donoho. Motivated by the work of Colona, Easley and Labate, we solve the same problem by shearlets. It turns out that our method attains the mean square error convergence to $O(\log(\varepsilon^{-1})\varepsilon^{\frac{2}{\frac{3}{2}+\alpha}})$, as the noisy level $\varepsilon$ goes to zero. Although this converge rate is the same as Candès and Donoho's in the case $\alpha = \frac{1}{2}$, the shearlets possess affine systems and avoid more complicated structure of the curvelet constructure. This makes it a better candidate for theoretical and numerical applications.

**Keywords:** Inverse problem, Bessel operator, Shearlets.

## 1 Introduction and Preliminary

This paper studies the problem of noisy convolution inversion. Assume $\alpha > 0$, define two-dimensional Bessel Potential of order $\alpha$ , the kernel $b_\alpha(x)$ with Fourier transform

$$\hat{b}_\alpha(\xi) = (1 + |\xi|^2)^{-\frac{\alpha}{2}}.$$

Lin Hu
Department of Applied Mathematics, Beijing University of Technology,
Beijing 100124, China
Department of Mathematics, Beijing Union University, Beijing 100101, China
e-mail: hulin9803@yeah.net

Youming Liu
Department of Applied Mathematics, Beijing University of Technology,
Beijing 100124, China
e-mail: liuym@bjut.edu.cn

S. Li (Eds.): Nonlinear Maths for Uncertainty and its Appli., AISC 100, pp. 419–426.
springerlink.com                                      © Springer-Verlag Berlin Heidelberg 2011

The Bessel operator $B_\alpha$ is the operator of convolution with $b_\alpha : B_\alpha f = b_\alpha * f$. Consider the problem of recovering an image $f$ in white noise:

$$Y = B_\alpha f + \varepsilon W \tag{1}$$

where $f$ is the object to be recovered which is compactly supported and $C^2$ away from a $C^2$ edge, $W$ denotes a Wiener sheet, $\varepsilon$ is a noisy level. This linear inverse problem has many applications. In 2002, Candès and Donoho ([1]) applied curvelets to this inverse problem of Bessel convolution with $\alpha = \frac{1}{2}$. They obtained the mean square error (MSE) convergence to $O(\log_2(\varepsilon^{-1})\varepsilon^{\frac{4}{5}})$, outperforming wavelet-based (which achieve only the $\varepsilon^{\frac{2}{3}}$ rate) and linear methods (which achieve only the $\varepsilon^{\frac{1}{2}}$ rate).

In 2010, Colonna ([2]) proposed a new technique for inverting the Radon transform using shearlet system. Motivated by their work, we apply shearlets to the problem (1) and receive the MSE convergence to $O(\log_2(\varepsilon^{-1})\varepsilon^{\frac{2}{3/2+2\alpha}})$. Our methods achieve the same rates of convergence as curvelets, and faster than similar competitive strategies based on any linear methods or wavelets in the case $\alpha = \frac{1}{2}$. Unlike the curvelet representation, the shearlet approach is based on the framework of affine systems, and avoids the more complicated structure of the curvelet construction.

There exists different constructions of discrete shearlets, we will follow mainly the one in [2, 4, 5]. Set

$$A_0 = \begin{pmatrix} 4 & 0 \\ 0 & 2 \end{pmatrix}, \quad B_0 = \begin{pmatrix} 1 & 1 \\ 0 & 1 \end{pmatrix},$$

and for any $\xi = (\xi_1, \xi_2) \in \mathbb{R}^2, \xi_1 \neq 0$, let

$$\hat{\psi}^{(0)}(\xi) = \hat{\psi}_1(\xi_1)\hat{\psi}_2\left(\frac{\xi_2}{\xi_1}\right),$$

where $\hat{\psi}_1, \hat{\psi}_2 \in C^\infty(\mathbb{R})$, $supp\ \hat{\psi}_1 \subset [-\frac{1}{2}, -\frac{1}{16}] \cup [\frac{1}{16}, \frac{1}{2}]$ and $supp\ \hat{\psi}_2 \subset [-1, 1]$. In addition, we assume that

$$\sum_{j \geq 0} |\hat{\psi}_1(2^{-2j}\omega)|^2 = 1 \qquad for\ |\omega| \geq \frac{1}{8},$$

and for each $j \geq 0$,

$$\sum_{l=-2^j}^{2^j-1} |\hat{\psi}_2(2^j\omega - l)|^2 = 1 \qquad for\ |\omega| \leq 1.$$

Let $D_0 = \{(\xi_1, \xi_2) \in \mathbb{R}^2 : |\xi_1| \geq \frac{1}{8}, |\frac{\xi_2}{\xi_1}| \leq 1\}$ and $L^2(D_0)^\vee = \{f \in L^2(\mathbb{R}^2) : supp\hat{f} \subset D_0\}$. Then the collection $\{\psi_{j,l,k}^{(0)} : j \geq 0, -2^j \leq l \leq 2^j - 1, k \in Z^2\}$ defined by

$$\psi^{(0)}_{j,l,k}(x) = 2^{\frac{3}{2}j}\psi^{(0)}(B_0^l A_0^j x - k)$$

is a Parseval frame for $L^2(D_0)^\vee$. Similarly, we can construct a Parseval frame for $L^2(D_1)^\vee$, where $D_1$ is the vertical cone $D_1 = \{(\xi_1, \xi_2) \in \mathbb{R}^2 : |\xi_2| \geq \frac{1}{8}, |\frac{\xi_1}{\xi_2}| \leq 1\}$. Let

$$A_1 = \begin{pmatrix} 2 & 0 \\ 0 & 4 \end{pmatrix}, \quad B_1 = \begin{pmatrix} 1 & 0 \\ 1 & 1 \end{pmatrix},$$

and

$$\hat{\psi}^{(1)}(\xi) = \hat{\psi}_1(\xi_2)\hat{\psi}_2(\frac{\xi_1}{\xi_2}).$$

Then the collection $\{\psi^{(1)}_{j,l,k} : j \geq 0, -2^j \leq l \leq 2^j - 1, k \in Z^2\}$ defined by

$$\psi^{(1)}_{j,l,k}(x) = 2^{\frac{3}{2}j}\psi^{(1)}(B_1^l A_1^j x - k)$$

is a Parseval frame for $L^2(D_1)^\vee$. Finally, let $\hat{\varphi} \in C_0^\infty(\mathbb{R}^2)$ be chosen to satisfy

$$1 = |\hat{\varphi}(\xi)|^2 + \sum_{d=0}^{1}\sum_{j\geq 0}\sum_{l=-2^j}^{2^j-1} |\hat{\psi}^{(d)}((B_d^{-l})^T A_d^{-j}\xi)|^2 X_{D_d}(\xi),$$

where $X_{D_i}(\xi)$ is the indicator function of the set $D_i$ ($i = 1, 2$). The following result[2] plays an important role in this paper:

**Theorem 1.** *Let $\varphi$ and $\psi^{(d)}_{j,l,k}$ be defined as above. Also, for $d = 0, 1$, let* $\hat{\tilde{\psi}}^{(d)}_{j,l,k}(\xi) = \hat{\psi}^{(d)}_{j,l,k}(\xi)X_{D_d}(\xi)$. *Then the collection of shearlets*

$$\{\varphi_k : k \in Z^2\} \cup \{\psi^{(d)}_{j,l,k}(x) : j \geq 0, -2^j + 1 \leq l \leq 2^j - 2, k \in Z^2, d = 0, 1\}$$

$$\cup \{\tilde{\psi}^{(d)}_{j,l,k}(x) : j \geq 0, l = -2^j, 2^j - 1, k \in Z^2, d = 0, 1\}$$

*is a Parseval frame for $L^2(\mathbb{R}^2)$.*

In the following, for brevity of notation, we introduce the index set $\mathcal{M} = N \cup M$, where $N = Z^2$, $M = \{\mu = (j, l, k, d) : j \geq 0, -2^j \leq l \leq 2^j - 1, k \in Z^2, d = 0, 1\}$. Hence, the shearlet system is denoted by $\{s_\mu : \mu \in \mathcal{M} = N \cup M\}$, where $s_\mu = \psi_\mu = \psi^{(d)}_{j,l,k}$ if $\mu \in M$, and $s_\mu = \varphi_\mu$ if $\mu \in N$. For $\{s_\mu : \mu \in M\}$, it is understood that the shearlet functions are modified as in Theorem 1.

This paper is organized as follows: In Section 2, we show two lemmas which will be used to prove the main theorem. In Section 3, we give the estimator performance of the shearlet-based inversion algorithm when the convolution data are corrupted by additive Gaussian noise.

## 2  Two Lemmas

In order to prove our main results in the next section, two lemmas are given in this part. We begin with introducing the companion representation of shearlet system.

**Definition 1.** Let $\alpha$ be a positive number. For $\mu \in \mathcal{M}$, define $\psi_\mu^+$ and $\varphi_\mu^+$ by their Fourier transforms:

$$\hat{\psi}_\mu^+(\xi) = 2^{-2\alpha j}(1 + |\xi|^2)^{\frac{\alpha}{2}}\hat{\psi}_\mu(\xi) \qquad for \ \mu \in M$$
$$\hat{\varphi}_\mu^+(\xi) = (1 + |\xi|^2)^{\frac{\alpha}{2}}\hat{\varphi}_\mu(\xi) \qquad for \ \mu \in N$$

Now, we construct an inversion formula of the Bessel operator based on them.

**Lemma 1.** *Let $\{s_\mu : \mu \in \mathcal{M}\}$ be the Parseval frame of shearlets and $\{s_\mu^+ : \mu \in \mathcal{M}\}$ be the companion representation of shearlets defined above. Then the following reproducing formula holds:*

$$f = \sum_{\mu \in \mathcal{M}} \langle B_\alpha f, s_\mu^+ \rangle 2^{2\alpha j} s_\mu$$

*for all $f \in L^2(\mathbb{R}^2)$.*

*Proof.* In order to prove this theorem, it is sufficient to derive

$$\langle f, s_\mu \rangle = 2^{2\alpha j} \langle B_\alpha f, s_\mu^+ \rangle$$

for all $\mu \in \mathcal{M}$, since $\{s_\mu : \mu \in \mathcal{M}\}$ is a Parseval frame for $L^2(\mathbb{R}^2)$. In fact, by the assumption $s_\mu = \psi_\mu$ for $\mu \in M$ and the Parseval equality, $2^{2\alpha j}\langle B_\alpha f, s_\mu^+ \rangle = 2^{2\alpha j}\langle \widehat{B_\alpha f}, \hat{s}_\mu^+ \rangle = 2^{2\alpha j}\int_{R^2}(1 + |\xi|^2)^{-\frac{\alpha}{2}}\hat{f}(\xi)2^{-2\alpha j}(1 + |\xi|^2)^{\frac{\alpha}{2}}\overline{\hat{s}_\mu(\xi)}d\xi = \langle f, s_\mu \rangle$ for $\mu \in M$. In the cases $\mu \in N$, $s_\mu = \varphi_\mu$, a similar computation yields $\langle B_\alpha f, s_\mu^+ \rangle = \langle \widehat{B_\alpha f}, \hat{\varphi}_\mu^+ \rangle = \int_{R^2}(1 + |\xi|^2)^{-\frac{\alpha}{2}}\hat{f}(\xi)(1 + |\xi|^2)^{\frac{\alpha}{2}}\overline{\hat{\varphi}_\mu(\xi)}d\xi = \langle f, \varphi_\mu \rangle$.  □

To exploit the shearlet system to the inverse problem, we give some necessary modification of the shearlet system defined in Section 1, by rescaling the coarse scale system. For a fixed $j_0 \in N$, let $\varphi \in C_0^\infty(\mathbb{R}^2)$ be such that

$$|\hat{\varphi}(2^{-j_0}\xi)|^2 + \sum_{d=0}^{1}\sum_{j \geq j_0}\sum_{l=-2^j}^{2^j-1} |\hat{\psi}^{(d)}((B_d^{-l})^T A_d^{-j})\xi|^2 X_{D_d}(\xi) = 1,$$

Therefore, the modified shearlet system:

$$\{2^{j_0}\varphi(2^{j_0}x - k) : k \in Z^2\} \cup \{\tilde{\psi}_{j,l,k}^{(d)}(x) : j \geq j_0, l = -2^j, 2^j - 1, k \in Z^2, d = 0, 1\}$$
$$\cup \{\psi_{j,l,k}^{(d)}(x) : j \geq j_0, -2^j + 1 \leq l \leq 2^j - 2, k \in Z^2, d = 0, 1\}$$

is a Parseval frame for $L^2(\mathbb{R}^2)$.

Similar to Section 1, we denote $\widetilde{M} = N \cup M^0$, where $N = \mathbb{Z}^2$, $M^0 = \{\mu = (j, l, k, d) : j \geq j_0, -2^j \leq l \leq 2^j - 1, k \in \mathbb{Z}^2, d = 0, 1\}$. Moreover, the new shearlet system $\{s_\mu : \mu \in \widetilde{M}\}$ is defined by $s_\mu = \psi_\mu = \psi_{j,l,k}^{(d)}$ if $\mu \in M^0$ and $s_\mu = 2^{j_0} \varphi(2^{j_0} x - \mu)$ if $\mu \in N$.

For our reconstruction, we need to introduce a collection $\mathcal{N}(\varepsilon)$ of significant shearlet coefficients, depending on the noise level $\varepsilon$. Let $j_0 = \frac{1}{\frac{9}{2} + 6\alpha} \log_2(\varepsilon^{-1})$, $j_1 = \frac{1}{\frac{3}{2} + 2\alpha} \log_2(\varepsilon^{-1})$. We define the set of significant coefficients $\mathcal{N}(\varepsilon) \subset \widetilde{M}$, which is given by the union $\mathcal{N}(\varepsilon) = M^0(\varepsilon) \cup N^0(\varepsilon)$, where

$$N^0(\varepsilon) = \{\mu = k \in \mathbb{Z}^2 : |k| \leq 2^{2j_0+1}\},$$
$$M^0(\varepsilon) = \{\mu = (j, l, k, d) : j_0 < j \leq j_1, |k| \leq 2^{2j+1}, d = 0, 1\}.$$

The following lemma gives properties of $\mathcal{N}(\varepsilon)$.

**Lemma 2.** *Let $\varepsilon$ denote the noise level, and $\mathcal{N}(\varepsilon)$ be the set of significant indices associated with the shearlet representation of $f$ given by*

$$f = \sum_{\mu \in \widetilde{M}} \langle f, s_\mu \rangle s_\mu.$$

*Then the following properties hold:*

*(i)    The neglected shearlet coefficients satisfy*

$$\sup_{f \in \varepsilon^2(A)} \sum_{\mu \in \mathcal{N}(\varepsilon)^c} |\langle f, s_\mu \rangle|^2 \leq C \varepsilon^{\frac{2}{\frac{3}{2} + 2\alpha}};$$

*(ii)   The risk proxy satisfies*

$$\sup_{f \in \varepsilon^2(A)} \sum_{\mu \in \mathcal{N}(\varepsilon)} \min(|\langle f, s_\mu \rangle|^2, 2^{4\alpha j} \varepsilon^2) \leq C \varepsilon^{\frac{2}{\frac{3}{2} + 2\alpha}};$$

*(iii)  The cardinality of $\mathcal{N}(\varepsilon)$ obeys*

$$\sharp \mathcal{N}(\varepsilon) \leq C \varepsilon^{-\frac{5}{\frac{3}{2} + 2\alpha}},$$

*where $C$ is positive constants independent of $f$ and $\varepsilon^2(A)$ is the set of functions supported inside $[0, 1]^2$ which are $C^2$ away from a $C^2$ edge.*

*Proof.* We shall omit the proof of (i) and (iii), since it is essentially the same as that of Theorem 4.1.1 in [2]. In order to show (ii), we introduce the set $R(j, \varepsilon) = \{\mu \in M_j : |c_\mu| > \varepsilon\}$ to denote the collection of large shearlet coefficients at a fixed scale $j$. According to Corollary 1.5 in [5] (which is valid both for coarse and fine scale shearlets), there is a constant $C > 0$ such that $\sharp R(j, \varepsilon) \leq C\varepsilon^{-\frac{2}{3}}$ for $\varepsilon > 0$. It follows by rescaling that

$$\sharp R(j, 2^{2\alpha j}\varepsilon) \leq C2^{-\frac{4}{3}\alpha j}\varepsilon^{-\frac{2}{3}}.$$

For $\mu = (j, l, k, d) \in M^0$ and $d = 0$, we have

$$|c_\mu| = |\langle f, \psi_\mu \rangle| = |\int_{R^2} f(x)2^{\frac{3}{2}j}\psi(B_0^l A_0^j x - k)dx| \leq 2^{-\frac{3}{2}j}\|f\|_\infty \|\psi\|_1 \leq C2^{-\frac{3}{2}j}.$$

Note that $2^{2\alpha j}\varepsilon > 2^{-\frac{3}{2}j}$, when $j > j_1 \geq \frac{1}{2\alpha + \frac{3}{2}}\log_2(\varepsilon^{-1})$. Thus, we can conclude $R(j, 2^{2\alpha j}\varepsilon) = \emptyset$ from the definition of $R(j, 2^{2\alpha j}\varepsilon)$. For the risk proxy, notice that

$$\sum_{\mu \in \mathcal{N}(\varepsilon)} \min(c_\mu^2, 2^{4\alpha j}\varepsilon^2) = S_1(\varepsilon) + S_2(\varepsilon),$$

where

$$S_1(\varepsilon) = \sum_{\{\mu \in \mathcal{N}(\varepsilon): |c_\mu| \geq 2^{2\alpha j}\varepsilon\}} \min(c_\mu^2, 2^{4\alpha j}\varepsilon^2),$$

and

$$S_2(\varepsilon) = \sum_{\{\mu \in \mathcal{N}(\varepsilon): |c_\mu| < 2^{2\alpha j}\varepsilon\}} \min(c_\mu^2, 2^{4\alpha j}\varepsilon^2).$$

By the assumption that $j_1 = \frac{1}{2\alpha + \frac{3}{2}}log_2(\varepsilon^{-1})$, the estimates in (ii) can be derived in a similar fashion as that of Theorem 4.1.1 (2) in [2]. $\qquad\square$

## 3   Main Result

This section is devoted to show the main theorem in this paper, based on Theorem 1, Lemma 1 and Lemma 2.

We return to the problem of noisy convolution inversion:

$$Y = B_\alpha f + \varepsilon W. \tag{2}$$

Projecting the data (2) onto the system $\{s_\mu^+ : \mu \in \mathcal{M}\}$, we obtain

$$y_\mu := 2^{2\alpha j}\langle Y, s_\mu^+ \rangle = 2^{2\alpha j}\langle B_\alpha f, s_\mu^+ \rangle + \varepsilon 2^{2\alpha j}\langle W, s_\mu^+ \rangle = \langle f, s_\mu \rangle + \varepsilon 2^{2\alpha j}n_\mu,$$

where $n_\mu$ is a Gaussian noise with zero mean and variance $\sigma_\mu^2 = \|s_\mu^+\|^2$.

To estimate $f$ from the noisy observation (2), we will introduce the soft thresholding function $T_s(y, t) = sgn(y)(|y| - t)_+$ and construct an estimator of the form:

$$\tilde{f} = \sum_{\mu \in \mathcal{N}(\varepsilon)} \tilde{c}_\mu s_\mu, \tag{3}$$

where the coefficients are estimated by the rule

$$
\tilde{c}_\mu =
\begin{cases}
T_s(y_\mu, \varepsilon\sqrt{2\log(\sharp\mathcal{N}(\varepsilon))}2^{2\alpha j}\sigma_\mu), & \mu \in \mathcal{N}(\varepsilon); \\
\\
0, & otherwise.
\end{cases}
\tag{4}
$$

where $\sigma_\mu = \|s_\mu^+\|$ and the term $\{\sigma_\mu : \mu \in \widetilde{\mathcal{M}}\}$ are uniformly bounded.

**Theorem 2.** *Let $f \in \varepsilon^2(A)$ be the solution of the inverse problem $Y = B_\alpha f + \varepsilon W$ and $\tilde{f}$ be the approximation to $f$ defined by (3)-(4). Then there exist a constant $C > 0$ such that*

$$
\sup_{f\in\varepsilon^2(A)} E\|\tilde{f} - f\|^2 \leq C\log(\varepsilon^{-1})\varepsilon^{\frac{2}{\frac{3}{2}+2\alpha}}
$$

*for small $\varepsilon > 0$, where $E$ is the expectation operator.*

*Proof.* Let $c_\mu = \langle f, s_\mu \rangle$ and $\tilde{c}_\mu$ be given by (4). From the Parseval frame property of the shearlet system $\{s_\mu : \mu \in \widetilde{\mathcal{M}}\}$, we know that

$$
E\|\tilde{f} - f\|^2 = E(\sum_{\mu\in\widetilde{\mathcal{M}}} |\tilde{c}_\mu - c_\mu|^2) = E(\sum_{\mu\in\mathcal{N}(\varepsilon)} |\tilde{c}_\mu - c_\mu|^2) + \sum_{\mu\in\mathcal{N}(\varepsilon)^c} |c_\mu|^2.
$$

Since $\sum_{\mu\in\mathcal{N}(\varepsilon)^c} |c_\mu|^2 \leq C\varepsilon^{\frac{2}{\frac{3}{2}+2\alpha}}$ due to Lemma 2 (i), it is sufficient to prove

$$
E(\sum_{\mu\in\mathcal{N}(\varepsilon)} |\tilde{c}_\mu - c_\mu|^2) \leq C\log(\varepsilon^{-1})\varepsilon^{\frac{2}{\frac{3}{2}+2\alpha}}.
\tag{5}
$$

According to the Oracle Inequality [3],

$$
E(\sum_{\mu\in\mathcal{N}(\varepsilon)} |\tilde{c}_\mu - c_\mu|^2) \leq L(\varepsilon)[\varepsilon^2 \sum_{\mu\in\mathcal{N}(\varepsilon)} \frac{2^{4\alpha j}\sigma_\mu^2}{\sharp\mathcal{N}(\varepsilon)} + \sum_{\mu\in\mathcal{N}(\varepsilon)} \min(c_\mu^2, \varepsilon^2 2^{4\alpha j}\sigma_\mu^2)],
\tag{6}
$$

where $L(\varepsilon) = 1 + 2\log(\sharp\mathcal{N}(\varepsilon))$. The assumption $j_1 = \frac{1}{\frac{3}{2}+\alpha}\log(\varepsilon^{-1})$ implies that

$$
\varepsilon^2 \sum_{\mu\in\mathcal{N}(\varepsilon)} \frac{2^{4\alpha j}\sigma_\mu^2}{\sharp\mathcal{N}(\varepsilon)} \leq C\varepsilon^2 2^{4\alpha j_1} \leq C\varepsilon^{\frac{3}{\frac{3}{2}+2\alpha}}.
$$

From the Lemma 2 (ii) and the uniform boundness of $\{\sigma_\mu : \mu \in \widetilde{\mathcal{M}}\}$, we have

$$
\sum_{\mu\in\mathcal{N}(\varepsilon)} \min(c_\mu^2, \varepsilon^2 2^{4\alpha j}\sigma_\mu^2) \leq C\varepsilon^{\frac{2}{\frac{3}{2}+2\alpha}}.
$$

Then for small $\varepsilon > 0$,

$$\varepsilon^2 \sum_{\mu \in \mathcal{N}(\varepsilon)} \frac{2^{4\alpha j}\sigma_\mu^2}{\sharp \mathcal{N}(\varepsilon)} + \sum_{\mu \in \mathcal{N}(\varepsilon)} \min(c_\mu^2, \varepsilon^2 2^{4\alpha j}\sigma_\mu^2) \le C\varepsilon^{\frac{2}{\frac{3}{2}+2\alpha}}.$$

On the other hand, Lemma 2 (iii) tells

$$\sharp \mathcal{N}(\varepsilon) \le C\varepsilon^{-\frac{5}{\frac{3}{2}+2\alpha}}.$$

All these with (6) lead to the desired (5). $\qquad\qquad\square$

When $\alpha = \frac{1}{2}$, the shearlet method achieves the near-optimal rate $\varepsilon^{\frac{4}{5}}$, outperforming wavelet-based (which achieve only the $\varepsilon^{\frac{2}{3}}$ rate) and linear methods (which achieve only the $\varepsilon^{\frac{1}{2}}$ rate).

**Acknowledgements.** This work is supported by the National Natural Science Foundation of China (No.10871012) and Natural Science Foundation of Beijing (No.1082003).

# References

1. Candés, E.J., Donoho, D.L.: Recovering edges in ill-posed inverse problems: optimality of curvelet frames. Ann. Stat. 30, 784–842 (2002)
2. Colonna, F., Easley, G., Labate, D.: Radon transform inversion using the shearlet representation. Appl. Comput. Harmon. Anal. 29, 232–250 (2010)
3. Donoho, D.L., Johnstone, I.M.: Ideal spatial adaptation via wavelet shrinkage. Biometrika 81, 425–455 (1994)
4. Easley, G., Lim, W., Labate, D.: Sparse directional image representations using the discrete shearlet transform. Appl. Comput. Harmon. Anal. 25, 25–46 (2008)
5. Guo, K., Labate, D.: Optimally sparse multidimensional representation using shearlets. SIAM J. Math. Anal. 39, 298–318 (2007)

# Globally Convergent Inexact Smoothing Newton Method for SOCCP

Jie Zhang and Shao-Ping Rui

**Abstract.** An inexact smoothing Newton method for solving second-order cone complementarity problems (SOCCP) is proposed. In each iteration the corresponding linear system is solved only approximately. Under mild assumptions, it is proved that the proposed method has global convergence and local superlinear convergence properties. Preliminary numerical results indicate that the method is effective for large-scale SOCCP.

**Keywords:** Second-order cone complementarity problems, Inexact methods, Large-scale problems.

## 1 Introduction

Let $\mathcal{K}^n$ be the second-order cone (SOC) in $R^n$, also called the Lorentz cone or ice-cream cone, defined by

$$\mathcal{K}^n = \{(x_1, x_2) \in R \times R^{n-1} | \, \|x_2\| \leq x_1\},$$

where $\| \cdot \|$ denotes the Euclidean norm. We are interested in complementarity problems involving the second-order cone in its constraints. In general, the second-order cone complementarity problems (SOCCP) has the follwing form:

Jie Zhang
School of Mathematical Science, Huaibei Normal University, Huaibei, 235000, People's Republic of China

Shaoping Rui
Faculty of Science, Xi'an Jiaotong University, Xi'an,
710049 and School of Mathematical Science, Huaibei Normal University, Huaibei, 235000, People's Republic of China
e-mail: rsp9999@163.com

S. Li (Eds.): Nonlinear Maths for Uncertainty and its Appli., AISC 100, pp. 427–434.
springerlink.com © Springer-Verlag Berlin Heidelberg 2011

$$\text{Find an } x \in \mathcal{K}, \text{ such that } F(x) \in \mathcal{K} \text{ and } x^T F(x) = 0, \tag{1}$$

where $F : R^n \to R^n$ is a continuously differentiable function, and $\mathcal{K} = \mathcal{K}^{n_1} \times \cdots \times \mathcal{K}^{n_m}$ with $m, n_1, \ldots, n_m \geq 1$ and $n_1 + \cdots + n_m = n$. Unless otherwise specified, in the following analysis we assume that $m = 1$ and $n_1 = n$.

Second-order cone complementarity problems have wide range of applications and, in particular, includes a large class of quadratically constrained problems as special cases [6]. Recently, SOCCP have attracted a lot of attention [4,5] and a number of smoothing methods for solving SOCCP have been proposed [1,3]. As in the smooth case, each iteration consists of finding a solution of linear system which may be cumbersome when solving a large-scale problem. The inexact method is one way to overcome this difficulty. Inexact Newton methods have been proposed for solving large scale complementarity problems [2,7]. In this paper, we extend the inexact Newton method to SOCCP. We propose a new inexact smoothing Newton algorithm for solving SOCCP under the framework of smoothing Newton method. We view the smoothing parameter as an independent variable. The forcing parameter of inexact Newton method links the norm of residual vector to the norm of mapping at the current iterate.

The paper is organized as follows: In the next Section, we introduce preliminaries and study a few properties of a vector-valued function. An inexact smoothing Newton method for solving the SOCCP (1) is proposed. Convergence results are analyzed in Section 3. In Section 4, numerical experiments are presented. Conclusions are given in Section 5.

The following notations will be used throughout this paper. " $:=$ " means "is defined as". $R_+$ and $R_{++}$ denote the nonnegative and positive reals. All vectors are column vectors, the superscript $T$ denotes transpose. For simplicity, we use $x = (x_1, x_2)$ instead of $x = (x_1, x_2^T)^T$. $\langle \cdot, \cdot \rangle$ represents the Euclidean inner product. The symbol $\| \cdot \|$ stands for the 2-norm. Landau symbols $o(\cdot)$ and $O(\cdot)$ are defined in usual way. Let int$\mathcal{K}$ denote the interior of $\mathcal{K}$. $x \succeq y$ $(x \succ y)$ means that $x - y \in \mathcal{K}$ $(x - y \in \text{int}\mathcal{K})$.

## 2 Preliminaries and Algorithm

First, we recall the Euclidean Jordan algebra associated with SOC. Next, we introduce a vector-valued function for SOCCP (1) and propose an inexact smoothing Newton method for SOCCP (1).

For any $x = (x_1, x_2)$, $y = (y_1, y_2) \in R \times R^{n-1}$, Jordan product associated with $\mathcal{K}$ is defined by

$$x \circ y = \begin{pmatrix} x^T y \\ y_1 x_2 + x_1 y_2 \end{pmatrix},$$

with $e = (1, 0, \ldots, 0) \in R^n$ being its unit element. We write $x^2$ to mean $x \circ x$. The linear mapping is defined by $L_x y = x \circ y, \quad \forall y \in R^n$.

We next introduce the spectral factorization of vectors in $R^n$ associated with $\mathcal{K}$. Let $x = (x_1, x_2) \in R \times R^{n-1}$. Then $x$ can be decomposed as

$$x = \lambda_1 u^{(1)} + \lambda_2 u^{(2)},$$

where $\lambda_1$, $\lambda_2$ and $u^{(1)}$, $u^{(2)}$ are the spectral values and the associated spectral vectors of $x$ given by

$$\lambda_i = x_1 + (-1)^i \|x_2\|,$$

$$u^{(i)} = \begin{cases} \frac{1}{2}\left(1, (-1)^i \frac{x_2}{\|x_2\|}\right) & \text{if } x_2 \neq 0, \\ \frac{1}{2}(1, (-1)^i w) & \text{if } x_2 = 0, \end{cases}$$

for $i = 1, 2$ with $w$ being any vector in $R^{n-1}$ satisfying $\|w\| = 1$.

**Lemma 1.** [3] *For any $(x_1, x_2) \in R \times R^{n-1}$ with spectral values $\lambda_1$, $\lambda_2$ and spectral vectors $u^{(1)}$, $u^{(2)}$ given as above, we have that*
*(i) $x \in \mathcal{K}$ if and only if $\lambda_i \geq 0$ and $x \in \text{int}\mathcal{K}$ if and only if $\lambda_i > 0, i = 1, 2$.*
*(ii)$x^2 = \lambda_1^2 u^{(1)} + \lambda_2^2 u^{(2)} \in \mathcal{K}$.*
*(iii) $x^{\frac{1}{2}} = \sqrt{\lambda_1} u^{(1)} + \sqrt{\lambda_2} u^{(2)} \in \mathcal{K}$ if $x \in \mathcal{K}$.*

**Lemma 2.** [10] *For any $a, b, u, v \in \mathcal{K}$. If $a \succ 0$, $b \succ 0$, $a \circ b \succ 0$, $\langle u, v \rangle \geq 0$ and $a \circ u + b \circ v = 0$, then $u = v = 0$.*

We introduce the following vector-valued function

$$\phi(\mu, x, y) = (1 + \mu)(x + y) - \sqrt{(x + \mu y)^2 + (y + \mu x)^2 + 2\mu^2 e}, \qquad (1)$$

where $(\mu, x, y) \in R_+ \times R^n \times R^n$. Let $z := (\mu, x, y) \in R_+ \times R^n \times R^n$ and

$$H(z) := \begin{pmatrix} \mu \\ F(x) - y \\ \phi(\mu, x, y) \end{pmatrix}. \qquad (2)$$

**Theorem 1.** *Let $\phi : R_+ \times R^n \times R^n \rightarrow R^n$ and $H : R_+ \times R^n \times R^n \rightarrow R_+ \times R^n \times R^n$ be defined by (1) and (2), respectively. Then*
*(i) $\phi(0, x, y) = 0 \Leftrightarrow x \circ y = 0, x \in \mathcal{K}, y \in \mathcal{K}$.*
*(ii) $\phi(\mu, x, y)$ is a smoothing function of $\phi(0, x, y)$ and $\phi(\mu, x, y)$ is semismooth on $R_+ \times R^n \times R^n$.*
*(iii) $H(z)$ is continuously differentiable at any $z = (\mu, x, y) \in R_{++} \times R^n \times R^n$ with its Jacobian*

$$\nabla H(z) = \begin{pmatrix} 1 & 0 & 0 \\ 0 & \nabla F & -I \\ \nabla \phi_\mu & \nabla \phi_x & \nabla \phi_y \end{pmatrix}, \qquad (3)$$

*where*

$$\nabla \phi_\mu = x + y - L_\omega^{-1}(L_{x+\mu y}y + L_{y+\mu x}x + 2\mu e),$$
$$\nabla \phi_x = (1+\mu)I - L_\omega^{-1}(L_{x+\mu y} + \mu L_{y+\mu x}),$$
$$\nabla \phi_y = (1+\mu)I - L_\omega^{-1}(L_{y+\mu x} + \mu L_{x+\mu y}),$$
$$\omega = \sqrt{(x+\mu y)^2 + (y+\mu x)^2 + 2\mu^2 e}.$$

*If $F(x)$ is monotone, then the Jacobian $\nabla H(z)$ is nonsingular for any $\mu > 0$.*

**Proof.** (i) By Proposition 2.1 in [3], we can obtain (i). Then solution set of

$$\Psi(z) := \|H(z)\|^2 = 0$$

coincide with the solution of SOCCP (1).

(ii) Now we prove that $\phi(\mu, x, y)$ is a smoothing function of $\phi(0, x, y)$. For any $x = (x_1, x_2), y = (y_1, y_2) \in R \times R^{n-1}$, it follows from the spectral factorization of

$$(x + \mu y)^2 + (y + \mu x)^2 + 2\mu^2 e$$

and Lemma 1 that

$$\phi(\mu, x, y) = (1+\mu)(x+y) - (\sqrt{\lambda_1(\mu)}u^{(1)}(\mu) + \sqrt{\lambda_2(\mu)}u^{(2)}(\mu)),$$

where

$$\lambda_i(\mu) = \|x+\mu y\|^2 + \|y+\mu x\|^2 + 2\mu^2 + 2(-1)^i\|c(\mu)\|, i = 1, 2,$$

$$u^{(i)}(\mu) = \begin{cases} \frac{1}{2}\left(1, (-1)^i \frac{c(\mu)}{\|c(\mu)\|}\right) & \text{if } c(\mu) \neq 0, \\ \frac{1}{2}(1, (-1)^i w) & \text{if } c(\mu) = 0, \end{cases}$$

$$c(\mu) = (x_1 + \mu y_1)(x_2 + \mu y_2) + (\mu x_1 + y_1)(\mu x_2 + y_2),$$

and $w \in R^{n-1}$ being an arbitrary vector satisfying $\|w\| = 1$. In a similar way,

$$\phi(0, x, y) = x + y - (\sqrt{\lambda_1}u^{(1)} + \sqrt{\lambda_2}u^{(2)}),$$

where

$$\lambda_i = \|x\|^2 + \|y\|^2 + 2(-1)^i\|v\|, i = 1, 2,$$

$$u^{(i)} = \begin{cases} \frac{1}{2}\left(1, (-1)^i \frac{v}{\|v\|}\right) & \text{if } v \neq 0, \\ \frac{1}{2}(1, (-1)^i w) & \text{if } v = 0, \end{cases}$$

$$v = x_1 x_2 + y_1 y_2.$$

If $v \neq 0$, then

$$\lim_{\mu \downarrow 0} \lambda_i(\mu) = \lambda_i,$$
$$\lim_{\mu \downarrow 0} u^{(i)}(\mu) = u^{(i)},$$

$i = 1, 2$. Hence, $\lim_{\mu \downarrow 0} \phi(\mu, x, y) = \phi(0, x, y)$. If $v = 0$, then

$$v(\mu) = 0,$$
$$\lambda_i = \|x\|^2 + \|y\|^2,$$
$$u^{(i)} = \frac{1}{2}(1, (-1)^i w),$$

$i = 1, 2$. Then

$$\lim_{\mu \downarrow 0} (\sqrt{\lambda_1(\mu)} u^{(1)}(\mu) + \sqrt{\lambda_2(\mu)} u^{(2)}(\mu))$$
$$= \sqrt{\|x\|^2 + \|y\|^2} u^{(1)}(\mu) + \sqrt{\|x\|^2 + \|y\|^2} u^{(2)}(\mu)$$
$$= \sqrt{\|x\|^2 + \|y\|^2} e$$
$$= \sqrt{\|x\|^2 + \|y\|^2} u^{(1)} + \sqrt{\|x\|^2 + \|y\|^2} u^{(2)}$$
$$= \sqrt{\lambda_1} u^{(1)} + \sqrt{\lambda_2} u^{(2)}.$$

Therefore, $\phi(\mu, x, y)$ is a smoothing function of $\phi(0, x, y)$. Semismoothness of the function $\phi(\mu, x, y)$ on $R_+ \times R^n \times R^n$ which can be obtained by Theorem 3.2 in [9].

(iii) It is easy to show that $H(z)$ is continuously differentiable at any $(\mu, z, y) \in R_{++} \times R^n \times R^n$. Since for any $(\mu, z, y) \in R_{++} \times R^n \times R^n$, we have $\omega \succ 0$, then $L_\omega$ is invertible. So the computation of $\nabla H(z)$ is not difficult. Next we want to show the nonsingularity of $\nabla H(z)$. Assume that $\nabla H(z)\Delta z = 0$, we need to shown $\Delta z = 0$. From $\nabla H(z)\Delta z = 0$ we have

$$\Delta \mu = 0, \tag{4}$$
$$\nabla F(x)\Delta x - \Delta y = 0, \tag{5}$$
$$\nabla \phi_\mu \Delta \mu + \nabla \phi_x \Delta x + \nabla \phi_y \Delta y = 0. \tag{6}$$

Form (4) and (6), we obtain

$$((1+\mu)L_\omega - L_{x+\mu y} - \mu L_{y+\mu x})\Delta x + ((1+\mu)L_\omega - L_{y+\mu x} - \mu L_{x+\mu y})\Delta y = 0. \tag{7}$$

Then, from (7), using Jordan product " $\circ$ " yield

$$((1+\mu)\omega - (x+\mu y) - \mu(y+\mu x)) \circ \Delta x + ((1+\mu)\omega - (y+\mu x) - \mu(x+\mu y)) \circ \Delta y = 0. \tag{8}$$

It is easy to see from the definition of $\omega$ that

$$(1+\mu)\omega - (x+\mu y) - \mu(y+\mu x) \succ 0, \, (1+\mu)\omega - (y+\mu x) - \mu(x+\mu y) \succ 0 \tag{9}$$

and

$$((1+\mu)\omega - (x+\mu y) - \mu(y+\mu x)) \circ ((1+\mu)\omega - (y+\mu x) - \mu(x+\mu y))$$
$$= 1/2((1+\mu)\omega - x - \mu y - \mu(y+\mu x) - y - \mu x - \mu(x+\mu y))^2$$
$$+ \mu((x+\mu y) - (y+\mu x))^2 + (1+\mu)^2 \mu^2 e \succ 0. \tag{10}$$

From (5) and the monotonicity of $F$, we can get

$$\langle \Delta x, \Delta y \rangle = \langle \Delta x, \nabla F(x) \Delta x \rangle \geq 0. \tag{11}$$

Then, by combining (9)-(11) with Lemma 2 we obtain that $\Delta x = 0$ and $\Delta y = 0$. The proof is completed.

1 Given constants $\delta \in (0,1)$ and $\sigma \in (0,1)$. Let $(\mu_0, x_0, y_0) \in R_{++} \times R^n \times R^n$ with $(\mu_0, x_0)$ arbitrary and $y_0 = F(x_0)$. Choose $\gamma \in (0,1)$ such that $\gamma \mu_0 < 1/2$ and a sequence $\{\eta_k\}$ such that $\eta_k \in [0, \eta]$, where $\eta \in [0, 1 - \gamma \mu_0]$ is a constant. Set $k := 0$.
2 *Step 1.* If $\Psi(z_k) = 0$, stop.
3 *Step 2.* Compute $\Delta z_k = (\Delta \mu_k, \Delta x_k, \Delta y_k)$ by

$$H(z_k) + \nabla H(z_k) \Delta z_k = R_k, \tag{12}$$

where $R_k = (\rho_k \mu_0, 0, r_k) \in R_{++} \times R^n \times R^n, \rho_k = \rho(z_k) = \gamma min\{1, \Psi(z_k)\}$ and $\|r^k\| \leq \eta_k \|H(z_k)\|$.
4 *Step 3.* Let $\theta_k$ be the maximum of the values $1, \delta, \delta^2, \ldots$ such that

$$\Psi(z_k + \theta_k \Delta z_k) \leq [1 - \sigma(1 - \gamma \mu_0 - \eta_k)\theta_k]\Psi(z_k). \tag{13}$$

5 *Step 4.* Set $z_{k+1} = z_k + \theta_k \Delta z_k$ and $k = k + 1$, go to Step 1.

We now prove that Algorithm 5 is well-defined. First, defined the set $\Omega = \{z = (\mu, x, y) \in R_{++} \times R^n \times R^n \mid \mu \geq \rho(z)\mu_0\}, \rho(z) = \gamma min\{1, \Psi(z)\}$.

**Theorem 2.** *Suppose that $F(x)$ is a monotone function. Then Algorithm 5 is well defined and infinite sequence $\{z_k = (\mu_k, x_k, y_k)\}$ generated by Algorithm 5 satisfy $\mu_k > 0$ and $z_k \in \Omega$ for all $k \geq 0$.*

**Proof.** Proof of the theorem similarly to Lemma 2.3 and Theorem 2.1 in [8]. For brevity, we omit the details here.

## 3   Convergence Analysis

In this section, we consider the global convergence and local superlinear convergence of Algorithm 5. We need the following assumption:

**Assumption 3.1.** *The level sets $\mathcal{L}(z_0) = \{z \in R^{2n+1} | \Psi(z) \leq \Psi(z_0)\}$ of $\Psi(z)$ are bounded.*

**Theorem 3.** *Suppose that $F(x)$ is a monotone function. If Assumption 3.1 holds, then each accumulation point $z^*$ of an infinite sequence $\{z_k\}$ generated by the Algorithm 5 is a solution of $H(z) = 0$.*

**Proof.** It follows from Theorem 2 that an infinite sequence $\{z_k\}$ is generated such that $z_k \in \Omega$. Without loss of generality, we assume that $\{z_k\}$ convergence to $\{z^*\}$. From the design of Algorithm 5, the sequence $\{\Psi(z_k)\}$

is monotonically decreasing. Then from the continuity of $\Psi(z)$, we have $\{\Psi(z_k)\} \rightarrow \Psi(z^*) \geq 0$. If $\Psi(z^*) = 0$, then we obtain the desired result. Suppose $\Psi(z^*) > 0$, from $z_k \in \Omega$, i.e., $\mu_k \geq \rho(z_k)\mu_0 = \gamma min\{1, \Psi(z_k)\}\mu_0$, we have $\mu^* \in R_{++}$. Then, $\nabla H(z^*)$ exists and is invertible. Hence, there exists a closed neighborhood $\mathcal{N}(z^*)$ of $z^*$ and a positive number $\bar{\alpha} \in (0, 1]$ such that for any $z = (\mu, x, y) \in \mathcal{N}(z^*)$, and all $\alpha \in [0, \bar{\alpha}]$

$$\Psi(z_k + \alpha\Delta z_k) \leq [1 - \sigma(1 - \gamma\mu_0 - \eta_k)\alpha]\Psi(z_k)$$

holds, which shows that for a nonnegative integer $l$ such that $\delta^l \in (0, \bar{\alpha}]$,

$$\Psi(z_k + \delta^l\Delta z_k) \leq [1 - \sigma(1 - \gamma\mu_0 - \eta_k)\delta^l]\Psi(z_k)$$

holds for all sufficiently large $k$. By the design of Algorithm 5, $\theta_k \geq \delta^l, \eta_k \leq \eta$ for all sufficiently large $k$. Then

$$\Psi(z_{k+1}) \leq [1 - \sigma(1 - \gamma\mu_0 - \eta)\delta^l]\Psi(z_k) \tag{14}$$

for all sufficiently large $k$. Taking the limit $k \rightarrow \infty$ in the both sides of inequality (14) generates

$$\Psi(z^*) \leq [1 - \sigma(1 - \gamma\mu_0 - \eta)\delta^l]\Psi(z^*).$$

This contradicts the assumption of $\Psi(z^*) > 0$. This completes the proof.

**Theorem 4.** *Suppose that $F(x)$ is a monotone function and $z^* = (\mu^*, x^*, y^*)$ is an accumulation point of the iteration sequence $\{z_k\}$ generated by Algorithm 5. If all $V \in \partial H(z^*)$ are nonsingular, Assumption 3.1 holds and $\eta_k \rightarrow 0$. Then, $\{z_k\}$ converges to $\{z^*\}$ superlinearly, that is, $\|z_{k+1} - z^*\| = o(\|z_k - z^*\|)$. Moreover, $\mu_{k+1} = o(\mu_k)$.*

**Proof.** From Theorem 3, $H(z)$ is semismooth at $z^*$, and we can prove the theorem similarly to Theorem 3.2 in [8]. For brevity, we omit the details here.

## 4   Numerical Experiments

In this section, we have implemented some numerical experiments on the SOCCP using the Algorithm 5 described in this paper. All programs are written in Matlab code, numerical test in PC, CPU Main Frequency 1.73GHz 1G run circumstance Matlab 7.1. In experiments, the function $F(x)$ in SOCCP (1) is $Mx + q$, $M \in R^n \times R^n, q \in R^n$. The matrix $M$ is obtained by setting $M = N^T N$, where $N$ is a square matrix. Elements of $N$ and $q$ are chosen randomly from the interval $[-1, 1]$. The parameters in algorithm 5 we choose as: $\sigma = 0.8, \gamma = 0.001, \delta = 0.8, \eta_k = 2^{-k}$. The initial $\mu_0$ is chosen randomly from interval $(0, 2)$ and $x_0$ is generated from a uniform distribution in the interval $(0, 1)$. We use $\|H(z)\| \leq 10^{-10}$ as the stoping criterion. The numerical results is listed in Table 1, where No.it denotes the numbers of iterations, the

**Table 1** Numerical results of Algorithm 5 for SOCCP of various problems size (n)

| n | No.it | CPU(sec) | $x^T F(x)$ |
|------|-----|----------|------------|
| 1000 | 14 | 70.8418 | 5.3340e-6 |
| 1500 | 19 | 264.639 | 1.5520e-3 |
| 2000 | 25 | 509.478 | 7.8776e-4 |

CPU time is in seconds. From Table 1, we see the inexact smoothing Newton Algorithm proposed in the paper needs few iterations and CPU. Moreover, Algorithm 5 can deal with large-scale second-order cone complementarity problems.

# References

1. Chen, J.S., Chen, X., Tseng, P.: Analysis of nonsmooth vector-valued functions associated with second-order cones. Mathematical Programming 101, 95–117 (2004)
2. Facchinei, F., Kanzow, C.: A nonsmooth inexact Newton method for the solution of large-scale nonlinear complementarity problems. Mathematical Programming 76, 493–512 (1997)
3. Fukushima, M., Luo, Z.Q., Tseng, P.: Smoothing functions for second-order-cone complementarity problems. SIAM Journal on Optimization 12, 436–460 (2001)
4. Hayashi, S., Yamashita, N., Fukushima, M.: A combined smoothing and regularization method for monotone second-order cone complementarity problems. SIAM Journal on Optimization 15, 593–615 (2004)
5. Kanzow, C., Ferenczi, I., Fukushima, M.: On the local convergence of semismooth Newton methods for linear and nonlinear second-order cone programs without strict complementarity. SIAM Journal on Optimization 20, 297–320 (2009)
6. Lobo, M.S., Vandenberghe, L., Boyd, S., Lebret, H.: Applications of second-order cone programming. Linear Algebra and its Applications 284, 193–228 (1998)
7. Rui, S.P., Xu, C.X.: Inexact non-interior continuation method for solving large-scale monotone SDCP. Applied Mathematics and Computation 215, 2521–2527 (2009)
8. Rui, S.P., Xu, C.X.: A smoothing inexact Newton method for nonlinear complementarity problems. Journal of Computational and Applied Mathematics 233, 2332–2338 (2010)
9. Sun, D., Sun, J.: Strong semismoothness of the fischer-burmeister SDC and SOC complementarity functins. Mathematical Programming 103, 575–581 (2005)
10. Yoshise, A.: Interior point trajectories and a homogeneous model for nonliear complementarity problems over symmetric cones. SIAM Journal on Optimization 17, 1129–1153 (2006)
11. Zhang, X.S., Liu, S.Y.: A smoothing method for second order cone complementarity problem. Journal of Computational and Applied Mathematics 228, 83–91 (2009)

# Existence of Positive Solution for the Cauchy Problem for an Ordinary Differential Equation

Toshiharu Kawasaki and Masashi Toyoda

**Abstract.** In this paper we consider the existence of positive solution for the Cauchy problem of the second order differential equation $u''(t) = f(t, u(t))$.

**Keywords:** Ordinary differential equation, Cauchy problem, Fixed point.

## 1 Introduction

The following ordinary differential equations arise in many different areas of applied mathematics and physics; see [2, 4]. In [3] Knežević-Miljanović considered the Cauchy problem

$$\begin{cases} u''(t) = P(t)t^a u(t)^\sigma, \ t \in [0,1], \\ u(0) = 0, \ u'(0) = \lambda, \end{cases} \tag{1}$$

where $a, \sigma, \lambda \in \mathbf{R}$ with $\sigma < 0$ and $\lambda > 0$, and $P$ is a continuous mapping of $[0,1]$ such that $\int_0^1 |P(t)||t^{a+\sigma}dt < \infty$. On the other hand in [1] Erbe and Wang considered the equation

$$u''(t) = f(t, u(t)), \ t \in [0,1]. \tag{2}$$

In this paper we consider the second order Cauchy problem

$$\begin{cases} u''(t) = f(t, u(t)), \ t \in [0,1], \\ u(0) = 0, \ u'(0) = \lambda, \end{cases} \tag{3}$$

Toshiharu Kawasaki
College of Engineering, Nihon University, Fukushima 963–8642, Japan
e-mail: toshiharu.kawasaki@nifty.ne.jp

Masashi Toyoda
Faculty of Engineering, Tamagawa University, Tokyo 194–8610, Japan
e-mail: mss-toyoda@eng.tamagawa.ac.jp

S. Li (Eds.): Nonlinear Maths for Uncertainty and its Appli., AISC 100, pp. 435–441.
springerlink.com　　　　　　　　　　　　　© Springer-Verlag Berlin Heidelberg 2011

where $f$ is a mapping from $[0,1] \times [0,\infty)$ into $\mathbf{R}$ satisfying the Carathéodory condition and $\lambda \in \mathbf{R}$ with $\lambda > 0$.

## 2  Main Results

**Theorem 1.** *Suppose that a mapping $f$ from $[0,1] \times [0,\infty)$ into $\mathbf{R}$ satisfies the following.*

(a)     *The mapping $f$ satisfies the Carathéodory condition, that is, the mapping $t \longmapsto f(t,u)$ is measurable for any $u \in (0,\infty)$ and the mapping $u \longmapsto f(t,u)$ is continuous for almost every $t \in [0,1]$.*

(b)     $|f(t,u_1)| \geq |f(t,u_2)|$ *for almost every $t \in [0,1]$ and for any $u_1, u_2 \in [0,\infty)$ with $u_1 \leq u_2$.*

(c)     *There exists $\alpha \in \mathbf{R}$ with $0 < \alpha < \lambda$ such that*

$$\int_0^1 |f(t,\alpha t)|dt < \infty.$$

(d)     *There exists $\beta \in \mathbf{R}$ with $\beta > 0$ such that*

$$\left| \frac{\partial f}{\partial u}(t,u) \right| \leq \frac{\beta |f(t,u)|}{u}$$

*for almost every $t \in [0,1]$ and for any $u \in (0,\infty)$.*

*Then there exist $h \in \mathbf{R}$ with $0 < h \leq 1$ such that the Cauchy problem (3) has a unique solution in $X$, where $X$ is a subset*

$$X = \left\{ u \; \middle| \; \begin{array}{l} u \in C[0,h], u(0) = 0, u'(0) = \lambda \\ \text{and } \alpha t \leq u(t) \text{ for any } t \in [0,h] \end{array} \right\}$$

*of $C[0,h]$, which is the class of continuous mappings from $[0,h]$ into $\mathbf{R}$.*

*Proof.* It is noted that $C[0,h]$ is a Banach space by the maximum norm

$$\|u\| = \max\{|u(t)| \mid t \in [0,h]\}.$$

Instead of the Cauchy problem (3) we consider the integral equation

$$u(t) = \lambda t + \int_0^t (t-s)f(s,u(s))ds.$$

By the condition (c) there exists $h \in \mathbf{R}$ with $0 < h \leq 1$ such that

$$\int_0^h |f(t,\alpha t)|dt < \min\left\{\lambda - \alpha, \frac{\alpha}{\beta}\right\}.$$

Let $A$ be an operator from $X$ into $C[0, h]$ defined by

$$Au(t) = \lambda t + \int_0^t (t - s) f(s, u(s)) ds.$$

Since a mapping $t \longmapsto \lambda t$ belongs to $X$, $X \neq \emptyset$. Moreover $A(X) \subset X$. Indeed by the condition (a) $Au \in C[0, h]$, $Au(0) = 0$,

$$(Au)'(0) = \left[ \lambda + \int_0^t f(s, u(s)) ds \right]_{t=0} = \lambda$$

and by the condition (b)

$$Au(t) = \lambda t + \int_0^t (t - s) f(s, u(s)) ds$$

$$\geq \lambda t - t \int_0^h |f(s, u(s))| ds$$

$$\geq \lambda t - t \int_0^h |f(s, \alpha s)| ds$$

$$\geq \alpha t$$

for any $t \in [0, h]$. We will find a fixed point of $A$. Let $\varphi$ be an operator from $X$ into $C[0, h]$ defined by

$$\varphi[u](t) = \begin{cases} \frac{u(t)}{t}, & \text{if } t \in (0, h], \\ \lambda, & \text{if } t = 0, \end{cases}$$

and

$$\varphi[X] = \{\varphi[u] \mid u \in X\}$$
$$= \{v \mid v \in C[0, h], v(0) = \lambda \text{ and } \alpha \leq v(t) \text{ for any } t \in [0, h]\}.$$

Then $\varphi[X]$ is a closed subset of $C[0, h]$ and hence it is a complete metric space. Let $\Phi$ be an operator from $\varphi[X]$ into $\varphi[X]$ defined by

$$\Phi\varphi[u] = \varphi[Au].$$

By the mean value theorem for any $u_1, u_2 \in X$ there exists a mapping $\xi$ such that

$$\frac{f(t, u_1(t)) - f(t, u_2(t))}{u_1(t) - u_2(t)} = \frac{\partial f}{\partial u}(t, \xi(t))$$

and

$$\min\{u_1(t), u_2(t)\} \leq \xi(t) \leq \max\{u_1(t), u_2(t)\}$$

for any $t \in [0, h]$. By the conditions (b) and (d)

$$|f(t, u_1(t)) - f(t, u_2(t))| = \left| \frac{\partial f}{\partial u}(t, \xi(t))(u_1(t) - u_2(t)) \right|$$

$$\leq \left| \frac{\beta f(t, \xi(t))}{\xi(t)} \right| |u_1(t) - u_2(t)|$$

$$\leq \left| \frac{\beta f(t, \alpha t)}{\alpha t} \right| |u_1(t) - u_2(t)|$$

for almost every $t \in [0, h]$. Therefore

$$|\Phi\varphi[u_1](t) - \Phi\varphi[u_2](t)| = \left| \frac{1}{t} \int_0^t (t - s)(f(s, u_1(s)) - f(s, u_2(s))) ds \right|$$

$$\leq \int_0^h \left| \frac{\beta f(s, \alpha s)}{\alpha s} \right| |u_1(s) - u_2(s)| ds$$

$$\leq \frac{\beta}{\alpha} \int_0^h |f(s, \alpha s)| ds \| \varphi[u_1] - \varphi[u_2] \|$$

for any $t \in [0, h]$. Therefore

$$\| \Phi\varphi[u_1] - \Phi\varphi[u_2] \| \leq \frac{\beta}{\alpha} \int_0^h |f(s, \alpha s)| ds \| \varphi[u_1] - \varphi[u_2] \|.$$

By the Banach fixed point theorem there exists a unique mapping $\varphi[u] \in \varphi[X]$ such that $\Phi\varphi[u] = \varphi[u]$. Then $Au = u$. □

**Theorem 2.** *Suppose that a mapping $f$ from $[0, 1] \times [0, \infty)$ into $\mathbf{R}$ satisfies the following.*

(a)     *The mapping $f$ satisfies the Carathéodory condition, that is, the mapping $t \longmapsto f(t, u)$ is measurable for any $u \in (0, \infty)$ and the mapping $u \longmapsto f(t, u)$ is continuous for almost every $t \in [0, 1]$.*

(e)     $|f(t, u_1)| \leq |f(t, u_2)|$ *for almost every $t \in [0, 1]$ and for any $u_1, u_2 \in [0, \infty)$ with $u_1 \leq u_2$.*

(f)     *There exists $\alpha \in \mathbf{R}$ with $0 < \alpha < \lambda$ such that*

$$\int_0^1 |f(t, (2\lambda - \alpha)t)| dt < \infty.$$

(d)     *There exists $\beta \in \mathbf{R}$ with $\beta > 0$ such that*

$$\left| \frac{\partial f}{\partial u}(t, u) \right| \leq \frac{\beta |f(t, u)|}{u}$$

*for almost every $t \in [0, 1]$ and for any $u \in (0, \infty)$.*

*Then there exist $h \in \mathbf{R}$ with $0 < h \leq 1$ such that the Cauchy problem (3) has a unique solution in $X$, where $X$ is a subset*

$$X = \left\{ u \;\middle|\; \begin{array}{l} u \in C[0,h], u(0) = 0, u'(0) = \lambda \\ \text{and } \alpha t \le u(t) \le (2\lambda - \alpha)t \text{ for any } t \in [0,h] \end{array} \right\}$$

of $C[0,h]$.

*Proof.* By the condition (f) there exists $h \in \mathbf{R}$ with $0 < h \le 1$ such that

$$\int_0^h |f(t, (2\lambda - \alpha)t)| dt < \min\left\{\lambda - \alpha, \frac{\alpha}{\beta}\right\}$$

and let $A$ be an operator from $X$ into $C[0,h]$ defined by

$$Au(t) = \lambda t + \int_0^t (t - s) f(s, u(s)) ds.$$

Since a mapping $t \longmapsto \lambda t$ belongs to $X$, $X \ne \emptyset$. Moreover $A(X) \subset X$. Indeed by the condition (a) $Au \in C[0,h]$, $Au(0) = 0$,

$$(Au)'(0) = \left[\lambda + \int_0^t f(s, u(s)) ds\right]_{t=0} = \lambda$$

and by the condition (e)

$$\begin{aligned} Au(t) &= \lambda t + \int_0^t (t - s) f(s, u(s)) ds \\ &\ge \lambda t - t \int_0^h |f(s, u(s))| ds \\ &\ge \lambda t - t \int_0^h |f(s, (2\lambda - \alpha)s)| ds \\ &\ge \alpha t \end{aligned}$$

and

$$Au(t) = \lambda t + \int_0^t (t - s) f(s, u(s)) ds$$

$$\le \lambda t + t \int_0^h |f(s, u(s))| ds$$

$$\le \lambda t + t \int_0^h |f(s, (2\lambda - \alpha)s)| ds$$

$$\le (2\lambda - \alpha)t$$

for any $t \in [0, h]$. We will find a fixed point of $A$. Let $\varphi$ be an operator from $X$ into $C[0,h]$ defined by

$$\varphi[u](t) = \begin{cases} \frac{u(t)}{t}, & t \in (0, h], \\ \lambda, & t = 0, \end{cases}$$

and

$$\varphi[X] = \{\varphi[u] \mid u \in X\}$$
$$= \{v \mid v \in C[0, h], v(0) = \lambda \text{ and } \alpha \le v(t) \le (2\lambda - \alpha) \text{ for any } t \in [0, h]\}.$$

Then $\varphi[X]$ is a closed subset of $C[0, h]$ and hence it is a complete metric space. Let $\Phi$ be an operator from $\varphi[X]$ into $\varphi[X]$ defined by

$$\Phi\varphi[u] = \varphi[Au].$$

Then we can show just like Theorem 1 that by the Banach fixed point theorem there exists a unique mapping $\varphi[u] \in \varphi[X]$ such that $\Phi\varphi[u] = \varphi[u]$ and hence $Au = u$. □

## 3 Examples

In this section we give some examples to illustrrate the results above.

*Example 1.* In [3] the Cauchy problem (1) is considered. Since $f(t, u) = P(t)t^a u^\sigma$, $a, \sigma, \lambda \in \mathbf{R}$ with $\sigma < 0$ and $\lambda > 0$ and $P$ is a continuous mapping such that $\int_0^1 |P(t)|t^{a+\sigma} dt < \infty$, the conditions (a), (b), (c) and (d) are satisfied. Indeed (a), (b) and (c) are clear and since

$$\left|\frac{\partial f}{\partial u}(t, u)\right| = |P(t)t^a \sigma u^{\sigma-1}|$$

$$= \frac{|\sigma||f(t, u)|}{u},$$

(d) holds. By Theorem 1 the Cauchy problem (1) has a unique solution in

$$X = \left\{ u \,\middle|\, \begin{array}{l} u \in C[0, h], u(0) = 0, u'(0) = \lambda \\ \text{and } \alpha t \le u(t) \text{ for any } t \in [0, h] \end{array} \right\}.$$

*Example 2.* We consider the Cauchy problem

$$\begin{cases} u''(t) = a(t) + u(t), & t \in [0, 1], \\ u(0) = 0, \ u'(0) = \lambda, \end{cases} \tag{4}$$

where $a$ is positive and integrable, and $\lambda \in \mathbf{R}$ with $\lambda > 0$. Since $f(t, u) = a(t) + u$, the conditions (a), (e), (f) and (d) are satisfied. Indeed (a), (e) and (f) are clear and since

$$\left|\frac{\partial f}{\partial u}(t, u)\right| = 1 \le \frac{a(t) + u}{u} = \frac{|f(t, u)|}{u},$$

(d) holds. By Theorem 2 the Cauchy problem (4) has a unique solution in

$$X = \left\{ u \; \middle| \; \begin{array}{l} u \in C[0,h], u(0) = 0, u'(0) = \lambda \\ \text{and } \alpha t \leq u(t) \leq (2\lambda - \alpha)t \text{ for any } t \in [0,h] \end{array} \right\}.$$

*Example 3.* We consider the Cauchy problem

$$\begin{cases} u''(t) = a(t)u(t)^\sigma, \; t \in [0,1], \\ u(0) = 0, \; u'(0) = \lambda, \end{cases} \tag{5}$$

where $\int_0^1 |a(t)| t^\sigma dt < \infty$ and $\sigma, \lambda \in \mathbf{R}$ with $\lambda > 0$. Since $f(t, u) = a(t)u^\sigma$, the conditions (a), (b), (c) and (d) are satisfied if $\sigma < 0$ and the conditions (a), (e), (f) and (d) are satisfied if $\sigma \geq 0$. Indeed (a) is clear, (b) and (c) are clear if $\sigma < 0$, (e) and (f) are clear if $\sigma \geq 0$, and since

$$\left| \frac{\partial f}{\partial u}(t, u) \right| = \begin{cases} |a(t)\sigma u^{\sigma-1}|, & \text{if } \sigma \neq 0, \\ 0, & \text{if } \sigma = 0, \end{cases}$$

$$= \frac{|\sigma| |f(t, u)|}{u},$$

(d) holds. By Theorem 1 if $\sigma < 0$ and by Theorem 2 if $\sigma > 0$ the Cauchy problem (5) has a unique solution in

$$X = \left\{ u \; \middle| \; \begin{array}{l} u \in C[0,h], u(0) = 0, u'(0) = \lambda \\ \text{and } \alpha t \leq u(t) \text{ for any } t \in [0,h] \end{array} \right\}$$

and

$$X = \left\{ u \; \middle| \; \begin{array}{l} u \in C[0,h], u(0) = 0, u'(0) = \lambda \\ \text{and } \alpha t \leq u(t) \leq (2\lambda - \alpha)t \text{ for any } t \in [0,h] \end{array} \right\},$$

respectively.

**Acknowledgements.** The authors would like to thank Professor Naoki Shioji and anonymous referees for their valuable suggestions and comments. The authors would like to thank Mr. Naoya Anamizu, Mr. Takuya Kimura, Mr. Sousuke Sato and Ms. Tsubasa Mogami for their support.

# References

1. Erbe, L.H., Wang, H.: On the existence of positive solutions of ordinary differential equations. Proceedings of the American Mathematical Society 120(3), 743–748 (1994)
2. Davis, H.T.: Introduction to Nonlinear Differential and Integral Equations. Dover Publications, New York (1962)
3. Knežević-Miljanović, J.: On the Cauchy problem for an Emden-Fowler equation. Differential Equations 45(2), 267–270 (2009)
4. Wong, J.S.W.: On the generalized Emden-Fowler equation. SIAM Review 17(2), 339–360 (1975)

# Stability of a Two Epidemics Model

T. Dumrongpokaphan, W. Jaihonglam, and R. Ouncharoen

**Abstract.** An $SI_1I_2RS$ epidemic model is studied. We derive the sufficient conditions on the system parameters which guarantee that the equilibrium points of the system are locally asymptotically stable or globally asymptotically stable.

**Keywords:** Epidemic, Incidence rate, Stability.

## 1 Introduction

Recently, many diseases caused by virus have emerged. The danger level of the disease such as dengue fever and malaria is calculated by the spreading of the disease. We determine who is infected by virus by measuring the amount of the disease or pathogen in the patient. Therefore, the model of the germ's behavior plays an important role in finding the characteristics of the disease and its control.

The system proposed in [18] is based on the assumptions that the natural birth rate is not the same as the natural death rate and the recovered population may lose its immunity and become susceptible again. S.Mena-Lorca and H.W.Hcthcote [15] used the system propose in [18] as the basis model,and introduced the SIR model in which the death of infected population is possible and the recovered population can be re-infected. S.Mena Lorca and H.W.Hethcote also derived the parametric conditions to classify the equilibrium points in the similar manner as in [18].

T. Dumrongpokaphan, W. Jaihonglam, and R. Ouncharoen
Department of Mathematics, Chiang Mai University, Chiang Mai 50200, Thailand

T. Dumrongpokaphan and R. Ouncharoen
Centre of Excellence in Mathematics, CHE, Si Ayuttaya road,
Bangkok 10400, Thailand
e-mail: thongcha@chiangmai.ac.th, kookkai.15296@gmail.com,
rujira@chiangmai.ac.th

S. Li (Eds.): Nonlinear Maths for Uncertainty and its Appli., AISC 100, pp. 443–451.
springerlink.com &copy; Springer-Verlag Berlin Heidelberg 2011

La-di Wang and Jian-quan Li [17], on the other hend, adapted the system from [7] with the assumptions that the system has two diseases and the disease - related death of the infected population may occur. Their model incorporates two classes of infectious individuals with different infectivities and the incidence rate is nonlinear.They investigated local and global stability of the disease-free and endemic equilibrium points.

In this work two diseases are considered in the system and we assume that the recovered population can be re-infected. So,we obtain

$$\frac{dS(t)}{dt} = A - \mu S(t) - [\beta_{10}I_1(t)S(t) + \beta_{20}I_2(t)S(t)] + \delta_{10}R(t)$$

$$\frac{dI_1(t)}{dt} = p_1 [\beta_{10}I_1(t)S(t) + \beta_{20}I_2(t)S(t)] - (\mu + \alpha_{10} + \gamma_{10}) I_1(t) \qquad (1)$$

$$\frac{dI_2(t)}{dt} = p_2 [\beta_{10}I_1(t)S(t) + \beta_{20}I_2(t)S(t)] - (\mu + \alpha_{20} + \gamma_{20}) I_2(t)$$

$$\frac{dR(t)}{dt} = \gamma_{10}I_1(t) + \gamma_{20}I_2(t) - \mu R(t) - \delta_{10}R(t),$$

where $S(t)$is the density of susceptable population at time $t$,$I_i(t)(i = 1, 2)$ is the density of infective population in the class $i$ at time $t$,$R(t)$ is the density of recorvered population at time $t$, $N(t)$ is the density of total population at time $t$,where $N(t) = S(t) + I_1(t) + I_2(t) + R(t)$,$A$ is the recruitment rate of populations,$\mu$ is the natural death rate,$\alpha_{i0}(i = 1, 2)$ is the disease-related death rate in the class $i$,$\gamma_{i0}(i = 1, 2)$is the recovery rate in the class $i$,$p_i(i = 1, 2)$is the probability of incidence in the class $i$ , $p_1 + p_2 = 1$,$\beta_{i0}I_iS(i = 1, 2)$ is the incidence rate in the class $i$,$\delta_{10}$ is the rate that recovered individuals lose immunity and return to the susceptible class,and all parameters are positive.

## 2　Main Results

We investigate local asymptotical stability and gobal asymptotical stability of the disease-free equilibrium point and the endemic equilibrium point. Letting $\tau = \mu t$ , we obtain the following system analogous to (1):

$$\frac{dS}{d\tau} = \frac{A}{\mu} - S - [\beta_1 I_1 S + \beta_2 I_2 S] + \delta R$$

$$\frac{dI_1}{d\tau} = p_1 [\beta_1 I_1 S + \beta_2 I_2 S] - (1 + \alpha_1 + \gamma_1) I_1 \qquad (2)$$

$$\frac{dI_2}{d\tau} = p_2 [\beta_1 I_1 S + \beta_2 I_2 S] - (1 + \alpha_2 + \gamma_2) I_2$$

$$\frac{dR}{d\tau} = \gamma_1 I_1 + \gamma_2 I_2 - R - \delta R$$

where $\beta_1 = \dfrac{\beta_{10}}{\mu}, \beta_2 = \dfrac{\beta_{20}}{\mu}, \delta = \dfrac{\delta_{10}}{\mu}, \alpha_1 = \dfrac{\alpha_{10}}{\mu}, \alpha_2 = \dfrac{\alpha_{20}}{\mu}, \gamma_1 = \dfrac{\gamma_{10}}{\mu}, \gamma_2 = \dfrac{\gamma_{20}}{\mu}.$

## 2.1  Disease - Free Equilibrium Point

Letting $d_i = 1 + \alpha_i + \gamma_i (i = 1, 2)$, the system (2) becomes :

$$\frac{dS}{d\tau} = \frac{A}{\mu} - S - [\beta_1 I_1 S + \beta_2 I_2 S] + \delta R$$

$$\frac{dI_1}{d\tau} = p_1 [\beta_1 I_1 S + \beta_2 I_2 S] - d_1 I_1 \qquad (3)$$

$$\frac{dI_2}{d\tau} = p_2 [\beta_1 I_1 S + \beta_2 I_2 S] - d_2 I_2$$

$$\frac{dR}{d\tau} = \gamma_1 I_1 + \gamma_2 I_2 - R - \delta R.$$

Next consider the system (3) by letting $N = S + I_1 + I_2 + R$.
Then, $\dfrac{dN}{d\tau} = \dfrac{A}{\mu} - N - (\alpha_1 I_1 + \alpha_2 I_2)$

Next we will prove that all solutions of the system (3) with positive data will remain positive for all times $\tau > 0$.

**Lemma 1.** *Let the initial solutions be* $S(0) > 0$, $I_1(0) > 0$, $I_2(0) > 0$, *and* $R(0) > 0$. *Then solutions* $S(\tau), I_1(\tau), I_2(\tau)$ *and* $R(\tau)$ *of the system* (3) *are positive for all* $\tau > 0$.

Thus the total population size $N$ may vary in time. In the absence of disease, the population size $N$ converges to the equilibrium $\dfrac{A}{\mu}$.

For biological considerations, we study the system (3) in the closed set

$$\Omega = \left\{ (S, I_1, I_2, R) \in \mathbb{R}_+^4 ; 0 \le S + I_1 + I_2 + R = N \le \frac{A}{\mu} \right\}$$

where $\mathbb{R}_+^4$ denotes the non - negative cone of $R^4$ including its lower dimensional faces. It can be verified by Lemma 1 that $\Omega$ is a positive invariant set with respect to the system (3). $\partial \Omega$ and $\dot{\Omega}$ denote the boundary and the interior of $\Omega$, respectively. The disease - free equilibrium point of the system (3) is $E_0 = \left( \dfrac{A}{\mu}, 0, 0, 0 \right) \in \Omega$, and it exists for all nonegative values of the parameters. Any equilibrium in $\dot{\Omega}$ corresponds to the disease being endemic and is called an endemic equilibrium. From the system (3) we obtain

$$I_1 = \frac{p_1 d_2 I_2}{d_1 p_2}. \qquad (4)$$

The disease-free equilibrium point corresponds to $I_1 = 0, I_2 = 0$ and $R = 0$.
Substituting $I_1 = 0, I_2 = 0$ and $R = 0$ into system (3), we obtain $S = \dfrac{A}{\mu}$.

### 2.1.1 Local Asymptotic Stability of the Disease-Free Equilibrium Point

Here, we derive the conditions which guarantee that the disease-free equilibrium point is locally asymptotically stable. Let $\Re_0 = \dfrac{p_1\beta_1 A}{d_1\mu} + \dfrac{p_2\beta_2 A}{d_2\mu})$ which will be called the basic reproduction number, that is the number of secondary infectious cases produced by an infectious individual during his or her effective infectious period when introduced into the population of susceptibles.

**Theorem 1.** *If $\Re_0 < 1$ , then the disease-free equilibrium point $E_0$ is locally asymptotically stable, and $E_0$ is unstable if $\Re_0 > 1$ .*

**Proof.** The characteristic equation is

$$0 = (\lambda + 1)(\lambda + 1 + \delta) \times$$
$$\left[ \lambda^2 + \left( d_1 + d_2 - \left( \frac{p_1\beta_1 A}{\mu} + \frac{p_2\beta_2 A}{\mu} \right) \right) \lambda + d_1 d_2 \left( 1 - \left( \frac{p_1\beta_1 A}{\mu} + \frac{p_2\beta_2 A}{\mu} \right) \right) \right].$$

Consider

$$0 = \lambda^2 + \left( d_1 + d_2 - \left( \frac{p_1\beta_1 A}{\mu} + \frac{p_2\beta_2 A}{\mu} \right) \right) \lambda + d_1 d_2 \left( 1 - \left( \frac{p_1\beta_1 A}{d_1\mu} + \frac{p_2\beta_2 A}{d_2\mu} \right) \right).$$

Equivalently,

$$0 = \lambda^2 + \left( d_1 + d_2 - \left( \frac{p_1\beta_1 A}{\mu} + \frac{p_2\beta_2 A}{\mu} \right) \right) \lambda + d_1 d_2 \left( 1 - \Re_0 \right) \qquad (5)$$

If $\Re_0 < 1$, then $\dfrac{p_1\beta_1 A}{d_1\mu} < 1$ and $\dfrac{p_2\beta_2 A}{d_2\mu} < 1$, which means $d_1 > \dfrac{p_1\beta_1 A}{\mu}$ and $d_2 > \dfrac{p_2\beta_2 A}{\mu}$. Therefore, $d_1 + d_2 - \left( \dfrac{p_1\beta_1 A}{\mu} + \dfrac{p_2\beta_2 A}{\mu} \right) > 0$.

Thus, all roots of the equation (5) have negative real parts if $\Re_0 < 1$ and one of its root has positive real part if $\Re_0 > 1$. Therefore, if $\Re_0 < 1$ then the disease-free equilibrium point $E_0$ is locally asymptotically stable. $\qquad \square$

### 2.1.2 Global Asymptotic Stability of the Disease-Free Equilibrium Point

We have seen that the disease-free equilibrium point is unstable if $\Re_0 > 1$. In this section, we shall prove the global stability of disease-free equilibrium point under the condition $\Re_0 \leq 1$.

**Theorem 2.** *If $\Re_0 \leq 1$, then the disease-free equilibrium point $E_0$ is globally asymptotically stable.*

**Proof.** Consider the Lyapunov function (constructed in [17])
$V(S, I_1, I_2, R) = \frac{\beta_1}{d_1} I_1 + \frac{\beta_2}{d_2} I_2$ . Its derivative along the solutions to the system
(3) is

$$\frac{dV}{d\tau}\Big|_{(3)} = \frac{\beta_1}{d_1}\frac{dI_1}{d\tau} + \frac{\beta_2}{d_2}\frac{dI_2}{d\tau} \leq (\Re_0 - 1)(\beta_1 I_1 + \beta_2 I_2).$$

Since $\Re_0 \leq 1$ and $S \leq \frac{A}{\mu}$, then $\frac{dV}{d\tau}\Big|_{(3)} \leq 0$.

When $\Re_0 < 1$ , the set $\bar{\Omega} = \left\{(S, I_1, I_2, R) \in \Omega : \frac{dV}{d\tau}\Big|_{(3)} = 0\right\}$, is the same as
the set $\{(S, I_1, I_2, R) \in \Omega : I_1 = I_2 = 0\}$. Therefore, from the first equation
of the system (3), we obtain $S = \frac{A}{\mu} - e^{-\tau}$.
Now,

$$\lim_{\tau \to +\infty} S(\tau) = \lim_{\tau \to +\infty} \left(\frac{A}{\mu} - e^{-\tau}\right) = \frac{A}{\mu}.$$

Next, we consider $\frac{dV}{d\tau} \leq (\Re_0 - 1)(\beta_1 I_1 + \beta_2 I_2) \leq (\Re_0 - 1) kV$
where $k = max\left\{\frac{\beta_1}{d_1}, \frac{\beta_2}{d_2}\right\}$. Therefore,

$$\frac{dV}{d\tau} \leq (\Re_0 - 1) kV$$
$$V \leq e^{(1-\Re_0)k\tau}.$$

If $\tau \to +\infty$ then $V(\tau) \to 0$, provided that $\Re_0 < 1$. Since $V(S, I_1, I_2, R) = \frac{\beta_1}{d_1} I_1 + \frac{\beta_2}{d_2} I_2$, we obtain

$$\lim_{\tau \to +\infty} I_1(\tau) = \lim_{\tau \to +\infty} I_2(\tau) = 0.$$

From the fourth equation of the system (3) if $I_1 = I_2 = 0$, we obtain $R = e^{-(1+\delta)\tau}$. Thus,
$$\lim_{\tau \to +\infty} R(\tau) = \lim_{\tau \to +\infty} e^{-(1+\delta)\tau} = 0.$$

Therefore, the disease - free equilibrium point $E_0$ is globally asymptotically
stable on the set $\Omega$ if $\Re_0 < 1$. When $\Re_0 = 1$, the set

$$\bar{\Omega} = \left\{(S, I_1, I_2, R) \in \Omega : \frac{dV}{d\tau}\Big|_{(3)} = 0\right\}$$

is one of the folowing four sets :

$\bar{\Omega}_1 = \{(S, I_1, I_2, R) \in \Omega : I_1 = I_2 = 0\},$

$\bar{\Omega}_2 = \left\{(S, I_1, I_2, R) \in \Omega : I_1 = 0, S = \dfrac{A}{\mu}\right\},$

$\bar{\Omega}_3 = \left\{(S, I_1, I_2, R) \in \Omega : I_2 = 0, S = \dfrac{A}{\mu}\right\},$

$\bar{\Omega}_4 = \left\{(S, I_1, I_2, R) \in \Omega : S = \dfrac{A}{\mu}\right\}.$

We are led to

$$\bigcup_{i=1}^{4} \overline{\Omega_i} = \overline{\Omega}_1 \cup \overline{\Omega}_4.$$

We consider set $\bar{\Omega}_1$ when $I_1 = I_2 = 0$ and substitute $I_1 = I_2 = 0$ into the system (3). We obtain $S = \dfrac{A}{\mu}$ and $R = 0$. Next, we consider the set $\bar{\Omega}_4$ when $S = \dfrac{A}{\mu}$ and substitute $S = \dfrac{A}{\mu}$ into the system (3). We obtain $I_1 = I_2 = R = 0$. Then, the largest compact invariant set of the system (3) on the set $\bar{\Omega}$ is the singleton set $\{E_0\}$. So the LaSalle's Invariance Principle [14] implies that $E_0$ is globally asymptotically stable on the set $\Omega$ if $\Re_0 = 1$.    □

## 2.2  Endemic Equilibrium Point

Now, we consider the endemic equilibrium point of the system (3) and its local stability and global stability. Consider at equilibrium point and using (4) we are led to

$$d_2 I_2 \left[\frac{\beta_1 p_1 S}{d_1} + \frac{\beta_2 p_2 S}{d_2} - 1\right] = 0.$$

At the endemic equilibrium point, $I_1 \neq 0$ and $I_2 \neq 0$. Then,

$$\frac{\beta_1 p_1 S}{d_1} + \frac{\beta_2 p_2 S}{d_2} - 1 = 0. \tag{6}$$

Let

$$F(S) = \frac{\beta_1 p_1 S}{d_1} + \frac{\beta_2 p_2 S}{d_2} - 1. \tag{7}$$

We obtain $\dfrac{dF(S)}{dS} = \dfrac{\beta_1 p_1}{d_1} + \dfrac{\beta_2 p_2}{d_2}.$

Since all parameters are positive, $\dfrac{dF(S)}{dS} > 0$. Thus, $F(S) > 0$ is an increasing function. At $S = 0$, we have $F(0) = -1 < 0$. At $S = \dfrac{A}{\mu}$, we have $F\left(\dfrac{A}{\mu}\right) = \Re_0 - 1$. If $\Re_0 > 1$, then $F\left(\dfrac{A}{\mu}\right) > 0$.

Since $F(S)$ is an increasing and continuous function, then there exists a positive root $S \in \left(0, \dfrac{A}{\mu}\right)$ by the intermediate value theorem.

Next, we consider the endemic equilibrium point $E^* = (S^*, I_1^*, I_2^*, R^*)$ by letting $\dfrac{dS}{d\tau} = \dfrac{dI_1}{d\tau} = \dfrac{dI_2}{d\tau} = \dfrac{dR}{d\tau} = 0$. We then have the endemic equilibrium point $E^* = (S^*, I_1^*, I_2^*, R^*)$, where $S^* = \dfrac{d_1 d_2}{\beta_1 p_1 d_2 + \beta_2 p_2 d_1}$, $I_1^* = \dfrac{(A - \mu S^*)(1+\delta) p_1 d_2}{d_1 d_2 + \delta(d_1 d_2 - \gamma_1 p_1 d_2 - \gamma_2 p_2 d_1)}$, $I_2^* = \dfrac{(A - \mu S^*)(1+\delta) p_2 d_1}{d_1 d_2 + \delta(d_1 d_2 - \gamma_1 p_1 d_2 - \gamma_2 p_2 d_1)}$, $R^* = \dfrac{(A - \mu S^*)(\gamma_1 p_1 d_2 + \gamma_2 p_2 d_1)}{d_1 d_2 + \delta(d_1 d_2 - \gamma_1 p_1 d_2 - \gamma_2 p_2 d_1)}$.

## 2.2.1  Local Asymptotic Stability of the Endemic Equilibrium Point

Here, we will derive for the conditions which ensure that the endemic equilibrium point is locally asymptotically stable.

**Theorem 3.** *If $\Re_0 > 1$, then the endemic equilibrium point $E^*(S^*, I_1^*, I_2^*, R^*)$ is locally asymptotically stable.*

**Proof.** Letting $B = \dfrac{d_2 p_1 \beta_1 S^*}{d_1} + \dfrac{d_1 p_2 \beta_2 S^*}{d_2}$, the characteristic equation of $J(E^*)$ is $\lambda^4 + a_1 \lambda^3 + a_2 \lambda^2 + a_3 \lambda + a_4 = 0$, where $a_1 = 2 + m + \delta + B > 0$, $a_2 = (1+\delta)(1+m) + m(d_1 + d_2) + (2+\delta)B > 0$, $a_3 = m\delta(2 + \alpha_1 + \alpha_2 + p_1\gamma_2 + p_2\gamma_1) + m(d_1 + d_2 + d_1 d_2) + (1+\delta)B > 0$, $a_4 = m d_1 d_2 + m\delta[1 + (1+\alpha_2)\gamma_1 p_2 + (1+\alpha_1)\gamma_2 p_1 + (1+\gamma_2)\alpha_1 + (1+\gamma_1)\alpha_2] > 0$.

By the Routh - Hurwitz criteria, the endemic equilibrium point is locally asymptotically stable if $a_1 > 0, a_3 > 0, a_4 > 0$ and $a_1 a_2 a_3 > a_3^2 + a_1^2 a_4$, we consider $a_1 a_2 a_3 - a_3^2 - a_1^2 a_4$ which yields

$$a_1 a_2 a_3 - a_3^2 - a_1^2 a_4$$
$$= m\delta B(1+\delta)[(2+\delta)(2+\alpha_1+\alpha_2) + (1+B)(p_1\gamma_2 + p_2\gamma_1) +$$
$$B(1+\delta)^2] + B^2[(1+\delta)(2+\delta)(1+\delta+B) + m\delta(p_1\gamma_2 + p_2\gamma_1)] +$$
$$m\delta(2 + \alpha_1 + \alpha_2 + p_1\gamma_2 + p_2\gamma_1)[B + mB(2+\delta) + m\delta(p_1\gamma_2 + p_2\gamma_1)] +$$
$$m(d_1 + d_2 + d_1 d_2)[m\delta(p_1\gamma_2 + p_2\gamma_1) + B + B^2(2+\delta)] +$$
$$mB^2(1+\delta)[(1+\delta)(2+\delta) + 1] + m\left(\dfrac{d_2^2 p_1 \beta_1 S^*}{d_1} + \dfrac{d_1^2 p_2 \beta_2 S^*}{d_2}\right) \times$$
$$[m\delta(2 + \alpha_1 + \alpha_2 + p_1\gamma_1 + p_2\gamma_1) + m(d_1 + d_2 + d_1 d_2) + B(1+\delta)].$$

We observe that $a_1 a_2 a_3 - a_3^2 - a_1^2 a_4 > 0$ if $\Re_0 > 1$. Thus, the endemic equilibrium point $E^*$ is locally asymptotically stable.                    □

## 3 Conclusions

This $SI_1 I_2 RS$ model incorporates the assumption that the recovered population can be re-infected. In this study, local asymptotic stability and global asymptotic stability are examined at the system equilibrium points which are disease-free and endemic. This research finds how these two equilibrium points could be stable. If the disease-free equilibrium point is stable, then the virus may be eradicated. On the other hand, if the endemic equilibrium point is stable, the infected population is reaches a constant level meaning that the infected rate equals the recovery rate. Consequently, we can predict the behavior of the spread of the disease and appropriate prevention program can be efficiently executed.

**Acknowledgements.** The research is partly supported by Centre of Excellence in Mathematics, CHI, Thailand and Faculty of Science Chiang Mai University, Thailand.

## References

1. Barnett, S., Cameron, R.G.: Introduction to Mathematical Control Theory. University of Bradford (1985)
2. Beltraml, E.: Mathematics for Dynamic Modeling. State University of New York (1998)
3. Beretta, E., Takeuchi, Y.S.: Global stability of an SIR epidemic model with time delays. J.Math. Biol. 40, 250–260 (1995)
4. Blyuss, K.B., Kyrychko, Y.N.: On a basic model of a two-disease epidemic. Appl. Math. Comput. 160, 177–187 (2005)
5. Cooke, K.L., vanden Driessche, P.: Analysis of an SEIRS epidemic model with two delays. J.Math. Biol. 35, 240–260 (1996)
6. De Leon, C.V.: Constructions of Lyapunov functions for classics SIS,SIR and SIRS epidemic model with vairable population size. UNAM, Mexico (2009)
7. Edelstein - Keshet, L.: Mathematical Models in Biology. Duke University, Random House, New York (1988)
8. Hethcote, H.W.: The mathematics of infectious diseases. SIAM Rev. 42, 599–653 (2000)
9. Kermack, W.O., McKendrick, A.G.: A contribution tu the Mathematical Theory of Epidemics. P. Soc. London 115, 700–721 (1927)
10. Khalil, H.K.: Nonlinear System. Upper Saddle River (1996)
11. Kyrychko, Y.N., Blyuss, K.B.: Global properties of a delayed SIR model with temporary immunity and nonlinear incidence rate. Nonlinear Anal. Real World Appl. 6, 495–507 (2005)
12. Li, M.Y., Graef, J.R., Wang, L., Karsai, J.: Global dynamics of a SEIR epidemic model with a varying total population size. Math. Biosci. 160, 191–213 (1999)

13. Li, M.Y., Smith, H.L., Wang, L.: Global dynamics of an SEIR epidemic model with vertical transmission. SIAM J. Appl. Math. 62 (2001)
14. LsSalle, J.P.: The Stabillity of Dynamic Systems, Philadelphia.PA (1976)
15. Mena-Lorca, Hethcote, H.W.: Dynamic - models of infections diseases as regulators of population size. J. Math. Biosci. 30, 693–716 (1992)
16. Thieme, H.R., Castillo-Chavez, C.: The role of variable infectivity in the human immumodeficiency virus epidemic. In: Engineering materials, pp. 157–177. Springer, Heidelberg (1989)
17. Wang, L.D., Li, J.Q.: Global stability of an epidemic model with nonlinear incidence rate and differential infectivity. Appl. Math. Comput. 161, 769–778 (2005)
18. Weber, A., Weber, M., Milligan, P.: Modeling epidemics caused by respiratory syncytial virus (RSV). J.Math. Biosci. 172, 95–113 (2001)

# The Impulsive Synchronization for m-Dimensional Reaction-Diffusion System

Wanli Yang and Suwen Zheng

**Abstract.** In this paper, an m-dimensional impulsive reaction-diffusion system is studied. Sufficient conditions are obtained for the global existence of solution for the impulsive system. By considering the equi-attractivity property of the impulsive error system, the impulsive synchronization of the m-dimensional reaction-diffusion system is investigated, and the sufficient conditions leading to the equi-attractivity property are obtained.

**Keywords:** Impulsive system, Reaction-diffusion systems, Global solution, Synchronization, Equi-attractivity.

## 1 Introduction

In this paper, the following m-dimensional impulsive predator-prey reaction-diffusion system is considered.

$$
\begin{cases}
\frac{\partial u_i}{\partial t} = d_i \Delta u_i + u_i(a_i - b_i u_i^2 - \sum\limits_{\substack{j=1 \\ j \neq i}}^{m} \frac{c_{ij}}{1+u_j}), t \neq t_k, \\
u_i(t_k + 0, x) = p_{ik}(x), \\
u_i(0, x) = u_{i0}(x), \\
u_i(t, x) = 0, \quad x \in \partial\Omega, \quad i = 1, 2, \cdots, m.
\end{cases}
\tag{P}
$$

where $x \in \Omega$, and $\Omega \in \mathbb{R}^n$ is a fixed bounded domain with smooth boundary $\partial\Omega$. $d_i, a_i, b_i, c_{ij}(i, j = 1, 2, \cdots, m)$ are positive constants, $u_{i0}, p_{ik}(i = 1, 2, \cdots, m)$ are given nonnegative functions.

Wanli Yang and Suwen Zheng
Institute of Nonlinear Science, Academy of Armored Forces Engineering,
Beijing, China
e-mail: `ywl-math@21cn.com,deehui@sina.com`

S. Li (Eds.): Nonlinear Maths for Uncertainty and its Appli., AISC 100, pp. 453–461.
springerlink.com          © Springer-Verlag Berlin Heidelberg 2011

We take the following assumptions

$$(\mathbf{H_0}): \begin{cases} u_{i0}, p_{ik} \in H^{1+\varepsilon}(\Omega) \cap L^{\infty}(\Omega), 0 < \varepsilon < 1, \\ u_{i0}(x) > 0, p_{ik}(x) > 0, x \in \Omega, \\ u_{i0} = p_{ik} = 0, x \in \partial\Omega, \\ i = 1, 2, \cdots, m; k = 1, 2, \cdots \end{cases}$$

The moments of impulse satisfy $0 < t_1 < \ldots < t_k < t_{k+1} < \ldots$, and $t_k \to \infty$, as $k \to \infty$. And for any $i = 1, 2, \cdots m$,

$$u_i(t_k + 0, x) = \lim_{t \to t_k^+} u_i(t, x), u_i(t_k - 0, x) = \lim_{t \to t_k^-} u_i(t, x) \triangleq u_i(t_k, x)$$

The system (P) arises in mathematical biology as a model of some competitive species which interact each other and migrate under self and cross-diffusion effects [7] [1]. In such a model, $u_i(i = 1, 2, \cdots, m)$ are population densities of some species.

The theory of impulsive ordinary differential equations and its applications to the fields of science and engineering have been very active research topics [16] [8] [9] [10] [14] [15]. Extending the theory of impulsive differential equations to partial differential equations has also gained considerable attention recently [5] [2] [3] [4]. Several differential inequalities are obtained, and asymptotic stability results, comparison results and uniqueness results involving first order PDE's and first order partial differential-functional equations are established using the method of Lyapunov functional equation. Unfortunately, there has been no theoretical analysis of the impulsive synchronization for impulsive reaction-diffusion systems, such as systems (P).

In this paper, we shall investigate the global existence of the solution for system (P) and the impulsive synchronization. Here, system (P) is the driven system, and the driving system is described as follows:

$$(\tilde{P}) \begin{cases} \frac{\partial \tilde{u}_i}{\partial t} = d_i \Delta \tilde{u}_i + \tilde{u}_i(a_i - b_i \tilde{u}_i^2 - \sum_{\substack{j=1 \\ j \neq i}}^{m} \frac{c_{ij}}{1+\tilde{u}_j}), t \neq t_k, \\ \tilde{u}_i(0, x) = \tilde{u}_{i0}(x), \\ \tilde{u}_i(t, x) = 0, \quad x \in \partial\Omega, \quad i = 1, 2, \cdots, m. \end{cases}$$

Let $e_i(t, x) = \tilde{u}_i(t, x) - u_i(t, x)$, $e_{i0}(x) = \tilde{u}_{i0}(x) - u_{i0}(x)$, then, the error system between reaction-diffusion system ($\tilde{P}$) and the impulsive reaction-diffusion system (P) can be described by the following equations:

$$(E) \begin{cases} \frac{\partial e_i}{\partial t} = d_i \Delta e_i + \varphi_i(u_i, \tilde{u}_i), t \neq t_k, \\ e_i(t_k + 0, x) = \tilde{u}_i(t_k, x) - p_{ik}(x), \\ e_i(0, x) = e_{i0}(x), \\ e_i(t, x) = 0, \quad x \in \partial\Omega, \quad i = 1, 2, \cdots, m. \end{cases}$$

where

$$\varphi_i(u_i, \tilde{u}_i) = \tilde{u}_i(a_i - b_i\tilde{u}_i^2 - \sum_{\substack{j=1 \\ j \neq i}}^m \frac{c_{ij}}{1 + \tilde{u}_j}) - u_i(a_i - b_iu_i^2 - \sum_{\substack{j=1 \\ j \neq i}}^m \frac{c_{ij}}{1 + u_j})$$

$$= a_ie_i - b_ie_i(\tilde{u}_i^2 + \tilde{u}_iu_i + u_i^2) - \sum_{\substack{j=1 \\ j \neq i}}^m c_{ij}\frac{e_i + e_iu_j - u_ie_j}{(1 + \tilde{u}_j)(1 + u_j)}.$$

In order to state the main results, the definition of the uniformly equi-attractive in the large for the solutions of the impulsive error systems (E) is introduced as follows.

**Definition 1.1.** [1] Solution of the error system (E) is said to be uniformly equi-attractive in the large if for each $\epsilon > 0, \delta > 0$, there exists a number $T = T(\epsilon, \delta) > 0$ such that $\|(e_1(0, x), e_2(0, x), \cdots, e_m(0, x))\|_2 < \delta$ implies $\|(e_1(t, x), e_2(t, x)), \cdots, e_m(t, x))\|_2 < \epsilon$ for $t \geq T$. where

$$\|(e_1, e_2, \cdots, e_m)\|_2 \triangleq \left( \int_\Omega \sum_{i=1}^m e_i^2 dx \right)^{\frac{1}{2}}.$$

## 2 The Global Existence of the Solution

Consider the following logistic differential equation with impulsive actions $(k = 1, 2, \cdots)$:

$$\begin{cases} \frac{dw}{dt} = w(a_0 - b_0w^r), t \neq t_k, \\ w(t_k + 0) = w_k, \\ w(0) = w_0. \end{cases} \tag{1}$$

where $a_0, b_0, r > 0$, and $w_0 > 0, w_k > 0(k = 1, 2, \cdots)$.

The following assertion is significant as an auxiliary result of our paper.

**Lemma 2.1.** Every solution $w(t) = w(t, 0, w_0; t_k, w_k)$ of the system (1) satisfies

$$0 < w(t) \leq \max_{k=0,1,2,\cdots} \left\{ \left( \frac{a_0}{b_0} \right)^{\frac{1}{r}}, w_k \right\}.$$

**Proof.** For $t \in (0, t_1]$, we can obtain

$$w^r(t) = \frac{a_0w_0^r}{a_0e^{-a_0rt} + b_0w_0^r(1 - e^{-a_0rt})} \leq \begin{cases} \frac{a_0}{b_0}, a_0 - b_0w_0^r \geq 0, \\ w_0^r, a_0 - b_0w_0^r < 0. \end{cases} \tag{2}$$

It implies that

$$w(t) \leq \begin{cases} (\frac{a_0}{b_0})^{\frac{1}{r}}, a_0 - b_0w_0^r \geq 0, \\ w_0, \quad a_0 - b_0w_0^r < 0. \end{cases} \tag{3}$$

For $t \in (t_k, t_{k+1}]$, we can obtain that

$$w^r(t) = \frac{a_0 w_k^r}{a_0 e^{-a_0 r(t-t_k)} + b_0 w_k^r[1 - e^{-a_0 r(t-t_k)}]} \leq \begin{cases} \frac{a_0}{b_0}, & a_0 - b_0 w_k^r \geq 0, \\ w_k^r, & a_0 - b_0 w_k^r < 0. \end{cases} \quad (4)$$

It implies that

$$w(t) \leq \begin{cases} (\frac{a_0}{b_0})^{\frac{1}{r}}, & a_0 - b_0 w_k^r \geq 0, \\ w_k, & a_0 - b_0 w_k^r < 0. \end{cases} \quad (5)$$

The following comparison theorems will be needed throughout the paper.

**Theorem 2.2.** (Walter) [13] Suppose that vector-functions

$$v(t,x) = (v_1(t,x), \cdots, v_m(t,x)), w(t,x) = (w_1(t,x), \cdots, w_m(t,x)), m \geq 1,$$

satisfy the following conditions:

(T1) they are of class $C^2$ in $x \in \Omega$ and of class $C^1$ in $(t,x) \in [a,b] \times \bar{\Omega}$, where $\Omega \subset R^n$ is a bounded domain with smooth boundary;

(T2) $v_t - \mu \Delta v - g(t,x,v) \leq w_t - \mu \Delta w - g(t,x,w)$, where $(t,x) \in [a,b] \times \Omega, \mu = (\mu_1, \ldots, \mu_m) > 0$ (inequalities between vectors are satisfied coordinate-wise), vector-function $g(t,x,u) = (g_1(t,x,u), \ldots, g_m(t,x,u))$ is continuously differentiable and quasi-monotonically increasing with respect to $u = (u_1, \cdots, u_m)$:

$$\frac{\partial g_i(t,x,u_1, \ldots, u_m)}{\partial u_j} \geq 0, i,j = 1, \ldots, m, \ i \neq j;$$

(T3) $v = w = 0, (t,x) \in [a,b] \times \partial \Omega$.
Then, $v(t,x) \leq w(t,x)$ for $(t,x) \in [a,b] \times \bar{\Omega}$.

**Theorem 2.3.** (Smith) [12] Assume that $T$ and $d$ are positive real numbers, the function $u(t,x)$ is continuous on $(0,T] \times \Omega$, continuously differentiable in $x \in \bar{\Omega}$, with continuous derivatives $\partial^2 u / \partial x_i \partial x_j$ and $\partial u / \partial t$ on $(0,T] \times \Omega$, and $u(t,x)$ satisfies the following inequalities:

$$\begin{cases} \frac{\partial u}{\partial t} - d\Delta u + c(t,x)u \geq 0, & (t,x) \in (0,T] \times \Omega, \\ u \geq 0, & (t,x) \in (0,T] \times \partial \Omega, \\ u(0,x) \geq 0, & x \in \Omega. \end{cases}$$

where $c(t,x)$ is bounded on $(0,T] \times \Omega$. Then $u(t,x) \geq 0$ on $(0,T] \times \bar{\Omega}$.
The local existence of the solution for the system (P) is well-known:

**Theorem 2.4.**[15-17] The system (P) admits a unique nonnegative solution $(u_1, u_2, \cdots, u_m) : \Omega \times [0,T^*) \to R_+^2$ where $0 < T^* \leq \infty$. And, if $T^* < \infty$, then

$$\sum_{i=1}^m \|u_i(t,0)\|_{\infty,\Omega} \to \infty, \quad \text{as } t \to T^*.$$

On the global existence of the solution for impulsive system (P), we have the following results:

**Theorem 2.5.** Under the assumption $(\mathbf{H_0})$, the solution of (P) is global existence and ultimately bounded. And

$$0 < u_i(t, x) \leq M_i, \quad i = 1, 2, \cdots, m$$

where

$$M_i = \max_{k=1,2,\cdots} \left\{ \sqrt{\frac{a_i}{b_i}}, \|u_{i0}(x)\|_\infty, \|p_{ik}(x)\|_\infty \right\}.$$

**Proof.** Let $u_i$ be the solution of system (P) in $(0, T^*)$. If $T^* < \infty$, then, there exists $k_0$ such that

$$t_{k_0} \leq T^* < t_{k_0+1}.$$

(1) $T^* \in (t_{k_0}, t_{k_0+1})$.

Let $v_i$ be the solution of the following impulsive differential equations respectively:

$$\begin{cases} \frac{dv_i}{dt} = v_i(a_i - b_i v_i^2), \\ v_i(t_k + 0) = \|p_{ik}(x)\|_\infty, \\ v_i(0) = \|u_{i0}(x)\|_\infty. \end{cases} \tag{6}$$

By the lemma 2.1, for $t \in [0, T^*)$, we have that

$$0 < v_i(t) \leq \max_{1 \leq k \leq k_0} \left\{ \sqrt{\frac{a_i}{b_i}}, \|u_{i0}(x)\|_\infty, \|p_{ik}(x)\|_\infty \right\} \tag{7}$$

From the assumption $(\mathbf{H_0})$, we obtain that

$$-v_i(a_i - b_i v_i^2) \leq -v_1(a_i - b_i v_i^2 - \frac{c_i}{1+v_i}) \tag{8}$$

It implies that

$$\begin{aligned} 0 &= \frac{du_i}{dt} - d_i \Delta u_i - u_i(a_i - b_i u_i^2 - \sum_{\substack{j=1 \\ j \neq i}}^{m} \frac{c_{ij}}{1+u_j}) \\ &= \frac{dv_i}{dt} - v_i(a_i - b_i v_i^2) \\ &\leq \frac{dv_i}{dt} - v_i(a_i - b_i v_i^2 - \sum_{\substack{j=1 \\ j \neq i}}^{m} \frac{c_{ij}}{1+v_j}) \end{aligned} \tag{9}$$

Using the comparison theorem 2.2, we have that

$$0 < u_i(t, x) \leq v_i(t), x \in \Omega, t \in [0, T^*), i = 1, 2, \cdots, m. \tag{10}$$

From (7) and theorem 2.4, we know that $T^* = \infty$.

(2) $T^* = t_{k_0}$.

Similarity to (10), for $t \in [0, T^*)$, we can obtain that

$$0 < u_i(t) \leq \max_{1 \leq k \leq k_0 - 1} \left\{ \sqrt{\frac{a_i}{b_i}}, \|u_{i0}(x)\|_\infty, \|p_{ik}(x)\|_\infty \right\} \tag{11}$$

From theorem 2.4, we also have $T^* = \infty$.

For the system $(\tilde{P})$, we have that

**Theorem 2.6.** The solution of the system $(\tilde{P})$ is global existent and ultimately bounded, and for all $t \in (0, \infty), x \in \Omega$, the following inequalities hold:

$$0 < \tilde{u}_i(t) \leq \max_{i=1,2,\cdots,m} \left\{ \sqrt{\frac{a_i}{b_i}}, \|u_{i0}(x)\|_\infty \right\} \triangleq N_i.$$

**Proof.** For all $T \in \mathrm{R}^1, t \in (0, T)$, let $\tilde{u}_i$ be the solution of $(\tilde{P})$. Similarity to theorem 2.5, we have

$$0 \leq \tilde{u}_i(t) \leq \tilde{u}_{i1}(t).$$

where $\tilde{u}_{i1}(t)$ is the solution of the following differential equations respectively:

$$\begin{cases} \frac{d\tilde{u}_{i1}}{dt} = \tilde{u}_{i1}(a_i - b_i \tilde{u}_{i1}^2), & t > 0, \\ \tilde{u}_{i1}(0) = \|u_{i0}(x)\|_\infty. \end{cases}$$

From (2), we obtain that

$$\begin{aligned} \tilde{u}_{i1}^2(t) &= \frac{a_i \|u_{i0}(x)\|_\infty^2}{a_i e^{-2a_i t} + b_i \|u_{i0}(x)\|_\infty^2 (1 - e^{-2a_i t})} \\ &\leq \begin{cases} \frac{a_i}{b_i}, & a_i - b_i \|u_{i0}(x)\|_\infty^2 \geq 0, \\ \|u_{i0}(x)\|_\infty^2, & a_i - b_i \|u_{i0}(x)\|_\infty^2 < 0. \end{cases} \end{aligned}$$

Therefore, for all $t \in (0, \infty)$, it follows that

$$0 < \tilde{u}_i(t) \leq \max_{i=1,2,\cdots,m} \left\{ \sqrt{\frac{a_i}{b_i}}, \|u_{i0}(x)\|_\infty \right\}.$$

# 3 Impulsive Synchronization

In this section, we will investigate the impulsive synchronization of the system (P) and $(\tilde{P})$. For the impulsive error system (E), we have the following result:

**Theorem 3.1.** Let

$$p_{ik} = \alpha_{ik} \tilde{u}_i(t_k) + (1 - \alpha_{ik}) u_i(t_k), \alpha_{ik} \in [0, 1],$$

and $c_0$ is the pioncare constant such that

$$\int_\Omega \varphi^2 \le c_0^2 \int_\Omega |\nabla\varphi|^2, \quad \forall \varphi \in H_0^1(\Omega).$$

If

$$\left[ \sum_{j=1}^k (\lambda_0 - \min_{i=1,2,\cdots,m} \frac{d_i}{c_0^2})(t_{j+1} - t_j) + \ln\lambda_k \right] \to -\infty, k \to \infty,$$

where

$$\lambda_k = \max_{i=1,2,\cdots,m} \{(1-\alpha_{ik})^2\}, k = 1,2,\cdots$$

$$\lambda_0 = \sup_{x_i \ne 0} \frac{\sum_{i=1}^m A_i x_i^2 + \sum_{i=1}^m \sum_{\substack{j=1 \\ j\ne i}}^m c_{ij}u_i x_i x_j}{\sum_{i=1}^m x_i^2}$$

with

$$A_i = \sup_{\substack{0 \le x \le N_i \\ 0 \le y \le M_i}} [a_i - b_i(x^2 + xy + y^2) - \sum_{\substack{j=1 \\ j\ne i}}^m c_{ij}\frac{1}{1+x}].$$

Then, the error system (E) is uniformly equi-attractive in the large.

**Proof.** Let

$$V(\mathbf{e}) = \frac{1}{2}\int_\Omega \sum_{i=1}^m e_i^2 dx. \tag{12}$$

It implies that

$$D_t' V(\mathbf{e}) = \int_\Omega \sum_{i=1}^m e_i \frac{\partial e_i}{\partial t} = \int_\Omega \left[ \sum_{i=1}^m d_i e_i \Delta e_i + \sum_{i=1}^m e_i \varphi_i(u_i, \tilde{u}_i) \right]$$

$$= \int_\Omega \left[ -\sum_{i=1}^m d_i |\nabla e_i|^2 + \sum_{i=1}^m e_i \varphi_i(u_i, \tilde{u}_i) \right]. \tag{13}$$

In the above equation (13),

$$e_i\varphi(u_i, \tilde{u}_i)$$

$$= a_i e_i^2 - b_i e_i^2(\tilde{u}_i^2 + \tilde{u}_i u_i + u_i^2) - \sum_{\substack{j=1 \\ j\ne i}}^m c_{ij}\frac{e_i^2 + e_i^2 u_j - u_i e_i e_j}{(1+\tilde{u}_j)(1+u_j)}$$

$$= e_i^2[a_i - b_i(\tilde{u}_i^2 + \tilde{u}_i u_i + u_i^2) - \sum_{\substack{j=1 \\ j\ne i}}^m c_{ij}\frac{1}{1+\tilde{u}_j}]$$

$$+ \sum_{\substack{j=1 \\ j\ne i}}^m \frac{c_{ij}u_i}{(1+\tilde{u}_j)(1+u_j)}e_i e_j. \tag{14}$$

Thus

$$\sum_{i=1}^{m} e_i \varphi(u_i, \tilde{u}_i) = \sum_{i=1}^{m} e_i^2 [a_i - b_i(\tilde{u}_i^2 + \tilde{u}_i u_i + u_i^2)] - \sum_{\substack{j=1 \\ j \neq i}}^{m} c_{ij} \frac{1}{1 + \tilde{u}_j}]$$

$$+ \sum_{i=1}^{m} \sum_{\substack{j=1 \\ j \neq i}}^{m} \frac{c_{ij} u_i}{(1 + \tilde{u}_j)(1 + u_j)} e_i e_j$$

$$\leq \sum_{i=1}^{m} A_i e_i^2 + \sum_{i=1}^{m} \sum_{\substack{j=1 \\ j \neq i}}^{m} c_{ij} u_i e_i e_j] \leq \lambda_0 \sum_{i=1}^{m} e_i^2. \tag{15}$$

Therefore, we have that

$$D_t' V(\mathbf{e}) \leq \int_{\Omega} \lambda_0 \sum_{i=1}^{m} e_i^2 - \sum_{i=1}^{m} \frac{d_i}{c_0^2} e_i^2 \leq \int_{\Omega} (\lambda_0 - \min_{i=1,2,\cdots,m} \frac{d_i}{c_0^2}) \sum_{i=1}^{m} e_i^2$$

$$= (\lambda_0 - \min_{i=1,2,\cdots,m} \frac{d_i}{c_0^2}) V(\mathbf{e}). \tag{16}$$

Thus

$$V(\mathbf{e}(t_{k+1}, x)) \leq \exp(\lambda_0 - \min_{i=1,2,\cdots,m} \frac{d_i}{c_0^2})(t_{k+1} - t_k) V(\mathbf{e}(t_k + 0, x)).$$

By the structure of the error system (E), we have that

$$e_i(t_k + 0, x) = \tilde{u}(t_k) - p_{ik}(x) = (1 - \alpha_{ik}) e_i(t_k, x)$$

Then,

$$V(\mathbf{e}(t_k + 0, x)) = \max_{i=1,2,\cdots,m} \{(1 - \alpha_{ik})^2\} \int_{\Omega} \sum_{i=1}^{m} e_i^2(t_k, x)$$

$$= \lambda_k V(\mathbf{e}(t_k, x)). \tag{17}$$

It implies

$$V(\mathbf{e}(t_{k+1}, x))$$

$$\leq \lambda_k \exp\left[(\lambda_0 - \min_{i=1,2,\cdots,m} \frac{d_i}{c_0^2})(t_{k+1} - t_k)\right] V(\mathbf{e}(t_k, x))$$

$$\leq \exp\left(\sum_{j=1}^{k} [(\lambda_0 - \min_{i=1,2,\cdots,m} \frac{d_i}{c_0^2})(t_{j+1} - t_j) + \ln \lambda_k]\right) V(\mathbf{e}(t_1, x)).$$

Therefore,

$$\lim_{k \to \infty} V(\mathbf{e}(t_{k+1}, x)) = 0. \tag{18}$$

# References

1. Akhmet, M.V., Beklioglu, M., Ergenc, T., Jkochenko, V.J.: An impulsive radio-dependent predator-prey system with diffusion. Nonlinear Analysis: Real World Applications 7, 1255–1267 (2006)
2. Bainov, D., Kamont, Z., Minchev, E.: On the impulsive partial differential functional inequalities of first order. Util Math. 48, 107–128 (1995)
3. Bainov, D., Kamont, Z., Minchev, E.: On the stability of solutions of impulsive partial differential equations of first order. Adv. Math. Sci. Appl. 6(2), 589–598 (1996)
4. Bainov, D., Minchev, E.: On the stability of solutions of impulsive partial differential-functional equations of first order via Lyapunov functions. Non-Linear Phenom Complex Syst 30(2), 109–116 (2000)
5. Khadra, A., Liu, X.Z., Shen, X.M.: Impulsive control and synchronization of spatiotemporal chaos. Chaos, solitons and Fractals 26, 615–636 (2005)
6. Haraus, A., Youkana, A.: On a result of K. Masada Concerning reaction-diffusion equations. Tohoku Math. J. 40, 159–163 (1988)
7. Lakshmikantham, V.: Theory of impulsive differential equations. World Scientific, Singapore (1989)
8. Liu, X.Z.: Stability results for impulsive differential systems with applications to population growth models. Dynam. Stabil. Syst. 9(2), 163–174 (1994)
9. Liu, X.Z.: Impulsive stabilization and control of chaotic systems. Non-Linear Anal. 47, 1081–1092 (2001)
10. Liu, X.Z., Willms, A.R.: Impulsive controllability of linear dynamical systems with applications to maneuvers of spacecraft. MPE 2, 277–299 (1996)
11. Pazy, A.: Semigroups of linear operators and applications to partial differential equations. Springer, Berlin (1883)
12. Smith, L.H.: Dynamics of competition. Lecture Notes in Mathematics, vol. 1714, pp. 192–240. Springer, Berlin (1999)
13. Walter, W.: Differential inequalities and maximum principles. Nonlinear Analysis: Theory, Methods and Applications 30, 4695–4771 (1997)
14. Yang, T., Yang, C.M., Yang, L.B.: Control of Rossler system to periodic motions using impulsive control methods. Phys. Lett. A 232, 356–361 (1997)
15. Yang, T., Yang, L.B., Yang, C.M.: Impulsive control of Lorenz system. Phys. D 110, 18–24 (1997)
16. Yang, T.: Impulsive control theory. Springer, Berlin (2001)
17. Yang, W., Wang, B.: On the question of global existence for the two-component reaction-diffusion systems with mixed boundary conditions. Nonliear Analysis 39, 755–766 (2000)

# A New Numerical Method for Solving Convection-Diffusion Equations

Hengfei Ding and Yuxin Zhang

**Abstract.** In this paper, we using semi-discrete method, transformed convection-diffusion equation into a ODEs: $\frac{dU(t)}{dt} = AU(t)$, then we get the solution of the ODEs: $U(t) = e^{tA}U_0$. Furthermore, we give a numerical approximation for $e^{tA}$ and get a special difference scheme for solving the convection-diffusion equation which improve the accuracy order and stability condition greatly. The accuracy order is fourth order and second order in space and time direction respectively. Finally, numerical result shows that this method is effective.

**Keywords:** Convection-diffusion equation, Difference scheme, High accuracy, System of ODEs.

## 1 Introduction

The convection-diffusion equation is commonly encountered in physical sciences governing the transport of a quantity such as mass, momentum, heat and energy. So, it is obvious very important to find a quick, stable and practical numerical method. we now consider one dimensional convection-diffusion equation [1]

$$\frac{\partial u}{\partial t} + \varepsilon \frac{\partial u}{\partial x} = \gamma \frac{\partial^2 u}{\partial x^2}, \quad 0 \le x \le 1, \quad t \ge 0. \tag{1}$$

subject to the initial condition

$$u(x, 0) = g(x), \quad 0 \le x \le 1.$$

Hengfei Ding and Yuxin Zhang,
School of Mathematics and statistics, Tianshui Normal University,
Tianshui 741000, P.R. China
e-mail: dinghf05@163.com, zhangyuxin2006@163.com

S. Li (Eds.): Nonlinear Maths for Uncertainty and its Appli., AISC 100, pp. 463–470.
springerlink.com                    © Springer-Verlag Berlin Heidelberg 2011

and boundary conditions

$$u(0,t) = 0, \quad t \geq 0.$$

$$u(1,t) = 0, \quad t \geq 0.$$

The parameter $\gamma$ is the viscosity coefficient and $\varepsilon$ is the phase speed, and both are assumed to positive.

At present, the accuracy order of some difference schemes for solving Eq.(1) are not very high. Generally, they are first order or second order in time direction and space direction respectively, furthermore, they are almost condition stable [1] [4] [2]]. In this paper, we using semi-discrete method, get a new high accuracy order difference scheme for solving Eq.(1), which is fourth accuracy order in space and second accuracy order in time direction respectively, and is uncondition stability. Finally, the numerical examples are presented which are fully matched with theory analysis.

The present paper is organized as follows. In section 2, we proposition of the difference scheme. In section 3, we discuss the accuracy order and stability. Some numerical examples are presented in section 4 and concluding remarks are given in section 5.

## 2  Proposition of the Difference Scheme

We subdivide the interval $0 \leq x \leq 1$ into $M$ equal subintervals by the grid points $x_j = jh$, $j = 0(1)M$, where $h = 1/M$, assume $\tau$ is time-size, mesh function $u(jh, k\tau)$ is written as $u_j^k$.

We extend interval $[0, 1]$ to $[-h, 1 + h]$, expanding $u(-h, t)$ and $u(h + 1, t)$ in a Taylor series expansion about $x = 0$ and $x = 1$, respectively. We get:

$$u(-h,t) = u(0,t) - hu_x(0,t) + \frac{h^2}{2}u_{xx}(0,t) + O(h^3) = u(0,t) + O(h) \quad (2)$$

$$u(1+h,t) = u(1,t) + hu_x(1,t) + \frac{h^2}{2}u_{xx}(1,t) + O(h^3) = u(1,t) + O(h) \quad (3)$$

neglecting the term of $O(h)$ from the (2) and (3) and we then obtain:

$$u(-h,t) = u(0,t), \quad u(1+h,t) = u(1,t) \quad (4)$$

In the same way, we expanding $u_t(-h,t)$ and $u_t(1 + h, t)$ in a Taylor series expansion about $x = 0$ and $x = 1$, respectively, and get

$$u_t(-h,t) = u_t(0,t), \quad u_t(1+h,t) = u_t(1,t) \quad (5)$$

Definite difference operator as follows:

$$\delta_x : \qquad \delta_x u_j = u_{j+\frac{1}{2}} - u_{j-\frac{1}{2}}$$

$$\mu_x : \qquad \mu_x u_j = \frac{1}{2}(u_{j+\frac{1}{2}} + u_{j-\frac{1}{2}})$$

$$\delta_x^2 : \qquad \delta_x^2 = u_{j+1} - 2u_j + u_{j-1}$$

$$\mu_x \delta_x : \qquad \mu_x \delta_x u_j = \frac{1}{2}(u_{j+1} - u_{j-1})$$

Write down Eq.(1) at every mesh point $x_j = jh$, $j = 1(1)M - 1$. Along time level $t$, then, we substitute compact difference approximation [7]

$$\frac{1}{h}\frac{\mu_x \delta_x u_j}{1 + \frac{1}{6}\delta_x^2} + O(h^4) \qquad (6)$$

and

$$\frac{1}{h^2}\frac{\delta_x^2 u_j}{1 + \frac{1}{12}\delta_x^2} + O(h^4) \qquad (7)$$

for $\frac{\partial u(x_j,t)}{\partial x}$ and $\frac{\partial^2 u(x_j,t)}{\partial x^2}$ in Eq.(1), respectively. By neglecting the terms of $O(h^4)$, it then follows that the values $u_j(t)$ approximating $u(x_j,t)$ will be the exact solution values of the system of the $M - 1$ ordinary differential equations

$$\frac{du_j(t)}{dt} = -\frac{\varepsilon}{h}\frac{\mu_x \delta_x u_j(t)}{1 + \frac{1}{6}\delta_x^2} + \frac{\gamma}{h^2}\frac{\delta_x^2 u_j(t)}{1 + \frac{1}{12}\delta_x^2} \qquad (8)$$

Let

$$\frac{du_j(t)}{dt} = v_j(t) \qquad (9)$$

Eq.(1) becomes

$$v_j(t) = -\frac{\varepsilon}{h}\frac{\mu_x \delta_x u_j(t)}{1 + \frac{1}{6}\delta_x^2} + \frac{\gamma}{h^2}\frac{\delta_x^2 u_j(t)}{1 + \frac{1}{12}\delta_x^2} \qquad (10)$$

simply (10) and yield to

$$\frac{1}{72}h^2 v_{j+2}(t) + \frac{7}{36}h^2 v_{j+1}(t) + \frac{7}{12}h^2 v_j(t) + \frac{7}{36}h^2 v_{j-1}(t) + \frac{1}{72}h^2 v_{j-2}(t)$$
$$= (\frac{1}{6}\gamma - \frac{1}{24}\varepsilon h)u_{j+2}(t) + (\frac{1}{3}\gamma - \frac{5}{12}\varepsilon h)u_{j+1}(t) - \gamma u_j(t) +$$
$$(\frac{1}{3}\gamma + \frac{5}{12}\varepsilon h)u_{j-1}(t) + (\frac{1}{6}\gamma + \frac{1}{24}\varepsilon h)u_{j-2}(t)$$

$$(11)$$

In matrix notation, (11) can be written as:

$$\begin{cases} \mathbf{A}V(t) = \mathbf{B}U(t) \\ \mathbf{U}(0) = \mathbf{U_0} \end{cases} \qquad (12)$$

since $\frac{du_j(t)}{dt} = v_j(t)$, (12) rewritten as

$$\begin{cases} \mathbf{A}\frac{d\mathbf{U}(t)}{dt} = \mathbf{B}\mathbf{U}(t) \\ \mathbf{U}(0) = \mathbf{U_0} \end{cases} \tag{13}$$

where

$$\mathbf{V}(t) = [v_1(t), v_2(t), \dots, v_{M-2}(t), v_{M-1}(t)]^T$$

$$\mathbf{U}(t) = [u_1(t), u_2(t), \dots, u_{M-2}(t), u_{M-1}(t)]^T$$

$$\mathbf{U}(0) = [g(x_1), g(x_2), \dots, g(x_{M-2}), g(x_{M-1})]^T$$

and $A$, $B$ are the five-diagonal matrix of order $M-1$ as below

$$A = Fivediag(\frac{1}{72}h^2, \frac{7}{36}h^2, \frac{7}{12}h^2, \frac{7}{36}h^2, \frac{1}{72}h^2),$$

$$B = Fivediag(\frac{1}{6}\gamma + \frac{1}{24}\varepsilon h, \frac{1}{3}\gamma + \frac{5}{12}\varepsilon h, -\gamma, \frac{1}{3}\gamma - \frac{5}{12}\varepsilon h, \frac{1}{6}\gamma - \frac{1}{24}\varepsilon h).$$

**Definition 1** [6]. An $(M-1) \times (M-1)$ complex matrix $D = [d_{ij}]$ is diagonally dominant if

$$|d_{ii}| \geq \sum_{j=1, \, j \neq i}^{M-1} |d_{ij}| \tag{14}$$

for all $1 \leq i \leq M-1$. An $(M-1) \times (M-1)$ matrix $D$ is strictly diagonally dominant if strict inequality in (14) is valid for all $1 \leq i \leq M-1$.

**Lemma 1** [6]. Let $D = [d_{ij}]$ be an $(M-1) \times (M-1)$ strictly or irreducibly diagonally dominant complex matrix, then, the matrix D is nonsingular.

From above, we know $A$ is nonsingular, so we easily get the solution of (13):

$$\mathbf{U}(t) = e^{tA^{-1}B}\mathbf{U_0} \tag{15}$$

As $t_k = k\tau, k = 0, 1, \dots$, so

$$\mathbf{U}(t_{k+1}) = e^{(k+1)\tau A^{-1}B}\mathbf{U_0}, \quad \mathbf{U}(t_k) = e^{k\tau A^{-1}B}\mathbf{U_0}$$

then we have the following scheme:

$$\mathbf{U}(t_{k+1}) = e^{\tau A^{-1}B}\mathbf{U}(t_k) \tag{16}$$

now, the problem is how to approximate $e^{\tau A^{-1}B}$ to get the numerical solution. A good approximation to $e^Z$ is the $[1, 1]$ padé approximation which have the form [3]

$$e^Z = \frac{2+Z}{2-Z} + O(Z^3) \tag{17}$$

neglecting the high order term of $O(Z^3)$, yield to

$$e^Z \doteq \frac{2+Z}{2-Z} \tag{18}$$

so we give a approximation to $e^{\tau A^{-1}B}$ with $(2I - \tau A^{-1}B)^{-1}(2I + \tau A^{-1}B)$ and applied it into (16), then get a new difference scheme for solving Eq.(1) as follow:

$$\mathbf{U}(t_{k+1}) = (2I - \tau A^{-1}B)^{-1}(2I + \tau A^{-1}B)\mathbf{U}(t_k) \tag{19}$$

writing (19) in the component form, we can yield to the difference scheme:

$$
\begin{aligned}
&[2h^2 - \tau(12\gamma + 3\varepsilon h)]u_{j-2}^{k+1} + [28h^2 - \tau(24\gamma + 30\varepsilon h)]u_{j-1}^{k+1} \\
&+(84h^2 + 72\tau\gamma)u_j^{k+1} + [28h^2 - \tau(24\gamma - 30\varepsilon h)]u_{j+1}^{k+1} \\
&+[2h^2 - \tau(12\gamma - 3\varepsilon h)]u_{j+2}^{k+1} = [2h^2 + \tau(12\gamma + 3\varepsilon h)]u_{j-2}^k \\
&+[28h^2 + \tau(24\gamma + 30\varepsilon h)]u_{j-1}^k + (84h^2 - 72\tau\gamma)u_j^k \\
&+[28h^2 + \tau(24\gamma - 30\varepsilon h)]u_{j+1}^k + [2h^2 + \tau(12\gamma - 3\varepsilon h)]u_{j+2}^k
\end{aligned} \tag{20}
$$

## 3  Accuracy Order and Stability Analysis

From the analysis and computation process above, we know that the leading error term of the (1) padé approximate to $e^{\tau A^{-1}B}$ is $O(\tau^3)$, but Equ.(16) was derived by interating the ordinary differential for $U(t)$ with respect to $t$, so the time direction is second accuracy order, the space direction is still fourth accuracy order, then we know the accuracy order of the difference scheme (19) is $O(\tau^2 + h^4)$.

Consider a function $u$ defined in a discrete set of points $x_j = jh$, $u_j = u(jh)$, the Euclid or $l_2$-norm is defined to be $||u||_2 = \sqrt{\sum_{j=1}^N |u_j|^2 h}$.

**Lemma 2** [5]. The difference scheme

$$\mathbf{U}^{k+1} = Q\mathbf{U}^k \tag{21}$$

is stable with respect to the $l_2$-norm if and only if there exists positive constants $\tau_0$, $h_0$ and $K$ so that

$$|\rho(\xi)| < 1 + K\tau \tag{22}$$

for $0 < \tau < \tau_0$, $0 < h < h_0$, and all $\xi \in [-\pi, \pi]$, and where $\rho$ is the symbol of difference scheme (21), then it is said that $\rho$ satisfies the Von Neumann condition. But when we consider a difference scheme with initial-boundary value problem, the Von Neumann condition is a necessary condition for stability.

We taking the discrete Fourier transform of Eq.(20) and yield to

$$\hat{u}^{n+1} = \rho(\xi)\hat{u}^n \tag{23}$$

where

$$\rho(\xi) = \{[(2h^2 + 12\tau\gamma)\cos 2\xi + (28h^2 + 24\tau\gamma)\cos\xi + (42h^2 - 36\tau\gamma)]$$

$$-i[(3\sin 2\xi + 30\sin\xi)\tau\varepsilon h]\}\{[(2h^2 - 12\tau\gamma)\cos 2\xi + (28h^2 - 24\tau\gamma)\cos\xi$$

$$+(42h^2 + 36\tau\gamma)] + i[(3\sin 2\xi + 30\sin\xi)\tau\varepsilon h]\}^{-1}$$

$$(24)$$

then, from the lemma 2 we get a necessary condition for stability is obtained by

$$|\rho(\xi)| \leq 1 \tag{25}$$

take (23) into (24) and yield to

$$\cos^4\xi + 8\cos^3\xi + 15\cos^2\xi - 4\cos\xi - 20 \leq 0 \tag{26}$$

**Lemma 3.** For $\forall \cos\xi \in [-1,1]$, (26) holds.
Proof :     since

$$\cos^4\xi + 8\cos^3\xi + 15\cos^2\xi - 4\cos\xi - 20$$
$$= \cos^4\xi + 8\cos^3\xi + 16\cos^2\xi - \cos^2\xi - 4\cos\xi - 20$$
$$= (\cos^2\xi + 4\cos\xi)^2 - (\cos^2\xi + 4\cos\xi) - 20$$
$$= (\cos^2\xi + 4\cos\xi + 4)(\cos^2\xi + 4\cos\xi - 5)$$
$$= (\cos\xi + 2)^2[(\cos\xi + 2)^2 - 9]$$

and           $(\cos\xi + 2)^2 > 0$, $(\cos\xi + 2)^2 - 9 \leq 0$.
so           $\cos^4\xi + 8\cos^3\xi + 15\cos^2\xi - 4\cos\xi - 20 \leq 0$
From above we know

**Theorem 1.** Difference scheme (20) is uncondition stability.

## 4  Numerical Examples

For the constant coefficient convection-diffusion equation:

$$\begin{cases} \frac{\partial u}{\partial t} + \varepsilon\frac{\partial u}{\partial t} = \gamma\frac{\partial^2 u}{\partial x^2}, & 0 \leq x \leq 1, \quad t \geq 0 \\ u(x,0) = e^{\frac{\varepsilon}{2\gamma}x}\sin\pi x, & 0 \leq x \leq 1 \\ u(0,t) = u(1,t) = 0, & t \geq 0 \end{cases} \tag{27}$$

Let

$$(1): \quad \varepsilon = \tfrac{1}{10}; \quad \gamma = \tfrac{1}{50}; \quad h = \tfrac{1}{100}; \quad \tau = \tfrac{1}{5000}$$

$$(2): \quad \varepsilon = \tfrac{1}{10}; \quad \gamma = \tfrac{1}{100}; \quad h = \tfrac{1}{200}; \quad \tau = \tfrac{1}{10000}$$

$$(3): \quad \varepsilon = \tfrac{1}{50}; \quad \gamma = \tfrac{1}{100}; \quad h = \tfrac{1}{200}; \quad \tau = \tfrac{1}{5000}$$

For convenience, we can use the exact solution to calute the value of the first state $u_j^1$, then according to the above scheme, we can get the values of the 20000th state, we compute its absolute error(which equals the absolute value of numerical solution minus exact solution). The results are shown in the following table 1, table 2 and table 3.

**Table 1** The absolute error at the condition (1)

| x | absolute error |
|---|---|
| 0.1 | 1.12456334137e-003 |
| 0.3 | 1.51081896755e-003 |
| 0.5 | 2.18596452693e-003 |
| 0.7 | 4.18620943296e-003 |
| 0.9 | 8.54100403283e-003 |

**Table 2** The absolute error at the condition (2)

| x | absolute error |
|---|---|
| 0.1 | 1.73175381084e-003 |
| 0.3 | 1.08837660703e-003 |
| 0.5 | 5.78126687751e-004 |
| 0.7 | 8.52338492887e-003 |
| 0.9 | 1.02787202415e-003 |

**Table 3** The absolute error at the condition (3)

| x | absolute error |
|---|---|
| 0.1 | 3.79786505083e-003 |
| 0.3 | 1.06744955984e-003 |
| 0.5 | 2.49066221474e-004 |
| 0.7 | 1.62952105685e-003 |
| 0.9 | 8.65239009388e-003 |

## 5   Concluding Remarks

In this work, we proposed a method to find the solution of the system of ordinary differential equations arisen from discing the convection-diffusion equation with respect to the space variable, the method is fourth order in space and second order in time direction, it is shown through a discrete Fourier analysis that it is unconditionally stable, numerical experiments are conducted to test its high accuracy and to shown its reasonableness.

**Acknowledgements.**  The project Supported by 'QingLan' Talent Engineering Funds and SRF(TSA0928) by Tianshui Normal University.

# References

1. Ismail, H.N.A., Elbarbary, E.M.E. Salem, G.S.E.: Restrictive Taylor's approximation for solving convection-diffusion equation. Appl. Math. Comput. 147, 355-363 (2004)
2. Salkuyeh, D.K.: On the finite difference approximation to the convection-diffusion equation. Appl. Math. Comput. 179, 79–86 (2006)
3. Smith, G.D.: Numerical Solution of Partial Differential Equations(finite difference method). Oxford University Press, Oxford (1990)
4. Sousa, E.: The controversial stability analysis. Appl. Math. Comput. 145, 777–794 (2003)
5. Thomas, J.W.: Numerical Partial Differential Equations(finite difference methods). Springer, New York (1995)
6. Varga, R.S.: Matrix Iterative Analysis. Springer, Berlin (2000)
7. Zhang, W.S.: Finite Difference Methods for Partial Differential Equation in Science Computation. Higher Education Press, Beijing (2006)

# Pullback Attractor for Non-autonomous P-Laplacian Equation in Unbounded Domain

Guangxia Chen

**Abstract.** By applying extended asymptotic a priori estimate method, we are concerned with the existence of $(L^2(\mathbb{R}^n), L^p(\mathbb{R}^n))$–pullback attractor for non-autonomous p-Laplacian equation defined in $\mathbb{R}^n$, where the external force $g(t,x)$ satisfies only a certain integrability condition.

**Keywords:** P-Laplacian equation, Pullback attractor, Unbounded domain.

## 1 Introduction

We consider the following non-autonomous p-Laplacian equation

$$\begin{cases} u_t - div(|\nabla u|^{p-2}\nabla u) + \lambda|u|^{p-2}u + f(u) = g(t,x) & \text{in } \mathbb{R}^n \times [\tau,\infty), \\ u(x,\tau) = u_\tau, \end{cases} \quad (1)$$

where $p > 2$, $\lambda > 0$. Assume that $f(\cdot) \in C^1(\mathbb{R})$ satisfies the conditions

$$\alpha_1|u|^2 - \beta_1|u|^p \le f(u)u \le \alpha_2|u|^2 + \beta_2|u|^p, \quad (2)$$

$$f'(u) > -l, \quad l > 0, \quad (3)$$

for some positive constants $\alpha_i, \beta_i$ $(i = 1,2)$ and $\lambda > \beta_1$. Furthermore, $g(t,x) \in L^2_{loc}(\mathbb{R}, L^2(\mathbb{R}^n))$, such that for some $\sigma \in (0,\alpha_1)$,

$$\int_{-\infty}^{t} e^{\sigma s}|g(s)|_2^2 ds < \infty, \quad \forall t \in \mathbb{R}. \quad (4)$$

Guangxia Chen

School of Mathematics and Informatics, Henan Polytechnic University,
Jiaozuo 454003, People's Republic of China
e-mail: we962@163.com

S. Li (Eds.): Nonlinear Maths for Uncertainty and its Appli., AISC 100, pp. 471–478.
springerlink.com &copy; Springer-Verlag Berlin Heidelberg 2011

The existence of global attractor for p-Laplacian equation has been studied extensively in many papers and monographs, see e.g., [1, 2, 5, 7, 8]. When $g$ is time dependent, [4] considered the existence of uniform attractors for p-Laplacian equation in bounded domains, and recently, in [3], the existence of pullback attractor in $L^2(\mathbb{R}^n)$ has been proved.

In this paper we will prove the existence of $(L^2(\mathbb{R}^n), L^p(\mathbb{R}^n))$–pullback attractor for problem (1). Instead of the hypothesis that $g(t)$ is translation bounded, we assume the external force satisfies only a certain integrability condition (4) which is less restrictive than translation bounded. At the same time, when consider the problem in whole space $\mathbb{R}^n$, the Sobolev embeddings are not compact. In fact, these are the two major difficulties when we discuss the existence of pullback attractor for problem (1). To attain our goal, in abstract framework, we extend the idea of asymptotic a priori estimate to non-autonomous dynamical systems, as application, we obtain the $(L^2(\mathbb{R}^n), L^p(\mathbb{R}^n))$–pullback asymptotic compactness of the process associated with problem (1).

Now, we recall the existence theorem for unique and global in time weak solution for problem (1)(see [1]).

**Theorem 1.** *Assume that $f$ satisfies (2) (3), and $g \in L^2_{\mathrm{loc}}(\mathbb{R}, L^2(\mathbb{R}^n))$. Then for any $\tau \in \mathbb{R}$, any initial data $u_\tau \in L^2(\mathbb{R}^n)$ and any $T \geq \tau$, there exists a unique solution $u \in C([\tau, T]; L^2(\mathbb{R}^n)) \cap C_w((\tau, T); W^{1,p}(\mathbb{R}^n))$ for equation (1), and the mapping $u_\tau \to u(t, \tau; u_\tau)$ is continuous in $L^2(\mathbb{R}^n)$.*

In view of Theorem 1, we define a continuous process $\{U(t, \tau) : -\infty < \tau \leq t < \infty\}$ in $L^2(\mathbb{R}^n)$ such that for all $t \geq \tau$, $U(t, \tau)u_\tau = u(t)$, where $u(t)$ is the solution of equation(1) with initial data $u(\tau) = u_\tau \in L^2(\mathbb{R}^n)$.

Our main results read as follows.

**Theorem 2.** *Assume that $f$ satisfies (2) (3) and $g(t) \in L^2_{\mathrm{loc}}(\mathbb{R}, L^2(\mathbb{R}^n))$ satisfies (4) for some $0 < \sigma < \alpha_1$. Then the process $U(t, \tau)$ associated with problem (1) possesses a $(L^2(\mathbb{R}^n), L^p(\mathbb{R}^n))$–pullback attractor $\mathcal{A} = \{A(t)\}_{t \in \mathbb{R}^n}$, in which*

$$A(t) = \bigcap_{s \leq t} \overline{\bigcup_{\tau \leq s} U(t, \tau)B(\tau)}^{L^p}$$

*and $\{B(t)\}$ is $(L^2(\mathbb{R}^n), L^p(\mathbb{R}^n))$–pullback absorbing for $U(t, \tau)$.*

## 2 Preliminaries and Abstract Results

### 2.1 Preliminaries

In this subsection, we recall some basic definitions and abstract results about bi-space pullback attractor, see [6] for details.

**Definition 1.** Let $X, Y$ be two Banach spaces, $U(t, \tau)$ be a process defined on $X$ and on $Y$. A family of sets $\mathcal{A} = \{A(t)\}_{t \in \mathbb{R}} \subset X$ is said to be a (X,Y)–pullback attractor for the process $U$, if

1.) for any $t \in \mathbb{R}$, $A(t)$ is closed in $X$ and compact in $Y$;

2.) $U(t, \tau) A(\tau) = A(t)$ for any $\tau \leq t$;

3.) it pullback attracts bounded subsets of $X$ in the topology of $Y$, i.e., for any bounded subset $D \subset X$, and any $t \in \mathbb{R}$, $\lim_{\tau \to -\infty} dist_Y(U(t, \tau)D, A(t)) = 0$.

**Theorem 3.** *( [6])Let $X, Y$ be two Banach spaces, $U(t, \tau)$ be a norm-to-weak continuous process defined on $X$ and on $Y$. Assume the family $\mathcal{B} = \{B(t)\}_{t \in \mathbb{R}}$ is $(X, X)$–pullback absorbing and $\mathcal{B}_1 = \{B_1(t)\}_{t \in \mathbb{R}}$ is $(X, Y)$–pullback absorbing for $U(t, \tau)$, and for any $t \in \mathbb{R}$, any sequence $\tau_n \to -\infty$, $x_n \in B(\tau_n)$, the sequence $\{U(t, \tau_n)x_n\}$ is precompact in $Y$ $((X, Y)$–pullback asymptotic compactness). Then the family $\mathcal{A} = \{A(t)\}_{t \in \mathbb{R}}$, where $A(t) = \bigcap_{s \leq t} \overline{\bigcup_{\tau \leq s} U(t, \tau)(B(\tau) \cap B_1(\tau))}^X = \bigcap_{s \leq t} \overline{\bigcup_{\tau \leq s} U(t, \tau)(B(\tau) \cap B_1(\tau))}^Y$ is a $(X, Y)$–pullback attractor for $U(t, \tau)$.*

## 2.2  Abstract Results

In this subsection, we extend the ideas in [7,8] to non-autonomous dynamical systems, it is useful for us to prove the existence of pullback attractor in $L^p(\mathbb{R}^n)$ $(p \geq 2)$.

**Lemma 1.** *(see [7]) Let $D \subset L^p(\mathbb{R}^n) \cap L^2(\mathbb{R}^n)$ be bounded in both $L^p(\mathbb{R}^n)$ and $L^2(\mathbb{R}^n)$. Then for any $\varepsilon > 0$, $D$ has a finite $\varepsilon$–net in $L^p(\mathbb{R}^n)$ if there exists a positive constant $M = M(\varepsilon)$, such that*

*1.) $D$ has a finite $(3M)^{(2-p)/2}(\varepsilon/2)^{p/2}$–net in $L^2(\mathbb{R}^n)$.*

*2.) For all $u \in D$, $(\int_{\mathbb{R}^n(|u| \geq M)} |u|^p dx)^{1/p} < 2^{-(2p+2)/p} \varepsilon$.*

**Lemma 2.** *Let $U$ be a process on $L^p(\mathbb{R}^n)$ $(p \geq 1)$, $\{B(t)\}$ is bounded and $(L^2(\mathbb{R}^n), L^p(\mathbb{R}^n))$–pullback absorbing for $U$, which satisfies*

$$for \; \forall t \in \mathbb{R}, \exists \tau^* = \tau^*(t), s.t. U(t, \tau)B(\tau) \subset B(t) \; as \; \tau < \tau^*. \tag{5}$$

*Then for any $t \in \mathbb{R}$ and $\varepsilon > 0$, there exist constants $\tau_0 = \tau_0(t, \varepsilon)$ and $M = M(t, \varepsilon)$ such that $m(\mathbb{R}^n(|U(t, \tau)u_\tau| \geq M)) < \varepsilon$ for any $u_\tau \in B(\tau)$ and $\tau \leq \tau_0$.*

*Proof.* Assume that $\{B(t)\}$ can be bounded by $M_1(t)$. Then for any $u_\tau \in B(\tau)$, we can deduce from (5) that $|U(t, \tau)u_\tau|_p^p \leq M_1(t)$ as $\tau \leq \tau^*$, then

$$M_1(t) \geq \int_{\mathbb{R}^n} |U(t, \tau)u_\tau|^p dx \geq \int_{\mathbb{R}^n(|U(t,\tau)u_\tau| \geq M(t))} |U(t, \tau)u_\tau|^p dx$$

$$\geq \int_{\mathbb{R}^n(|U(t,\tau)u_\tau| \geq M(t))} M^p(t) dx \geq M^p(t) \cdot m(\mathbb{R}^n(|U(t, \tau)u_\tau| \geq M(t))),$$

take $M(t) \geq (M_1(t)/\varepsilon)^{1/p}$, then we have $m(\mathbb{R}^n(|U(t, \tau)u_\tau| \geq M)) < \varepsilon$. $\qquad \square$

**Lemma 3.** *(see [7]) Let $D$ be a bounded subset in $L^p(\mathbb{R}^n)$ $(p \geq 1)$. If $D$ has a finite $\varepsilon$–net in $L^p(\mathbb{R}^n)$, then there exists an $M = M(D, \varepsilon)$ such that for any $u \in D$, the estimate $\int_{\mathbb{R}^n(|u| \geq M)} |u|^p dx \leq 2^{p+1} \varepsilon^p$ is valid.*

**Theorem 4.** *Let $U$ be a norm-to-weak continuous process on $L^2(\mathbb{R}^n)$ and on $L^p(\mathbb{R}^n)$, where $2 \leq p < \infty$. Suppose that $U$ possesses a compact $(L^2(\mathbb{R}^n), L^2(\mathbb{R}^n))$–pullback attracting set. Then $U$ has a $(L^2(\mathbb{R}^n), L^p(\mathbb{R}^n))$– pullback attractor provided the following conditions hold:*

*1.) $U(t, \tau)$ has a bounded $(L^2(\mathbb{R}^n), L^p(\mathbb{R}^n))$–pullback absorbing set $\mathcal{B} = \{B(t)\}$;*

*2.) for any $t \in \mathbb{R}$ and $\varepsilon > 0$, there exist positive constants $M = M(\varepsilon, t)$ and $\tau_0 = \tau(\varepsilon, t)$ such that the estimate $\int_{\mathbb{R}^n(|U(t,\tau)u_\tau| \geq M)} |U(t, \tau)u_\tau|^p dx \leq \varepsilon$ is valid for any $u_\tau \in B(\tau)$ and $\tau \leq \tau_0$.*

*Proof.* From Theorem 3, we need to verify that for any $\tau_n \to -\infty$ and $u_{\tau_n} \in B(\tau_n)$, $\{U(t, \tau_n)u_{\tau_n}\}$ is precompact in $L^p(\mathbb{R}^n)$. Let $\{P(t)\}$ be the compact $(L^2(\mathbb{R}^n), L^2(\mathbb{R}^n))$–pullback attracting set, then any $\varepsilon$–neighborhood of $P(t)$ in $L^2(\mathbb{R}^n)$, we denote it by $\{B_0(t)\}$, is $(L^2(\mathbb{R}^n), L^2(\mathbb{R}^n))$–pullback absorbing, take $B_1(t) = B(t) \cap B_0(t)$, then $\{B_1(t)\}$ is not only $(L^2(\mathbb{R}^n), L^2(\mathbb{R}^n))$– pullback absorbing, but also $(L^2(\mathbb{R}^n), L^p(\mathbb{R}^n))$–pullback absorbing. Therefore, it is sufficient to prove that for any $\tau_n \to -\infty$, and $u_{\tau_n} \in B_1(\tau_n)$, the sequence $\{U(t, \tau_n)u_{\tau_n}\}_n$ is precompact in $L^p(\mathbb{R}^n)$.

In fact, from Theorem 3, it is easy to see that there exists $\tau_1(\varepsilon, t)$, such that $\{U(t, \tau_n)u_{\tau_n} | \tau_n \leq \tau_1\}$ has a finite $(3M)^{(2-p)/2}(\varepsilon/2)^{p/2}$–net in $L^2(\mathbb{R}^n)$. Taking $\tau^* = \min\{\tau_0, \tau_1\}$, then Lemma 3 implies that $\{U(t, \tau_n)u_{\tau_n} | \tau_n \leq \tau^*\}$ has a finite $\varepsilon$–net in $L^p(\mathbb{R}^n)$, from the arbitrariness of $\varepsilon$, we obtain $\{U(t, \tau_n)u_{\tau_n}\}$ is precompact in $L^p(\mathbb{R}^n)$. □

## 3  Proof of Theorem 2

**Theorem 5.** *( [3]) Assume $f$ satisfies (2) (3), $g \in L^2_{loc}(\mathbb{R}, L^2(\mathbb{R}^n))$ satisfies (4) for some $\sigma \in (0, \alpha_1)$. Let $U$ be the process associated with problem (1). Then for any bounded subset $D \subset L^2(\mathbb{R}^n)$ and any $t \in \mathbb{R}$, there exists $\tau_1(D, t) \leq t$ such that for $\forall \tau \leq \tau_1(D, t)$*

$$\int_{\mathbb{R}^n} |u|^2 + |u|^p + |\nabla u|^p \leq C(e^{\sigma(\tau-t)}|u(\tau)|_2^2 + e^{-\sigma t} \int_{-\infty}^t e^{\sigma s}|g(s)|_2^2 ds). \quad (6)$$

*Furthermore, there exists a $(L^2(\mathbb{R}^n)), L^2(\mathbb{R}^n))$–pullback attractor for $U$.*

Denote $\{B(t)\}_{t \in \mathbb{R}}$ by

$$B(t) = \{u \in L^2(\mathbb{R}^n) \cap W^{1,p}(\mathbb{R}^n) : |u|_2^2 + |u|_p^p + |\nabla u|_p^p \leq R(t)\}, \quad (7)$$

where $R(t) := C(1 + e^{-\sigma t} \int_{-\infty}^t e^{\sigma s}|g(s)|_2^2 ds)$. Then $\{B(t)\}$ is $(L^2(\mathbb{R}^n), L^2(\mathbb{R}^n))$– pullback absorbing and $(L^2(\mathbb{R}^n), W^{1,p}(\mathbb{R}^n))$–pullback absorbing for $U$ associated with problem (1). From (6), we know that

$$e^{\sigma t} R(t) \to 0 \text{ as } t \to -\infty. \tag{8}$$

Next, we give an asymptotic a priori estimate for the unbounded part of the modular $|u|$ in $L^p(\mathbb{R}^n)$.

**Lemma 4.** *Assume that (2) (3) hold and $g(t) \in L^2_{loc}(\mathbb{R}, L^2(\mathbb{R}^n)$ satisfies (4) with $0 < \sigma < \alpha_1$, let $\{B(t)\}$ be the $(L^2(\mathbb{R}^n), L^p(\mathbb{R}^n))$–pullback absorbing set defined by (7). Then for any $t \in \mathbb{R}$, $\varepsilon > 0$, there exist constants $\tau_0 = \tau_0(\varepsilon, t) \le t$ and $M = M(\varepsilon, t) > 0$, such that*

$$\int_{\mathbb{R}^n(|u| \ge M)} |U(t, \tau) u_\tau|^p dx \le C\varepsilon \quad \text{for any } \tau \le \tau_0 \text{ and } u_\tau \in B(\tau).$$

*Proof.* For any fixed $\varepsilon > 0$, there exists a $\delta > 0$, such that if $e \subset \mathbb{R}^n$ and $m(e) < \delta$, then

$$\int_{-\infty}^{t} e^{\sigma r} \int_e |g(r)|^2 dx dr < \varepsilon. \tag{9}$$

On the other hand, from Lemma 1, Lemma 2 and Theorem 5, we know that there exist $\tau_1 = \tau_1(\varepsilon, t) < t$ and $M_1 = M_1(\varepsilon, t)$, such that for any $u_\tau \in B(\tau)$ and $\tau \le \tau_1$, it yields

$$m(\mathbb{R}^n(|u(t)| \ge M_1)) \le min\{\varepsilon, \delta\}. \tag{10}$$

Moreover, from (2) we can choose $M_0$ large enough such that

$$\alpha_1 |u| - \beta_1 |u|^{p-1} \le f(u) \le \alpha_2 |u| + \beta_2 |u|^{p-1} \text{ in } \mathbb{R}^n(u \ge M_0). \tag{11}$$

Take $M_2 = max\{M_0, M_1\}$ and $\tau \le \tau_1$, let $(u - M_2)_+$ be the positive part of $u - M_2$, that is

$$(u - M_2)_+ = \begin{cases} u - M_2, & u \ge M_2 \\ 0, & u \le M_2. \end{cases}$$

Multiplying (1) by $(u - M_2)_+$ and integrating on $\mathbb{R}^n$, we have

$$\frac{1}{2}\frac{d}{dt}\int_{\Omega_2} |(u - M_2)_+|^2 dx + \int_{\Omega_2} |\nabla u|^p dx + \lambda \int_{\Omega_2} |u|^{p-2} u(u - M_2)_+ dx$$

$$+ \int_{\Omega_2} f(u)(u - M_2)_+ dx = \int_{\Omega_2} g(t)(u - M_2)_+ dx,$$

where $\Omega_2 \triangleq \mathbb{R}^n(u \ge M_2)$, then

$$\frac{1}{2}\frac{d}{dt}\int_{\Omega_2} |(u - M_2)_+|^2 dx + \int_{\Omega_2} |\nabla u|^p dx + \lambda \int_{\Omega_2} |u|^p dx + \int_{\Omega_2} f(u) u dx$$

$$\leq \int_{\Omega_2} g(t)(u - M_2)_+ dx + M_2 \left[ \lambda \int_{\Omega_2} |u|^{p-1} dx + \int_{\Omega_2} f(u) dx \right].$$

From (2), (10), (11) and Young's inequality, we have

$$\frac{1}{2} \frac{d}{dt} \int_{\Omega_2} |(u - M_2)_+|^2 dx + \int_{\Omega_2} |\nabla u|^p dx + \frac{\lambda - \beta_1}{2} \int_{\Omega_2} |u|^p dx + \alpha_1 \int_{\Omega_2} |u|^2 dx$$

$$\leq C(\delta) \int_{\Omega_2} |g(t)|^2 dx + \delta \int_{\Omega_2} |u|^2 + C\varepsilon, \tag{12}$$

for some $\delta > 0$ small enough, such that $0 < \sigma < \alpha_1 - \delta$. Multiplying (12) by $e^{\sigma t}$, we get that

$$\frac{d}{dt} e^{\sigma t} \int_{\Omega_2} |(u - M_2)_+|^2 dx + e^{\sigma t} \int_{\Omega_2} \left( 2|\nabla u|^p + (\lambda - \beta_1)|u|^p + 2\alpha_1 |u|^2 \right) dx$$

$$- \sigma e^{\sigma t} \int_{\Omega_2} |(u - M_2)_+|^2 dx \leq C e^{\sigma t} \int_{\Omega_2} |g(t)|^2 dx + C e^{\sigma t} \varepsilon,$$

then we obtain that

$$\frac{d}{dt} e^{\sigma t} \int_{\Omega_2} |(u - M_2)_+|^2 dx + e^{\sigma t} \int_{\Omega_2} \left( 2|\nabla u|^p + (\lambda - \beta_1)|u|^p + \alpha_1 |u|^2 \right) dx$$

$$+ (\alpha_1 - \sigma) e^{\sigma t} \int_{\Omega_2} |(u - M_2)_+|^2 dx \leq C e^{\sigma t} \int_{\Omega_2} |g(t)|^2 dx + C e^{\sigma t} \varepsilon. \tag{13}$$

Therefore, Gronwall's inequality yields that

$$\int_{\Omega_2} |(u(t) - M_2)_+|^2 dx \leq e^{-\sigma t} e^{\sigma \tau} \int_{\Omega_2} |(u(\tau) - M_2)_+|^2 dx$$

$$+ C e^{-\sigma t} \int_\tau^t e^{\sigma s} \int_{\Omega_2} |g(s)|^2 dx ds + C\varepsilon,$$

combining this inequality with (8) and (9), there exists $\tau_3(t)$, such that

$$\int_{\Omega_2} |(u(t) - M_2)_+|^2 dx \leq C\varepsilon, \quad \text{for } \forall\, u_\tau \in B(\tau) \text{ and } \tau \leq \tau_3.$$

Now, for any $\tau \leq \min\{\tau_3, \tau_1\}$, integrating (13) from $\tau$ to $t$, we get that

$$e^{\sigma t} \int_{\Omega_2} |(u(t) - M_2)_+|^2 dx + C \int_\tau^t e^{\sigma t} \int_{\Omega_2} \left( |\nabla u|^p + |u|^p + |u|^2 \right) dx dt$$

$$\leq e^{\sigma \tau} \int_{\Omega_2} |(u(\tau) - M_2)_+|^2 dx + C \int_\tau^t e^{\sigma t} \int_{\Omega_2} |g(t)|^2 dx dt + C\varepsilon \int_\tau^t e^{\sigma t} dt$$

$$\leq C\varepsilon. \tag{14}$$

Denote $F(u) = \int_0^u f(s)ds$, we can deduce from (2) that

$$\frac{\alpha_1}{2}|u|_2^2 - \frac{\beta_1}{p}|u|_p^p \leq \int_{\Omega_2} F(u) \leq \alpha_3|u|_2^2 + \beta_3|u|_p^p, \tag{15}$$

for some $\alpha_3, \beta_3 > 0$. Then (14) implies that

$$\int_\tau^t e^{\sigma t} \int_{\Omega_2} (|\nabla u|^p + |u|^p + F(u))\, dxdt \leq C\varepsilon. \tag{16}$$

On the other hand, multiplying (1) by $(u - M_2)_{+t}$ and integrating on $\mathbb{R}^n$, we have

$$\int_{\mathbb{R}^n} |(u - M_2)_{+t}|^2 dx + \int_{\mathbb{R}^n} |\nabla u|^{p-2}\nabla u \nabla (u - M_2)_{+t} dx$$

$$+ \lambda \int_{\mathbb{R}^n} |u|^{p-2} u \cdot (u - M_2)_{+t} dx + \int_{\mathbb{R}^n} f(u) \cdot (u - M_2)_{+t} dx$$

$$= \int_{\mathbb{R}^n} g(t) \cdot (u - M_2)_{+t} dx.$$

Then we have

$$\frac{1}{2} \int_{\mathbb{R}^n} |(u - M_2)_{+t}|^2 dx + \frac{d}{dt} \int_{\Omega_2} \left(\frac{1}{p}|\nabla u|^p + \frac{\lambda}{p}|u|^p + F(u)\right) dx$$

$$\leq \frac{1}{2} \int_{\Omega_2} |g(t)|^2 dx,$$

after some simple calculation, it becomes

$$\frac{d}{dt} e^{\sigma t} \int_{\Omega_2} \left(\frac{1}{p}|\nabla u|^p + \frac{\lambda}{p}|u|^p + F(u)\right) dx$$

$$\leq \sigma e^{\sigma t} \int_{\Omega_2} \left(\frac{1}{p}|\nabla u|^p + \frac{1}{p}\lambda|u|^p + F(u)\right) dx + \frac{1}{2}e^{\sigma t} \int_{\Omega_2} |g(t)|^2 dx,$$

in which $0 < \sigma < \alpha_1$, then from uniform Gronwall's inequality, we have

$$\int_{\Omega_2} \left(\frac{1}{p}|\nabla u|^p + \frac{\lambda}{p}|u|^p + F(u)\right) dx \leq C\varepsilon,$$

by applying (15) once again, it is easy to get that

$$\int_{\Omega_2} (|\nabla u|^p + |u|^p + |u|^2)\, dx \leq C\varepsilon. \tag{17}$$

Just taking $|(u + M_2)_-|$ instead of $(u - M_2)_+$, and replacing $(u - M_2)_{+t}$ with $(u + M_2)_{-t}$, after repeating the same steps above, we can deduce that

$$\int_{\mathbb{R}^n (u \leq -M_2)} \left( |\nabla u|^p + |u|^p + |u|^2 \right) dx \leq C\varepsilon, \tag{18}$$

combining (17) with (18), we conclude that for any $M \geq M_2$,

$$\int_{\mathbb{R}^n (|u| \geq M)} \left( |\nabla u|^p + |u|^p + |u|^2 \right) dx \leq C\varepsilon,$$

and the proof is completed.                                                  □

**Proof of Theorem 2:** First of all, it is obvious that $U$ is norm-to-weak continuous process on $L^p(\mathbb{R}^n)$. Collecting Theorem 4, Theorem 5, and Lemma 4, the $(L^2(\mathbb{R}^n), L^p(\mathbb{R}^n))$–pullback attractor for $U$ exists.                □

**Acknowledgements.** This work is supported by the Doctor Funds of Henan Polytechnic University (No. B2009-3 and No.B2008-56) and the NSF of Henan Provincial education department (No.2011B110015).

# References

1. Babin, A.V., Vishik, M.I.: Attractors of Evolution Equations. North-Holland, Amsterdam (1992)
2. Carcalho, A.N., Gentile, C.B.: Asymptotic behaviour of nonlinear parabolic equations with monotone principal part. J. Math. Anal. Appl. 280, 252–272 (2003)
3. Chen, G., Liu, Z.: Pullback attractor for non-autonomous p-Laplacian equation in unbounded domain. Journal of XinYang Normal University (national science edition) 24, 162–166 (2011)
4. Chen, G., Zhong, C.: Uniform attractor for non-autonomous p-Laplacian equation. Nonlinear Anal. 68, 3349–3363 (2008)
5. Khanmanedov, A.K.: Existence of a global attractor for the degenerate parabolic equation with p-Laplacian principal part in unbounded domain. J. Math. Anal. Appl. 316, 601–615 (2006)
6. Wang, Y: On the existence of pullback attractors for non-autonomous infinite dimensional dynamical systems. Doctoral dissertation, Collected in library of Lanzhou University, http://218.196.244.89:8088/D/Thesis_Y1332668.aspx
7. Yang, M., Sun, C., Zhong, C.: Existence of a global attractor for a p-Laplacian equation in $\mathbb{R}^n$.. Nonlinear Anal. 66, 1–13 (2007)
8. Yang, M., Sun, C., Zhong, C.: Global attractor for $p$-Laplacian equation. J. Math. Anal. Appl. 327, 1130–1142 (2007)

# Monte-Carlo Simulation of Error Sort Model

Ryoji Fukuda and Kaoru Oka

**Abstract.** We propose a novel quantization method for human likelihood or preference using rank data. We estimate an evaluation function by assuming that the data is sorted on the basis of observed function values and that the observation includes some errors. We use the difference in the rank order to estimate the corresponding function. Then, its performance will vary depending on the settings. Our data are normally distributed random numbers. First, function values are given as normally distributed random numbers, and after that, observed values are given by adding normally distributed random errors. The performance will depend on the difference in the variances of two normal distributions: for the function values and for the error values. We present a test to evaluate the proposed method by using random data.

**Keywords:** Monte-Carlo Simulation, Evaluation function, Rank data.

## 1 Introduction

Service providers struggle with the evaluation of human feelings to create good industrial products or to render quality services. For example, computer software may take over some human task if certain human feelings are quantified. Such features are important for supporting handicapped persons who use assistive software.

Ryoji Fukuda
Faculty of Engineering Oita University 870-1192 Oita Japan
e-mail: rfukuda@oita-u.ac.jp

Kaoru Oka
Faculty of Computer Science and Systems Engineering Kyushu Institute of Technology 820-8502 Iizuka Japan
e-mail: oka@oz.ces.kyutech.ac.jp

S. Li (Eds.): Nonlinear Maths for Uncertainty and its Appli., AISC 100, pp. 479–486.
springerlink.com &copy; Springer-Verlag Berlin Heidelberg 2011

In general, the quantification of human feelings is not easy. Questionnaires, which are typically used to quantify human feelings, do not provide sufficient information. The format of the questionnaire is important. If the questions are too simple, the structure of the data is also simple, and we need a large number of respondents. Conversely, complex questions will wear down the respondents. From this point of view, we believe that the ranking of data is useful. Arranging the data is a primitive task, and people may be accustomed to it. Indeed, there are many TV programs and magazine articles on ranking data.

In this report, we deal with incomplete rank data, that is, data of integer sequences $\{\{r_j^{(k)}\}_{j\leq n}\}_{k\leq K}$, where $\{r_j^{(k)}\}_{j\leq n}$ consists of $n$ different integers for each $k \leq K$, and these sets may be different from each other. Moreover, we assume that these ranks are determined by an unknown estimation function $f(x)$, and that these values are observed with independent $N(0,1)$-errors. This model was defined as error sort model in [2]. Our objective is to develop an estimation method for the function $f(x)$ using the above rank data. Another try for rank data is found, for example, in [1].

Simulations for complete rank data are given in [2] (which is submitted to FSS and is written in Japanese), and these outlines will be explained in the following sections. This is the case where $\{r_j^{(k)}\}_{j\leq n}$ are permutations of $\{1, 2, \ldots, n\}$. We use data of the difference in the ranks in these estimations; we cannot estimate the function when all rank data are the same. The performance of the estimation may depends on the difference in the function $f(x)$. We will analyze them using Monte-Carlo simulation.

## 2 Error Sort Model

The definition of this model is given in [2] (in Japanese), and some results for complete rank data (in the case $N = n$) are given. This section includes the definition and some results of [2], which are explained for self-containedness.

### 2.1 Definition of the Model

Let $n, K$ be integers with $n < N$, and $\{\{r_j^{(k)}\}_{j\leq n}\}_{k\leq K}$ be a family of integer sequences such that

$$1 \leq r_j^{(k)} \leq N, \quad r_i^{(k)} \neq r_j^{(k)} \quad \text{if } i \neq j \text{ for each } k \leq K.$$

Let $f(x)$ be a function defined on $I = 1, 2, \ldots, N$. This is an ideal evaluation function and all rank data are sorted according to this function. Differences among the rank data arise from errors of observation.

Let $\{g_j^{(k)}\}_{j\leq n, k\leq K}$ be a sequence of independent random variables, and assume that the distribution of $g_j^{(k)}$ is $N(0, \sigma^2)$ for each $j$ and $k$. We assume that the rank data $\{r_j^{(k)}\}_{j\leq n}$ $(k \leq K)$ is a sorting result of

$$\{f(j) + g_j^{(k)}\}_{j\leq n}.$$

This was defined as error sort model in [2].

## 2.2 Reverse Count

When all the rank data $\{r_j^{(k)}\}_{j\leq n}$ $(k \leq K)$ are of the same order, we can only determine the order but never estimate the function values.

Assume that there exists a standard rank order $\{r_j^*\}_{j\leq n}$. We do not assume that "$f(r_j^*) < f(r_k^*)$ if $j < k$" is always true; in the actual analysis we obtain this rank order using the average of individual ranks. First, we consider the case $n = N$. Let $R_{i,j}$ be the number of the reverse order for two elements $i, j \in I$, as compared to the standard order, that is,

$$R_{i,j} = \#\{k : (i^* - j^*)(i^{(k)} - j^{(k)}) < 0\},$$

where $i^*, i^{(k)}$ are integers that satisfy $r_{i^*}^* = i, r_{i^{(k)}}^{(k)} = i$ and $\#\{\ldots\}$ denotes the cardinality of the set $\{\ldots\}$. The number $R_{i,j}$ is defined as the *reverse count* for $i, j$

Next, we consider general cases $(n \leq N)$. Let $I'$ be a set of all integers in $\{r_j^{(k)}\}_{j\leq n}$ $(k \leq K)$, that is,

$$I' = \bigcup_{k\leq K} \{r_j^{(k)}\}_{j\leq n}.$$

Assume that the given standard rank order is a permutation of $I'$. Then, we can extend $R_{i,j}$ as

$$R_{i,j} = \#\left\{ k : \begin{array}{ll} (i^* - j^*)(i^{(k)} - j^{(k)}) < 0 \\ (i^* - j^*) < 0 & i^{(k)} \text{ does not exist} \\ (i^* - j^*) > 0. & j^{(k)} \text{ does not exist} \end{array} \right\},$$

Next, we define

$$T_{i,j} = \#\left\{ k : \{i,j\} \cap \{r_j^{(k)}\}_{j\leq n} \neq \emptyset \right\}.$$

Then, the distribution of the random variable $R_{i,j}$ is binomial $B(T_{i,j}, p)$. Without loss of generality, we assume that $i^* < j^*$; then the probability $p$ is given by

$$p = \frac{1}{\sqrt{2\pi\sigma^2}} \int_{f(j)-f(i)}^{\infty} e^{-\frac{x^2}{2\sigma^2}} \, dx.$$

The difference $f(j) - f(i)$ can be negative if the standard order is different from the ascending order of $\{f(i)\}$. All distributions for reverse counts are determined by the function $f(i)$. Then, we are able to estimate the function by the maximum likelihood method.

# 3  Simulations for Reverse Counts

## 3.1  For Complete Data

Let $n = N$ and $K$ be a positive integers and $\{\{r_j^{(k)}\}_{j \leq n}\}_{k \leq K}$ be a family of rank data. We define a tentative evaluation function $f^*$ on $I(= I')$.

$$f^*(i) = \sum_{k=1}^{K} i^{(k)},$$

where $i^{(k)}$ denotes the rank of the integer '$i$' in $\{r_j^{(k)}\}_{j \leq n}$ (see also section 2.2). We define a standard rank order $\mathbf{r}^* = \{r_j^*\}_{j \leq n'}$ as the sorting result according to the function $f^*$. Then we obtain the reverse count $\{R_{i,j}\}$ with respect to the standard order $\mathbf{r}^*$.

We define that a pair $i, j$ is $m$-connected if $R_{i,j} \geq m$, and that $I$ is $m$-connected if $I$ consists of one equivalent class with respect to the equivalence relation "$m$-connected". The maximum connected reverse count (MCRC) is the maximum $m$ with $I$ is $m$-connected. We believe that this value has certain dependence with the estimate accuracy by our method. We try to obtain an average values for MCRCs under several situations, according to the table 1.

In general, we use $N(0, SD^2)$-random numbers and obtain the function values $\{f(1), f(2), \ldots, f(N)\}$ Let $\{g_j\}_{j=1}^{N}$ be a sequence of independent $N(0, sd^2)$-random numbers ($sd < SD$). ('Target Num.' stands for $N$, 'Base Sd' stands for $SD$, 'Error Sd' stands for $sd$.) One datum $\{r_j\}_{j=1}^{n}$ is a sorting result of top $n$ indices ($f(r_1) + g_{r_1} \leq f(r_2) + g_{r_2} \leq \cdots$, and 'Select Num' stands for $n$). The left graph in Fig. 1 is a graph of 1000 trials' average of the MCRCs.

## 3.2  For Incomplete Data

Let $I'(\subset I = \{1, 2, \ldots, N\}$ be a set of all appeared indices. The number of $I'$ is larger than the select num $n$. We select $n$ reliable indices via following steps.

1. Set the tentative evaluation function $f^*(j)$ on $I'$, let $\{r_j^*\}_{j=1}^{n'}$ be the sort result with respect to $f^*$.

**Table 1** Simulation 1

| | |
|---|---|
| Target Num. | 15 |
| Select Num. | 15 |
| Sample Num. | 100 |
| Base SD | 1.0 |
| Error SD | 0.1 ∼ 0.5 (step 0.1) |
| Try Num | 1000 |

2. Let $C_1$ be the 1-connected component including $r_1$, and stop this procedure if $|C_1|$ is larger than $n$.
3. Let $C_2$ be the 1-connected component including $r_j$, where $j$ is the minimal number with $r_j \notin C_1$. Stop this procedure if $|C_1 \cup C_2| \geq n$.
4. Iterate similar steps until $|\bigcup_{j=1}^{J} C_j| \geq n$.
5. Sort all indices in $\bigcup_{j=1}^{J} C_j$ with respect to $f^*$ and select the top $n$ elements.

Let $I^*$ be the set of $n$ selected indices by the above steps. In the next section we estimate the evaluation function $f$ on $I^*$. We try the test of MCRC for the random data in Table 2, and the right graph in Fig. 1 is its result, Each MCRC's is caluculated for the selected set $I^*$.

**Table 2** Simulation 2

| | |
|---|---|
| Target Num. | 150 |
| Select Num. | 15 |
| Sample Num. | 100 |
| Base SD | 1.0 |
| Error SD | 0.01 ∼ 0.10 (step 0.01) |
| Try Num | 300 |

## 4  Simulations for Estimations

Fix a pair $i_1, i_2 \in I^*$ and assume that $i_2$ appears after $i_1$ in $\{r_j^*\}$. The reverse count $R_{i_1,i_2}$ is the number of data $\{r_j^{(k)}\}_j$ in which $i_1$ appears after $i_2$(see Section 2). The probability of "$i_1$ appears after $i_2$" is given by

$$E(f(i_2) - f(i_1)) = \frac{1}{\sqrt{4\pi\sigma^2}} \int_{f(i_2)-f(i_1)}^{\infty} e^{-\frac{x^2}{4sd^2}} \, dx,$$

where the distribution of the errors $g_{i_1}$ and $g_{i_2}$ is $N(0, sd^2)$. Let $T(i_1, i_2)$ be the number of appearance of the pair $i_1, i_2$, then the distribution of the reverse count $R_{i_1,i_2}$ is binomial $B(T(i_1, i_2), E(f(i_2) - f(i_1)))$. Then, by the

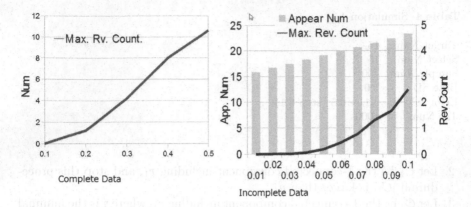

**Fig. 1** Average of Reverse Count

maximum likelihood estimation, we obtain the difference $f(i_2) - f(i_1)$ as follows:

$$f(i_2) - f(i_1) = E^{-1}(R(i_1, i_2)/T(i_1, i_2)).$$

Note that we cannot obtain this value when $R(i_1, i_2) = 0$ or $R(i_1, i_2) = T(i_1, i_2)$). In the case of complete data, $T(i_1, i_2) = n$, and for general case, $T(i_1, i_2)$ is the numbers of $\{r_j^{(k)}\}_j$ satisfying $\{i_1, i_2\} \cap \{r_j^{(k)}\}_j \neq \emptyset$.

If $I^*$ is 1-connected, all values of $f(j)$ ($j \in I^*$) can be defined using the difference of the function. However, the function cannot be uniquely defined. The following steps are our suggested method.

1. Calclate the maximum connected reverse count $m$ for $I^*$.
2. Sort $I^*$ with respect to $\{r_j^*\}_j$.
3. Select the top index $i_0 \in I^*$ in the above rank, and seek a $m$-connected index $i_1 \in I^*$. (The seek order is ascending order of $\{r_j^*\}_j$)
4. For $i_1$, seek a $m$-connected index $i_2$ and iterate this procedure until there is no connected element. Then we obtain a $m$-connected path. Define $f(i_0) = 0$, and define function values for other indices in this path using the above method.
5. Next we construct another path starting from top index among remaining indices, and define the function values.
6. Seek $m$-connected index pair between the new path and the old path (or the union of all other paths). In the case where there exist a connected pair, obtain a difference of corresponding function values using the above method, and if there is no connected pair, give a fixed large value for the nearest rank pair. Then arrange the function values for indices of new path.
7. Iterate this procedure until we obtain all function values.

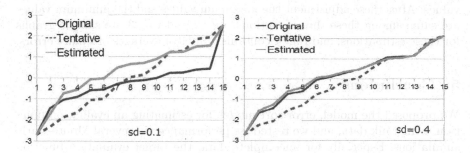

**Fig. 2** Estimations for Complete Data

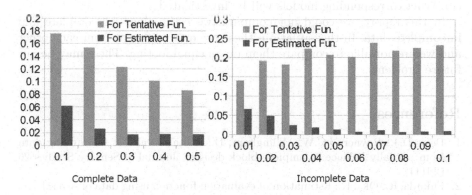

**Fig. 3** Estimations for Incomplete Data

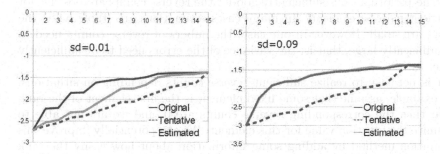

**Fig. 4** Sq. Errors for Estimations

For the two settings in Table 1 and Table 2, we try an estimation tests by the above procedure. Our purpose is to obtain evaluation functions, and assume that the role or the meaning of these functions never changes by linear transformations. Then, in the comparison among a original function, and the corresponding tentative and estimated function; we give linear transformations for the tentative and estimated function, to adjust these function values. After these adjustment the maximum values and the minimum values are same among these three functions. Fig. 2 and Fig. 3 are sample graphs for these estimations, and Fig. 4 contains graphs of averages of square errors.

## 5   Conclusions

We proposed the model, error sort model, for estimating an evaluation function using rank data, and we tested its performance via several Monte-Carlo simulations. Especially for imcomplete data, the target evaluation function was not clear and we set a reliable domain for this function.

Reverse counts, the numbers of different orders, constitute key information and the performance is sufficiently good if the reverse counts are positive for sufficient pairs of elements. In many cases estimations are stable when the errors are small. However in our case, the nubers of reverse counts need to be sufficiently large, that is, the variance of the errors need to be sufficiently large.

It is not easy to realize sufficient reverse counts depending on situations. For example, assume that some fixed element is the first element of the rank data; then the corresponding reverse counts are 0, and we are not able to estimate the function value for this element. We can potentially improve this estimation method by adding some information about how easily the rank is determined. However, simplicity and easy-to-get are the major merits of this method. In the future, ways to obtain effective information easily and to construct corresponding models will be investigated.

In this report, we could not compare two estimations for complete and imcomplete data. In this situation, these are quite different from each other and we are not able to compare them on an equal footing. These may be our future problems.

## References

1. Best, D.J., Rayner, J.C.W., Allingham, D.: A statistical test for ranking data from partially balance incomplete block designs. Journal of Sensory Studies 26 (2011)
2. Fukuda, R., Oka, K.: Estimation of evaluation function using data of several sort results. Submited to Journal of Japan Society for Fuzzy Theory and Intelligent Informatics (2011)

# A Fuzzy Estimation of Fuzzy Parameters with Rational Number Cuts

Dabuxilatu Wang and Jingjing Wang

**Abstract.** Based on the distribution function of an interval-valued random variable, we propose the parametric set for random sets, and extend the concept of fuzzy parameters proposed by Kruse and Meyer [5] and Wu [10]. Under the assumption of "inducibility from the original", we consider some novel point estimation methods for the proposed fuzzy parameters.

**Keywords:** Fuzzy estimator, Fuzzy parameters, Fuzzy random variables

## 1 Introduction

Point estimation for fuzzy parameters proposed by Kruse and Meyer [5] and Wu [10] is an important aspect of statistics with fuzzy data. The fuzzy parameters here are associated with fuzzy random variables (FRVs, [6] [5] [2] [7]) that model the fuzzy perception of a crisp but unknown random variable. For example, in an acceptance sampling by attributes, the number of conformity items in the sequence of a $n$ Bernoulli trials is governed by the binomial distribution $b(n, p)$. However, often in practical case, it is difficult to classify inspected items as "conformity" or "nonconformity". When we can not distinguish clearly a conformity item from a nonconformity item, a fuzzy observation method have to be considered, in this case, the percent defective $p$ could

Dabuxilatu Wang
School of Mathematics and Information Sciences, Guangzhou University,
No. 230 WaiHuan XiLu, Higher Education Mega Center, Guangzhou,
510006, P.R. China
e-mail: dbxlt0@yahoo.com

Jingjing Wang
School of Journalism and Communication, Jinan University,
No. 601 Huangpu Dadaoxi, Guangzhou, 510632, P.R. China
e-mail: jjwang@gmail.com

S. Li (Eds.): Nonlinear Maths for Uncertainty and its Appli., AISC 100, pp. 487–494.
springerlink.com                      © Springer-Verlag Berlin Heidelberg 2011

be considered as a vague parameter. Estimation of fuzzy parameters plays
a key role in modelling the complicated uncertainties in systems. Wang [8]
extended the fuzzy parameter by Kruse and Meyer [5] to $n$-dimensional case
with considering the notion of selector of a set-valued function under the
generalized Hausdorff distance $d_\infty$. However, there has been some tweaking
of the original two definitions of fuzzy parameters. It is reasonable to propose a unified standard definition for fuzzy parameters under the structure
of the distribution function of random sets [1]. The aim of this note is to
propose some novel fuzzy parameters and a novel point estimation approach
by introducing the notion "inducibility from the original".

## 2  Preliminary Concepts

Let $U$ be a universal set. A mapping $A : U \to [0,1]$ is said to be a fuzzy set on
$U$. $A(\cdot)$ is called the membership function of fuzzy set $A$. For each $\alpha \in [0,1]$,
the set $A_\alpha := \{x \in U | A(x) \geq \alpha\}$ ($A_{s\alpha} := \{x \in U | A(x) > \alpha\}$ ) is said to be
the $\alpha$-cut (strong $\alpha$-cut ) of fuzzy set $A$. The family $\{A_\alpha : \alpha \in [0,1]\}$ is a
set representation of fuzzy set $A$. Based on the resolution identity, we have
$A(x) = \sup_{\alpha \in [0,1]} \{\alpha | x \in A_\alpha\}, x \in U$. In the sequel, we assume $U = \mathbb{R}$ (or
more general $\mathbb{R}_0 \subset \mathbb{R}$), where $\mathbb{R}$ is the set of all real numbers.

**Definition 2.1.** A mapping $a : \mathbb{R} \to [0,1]$ is said to be a fuzzy number if it
satisfies (1)$\{x | a(x) = 1, x \in \mathbb{R}\} \neq \varnothing$. (2)$\forall \alpha \in [0,1]$, the $\alpha$−cuts of $a$, $a_\alpha$ is a
closed interval,i.e. $a_\alpha = [a_\alpha^-, a_\alpha^+]$, where $a_\alpha^- \leq a_\alpha^+$, $a_\alpha^-, a_\alpha^+ \in \mathbb{R}$.
By $\mathcal{F}(\mathbb{R})$ we denote the set of all fuzzy numbers.

**Definition 2.2.** A fuzzy number $a$ of $\mathbb{R}$ is said to be a fuzzy real number
if and only if (1)$a$ is strictly increasing on the interval $[a_0^-, a_1^-]$ and strictly
decreasing on the interval $[a_1^+, a_0^+]$ (2)$a_1^- = a_1^+$. By $\mathcal{F}_\mathbb{R}$ we denote the set of
all fuzzy real numbers.

Let $\mathbb{Q}$ be the set of all rational numbers. The family $\{a_{s\alpha} : \alpha \in [0,1] \cap \mathbb{Q}\}$ is
a set representation of fuzzy number $a$, i.e. $a(x) = \sup_{\alpha \in [0,1] \cap \mathbb{Q}} \{\alpha | x \in a_{s\alpha}\}$,
$x \in \mathbb{R}$.

We describe a random experiment by a probability space, $(\Omega, \mathcal{A}, P)$, where
$\Omega$ is the set of all possible outcomes of the experiment, $\mathcal{A}$ is a $\sigma-$ algebra of
subsets of $\Omega$ and the set function $P$, defined on $\mathcal{A}$, is a probability measure.

We will represent by a random variable, $\xi : \Omega \to \mathbb{R}$, the observation of
some attribute of the elements in the referential set, $\Omega$. When our observation can not be totally precise, we do not know the exact value, $\xi(\omega)$,
of the characteristics for the individual $\omega$. Hence, we can define a fuzzy-
valued mapping, $X : \Omega \to \mathcal{F}(\mathbb{R})$ (or $\mathcal{F}_\mathbb{R}$), that represents the fuzzy perception of $\xi$, all we can observe about the point $\xi(\omega)$ is that it belongs to
sets $X_\alpha(\omega) = [X_\alpha^-(\omega), X_\alpha^+(\omega)], \forall \alpha \in [0,1]$. We will call $\xi$ the original random variable of $X$ (or the original of $X$)(see. [5]). A fuzzy-valued mapping
$X$ is said to be FRV of Kruse and Meyer's sense if (1)$\{X_\alpha(\omega) : \alpha \in [0,1]\}$
is a set representation of $X(\omega)$ for all $\omega \in \Omega$. (2).For each $\alpha \in [0,1]$, both

$X_\alpha^-(\omega) := \inf X_\alpha(\omega)$ and $X_\alpha^+(\omega) := \sup X_\alpha(\omega)$ are real valued random variables on $(\Omega, \mathcal{A}, P)$(see. [5]).

By definition 3.3 and proposition 3.1 of [2], $X_\alpha(\omega) = [X_\alpha^-(\omega), X_\alpha^+(\omega)]$ , $\alpha \in [0, 1], \omega \in \Omega$, are random sets with respect to an appropriate measurable structure. A random variable $\xi$ is said to be a measurable selection of the random set $X_\alpha$ if there $\xi(\omega) \in X_\alpha(\omega) = [X_\alpha^-(\omega), X_\alpha^+(\omega)]$, $\forall \omega \in \Omega$. It is easy to show that the both random variables $X_\alpha^-, X_\alpha^+$, $\alpha \in [0, 1]$ are measurable selections of the random set $X_\alpha$. By $S(X_\alpha)$, we denote the set of all measurable selections of the random set $X_\alpha$, i.e.

$$S(X_\alpha) = \{\eta | \eta : \Omega \to \mathbb{R} \quad \text{is random variable and}$$

$$\eta(\omega) \in X_\alpha(\omega), \forall \omega \in \Omega\}, \alpha \in [0, 1].$$

The set of all measurable selections of $X_\alpha, \alpha \in [0, 1]$ is denoted by $\chi$, which is also called the set of all measurable selections of the FRV $X$. It is easy to verify that $\chi = \cup_{\alpha \in [0,1]} S(X_\alpha) = S(X_0)$. Note that, the $\alpha$ -level here can be viewed as a membership value which stands for the acceptance degree of the measurable selection being the original in values it takes on. Thus, we can further define a special fuzzy set $v$ on $\chi$, $v : \chi \to [0, 1]$, for a fixed $\eta \in \chi$, $v(\eta) = \sup_{\alpha \in [0,1]} \{\alpha | \eta \in S(X_\alpha)\}$.

According to I.Couso's distribution function of a random set (see. [1]),

$$F(X_\alpha) = \{F_\eta : \mathbb{R} \to [0, 1] | \eta \in S(X_\alpha)\}, \alpha \in [0, 1].$$

and for a fixed $x \in \mathbb{R}$

$$F_{X_\alpha}(x) = \{F_\eta(x) | \eta \in S(X_\alpha)\}, \alpha \in [0, 1].$$

The union $\cup_{\alpha \in [0,1]} F(X_\alpha)$ must contain the distribution $F_\xi$ of the original $\xi$. It is easy to show that $\cup_{\alpha \in [0,1]} F(X_\alpha) = F(X_0)$, and $\cup_{\alpha \in [0,1]} F_{X_\alpha}(x) = F_{X_0}(x)$.

We now assume that the type of the distribution of the original $\xi$ is known as $F_\xi(x; \theta)$, where $\theta \in \Theta$ is unknown, $\Theta$ is the parameter space of $\theta$. And each measurable selection $\eta$ of $X$ has its own unknown parameter $\theta_\eta \in \mathbb{R}$ and the distribution function $F_\eta(x; \theta_\eta)$. The distribution of $X_\alpha$ with unknown parameters is denoted by $F_{X_\alpha}(\theta(X_\alpha)) := \{F_\eta(x; \theta_\eta) | \eta \in S(X_\alpha)\}$, where $\theta(X_\alpha) := \{\theta_\eta | \eta \in S(X_\alpha), \eta \sim F_\eta(x; \theta_\eta)\}, \theta_\eta \in \mathbb{R}$, is called the parametric set of the random set $X_\alpha$, their union

$$\cup_{\alpha \in [0,1]} \theta(X_\alpha) = \theta(X_0) = \{\theta_\eta | \eta \in S(X_\alpha), \eta \sim F_\eta(x; \theta_\eta), \alpha \in [0, 1]\}, \theta_\eta \in \mathbb{R}$$

contains parameter $\theta$. Therefore, this union implies all the available information about $\theta$. On the other hand, this union can be viewed as a union of all $\alpha$-cuts of the so-called fuzzy parameter $\tilde{\theta}$ of the FRV $X$, here $\tilde{\theta} : \theta(X_0) \to [0, 1]$ is defined as

$$\tilde{\theta}(\theta_\eta) = \sup_{\alpha \in [0,1]} \left\{ \alpha | \eta \in S(X_\alpha), F_\eta(x; \theta_\eta) \in F_{X_\alpha}(\theta(X_\alpha)) \right\}, \theta_\eta \in \theta(X_0).$$

Note that $\tilde{\theta}(\theta_\eta) = v(\eta)$, and $\Theta, \theta(X_0) \subset \mathbb{R}$.

If all members $\theta_\eta \in \theta(X_0)$ are unknown, then $\theta(X_0)$ will be completely unknown, so that we have no any information about $\theta$. We need a procedure of point estimation to give an estimator for the parametric set $\theta(X_0)$ depending on the available fuzzy sample $X_1, \cdots, X_n$ from the population $X$.

**Remark 2.1**

(1)$\chi \neq \varnothing$. It is reasonable to put restriction on the fuzzy set $v$ that the identically distributed measurable selections should take the same membership value. i.e., if the set $\chi$ can be divided into different identically distributed groups $[F_1], [F_2], \cdots$, then choose a representative $\eta^{(i)}, i = 1, 2, \cdots$ from each group and calculate the membership value $v(\eta^{(i)}) = \sup_{\alpha \in [0,1]} \{\alpha | \eta^{(i)} \in S(X_\alpha)\}, i = 1, 2, \cdots$, where $\eta^{(i)}$ has the distribution $F_i$ and for any $\eta \in [F_i]$, $v(\eta) = v(\eta^{(i)}), i = 1, 2, \cdots$.

(2)The fuzzy parameter $\tilde{\theta}$ proposed by Kruse and Meyer [5] is defined as a fuzzy set on the parameter space $\Theta$ of $\theta$ with considering the parameter family $\{F_\xi(x; \theta) : \theta \in \Theta\}$ of the original $\xi$, i.e.,

$$\tilde{\theta}(t) = \sup \left\{ v(\eta) : \eta \in \chi_0, \theta_\eta = t \right\}, t \in \Theta.$$

where $\chi_0$ is the set of all measurable selections of $X$ with same type of distribution as the original $\xi$, and the function $v$ is defined above. Thus, the fuzzy parameter by Kruse and Meyer is a special case of the fuzzy parameter of the FRV $X$ defined above.

(3)The fuzzy parameter $\tilde{\theta}$ proposed by Wu [10] for FRV $X$ is directly defined as a fuzzy real number on the parameter space $\Theta$, where $\Theta$ is assumed to be an interval of $\mathbb{R}$, and the distributions $F_{X_\alpha^-}, F_{X_\alpha^+}$ of the random variables $X_\alpha^-, X_\alpha^+$ ($\alpha \in [0,1]$) satisfy that

$$F_{X_\alpha^-}(x) = F_\xi(x; \tilde{\theta}_\alpha^-), F_{X_\alpha^+}(x) = F_\xi(x; \tilde{\theta}_\alpha^+).$$

This parameter is also a special case of the fuzzy parameter of the FRV $X$.

## 3   Fuzzy Parameters Estimation under Inducibility

In this section, we propose a novel point estimation approach for the proposed fuzzy parameter in Section 2. The population $X$ is a FRV with the original $\xi$ behind it. Assume that only fuzzy observations (fuzzy data) $X_1, \cdots, X_n$ (the realization of $X$) on the unknown original $\xi$ as well as the distribution $F_\xi(x; \theta)$ with an unknown parameter $\theta \in \Theta \subset \mathbb{R}$ are available. It is clear that the estimation on $\tilde{\theta}$ can be reduced to the estimation on the parametric set $\theta(X_\alpha), \alpha \in [0,1]$, which means that we need to obtain a set-valued estimator

$\bigcup_{\eta \in X_\alpha} \hat{\theta}_\eta(X_\alpha^{(s)})$, where $X_\alpha^{(s)}$ represents the sample from the population $X_\alpha$, so that the estimator of the fuzzy parameter $\tilde{\theta}$ can be obtained formally as $\bigcup_{\alpha \in [0,1]} \bigcup_{\eta \in X_\alpha} \hat{\theta}_\eta(X^{(s)})$, where $X^{(s)}$ represents the fuzzy sample from the population $X$. However, in general, there are infinite number of distribution types in $F_{X_\alpha}(\theta(X_\alpha)) := \{F_\eta(x; \theta_\eta) | \eta \in S(X_\alpha)\}, \alpha \in [0,1]$, so it is impossible to estimate $\theta(X_\alpha), \alpha \in [0,1]$ by distribution-wise way.

We have to consider to put some restriction on the family $F_{X_\alpha}(\theta(X_\alpha)), \alpha \in [0,1]$.

**Definition 3.1.** Let $X$ be a FRV with the original $\xi$. $X$ is said to be *inducible from the original* if all distributions $\left\{ F_{X_r^-}(x; \theta_{X_r^-}), F_{X_r^-}(x; \theta_{X_r^-}) : r \in [0,1] \cap \mathbb{Q}, X_r^\pm \in \chi \right\}, \theta_{X_r^\pm} \in \mathbb{R}$, are in the parametric family $\left\{ F_\xi(x; \theta), \theta \in \Theta \right\}$.

Obviously, if the distributions of random variables $X_\alpha^-, X_\alpha^+$ are in the parametric family of the original for all $\alpha \in [0,1]$ (this is the case of [5] [10]), then $X$ is inducible from the original. If $X$ is degenerated to a random variable then it naturally inducible; If $X$ is valued in the set of all identically shaped fuzzy real numbers, then it is inducible from the original.
A FRV $X$ is not inducible from the original if there exists rational number $r_0 \in [0,1]$ such that $X_{r_0}^-$ or $X_{r_0}^+$ has no distribution in the parameter family of the original.

Let $X$ be inducible from the original. Then, for any $\alpha \in [0,1]$, $X_\alpha^\pm$ are in the parametric family of the original if and only if $X_{r_n}^- \to^P X_\alpha^-, X_{r_n}^+ \to^P X_\alpha^+$ when $r_n \to \alpha, (n \to \infty)$, where $\{r_n\} \subset [0,1] \cap \mathbb{Q}$. Definition 3.1 indicates that an inducible FRV has a set of measurable selections reduced to the original.

Now we consider the estimation of fuzzy parameter of the FRV $X$ which is inducible from the original $\xi$ with a parametric family $\left\{ F_\xi(x; \theta), \theta \in \Theta \right\}$. Since

$$\left\{ F_{X_r^-}(x; \theta_{X_r^-}), F_{X_r^+}(x; \theta_{X_r^+}); \ r \in [0,1] \cap \mathbb{Q}, X_r^\pm \in \chi \right\}$$
$$\subset \left\{ F_\xi(x; \theta), \theta \in \Theta \right\}, \theta_{X_r^+}, \theta_{X_r^-} \in \mathbb{R},$$

and the set $[0,1] \cap \mathbb{Q}$ is dense in the interval $[0,1]$, the parametric set $\{\theta_{X_r^-}, \theta_{X_r^+}; r \in [0,1] \cap \mathbb{Q}, X_r^\pm \in \chi\} \approx \Theta \subset \theta(X_0)$. Given a fuzzy sample $X_1, \ldots, X_n$, then the crisp samples $X_{1r}^-, X_{1r}^+, \ldots, X_{nr}^-, X_{nr}^+$ ($r \in [0,1] \cap \mathbb{Q}$) are given, then the unknown parameter $\theta_{X_r^-}$ ($\theta_{X_r^+}$) of $F_{X_r^-}(x; \theta_{X_r^-})$ ($F_{X_r^+}(x; \theta_{X_r^+})$) can be estimated by moment method or maximum likelihood method. The estimator $\hat{\theta}_{X_r^-} = \hat{\theta}_{X_r^-}(x_{1r}^-, x_{2r}^-, \ldots, x_{nr}^-)$ (or $\hat{\theta}_{X_r^+} = \hat{\theta}_{X_r^+}(x_{1r}^+, x_{2r}^+, \ldots, x_{nr}^+)$) can be viewed as an estimator of $\theta$, i.e.

$$\hat{\theta} = \hat{\theta}(x_1, x_2, \ldots, x_n) := \hat{\theta}_{X_r^-}(x_{1r}^-, x_{2r}^-, \ldots, x_{nr}^-)$$

(or $\hat{\theta}_{X_r^+}(x_{1r}^+, x_{2r}^+, \ldots, x_{nr}^+)$), where $\{x_1, x_2, \ldots, x_n\} \subset E$,

$$E := \left\{ X_r^-(\omega), X_r^+(\omega), r \in [0,1] \cap \mathbb{Q}, \omega \in \Omega \right\}.$$

We can extend the estimator $\hat{\theta}(x_1, x_2, \ldots, x_n)$ to a fuzzy set $\hat{\theta}(X_1, X_2, \ldots, X_n)$ within $E$ by means of Zadel's extension principle. This fuzzy set is a kind of fuzzy estimator of fuzzy parameter $\tilde{\theta}$ of the FRV $X$ when $X$ is inducible from the original, i.e.

$$\hat{\tilde{\theta}}(t) = \hat{\theta}(X_1, X_2, \ldots, X_n)(t)$$
$$= \sup_{t = \hat{\theta}(x_1, x_2, \ldots, x_n), \{x_1, x_2, \ldots, x_n\} \subset E} \min \left\{ X_1(x_1), \ldots, X_n(x_n) \right\}.$$

equivalently, $\hat{\tilde{\theta}}(t) = \sup_{\alpha \in [0,1]} \left\{ \alpha | t \in \hat{\tilde{\theta}}_\alpha \right\}$, $t \in \Theta$, where

$$\hat{\tilde{\theta}}_\alpha = \left\{ \hat{\theta}_s(x_1, x_2, \ldots, x_n), s = X_r^+, X_r^-, \right.$$
$$\left. r \in [0,1] \cap \mathbb{Q}, \{x_1, x_2, \ldots, x_n\} \subset E \right\}.$$

If $\hat{\theta}(x_1, x_2, \ldots, x_n)$ is a maximum likelihood (or moment) estimator of $\theta$, then the fuzzy set $\hat{\theta}(X_1, X_2, \ldots, X_n)$ is said to be a *fuzzy maximum likelihood estimator (or fuzzy moment estimator)* $\hat{\tilde{\theta}}$ of the fuzzy parameter $\tilde{\theta}$ of the FRV $X$.

**Example 3.1.** Assume that some risk variable $\xi$ about environmental contamination in city area can be governed by a Gaussian distribution $N(\mu, \sigma^2)$. The environmental contamination risks are perceived by the inhabitants in different senses. Assume that the fuzzy perception of the inhabitants on the risk variable $\xi$ can be expressed with a triangular FRV $X = (\xi, l, s)_{LR}$ [9], where $L(x) = R(x) = \max\{0, 1 - x\}$, $l \sim N(\mu_l, \sigma_l^2)$, $s \sim N(\mu_s, \sigma_s^2)$, and $\xi, l, s$ are independent. Then, for any $\alpha \in [0,1]$,

$$X_\alpha^- = \xi - (1 - \alpha)l \sim N\big(\mu - (1 - \alpha)\mu_l, \sigma^2 + (1 - \alpha)^2 \sigma_l^2\big),$$

$$X_\alpha^+ = \xi + (1 - \alpha)s \sim N\big(\mu + (1 - \alpha)\mu_s, \sigma^2 + (1 - \alpha)^2 \sigma_s^2\big),$$

which means

$$F_{X_\alpha^-} F_{X_\alpha^+} \in \left\{ N(\mu, \sigma^2); -\infty < \mu < \infty, \sigma^2 > 0 \right\},$$

therefore, the FRV $X$ is inducible from the original $\xi$, and here $E = \left\{ X_r^-(\omega), X_r^+(\omega), r \in [0,1], \omega \in \Omega \right\}$. Since $X$ is a fuzzy perception of $\xi$, then there are also fuzzy perceptions on the two unknown parameters $\mu, \sigma^2$, they are fuzzy parameters denoted by $\tilde{\mu}_X, \tilde{\sigma}_X^2$, and

$$\tilde{\mu}_X(t) = \sup_{\alpha \in [0,1]} \left\{ \alpha | t \in \big[ \mu - (1 - \alpha)\mu_l, \mu + (1 - \alpha)\mu_s \big] \right\}, t \in \mu(X_0),$$

$$\tilde{\sigma}_X^2(t) = \sup_{\alpha \in [0,1]} \left\{ \alpha | t \in \left\{ \sigma^2 + (1-\beta)^2 \sigma_l^2, \sigma^2 + (1-\beta)^2 \sigma_s^2, \beta \in [\alpha, 1] \right\} \right\}, t \in \sigma^2(X_0).$$

Assume that we have a sample of size $n$ from the inhabitants by questionnaire, $X_i = (\xi_i, l_i, s_i)_{LR}$, $i = 1, \ldots, n$, based on which we attempt to estimate the fuzzy parameters. From the $\alpha$-cut of each fuzzy sample data, we have the crisp samples $X_{i\alpha}^- = \xi_i - (1-\alpha)l_i, X_{i\alpha}^+ = \xi_i + (1-\alpha)s_i, i = 1, \ldots, n$ for each $\alpha \in [0,1]$, and the estimates of the parameters

$$\mu_{X_\alpha^-} = \mu - (1-\alpha)\mu_l, \mu_{X_\alpha^+} = \mu + (1-\alpha)\mu_s,$$

$$\sigma_{X_\alpha^-}^2 = \sigma^2 + (1-\alpha)^2 \sigma_l^2, \sigma_{X_\alpha^+}^2 = \sigma^2 + (1-\alpha)^2 \sigma_s^2,$$

can be obtained easily as

$$\hat{\mu}_{X_\alpha^-} = \bar{\xi} - (1-\alpha)\bar{l}, \hat{\mu}_{X_\alpha^+} = \bar{\xi} + (1-\alpha)\bar{s},$$

$$\hat{\sigma}_{X_\alpha^-}^2 = \frac{1}{n} \sum_{i=1}^n \left[ (1-\alpha)(l_i - \bar{l}) + (\xi_i - \bar{\xi}) \right]^2,$$

$$\hat{\sigma}_{X_\alpha^+}^2 = \frac{1}{n} \sum_{i=1}^n \left[ (1-\alpha)(s_i - \bar{s}) + (\xi_i - \bar{\xi}) \right]^2.$$

By Zadel's extension principle, we have

$$\hat{\tilde{\mu}}_X(t) = \sup_{\alpha \in [0,1]} \left\{ \alpha | t \in \left[ \bar{\xi} - (1-\alpha)\bar{l}, \bar{\xi} - (1-\alpha)\bar{s} \right] \right\}, t \in \left\{ \hat{\mu}_{X_\alpha^-}, \hat{\mu}_{X_\alpha^+}, \alpha \in [0,1] \right\},$$

i.e. $\hat{\tilde{\mu}}_X = (\bar{\xi}, \bar{l}, \bar{s})_{LR}$.

$$\hat{\tilde{\sigma}}_X^2(t) = \sup_{\alpha \in [0,1]} \left\{ \alpha | t \in \left\{ \frac{1}{n} \sum_{i=1}^n \left[ (1-\beta)(s_i - \bar{s}) + (\xi_i - \bar{\xi}) \right]^2, \right. \right.$$

$$\left. \left. \frac{1}{n} \sum_{i=1}^n \left[ (1-\beta)(l_i - \bar{l}) + (\xi_i - \bar{\xi}) \right]^2, \beta \in [\alpha, 1] \right\} \right\},$$

$$t \in \left\{ \hat{\sigma}_{X_\alpha^-}^2, \hat{\sigma}_{X_\alpha^+}^2, \alpha \in [0,1] \right\}.$$

The estimator $\hat{\tilde{\mu}}_X, \hat{\tilde{\sigma}}_X^2$ is a fuzzy maximum likelihood estimator of $\tilde{\mu}_X$, $\tilde{\sigma}_X^2$, respectively.

## 4   Conclusions

We have proposed a parametric set , a general fuzzy parameter and the notion "inducibility from the original " for fuzzy random variable of Kruse and Meyer. Based on the fuzzy observations and the given parameters model of the original, we present a fuzzy estimation on the fuzzy parameters by using extension principle with simple restriction on the set of parameters of selections. The proposed method can be used to all inducible fuzzy random variables with known parameters models.

# References

1. Couso, I., Sánchez, L., Gil, P.: Imprecise distribution function associated to a random set. Information Sciences 159, 109–123 (2004)
2. Gil, M.A., López-Díaz, M., Ralescu, D.A.: Overview on the development of fuzzy random variables. Fuzzy Sets and Systems 157, 2546–2557 (2006)
3. Grzegorzewski, P.: Testing statistical hypotheses with vague data. Fuzzy Sets and Systems 112, 501–510 (2000)
4. Nguyen, H.T.: Fuzzy and random sets. Fuzzy Sets and Systems 156(3), 349–356 (2005)
5. Kruse, R., Meyer, K.D.: Statistics with Vague Data. D. Reidel Publishing Company, Dordrecht (1987)
6. Kwakernaak, H.: Fuzzy random variables, Part-1: Definitions and theorems. Informatioin sciences 15, 1–15 (1978); Part-2: Algorithms and examples for the discrete case. Information Sciences 17, 253-278 (1978)
7. Puri, M.D., Ralescu, D.: Fuzzy Random Variables. Journal of Mathematical Analysis and Applications 114, 409–422 (1986)
8. Wang, D.: A note on consistency and unbiasedness of point estimation with fuzzy data. Metrika 60, 93–104 (2004)
9. Wang, D.: A cusum control chart for fuzzy quality data. Advanced soft computing 37, 357–364 (2006)
10. Wu, H.C.: The fuzzy estimators of fuzzy parameters based on fuzzy random variables. European Journal of Operational Research 146, 101–114 (2003)

# Regularized REML for Estimation in Heteroscedastic Regression Models

Dengke Xu and Zhongzhan Zhang

**Abstract.** In this paper, we propose a regularized restricted maximum likelihood(REML) method for simultaneous variable selection in heteroscedastic regression models. Under certain regularity conditions, we establish the consistency and asymptotic normality of the resulting estimator. A simulation study is conducted to illustrate the performance of the proposed method.

**Keywords:** Heteroscedastic regression models, Variable selection, REML, Regularization.

## 1  Introduction

Many different approaches have been suggested to the problem of flexibly modeling of the mean. Less attention, however, has been devoted to the problem of modeling of the variance compared with that of the mean in statistical literature. In many applications, particularly in the econometric area and industrial quality improvement experiments, modeling the variance will be of direct interest in its own right, to identify the source of variability in the observations. On the other hand, modeling the variance itself may be of scientific interest. Thus, modeling of the variance can be as important as that of the mean.

Heteroscedastic regression models for normal data have been received a lot of attention in recent years. For example, Park [9] proposed a log linear model for the variance parameter and described the Gaussian model using a two stage process to estimate the parameters. Harvey [5] discussed maximum likelihood (ML) estimation of the mean and variance effects and the

Dengke Xu and Zhongzhan Zhang
College of Applied Sciences, Beijing University of Technology,
100 Pingleyuan, Chaoyang District, Beijing, 100124, P.R. China
e-mail: xudengke1983@emails.bjut.edu.cn, zzhang@bjut.edu.cn

S. Li (Eds.): Nonlinear Maths for Uncertainty and its Appli., AISC 100, pp. 495–502.
springerlink.com                    © Springer-Verlag Berlin Heidelberg 2011

subsequent likelihood ratio test under general conditions. Aitkin [1] provided ML estimation for the joint mean and variance models and applied it to the commonly cited Minitab tree data. Verbyla [13] estimated the parameters using restricted maximum likelihood (REML) and provided leverage and influence diagnostics for ML and REML. Engel and Huele [2] applied a similar model to Taguchi-type experiments for robust design. Taylor and Verbyla [11] proposed joint modeling of location and scale parameters of the t distribution to accommodate possible outliers. More general distributions from the family of generalized linear models are considered by Nelder and Lee [8], Lee and Nelder [7], Smyth and Verbyla [10] and Wang and Zhang [14].

To the best of our knowledge, most existing variable selection procedures are limited to only select the mean explanation variables, e.g., Fan and Lv [4] and references therein. In practice, it is also important that we can find which variables drive the variance by variable selection procedures. However, little work has been done to select the variance explanation variables. Wang and Zhang [14] proposed an only variable selection of the mean explanation variable criterion EAIC based on the extended quasi-likelihood which is for joint generalized linear models with structured dispersions.

The main objective of this paper is to develop an efficient regularized REML based method to select explanatory variables that make a significant contribution to the joint mean and variance models. We propose a unified procedure that simultaneously selects significant variables in joint mean and variance models. Furthermore, with proper choice of tuning parameters, we show that this variable selection procedure is consistent, and the estimators of regression coefficients have oracle property. This indicates that the regularized estimators work as well as if the subset of true zero coefficients were already known. A simulation study is used to illustrate the proposed methodologies.

The rest of this paper is organized as follows. In Section 2 we first describe heteroscedastic regression models. Then, we propose a regularized method for the joint models via regularized REML method. Asymptotic properties of the resulting estimators are considered in Section 3. In Section 4 we carry out simulation studies to assess the finite sample performance of the method.

## 2 Variable Selection in Heteroscedastic Regression Models

### 2.1 Heteroscedastic Regression Models

Let $Y = (y_1, y_2, \cdots, y_n)^T$ be a vector of $n$ independent responses, where $n$ is the sample size; $X = (x_1, x_2, \cdots, x_n)^T$ be an $n \times p$ matrix whose $i$th row $x_i^T = (x_{i1}, \cdots, x_{ip})$ is the observation of explanatory variables associated with the mean of $y_i$; and $Z = (z_1, z_2, \cdots, z_n)^T$ be an $n \times q$ matrix whose $i$th row $z_i^T = (z_{i1}, \cdots, z_{iq})$ is the observation of explanatory variables associated with the variance of $y_i$. There might be some $z$s which coincide with some $x$s. The heteroscedastic regression models are then written as

$$\begin{cases} y_i \sim N(\mu_i, \sigma_i^2) \\[2mm] \mu_i = x_i^T \beta \\[2mm] \sigma_i^2 = h^2(z_i^T \gamma) \end{cases} \tag{1}$$

where $\beta = (\beta_1, \cdots, \beta_p)^T$ is a $p \times 1$ vector of unknown regression coefficients in the mean model, and $\gamma = (\gamma_1, \cdots, \gamma_q)^T$ is a $q \times 1$ vector of regression coefficients in the variance model. For the identifiability of the model, we always suppose that $h(\cdot)$ is a monotone function. We first assume all the explanatory variables of interest, and perhaps their interactions as well, are already included into the initial models. We then aim to remove the unnecessary explanatory variables from the models.

## 2.2  Regularized REML Estimation

Our variable selection method is built upon standard methods of estimation in the heteroscedastic regression models, specifically, maximum likelihood (ML) and restricted maximum likelihood (REML) methods. Under model (1) and up to a constant, the (full) log-likelihood for the data is

$$\ell_F(\beta, \gamma) = -\frac{1}{2} \log |\Sigma| - \frac{1}{2}(Y - X^T \beta)^T \Sigma^{-1}(Y - X^T \beta), \tag{2}$$

where $\Sigma = \mathrm{diag}\{h^2(z_1^T \gamma), \cdots, h^2(z_n^T \gamma)\}$, and the ML estimates of parameters $\beta, \gamma$ can be obtained by maximizing the log-likelihood function (2). Note that when $\gamma$ is known, the MLE for $\beta$ is given by

$$\hat{\beta}(\gamma) = \arg\min_{\beta} \frac{1}{2}(Y - X^T \beta)^T \Sigma^{-1}(Y - X^T \beta). \tag{3}$$

One well-known criticism on the ML estimation is that for the variance components (i.e. $\gamma$), there is a downward finite-sample bias due to the fact that the ML method does not take into account the loss in degrees of freedom from the estimation of $\beta$. The restricted maximum likelihood estimator (REML) corrects for this bias by defining estimators of the variance components as the maximizers of the log-likelihood based on $n - p$ linearly independent error contrasts. This log-likelihood, according to Harville [6], is

$$\ell_R(\gamma) = -\frac{1}{2} \log |\Sigma| - \frac{1}{2} \log |X \Sigma^{-1} X^T| - \frac{1}{2}(Y - X^T \hat{\beta}(\gamma))^T \Sigma^{-1}(Y - X^T \hat{\beta}(\gamma)), \tag{4}$$

where $\hat{\beta}(\gamma)$ is given by (3).

One way to obtain the estimate of $(\beta, \gamma)$ is to solve (3) and (4) iteratively until convergence. Joining the estimator (3) and the REML (4), we may write a modified log-likelihood as

$$\ell_n(\beta, \gamma) = -\frac{1}{2} \log |\Sigma| - \frac{1}{2} \log |X \Sigma^{-1} X^T| - \frac{1}{2} (Y - X^T \beta)^T \Sigma^{-1} (Y - X^T \beta). \quad (5)$$

Clearly, the MLE of $\beta$ and the REML of $\gamma$ can be obtained by jointly maximizing (5). In order to obtain the desired sparsity in the final estimators, we propose the regularized likelihood function

$$\mathcal{L}(\beta, \gamma) = \ell_n(\beta, \gamma) - n \sum_{j=1}^{p} p_{\lambda_j^{(1)}} (|\beta_j|) - n \sum_{k=1}^{q} p_{\lambda_k^{(2)}} (|\gamma_k|). \quad (6)$$

For notational simplicity, we rewrite (6) in the following

$$\mathcal{L}(\theta) = \ell_n(\theta) - n \sum_{j=1}^{p} p_{\lambda_j^{(1)}} (|\beta_j|) - n \sum_{k=1}^{q} p_{\lambda_k^{(2)}} (|\gamma_k|), \quad (7)$$

where $\theta = (\theta_1, \cdots, \theta_s)^T = (\beta_1, \cdots, \beta_p; \gamma_1, \cdots, \gamma_q)^T$ with $s = p + q$ and $p_{\lambda^{(l)}}(\cdot)$ is a given penalty function with the tuning parameter $\lambda^{(l)} (l = 1, 2)$. The tuning parameters can be chosen by a data-driven criterion such as cross validation (CV), generalized cross-validation (GCV) (Tibshirani [12]), or the BIC-type tuning parameter selector (Wang et al. [15]). Here we use the same penalty function $p(\cdot)$ for all the regression coefficients but with different tuning parameters $\lambda^{(1)}$ and $\lambda^{(2)}$ for the mean parameters and the variance parameters, respectively. Note that the penalty functions and tuning parameters are not necessarily the same for all the parameters. For example, we wish to keep some important variables in the final model and therefore do not want to penalize their coefficients. In this paper, we use the smoothly clipped absolute deviation (SCAD) penalty whose first derivative satisfies

$$p_\lambda'(t) = \lambda \left\{ I(t \leq \lambda) + \frac{(a\lambda - t)_+}{(a - 1)\lambda} I(t > \lambda) \right\}$$

for some $a > 2$ (Fan and Li [3]). Following the convention in Fan and Li [3], we set $a = 3.7$ in our work. The SCAD penalty is a spline function on an interval near zero and constant outside, so that it can shrink small value of an estimate to zero while having no impact on a large one.

The regularized REML estimator of $\theta$, denoted by $\hat{\theta}$, maximizes the function $\mathcal{L}(\theta)$ in (7). With appropriate penalty functions, maximizing $\mathcal{L}(\theta)$ with respect to $\theta$ leads to certain parameter estimators vanishing from the initial models so that the corresponding explanatory variables are automatically removed. Hence, through maximizing $\mathcal{L}(\theta)$ we achieve the goal of selecting important variables and obtaining the parameter estimators, simultaneously.

## 3  Asymptotic Properties

We next study the asymptotic properties of the resulting regularized REML estimators. We first introduce some notations. Let $\theta_0$ denote the true values of $\theta$. Furthermore, let $\theta_0 = (\theta_{01}, \cdots, \theta_{0s})^T = ((\theta_0^{(1)})^T, (\theta_0^{(2)})^T)^T$. For ease of presentation and without loss of generality, it is assumed that $\theta_0^{(1)}$ consists of all nonzero components of $\theta_0$ and that $\theta_0^{(2)} = 0$. Denote the dimension of $\theta_0^{(1)}$ by $s_1$. Let

$$a_n = \max_{1 \le j \le s} \{p'_{\lambda_n}(|\theta_{0j}|), \theta_{0j} \ne 0\}$$

and

$$b_n = \max_{1 \le j \le s} \{|p''_{\lambda_n}(|\theta_{0j}|)| : \theta_{0j} \ne 0\},$$

where $\lambda_n$ is equal to either $\lambda_n^{(1)}$ or $\lambda_n^{(2)}$, depending on whether $\theta_{0j}$ is a component of $\beta_0$ or $\gamma_0 (1 \le j \le s)$.

To obtain the theorems in the paper, we require the following regularity conditions:

(C1): The parameter space is compact and the true value $\theta_0$ is in the interior of the parameter space.

(C2): The design matrices $x_i$ and $z_i$ in the joint models are all bounded, meaning that all the elements of the matrices are bounded by a single finite real number.

(C3): $\lim_{n \longrightarrow \infty} (\frac{1}{n} X \Sigma^{-1} X^T) = \mathcal{I}_\beta$, $\lim_{n \longrightarrow \infty} (-\frac{1}{n} \frac{\partial^2 \ell_R}{\partial \gamma \partial \gamma^T}) = \mathcal{I}_\gamma$, where $\mathcal{I}_\beta$ and $\mathcal{I}_\gamma$ are positive definite matrices.

**Theorem 1.** *Assume* $a_n = O_p(n^{-\frac{1}{2}})$, $b_n \to 0$ *and* $\lambda_n \to 0$ *as* $n \to \infty$. *Under the conditions* $(C1) - (C3)$ , *with probability tending to 1 there must exist a local maximizer* $\hat{\theta}_n$ *of the regularized likelihood function* $\mathcal{L}(\theta)$ *in (7) such that* $\hat{\theta}_n$ *is a* $\sqrt{n}$-*consistent estimator of* $\theta_0$.

The following theorem gives the asymptotic normality property of $\hat{\theta}_n$. Let

$$A_n = \text{diag}(p''_{\lambda_n}(|\theta_{01}^{(1)}|), \cdots, p''_{\lambda_n}(|\theta_{0s_1}^{(1)}|)),$$

$$c_n = (p'_{\lambda_n}(|\theta_{01}^{(1)}|)\text{sgn}(\theta_{01}^{(1)}), \cdots, p'_{\lambda_n}(|\theta_{0s_1}^{(1)}|)\text{sgn}(\theta_{0s_1}^{(1)}))^T,$$

where $\theta_{0j}^{(1)}$ is the $j$th component of $\theta_0^{(1)}$ $(1 \le j \le s_1)$. Denote the Fisher information matrix of $\theta$ by $\mathcal{I}_n(\theta)$.

**Theorem 2.** *Assume that the penalty function* $p_{\lambda_n}(t)$ *satisfies*

$$\liminf_{n \to \infty} \liminf_{t \to 0^+} \frac{p'_{\lambda_n}(t)}{\lambda_n} > 0$$

*and* $\bar{\mathcal{I}}_n = \mathcal{I}_n(\theta_0)/n$ *converges to a finite and positive definite matrix* $\mathcal{I}_\theta(\theta_0)$ *as* $n \to \infty$. *Under the same mild conditions as these given in Theorem 1, if*

$\lambda_n \to 0$ and $\sqrt{n}\lambda_n \to \infty$ as $n \to \infty$, then with probability tending to 1, the $\sqrt{n}$-consistent estimator $\hat{\theta}_n = ((\hat{\theta}_n^{(1)})^T, (\hat{\theta}_n^{(2)})^T)^T$ in Theorem 1 must satisfy

(i) $\hat{\theta}_n^{(2)} = 0$.

(ii) $\sqrt{n}(\bar{\mathcal{I}}_n^{(1)})^{-1/2}(\bar{\mathcal{I}}_n^{(1)} + A_n)\{(\hat{\theta}_n^{(1)} - \theta_0^{(1)}) + (\bar{\mathcal{I}}_n^{(1)} + A_n)^{-1}c_n\} \xrightarrow{D} \mathcal{N}_{s_1}(0, I_{s_1})$.

where $\bar{\mathcal{I}}_n^{(1)}$ is the $(s_1 \times s_1)$ submatrix of $\bar{\mathcal{I}}_n$ corresponding to the nonzero components $\theta_0^{(1)}$ and $I_{s_1}$ is the $(s_1 \times s_1)$ identity matrix.

Remark: Proofs of Theorem 1 and Theorem 2 are essentially along the same line as Fan and Li [3]. This will lead to the conclusions. To save space, the proofs are omitted.

## 4  Simulation Study

In this section we conduct a simulation study to assess the small sample performance of the proposed procedures. We consider the sample size $n=50$, 75 and 100 respectively. We choose the true values of the parameters in the model (1) to be $\beta_0 = (\beta_1, \beta_2, \cdots, \beta_7)^T$ with $\beta_1 = 1, \beta_2 = 1, \beta_3 = 1$, and $\gamma_0 = (\gamma_1, \gamma_2, \cdots, \gamma_7)^T$ with $\gamma_1 = 1, \gamma_2 = 1, \gamma_3 = 1$, respectively, while the remaining coefficients, corresponding to the irrelevant variables, are given by zeros.

In the models, the covariates $x_i$ and the covariates $z_i$ are generated by drawing random samples from uniform distribution on [-1,1]. The responses $y_i$ are then drawn from the normal distribution $N(\mu_i, \sigma_i^2)(i = 1, \cdots, n)$.

In the simulation study, 1000 repetitions of random samples are generated. For each random sample, the proposed variable selection procedure for finding out regularized REML estimators with SCAD penalty function is considered, says RSCAD. The unknown tuning parameters $\lambda^{(l)}, (l = 1, 2)$ for the penalty function are chosen by BIC criterion in the simulation. The average number of the estimated zero coefficients with the 1000 simulation runs is reported in Table 1. Note that "Correct" in Table 1 means the average number of zero regression coefficients that are correctly estimated as zero, and "Incorrect" depicts the average number of non-zero regression coefficients that are erroneously set to zero. The performance of estimator $\hat{\beta}$ and $\hat{\gamma}$ will be assessed by the generalized mean square error(GMSE), defined as

$$\text{GMSE}(\hat{\beta}) = (\hat{\beta} - \beta_0)^T E(XX^T)(\hat{\beta} - \beta_0), \text{GMSE}(\hat{\gamma}) = (\hat{\gamma} - \gamma_0)^T E(ZZ^T)(\hat{\gamma} - \gamma_0).$$

We compare the performance of the RSCAD variable selection procedure, proposed by this paper, with the SCAD variable selection procedure based the log-likelihood (2), says NSCAD.

From Table 1, we can make the following observations. Firstly, the performances of both variable selection procedures become better and better as $n$ increases. For example, the values in the column labeled 'Correct' become

**Table 1** Variable selections for the parametric component with different methods

| $\beta$ Method | $n = 50$ | | | $n = 75$ | | | $n = 100$ | | |
|---|---|---|---|---|---|---|---|---|---|
| | GMSE | Correct | Incorrect | GMSE | Correct | Incorrect | GMSE | Correct | Incorrect |
| RSCAD | 0.1011 | 3.8340 | 0.1130 | 0.0393 | 3.9680 | 0.0150 | 0.0238 | 3.9940 | 0.0070 |
| NSCAD | 0.1279 | 3.7340 | 0.1400 | 0.0403 | 3.9510 | 0.0150 | 0.0241 | 3.9890 | 0.0030 |

| $\gamma$ Method | GMSE | Correct | Incorrect | GMSE | Correct | Incorrect | GMSE | Correct | Incorrect |
|---|---|---|---|---|---|---|---|---|---|
| RSCAD | 0.4357 | 3.1530 | 0.5700 | 0.2288 | 3.5780 | 0.3190 | 0.1301 | 3.7710 | 0.1490 |
| NSCAD | 0.6402 | 2.5690 | 0.4580 | 0.2558 | 3.3570 | 0.2480 | 0.1399 | 3.6660 | 0.1170 |

more and more closer to the true number of zero regression coefficients in the models. Secondly, the performance of RSCAD is significantly better than that of NSCAD. The latter cannot eliminate some unimportant variables and gives larger model errors. Thirdly, the performance of the variable selection procedure in mean model outperforms that in variance model.

**Acknowledgements.** This work is supported by grants from the National Natural Science Foundation of China(10971007); Funding Project of Science and Technology Research Plan of Beijing Education Committee(JC006790201001).

# References

1. Aitkin, M.: Modelling variance heterogeneity in normal regression using GLIM. Appl. Statist. 36, 332–339 (1987)
2. Engel, J., Huele, A.F.: A generalized linear modeling approach to robust design. Technometrics 38, 365–373 (1996)
3. Fan, J.Q., Li, R.: Variable selection via nonconcave penalized likelihood and its oracle properties. Journal of American Statistical Association 96, 1348–1360 (2001)
4. Fan, J.Q., Lv, J.C.: A selective overview of variable selection in high dimensional feature space. Statistica Sinica 20, 101–148 (2010)
5. Harvey, A.C.: Estimating regression models with multiplicative heteroscedasticity. Econometrica 44, 460–465 (1976)
6. Harville, D.A.: Bayesian Inference for Variance Components Using Only Error Contrasts. Biometrika 61, 383–385 (1974)
7. Lee, Y., Nelder, J.A.: Generalized linear models for the analysis of quality improvement experiments. The Canadian Journal of Statistics 26(1), 95–105 (1998)
8. Nelder, J.A., Lee, Y.: Generalized linear models for the analysis of Taguchi-type experiments. Applied Stochastic Models and Data Analysis 7, 107–120 (1991)
9. Park, R.E.: Estimation with heteroscedastic error terms. Econometrica 34, 888 (1966)
10. Smyth, G.K., Verbyla, A.P.: Adjusted likelihood methods for modelling dispersion in generalized linear models. Environmetrics 10, 696–709 (1999)

11. Taylor, J.T., Verbyla, A.P.: Joint modelling of location and scale parameters of the t distribution. Statistical Modelling 4, 91–112 (2004)
12. Tibshirani, R.: Regression shrinkage and selection via the LASSO. Journal of Royal Statistical Society Series B 58, 267–288 (1996)
13. Verbyla, A.P.: Variance heterogeneity: residual maximum likelihood and diagnostics. Journal of the Royal Statistical Society Series B 52, 493–508 (1993)
14. Wang, D.R., Zhang, Z.Z.: Variable selection in joint generalized linear models. Chinese Journal of Applied Probability and Statistics 25, 245–256 (2009)
15. Wang, H., Li, R., Tsai, C.: Tuning parameter selectors for the smoothly clipped absolute deviation method. Biometrika 94, 553–568 (2007)

# Testing of Relative Difference under Inverse Sampling

Shaoping Jiang and Yanfang Zhao

**Abstract.** Inverse sampling is one of the most hot topic in the continuous sampling process. It is considered to be a more appropriate sampling scheme than the usual binomial sampling scheme when the subjects arrive sequentially and the underlying response of interest is acute. In this article, we have studied various test statistics for testing relative difference in case-control studies under inverse sampling. We utilized the way of Fisher-score to get the variance of interest parameter. Then we got type I error and power by Monte Carlo simulation. In general, Score statistic is the best, which can get the small type I error and large power.

**Keywords:** Inverse sampling, Risk difference, Score-statistic.

## 1 Introduction

Inverse sampling scheme, first proposed by Haldane [1], suggests that one continues to sample subjects until a pre-specified number of index subjects with the rare event of interest is observed. This inverse sampling scheme is referred as standard inverse sampling scheme in all subsequent discussion. In contrast with binomial sampling, it is more appropriate in detecting the difference between two treatments for a rare disease by avoiding sparse data structure due to the low incidence of the disease. Moreover, inverse sampling is preferred to binomial sampling when subjects arrive sequentially and when maximum likelihood estimators of some epidemiological measures do not exist under binomial sampling [3].

Shaoping Jiang and Yanfang Zhao
School of mathematics & computer science, Yunnan Nationalities University,
131 Yi Er Yi Avenue, Kunming, Yunnan, 650031, P.R. China
e-mail: jiang2005124@hotmail.com, jiang2005124@163.com

S. Li (Eds.): Nonlinear Maths for Uncertainty and its Appli., AISC 100, pp. 503–508.
springerlink.com   © Springer-Verlag Berlin Heidelberg 2011

In reality, inverse sampling has important practical value, many scholars have made a lot of research about this problem. For instance, Kikuchi [2] studied the association between congenital heart disease and low birth weight under inverse sampling. Smith [6] used inverse sampling to study the level of HIV-1 mRNA-expressing (positive) mononuclear cells within the esophageal mucosa of patients with acquired immune deficiency syndrome (AIDS) and esophageal disease under inverse sampling. Lui [4] considered the asymptotic conditional test procedures for relative difference under inverse sampling. In these paper they haven't used the way of Fisher-score to get the variance of interest parameter. According to this method we consider various test statistics(Wald, Score, Wald-score and likelihood ratio test statistic) for testing the hypothesis of equality of the relative difference. Relative difference is introduced in section 2. In section 3, parameters estimate and four test statistics are given. In section 4, simulation studies are constructed to investigate to performance of the above test statistic. Finally, some concluding remarks are given in section 5.

## 2  Definition of Relative Difference

Consider a study involving two comparative groups, with group 0 denoting exposed and group 1 unexposed. The key feature of the inverse sampling design is that the number of events $r_i$ (i=0,1) is pre-specified. Let $y_i$ denote the number of non-events to ensure that pre-specified events are observed. The probability mass function is given by

$$f(Y_i = y_i | p_i) = \binom{y_i + r_i - 1}{y_i} p_i^{r_i} (1 - p_i)_i^y, (i = 0, 1) \qquad (1)$$

Consider the two samples is independent, the probability density function of $(y_0, y_1)$ is given by:

$$
\begin{aligned}
&f(Y_1 = y_1, Y_0 = y_0 | p_1, p_0) \\
&= \binom{y_1 + r_1 - 1}{y_1} p_1^{r_1} (1 - p_1)^{y_1} \binom{y_0 + r_0 - 1}{y_0} p_0^{r_0} (1 - p_0)^{y_0} \\
&= \binom{y_1 + r_1 - 1}{y_1} \binom{y_0 + r_0 - 1}{y_0} p_1^{r_1} q_1^{y_1} p_0^{r_0} q_0^{y_0}.
\end{aligned} \qquad (2)
$$

where $q_i = 1 - p_i, i = 0, 1$.

Following the arguments of Sheps [5], we define relative difference as:

$$\delta = \frac{p_1 - p_0}{1 - p_0} = 1 - \Phi. \qquad (3)$$

where $\Phi = \frac{q_1}{q_0}$, and we can get that relative difference $\delta \leq 1$.

Here, our main interest is to test the following hypothesis:

$$H_0 : \delta = \delta_0 \longleftrightarrow H_1 : \delta \neq \delta_0$$

Where $\delta_0$ is a fixed constant. According to Eq.(2), the log-likelihood function of the observed frequencies $(y_0, y_1)$ is given by:

$$L(\delta, p_0) = r_1 \log[p_0 + \delta(1 - p_0)] + y_1 \log[(1 - \delta)(1 - p_0)]$$
$$+ r_0 \log p_0 + y_0 \log(1 - p_0) + C. \tag{4}$$

where C is a constant which does not depend on the parameters $\delta$ and $p_0$, $\delta$ is the parameter of interest, and $p_0$ is the nuisance parameter.

## 3 Parameter Estimate and Statistic

### 3.1 Parameter Estimate

We can get the MLE of parameter by solve the efficient score function,

$$U(\theta) = (U_\delta(\theta), U_{P_0}(\theta))^T = (\frac{\partial L(y_1, y_0)}{\partial \delta}, \frac{\partial L(y_1, y_0)}{\partial p_0})^T = 0 \tag{5}$$

The MLE of $\theta$ denoted by $\hat{\theta} = (\hat{\delta}, \hat{p}_0)$, where $\hat{\delta} = \frac{r_1 y_0 - r_0 y_1}{y_0 (r_1 + y_1)}$ and $\hat{p}_0 = \frac{r_0}{y_0 + r_0}$.

We can get the MLE of parameter under $H_0$ by solve the $U_{p_0}(\theta) = \frac{\partial L(y_1, y_0)}{\partial p_0} = 0$, the the estimate is given by $\tilde{\theta} = (\delta_0, \tilde{p}_0)$, and $\tilde{p}_0 = \frac{-B + \sqrt{B^2 - 4AC}}{2A}$, where $A = (1 - \delta_0)(y_1 + y_0 + r_1 + r_0)$, $B = \delta_0(y_0 + y_1 + r_1 + 2r_0) - (r_1 + r_0)$, $C = -r_0\delta_0$.

### 3.2 Statistic

In past research we use the delta method to get the expectations and variance of interest parameters. But the delta way is a method of approximate solution, the result with a certain degree of deviation; To avoid bias, we use the Fisher-score approach to get the variance of parameter. The Fisher information matrix as follows:

$$I = \begin{pmatrix} I_{11} & I_{12} \\ I_{21} & I_{22} \end{pmatrix} = \begin{pmatrix} E(-\frac{\partial^2 L}{\partial \delta^2}) & E(-\frac{\partial^2 L}{\partial \delta \partial p_0}) \\ E(-\frac{\partial^2 L}{\partial \delta \partial p_0}) & E(-\frac{\partial^2 L}{\partial p_0^2}) \end{pmatrix}$$

We must know that the parameter $y_i (i=0, 1)$ follow the negative binomial distribution, the expectation of parameter given by: $E(y_0) = \frac{r_0(1 - p_0)}{p_0}$, $E(y_1) = \frac{r_1(1 - p_1)}{p_1}$. And the variance of interest parameter is given by:

$$\mathrm{Var}(\hat{\delta}) = \sum(\theta)$$
$$= (I_{11} - I_{12}^2 I_{22}^{-1})^{-1} = \frac{(1-\delta)\{r_1 p_0^2 (1-\delta) + r_0[p_0 + \delta(1-p_0)]^2\}}{r_1 r_0 (1-p_0)}$$

We consider the following test statistics,

(1) Wald statistic:

$$T_1 = (\hat{\delta} - \delta_0)^T (\sum(\hat{\theta}))^{-1} (\hat{\delta} - \delta_0). \tag{6}$$

(2) Score statistic:

$$T_2 = U_\delta(\tilde{\theta})^T \sum(\tilde{\theta}) U_\delta(\tilde{\theta}). \tag{7}$$

(3) Wald-score statistic:

$$T_3 = (\hat{\delta} - \delta_0)^T (\sum(\tilde{\theta}))^{-1} (\hat{\delta} - \delta_0). \tag{8}$$

(4) Likelihood statistic:

$$T_4 = -2[L(\tilde{\theta}) - L(\hat{\theta})]. \tag{9}$$

which are asymptotically distributed as the chi-square distribution with one degree of freedom under $H_0 : \delta = \delta_0$.

## 4  Simulation Studies

In this section, several simulation studies are conducted to investigate the performance of various test statistics. In simulation studies, we set $\delta = -0.1, 0.0, 0.1$, $p_0 = 0.2, 0.3, 0.5$, and $r_0 = r_1 =$ 30,50,100,150. We evaluate the type I error and power under the statistics in the Table1.

After observed the value of Table1, we get the conclusion as follows:

Score statistic is optimal because it can be ensure that committed the type I error is the smallest and power is large in the same condition with other statistics. Especially when $r$ below 50, the performance is perfect, so we can use the score statistic to deal with small samples. By observing the type I error we found the $T_4$ statistics has stable nature. But compare with score statistic, the $T_4$ statistics is worse, because it has big type I error than score statistic.

The Wald statistic $T_1$ applied to the large samples. The type I error become smaller and smaller as the sample size n increases, and tend to critical level. Compare with $T_3$ statistic, $T_1$ statistic is stable because the type I error of $T_3$ statistic has big wavy.

**Table 1** Empirical Type I error rates and power for testing hypothesis $H_0 : \delta = \delta_0$ based on 10000 trials at $\alpha = 5\%$

| $\delta_0$ | $p_0$ | $r_0 = r_1$ | $T_1$ | $T_2$ | $T_3$ | $T_4$ |
|---|---|---|---|---|---|---|
| 0.1 | 0.2 | 10 | 6.03(100) | 5.00(100) | 7.17(100) | 5.64(100) |
| | | 30 | 5.55(100) | 4.29(99.9) | 6.11(100) | 5.25(99.9) |
| | | 50 | 5.11(100) | 4.84(100) | 5.48(100) | 4.95(100) |
| | | 80 | 5.17(100) | 4.92(100) | 5.56(100) | 5.14(100) |
| | | 100 | 5.07(100) | 4.89(100) | 5.37(100) | 5.06(100) |
| | 0.3 | 10 | 5.43(100) | 4.65(100) | 6.53(100) | 5.13(100) |
| | | 30 | 5.39(100) | 5.03(100) | 6.36(100) | 5.29(100) |
| | | 50 | 5.15(99.9) | 4.74(99.2) | 5.58(99.7) | 5.12(99.8) |
| | | 80 | 5.00(100) | 4.93(99.9) | 5.43(100) | 5.01(99.7) |
| | | 100 | 5.14(100) | 4.90(100) | 5.38(100) | 5.03(100) |
| | 0.5 | 10 | 5.12(100) | 4.74(100) | 7.15(100) | 5.54(100) |
| | | 30 | 4.87(100) | 4.77(100) | 6.63(100) | 5.29(100) |
| | | 50 | 5.03(100) | 5.04(100) | 5.86(100) | 5.30(100) |
| | | 80 | 4.68(95.6) | 4.80(84.4) | 5.48(75.9) | 4.75(94.6) |
| | | 100 | 4.81(99.6) | 4.62(96.0) | 5.12(96.9) | 4.73(99.5) |
| 0.0 | 0.2 | 10 | 6.28(100) | 5.05(99.9) | 7.37(99.9) | 5.56(100) |
| | | 30 | 6.00(100) | 5.17(99.9) | 6.70(100) | 5.53(99.9) |
| | | 50 | 5.65(100) | 5.20(100) | 6.25(100) | 5.46(100) |
| | | 80 | 5.44(100) | 5.07(100) | 5.78(100) | 5.25(100) |
| | | 100 | 4.77(100) | 4.58(100) | 4.99(100) | 4.69(100) |
| | 0.3 | 10 | 5.37(100) | 4.65(100) | 6.95(100) | 5.24(100) |
| | | 30 | 5.22(100) | 4.89(100) | 6.53(100) | 5.23(100) |
| | | 50 | 4.70(100) | 4.61(99.8) | 5.66(99.9) | 4.86(99.9) |
| | | 80 | 5.15(100) | 5.02(100) | 5.69(100) | 5.16(100) |
| | | 100 | 4.65(100) | 4.55(100) | 4.94(100) | 4.66(100) |
| | 0.5 | 10 | 5.09(100) | 4.50(100) | 8.24(100) | 5.72(100) |
| | | 30 | 4.81(100) | 4.73(100) | 6.83(100) | 5.23(100) |
| | | 50 | 4.72(100) | 4.87(100) | 5.97(100) | 5.02(100) |
| | | 80 | 4.64(98.1) | 4.60(90.6) | 5.47(83.0) | 4.80(97.1) |
| | | 100 | 4.77(99.8) | 4.97(98.2) | 5.71(98.7) | 5.10(99.8) |
| -0.1 | 0.2 | 10 | 6.39(100) | 4.85(100) | 8.36(100) | 5.58(100) |
| | | 30 | 6.12(100) | 5.15(100) | 7.07(100) | 5.65(99.9) |
| | | 50 | 5.68(100) | 4.94(100) | 6.62(100) | 5.23(100) |
| | | 80 | 5.43(100) | 5.01(100) | 5.82(100) | 5.26(100) |
| | | 100 | 5.08(100) | 4.90(100) | 5.51(100) | 4.95(100) |
| | 0.3 | 10 | 5.58(100) | 4.92(100) | 8.00(100) | 5.74(100) |
| | | 30 | 5.03(100) | 4.65(100) | 6.51(100) | 5.10(100) |
| | | 50 | 5.25(100) | 4.91(99.9) | 6.23(100) | 5.23(99.9) |
| | | 80 | 5.27(100) | 5.17(100) | 5.82(100) | 5.16(100) |
| | | 100 | 5.00(100) | 4.94(100) | 5.62(100) | 5.05(100) |
| | 0.5 | 10 | 4.34(100) | 4.74(100) | 8.23(100) | 5.61(100) |
| | | 30 | 4.36(100) | 5.11(100) | 7.30(100) | 5.60(100) |
| | | 50 | 4.50(100) | 4.76(100) | 6.13(100) | 4.95(100) |
| | | 80 | 4.73(99.1) | 4.87(93.9) | 5.76(87.9) | 4.86(98.3) |
| | | 100 | 4.76(99.9) | 4.92(99.1) | 5.72(99.5) | 4.90(99.9) |

[a] $T_1$, $T_2$, $T_3$ and $T_4$ are correspond to Wald statistics, Score statistic, Wald-score statistic and Likelihood statistic.

## 5   Discussion

Inverse samples are often involved in our life. In this paper we use the Fisher-Score method to get the variance of interesting parameter. This method can be more accurate than delta way. we constructed four statistics and got the priority one based on the evaluation of type I error and power. In the future, we can use the same method to discuss the test about risk ratio.

**Acknowledgements.** This work is supported by grants from the Foundation of Yunnan Nationalities University (09QN11).

## References

1. Haldane, J.B.S.: On a method of estimating frequencies. Biometrika 33, 222–225 (1945)
2. Kikuchi, D.A.: Inverse sampling in case control studies involving a rare exposure. Biom. J. 29, 243–246 (1987)
3. Lui, K.J.: A note on the use ofinverse sampling: point estimation between successive infections. J. Off. Stat. 16, 31–37 (2000)
4. Lui, K.J.: Asymptotic conditional test procedures for relative difference under inverse sampling. Computational Statistics and Data Analysis 34, 335–343 (2000)
5. Sheps, M.C.: An examination of some methods of comparing several rates or proportions. Biometrics 15, 87–97 (1959)
6. Smith, P.D., Fox, C.H., Masur, H., Winter, H.S., Alling, D.W.: Quantitative analysis of mononuclear cells expressing human immunodeficiency virus type 1 RNA in esophageal mucosa. J. Exp. Med. 180, 1541–1546 (1994)

# Adaptive Elastic-Net for General Single-Index Regression Models

Xuejing Li, Gaorong Li, and Suigen Yang

**Abstract.** In this article, we study a general single-index model with diverging number of predictors by using the adaptive Elastic-Net inverse regression method. The proposed method not only can estimate the direction of index and select important variables simultaneously, but also can avoid to estimate the unknown link function through nonparametric method. Under some regularity conditions, we show that the proposed estimators enjoy the so-called oracle property.

**Keywords:** Elastic-Net, Single-index Model, Inverse regression, High dimensionality, Dimension reduction, Variable Selection, Oracle property.

## 1 Introduction

The single-index models combine flexibility of modelling with interpretability of linear models, and have more advantages than the linear or nonparametric regression models. One advantage of the single index model is that it overcomes the risk of misspecifying the link function. Another advantage of the single index model is that it avoids the so-called curse of dimensionality. Therefore, single index models have been widely studied by many statistician. In this paper, we consider the general single-index model proposed by Li and Duan (1989) and Li (1991) as follows

$$Y_i = G(\beta^T X_i, e_i), \quad i = 1, \ldots, n, \tag{1}$$

where $X_i = (X_{i1}, \ldots, X_{ip_n})^T$ is the linearly independent predictors, $\beta$ is a $p_n \times 1$ interest index, the error terms $e_i$ are assumed to be independent of $X_i$,

Xuejing Li, Gaorong Li and Suigen Yang
College of Applied Sciences, Beijing University of Technology,
Beijing 100124, China
e-mail: lxj@bjut.edu.cn, ligaorong@bjut.edu.cn,
yangsuigen@emails.bjut.edu.cn

S. Li (Eds.): Nonlinear Maths for Uncertainty and its Appli., AISC 100, pp. 509–516.
springerlink.com                    © Springer-Verlag Berlin Heidelberg 2011

and $G(\cdot)$ is an unknown link function, which relates the response variable $Y$ to the the predictor vector $X$ through a linear combination $\beta^T X$ and the error term $e$. Equivalently, $Y$ is independent of $X$ when $\beta^T X$ is given (Cook 1998). Clearly, model (1) is very general and covers the heteroscedastic model as a special case. Note that when the link function $G(\cdot)$ is unspecified, the slope vector $\beta$ is identified only up to a multiplicative scalar because any location-scale change in $\beta^T X$ can be absorbed into the link function. If we can identify the direction of $\beta$ in model (1), then we reduce the dimension $p_n$ of predictors to one dimension. In the literature the theory of *sufficient dimension reduction* (SDR) provides an effective starting point to estimate $\beta$ without loss of regression information of $Y$ on $X$ and without assuming the specific link function. The promising methods include sliced inversion regression (SIR, Li 1991, Cook and Ni 2005), sliced average variance estimation (Cook and Weisberg 1991), contour regression (Li, Zha and Chiaromonte 2005), nonconcave penalized inverse regression (Zhu and Zhu 2009) and references therein.

In practice, $p_n$ is usually very large compared to the sample size $n$, and thus $p_n$ can be assumed to depend on $n$ at some rate (Donoho 2000, Li, Peng and Zhu 2011). To enhance the prediction performance of the fitted model, there are two important issues in regressions: reducing dimensionality and fitting parsimonious model through excluding the unimportant variables. To solve the problems above, various penalized methods have been proposed and shown a better performance, such as ridge regression (Hoerl and Kennard 1970), Lasso (Tibshirani 1996) and SCAD (Fan and Li 2001). Recently, Zou (2006) has showed explicitly that Lasso could be inconsistent in certain situation. Zou (2006) proposed an modified version of Lasso called adaptive Lasso and explicitly proved its consistency in terms of oracle properties. Through combining both ridge ($L_2$) and lasso ($L_1$) penalty together, Zou and Hastie (2005) proposed the Elastic-Net, which also has the property of sparsity, to solve the collinearity problems. Zou and Hastie (2005) and Zou and Zhang (2009) further showed that the Elastic-Net can significantly improve the prediction accuracy of the Lasso when the correlations among the predictors become high.

In this paper we study the variable selection when the assumed link function might be incorrect. Model (1) is assumed to be "sparse", i.e. most of the regression coefficients $\beta$ are exactly zero corresponding to predictors that are irrelevant to the response. Let $\mathcal{A} = \{j : \beta_j \neq 0, j = 1, \ldots, p_n\}$ denote the true model set and $|\mathcal{A}| = k_n < p_n$. We intend to understand the model selection performance of the adaptive Elastic-Net inverse regression method. The proposed method not only can estimate the direction of index and select important variables simultaneously, but also can avoid to estimate the unknown link function through nonparametric method. Under some regularity conditions, we demonstrate that the proposed estimators enjoy the so-called oracle property.

The rest of the article is organized as follows. In Section 2, we introduce the adaptive Elastic-Net inverse regression for general single-index model with high dimensional predictors. In Section 3, some statistical properties, including the oracle property, are established. The technical details are available upon request.

# 2   Methodology

## 2.1   Identification of $\beta$

In this subsection, we will define a population identification of $\beta$ for single-index models (1) without introducing the penalty. In the SDR context, we will define a criterion to identify $\beta$ in model (1) so that we can have a sufficient recovery of the direction of $\beta$ in the sense that the projection direction under our criterion identifies $\beta$ only up to a multiplicative scalar. To be precise, denote by $F(y) = \text{Prob}(Y \leq y)$ the distribution function of the continuous response $Y$. Let the loss function take the form of

$$\ell(\beta^T X, F(Y)) = -F(Y)\beta^T X + \psi(\beta^T X), \tag{2}$$

where $\psi(\beta^T X)$ is convex in $\beta^T X$. The loss function $\ell(\beta^T X, F(Y))$ defined in (2) covers the least square measure as a special case, namely,

$$\ell(\beta^T X, F(Y)) = (F(Y) - \beta^T X)^2/2. \tag{3}$$

by letting $\psi(\beta^T X) = [\beta^T X X^T \beta + F^2(Y)]/2$ in (2). The the estimate of $\beta$ is a solution of the following minimization problem:

$$\beta_0 = \arg\min_{\beta} E[\ell(\beta^T X, F(Y))], \tag{4}$$

The minimizer is very general and includes least squares, $M$-estimates with nondecreasing influence functions, etc. When the loss function is defined by (3), then the least squares estimation is

$$\beta_{n0} =: \arg\min_{\beta} E[\ell(\beta^T X, F(Y))] = \arg\min_{\beta} E[F(Y) - \beta^T X]^2$$
$$= \Sigma^{-1}\text{Cov}(X, F(Y)). \tag{5}$$

Note that we do not restrict $\|\beta\| = 1$, which indicates only the direction of $\beta$ is of our concern. Without loss of generality, we assume the predictor variables are centered and satisfy $E(X) = 0$. The identification of $\beta$ is stated in the following proposition.

**Proposition 1.** *Under model* (1) *and the criterion function* (2). *Further assume:*

(C1) $\psi(\beta^T X)$ is convex in $\beta^T X$;

(C2) The expected criterion function $E[\ell(\beta^T X, F(Y))]$ has a proper minimizer;

(C3) The linearity condition: $E(X|\beta^T X) = \Sigma\beta(\beta^T \Sigma\beta)^{-1}\beta^T X$, where $\Sigma = \text{Cov}(X)$.

Then the minimizer $\beta_0$ of (4) using the objective function (2) is proportional to $\beta$.

In this proposition, condition (C1) is satisfied for many important estimation methods, including least squares, $M$-estimates with non-decreasing influence functions, etc. The convexity property of the criterion is crucial here. Without the convexity, we may have inconsistency. When the convexity of $\psi(\cdot)$ is not strict, we need some additional assumptions to reach the same conclusion. The study of this issue could be parallel to Theorem 2.2 in Li and Duan (1989). Condition (C2) is quite mild and can usually be satisfied. All elliptical contoured distributions satisfy the linearity condition. The linearity condition (C3) is widely assumed in the dimension reduction context such as Li (1991) and Cook (1998, proposition 4.2, page 57). Hall and Li (1993) showed that, as $p_n \to \infty$, this linearity condition holds to a good approximation in model (1).

This proposition implies that, if $\psi(\cdot)$ in the objective function (2) is strictly convex, then all minimizers of (4) must fall along the directions of $\beta$ even though the link function $G(\cdot)$ is unspecified. Thus any regression slope estimate based on minimizing the criterion function (4) is proportional to $\beta$ up to a multiplicative scalar.

## 2.2   Adaptive Elastic Net

Suppose that $(X_i^T, Y_i)^T, i = 1, \ldots, n$, are independent copies of $(X^T, Y)^T$ and come from model (1). Let $\mathbf{Y} = (Y_1, \ldots, Y_n)^T$ is the response vector, $\mathbf{X}_j = (X_{1j}, \ldots, X_{nj})^T, j = 1, \ldots, p_n$, are the linearly independent predictors, $\beta$ is a $p_n \times 1$ interest index. Let $\mathbf{X} = [\mathbf{X}_1, \ldots, \mathbf{X}_{p_n}]$ be the predictor matrix. From now on, we assume that $\mathbf{X}$ is centered so that each column has mean 0. Due to the diverging number of parameters, we cannot assume that the least square functions are invariant in our study. Let $\mathbf{F}_n(\mathbf{Y}) = (F_n(Y_1), \ldots, F_n(Y_n))^T$, where $F_n(Y) = \frac{1}{n}\sum_{i=1}^n \mathbf{1}_{\{Y_i \leq Y\}}$ is the empirical distribution function and is estimable from the response sample $Y$. Thus, the sample version of the least square measure in (3) becomes

$$L_n(\beta) = [\mathbf{F}_n(\mathbf{Y}) - \mathbf{X}\beta]^T[\mathbf{F}_n(\mathbf{Y}) - \mathbf{X}\beta]. \tag{6}$$

Now we consider the following optimization problem to get the naive Elastic-Net estimation of $\beta$

$$\hat{\beta}_{\text{naive}} = \arg\min_{\beta} \left\{ L_n(\beta) + \lambda_2 \sum_{j=1}^{p_n} \beta_j^2 + \lambda_1 \sum_{j=1}^{p_n} |\beta_j| \right\}. \tag{7}$$

As discussed in Zou and Zhang (2009), the Elastic-Net estimator which is a scaled version of the naive Elastic-Net estimator is defined as

$$\hat{\beta}_{\text{enet}} = \left(1 + \frac{\lambda_2}{n}\right)\hat{\beta}_{\text{naive}}. \tag{8}$$

Parameters $\lambda_1$ and $\lambda_2$ in (7) control the amount of regularization applied to the estimate. $\lambda_2 = 0$ leads the naive Elastic-Net estimate $\hat{\beta}_{\text{naive}}$ back to the Lasso estimate. Although the Elastic-Net can do both continuous shrinkage and automatic variable selection simultaneously, and handle the collinearity for high dimensional data, it is easy to see that the Elastic-Net is lack of the oracle property.

Next we propose a revised version of the Elastic-Net by incorporating adaptive weights in the Lasso penalty of equation (7). The adaptive Elastic-Net can be viewed as a combination of the Elastic-Net and the adaptive Lasso. Suppose that $\hat{\beta}$ is a root $n$-consistent estimator of $\beta$. We can choose the naive Elastic-Net estimator $\hat{\beta}_{\text{enet}}$ given by (8), then we define the adaptive weights by

$$\hat{\omega}_j = (|\hat{\beta}_{\text{enet},j}|)^{-\gamma}, \quad j = 1, \ldots, p_n, \tag{9}$$

where $\hat{\beta}_{\text{enet},j}$ denotes the $j$-th component of naive Elastic-Net estimator $\hat{\beta}_{\text{enet}}$, and $\gamma$ is a positive constant. The adaptive Elastic-Net estimate is then defined as

$$\hat{\beta}_n = \left(1 + \frac{\lambda_2}{n}\right)\arg\min_{\beta}\left\{L_n(\beta) + \lambda_2 \sum_{j=1}^{p_n} \beta_j^2 + \lambda_1 \sum_{j=1}^{p_n} \hat{\omega}_j|\beta_j|\right\}. \tag{10}$$

$\lambda_2 = 0$ leads the adaptive Elastic-Net estimate back to the adaptive Lasso estimate. We would also like to define $\mathcal{A}_n = \{j : \hat{\beta}_{n,j} \neq 0\}$. The adaptive Elastic-Net variable selection is consistent if and only if $\lim_{n\to\infty} P(\mathcal{A}_n = \mathcal{A}) = 1$.

## 3    Theoretical Results

In this section, we are interested in the oracle properties of adaptive Elastic-Net expressed in (10).

Throughout the paper, we denote $\gamma_{\min}(A)$ and $\gamma_{\max}(A)$ as the minimum and maximum eigenvalues and $\text{tr}(A)$ as the trace operator of a matrix $A$, respectively. For the convenience and simplicity, we shall employ $c > 0$ to denote some constants not depending on $n$ and $p_n$ but may take difference values at each appearance. In order to establish the theoretical results, we introduce the following regularity conditions.

(A) We assume that $b \leq \gamma_{\min}\left(\frac{1}{n}\mathbf{X}^T\mathbf{X}\right) \leq \gamma_{\max}\left(\frac{1}{n}\mathbf{X}^T\mathbf{X}\right) \leq B$, where $b$ and $B$ are two positive constants.

(B) $0 < C_1 < \gamma_{\min}\{E(XX^T)^2\} \leq \gamma_{\max}\{E(XX^T)^2\} < C_2 < \infty$ for all $p_n$.

(C) $\lim\limits_{n\to\infty} \dfrac{\lambda_1}{\sqrt{n}} = 0$, $\qquad \lim\limits_{n\to\infty} \dfrac{\lambda_2}{n} = 0$, $\qquad \lim\limits_{n\to\infty} \dfrac{\lambda_1}{\sqrt{n}} n^{\gamma/2} p_n^{-(1+\gamma)/2} = \infty$.

(D) $\lim\limits_{n\to\infty} \dfrac{\lambda_2}{n}\sqrt{\sum\limits_{j\in\mathcal{A}}\beta_j^2} = 0$ $\qquad \lim\limits_{n\to\infty} \left(\dfrac{\sqrt{n}}{\sqrt{p_n}\lambda_1}\right)^{1/\gamma}\left(\min\limits_{j\in\mathcal{A}}|\beta_j|\right) = \infty$.

Condition (A) and (B) assume the predictor matrix has a reasonably good behavior. Similar conditions were considered in Zou and Zhang (2009) and Zhu and Zhu (2009). Conditions (C) and (D) is similar to conditions (A5) and (A6) in Zou and Zhang (2009). Note that $\frac{\lambda_1}{\sqrt{n}}n^{\gamma/2}p_n^{-(1+\gamma)/2} \to \infty$ is reduced to $\frac{\lambda_1}{\sqrt{n}}n^{\gamma/2} \to \infty$ in the finite dimension setting, which agree with the condition in Ghosh (2007).

**Theorem 1.** *Suppose that the data set is given by* $(\mathbf{Y},\mathbf{X})$, *and let* $\hat{\omega} = (\hat{\omega}_1,\ldots,\hat{\omega}_{p_n})$ *be a vector whose components are all non-negative and can depend on* $(\mathbf{Y},\mathbf{X})$. *Define*

$$\hat{\beta}_{\hat{\omega}}(\lambda_2,\lambda_1) = \arg\min_b \left\{ L_n(\beta) + \lambda_2 \sum_{j=1}^{p_n}\beta_j^2 + \lambda_1 \sum_{j=1}^{p_n}\hat{\omega}_j|\beta_j| \right\},$$

*for non-negative parameters* $\lambda_2$ *and* $\lambda_1$. *If* $\hat{\omega}_j = 1$ *for all* $j$, *we denote* $\hat{\beta}_{\hat{\omega}}(\lambda_2,\lambda_1)$ *by* $\hat{\beta}(\lambda_2,\lambda_1)$ *for convenience.*

*If we assume the model (1) and condition (A), then we have*

$$E(\|\hat{\beta}_{\hat{\omega}}(\lambda_2,\lambda_1) - \beta\|^2)$$

$$\leq 4 \frac{\frac{3}{2}\lambda_2^2\|\beta\|^2 + 4C_2np_n + 3C_2n\|\beta\|^2 + \lambda_1^2 E\left[\sum\limits_{j=1}^{p_n}\hat{\omega}_j^2\right]}{(bn+\lambda_2)^2} + o\left(\frac{p_n\log^2 n}{(bn+\lambda_2)^2}\right). \tag{11}$$

*In particular, when* $\hat{\omega}_j = 1$ *for all* $j$, *we have*

$$E(\|\hat{\beta}(\lambda_2,\lambda_1) - \beta\|^2)$$

$$\leq * 4 \frac{\frac{3}{2}\lambda_2^2\|\beta\|^2 + 4C_2np_n + 3C_2n\|\beta\|^2 + \lambda_1^2 p_n}{(bn+\lambda_2)^2} + o\left(\frac{p_n\log^2 n}{(bn+\lambda_2)^2}\right).$$

Theorem 1 implies that $\hat{\beta}(\lambda_2,\lambda_1)$ is a root-$(n/p_n)$ consistent estimator. This consistent rate is the same as the result of SCAD (Zou and Zhang 2009, and Fan and Peng 2004). Therefore, the root-$(n/p_n)$ consistency result suggests that it is appropriate to use the Elastic-Net estimator to construct the adaptive weights.

**Theorem 2.** *(Oracle properties) Suppose that conditions (B)-(D) hold, if $p_n^3/n \to 0$ as $n \to \infty$, then the adaptive Elastic-Net estimator $\hat{\beta}_n$ defined by (10) must satisfy the following:*

1. Consistency in variable selection: $\lim\limits_{n\to\infty} P\left(A_n = \mathcal{A}\right) = 1$.

2. Asymptotic normality: *Further assume that* $0 < C_1 < \gamma_{\min}\{E(XX^T)^4\} \le \gamma_{\max}\{E(XX^T)^4\} < C_2 < \infty$ *holds uniformly for* $p_n$. *Then* $\sqrt{n}A_n\Sigma_{\mathcal{A}}(\hat{\beta}_{\mathcal{A}} - \beta_{\mathcal{A}}) \to N(0, G)$, *where* $\beta_{\mathcal{A}}$ *is the first* $k_n$ *nonzero elements of* $\beta$, $A_n$ *is a* $q \times k_n$ *matrix such that* $\gamma_{\max}(A_n A_n^T) < \infty$ *and* $A_n\text{Cov}\left(X_{\mathcal{A}}^T[F(Y) - X_{\mathcal{A}}^T\beta_{\mathcal{A}}]\right)A_n^T \to G$, *and* $G$ *is a* $q \times q$ *nonnegative symmetric matrix.*

By Theorem 2, the selection consistency and the asymptotic normality are still valid when the number of parameters diverges.

**Acknowledgements.** Xuejing Li's research was supported by PHR (No. 201006102) and Beijing Natural Science Foundation(Stochastic Analysis with uncertainty and applications in finance). Gaorong Li's research was supported by Ph.D. Program Foundation of Ministry of Education of China (20101103120016), Funding Project for Academic Human Resources Development in Institutions of Higher Learning Under the Jurisdiction of Beijing Municipality (PHR20110822), Training Programme Foundation for the Beijing Municipal Excellent Talents (2010D005015000002) and Doctor Foundation of BJUT (X0006013201101).

# References

1. Cook, R.D.: Regression graphics: Ideas for Studying Regressions through Graphics. Wiley & Sons, New York (1998)
2. Cook, R.D., Ni, L.: Sufficient dimension reduction via. J. Amer. Statist. Assoc. 100, 410–428 (2005)
3. Cook, R.D., Weisberg, S.: Discussion to "Sliced inverse regression for dimension reduction". J. Amer. Statist. Assoc. 86, 316–342 (1991)
4. Donoho, D.L.: High-dimensional data analysis: The curses and blessings of diemsnionality. In: Aide-Memoire of a Lecture at AMS Conference on Math Challenges of the 21st Century (2000)
5. Fan, J.Q., Li, R.: Variable selection via nonconcave penalized likelihood and its oracle properties. J. Amer. Statist. Assoc. 96, 1348–1360 (2001)
6. Fan, J.Q., Peng, H.: Nonconcave penalized likelihood with a diverging number of parameters. Ann. Statist. 32, 928–961 (2004)
7. Ghosh, S.: Adaptive Elastic Net: an improvement of Elastic Net to achieve oracle properties. Indiana Univ.–Purdue Univ. Tech. report (2007
8. Hall, P., Li, K.C.: On almost linearity of low dimensional projection from high dimensional data. Ann. Statist. 21, 867–889 (1993)
9. Hoerl, A.E., Kennard, R.W.: Ridge regression: biased estimation for nonorthogonal problems. Technometrics 12, 55–67 (1970)

10. Li, B., Zha, H., Chiaromonte, F.: Contour regression: A general approach to dimension reduction. Ann. Statist. 33, 1580–1616 (2005)
11. Li, G.R., Peng, H., Zhu, L.X.: Nonconcave penalized M-estimation with diverging number of parameters. Statistica Sinica 21, 391–419 (2011)
12. Li, K.C.: Sliced inverse regression for dimension reduction (with discussion). J. Amer. Statist. Assoc. 86, 316–342 (1991)
13. Li, K.C., Duan, N.H.: Regression analysis under link violation. Ann. Statist. 17, 1009–1052 (1989)
14. Tibshirani, R.: Regression shrinkage and selection via the lasso. J. Roy. Statist. Soc. B. 58(1), 267–288 (1996)
15. Zhu, L.P., Zhu, L.X.: Nonconcave penalized inverse regression in single-index models with high dimensional predictors. J. Multivariate Anal. 100, 862–875 (2009)
16. Zou, H.: The adaptive Lasso and its oracle properties. J. Amer. Statist. Assoc. 101(476), 1418–1429 (2006)
17. Zou, H., Hastie, T.: Regularization and variable selection via the elastic net. J. Roy. Statist. Soc. B 67, 301–320 (2005)
18. Zou, H., Zhang, H.H.: On the adaptive elastic-net with a diverging number of parameters. Ann. Statist. 37(4), 1733–1751 (2009)

# Some Remark about Consistency Problem of Parameter Estimation

Mingzhong Jin, Minqing Gong, and Hongmei Liu

**Abstract.** Let $Y_i = x_i'\beta + e_i, 1 \leq i \leq n, n \geqslant 1$, be a linear regression model. Denote by $\lambda_n$ and $\mu_n$ the smallest and largest eigenvalues of $\sum_{i=1}^{n} x_i x_i'$. Assume that the random errors $e_1, e_2, \cdots$ are iid, $Ee_1 = 0$ and $E|e_1| < \infty$. Under the restriction that $\mu_n = O(\lambda_n)$, this paper obtains the necessary and sufficient condition for the LS estimate of $\beta$ to be strongly consistent.

**Keywords:** Linear regression model, Least squares estimate, Strong consistency.

## 1 Introduction

Consider the linear regression model

$$Y_i = x_i'\beta + e_i, 1 \leq i \leq n, n \geqslant 1. \tag{1}$$

We shall always assume that $x_1, x_2, \cdots$ are known non-random $p$-vectors. The least squares estimate of $\beta$, the $p$-vector of regression coefficients, will be denoted by $\hat{\beta}_n$.

Many statisticians studied the problem of strong consistency (SC) of $\hat{\beta}_n$. In earlier days this problem was studied under the assumption that the random errors $e_1, e_2, \cdots$ possess finite variance. This case was finally solved by Lai and others in an important work [6] published in 1979. In that paper they

Mingzhong Jin and Hongmei Liu
Guizhou Universicy for Nationalities.Guiyang Huaxi, 550025, China
e-mail: jmz6899@163.com

Minqing Gong
Guizhou Universicy, Guiyang Huaxi, 550025, China

S. Li (Eds.): Nonlinear Maths for Uncertainty and its Appli., AISC 100, pp. 517–524.
springerlink.com &copy; Springer-Verlag Berlin Heidelberg 2011

showed that if the random errors $e_1, e_2, \cdots$ are iid, with $Ee_1 = 0$ and $0 < Ee_1^2 < \infty$(these assumptions can be considerably weakened), then $S_n^{-1} \equiv (\sum\limits_{i=1}^{n} x_i x_i')^{-1} \to 0$ is a sufficient condition for SC of $\hat{\beta}_n$. Since it was known earlier [5]that this condition is also necessary, so $S_n^{-1} \to 0$ is the necessary and sufficient (NS) condition. Later the research effort turned to the case where a lover-order moment for $e_i$ is assumed. A typical formulation is that $e_1, e_2, \cdots$ are iid, $Ee_1 = 0$ and $E|e_1|^r < \infty$ for some $r \in [1,2)$. In 1981, Chen [2]showed that $\hat{\beta}_n$ is SC if $S_n^{-1} = O(n^{-(2-r)/r}(\log n)^{-a})$ for some $a > 1$ and some other conditions are satisfied. At almost the same time Chen et al. [1]obtained a single sufficient condition $S_n^{-1} = O(n^{-(2-r)/r}(\log n)^{-2/r-\varepsilon})$ for some $\varepsilon > 0$. In 1989, Zhu in his doctorial dissertation [7]made a substantial improvement by showing that $S_n^{-1} = O(n^{-(2-r)/r})$ is sufficient. Using a result recently published [3], it can be shown that Zhu's result cannot be further improved. For any $c_n \downarrow 0$ such that $\limsup c_n n^{(2-r)/r} = \infty$, the condition $S_n^{-1} = O(c_n)$ is no longer sufficient even for the weak consistency of $\hat{\beta}_n$.

Zhu also attempted to find the NS condition for $\hat{\beta}_n$ to be SC in the special case $p = 1$, which was later completely solved by Chen and others [4]. For general $p$ the NS condition is still unsolved. The problem seems more difficult than earlier expected.

The purpose of this paper is to give, for general $p$ but under some additional restriction , a NS condition for $\hat{\beta}_n$ to be SC. We need some notations. Assume that $S_n^{-1}$ exists for large $n$ and write

$$a_i = S_i^{-1} x_i, i \geq 1.$$

$a_i$ can be arbitrarily defined if $S_i^{-1}$ does not exist. Define

$$N(K) = \#\{i : i \geq 1, \| a_i \| \geq K^{-1}\}.$$

Let $((n,1), (n,2), \cdots, (n,n))$ be a permutation of $(1, 2, \cdots, n)$ such that

$$\| a_{(n,1)} \| \geq \| a_{(n,2)} \| \geq \cdots \geq \| a_{(n,n)} \|.$$

Dfine

$$V(n,j) = S_n^{-1} \sum_{i=1}^{n} x_i I(\| a_i \| \geq \| a_{(n,j)} \|), 1 \leq j \leq n.$$

$$V(n) = \max_{1 \leq j \leq n} \| (n,j) \|,$$

Where $I(\cdot)$ is the indicator. Denote by $\lambda_n$ and $\mu_n$ the smallest and largest eigenvalues of $S_n$. Finally, write $|A|$ for $\max\limits_{i,j}|a_{ij}|$, where $A = (a_{ij})$.

Now we can formulate the main result of this paper:

**Theorem 1.** *Suppose that* $e_1, e_2, \cdots$ *in model(1) are iid,* $Ee_1 = 0$ *and* $E|e_1|^r < \infty$, *and that*

$$\mu_n = O(\lambda_n) \tag{2}$$

*Then the NS condition for* $\hat{\beta}_n$ *to be SC is that*

$$
\begin{aligned}
&\text{For } 1 < r < 2: S_n^{-1} \to 0, N(K) = O(K^r) \text{ as } K \to \infty. \\
&\text{For } r = 1: S_n^{-1} \to 0, N(K) = O(K^r) \text{ as } K \to \infty, V(n) = O(1).
\end{aligned} \tag{3}
$$

*Remark 1.* The necessity means that is one of these conditions is not satisfied, then there exists $\{e_i\}$ satisfying the conditions specified in the theorem such that if the model (1) has this $\{e_i\}$, then $\hat{\beta}_n$ does not converge a.s. to $\beta$. Thus the necessary condition is not related to individual $\{e_i\}$, but rather to the whole class $\mathcal{F}$ of $\{e_i\}$ satisfying the conditions specified in the theorem. It would be ideal if we find a condition C such that is C is(not) satisfied, the $\hat{\beta}_n$ converges(does not converge) a.s. to $\beta$ for any $\{e_i\} \in \mathcal{F}$. But it can easily be shown that such a condition C does not exist when $r < 2$.

We divide the proof of this theorem into several sections(from theorem 2 to theorem 4).

## 2  Necessity of $S_n^{-1} \to 0$

This follows from a lemma proved in [1]: Let $\{e_i\}$ be a sequence of independent random variables containing no asymptotically degenerate subsequence, and $\{C_{ni}, 1 \leq i \leq n, n \geq 1\}$ be an array of constants such that $\sum\limits_{i=1}^{n} C_{ni} e_i \to 0$ in pr. Then $\sum\limits_{i=1}^{n} C_{ni}^2 \to 0$.

## 3  Necessity of $N(K) = O(K^r)$

**Theorem 2.** *If* $\mu_n = O(\lambda_n)$, *then*

$$|S_n^{-1} S_{n-1}| = O(1). \tag{4}$$

*Proof.* We have

$$S_n^{-1} = (S_{n-1} + x_n x_n')^{-1} = S_{n-1}^{-1} - S_{n-1}^{-1} x_n x_n' S_{n-1}^{-1} / (1 + x_n' S_{n-1}^{-1} x_n),$$

hence

$$S_n^{-1} S_{n-1} = I - S_{n-1}^{-1} x_n x_n' / (1 + x_n' S_{n-1}^{-1} x_n).$$

Since $\mu_n = O(\lambda_n)$, we have $|S_{n-1}^{-1}| = O(\mu_{n-1}^{-1})$, thus

$$|S_{n-1}^{-1} x_n x_n'| = O(\| x_n \|^2 / \mu_{n-1}).$$

On the other hand, $x_n' S_{n-1}^{-1} x_n \geq \| x_n \|^2 / \mu_{n-1}$. Hence (4) follows.

**Theorem 3.** *If $S_{n-1}^{-1} \to 0$, then $a_n \to 0$ and $\max\limits_{1 \leq i \leq n} \| S_n^{-1} x_i \| \to 0$.*

*Proof.* Write $S_n = P_n' \Lambda_n P_n$, where $P_n$ is orthogonal and $\Lambda_n$ is diagonal. Put

$$z_{ni} \equiv (z_{ni1}, \cdots, z_{nip})' = P_n x_i, 1 \leq i \leq n.$$

Then $a_n = P_n' \Lambda_n^{-1} z_{nn}$ and $\Lambda_n = \sum\limits_{i=1}^{n} z_{ni} z_{ni}'$. Hence the $j$-th component of

$\Lambda_n^{-1} z_{nn}$ is $z_{nnj} / \sum\limits_{i=1}^{n} z_{nij}^2$. $S_n^{-1} \to 0$ entails $\lim\limits_{n \to \infty} \sum\limits_{i=1}^{n} z_{nij}^2 = \infty$ and hence

$\lim\limits_{n \to \infty} z_{nnj} / \sum\limits_{i=1}^{n} z_{nij}^2 = 0$. This proves $\Lambda_n^{-1} z_{nn} \to 0$ and hence $a_n \to 0$. The

second assertion can be proved likewisely.

Since the length restriction, we omit the proof detailedly.

## 4  Sufficiency, $1 < r < 2$

**Theorem 4.** *(Matrix Form of Kronecker Theorem). If $\mu_n = O(\lambda_n)$ and $S_n^{-1} \to 0$, then the convergence of $\sum\limits_{i=1}^{\infty} a_i e_i$ entails the convergence to zero of $S_n^{-1} \sum\limits_{i=1}^{n} x_i e_i$.*

*Proof.* Write $B_i = x_{i+1} x_{i+1}', i = 0, 1, \cdots, T_0 = 0, T_n = \sum\limits_{i=1}^{n} a_i e_i, n \geq 1$. Simple manipulations give

$$S_n^{-1} \sum_{i=1}^{n} x_i e_i = S_n^{-1} (\sum_{i=1}^{n_0} x_i e_i - S_{n_0+1} T_{n_0}) + S_n^{-1} \sum_{j=0}^{n-1} B_j (T_n - T_j) \equiv J_1 + J_2.$$

where $n_0$ is fixed so that $S_{n_0}^{-1}$ exists. We have $J_1 \to 0$ in view of $S_n^{-1} \to 0$. To deal with $J_2$, denote by $s_{iuv}$ and $b_{juv}$ the $(u, v)$-element of $S_i$ and $B_j$. Then $s_{iuv} = \sum\limits_{j=0}^{-1} b_{juv}$ and $|b_{juv}| \leq (b_{juu} + b_{jvv})/2$. Denote by $t_{njl}$ the $l$-th component of $T_n - T_j$, and

$$t_{nj} = \max_{1 \leq i \leq pj} \max_{\leq k \leq n} |t_{nkl}|.$$

$t_{nj}$ does not increase as $j$ increases. We have

$$|u - th\ component\ of\ \sum_{j=k}^{n-1} B_j(T_n - T_j)| \leq \sum_{j=k}^{n-1} \sum_{v=1}^{p} |b_{juv}||t_{njv}|$$

$$\leq t_{nk} \sum_{j=k}^{n-1} \sum_{v=1}^{p} (b_{juu} + b_{jvv})/2$$

$$\leq t_{nk}(p \sum_{j=k}^{n-1} (b_{juu} + \sum_{v=1}^{p} \sum_{j=k}^{n-1} b_{jvv}))$$

$$\leq t_{nk}(ps_{nuu} + \sum_{v=1}^{p} s_{nuu})$$

$$\leq ct_{nk}\mu_n,$$

$$(5)$$

where $c$ is a constant. Since $\mu_n = O(\lambda_n), |S_n^{-1}| \leq c_0/\mu_n$ for some constant $c_0$. Hence by (5) we have

$$|S_n^{-1} \sum_{j=k}^{n-1} b_j(T_n - T_j)| \leq cc_0pt_{nk} \to 0, as\ n > k \to \infty. \qquad (6)$$

Here we use the fact that $t_{nk} \to 0$ as $n > k \to \infty$, which follows from the convergence of $T_n$ as $n \to \infty$. On the other hand, since $S_n^{-1} \to 0$, we have $S_n^{-1} \sum_{j=0}^{k-1} b_j(T_n - T_j) \to 0$ as $n \to \infty$. (Here again we use the fact that $T_n$ converges as $n \to \infty$). Combining this with (6), we obtain $J_2 \to 0$, and the theorem is proved.

*Remark 2.* Simple counter-example shows that the condition $\mu_n = O(\lambda_n)$ is essential. This is the reason why the restriction $\mu_n = O(\lambda_n)$ is imposed.

Now turn to the main task of this section. According to Theorem 4, it suffices to show that $\sum_{i=1}^{\infty} d_i e_i$ converges a.s., where $d_i$ is the $j$-th component of $a_i, 1 \leq j \leq p$. Put $e_i' = e_i I(|e_i| < |d_i|^{-1}), i \geq 1$. $N(K) = O(K^r)$ entails

$$\tilde{N}(K) \equiv \#\{i : i \geq 1, |d_i| \geq K^{-1}\} = O(K^r).$$

Employing this fact and $E|e_1|^r < \infty$, it can easily be shown that $P(e_i \neq e_i', i.o.) = 0$. Hence we have only to show that

$$\sum_{i=1}^{\infty} d_i e_i' \quad converges\ a.s.. \qquad (7)$$

(7) is the true if the following two assertions are true:

$$\sum_{i=1}^{\infty} d_i E(e_i') \ converges \tag{8}$$

$$\sum_{i=1}^{\infty} d_i^2 E(e_i'^2) < \infty. \tag{9}$$

We proceed to prove (8). The proof of (9) is similar and will be omitted. Put $q_i = P(i-1 \le |e_1| < i), i = 1, 2, \cdots$. Since $Ee_i = 0$, we have $Ee_i' = -E(e_i I(|e_i| \ge |d_i|^{-1}))$. If $k-1 \le |d_i|^{-1} < k$, then

$$|Ed_i e_i'| \le |d_i| E|e_i I(|e_i| \ge |d_i|^{-1})| \le (k-1)^{-1} \sum_{j=1}^{\infty} j q_i, k \ge 2.$$

Further,

$$\#\{i : i \ge 1, k-1 < |d_i|^{-1} \le k\} = \tilde{N}(k) - \tilde{N}(k-1).$$

Hence

$$\sum_{i=1}^{\infty} |Ed_i e_i'| \le \tilde{N}(1)\sup_{i \ge 1}|d_i| E|e_1| + \sum_{k=2}^{\infty} (\tilde{N}(k) - \tilde{N}(k-1))(k-1)^{-1} \sum_{j=k-1}^{\infty} j q_j$$

$$\equiv J_1 + J_2.$$

Since $S_n^{-1} \to 0$, we have $a_n \to 0$ by Theorem 3. Hence $J_1$ remains finite as $n \to \infty$. On the other hand

$$J_2 = \sum_{j=1}^{\infty} (j^{-1}\tilde{N}(j+1) - \tilde{N}(1)) + \sum_{k=2}^{\infty} \tilde{N}(k)((k-1)^{-1} - k^{-1})j q_j \equiv H_1 - H_2 + H_3.$$

$H_1 < \infty$ in view of $\tilde{N}(j+1) \le c(j+1)^r$ and $E|e_1|^r < \infty$. $H_2 < \infty$ follows from $E|e_1| < \infty$. As for $H_3$, since $\tilde{N}(k) \le ck^r$ and $r > 1$, we see that

$$\sum_{k=2}^{j} \tilde{N}(k)((k-1)^{-1} - k^{-1}) = O(j^{r-1}),$$

and $H_3 < \infty$ follows from $E|e_1|^r < \infty$. Thus we obtain (7). The proof is concluded.

## 5  Sufficiency, $r = 1$

The argument in section 4 fails in the case $r = 1$. The trouble lies in dealing with $H_3$, since we have only $\sum_{k=2}^{\infty} \tilde{N}(k)((k-1)^{-1} - k^{-1}) = O(\log j)$ and not

$O(j^{r-1})$ or $O(1)$. This is why we need the additional condition $V(n) = O(1)$ in this case.

In order to make an easy use of the condition $V(n) = O(1)$, we slightly change the argument in Sect 4. Define $e_i'' = e_i I(|e_i| < \| a_i \|^{-1})$. Then $P(e_i \neq e_i'', i.o) = 0$. Since $S_n^{-1} \to 0$, in order to show $S_n^{-1} \sum_{i=1}^{n} x_i e_i \to 0, a.s..$, we have only to show that $S_n^{-1} \sum_{i=1}^{n} x_i e_i'' \to 0, a.s..$, which is true if the following tow assertions are true:

$$S_n^{-1} \sum_{i=1}^{n} x_i E e_i'' \to 0,$$

$$S_n^{-1} \sum_{i=1}^{n} x_i (e_i'' - E e_i'') \to 0, a.s.. \tag{10}$$

According to Theorem 4, (10) is true if $\sum_{i=1}^{n} a_i (e_i'' - E e_i'')$ converges a.s., and the latter is true if

$$\sum_{i=0}^{n} d_i^2 E(e_i''^2) < \infty. \tag{11}$$

where $d_i$ is the same as in Sect 4.

Since the length restriction ,we omit the proof detailedly.


## 6   Necessity of $V(n) = O(1)$ in Case $r = 1$

Suppose that $\hat{\beta}_n$ is SC, so $S_n \sum_{i=1}^{n} x_i e_i \to 0, a.s..$ From Sect 2 and Sect 3 we have $S_n^{-1} \to 0$ and $N(k) = O(K)$. In Sect 5 we pointed out that these two facts entail (11) which in turn entails $S_n^{-1} \sum_{i=1}^{n} x_i (e_i'' - E e_i'') \to 0, a.s..$ Therefore

$$S_n^{-1} \sum_{i=1}^{n} x_i E e_i'' \to 0. \tag{12}$$

Thus, we have only to show that if $V(n)$ is not $O(1)$, Then we can construct an iid. sequence $\{e_i\}$ with $E e_i = 0, E|e_i| \to \infty$, such that (12) fails.

Since the length restriction ,we omit the proof detailedly. Therefore, if in model(1) the random errors $e_1, e_2, \cdots$ are iid. with the common distribution $F_0$, then $\hat{\beta}_n$ does not converge a.s. to $\beta$, and the necessity of $V(n) = O(1)$ is established.

**Acknowledgements.** This research is partially supported by Guizhou province government([2010] No.04); The Dept. of Guizhou Education(2011); Guizhou Province ([2010] No.2136, [2010] No.7011, [2009] No.21); The Committee of National Work of Guizhou Province; The Committee of National Work of CHINA([2010] No.10gz08).

# References

1. Chen, G., Lai, T.L., Wei, C.Z.: Convergence systems and strong consistency of least squares estimates in liear models. J. Multivariate. Anal. 11, 319–333 (1981)
2. Chen, X.: Again on the consistency of least squares estimates in multiple regression. Acta. Math. Sinica. 24, 34–36 (1981)
3. Chen, X.: Necessary and sufficient conditions for the weak consistency of LS estimates in linear regression under a low-order moment condition. Science in China 25, 349–358 (1995)
4. Chen, X., Zhu, L., Fang, K.: Convergence o weighted sum of random variables. Statistica Sinica 6, 2 (1996)
5. Drygas, H.: Weak and strong consistency of the least squares estimators in regression model. Z. Wahrsch. Verw. Gebiete. 34, 119–127 (1976)
6. Lai, T.L., Robbins, H., Wei, C.Z.: Strong consistency of least squares estimates in multiple regression II. J. Multivariate. Anal. 9, 343–362 (1979)
7. Zhu, L.: Doctorial disertation. Inst. Systems. Science Academic. Sinica (1989); J. Multivariate. Anal. 9, 343–362 (1979)

# Variable Selection for Semiparametric Isotonic Regression Models

Jiang Du, Zhongzhan Zhang, and Tianfa Xie

**Abstract.** In this paper, we propose a penalized constrained least squares method for variable selection in semiparametric isotonic regression model. Under certain regularity conditions, asymptotic properties of the proposed estimators are established. A simulation study is presented for illustrations.

**Keywords:** Semiparametric isotonic regression models, Variable selection, Tuning parameter.

## 1 Introduction

In practice, the number of potential explanatory variables is often large, however, only a subset of them are predictive to the response. Variable selection is necessary to improve prediction accuracy and model interpretability of final models. For classical linear regression models, many variable selection procedures have been proposed since the 1970s such as the Akaike information criterion, Bayesian information criterion, Risk information criterion, least absolute selection and shrinkage operator (LASSO), smooth clipped absolute deviation (SCAD), least angle regression, adaptive lasso. An earlier review on variable selection for linear regression is given by Linhart and Zucchini [8] and a recent review is of Fan and Lv [7]. Though there is a vast amount of work on variable selection for linear model, limited works have been done on model selection for semiparemetric isotonic regression model. Model selection for semiparametric regression models is challenging, since it consists of several interrelated estimation and selection problems: nonparametric estimation, smoothing parameter selection, and variable selection and estimation of linear component.

Jiang Du, Zhongzhan Zhang, and Tianfa Xie
College of Applied Sciences, Beijing University of Technology, P.R. China, 100124
e-mail: dujiang84@163.com, zzhang@bjut.edu.cn, xietf@bjut.edu.cn

S. Li (Eds.): Nonlinear Maths for Uncertainty and its Appli., AISC 100, pp. 525–532.
springerlink.com                                      © Springer-Verlag Berlin Heidelberg 2011

Consider the following semiparametric regression model

$$Y = X\beta + g(Z) + \varepsilon, \tag{1}$$

where $Y$ is a response, $X$ is a $1 \times d$ vector consisting of explanatory variables of primary interest, $\beta = (\beta_1, \cdots, \beta_d)^T$ is a $d \times 1$ vector of unknown parameters, $g(\cdot)$ is an unknown smooth function of auxiliary covariate $Z$, $\varepsilon$ is a model error with $E(\varepsilon|X, Z) = 0$. In this model, the mean response is linearly related to $X$, while its relation with $Z$ is not specified up to any finite number of parameters. This model combines the flexibility of nonparametric regression and parsimony of linear regression. An advantage of the semiparamtric regression model is when the relation between $Y$ and $X$ is of main interest and can be approximated by a linear function, it offers more interpretability than a purely nonparametric model. Another advantage of the semiparamtric regression model is that it provides a convenient way. In this paper, to avoid the curse of dimensionality, we assume that $Z$ is univariate that ranges over a nondegenerate compact interval. Without loss of generality, it is assumed to be the unit interval $W = [0, 1]$.

Model (1) was introduced by Engle et al [5]. Variable selection for this model has been studied by many authors recently; see for example, Bunea [1], Bunea and Wegkamp [2], Li and Liang [10]. Semiparametric model (1) with $g$ being assumed as a monotone function is studied by Huang [9] and Cheng [4].To the best of our knowledge, most existing estimation method are limited to directly estimate the parameters of linear part and most existing variable selection procedures of semiparametric regression model are limited to smooth function, which is not constrained by isotonic property. Variable selection of semiparametric isotonic regression model imposes challenges for many practical statisticians. Variable selection for semiparametric isotonic regression model is not investigated yet. Therefore, to solve this problem, in this paper, we proposed a variable selection method for scmiparametric regression model with function $g$ being increasing. With proper choices of the penalty functions and the tuning parameter, the consistency and asymptotic normality of the resulting estimators of both parametric component and nonparametric component are established.

The rest of this article is organized as follows. In Section 2,we propose the variable selection procedures for semiparametric isotonic regression models via penalized constrained least squares. The asymptotic properties of the resulting estimators are established in the Section 3. Simulation studies are given in Section 4. Regularity conditions is relegated to the Appendix.

## 2    Penalized Constrained Least Squares Method

Consider the semiparametric isotonic regression model

$$Y = X\beta + g(Z) + \varepsilon, \tag{2}$$

where $X = (X_1, \cdots, X_d)$ is a $1 \times d$ dimensional covariate, $\beta \in \Theta$ is a $d \times 1$ dimensional parameter of interest, and $g$ is an unknown increasing function, $\varepsilon$ is the model error with $E(\varepsilon|X, Z) = 0$, $\mathrm{Var}(\varepsilon) = \sigma^2$. In this article, we only consider univariate $Z$. The proposed method is applicable to multivariate $Z$. The extension to the multivariate $Z$ might be practically less useful due to the "curse of dimensionality."

Let $(\beta_0, g_0)$ be the true value of the parameters $(\beta, g)$. Assume that $\beta_0$ belongs to a subset $\Theta \subset R^d$ and to be definitive, $\beta_0$ also has sparse representation. Without loss of generality, assume that $\Theta$ is a convex set. If it is not, then a larger convex set containing $\Theta$ can be used. Suppose that $\{(X_i, Y_i, Z_i), i = 1, 2, \cdots, n\}$ is a random sample from model (2). Let $Z_{(1)} \le Z_{(2)} \le \cdots \le Z_{(n)}$ be the ordered values of $Z_i$'s. For the sake of simplicity of notation, let $g_i = g(Z_{(i)}), 1 \le i \le n$. The constrained penalized least squares estimator $(\hat{\beta}, \hat{g}_n)$ of $(\beta_0, g_0)$ is defined to be the minimizer of $M_n(\beta, g)$ subjected to $\beta \in \Theta$ and $g_1 \le g_2 \le \cdots \le g_n$, where

$$M_n(\beta, g) = \sum_{i=1}^{n} (Y_i - X_i\beta - g(Z_i))^2 + n \sum_{j=1}^{d} p_{\lambda_{jn}}(|\beta_j|), \qquad (3)$$

$p_{\lambda_{jn}}(|\beta_j|)$ is a penalty function with a tuning parameter $\lambda_{jn}$, which may be chosen using BIC proposed by Wang et al. [11], for $j = 1, \cdots, d$. The tuning parameters are not necessarily the same for all $j$. For example, we wish keep some important variables in the final model, and therefore we should not penalize their coefficients. For the sake of simplicity of notation, we use $\lambda_j$ to stand for $\lambda_{jn}$ throughout this article.

Since $g(Z) = E(Y|Z) - E(X|Z)\beta$, then the objective function can be rewritten as

$$M_n(\beta) = \sum_{i=1}^{n} [Y_i - E(Y_i|Z_i) - \{X_i - E(X_i|Z_i)\}\beta]^2 + n \sum_{j=1}^{d} p_{\lambda_j}(|\beta_j|).$$

Denote $m_y(Z) = E(Y|Z)$ and $m_x(Z) = E(X|Z)$. Let $\hat{m}_y(\cdot)$ and $\hat{m}_x(\cdot)$ be estimates of $m_y(\cdot)$ and $m_x(\cdot)$, respectively. In this section, we use local linear regression to estimate both $m_y(\cdot)$ and $m_x(\cdot)$. Let $\hat{Y}_i = Y_i - \hat{m}_y(Z_i)$ and $\hat{X}_i = X_i - \hat{m}_x(Z_i)$. Thus, $M_n(\beta)$ can be approximated as

$$\hat{M}_n(\beta) = \sum_{i=1}^{n} \left[\hat{Y}_i - \hat{X}_i\beta\right]^2 + n \sum_{j=1}^{d} p_{\lambda_j}(|\beta_j|).$$

This suggests us that we can define an estimator of $\beta$ as

$$\hat{\beta}_n = \arg\min_{\beta} \left\{ \sum_{i=1}^{n} [\hat{Y}_i - \hat{X}_i\beta]^2 + n \sum_{j=1}^{d} p_{\lambda_j}(|\beta_j|) \right\}.$$

The isotonic least squares estimator of $g_0(\cdot)$ is defined as

$$\hat{g}_n = \arg\min_{h \in \mathscr{H}} \left\{ \sum_{i=1}^{n} (Y_i - X_i \hat{\boldsymbol{\beta}}_n - h(Z_i))^2 \right\}, \tag{4}$$

where $\mathscr{H}$ is the class of all non-decreasing functions on $W$. Note that since $\mathscr{H}$ forms a closed convex cone, the optimization problem is that of minimizing a convex function over a convex cone, therefore, $\hat{g}_n$ is well defined. (4) gives the isotonic least squares estimator proposed by Brunk [4] that is

$$\hat{g}_n(z) = \max_{u \le z} \min_{z \le t} \frac{\sum_{\{i:u \le Z_i \le t\}} (Y_i - X_i \hat{\boldsymbol{\beta}}_n)}{N_n([u,t])}, \tag{5}$$

where $z \in W$, $N_n([u,t]) = \sharp\{i : u \le Z_i \le t, 1 \le i \le n\}$, $\sharp$ denotes the counting measure.

## 3  Sampling Properties

In this section, we establish the asymptotic behavior of the estimators. Let

$$\boldsymbol{\beta}_0 = (\beta_{10}, \beta_{20}, \cdots, \beta_{d0})^{\tau} = (\boldsymbol{\beta}_{10}^{\tau}, \boldsymbol{\beta}_{20}^{\tau})^{\tau}.$$

Without loss of generality, assume that $\boldsymbol{\beta}_{10}$ consists of all nonzero components of $\boldsymbol{\beta}_0$, and $\boldsymbol{\beta}_{20} = 0$. Let $s$ denote the dimension of $\boldsymbol{\beta}_{10}$. Denote

$$a_n = \max_{1 \le j \le d}\{|p'_{\lambda_j}(|\beta_{j0}|)|, \beta_{j0} \ne 0\}, \qquad b_n = \max_{1 \le j \le d}\{|p''_{\lambda_j}(|\beta_{j0}|)|, \beta_{j0} \ne 0\},$$

$$\boldsymbol{b}_n = \{p'_{\lambda_1}(|\beta_{10}|)\operatorname{sgn}(\beta_{10}), p'_{\lambda_2}(|\beta_{20}|)\operatorname{sgn}(\beta_{20}), \cdots, p'_{\lambda_s}(|\beta_{s0}|)\operatorname{sgn}(\beta_{s0})\}^{\tau},$$

$$\Sigma_\lambda = \operatorname{diag}\{p''_{\lambda_1}(|\beta_{10}|, p''_{\lambda_2}(|\beta_{30}|, \cdots, p''_{\lambda_s}(|\beta_{s0}|)\},$$

In what follows, denote $\mathbf{A}^{\otimes 2} = \mathbf{A}\mathbf{A}^{\tau}$ for any vector or matrix $\mathbf{A}$. $\|v\|$ denotes the Euclidean norm for the vector $v$. We only state our theorems here, and put the conditions for the theorems into the appendix. The proofs of the theorems are omitted for the sake of space.

**Theorem 1.** *Suppose that $a_n = O(n^{-1/2}), b_n \to 0$, and the regularity conditions $A_1 - A_6$ in the appendix hold. Then we have the following conclusions.*

(1) *With probability approaching one, there exists a local minimizer $\hat{\boldsymbol{\beta}}$ of $\hat{M}_n(\boldsymbol{\beta})$ such that $\|\hat{\boldsymbol{\beta}} - \boldsymbol{\beta}_0\| = O_p(n^{-1/2})$.*

(2) *Further assume that all $\lambda_{jn} \to 0, n^{1/2}\lambda_{jn} \to \infty$, and*

$$\liminf_{n \to \infty} \liminf_{t \to 0^+} \frac{p'_{\lambda_{jn}}(t)}{\lambda_{jn}} > 0, \tag{6}$$

*the root $n$ consistent estimator $\hat{\beta} = (\hat{\beta_1}^\tau, \hat{\beta_2}^\tau)$ in conclusion (1) satisfies (a) $\hat{\beta_2} = 0$, and (b) $\hat{\beta_1}$ has an asymptotic normal distribution, i.e.*

$$\sqrt{n}(H_{11} + \Sigma_\lambda)\{(\hat{\beta_1} - \beta_{10}) + (H_{11} + \Sigma_\lambda)^{-1}\boldsymbol{b}_n\} \to N_s(0, I_{11}),$$

*in distribution, where $I_{11}$ is $d \times d$ unit matrix and $H_{11}$ is given in the appendix.*

The following theorem gives the sampling property of the nonparametric isotonic function of $g(\cdot)$.

**Theorem 2.** *Under the regularity conditions $A_1 - A_6$, if $g(\cdot)$ is continuously differentiable at $z_0$, and the derivative $\dot{g}(z_0) > 0$, we have*

$$n^{1/3}\left[\frac{2f_Z(z_0)}{\sigma^2\dot{g}(z_0)}\right]^{1/3}(\hat{g}(z_0) - g(z_0)) \longrightarrow \zeta, n \longrightarrow \infty,$$

*in distribution, where $\zeta$ is the slope at zero of the greatest convex minorant of $B(t) - t^2$. $B(t)$ is a two-sided standard Brownian motion.*

Theorem 2 can be shown along the same lines as Theorem 2.1 of Huang [9].

## 4  Simulation Studies

To demonstrate the finite sample performance of the proposed penalized constrained least squares method, we consider the following semiparametric isotonic regression model

$$Y = X\beta + g(Z) + \varepsilon, \tag{7}$$

where $X$ is generated from a multivariate normal distribution with zero mean and covariance

$$Cov(x_{ij_1}, x_{ij_2}) = \rho^{|j_1 - j_2|}, 1 \leq j_1, j_2 \leq d, 1 \leq i \leq n, \tag{8}$$

with $\rho = 0.5, 0.75$. In our simulation, the sample size $n$ is 50,100 and 200,respectively, $d = 10$ and parameter $\beta_{10} = (3, 1.5, 0.75)$, $g(t) = t^3$, $Z$ is generated from uniform distribution [-1,1], $\varepsilon$ is generated from standard normal distribution. For each case, we repeat the simulation 1,000 times. We use SCAD, Lasso and ALasso penalty function to chose significant variables in parametric part. The unknown tuning parameters for the penalty function in the simulation are chosen by BIC (Wang et al. [11]). For the sake of computational convenience, we employ the algorithm which is proposed by Fan and Li [6], and take $\varepsilon_0 = 10^{-6}$ as threshold value. For each parametric estimator $\hat{\beta}$, its estimation accuracy is measured by mean squared error defined as $MSE = E[(\hat{\beta} - \beta_0)'E(X'X)(\hat{\beta} - \beta_0)]$. The variable selection performance is gauged by "C" , and "IC", where "C" is the number of zero coefficients those are correctly estimated by zero, "IC" is the number of nonzero coefficients incorrectly estimated to be zero. For nonparametric component, the estimation accuracy is measured by mean integrated square error defined as

$$MISE = E\left\{ \int_{-1}^{1} [\hat{g}(t) - g(t)]^2 \, \mathrm{d}t \right\}. \tag{9}$$

The results are summarized in the following tables.

We now summarize the main findings from this study. Firstly, from the MSE of the parametric estimators, we can conclude that the parametric estimator has little impact on nonparametric estimator, which is in accordance with theory, estimator of the regression parameter is root-n consistent and the isotonic estimator of the functional component, at a fixed point, is cubic root-n consistent. Secondly, for small sample size, the performance of the estimators is good, especially when the correlation coefficient of covariates is small. Last but not least, the correlation coefficient of covariates is important for the resulting estimators. With increase of the correlation coefficient of covariates, MSE and MISE is becoming increasing.

**Table 1** The sample size is $n = 50$

| Methods | $\rho = 0.50$ | | | | $\rho = 0.75$ | | | |
|---------|-----|-------|--------|-------|-------|-------|--------|-------|
|         | IC  | C     | MSE    | MISE  | IC    | C     | MSE    | MISE  |
| SCAD    | 0.963 | 5.687 | 2.906 | 2.313 | 1.263 | 4.362 | 17.020 | 2.853 |
| Lasso   | 0.708 | 5.650 | 2.181 | 2.238 | 1.444 | 5.676 | 7.153 | 2.596 |
| ALasso  | 0.972 | 5.541 | 3.146 | 2.401 | 1.563 | 5.446 | 11.300 | 2.772 |
| Oracle  | 0.000 | 7.000 | 0.881 | 2.053 | 0.000 | 7.000 | 3.246 | 2.378 |

**Table 2** The sample size is $n = 100$

| Methods | $\rho = 0.50$ | | | | $\rho = 0.75$ | | | |
|---------|-----|-------|--------|-------|-------|-------|--------|-------|
|         | IC  | C     | MSE    | MISE  | IC    | C     | MSE    | MISE  |
| SCAD    | 0.781 | 6.576 | 0.923 | 1.968 | 1.372 | 6.236 | 3.268 | 2.149 |
| Lasso   | 0.332 | 6.565 | 0.498 | 1.967 | 1.130 | 6.598 | 1.288 | 2.094 |
| ALasso  | 0.766 | 6.559 | 0.977 | 2.006 | 1.437 | 6.413 | 2.554 | 2.157 |
| Oracle  | 0.000 | 7.000 | 0.318 | 1.915 | 0.000 | 7.000 | 1.045 | 2.042 |

## 5  Appendix: Assumptions

In order to proof the theorems, we need the following conditions.

($A_1$) The distribution of $\varepsilon$ satisfies the moment condition $E|\varepsilon|^2 < \infty$.

($A_2$)

$$E(X - E(X|Z))^{\otimes 2} = H = \begin{pmatrix} H_{11} & H_{12} \\ H_{21} & H_{22} \end{pmatrix}$$

**Table 3** The sample size is $n = 200$

| Methods | $\rho = 0.50$ | | | | $\rho = 0.75$ | | | |
|---|---|---|---|---|---|---|---|---|
| | C | IC | MSE | MISE | C | IC | MSE | MISE |
| SCAD | 0.458 | 6.795 | 0.435 | 2.085 | 1.149 | 6.870 | 0.998 | 2.060 |
| Lasso | 0.078 | 6.805 | 0.157 | 2.058 | 0.707 | 6.776 | 0.481 | 2.048 |
| ALasso | 0.415 | 6.855 | 0.404 | 2.085 | 1.185 | 6.790 | 1.071 | 2.080 |
| Oracle | 0.000 | 7.000 | 0.104 | 2.038 | 0.000 | 7.000 | 0.336 | 2.023 |

$H_{11}$ is a $s \times s$ positive definite matrix.

($\mathbf{A_3}$) The function $\zeta(Z) \equiv E(X|Z = z)$ satisfies the Lipschitz condition

$$\| \zeta(z_1) - \zeta(z_2) \| \leq C|z_1 - z_2|$$

for all $z_1, z_2 \in W$, $W$ is compact interval of $R$ and $C$ is a constant.

($\mathbf{A_4}$) The support of $X$ is a bounded subset of $R^d$, and the density of $Z$, denoted as $f_Z$, is continuous.

($\mathbf{A_5}$) $g(\cdot)$ is differentiable and its derivatives are bounded on $W$. For some constants $C, \gamma > 0$, it holds that $\inf_{|u-v|>\delta} | g(u) - g(v) | \geq C\delta^{\gamma}$ for all $\delta > 0$.

($\mathbf{A_6}$) $g$ is strictly increasing and is bounded, that is, $\sup_z |g(z)| < C$ for a finite constant C; and the parametric space $\Theta$ is bounded.

**Acknowledgements.** This work is supported by grants from the National Natural Science Foundation of China(10971007); Funding Project of Science and Technology Research Plan of Beijing Education Committee(JC006790201001) and research fund of BJUT (X1006013201001).

# References

1. Breiman, L.: Better subset selection using nonnegative garrote. Techonometrics 37, 373–384 (1995)
2. Breiman, L.: Heuristics of instability and stabilization in model selection. The Annals of Statistics 24, 2350–2384 (1996)
3. Brunk, H.D.: On the estimation of parameters restricted by inequalities. Ann Math Statist 29, 437–454 (1958)
4. Cheng, G.: Semiparametric additive isotonic regression. Journal of Statistical Planning and Inference 139, 1980–1991 (2009)
5. Engle, R., Granger, C., Rice, J., Weiss, A.: Nonparametric estimates of the relation between weather and electricity sales. Journal of American Statistical Association 81, 310–386 (1986)
6. Fan, J.Q., Li, R.: Variable selection via nonconcave penalized likelihood and its oracle properties. Journal of American Statistical Association 96, 1348–1360 (2001)
7. Fan, J., Lv, J.: A selective overview of variable selection in high dimensional feature space. Statistica Sinica 20, 101–148 (2010)

8. Linhart, H., Zucchini, W.: Model Selection. John Wiley and Sons, New York (1986)
9. Huang, J.: A note on estimating a partly linear model under monotonicity constraints. Journal of Statistical Planning and Inference 107, 345–351 (2002)
10. Li, R., Liang, H.: Variable selection in semiparametric regression modeling. Annals of Statistics 36, 261–286 (2008)
11. Wang, H., Li, R., Tsai, C.: Tuning parameter selectors for the smoothly clipped absolute deviation method. Biometrika 94, 553–568 (2007)
12. Wang, H., Li, B., Leng, C.: Shrinkage tuning parameter selection with a diverging number of parameters. Journal of Royal Statistical Society Series B 71, 671–683 (2009)

# Study of Prognostic Factor Based on Factor Analysis and Clustering Method

Zheng Liu, Liying Fang, Mingwei Yu, and Pu Wang

**Abstract.** Relevance exists in Traditional Chinese Medicine(TCM) clinical symptoms. Their different combinations reflect different effects. Focusing on these characteristics, an univariate analysis method based on the factor analysis and clustering(FACUA) is proposed. First, the independent common factors extracted from the correlative multivariable are used to establish the eigenvectors of symptoms for patients. Then, the symptom patterns are discovered from the gathered similar symptoms combination. The method is verified by the patients with advanced NSCLC(non-small cell lung cancer) from Beijing Hospital of Traditional Chinese Medicine. The experimental result shows that the FACUA method can deal with the TCM clinical symptoms and analyze the relationship between the TCM clinical symptoms and the tumor progression. The FACUA method can improve the universal applicability of the univariate analysis in TCM clinical symptoms.

**Keywords:** Factor Analysis, Clustering, Univariate Analysis, Clinical Symptoms.

## 1 Introduction

More attention has been paid on the quality of life since a new concept of health was proposed by the WHO [4]. Clinical experience shows that effective TCM treatment for patients with advanced cancer can improve clinical

Zheng Liu, Liying Fang, and Pu Wang
College of Electronic Information and Control Engineering,
Beijing University of Technology, Beijing 100124, China
e-mail: catherine5656@163.com, fangliying@bjut.edu.cn, wangpu@bjut.edu.cn

Mingwei Yu
Beijing Hospital of Traditional Chinese Medicine, CPUMS, Beijing 100124, China
e-mail: yumingwei1120@163.com

S. Li (Eds.): Nonlinear Maths for Uncertainty and its Appli., AISC 100, pp. 533–540.
springerlink.com                    © Springer-Verlag Berlin Heidelberg 2011

symptoms, enhance the quality of life, and prolong survival time. Therefore, more and more researchers focus on tumor therapy with Traditional Chinese Medicine(TCM), however, the study is still in the stage of exploration [3]. Some researchers have analyzed the relationship between the single clinical symptom or TCM syndrome and the tumor progression by directly using a survival analysis method directly [5], which not only neglects the relevance of clinical symptoms but also affects the experimental result obtained by the visual observation and experience of doctors. Thus, the medical significance and reproducibility of the experimental result are not convincible [7].

To compensate for these shortages, this paper takes the factor analysis and clustering method into analyz the prognostic factors of tumor progression by considering the relevance between the symptoms and the similarity of the comprehensive effect. The reserch allows the practical research of an appropriate method for clinical symptoms and can provide a reference for advanced NSCLC with the TCM treatment.

## 2 Object and Implementation Framework

### 2.1 Object of Study

Symptom refers to the information that is relevant to the disease, which is the main contents of Chinese interrogation and basis of TCM syndrome [1]. The characteristics of TCM symptoms are more complicated. On one hand, some symptoms are caused by the common potential risk factor. On the other hand, some different symptoms combinations reflect the same type of comprehensive effect. For example, from the Chinese perspective, the spontaneous sweat and night sweat are all caused by asthenic-syndrome. When dry throat, fever and other symptoms occur together, the risk of tumor progression will be higher.

### 2.2 Implementation Framework

To solve the above problems, this paper proposes a univariate analysis method based on the factor analysis and clustering(FACUA) to analyze the prognostic symptoms factors of tumor progression. Fig. 1 shows the steps of data processing and the framework of the FACUA system. The framework mainly includes four modules.

#### 2.2.1 Preprocessing Module

(1) Calculation of PFS time

Progression-free survival(PFS) is a term used to describe the length of time during and after medication or treatment during which the disease being treated(usually cancer) does not get worse [6].The PFS time is calculated as $T = T_e - T_s$, where $T_s$ is the time of patients enrolled and $T_e$ is the first

**Fig. 1** The framework
of the FACUA system.

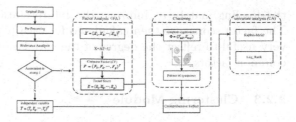

time when tumor progression occurs(include death). During the research, $T_e$
will be the deadline of the research with no progression and to be the last
follow-up time if patients withdraw.

(2) Calibration of tumor progression

The PFS time is only partially known once the patients withdraw or no progression, thus a statue variable must be added to make a distinction between
accurate data and censored data. In this paper, 1 is for tumor progression
and 0 is for another statue.

(3) Processing of missing data

The symptom of cold extremities has a little information and more obstruction for the next procedures. Therefore, we eliminate the symptom index from
the data set.

### 2.2.2 Factor Analysis Module

Factor analysis(FA) is a statistical method, which describes variability among
observed variables in terms of a potentially lower number of unobserved variables called common factor(CF). In this paper, FA is used to uncover the
underlying structure of the TCM symptoms and to seek the least number of
CF that can account for the majority of the information of the set of symptoms variables, i.e., the FA module mainly reduces dimension and extracts
CF for the clinical symptoms.

Firstly, a patient-symptoms original data matrix is established based
on indicators samples. The FA model is establish by using the correlated
symptoms.

$$
\begin{cases}
X_1 = a_{11}F_1 + a_{12}F_2 + \cdots + a_{1p}F_p + U_1 \\
X_2 = a_{21}F_1 + a_{22}F_2 + \cdots + a_{2p}F_p + U_2 \\
\vdots \\
X_m = a_{m1}F_1 + a_{m2}F_2 + \cdots + a_{mp}F_p + U_m
\end{cases}
\tag{1}
$$

Where $X = (X_1, X_2, \cdots, X_m)^T$ is an observed clinical symptoms vector,
$F = (F_1, F_2, \cdots, F_p)^T$ is a CF vector, $A = (a_{ij})_{m \times p}$ is the factor loading
matrix. In matrix notation, we have $X_{m \times 1} = A_{m \times p} \cdot F_{p \times 1} + U_{m \times 1}$

The principle component analysis(PCA) is chosen to extract the CF and
estimate the factor loading matrix. Varimax orthogonal rotation is used on

the loading matrix to improve the interpretation of the common factors. The factor scores are then calculated for each patient on every CF, which is very useful to reflect the degree of the correlation between the patients and common factors.

### 2.2.3 Clusternig Module

Clustering is a method for unsupervised learning and best known as a useful technique of data automatic classification according to their similarities to each other. The symptom eigenvector is used to describe the combination of the multiple symptoms. The main objective of clustering module is to assign patients into subsets according to the similar of symptom eigenvectors, which can be considered as the pattern of symptoms.

Sample properties including relatively independent symptoms index and scores of patients in common factors, which separately belong to categorical variables and numerical variables. K - prototypes clustering algorithm is chosen to deal with the two types of variables. Mean value and Euclidean distance are used when dealing with numerical variables. Mode and principle of minimum difference degree are used when dealing with categorical variables.

### 2.2.4 Univariate Analysis Module

Kaplan-Meier is a non-parametric estimate method for the survival analysis [2]. An important advantage of the Kaplan-Meier curve is that the method can consider some types of censored data that always occurs in medicine. A univariate analysis(UA) method, including Kaplan-Meier estimate and Log-Rank test, will be used for analyzing the relationship between the TCM clinical symptoms and the tumor progression.

Comprehensive effect(CE) of the clinical symptoms can be defined as a variable used to describe the combination effect to health. PFS survival function will be illustrated by Kaplan-Meier curve. The null hypothesis of no difference between the survival functions will be test by Log-Rank test for each level of the CE. It can be judged by P-value whether CE can represent highly statistically significant predictors of PFS time or not.

## 3 Experiments and Result Analysis

For further discussing the relationship between the TCM clinical symptoms and the tumor progression, we use the conventional UA and FACUA method to compare the following experiments.

51 patients in advanced NSCLC were collected from the Beijing Chinese Medicine Hospital. Their symptoms include cough, chest pain, appetite and other 17 indicators. Each symptom is divided into four grades marked as 0,1,2, and 3 according to the severiry is. After preprocessed, the PFS time is

calculated, the state of tumor progression is marked, and the useless symptoms are excluded.

## 3.1   Conventional UA Experimental

Some experiments were carried out with the UA method for analyzing the relationship between the single symptom and the tumor progression directly. The significant effect of each symptom is tested by Log-Rank. Results are listed in Table 1.

**Table 1** Results of conventional single-factor analysis

| Factor | P-value | Factor | P-value | Factor | P-value |
|---|---|---|---|---|---|
| Age | 0.375 | chest tightness | 0.122 | constipation | 0.012 |
| Sex | 0.216 | mouth parched | 0.720 | fever | 0.626 |
| Time | 0.053 | asthenic fever | 0.016 | nocturia | 0.012 |
| ECOG | 0.004 | expectoration | 0.435 | night sweating | 0.014 |
| TNM | 0.000 | inappetency | 0.236 | weariness | 0.017 |
| chest pain | 0.675 | diarrhoea | 0.984 | blood sputum | 0.300 |
| cough | 0.169 | insomnia | 0.053 | breathe hard | 0.378 |

It can be seen that the symptoms of weariness, constipation, nocturia, night sweating and asthenic fever, and TNM stage and ECOG subtypes represent highly statistically significant effects of PFS time ($P < 0.05$). However, the sample data indicate that, more than 80%, these 5 symptoms are appeared combining with others. Thhus, the results are easy to be effected by the overlap influence of multi-avaliables.

## 3.2   FACUA Experimental

The correlation test is performed for the 10 related symptoms selected from the medical point of view, KMO=0.811 ($KMO > 0.5$) and refuses to Bartlett's hypothesis ($P < 0.001$). The results indicated that the FA method is appropriate for the selected symptoms set.

At first, extracting 3 CF with the eigenvalue $\lambda > 1$ according with the results of the principle component analysis which can explain 71.9% of the total variance with less lost, partial results are shown in Table 2.

The rotated component matrix showed in Table 3 illustrates the correlation between the observed symptoms and the CF. For example, the CF1 is characterized by very high loadings of the symptoms of night sweating, nocturia, constipation and asthenic fever. Thus, the CF1 is accordingly defined as the asthenia-syndrome factor.

**Table 2** Total Variance Explained

| cmpnt | Initial Eigenvalues | | | Extraction Sums of Squared Loadings | | |
|---|---|---|---|---|---|---|
| | Total | %of Variance | Cumulative% | Total | %of Variance | Cumulative% |
| 1 | 4.817 | 48.175 | 48.175 | 4.817 | 48.175 | 48.175 |
| 2 | 1.299 | 12.991 | 61.165 | 1.299 | 12.991 | 61.165 |
| 3 | 1.074 | 10.741 | 71.907 | 1.074 | 10.741 | 71.907 |
| 4 | 0.726 | 7.262 | 79.169 | | | |
| 5 | 0.576 | 5.757 | 84.925 | | | |

**Table 3** Rotated Component Matrix

| | Component | | |
|---|---|---|---|
| | 1 | 2 | 3 |
| cough | .049 | .857 | -.017 |
| chest tightness | .186 | .707 | .311 |
| chest pain | .337 | .396 | .618 |
| mouth parched | .015 | -.045 | .860 |
| weariness | .369 | .582 | .449 |
| inappetency | .399 | .634 | -.051 |
| night sweating | .756 | .346 | .207 |
| constipation | .517 | .150 | .628 |
| nocturia | .898 | .039 | .251 |
| asthenic fever | .889 | .322 | .070 |
| TCM signification | asthenia-syndrome | lung tumor-syndrome | sthenia-syndrome |

Similarly, the CF2 and CF3 are defined as the lung tumor-syndrome and sthenia-syndrome which is consistent with TCM conclusion.

The eigenvectors of symptoms for patients can be establihed through the factor scores and other independent symptoms. Symptom patterns can be discovered by the cluster analysis through the eigenvectors. As shown in Table 4, the test indicates that the criterion for the K-prototypes algorithm is satisfied and the symptom patterns of TCM are clear when the clustering group is 4. Using the second group as an example, the symptom pattern is a combination result thst includes expectoration, breathe hard, and lung tumor symptoms(CF2).

CE can be divided into four levels corresponding to these four symptom patterns and the UA can be done on each level. Fig. 2 shows the curve of PFS survival function. The results show that CE represented for accordingly pattern of the symptoms is the highly statistically significant predictor of

**Table 4** Final Cluster Centers

| | Cluster | | | |
|---|---|---|---|---|
| | 1 | 2 | 3 | 4 |
| expectoration | 1 | 1 | 1 | 1 |
| Blood sputum | 0 | 0 | 0 | 0 |
| breathe hard | 0 | 1 | 1 | 0 |
| fever | 0 | 0 | 0 | 0 |
| insomnia | 0 | 0 | 0 | 0 |
| diarrhoea | 0 | 0 | 0 | 0 |
| factor score 1 | .03793 | -.24058 | -.27493 | 4.35865 |
| factor score 2 | -.47556 | .18155 | -.03761 | 1.57654 |
| factor score 3 | -.44524 | -.07984 | .41887 | .34146 |

PFS ($P = 0.001, P < 0.05$). In the point of medicine, it can be interpreted that the less CF the patients have, the longer PFS time the patients get.

**Fig. 2** The PFS survival function

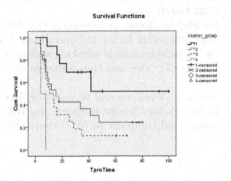

## 4 Discussion and Conclusions

Because of ignoring the correlation between the different TCM clinical symptoms, the results of conventional UA are not comprehensive and superimposed which will affect the overall results. This paper proposes the FACUA method to deal with the multiple types and strong correlation characteristics of TCM clinical symptoms. The results show that FACUA can extract CF represented comprehensive information and discover the symptom patterns from multiple symptom eigenvectors.

The CE represented according to different symptom patterns are highly statistically significant predictors of PFS. The FACUA has proven itself to be an effective method for processing clinical symptoms for TCM.

**Acknowledgements.** This paper is supported by 2010 Program for Excellent Talents in Beijing Municipal Organization Department (2010D005015000001), the New Centaury National Hundred, Thousand and Ten Thousand Talent Project, and got the cooperation with Beijing Hospital of Traditional Chinese Medicine Affiliated to CPUMS. Special thanks have been given there.

# References

1. Ap, L., Lukman, S., Yy, W.: To explore the scientific foundation of classification of zheng from the objective laws of symptoms. Journal of Traditional Chinese Medicine 46, 5–7 (2005)
2. Kaplan, E.L., Meier, P.: Nonparametric estimation from incomplete observations. J. Am. Stat. Assoc. 53, 457–481 (1958)
3. Lin Lz Fau Zhou, D., Zhou Dh Fau Zheng, X., Zheng, X.T.: Effect of traditional Chinese medicine in improving quality of life of patients with non-small cell lung cancer in late stage. Zhongguo Zhong Xi Yi Jie He Za Zhi 26, 389–393 (2006)
4. Movsas, B., Moughan, J., Sarna, L., Langer, C., Werner-Wasik, M., Nicolaou, N., Komaki, R., Machtay, M., Wasserman, T., Bruner, D.W.: Quality of life supersedes the classic prognosticators for long-term survival in locally advanced non-small-cell lung cancer: An Analysis of Rtog 9801. J. Clin. Oncol. 27, 5816–5822 (2009)
5. Ou Sh Fau Ziogas, A., Ziogas A Fau Zell, J.A., Zell, J.A.: ian ethnicity is a favorable prognostic factor for overall survival in non-small cell lung cancer (Nsclc) and is independent of smoking status. J. Thorac. Oncol. 4, 1083–1093 (2009)
6. Raftery, J.: Nice and the challenge of cancer drugs. Brit Med. J 67, 1468–5833 (2009)
7. Wang, Y.: Progress and prospect of objectivity study on four diagnostic methods in traditional Chinese medicine. In: IEEE International Conference on Bioinformatics and Biomedicine Workshops, BIBMW (March 2010)

# Applying Factor Analysis to Water Quality Assessment: A Study Case of Wenyu River

Chen Gao, Jianzhuo Yan, Suhua Yang, and Guohua Tan

**Abstract.** This paper takes the mean monitoring data of Wenyu river basin in Beijing during 2006-2010 as an example. Based on the characteristics of origin data, the water quality assessment (WQA) is analyzed by factor analysis, which focuses on five aspects: data standardization, applicability evaluation, principle factor extraction, principle factor interpretation and factor scores. Results show that, by objectively and reasonably using the factor analysis to evaluate water quality, the development tendency and variation law of regional water quality can be further understood, which can be a reference for contamination control planning of a region environmental system.

**Keywords:** Water quality assessment, Factor analysis, Wenyu river basin, Beijing.

## 1 Introduction

According to several water quality indicators, water quality assessment(WQA) aims at comprehensively evaluating water quality with establishing a mathematical model. The results can further understands of the development tendency and variation law of regional water quality, which can be a reference for contamination control planning of a region environmental system [6].

WQA with multiple factors mainly analyzes the interaction between the water quality assessment factors with the minimum human factors. In recent years, with the application and popularization of multivariate statistical methods and computers, factor analysis(FA) has been widely used in

Chen Gao, Jianzhuo Yan, Suhua Yang, and Guohua Tan
College of Electronic Information and Control Engineering,
Beijing University of Technology, Beijing, China
e-mail: gchenmail@126.com, yanjianzhuo@emails.bjut.edu.cn,
yangsuhua@emails.bjut.edu.cn, tgh@emails.bjut.edu.cn

S. Li (Eds.): Nonlinear Maths for Uncertainty and its Appli., AISC 100, pp. 541–547.
springerlink.com

WQA [5]. However, there are lots of problems about the application and selection of methods when water quality is evaluated by using FA, which lead to
deficient explanation and persuasion of an empirical analysis. Thus, properly
applying FA in WQA is very important.

## 2  Basic Principle and Mathematical Model

FA is a multivariate statistical analysis method. It reduces the complexity of
large-scale data set and eliminates redundant information. Based on Eigen-
analysis of the correlation or a covariance matrix, FA converts many indicators to a few irrelevant composite indicators and individuates the meaning of
each principle factor(PF) [7].

Assume that the number of cross-sections is m, and each cross-section is
described by a total of p strong correlation variables. Thus, a mathematical
model of FA is [8]:

$$\begin{cases} x_1 = a_{11}f_1 + a_{12}f_2 + \cdots + a_{1p}f_p + \varepsilon_1 \\ x_2 = a_{21}f_1 + a_{22}f_2 + \cdots + a_{2p}f_p + \varepsilon_2 \\ \vdots \\ x_m = a_{m1}f_1 + a_{m2}f_2 + \cdots + a_{mp}f_p + \varepsilon_m \end{cases} \tag{1}$$

where $X = (x_1, x_2, \cdots, x_m)^T$ is standardized monitoring data vector; $F = (f_1, f_2, \cdots, f_p)^T$ is PF vector; $a_{ij}$ is factor loading, which reflects correlation
between $x_i$ and $f_j$; $\varepsilon_i$ stands for other affection factors, which is not included
in ahead p factors and often be ignored in actual analysis.

## 3  Study Case

Based on the mean monitoring data of Wenyu river basin in Beijing during
2006-2010, this section objectively and reasonably applies factor analysis to
WQA.

### 3.1  Data Standardization

Based on analysis of original data, the correlation matrix is applied to FA,
therefore the original data must be converted into normalized format. Generally, data standardization keeps the correlation of variables and tends to
lessen the influence of large magnitude gap. Thus, the resulting correlation
matrix is more suitable for FA in WQA. The equation of standardization is
as follows [4]:

$$x'_{ij} = \frac{x_{ij} - \overline{x_j}}{S_j}, (i = 1, 2, \cdots, n; j = 1, 2, \cdots, k) \qquad (2)$$

where $\overline{x_j} = \frac{1}{n} \sum x_{ij}$ and $S_j = \sqrt{\frac{\sum (x_{ij} - \overline{x_j})^2}{n-1}}$.

## 3.2 Applicability Evaluation

The correlation coefficient matrix is shown in Table 1. It can be seen that most values are more than 0.3, which means that there is strong correlation between indicators [2]. Thus, utilizing PFs to evaluate Wenyu river water quality is objective and reasonable.

**Table 1** Correlation Coefficient Matrix

| Corre-lation | Turbi-dity | DO | $NO_2^-$ | $NO_3^-$ | PH | TN | $NH_3$-N | TP | $COD_{Mn}$ | $BOD_5$ |
|---|---|---|---|---|---|---|---|---|---|---|
| Turbidity | 1.000 | -0.490 | -0.529 | -0.258 | -0.138 | 0.670 | 0.675 | 0.622 | 0.570 | 0.494 |
| DO | -0.490 | 1.000 | 0.715 | 0.360 | 0.535 | -0.638 | -0.622 | -0.506 | -0.599 | -0.786 |
| $NO_2^-$ | -0.529 | 0.715 | 1.000 | 0.454 | 0.373 | -0.698 | -0.704 | -0.505 | -0.572 | -0.697 |
| $NO_3^-$ | -0.258 | 0.360 | 0.454 | 1.000 | 0.057 | -0.324 | -0.329 | -0.186 | -0.369 | -0.379 |
| PH | -0.138 | 0.535 | 0.373 | 0.057 | 1.000 | -0.403 | -0.368 | -0.339 | -0.413 | -0.625 |
| TN | 0.670 | -0.638 | -0.698 | -0.324 | -0.403 | 1.000 | 0.995 | 0.917 | 0.942 | 0.891 |
| $NH_3$-N | 0.675 | -0.622 | -0.704 | -0.329 | -0.368 | 0.995 | 1.000 | 0.930 | 0.945 | 0.869 |
| TP | 0.622 | -0.506 | -0.505 | -0.186 | -0.339 | 0.917 | 0.930 | 1.000 | 0.927 | 0.762 |
| $COD_{Mn}$ | 0.570 | -0.599 | -0.572 | -0.369 | -0.413 | 0.942 | 0.945 | 0.927 | 1.000 | 0.831 |
| $BOD_5$ | 0.494 | -0.786 | -0.697 | -0.379 | -0.625 | 0.891 | 0.869 | 0.762 | 0.831 | 1.000 |

## 3.3 Principle Factor Extraction

The main purpose of FA is to use a few irrelevant composite indicators to interpret FA model. Too many PFs can not achieve the purpose, however, too little PFs mean more information lost from origin dataset [1]. Based on correlation coefficient matrix listed in Table 1, each PF's eigenvalue and cumulative proportion in anova can be calculated in SPSS software. In the paper, factors are extracted as PFs when the factor's eigenvalue more than 1 and cumulative proportion in anova beyond 85%.

Total variance explained result is shown in Table 2. Three PFs are extracted. Nearly 85.911% information of primitive variables can be explained by the only three PFs, which can conveniently and efficiently interpret the contamination type and contamination level among 19 cross-sections of Wenyu river basin.

**Table 2** Total Variance Explained

| Factor | Initial Eigenvalues Extraction | | | Sums of Squared Loadings | | | Rotation Sums of Squared Loadings | | |
|---|---|---|---|---|---|---|---|---|---|
| | Total | %of Variance | Cumulative% | Total | %of Variance | Cumulative% | Total | %of Variance | Cumulative% |
| 1 | 6.493 | 64.934 | 64.934 | 6.493 | 64.934 | 64.934 | 4.681 | 46.808 | 46.808 |
| 2 | 1.075 | 10.753 | 75.686 | 1.075 | 10.753 | 75.686 | 2.209 | 22.086 | 68.893 |
| 3 | 1.022 | 10.225 | 85.911 | 1.022 | 10.225 | 85.911 | 1.702 | 17.018 | 85.911 |
| 4 | 0.598 | 5.983 | 91.894 | | | | | | |
| 5 | 0.365 | 3.654 | 95.548 | | | | | | |
| 6 | 0.268 | 2.680 | 98.228 | | | | | | |
| 7 | 0.110 | 1.096 | 99.624 | | | | | | |
| 8 | 0.050 | 0.500 | 99.824 | | | | | | |
| 9 | 0.015 | 0.148 | 99.972 | | | | | | |
| 10 | 0.003 | 0.028 | 100.00 | | | | | | |

## 3.4 Principle Factor Interpretation

A FA model is established not only for extracting PFs but also for the actual meaning of each PF, which can be really helpful for a real situation analysis. When factor loading values are relatively average, reasons of factor scores' difference about assessment objects in each factor can not be obtained from original indicators. Thus, factor rotation is necessary. Factor rotation mainly includes oblique rotation and orthogonal rotation. Oblique rotation depends on parameters defined by users and factors still have a little correlation. The factors of orthogonal rotation are irrelevant without information overlaps, which is exactly needed in this study case [4].

As a popular method of orthogonal rotation, a varimax is applied to define the three extracted factors listed in Table 2. The component matrix is shown in Table 3. The rotated component matrix is shown in Table 4. Compared with Table 3, factor loading values of $NO_3^-$, PH and DO listed in Table 4 apparently approach to extremes by rotating without information loss. The loading absolute values of variables are greater than 0.6 because the loading absolute value is an indicator of the participation of the variable in each PF. Thus the implications of factors are distinct and each PF can be defined.

PF1 accounts for 64.964% of the total variance, and it is characterized by very high loadings of Turbidity, TN, $NH_3$-N, TP, $COD_{Mn}$ and $BOD_5$, so PF1 is defined as Organic Pollution Factor(OPF). PF2 accounts for 10.753% of the total variance, and it is characterized by very high loadings of PH and DO. Therefore, PF2 is defined as Eutrophication Pollution Factor(EPF).

**Table 3** Component Matrix

| Indicator | Factor | | |
|---|---|---|---|
| | 1 | 2 | 3 |
| Turbidity | 0.693 | 0.355 | -0.153 |
| DO | -0.777 | 0.408 | 0.085 |
| $NO_2^-$ | -0.779 | 0.224 | 0.292 |
| $NO_3^-$ | -0.424 | 0.171 | 0.798 |
| PH | -0.515 | 0.656 | -0.435 |
| TN | 0.967 | -0.182 | -0.080 |
| $NH_3$-N | 0.963 | -0.218 | -0.060 |
| TP | 0.878 | -0.339 | -0.228 |
| $COD_{Mn}$ | 0.925 | -0.182 | -0.098 |
| $BOD_5$ | 0.931 | -0.209 | -0.094 |

**Table 4** Rotated Component Matrix

| Indicator | Factor | | |
|---|---|---|---|
| | 1 | 2 | 3 |
| Turbidity | 0.737 | 0.025 | -0.291 |
| DO | -0.394 | 0.646 | 0.453 |
| $NO_2^-$ | -0.464 | 0.418 | 0.594 |
| $NO_3^-$ | -0.122 | 0.009 | 0.912 |
| PH | -0.127 | 0.932 | -0.044 |
| TN | 0.903 | -0.332 | -0.222 |
| $NH_3$-N | 0.917 | -0.294 | -0.229 |
| TP | 0.941 | -0.226 | -0.017 |
| $COD_{Mn}$ | 0.872 | -0.320 | -0.191 |
| $BOD_5$ | 0.659 | -0.631 | -0.295 |

PF3 accounts for 10.225% of the total variance, and it is characterized by $NO_3^-$ and $NO_2^-$. So PF3 is defined as Nitrate Pollution Factor(NPF).

## 3.5 Factor Scores

Factor scores can be calculated in SPSS by a regression method and the results are shown in Table 5. The comprehensive evaluation function is [9]:
$$F = (64.934/85.911) * PF1 + (10.753/85.911) * PF2 + (10.225/85.911) * PF3.$$

As shown in Table 5, section NO.14 possesses an extremely high level of PF1 and F, which indicates this section has the worst pollution situation of water quality among 19 cross-sections and is mainly affected by organic pollution factor. Section NO.13 possesses a high level of PF2, which indicates that eutrophication is the main pollution factor, and section NO.15 possesses the highest content of nitrate. According to Environmental Quality Standards for Surface Water [3], the water quality situation is tally with the actual water quality.

**Table 5** Factor Scores

| NO. | Section | OPF | EPF | NPF | F |
|---|---|---|---|---|---|
| 1 | Wenquan | 0.162483433 | -1.017321597 | -0.363872021 | -0.047830328 |
| 2 | Qianshajian | -0.521036617 | -2.089409411 | -1.127792864 | -0.789562374 |
| 3 | Daoxianghu | -0.613710586 | -0.397753953 | 0.245329048 | -0.484446019 |
| 4 | Shangzhuang | -0.669855934 | 0.331661442 | -0.641677277 | -0.541155614 |
| 5 | Shaheqiao | 0.189568346 | 0.687050879 | -0.568459412 | 0.161618321 |
| 6 | Nanshahe | -0.587717586 | -1.567742935 | -0.790908445 | -0.734572201 |
| 7 | Beishahe | 1.060608301 | 0.019090927 | 0.698329615 | 0.887141862 |
| 8 | Chaozhong | -0.463883625 | -0.802269832 | -1.055451513 | -0.576650470 |
| 9 | Longtan | -0.864281169 | 0.853902816 | -0.153707709 | -0.564664336 |
| 10 | Yangtaizidong | -0.544041961 | 0.705033686 | 1.025133914 | -0.200947483 |
| 11 | Yangtaizixi | -0.732899027 | 0.786212003 | 0.230557159 | -0.428099787 |
| 12 | Yangtaizi | -0.618138163 | 1.027967046 | 0.530118742 | -0.275447727 |
| 13 | Nanzhuang | 0.531468891 | 2.053247353 | -1.448513707 | 0.486292990 |
| 14 | Xiaocunqiao | 3.540005135 | -0.392162463 | 0.106218841 | 2.639194726 |
| 15 | Shagouqiaoxi | -0.168949057 | -0.355103222 | 2.345284569 | 0.106989462 |
| 16 | Shagouqiaodon | -0.407269783 | -0.062797437 | 1.389940242 | -0.150257569 |
| 17 | Qintunhe | 0.160176810 | 0.134984289 | -0.996825747 | 0.019320736 |
| 18 | Tugoucunqiao | 0.341260805 | 0.859640080 | -0.567194752 | 0.298024381 |
| 19 | Xinbaozha | 0.206211789 | -0.774229672 | 1.143491317 | 0.195051430 |

## 4 Conclusions

Combined with characteristics of Wenyu river water quality monitoring data, factor analysis is applied to water quality assessment of Wenyu river basin by focusing on five aspects: data standardization, applicability evaluation, principle factor extraction, principle factor interpretation and factor scores.

Standardization equation is utilized to normalize the origin monitoring data to ensure the comparability among data. Based on the correlation coefficient matrix, the correctness and effectiveness of water quality assessment with factor analysis is guaranteed. To ensure that each extracted PF can completely interpret at least one indicator's variance and all PFs can interpret at least 85% of total information, factor is extracted as PF on condition that its eigenvalue over 1 and cumulative proportion in anova beyond 85%.

Three comprehensive factors are rotated by varimax with Kaiser Normalization, and can be defined as organic pollution factor, eutrophication pollution factor and nitrate pollution factor, which have actual evaluation significance. Based on factor scores and comprehensive factor scores of each cross-section, the pollution situation and level of water quality about Wenyu river basin is analyzed.

With tentative exploration for research and innovation of water quality assessment, this paper provides a scientific, reasonable and practical guiding for reference.

# References

1. Bramha, S., Panda, U.C., Rath, P., Sahu, K.C.: Application of factor analysis in geochemical speciation of heavy metals in the sediments of a lake system-chilika (India): a case study. Journal of Coastal Research 26(5), 860–868 (2010)
2. Farnham, I.M., Hodge, V.F., Johannesson, K.H., Singh, A.K., Stetzenbach, K.J.: Factor analytical approaches for evaluating groundwater trace element chemistry data. Analytica Chimica Acta 490, 123–138 (2003)
3. GB3838-2002: Environmental Quality Standards for Surface Water (2002)
4. Li, Z.H., Luo, P., Tong, J., Fen, X., Jiao, C.: SPSS for Windows. Publication House of Electronics Industry, Beijing (2005)
5. Liao, S.W., Sheu, J.Y., Chen, J.J., Lee, C.G.: Water quality assessment and apportionment source of pollution from neighbouring rivers in the Tapeng Lagoon (Taiwan) using multivariate analysis: A case study. Water Science and Technology 54(11-12), 47–55 (2006)
6. Liu, L., Chen, L., Gao, P., Chen, G.: Study on Water Quality Assessment of Urban River. In: Proceedings 2011 International Conference on Computer Distributed Control and Intelligent Environmental Monitoring (CDCIEM 2011), pp. 2244–2247 (2011)
7. Shrestha, S., Kazama, F., Nakamura, T.: Use of principal component analysis, factor analysis and discriminant analysis to evaluate spatial and temporal variations in water quality of the Mekong River. Journal of Hydroinformatics 10(1), 43–56 (2008)
8. Wang, L.B.: Multivariate statistical analysis. Ecnomic Science Press, Beijing (2010)
9. Wang, S., Yu, J., Zhang, S.X.: Analysis of Water Quality Evolution and its Influence Factor in Qingdao Inshore Area. Advances in Management of Technology PT 2, 310–314 (2008)

# On the Profit Manipulations of Chinese Listed Companies

Shuangjie Li and Xingxing Chen

**Abstract.** This paper proposes a recognizing model of profit manipulations based on BP neural network model. Then it improves the model by adding DEA efficiency index according to the empirical result of Chinese listed company's data in 2005-2009. The Error type II of the model is reduced rapidly and the discriminate rate of the model is successfully improved to 90%.

**Keywords:** Listed company, Profit manipulation, BP neural network model, DEA efficiency.

## 1 Introduction

Chinese security market made a rapid development in 1990s. In 2008, 4701 enterprises were punished by the ministry of finance departments, and the punished enterprises were required to adjust their financial records, repay taxes or pay fines. By regulating behaviors of listed companies, China's national treasury gathered taxes of 1.03 billion Yuan, and captured fines of 55.96 million Yuan[1]. Therefore,profit manipulations in China are serious.

Earnings management has been studied by western countries since 1960s [7]. Healy and Wahlen (1999) studied impacts of resource distribution made

Shuangjie Li
School of Economics and Management, Beijing University of Technology,
Beijing, China
e-mail: lishuangjie@bjut.edu.cn

Xingxing Chen
Institute of Quantitative & Technical Economics, Chinese Academy of
Social Sciences, Beijing, China
e-mail: xingstar56@126.com

[1] Data comes from "government information" in Treasury website at
www.mof.gov.cn/zhengwuxinxi/bulinggonggao/tongzhitonggao/201002/
t20100203_267994.html

S. Li (Eds.): Nonlinear Maths for Uncertainty and its Appli., AISC 100, pp. 549–556.
springerlink.com                                    © Springer-Verlag Berlin Heidelberg 2011

by earning management [3], pioneering a new chapter of researches in earning management. In the early of the 21st century, Syed did some researches on improved Jones model [1]. Since then, studies of earning management tended to quantitative and modeling. Researches of profit manipulations in China begun by Li Xin (2003) [5], and were about 40 years later than western country. Li Yanxi (2006) set up a recognizing model of profit manipulation by using statistical methods [6]. This thesis builds a model for recognizing profit manipulations of listed companies in China, and successfully combines knowledge of artificial intelligence and profit manipulation for the first time.

## 2  Theoretical Bases

### 2.1  Definition of Profit Manipulation

Definitions of earnings management can be classified as white, gray and black. White earnings management is beneficial, the black is pernicious, and the gray is manipulation of reports within the boundaries of compliance with bright-line standards [2]. The concept of profit manipulation is related to the definition of earnings management and the understanding of the word "manipulation". When "manipulation" is derogatory, earnings management is considered to be neutral, and contains the concept of profit manipulation. While regarding "manipulation" as neutral, profit manipulation contains the concept of earnings management [8].

This paper identifies the concept of manipulation as neutral and considers profit manipulation to be a behavior of an enterprise which controls its financial indicators and to dominate profit of the company. The study analyzes conditions of profit manipulations of Chinese listed companies. By using the special concept of profit manipulation according to domestic researches, this article successfully collects data and builds a recognizing model.

### 2.2  BP Neural Network

BP neural network system is a neural network that its error propagates back and forward with multilayer, which includes an input layer, an output layer and one or more hidden layers. It is said that BP neural network theory offers a good prediction on the area of financial warning for enterprises. Although there aren't any researches on profit manipulation by using BP neural network, the method which is used in identifying profit manipulations of enterprises can be theoretically demonstrated as feasible. In order to determine the inputs, outputs and the cell numbers of hidden layers, BP neural network processes the training sample data, finds out the data decision-making threshold, and controls the error of the model. Finally, BP neural network model is created.

## 2.3 DEA Efficiency

In 1978, Charnes, A., Cooper, W.W., Rhodes, E. proposed a method called Data Envelopment Analysis (DEA) to evaluate relative efficiency among each Decision Making Unit (DMU) for the first time. DEA is a kind of nonparametric analytical methods that evaluates efficiencies. It measures relative efficiency of a group of inputs and outputs by using the method of linear programming [4]. According to whether the returns to scale changed, DEA model can be divided into Constant Returns to Scale (CRS) model and Variable Returns to Scale (VRS) model.

# 3 BP Neural Network Model

## 3.1 Data Gathering and Preprocessing

This paper uses some financial indicators in literatures for reference[2], and the formulas of indicators are shown in Table 1.

The symbol $\Delta$ represents the value change of a variable form year $t$-1 to year $t$, and year $t$ is the year when listed companies are punished by CSRC because of their profit manipulations.

Data of listed corporations of profit manipulation from 2005 to 2009 are collected from the illegal database, a sub-database in China Center for Economic Research (CCER). Companies that have missing values are deleted, and finally 19 corporations are obtained, which include 3 companies in 2005, 5 companies in 2006, 7 companies in 2007, and 4 companies in 2008. In addition, another group of companies with profit manipulation is set up as a control group, so the number of corporations is 38. The companies are divided into 28 training samples and 10 testing samples. As samples of $t$-1, $t$-2 and $t$-3 periods shall be gathered to study the financial trends for each company, so the number of training sample and test sample is respectively 84 (28×3) and 30 (10×3), and the total number of samples is 114. The article uses paired samples T-test to test significant differences between the two groups, and the data is divided into $t$-1, $t$-2 and $t$-3 periods. Then T value and P value of 26 indicators in the three periods are gotten. Selecting 0.1 as significance, the chosen financial indicators are $x3$, $x4$, $x6$, $x7$, $x8$, $x17$, $x18$, $x19$, $x22$, and $x25$. Correlation analysis in the three periods is also used, and the results shows the 10 indicators are uncorrelated at the 0.05 significant level.

---

[2] Yao Hong, Li Yanxi, Gao Rui: A Recognition Model of Aggressive Earnings Management in Chinese Listed Companies Based on the Principal Components Method. Journal of Management Science. Oct. vol. 20(5), 83-91 (2007).

**Table 1** Formula for indexes

| Index | Formula | Index | Formula |
|-------|---------|-------|---------|
| $x1$ | Net cash flow from operating activities/Gross profit | $x14$ | Accounts payable/$\Delta$Revenue |
| $x2$ | Net cash flow from operating activities/Net profit | $x15$ | Accounts payable/Total assets |
| $x3$ | Net cash flow from investing activities/Net profit | $x16$ | Cost of sales/Revenue |
| $x4$ | Net cash flow from operating activities/Revenue | $x17$ | Finance expense/Revenue |
| $x5$ | Net cash flow from operating activities per share/Earnings per share[a] | $x18$ | Ending balance of cash/(Account receivable +Notes receivable) |
| $x6$ | Outflows of operating cash/Total cost | $x19$ | Ending balance of cash and cash equivalents/Total assets |
| $x7$ | $\Delta$Account receivable/$\Delta$Revenue | $x20$ | Impairment for fixed assets/Total assets |
| $x8$ | Bad debt provision/Account receivable | $x21$ | Tax payable/Revenue |
| $x9$ | $\Delta$Commodity stocks/$\Delta$Revenue | $x22$ | Gross profit margin/Gross margin in the same industry |
| $x10$ | $\Delta$Commodity stocks/$\Delta$Cost of sales | $x23$ | Gross profit margin this year/Average gross margin[b] |
| $x11$ | $\Delta$Commodity stocks/$\Delta$Accounts payable | $x24$ | (Profit before tax-Nonoperating expense/Profit before tax |
| $x12$ | Provision for inventory/Inventory | $x25$ | Cash gross margin/Cash gross margin in the same industry |
| $x13$ | Impairment for fixed assets/Depreciation of fixed assets | $x26$ | Cash gross margin/Sales margin |

[a] Fully diluted method; [b] Over the past five years.

## 3.2 *Normalizes Input Variables*

Input variables of the model for recognizing profit manipulations of listed corporations shall be normalized before setting up the model, which means the data should be transformed into [0.001, 0.999]. However, when synthetically analyzing multiple indicators, some indicators are treated as positive indicators, others indicators are reverse indicators, and still others are moderate indicators[3]. Based on the recognizing model, positive indicators mean that identification of profit manipulations will be easier when the value of indicators are large; Reverse indicators mean the identification will be easier when the value of indicators is small; Moderate indicators are defined as behaviors of profit manipulations will be easily recognized when the value of indicators is too large or too small. Therefore, this paper will ultimately divides 10 financial indicators into three categories, in which $x6$, $x18$, $x22$ are

---

[3] The specific definition will be look up at Ye Zongyu: Methods of Dealing indicators with Positive and Dimensionless Methods through Synthetical Evaluation of Multi-indicators. Zhejiang Statistics. 4, 24-25 (2003).

positive indicators, $x3$, $x4$, $x8$, $x17$, $x19$, $x25$ are reverse indicators, and $x7$ is moderate indicator. For different types of indicators, different formulas are used to normalize its value. The specific methods are as follows.

1. For positive indicators and moderate indicator, Eq.1 is used. As for reverse indicators, Eq.2 is used.

$$y_i = 0.001 + \frac{x_i - min\, x_i}{max\, x_i - min\, x_i} \times 0.998 \qquad i = 1, 2, \ldots, n \qquad (1)$$

$$y_i = 0.001 + \frac{max\, x_i - x_i}{max\, x_i - min\, x_i} \times 0.998 \qquad i = 1, 2, \ldots, n \qquad (2)$$

2. 114 (57 pairs) samples of $t$-1, $t$-2, $t$-3 periods are divided into 84 (42 pairs) training samples and 30 (15 pairs) test samples at random.

## 3.3 Construction of BP Neural Network Model

MATLAB program is compiled by using the reduced samples and BP neural network model is set up. The program is run 30566 times and the error of the model reaches 0.02, consuming 5 minutes 28 seconds. The gradient of the model is 0.00873 when the performance goal is met.

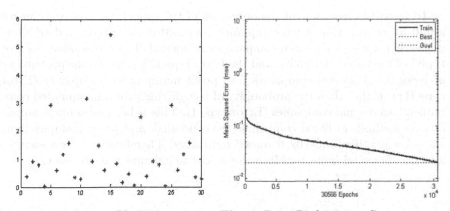

**Fig. 1** Error Scatter Plot        **Fig. 2** Error Performance Curve

Figure.1 shows that scatter of error is almost around the original point, so there is only a little abnormal value. Also, the model is convergent according to the error performance curve of Fig.2. The error of the model is smaller than the expected error of 0.02. Therefore, the model is reasonable.

## 3.4  Discriminate Rate of the Model

Through three-layer BP neural network, samples are trained, and weights and thresholds of each layer are gained. Final output of the model is gotten by MATLAB, and it is close to the expected values. So if financial indicators are acquired, whether listed corporations have behaviors of profit manipulations or not can be predicted by the three-layer BP neural network model. Also, there is a conclusion that it is feasible to apply BP neural network model to researches of profit manipulation. The result shows that 22 companies can be correctly recognized by using the three-layer BP neural network model, and the discriminate rate of the model reaches 73.33%.

## 3.5  Results of the Model

Results of 30 test samples reflect that in the 15 manipulated companies, 13 companies can be successfully recognized as manipulated, while 2 companies are treated as normal. And the discriminate rate of the 15 manipulated companies is 86.67%. At the same time, in 15 normal companies, there are 9 companies recognized as normal and 6 as manipulated. Thus the discriminate rate of the 15 normal companies is only 60%, and the average discriminate rate of the model is 73.33%. So although the predicted result of the model is good, there is still room for improvement.

To have a better view of the outcomes of the model, this research makes a null hypothesis that listed companies are profit manipulated, and an alternative hypothesis that listed companies are normal. The probability of Error type I of the model is 13.33%, and the Error type II is 40%. So the probability of recognizing normal companies into profit manipulated companies (Error type II) is higher than the probability of recognizing profit manipulated companies into normal companies (Error type I). This is because profit manipulations methods of listed companies are concealed, and profit manipulations can't be identified only by financial indicators. Therefore, it is necessary to add non-financial indicators into the model to enhance the efficiency.

## 4  Improved Model Based on DEA Efficiency

A DEA efficiency indicator is added into the model to set up an improved model. The thesis uses total employees $x\_1$ (People), total assets $x\_2$ (100 million Yuan), main business cost $x\_3$ (100 million Yuan) as input indicators, and net profit $y\_1$ (10 million Yuan), main business income $y\_2$ (100 million Yuan) as output indicators to establish a DEA efficiency model.

## 4.1 Improved Model

Constant Scale Efficiency (CRS) model is used, and efficiencies of paired samples for each period are calculated by Efficiency Measurement System (EMS). The indicator calculated by the DEA efficiency model is marked as $x27$. Then paired sample T-test is made by SPSS. From the correlation analysis of each period, variable $x17$, $x19$ and $x25$ that are significantly correlated are deleted, and the rest of the variables are uncorrelated under the significant level of 0.001. So the input variables of the improved model are $x3$, $x4$, $x6$, $x7$, $x8$, $x18$, $x22$, and $x27$.

An improved model is created based on the reduced samples and the DEA efficiency indicators by compiling MATLAB program. After operating 44994 times, the improved model reaches the expected error of 0.03. As the model is convergent according to the error performance curve, the model is reasonable.

## 4.2 Results of the Improved Model

Comparing with the outcome of the model before, the improved model has a better prediction on profit manipulations. 30 test samples as mentioned before are adopted again. By using the improved model, 13 companies can be successfully recognized as manipulated in the 15 manipulated companies, while 2 companies are marked as normal. So the discriminate rate of the manipulated companies is still 86.67%. But in 15 normal companies, 14 companies are recognized as normal and only one company as manipulated. Thus the discriminate rate of the 15 normal companies is 93.33%, and the average discriminate rate of the model is successfully improved to 90%. Therefore, although the probability of Error type I of the improved model is still 13.33%, the probability of recognizing normal companies into profit manipulated companies (Error type II) decreases from 40% to 6.67%. This is because efficiencies of listed corporations with profit manipulations are lower than that of normal corporations. In a word, it is rigorous to distinguish whether there are profit manipulations in companies only by using financial indicators.

In order to study whether behaviors of profit manipulations in listed companies can be effectively recognized by DEA efficiency indicator, this paper does some researches on changes of efficiencies in listed corporations. Hypothesis that "Efficiencies of profit manipulated companies decline and normal companies increase", and it is found that efficiencies of profit manipulated corporations will decline with a probability of 63.16% and efficiencies of normal companies increase with a probability of 94.74%. On the one hand, normal companies can be successfully recognized by the improved model after entering the DEA efficiency indicator, and this can decrease the probability of recognizing normal companies into profit manipulated companies (Error type II). On the other hand, the operating situations of some listed companies of profit manipulations don't deteriorate, which means motivations of these companies are misappropriating, raising equity, smoothing profits and so on. Therefore, motivations of profit manipulations in the improved model include not only financial crises of companies, but also contain a broader way.

# 5  Conclusions

Profit manipulations of listed companies damage the interests of stakeholders and disturb the order of the security market. Based on financial indicators in previous researches, the article studies indicators of profit manipulations and creates a recognizing model. The outcome of the model is influenced by error type II, which reduces the accuracy of the model. So the paper creatively introduces DEA efficiency index, and successfully reduces error of the second kind. In addition, the paper points out that motivations of profit manipulations are multiple and complicated, and behaviors of profit manipulations may not occur only in listed companies, whose operating condition deteriorate. To improve the accuracy of the model, non-financial indicators will continue to be explored in the future.

# References

1. Ali shah, S.Z., Ali Butt, S., Hasan, A.: Corporate governance and earnings management an empirical evidence form Pakistani listed companies. European Journal of Scientific Research, 624–638 (2009)
2. Demski, J.S.: Earnings Management: Emerging Insights in Theory, Practice, and Research. New York University, New York (2008)
3. Healy, P.M., Wahlen, J.M.: A review of the earnings management literature and its implications for standard setting. Boston, Accounting Horizons 365-383 allocation based on DEA model. Industrial Technology & Economy, 365–383 (1999)
4. Li, S., Wang, H., Liu, R.: Efficienct analysis of manufacturing technological innovation and resources
5. Li, X.: Evaluation and application on models of profit manipulation. Finance and Accounting Monthly 5, 6–8 (2003)
6. Li, Y., Yao, H., Gao, R.: Study on the recognizing model of aggressive earnings management on the stock market. Management Review 1, 3–9 (2006)
7. Lin, C.: Testing of the market reaction to earnings manipulation. Ph.D. Thesis, Xiamen University, Xiamen (2002)
8. Yan, W.: The relations between profit operating and surplus management. Shanxi Coking Coal Science & Technology 1, 40–41 (2006)

# Fuzzy Portfolio Optimization Model with Fuzzy Numbers

Chunquan Li and Jianhua Jin

**Abstract.** A new portfolio optimization model with triangular fuzzy numbers is proposed in this paper. The objective function is considered as maximizing fuzzy expected return of securities under the constraint that the risk will not be greater than a preset tolerable fuzzy number, where the expected return and risk of securities are described as triangular fuzzy numbers. By using the method of dominance possibility criterion, the portfolio optimization model is converted into its equivalent crisp linear programming problem. Finally an example is presented to illustrate the effectiveness of the proposed algorithm.

**Keywords:** Fuzzy portfolio optimization model, Fuzzy number, Fuzzy expected rate of return, Risk, Securities markets.

## 1 Introduction

Portfolio selection has been one of the important research topics in financial area since *Markowitz* [14] proposed the famous mean-variance theory in 1952. *Markowitz* initially investigated a mathematical way for analyzing portfolio selection problem. Thereafter, a variety of simplified mean-variance models have been constantly developed such as references [1, 8, 12]. In practice processing, the majority of portfolio selection models require a perfect

Chunquan Li
School of Sciences, Southwest Petroleum University, Chengdu, 610500, China
e-mail: spring19810418@yahoo.com.cn

Jianhua Jin
School of Sciences, Southwest Petroleum University, Chengdu, 610500, China
College of Mathematics and Econometrics, Hunan University, Changsha, 410082, China
e-mail: jjh2006ok@yahoo.com.cn

S. Li (Eds.): Nonlinear Maths for Uncertainty and its Appli., AISC 100, pp. 557–565.
springerlink.com                                    © Springer-Verlag Berlin Heidelberg 2011

knowledge of data. However, the data are often prone to errors since it may be difficult to obtain statistically meaningful estimates from available historical data, especially for the rate of security returns. Portfolio optimization based on inaccurate point estimates may be highly misleading. Black [5] said that the portfolio selection system cannot be tolerant of a little change appearing in the rate of return. The security market is sensitive to various economic and political factors, which can be affected by not only probabilistic factors but also complex uncertainty or vagueness factors.

After the inception of fuzzy set theory by Zadeh [17], many fuzzy portfolio selection models were constructed by scholars [3, 4, 7, 16], etc. Sometimes the security return or the rate of risk cannot be characterized approximately by fuzzy variables, the portfolio selection problem is solved in more complex uncertainty situations. When the security returns are considered as fuzziness with random parameters, Ammar [2] employed fuzzy random programming method to select the optimal portfolio. Regarding the security returns to be random but the expected returns of the securities as fuzziness, Huang [9] proposed random fuzzy mean-variance models. Recently, Huang [10] discussed fuzzy portfolio selection problem in the situation that each security return belongs to a certain class of fuzzy variables, and proposed two credibility-based minimax mean-variance models. In this paper, we will discuss portfolio selection optimization problem in the situation that the fuzzy expected return and risks of securities are predicted as the triangular fuzzy numbers, and propose a new fuzzy portfolio optimization model by maximizing the expected rate of return on portfolios under the constraint that the risks will not be greater than a preset tolerable fuzzy number.

The remainder part of the paper is organized in the following way. Some necessary knowledge about fuzzy set are presented in Section 2. In Section 3, a new portfolio selection model is proposed. The objective function is considered as maximizing fuzzy expected return of securities under the constraint that the risks will not be greater than a preset tolerable fuzzy number, where the expected rates of return and risk are described as fuzzy numbers. The resolution of the portfolio selection model is presented in Section 4. An numerical example is given to illustrate the effectiveness of the proposed algorithm in Section 5. The conclusions are presented in Section 6.

## 2  Preliminaries

In this section we review some definitions and results already known, and give some notations used throughout the paper. Let $R$ denote the set of real numbers and $F(X) = \{f : X \to [0,1] | f$ is a mapping from $X$ to the unit interval $[0,1]\}$ denote the family of all fuzzy subsets of the set $X$.

**Definition 1.** [13] Let $A \in F(R)$. Then the fuzzy subset $A$ is called a fuzzy number with the following conditions satisfied:

(i) $A$ is normal, that is, there is an element $x_0 \in R$ such that $A(x_0) = 1$;

(ii) the $\alpha$−cut set of $A$, denoted by $A_\alpha = \{x \in R | A(x) \geq \alpha\}$, is a closed interval in $R$, for any $\alpha \in (0, 1]$.

**Definition 2.** A fuzzy number $\tilde{A}$ is called a triangular fuzzy number if it has a triangular membership function which is given by

$$\tilde{A}(x) = \begin{cases} 0, & \text{if } x < \mathcal{B}a \text{ or } x > \bar{a} \\ \frac{x - \mathcal{B}a}{a_0 - \mathcal{B}a}, & \text{if } \mathcal{B}a \leq x \leq a_0 \\ \frac{\bar{a} - x}{\bar{a} - a_0}, & \text{if } a_0 \leq x \leq \bar{a} \end{cases}$$

We denote it by $\tilde{A} = [\mathcal{B}a, a_0, \bar{a}]$. For the triangular fuzzy numbers $\tilde{A} = [\mathcal{B}a, a_0, \bar{a}]$ and $\tilde{B} = [\mathcal{B}b, b_0, \bar{b}]$, the following assertions hold clearly: $\tilde{A} + \tilde{B} = [\mathcal{B}a + \mathcal{B}b, a_0 + b_0, \bar{a} + \bar{b}]$, $\tilde{A} - \tilde{B} = [\mathcal{B}a - \bar{b}, a_0 - b_0, \bar{a} - \mathcal{B}b]$, $k[\mathcal{B}a, a_0, \bar{a}] = [k\mathcal{B}a, ka_0, k\bar{a}]$, and $(\tilde{A} + \tilde{B})_\lambda = \tilde{A}_\lambda + \tilde{B}_\lambda, \forall k \geq 0, \forall \lambda \in [0, 1]$.

The comparison between triangular fuzzy numbers can be carried out using the dominance possibility criterion [15]. The above criterion is frequently applied in fuzzy programming and stochastic fuzzy programming [11]. Given two triangular fuzzy numbers $\tilde{A} = [\mathcal{B}a, a_0, \bar{a}]$ and $\tilde{B} = [\mathcal{B}b, b_0, \bar{b}]$, assume that a given relation is $\tilde{A} \geq \tilde{B}$, then the binary relation can be presented as follows through the dominance possibility criterion(denoted by $DPC$):

$$Poss(\tilde{A} \geq \tilde{B}) = \begin{cases} 1, & \text{if } a_0 \geq b_0 \\ \frac{\bar{a} - \mathcal{B}b}{\bar{a} - a_0 + b_0 - \mathcal{B}b}, & \text{if } b_0 \geq a_0, \bar{a} \geq \mathcal{B}b \\ 0, & \text{if } \mathcal{B}b \geq \bar{a} \end{cases}$$

It means the possibility that the maximum value of $\tilde{A}$ exceeds the maximum value of $\tilde{B}$.

# 3 Fuzzy Portfolio Optimization Model

Assume that an investor wants to allocate one's wealth among n risky assets and a non-risky assets. $r_{tj}$ represents the historical rate of return of risky asset $j$ at period $t$, $t = 1, 2, \cdots, T$, $j = 1, 2, \cdots, n$. Traditionally, researchers consider the arithmetic mean $r_j = \frac{1}{T} \sum_{t=1}^{T} r_{tj}$ as the expected return of risky asset $j$. However, this technique would have great defect since there are many uncertain factors in securities markets. Furthermore, the later historical date should contain more information than the earlier historical date. If the time horizon of history data of assets is long enough, the defect would be greater. Hence, it seems more reasonable and realistic to describe the expected returns in terms of fuzzy numbers, according to the developing trendy and performance of risky assets, or according to experts' knowledge and experience.

In this paper, the expected rate of return and risks are described to be triangular fuzzy numbers. Let $\tilde{r}_j = [\mathcal{B}r_j, r_{0j}, \bar{r}_j]$ be the fuzzy expected rate of return of risky asset $j$, $x_j$ be proportion of the total investment devoted to risky asset $j$, and $x_j \geq 0, j = 1, 2, \cdots, n$, $\sum_{j=1}^{n} x_j \leq 1$. Let a nonnegative real number $r_f$ denote the return of non-risky asset, the proportion of the total investment devoted to it be denoted by $x_{n+1}$. Then $x_{n+1} = 1 - \sum_{j=1}^{n} x_j$, and the fuzzy expected return $\tilde{r}$ of portfolio $x = (x_1, \cdots, x_{n+1})$ can be stated as

$$\tilde{r} = \sum_{j=1}^{n} \tilde{r}_j x_j + r_f x_{n+1} = [\sum_{j=1}^{n} \mathcal{B}r_j x_j + r_f x_{n+1}, \sum_{j=1}^{n} r_{0j} x_j + r_f x_{n+1}, \sum_{j=1}^{n} \bar{r}_j x_j + r_f x_{n+1}].$$

The risk of portfolio $x$ is represented by $\tilde{V}(x)$ in the following way.

$$\tilde{V}(x) = \frac{1}{T} \sum_{t=1}^{T} (\sum_{j=1}^{n} (r_{tj} - \tilde{r}_j)^2 x_j + (r_f - r_f)^2 x_{n+1}) = [\mathcal{B}V, V_0, \bar{V}],$$

where, $\mathcal{B}V = min\{\frac{1}{T} \sum_{t=1}^{T} \sum_{j=1}^{n} (r_{tj} - \mathcal{B}r_j)^2 x_j, \frac{1}{T} \sum_{t=1}^{T} \sum_{j=1}^{n} (r_{tj} - \bar{r}_j)^2 x_j\}, \bar{V} = max$

$\{\frac{1}{T} \sum_{t=1}^{T} \sum_{j=1}^{n} (r_{tj} - \mathcal{B}r_j)^2 x_j, \frac{1}{T} \sum_{t=1}^{T} \sum_{j=1}^{n} (r_{tj} - \bar{r}_j)^2 x_j\}, V_0 = \frac{1}{T} \sum_{t=1}^{T} \sum_{j=1}^{n} (r_{tj} - r_{0j})^2 x_j.$

Suppose that the triangular fuzzy number $\tilde{\beta}$ is the maximum variance level of risk the investors can tolerate. Then, for the purpose of pursuing profit, the investor should require that the variance of the portfolio risk in all cases should not be greater than the preset $\tilde{\beta}$ and then pursue maximum fuzzy expected return of securities in the worst case. To express the idea mathematically, the portfolio optimization model is constructed as follows:

$$(P1) \qquad\qquad max \quad \tilde{r}, \qquad\qquad\qquad (1)$$

$$\text{s.t.} \quad \begin{cases} \tilde{V} \leq \tilde{\beta} \\ \sum_{j=1}^{n} x_j + x_{n+1} = 1, x_j \geq 0, \ j = 1, 2, \cdots, n+1 \end{cases} \qquad (2)$$

Where, $\tilde{\beta} = [\mathcal{B}\beta, \beta_0, \bar{\beta}]$, $\mathcal{B}\beta < \bar{\beta}$. The constraint $\tilde{V} \leq \tilde{\beta}$ means that the rate of portfolio risk in any cases will not be greater than the given fuzzy number $\tilde{\beta}$, and the objective function (1) means that the optimal portfolio should be the one with maximum expected return in the worst case.

## 4  The Resolution of Fuzzy Portfolio Optimization Model

Firstly, the equivalent deterministic-crisp objective function for the fuzzy objective function (1) is considered to be presented based on the $\lambda-$cut of fuzzy numbers given in Section 2. For the given level $\lambda \in [0, 1]$, according to the definition of $\lambda-$cut, we have $\tilde{r}(x) \geq \lambda$. Then $\tilde{r}_\lambda = [\tilde{r}_\lambda^-, \tilde{r}_\lambda^+]$, where

$$\tilde{r}_\lambda^- = \sum_{j=1}^{n} (\lambda r_{0j} + (1 - \lambda)\mathcal{B}r_j)x_j + r_f x_{n+1}, \qquad\qquad (3)$$

$$\tilde{r}_\lambda^+ = \sum_{j=1}^{n} (\lambda r_{0j} + (1 - \lambda)\bar{r}_j)x_j + r_f x_{n+1}. \tag{4}$$

Assume that for the level $\lambda \in [0, 1]$, the investor maximize the expected return, then the fuzzy objective function(1) can be represented as

$$max \;\; \tilde{r}_\lambda^- = \sum_{j=1}^{n} (\lambda r_{0j} + (1 - \lambda)\mathcal{B}r_j)x_j + r_f x_{n+1}. \tag{5}$$

It is shown that the model $(P1)$ is an optimization problem with triangular fuzzy numbers in objective function and constraints. Therefore, techniques of classical programming can not be applied. In the following, we will get the equivalent deterministic crisp constraints for the fuzzy constraint in (2), according to the dominance possibility criterion as presented in Section 2. Assume that $\tilde{\beta} \geq \tilde{V}$, according to $DPC$, it can be represented as follows:

$$Poss(\tilde{\beta} \geq \tilde{V}) = \begin{cases} 1, & \text{if } \beta_0 \geq V_0 \\ \frac{\bar{\beta} - \mathcal{B}V}{\bar{\beta} - \beta_0 + V_0 - \mathcal{B}V}, & \text{if } V_0 \geq \beta_0, \bar{\beta} \geq \mathcal{B}V \\ 0, & \text{if } \mathcal{B}V \geq \bar{\beta} \end{cases}$$

Given level $\lambda \in [0, 1]$, if $Poss(\tilde{\beta} \geq \tilde{V}) \geq \lambda$, then the equivalent deterministic-crisp constraints for the fuzzy constraint in (2) can be stated as

$$\begin{cases} (1 - \lambda)\bar{\beta} + \lambda\beta_0 \geq (1 - \lambda)\mathcal{B}V + \lambda V_0 \\ \bar{\beta} \geq \mathcal{B}V \end{cases} \tag{6}$$

Let $\tilde{V}_\lambda^- = (1 - \lambda)\mathcal{B}V + \lambda V_0$. According to the $DPC$, $(P1)$ can be transformed equivalently to the following linear programming model $(P2)$:

$$(P2): \;\; max(\tilde{r})_\lambda^- = \sum_{j=1}^{n} (\lambda r_{0j} + (1 - \lambda)\mathcal{B}r_j)x_j + r_f x_{n+1},$$

$$s.t. \begin{cases} (1 - \lambda)\bar{\beta} + \lambda\beta_0 \geq (1 - \lambda)\mathcal{B}V + \lambda V_0, \bar{\beta} \geq \mathcal{B}V \\ \mathcal{B}V = min\{\frac{1}{T} \sum_{t=1}^{T} \sum_{j=1}^{n} (r_{tj} - \mathcal{B}r_j)^2 x_j, \frac{1}{T} \sum_{t=1}^{T} \sum_{j=1}^{n} (r_{tj} - \bar{r}_j)^2 x_j\} \\ \sum_{j=1}^{n} x_j + x_{n+1} = 1, x_j \geq 0, j = 1, 2, \cdots, n + 1 \end{cases}$$

Clearly, the model $(P2)$ can be transformed to the following equivalent model $(LP2)$, where $(\tilde{r}_\lambda^-)^*$ and $(\tilde{r}_\lambda^-)^{**}$ be the maximum values of the following models $(LP2)(a)$ and $(LP2)(b)$, respectively.

$$(LP2): \;\; max\{((\tilde{r}(x))_\lambda^-)^*, \;\; ((\tilde{r}(x))_\lambda^-)^{**}\},$$

$$(LP2)(a): \;\; max\tilde{r}_\lambda^- = \sum_{j=1}^{n} (\lambda r_{0j} + (1 - \lambda)\mathcal{B}r_j)x_j + r_f x_{n+1},$$

$$s.t. \begin{cases} (1-\lambda)\bar{\beta} + \lambda\beta_0 \geq (1-\lambda)\omega + \lambda\frac{1}{T}\sum_{t=1}^{T}\sum_{j=1}^{n}(r_{tj} - r_{0j})^2 x_j \\ \sum_{t=1}^{T}\sum_{j=1}^{n}(r_{tj} - \mathcal{B}r_j)^2 x_j \leq \sum_{t=1}^{T}\sum_{j=1}^{n}(r_{tj} - \bar{r}_j)^2 x_j \\ \omega = \frac{1}{T}\sum_{t=1}^{T}\sum_{j=1}^{n}(r_{tj} - \mathcal{B}r_j)^2 x_j, \bar{\beta} \geq \omega \\ \sum_{j=1}^{n} x_j + x_{n+1} = 1, x_j \geq 0, j = 1, 2, \cdots, n+1 \end{cases}$$

$$(LP2)(b): \quad max\tilde{r}_{\lambda}^{-} = \sum_{j=1}^{n}(\lambda r_{0j} + (1-\lambda)\mathcal{B}r_j)x_j + r_f x_{n+1},$$

$$s.t. \begin{cases} (1-\lambda)\bar{\beta} + \lambda\beta_0 \geq (1-\lambda)\omega + \lambda\frac{1}{T}\sum_{t=1}^{T}\sum_{j=1}^{n}(r_{tj} - r_{0j})^2 x_j \\ \sum_{t=1}^{T}\sum_{j=1}^{n}(r_{tj} - \mathcal{B}r_j)^2 x_j \geq \sum_{t=1}^{T}\sum_{j=1}^{n}(r_{tj} - \bar{r}_j)^2 x_j \\ \omega = \frac{1}{T}\sum_{t=1}^{T}\sum_{j=1}^{n}(r_{tj} - \bar{r}_j)^2 x_j, \bar{\beta} \geq \omega \\ \sum_{j=1}^{n} x_j + x_{n+1} = 1, x_j \geq 0, j = 1, 2, \cdots, n+1 \end{cases}$$

From the above analysis, the following assertion holds:

**Theorem 1.** $x^* = (x_1^*, x_2^*, \cdots, x_n^*, x_{n+1}^*)$ is the optimal portfolio of the model $(P2)$ if and only if $x^* = (x_1^*, x_2^*, \cdots, x_n^*, x_{n+1}^*)$ is the optimal portfolio of the model $(LP2)$.

One can see that both $(LP2)(a)$ and $(LP2)(b)$ are two standard linear programming problems. They can be solved by several algorithms of linear programming efficiently. Therefore, we can solve the original portfolio selection problem $(P1)$ by solving $(P2)$ or $(LP2)$.

## 5 An Example

In order to illustrate the proposed methods, we will consider an example introduced by Fang [6], which is shown in Table 1. Let $r_{0j}$ be the arithmetical mean of the history data of securities, $\mathcal{B}r_j = (1 - \varepsilon)r_{0j}$, $\bar{r}_j = (1 + \varepsilon)r_{0j}$, $\varepsilon = 0.1, j = 1, 2, \cdots, 8$. Assume that the return of non-risky security is the tenth column given in Table 1, denoted by $r_f = 0.03$. For the model $(LP2)$, given the level $\lambda = 0.8$ and the fuzzy tolerable risk in terms of a triangular fuzzy number $\tilde{\beta}$, the optimal portfolio strategies are obtained by using algorithms of linear programming. Some of the results are presented in Table 2. One can see that the results in Table 2 display the distributive investment.

**Table 1** Returns of nine securities

| year | 1 | 2 | 3 | 4 | 5 | 6 | 7 | 8 | 9 |
|---|---|---|---|---|---|---|---|---|---|
| 1937 | -0.305 | -0.173 | -0.318 | -0.477 | -0.457 | -0.065 | -0.319 | -0.435 | 0.03 |
| 1938 | 0.513 | 0.098 | 0.285 | 0.714 | 0.107 | 0.238 | 0.076 | 0.238 | 0.03 |
| 1939 | 0.055 | 0.2 | -0.047 | 0.165 | -0.424 | -0.078 | 0.381 | -0.295 | 0.03 |
| 1940 | -0.126 | 0.03 | 0.104 | -0.043 | -0.189 | -0.077 | -0.051 | -0.036 | 0.03 |
| 1941 | -0.28 | -0.183 | -0.171 | -0.277 | 0.637 | -0.187 | 0.087 | -0.24 | 0.03 |
| 1942 | -0.003 | 0.067 | -0.039 | 0.476 | 0.865 | 0.156 | 0.262 | 0.126 | 0.03 |
| 1943 | 0.428 | 0.3 | 0.149 | 0.225 | 0.313 | 0.351 | 0.341 | 0.639 | 0.03 |
| 1944 | 0.192 | 0.103 | 0.26 | 0.29 | 0.637 | 0.233 | 0.227 | 0.282 | 0.03 |
| 1945 | 0.446 | 0.216 | 0.419 | 0.216 | 0.373 | 0.349 | 0.352 | 0.578 | 0.03 |
| 1946 | −0.088 | −0.046 | −0.078 | −0.272 | −0.037 | −0.209 | 0.153 | 0.289 | 0.03 |
| 1947 | −0.127 | −0.071 | 0.169 | 0.144 | 0.026 | 0.355 | −0.099 | 0.184 | 0.03 |
| 1948 | −0.015 | 0.056 | −0.035 | 0.107 | 0.153 | −0.231 | 0.038 | 0.114 | 0.03 |
| 1949 | 0.305 | 0.038 | 0.133 | 0.321 | 0.067 | 0.246 | 0.273 | −0.222 | 0.03 |
| 1950 | −0.096 | 0.089 | 0.732 | 0.305 | 0.579 | −0.248 | 0.091 | 0.327 | 0.03 |
| 1951 | 0.016 | 0.09 | 0.021 | 0.195 | 0.04 | −0.064 | 0.054 | 0.333 | 0.03 |
| 1952 | 0.128 | 0.083 | 0.131 | 0.39 | 0.434 | 0.079 | 0.109 | 0.062 | 0.03 |
| 1953 | −0.01 | 0.035 | 0.006 | −0.072 | −0.027 | 0.067 | 0.21 | −0.048 | 0.03 |
| 1954 | 0.154 | 0.176 | 0.908 | 0.715 | 0.469 | 0.077 | 0.112 | 0.185 | 0.03 |

**Table 2** The optimal portfolio strategies

| $\tilde{\beta}$ | return | $\tilde{V}_\lambda^-$ | optimal portfolio $x^*$ |
|---|---|---|---|
| [0.005, 0.010, 0.015] | 0.0645 | 0.0110 | (0, 0, 0, 0, 0.01, 0, 0.348, 0, 0.642) |
| [0.010, 0.020, 0.030] | 0.0985 | 0.0220 | (0, 0, 0, 0, 0.01, 0.012, 0.7006, 0,0.2774) |
| [0.0200, 0.030, 0.040] | 0.1266 | 0.0320 | (0, 0, 0, 0.012, 0.0225, 0.008, 0.9575, 0, 0) |
| [0.030, 0.040, 0.050] | 0.1337 | 0.0420 | (0, 0, 0, 0.014, 0.1224, 0.006, 0.8576, 0, 0) |
| [0.05, 0.060, 0.070] | 0.1477 | 0.0620 | (0, 0, 0, 0.01, 0.3270, 0.0050, 0.6580, 0, 0) |
| [0.060, 0.070, 0.080] | 0.1541 | 0.0720 | (0, 0, 0, 0.1, 0.3667, 0.0100, 0.5233, 0, 0) |
| [0.080, 0.090, 0.100] | 0.1685 | 0.0920 | (0, 0, 0, 0.2, 0.5023, 0.0020, 0.2957, 0, 0) |
| [0.0093, 0.1123, 0.1153] | 0.1836 | 0.1129 | (0, 0, 0, 0.009, 0.8419, 0, 0.1491, 0, 0) |

**Fig. 1** The binary relation that the expected risk $\hat{V}_\lambda^-$ corresponding to the optimal return of portfolio is given in Table 2. It is believed that the more the tolerable risk $\tilde{\beta}$, the larger both the expected return and the corresponding risk of portfolios. The investor may choose investment strategy from the portfolios according one's attitude towards the expected return and risk of securities.

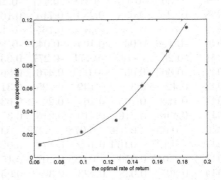

## 6 Conclusions

This paper proposes a fuzzy portfolio optimization model based on triangular fuzzy numbers. The expected returns and risks of securities are described as triangular fuzzy numbers. The proposed models are converted into equivalent crisp linear programming models by using the method of dominance possibility criterion. As is well known, the linear programming problem can be solved by several algorithms of linear programming efficiently. Finally, an example is presented to illustrate the effectiveness of the proposed models.

## References

1. Abdelaziz, F.B., Aouni, B., Fayedh, R.E.: Multi-objective stochastic programming for portfolio selection. Eur. J. Oper. Res. 177, 1811–1823 (2007)
2. Ammar, E.E.: On solutions of fuzzy random multiobjective quadratic programming with applications in portfolio problem. Inf. Sci. 178, 468–484 (2008)
3. Arenas-Parra, M., Bilbao-Terol, A., Rodríguez-Uña, M.V.: A fuzzy goal programming approach to portfolio selection. Eur. J. Oper. Rese. 133, 287–297 (2001)
4. Bilbao-Terol, A., Pérez-Gladish, B., Arenas-Parra, M., Rodríguez-Uña, M.V.: Fuzzy compromise programming for portfolio selection. Appl. Math. Comput. 173, 251–264 (2006)
5. Black, F., Litterman, R.: Asset allocation: combining investor views with market equilibrium. Journal of Fixed Income 1, 7–18 (1991)
6. Yong, F., Shouyang, W.: Fuzzy Portfolio Optimization—Theory and Method. Higher Education Press, Beijing (2005)
7. Tanaka, H., Guo, P., Turksen, B.: Portfolio selection based on fuzzy probabilities and possibility distributions. Fuzzy Sets and Systems 111, 387–397 (2000)

8. Hirschberger, M., Qi, Y., Steuer, R.E.: Randomly generatting portfolio-selection covariance matrices with specified distributional characteristics. European J. of Operational Research 177, 1610–1625 (2007)
9. Huang, X.: Two new models for portfolio selecion with stochastic returns taking fuzzy information. Eur. J. Oper. Res. 180, 396–405 (2007)
10. Huang, X.: Minimax mean-variance models for fuzzy portfolio selection. Soft Comput. 15, 251–260 (2011)
11. Iskander, M.: Using different dominance criteria in stochastic fuzzy linear multiobjective programming— a case of fuzzy weighted objective function. Mathematical and Computer Modelling 37, 167–176 (2003)
12. Konno, H., Yamazaki, H.: Mean-absolute deviation portfolio optimization model and its application to Tolyo Stock Market. Management Scicence 37, 519–531 (1991)
13. Yongming, L.: Analysis of Fuzzy System. Science Press, Beijing (2005)
14. Markowitz, H.: Portfolio selection. Journal of Finance 7, 77–91 (1952)
15. Negi, D.S., Lee, E.S.: Possibitlity programming by the comparison of fuzzy numbers. Computers Mate. Applic. 25, 43–50 (1993)
16. Xiuguo, W., Wanhua, Q., Jichang, D.: Portfolio Selection Model with Fuzzy Coefficients. Fuzzy Systems and Mathematics 20, 109–118 (2006)
17. Zadeh, L.A.: Fuzzy sets. Inform. and Contr. 8, 338–353 (1965)

# Soft Computing Techniques and Its Application in the Intelligent Capital Compensation

Kuo Chang Hsiao, Berlin Wu, and Kai Yao

**Abstract.** Measurement of the value of intellectual capital is increasingly receiving more and more attention. However, the evaluation of intellectual capital is complex, it involves many factors, such as peoples' utility (human subjective recognition) and the economic efficiency etc. that are very difficult to compute by traditional methods. In this paper we propose an integrated fuzzy evaluation procedure to measure intellectual capital. The main methods used in this research are fuzzy statistical analysis, fuzzy-weighting and fuzzy ranking. This integrated procedure is aimed at yielding appropriate and reasonable rankings and value of intellectual capital. Empirical study shows that fuzzy statistics with soft computing are more realistic and reasonable in the intellectual capital evaluation.

**Keywords:** Intellectual capital, Evaluation, Fuzzy statistics, Human thought, Fuzzy-weighting.

## 1 Introduction

The implication of intellectual property rights(IPR) reveals one of the most important competence ability of a country. Aside from traditional property owners, IP holders own their properties in intangible form. Therefore, intellectual property is easier to be violated than tangible one. The standard of compensation is also difficult to be judged upon once it happened.

Kuo Chang Hsiao
Dept. of Education Policy and Ad., National ChiNan University, Taiwan

Berlin Wu
Department of Mathematical Sciences, National Chengchi University, Taiwan
e-mail: Berlin@nccu.edu.tw

Kai Yao
Management of school, Fudan University, China

S. Li (Eds.): Nonlinear Maths for Uncertainty and its Appli., AISC 100, pp. 567–578.
springerlink.com                                    © Springer-Verlag Berlin Heidelberg 2011

According to Tort Law (TL), the compensation that IP holder demands relies on the testimony of how the property is violated and the "level of being damaged." The fact is that, many IP holders give up prosecuting violating cases since it's difficult to prove and measure the actual loss; therefore, IPR relief system is a nominal in reality.

We live in a knowledge economy society. Knowing how to evaluate intellectual property is an important issue. Therefore, establishing a method for objectively evaluating intellectual capital is not only important but also urgent. The difficulties of evaluating intellectual capital follow from (1) the involvement of too many influential variables; (2) inappropriate measurement techniques (3) vagueness perception and cognition of intellectual capital. Following are few researches that are available on evaluating intellectual capital evaluation, this paper proposes an integrated procedure to compute the intellectual capital via fuzzy statistical analysis.

This essay is aiming at exploring the intellectual property value(IPV) of the academic journals. The indicators applied in this articles are: author's academic background, author's position, ranking of journal he/she published, framework of the article, research method and whether the research meets applicative and academic needs and demands. These factors are based on the opinions of professional experts to gauge intellectual capital, and are thus highly subjective, uncertain as well as incomplete.

Since Zadeh (1965) developed fuzzy set theory, its applications are extended to traditional statistical inferences and methods in social sciences, including medical diagnosis or a stock investment system. For example, Dubois and Parde (1991) , demonstrated the approximate reasoning econometric methods one after another. Wu and Hsu (2004) developed fuzzy time series model to overcome the bias of stock market which might be explained unreasonable.

There has been considerable and increasing attention paid to the idea of fuzzy logic since it was introduced by Zadeh as a modification of conventional mathematical Set Theory (Zadeh 1965). The basic aim is that vagueness and ambiguity can be described and distinguished mathematically. Fuzzy theory has been widely used by a variety of authors. Tseng and Wu (2002) applied fuzzy regression models to business cycle analysis. Wu *et al.,* (2002) proposed new approaches on market research with fuzzy statistical analysis. Kostas Metaxiotis (2003) integrated fuzzy logic into a decision support system. Hong Zhang (2004) proposed the fuzzy discrete-event simulation to model the uncertain activity duration. Rajkumer Ohdar *et al.,* (2004) introduced the fuzzy-based approach to measure and evaluate the performance of suppliers in the supply chain. Jeng *et al.,* (2009) applied fuzzy forecasting techniques in DNA computing. Ho and Wu (2008) used integrated fuzzy statistics analysis in the valuation of intellectual capital. Sharon M. Ordoobadi (2008) used fuzzy logic to evaluate advanced technologies for decision makers and provided a model based on fuzzy logic for decision makers to help them with selection of appropriate suppliers in 2009.

## 2  How to Evaluate an Intelligent Capital

People's thinking is full of uncertainty and whose wisdom is difficult to measure or quantify as well. The value of intellectual property is nearly impossible to evaluate by traditional statistics. For this reason, researchers start to apply fuzzy statistics/soft computing to study intellectual property. Hence the result shall be more objective, and prone to meet the social expectation.

### 2.1  What Is the Discussion Domain for an Intellectual Capital?

An intellectual capital means all resources that determine the value and the competitiveness of a work. Edvisson and Malone(1999)suggested that the intellectual property rights(IPR)consist with knowledge capital, non-financial capital, concealed property capital, non-visible property, or intangible assets. Brooking, Board and Jones(1998) indicate that IPR include market assets, intellectual property assets, infrastructure assets, human-centered assets. Such intellectual property, being the difference between company market value and book value, is the counterpart of visible entity capital and finance capital.

EIU(1995), Smith and Parr(1994), Stewart(1994), Edvisson and Malone (1999), and Reilly and Schweihs(1998)propose that intellectual capital includes three basic classes, such as human capital, structural capital, and relationship capital. Human capital represents the individual skills applied to satisfy customers. Structural capital is the organizational capabilities of the enterprise, demanded by the market. Relationship capital is the strength of a franchise afforded by a business's ownership of rights. While Mar(2000)surmise the literature and suggests that the intellectual capital in a wide sense, including knowledge capital, management capital, and market capital.

### 2.2  Statistical Evaluation with Fuzzy Data

Many phenomena in the world, such as human language, thinking and decision-making, all possess non-quantitative characteristics. Human behavior is particularly difficult to quantize. The argument is about the principle of applying fuzzy scale and categorization into human's interaction with the dynamic environment, and to give a more concrete description and solution toward complicated/vague phenomenon.

Since mode, mean and median are essential statistics in analyzing the sampling survey. For instance, when people process a happy assessment, they classify the distraction into two categories: happy and unhappy. This kind of classification is not realistic, since that happiness is a fuzzy concept (degree) and can hardly be justified by the true-false logic. Therefore, to compute the information based on the fuzzy logic should be more reasonable. Calvo and

Mesiar (2001)] proposed the generalized median by discussing aggregation operators closely related to medians and to propose new types of aggregation operators appropriate, both for the cardinal and ordinal types of information.

## 3  Soft Computing in the Fuzzy Evaluation

If the researcher can use the membership function to express the degree of persons' feelings based on their own choices, the result presented will be closer to the people's real thinking. 3.1 shows how to compute sample mean for fuzzy data

### 3.1  The Role of Fuzzy Weight

How to decide the weights, called fuzzy weights, becomes a primary work before evaluating the intellectual capital. In this section we will demonstrate an integrated design via appropriate questionnaires of field study to reach a common agreement for weight of fuzzy factors for an object/event.

For example, to suppose that we are doing the sampling survey during an election to period. If the questionnaire is designed to forecast the percentage of popular ballot, then the factors to supporting the candidate are affected by people's judgment. Because the trend in voting intentions are crucial to understanding actual voting behavior, it is important to measure the weight of weight for forecasting the in preferable voting.

Therefore, a voter's views on a variety of issues determine his view of each candidate. Voters make choices by synthesizing a range of factors and then determines how these are related to each candidate. However, different factors affect a popular ballot with different degree. But each individual voter emphasizes these trends differently. Accordingly, consider firstly, the fuzzy weight by an individual.

Here, the calculating process of entity fuzzy weight is presented:

*Step1: First, determine the factors* $A = \{A_1, A_2, ...., A_k\}$ *for the intellectual capital*

*Step2: Ask each interviewee i to give the importance of factors set with an membership* $m_{ij}$ $\sum_{j=1}^{k} m_{ij} = 1$. *Let* $m_{ij}$ *be the membership of importance of factor j for the ith interviewee*

*Step3: Calculate the fuzzy weight* $w_j$ *of* $A_j$ *by* $w_j = \dfrac{\sum_{i=1}^{n} m_{ij}}{\sum_{j=1}^{k} \sum_{i=1}^{n} m_{ij}}$

**Example 3.1.** Suppose there are five interviewees rank a certain event with five factors for a discussion domain, Table 1 illustrates the result.

**Table 1** Memberships for five factors with five interviewees

| sample | 1 | 2 | 3 | 4 | 5 |
|---|---|---|---|---|---|
| 1 | 0.10 | 0.10 | 0.6 | 0.10 | 0.10 |
| 2 | 0.20 | 0.15 | 0.5 | 0.05 | 0.10 |
| 3 | 0.10 | 0.10 | 0.7 | 0.05 | 0.05 |
| 4 | 0.10 | 0.20 | 0.5 | 0.10 | 0.10 |
| 5 | 0.25 | 0.10 | 0.4 | 0.15 | 0.10 |
| Sum of memberships | 0.75 | 0.65 | 2.7 | 0.45 | 0.45 |
| w | 0.15 | 0.13 | 0.54 | 0.09 | 0.09 |

It shows that $w_1 = \dfrac{\sum\limits_{i=1}^{5} m_{i1}}{\sum\limits_{k=1}^{5}\sum\limits_{i=1}^{5} m_{ij}} = \dfrac{0.75}{5} = 0.15, \ldots, w_5 = \dfrac{\sum\limits_{i=1}^{5} m_{i5}}{\sum\limits_{k=1}^{5}\sum\limits_{i=1}^{5} m_{ij}} = \dfrac{0.45}{5} = 0.09$.

## 3.2 Distance among Fuzzy Data

Once such a transformation has been selected, instead of the original trapezoiddata, we have a new value y = f(x). In the ideal situation, this new quantity y is normally distributed. (In practice, a normal distribution for y may be a good first approximation ) When selecting the transformation, we must take into account that, due to the possibility of a rescaling, the numerical values of the quantity x is not uniquely determined.

**Definition 3.1.** *Scaling for a interval fuzzy number on R*

*Let $A = [a, b]$ be an interval fuzzy number on U with its center $(a - b)/2$. Then the defuzzification number RA of $A = [a, b]$ is defined as*

$$RA = cx + (1 - \frac{\ln(1 + \|A\|)}{\|A\|});$$

*where, $\|A\|$ is the length of the interval.*

However, there are few literatures and definitions appear on the measurement system. In this section, a well-defined distance for interval data will be presented.

**Definition 3.2.** *Let $A_i = [a_i, b_i]$ $(i=1,2,\ldots n)$ be a sequence of interval fuzzy number on U with its center $(a-b)/2$. Then the distance between the trapezoid fuzzy number $A_i$ and $A_j$ is defined as*

$$d(A_i, A_j) = |cx_i - cx_j| + \left| \frac{\ln(1 + \|A_i\|)}{\|A_i\|} - \frac{\ln(1 + \|A_j\|)}{\|A_j\|} \right|$$

# 4   An Integrated Fuzzy Evaluation Process

Analysis by traditional methods usually involves the following weaknesses: (a) the use of arithmetic in traditional questionnaires is often over-explanation. (b) experimental data are often overused just to cater to the need for apparent numerical accuracy. (c) for the sake of simplifying the evolutional model, however, will neglect the relationship of actual condition and dynamic characteristic. That is why we will prefer to apply fuzzy theory to handle the questions that involve human opinion.

## 4.1   Discussion Domain and Weight of Factors

It is appropriate to apply the membership function, a more precise mathematical techniques, in analyzing the fuzzy information. The value of the membership function, between 0 and 1, is derived from the characteristic function, to express the membership grade of each element in a set.

There are many types of membership function, such as Z- type, $\Lambda$- type, $\Pi$- type, S- type,and so on (Nguyen and Wu, 2006). In this research the $\Lambda$ – type membership functions is applied. It assesses the fuzzy interval of various evaluations, and then calculates the fuzzy value of an enterprise according to appraiser's fuzzy weighting.

The researcher also uses the $\Lambda$- type to reflect the value of the intelligent capital distribution. That is, we will give the value of intelligent capital into different linguistic terms, such as, *valueless, not too valuable, lightly valuable, valuable, very valuable, extremely valuable, hugely valuable, invaluable.* Each term will be correspondent with a real value, which will be determined by the sampling survey and fuzzy statistical analysis.

## 4.2   Intelligent Capital: Factors with Highly Co Integrated Property

After a detailed discussion from the above sections, an integrated process of fuzzy evaluation is shown following. The geometric average is used to instead for the weighted arithmetic average. The reason is that the factors are highly correlated, any extreme value of a certain factor will influence the real intelligent capital. For instance in evaluating the intelligent capital of an academic paper.

How to assess the value of a paper? According to an investigation on the scholars, the conclusion is drawn as that the five key items to judge a paper's value are: (1)*author's academic background; (2)author's position; (3)ranking of journal he/she published; (4)structure and creative of the article,* and whether the (5) *research meets applicative and academic needs.* When applying these five elements into paper-value judgement, subjective opinions are also involved. It makes a difference from the evaluative ratio, therefore we make a geometric average to get a more appropriate evaluation. That is, if to suppose the factor sets is $A = \{A_i = [a_i, b_i], \ i = 1..., l\}$ and which corresponding to the with a weight set $w = \{w_1, w_2, ...., w_l\}$, then the integrated *Intelligent Capital* will be

$$IC = [\prod_{i=1}^{l} a_i^{w_i}, \ \prod_{i=1}^{l} b_i^{w_i}] \tag{1}$$

## 4.3 How Much Does the Author Lost

When we calculate the lost for intelligent capital compensation, we may take two points into considerations: 1. Personal 2. Public. Form the personal point of view we will consider how long does the author work to complete this piece, how much does the median of author's monthly income. From the public point of view, the *rank of the journal*, the *degree of similarity to the original work* and the time from *the first publication of this work.*

We will put these two parts of compensation by addition. While inside these two factors, we would like to take it by the production. Since inside the factors, the variables are highly co-integrated. Model (2) illustrates the above description for *Lost Evaluation (LE)*.

$$LE = (M \cdot \sqrt{W} + R \cdot T \cdot e^{-Y}) \cdot S \tag{2}$$

*Where M=median of monthly income /salary in recent 10 years (U\$ dollar), $0 \leq M < \infty$*

*W=working months to complete the paper, $0 \leq T < \infty$.*

*R=the status of the journal that the paper published (R=1,2,...10. 1=the beginning rank,10=the highest rank, for example, top 10 journal of the field will be ranked as 10, rank 6 to rank 8 will be the international journal such as SSCI. SCI, EI, A&H, etc; rank 4 to rank 6 will be the international journal or core journal for local area, rank 1 to rank 3 will be the local journal, refereed conference or preprint report),*

*T= total spend on completing this work, Y=year from the first publication, $0 \leq Y < \infty$.*

*S= degree of similarity to the original paper, an interval data on the $[0,1]$.*

To explain the equation (2), the following points are necessary and important: (1) We use the median of monthly income in recent 10 years to evaluate the lost of intelligent capital. (2) We measure the working months to complete the work by square root function (since intelligent action needs kind of concentration working.) (3)We use the journal of rank to evaluate the status/quality of this paper. (4) We use the fuzzy expense evaluation to measure the total spend for completing the work. (5) We use the year with exponential decay function to measure the creative value of this work. (6)We use the degree of similarity to judge the ratio value of compensation.

For example, if a person has the median of salary in recent ten years is U$3,000. He worked 30 months to complete the paper. The rank of this paper published in the field is rank 8. Total spend in completing this work is U$5,000. The year from the first publication is one. The degree of similarity to the original paper is 60%, hence according to the formula (2) we evaluate the *Lost Evaluation as:*

$$LE = (3 \cdot \sqrt{30} + 8 \cdot 5 \cdot e^{-1}) \cdot [.6, .7] = [18.7, 21.8]$$

## 4.4  *Integrated Compensation Evaluation and Decision Making.*

Due to it is difficult to meet the common agreement of compensation for the both sides, in order to minimize the gap, it is necessary to propose an appropriate decision rule. Let $LE=[a,b]$  *and* $IC=[c,d]$ stand for the lost evaluation and intelligent capital respectively, we set the compensation as

$$Compensation = [l, u], \tag{3}$$

where $l = \min \sqrt{ad}, \sqrt{bc}$ and $u = \max \sqrt{ad}, \sqrt{bc}$.

The reason for the researcher to choose geometric average with endpoints cross production is to avoid people's unusual estimation in the lost evaluation as well as the intelligent capital. The advantage of this estimation is that we can use this two evaluated interval to reach an appropriate (common agreement) interval so that both sides of people would like to accept the result. While most traditional evaluation methods are base on the real value operation with the mean (arithmetic average) functions.

## 5  A Case Study about the Academic Paper

In 2010, from the news of web, there is a PhD candidate, works as a teacher, was founded plagiarism for 30% of the article he submitted. The author, his professor, has found it out and made his lawyer sending legal confirm letter to him. Taking his career future into consideration, the candidate thus took initiative move: providing U$5000 dollars as compensation, signing recognizance for confession and promising never to make the mistake again. Later, the PhD candidate and his professor reach a peacemaking result without further prosecution.

The case mentioned above is studied as an example to discuss the rationality of intellectual property compensation. This case has revealed two layers of the issue: plagiarism and compensation. The former is easy to identify: a published journal article is plagiarized. But how to decide the property value to the holder? There's no fair means for calculating, so as the difficulty to estimate the loss; consequentially rationality of compensation is hard to reach.

Firstly, we interviewed 20 experts in the academic field, including university professors, associate professors, assistant professors, lecturers and graduated students, to rating the indicators for valuating articles. Five items stand out after the analysis: author's academic background, author's position, ranking of journal he/she published, structure of the article, research method and whether the research meets applicative and academic needs. Table 2 illustrate the weight for five intelligent capital factors.

**Table 2** The weights of intelligent capital factors for an academic paper

|  | Education Background | Position Position | Ranking of Journals | Structure and Method | App. and Technicality |
|---|---|---|---|---|---|
| *Professor* | 0.2 | 0.2 | 0.3 | 0.1 | 0.2 |
| *Associate Prof* | 0.1 | 0.2 | 0.3 | 0.1 | 0.3 |
| *Assistant Prof* | 0.1 | 0.1 | 0.3 | 0.1 | 0.4 |
| *Lecturer* | 0.1 | 0.1 | 0.2 | 0.2 | 0.4 |
| *Grad. Student* | 0.2 | 0.2 | 0.2 | 0.1 | 0.3 |

Table 2 shows the fuzzy statistics for the *Intelligent Capital value* From Table 2 we can find that the fuzzy mode and fuzzy mean of the capital intelligent for four levels. Suppose the factors in the discussion domain include :$A_1$ =Educational Background, $A_2$= Position, $A_3$= Ranking of Journals, $A_4$= Structure and Method, $A_5$= App. and Technicality. For instance, the weight for the professor level is $w_1 = 0.2, w_2 = 0.2, w_3 = 0.3, w_4 = 0.1, w_5 = 0.2$. For the reason of comparison, by the equation (1), we compute the value of intellectual capital for the professor level as well as for the graduate student's level:

$$IC = [15^{0.2} \cdot 13^{0.2} \cdot 11^{0.3} \cdot 12^{0.1} \cdot 16^{0.2}, \ 35^{0.2} \cdot 25^{0.2} \cdot 27^{0.3} \cdot 21^{0.1} \cdot 28^{0.2} = [13,28]$$

$$IC = [12^{0.2} \cdot 8^{0.2} \cdot 11^{0.2} \cdot 12^{0.1} \cdot 13^{0.3}, \ 19^{0.2} \cdot 15^{0.2} \cdot 19^{0.2} \cdot 16^{0.1} \cdot 17^{0.3}] = [11,17]$$

Secondly, we calculate the author's lost by model (2) from his background to reach a reasonable compensation. For comparison, we illustrate it as a professorship and a graduate student.

*LE (professor)*= $(3 \cdot \sqrt{30} + 8 \cdot 5 \cdot e^{-1}) \cdot [.6, .7] = [18.7, 21.8]$
*LE(graduated student)*= $(1 \cdot \sqrt{30} + 8 \cdot 5 \cdot e^{-1}) \cdot [.6, .7] = [6.2, 7.2]$

Table 3 illustrates the LE, IC and the common agreement of the evaluate value for this case.

**Table 3** Intelligent Capital Evaluation for an academic paper Unit = U$ dollar (*thousand*)

| Factors | | Education Background | Position | Ranking of Journals | Structure and Method | Appl. and Technicality | Intelligent capital |
|---|---|---|---|---|---|---|---|
| Professor | Fuzzy Mode | [10,30] | [10,30] | [10,30] | [10,30] | [10,30] | [10,30] |
| | Fuzzy mean | [15,35] | [13,25] | [11,27] | [12,21] | [16,28] | [13,28] |
| Associate | Fuzzy Mode | [10,30] | [10,20] | [10,30] | [10,20] | [10,20] | [10,24] |
| Professor | Fuzzy mean | [17,30] | [12,26] | [14,25] | [11,18] | [13,22] | [13,24] |
| Assistant Prof. | Fuzzy Mode | [10,20] | [10,20] | [10,20] | [10,20] | [10,20] | [10,20] |
| | Fuzzy mean | [12,22] | [13,23] | [13,23] | [11,21] | [11,19] | [12,21] |
| Lecturer | Fuzzy Mode | [10,20] | [5,20] | [10,20] | [10,20] | [10,15] | [9,17] |
| | Fuzzy mean | [11,19] | [8,17] | [12,20] | [11,18] | [10,18] | [11,18] |
| Grad. student | Fuzzy Mode | [10,20] | [10,15] | [10,30] | [10,20] | [10,15] | [10,20] |
| | Fuzzy mean | [12,19] | [8,15] | [11,19] | [12,16] | [13,17] | [11,17] |

**Table 4** The LE, IC as well and the common agreement of the evaluate value

| | Evaluated value (*Professor*) | Evaluated value (*Graduate Student*) |
|---|---|---|
| LE | [19, 22] | [6, 7] |
| IC | [13, 28] | [11,17] |
| Compensation | [17,23] | [9,12] |

# 6  Conclusions

Intellectual capital is a wide-ranging and complex area, and its evaluation includes much dispute. The advantage of the fuzzy statistical analyzing techniques proposed in this article, lies in its method to handle human thought and recognition, improving the vague measurement. The presented integrated procedure differs from the traditional assessment method, and establishes the membership grade of evaluator's weight to better capture real values.

Moreover, if we survey an intelligent capital object. No matter how carefully we read the measuring process, we can never be certain of the exact value, but we can answer with more confident that the appropriate area lies within certain bounds. The fact is that intervals can be considered as a number or a according to the underlying applications. Though interval analysis and fuzzy set theory being as areas of active research in mathematics, numerical analysis and computer science began in the late 1950s and early 1960s. The application to statistical evaluations is just beginning.

Human mind is full of uncertainty. It's easier to catch the truth by applying fuzzy statistics rather than tradition ones. This research, applying fuzzy statistic on valuating articles and IPR compensation, hopes to understand better what contribute to article values and the rationality of compensation.

This paper also finds that (1) traditional methods use all equally weights for every assessment factor, but in reality, factors are variously important. This text proposes fuzzy weighting in accordance with real conditions. (2) This research provides a method for evaluating intellectual capital, using a $\Lambda$- type membership function to establish the value interval, according to the above weights and to determine a membership grade to calculate fuzzy value and rank.

The future development of this research will be: (1) to apply the soft computing technique to get the more appropriate evaluation (2) intellectual capital, wide ranging, complex expand the assessment of factors to include the type of enterprise, increasing the objectivity of the evaluation, and (3) using of the fuzzy regression methods, according to sub assessment factors, to determine the appropriate value of intellectual capital.

# References

1. Brooking, A., Board, P., Jones, S.: The predictive potential of intellectual capital. Int. J. Technology Management 16, 115–125 (1998)
2. Calvo, T., Mesiar, R.: Generalized median. Fuzzy Sets and Systems 124, 59–61 (2001)
3. Jeng, D.J.F., Watada, J., Wu, B.: Biologically inspired fuzzy forecasting: a new forecasting methodology. International Journal of Innovative Computing. Information and Control 5, 12(B), 4835–4844 (2009)
4. Dubois, D., Prade, H.: Fuzzy sets in approximate reasoning, Part 1: Inference with possibility distributions. Fuzzy Sets and Systems 40, 143 202 (1991)
5. Economist Intelligence Unit (EIU). The valuation of intangible assets of Accounting for Intangible Assets: A new Perspective on the True and Fair View. Addison-Wesley Publishers Co, Special Report 254 (January 1992)
6. Edvisson, L., Malone, M.: The Wood is Big to Permit the Intelligence Capital-How to Measure the Information Ages Immaterial Assets. City Culture, Taipei (1999)
7. Ho, S.-M., Wu, B.: Evaluating intellectual capital with integrated fuzzy statistical analysis: a case study on the CD POB. International Journal of Information and Management Sciences 19(2), 285–300 (2008)
8. Zhang, H., Li, H., Tam, C.M.: Fuzzy discrete-event simulation for modeling uncertain activity duration. Engineering, Construction and Architectural Management 11(6), 426–437 (2004)
9. Kostas, M., Psarras, J., Samouilidis, E.: Integrating fuzzy logic into decision support system: current research and future prospects. Information Management & Computer Security 1 10, 53–59 (2003)
10. Mar, S.: The Research of the Information Software Immaterial Assets Price Evaluation System, Taiwan Stock Exchange Incorporated Company (2000)
11. Ohdar, R., Ray, P.K.: Performance measurement and evaluation of suppliers in supply chain: an evolutionary fuzzy-based approach. Journal of Manufacturing Technology Management 15(8), 723–734 (2004)
12. Reilly, R.F., Schweihs, R.P.: Valuing Intangible Assets. McGraw-Hill, New York (1998)

13. Sharon, M., Ordoobadi: Fuzzy logic and evaluation of advanced technologies. Industrial Management & Data Systems 108(7), 928–946 (2008)
14. Smith, G.V., Parr, R.L.: Valuation of Intellectual Property and Intangible Assets, 2nd edn. John Wily & Sons, Inc., Canada (1994)
15. Stewart, T.: Your company's most valuable assets: intellectual capita. Fortune, 68-74 (1994)
16. Wu, B.-L.: Modren Statistics. Chien Chen, Taipei (2000)
17. Wu, B., Tseng, N.: A new approach to fuzzy regression models with application to business cycle analysis. Fuzzy Sets and System 130, 33–42 (2002)
18. Wu, B., Hsu, Y.-Y.: The use of kernel set and sample memberships in the identification of nonlinear time series. Soft Computing Journal 8, 207–216 (2004)

# Geometric Weighting Method and Its Application on Load Dispatch in Electric Power Market

Guoli Zhang and Chen Qiao

**Abstract.** The paper gives geometric weighting method(GWM), which can change the multi-objective programming into a single objective programming, prove that the optimal solution of the single objective programming is non-inferior solution of original multi-objective programming. We take three objective load dispatch problem as an example, and use the geometric weighting method to solve the load dispatch problem. Simulation results show that the geometric weighting method is feasible and effective.

**Keywords:** Multi-objective programming, Geometric weighting method, Electric power market, Load dispatch.

## 1 Introduction

A multi-objective optimization problem can be stated as follows:

$$\begin{cases} \min f(x) = (f_1(x), f_2(x), \cdots, f_K(x)) \\ s.t. \ x \in X \end{cases} \tag{1}$$

Where, $X = \{x = (x_1, x_2, \cdots, x_n)^T \in R^n | h_i(x) = 0, i = 1, 2, \cdots, p; g_j(x) \leq 0, j = 1, 2, \cdots, q\}$ is the constraint set, $f(x) = (f_1(x), f_2(x), \cdots, f_K(x))$ is a $K(K \geq 2)$ dimensional objective function. Any or all of the functions $f_k(x)$, $h_i(x)$ and $g_j(x)$ may be nonlinear. The multi-objective optimization problem is also known as a vector minimization problem. The multi-objective programming generally has no absolute optimal solution, so how to get its non-inferior solutions is very important issue. Several methods have been developed for solving a multi-objective optimization problem. The traditional

Guoli Zhang and Chen Qiao
North China Electric Power University
e-mail: zhangguoli@necpu.edu.cn, qiaochen069@163.com

S. Li (Eds.): Nonlinear Maths for Uncertainty and its Appli., AISC 100, pp. 579–586.
springerlink.com ⓒ Springer-Verlag Berlin Heidelberg 2011

methods have main utility (evaluation) function method, constraint method, interactive method, layered sequence method and goal programming method etc. In recent years, some new method such as fuzzy method [5], genetic algorithms, particle swarm optimization [11] [10] etc. have been proposed to solve the multi-objective optimization problem.

Economic dispatch (ED) is very important in power systems, with the basic objective of scheduling the committed generating unit outputs to meet the load demand at minimum operating cost, while satisfying all units and system constraints. Different models and techniques have been proposed in the literature [12] [6] [2] [9]. As more and more attention fixes on environmental protection, economic dispatch problem is no longer a pure problem of minimization of fuel consumption (operating costs), how to reduce the emissions of pollutants must be taken into consideration. In addition, in electric market environment, the expense of purchasing electricity is also should consider. Some new models and solving algorithms have been proposed to solve load dispatch problem in electric power market [7] [1] [4] [3].

The rest of this paper is organized as follows: Section 2 gives geometric weighting method which change multi-objective programming into a single objective programming, proves that the optimal solution of the single objective programming is the non-inferior solution of original multi-objective programming. Section 3 introduces a load dispatch model which has three objective functions, and uses geometric weighting method to solve the three objective load dispatch problem. The conclusion is drawn in section 4.

## 2    Geometric Weighting Method

Evaluation function method is one of effective methods for solving multi-objective programming problem, through evaluation function $u(f(x))$, the optimization problem (1) becomes a single objective optimization problem as follows:

$$\begin{cases} \min u(f(x)) \\ s.t. \ x \in X \end{cases} \tag{2}$$

Where $u : R^K \to R$ is evaluation function. The evaluation function takes commonly linear weighting function, maximum function, distance function etc.

As we all known, if the relative important degrees of the objective functions are known, the linear weighting method is commonly used to solve multi-objective optimization problem. Gives weight vector $(w_1, w_2, \cdots, w_K)^T$, linear weighting method takes evaluation function as follows:

$$u(f(x)) = \sum_{k=1}^{K} w_k f_k(x) \tag{3}$$

The original multi-objective optimization problem (1) can be changed into the following single objective optimization problem (4):

$$\begin{cases} \min \sum_{k=1}^{K} w_k f_k(x) \\ s.t. \ x \in X \end{cases} \qquad (4)$$

Specifically, when $w_k = 1/K (k = 1, 2, \cdots, K)$ the evaluation function is:

$$(1/K) \times f_1(x) + (1/K) \times f_2(x) + \cdots + (1/K) \times f_K(x) \qquad (5)$$

That is the arithmetic mean for $f_1(x), f_2(x), \cdots, f_K(x)$, similarly the geometric mean for the objective functions $f_1(x), f_2(x), \cdots, f_K(x)$ can be described by the formula (6):

$$f_1(x)^{1/K} \times f_2(x)^{1/K} \times \cdots \times f_m(x)^{1/K} \qquad (6)$$

In general, the important degrees of the objective functions are different, so the evaluation function is taken as the following form:

$$u(f(x)) = \prod_{k=1}^{K} f_k(x)^{w_k} \qquad (7)$$

and the multi-objective optimization problem (1) is formulated as:

$$\begin{cases} \min \prod_{k=1}^{K} f_k(x)^{w_k} \\ s.t. \ x \in X \end{cases} \qquad (8)$$

Where, $w_k (k = 1, 2, \cdots, K)$ is weight such that $w_k \geq 0$, $\sum_{k=1}^{K} w_k = 1$.
The problem (8) is a single objective optimization problem.

**Lemma 1.** *Let $X \subseteq R^n$, $f : X \to R^K$, $u : R^K \to R^1$, if $u(y)$ is strict increase function, $x^*$ is one optimal solution of the problem (8), then $x^*$ is one non-inferior solution of the problem (1).*

**Theorem 1.** *Let $X \subseteq R^n$, $f : X \to R^K$, $f(x) > 0 \ (\forall x \in X)$, $w = (w_1, w_2, \cdots, w_K) > 0$ is constant vector, if $x^*$ is one optimal solution of the problem (8), then $x^*$ is one non-inferior solution the problem (1).*

*Proof.* $\forall f', f'' \in R^K$, assume $\forall f' \leq f''$, denote $f' = (f_1', f_2', \cdots, f_K')$, $f'' = (f_1'', f_2'', \cdots, f_K'')$, then $f_k' \leq f_k'' (k = 1, 2, \cdots, K)$, and there exist at least one $k_0 \in \{1, 2, \cdots, K\}$ such that $f_{k_0}' < f_{k_0}''$.

Because $f(x) > 0$, $w > 0$, hence $f_k'^{w_k} \leq f_k''^{w_k} (k = 1, 2, \cdots, K)$ and $f_{k_0}'^{w_{k_0}} < f_{k_0}''^{w_{k_0}}$.

Therefore we have

$$\prod_{k=1}^{K} f_k'^{w_k} < \prod_{k=1}^{K} f_k''^{w_k} \qquad (9)$$

That is $u(f') = \prod_{k=1}^{K} f_k'^{w_k} < \prod_{k=1}^{K} f_k''^{w_k} = u(f'')$, in other words, $u(f(x))$ is strict increase function. According to the lemma 1, $x^*$ is one non-inferior solution of the problem (1). $\qquad \square$

# 3 Application of GWM on Load Dispatch in Electric Power Market

## 3.1 The Load Dispatch Model

In electric market environment, load dispatch problem is no longer a pure economic dispatch problem which minimizes only fuel consumption or operating costs. The reference [8] established multi-objective a load dispatch model which includes three objective functions, they are fuel consumption, harmful gas emissions and power purchase cost functions respectively. The multi-objective load dispatch model given in reference [8] can be described as follows:

$$\begin{cases} \min F = \{f_1(P_{G_r}), f_2(P_{G_r}), f_3(P_{G_r})\} \\ \sum_{r=1}^{M} P_{G_r}(t) - P_d(t) - P_{rc}(t) = 0 \\ P_{G_{rmin}} \leq P_{G_r} \leq P_{G_{rmax}} \end{cases} \tag{10}$$

Where,

$$f_1(P_{G_r}) = \sum_{t=1}^{T} \sum_{r=1}^{M} [a_r P_{G_r}^2(t) + b_r P_{G_r}(t) + c_r] \tag{11}$$

$$f_2(P_{G_r}) = \sum_{t=1}^{T} \sum_{r=1}^{M} [\alpha_r P_{G_r}^2(t) + \beta_r P_{G_r}(t) + \gamma_r] \tag{12}$$

$$f_3(P_{G_r}) = \sum_{t=1}^{T} \sum_{r=1}^{M} P_{G_r}(t)\rho_{G_r}(t)] \tag{13}$$

Where, $T$ is the number of time intervals per dispatch cycle, (usually taken 24, 48, 96), $M$ represents the number of committed units, $f_1(P_{G_r})$ is the fuel consumption of power generation, $a_r$, $b_r$, $c_r$ are the constant coefficients; $f_2(P_{G_r})$ is the function of the emission of harmful gases, $\alpha_r$, $\beta_r$, $\gamma_r$ are the constant coefficients; $f_3(P_{G_r})$ is function of the purchase cost; $P_{G_r}(t)$ is the output power of the r-th unit in the t-th time interval; $P_d(t)$ and $P_{rc}(t)$ are active load and active network loss in the t-th time interval respectively; $P_{G_{rmax}}$, $P_{G_{rmin}}$ are the maximum and minimum output power of the r-th unit respectively; $\rho_{G_r(t)}$ is the bid of the r-th unit in the t-th time interval.

## 3.2 Simulation Experiment

First, by the geometric weighting method, the load dispatch problem (10) becomes:

$$\begin{cases} \min F = f_1(P_{G_r})^{w_1} \times f_2(P_{G_r})^{w_2} \times f_3(P_{G_r})^{w_3} \\ \sum_{r=1}^{M} P_{G_r}(t) - P_d(t) - P_{rc}(t) = 0 \\ P_{G_{rmin}} \leq P_{G_r} \leq P_{G_{rmax}} \end{cases} \tag{14}$$

**Table 1** Unit parameters

| Unit No. | Coefficient of Fuel Consumption(t/h) | | | Coefficient of Nitrogen Oxides Emission($10^{-7}$t/h) | | | Output Constraints (MW) | | Bid(RMB /(Kwh)) |
|---|---|---|---|---|---|---|---|---|---|
| | $a_r$ | $b_r$ | $c_r$ | $\alpha_r$ | $\beta_r$ | $\gamma_r$ | $P_{G_{rmin}}$ | $P_{G_{rmax}}$ | |
| 1 | 0.000175 | 0.11 | 3.0 | 6.490 | -5.554 | 4.091 | 310 | 570 | 0.27 |
| 2 | 0.000230 | 0.15 | 5.0 | 5.638 | -6.047 | 2.543 | 250 | 425 | 0.20 |
| 3 | 0.000116 | 0.07 | 7.0 | 4.586 | -5.094 | 4.257 | 350 | 700 | 0.25 |
| 4 | 0.000150 | 0.13 | 3.0 | 3.380 | -3.550 | 5.326 | 300 | 610 | 0.23 |
| 5 | 0.000120 | 0.12 | 5.0 | 4.586 | -5.904 | 4.258 | 325 | 660 | 0.27 |

**Table 2** Objective function values for different weights

| Weight($w_1, w_2, w_3$) | | | Total Fuel Consumption $f_1$ | Total Emissions $f_2$ | Power Purchase Costs $f_3$ |
|---|---|---|---|---|---|
| 1 | 0 | 0 | 9.7937e+003 | 12.2672 | 1.3082e+007 |
| 0 | 1 | 0 | 1.0198e+004 | 11.0580 | 1.2891e+007 |
| 0 | 0 | 1 | 1.0413e+004 | 11.4211 | 1.2701e+007 |
| 0.3 | 0.1 | 0.6 | 1.0079e+004 | 11.2780 | 1.2825e+007 |
| 0.7 | 0.2 | 0.1 | 9.8795e+003 | 11.5387 | 1.2985e+007 |
| 0.4 | 0.3 | 0.4 | 1.0077e+004 | 11.1599 | 1.2869e+007 |
| 0.4 | 0.5 | 0.1 | 1.0100e+004 | 11.1256 | 1.2881e+007 |
| 0.2 | 0.7 | 0.1 | 1.0197e+004 | 11.0711 | 1.2846e+007 |
| 0.2 | 0.6 | 0.2 | 1.0206e+004 | 11.0687 | 1.2840e+007 |

This is a single objective nonlinear optimization problem, which can be solved by using classical single objective optimization methods.

Then consider the five power plants the system, each competitor as a single business unit. To simplify the calculation, $T$ takes 24, uses the same bid in each time intervals, neglects network losses, unit parameters shown in Table 1 come from literature [8].

Simulation program is realized by matlab, the results are shown from Table 2 to Table 5.

It can be seen that each objective function value how to change as weights changing from table 2 to table 5, it show also that multi-objective load dispatch model is suitable for power market, application still need to determine the weights.

**Table 3** Active power outputs for $w = (0.3, 0.1, 0.6)$

| t | Load | Unit 1 | Unit 2 | Unit 3 | Unit4 | Unit 5 |
|---|------|--------|--------|--------|-------|--------|
| 0 | 1880.80 | 310.0000 | 311.2572 | 493.0673 | 439.7836 | 326.6920 |
| 1 | 2012.01 | 310.0000 | 334.2215 | 529.2445 | 476.2384 | 362.3056 |
| 2 | 1852.12 | 310.0000 | 304.7820 | 482.8498 | 429.4881 | 325.0000 |
| 3 | 1786.95 | 310.0000 | 289.1142 | 458.1920 | 404.6438 | 325.0000 |
| 4 | 1810.95 | 310.0000 | 294.8788 | 467.2748 | 413.7964 | 325.0000 |
| 5 | 1873.01 | 310.0000 | 309.8063 | 490.7528 | 437.4509 | 325.0000 |
| 6 | 1849.90 | 310.0000 | 304.2362 | 482.0154 | 428.6484 | 325.0000 |
| 7 | 1898.98 | 310.0000 | 314.4389 | 498.0796 | 444.8352 | 331.6263 |
| 8 | 2011.98 | 310.0000 | 334.2166 | 529.2362 | 476.2299 | 362.2973 |
| 9 | 2055.99 | 310.0000 | 341.9189 | 541.3706 | 488.4574 | 374.2431 |
| 10 | 2366.64 | 329.3979 | 392.8826 | 621.6786 | 569.3833 | 453.2975 |
| 11 | 2231.75 | 310.0000 | 372.6878 | 589.8281 | 537.2853 | 421.9488 |
| 12 | 2195.77 | 310.0000 | 366.3838 | 579.9104 | 527.2929 | 412.1829 |
| 13 | 2272.76 | 314.4921 | 379.0714 | 599.8998 | 547.4340 | 431.8628 |
| 14 | 2302.72 | 319.2338 | 383.4911 | 606.8513 | 554.4415 | 438.7022 |
| 15 | 2302.72 | 319.2337 | 383.4905 | 606.8516 | 554.4416 | 438.7026 |
| 16 | 2260.72 | 312.5504 | 377.3056 | 597.1148 | 544.6280 | 429.1212 |
| 17 | 2298.72 | 318.5922 | 382.9432 | 605.9115 | 553.4920 | 437.7810 |
| 18 | 2693.79 | 401.7084 | 425.0000 | 700.0000 | 610.0000 | 557.0816 |
| 19 | 2723.50 | 413.9052 | 425.0000 | 700.0000 | 610.0000 | 574.5948 |
| 20 | 2699.50 | 404.0523 | 425.0000 | 700.0000 | 610.0000 | 560.4477 |
| 21 | 2576.55 | 364.6276 | 425.0000 | 673.0534 | 610.0000 | 503.8690 |
| 22 | 2446.66 | 342.1237 | 404.6596 | 640.2334 | 588.0804 | 471.5628 |
| 23 | 2190.94 | 310.0000 | 365.5376 | 578.5791 | 525.9515 | 410.8718 |

[a] t represents the t-th time interval.

**Table 4** The objective function values corresponding to optimal dispatch for $w = (0.3, 0.1, 0.6)$

| t | $f_1$ | $f_2$ | $f_3$ | s | t | $f_1$ | $f_2$ | $f_3$ | s |
|---|-------|-------|-------|---|---|-------|-------|-------|---|
| 0 | 343.7985 | 0.3419 | 4.5858e+005 | 1.2916e+004 | 12 | 419.4567 | 0.4631 | 5.3452e+005 | 1.5494e+004 |
| 1 | 374.4413 | 0.3897 | 4.9021e+005 | 1.3974e+004 | 13 | 438.8754 | 0.4959 | 5.5321e+005 | 1.6143e+004 |
| 2 | 337.0216 | 0.3315 | 4.5190e+005 | 1.2687e+004 | 14 | 446.4251 | 0.5090 | 5.6058e+005 | 1.6398e+004 |
| 3 | 321.7753 | 0.3087 | 4.3689e+005 | 1.2174e+004 | 15 | 446.4250 | 0.5090 | 5.6058e+005 | 1.6398e+004 |
| 4 | 327.3386 | 0.3169 | 4.4242e+005 | 1.2362e+004 | 16 | 435.8583 | 0.4908 | 5.5026e+005 | 1.6041e+004 |
| 5 | 342.0019 | 0.3391 | 4.5671e+005 | 1.2854e+004 | 17 | 445.4175 | 0.5072 | 5.5959e+005 | 1.6363e+004 |
| 6 | 336.4942 | 0.3307 | 4.5139e+005 | 1.2670e+004 | 18 | 545.7669 | 0.6980 | 6.5917e+005 | 1.9810e+004 |
| 7 | 347.9697 | 0.3483 | 4.6296e+005 | 1.3061e+004 | 19 | 553.3294 | 0.7136 | 6.6719e+005 | 2.0081e+004 |
| 8 | 374.4342 | 0.3897 | 4.9021e+005 | 1.3974e+004 | 20 | 547.2106 | 0.7010 | 6.6072e+005 | 1.9862e+004 |
| 9 | 384.9919 | 0.4065 | 5.0082e+005 | 1.4333e+004 | 21 | 517.3768 | 0.6368 | 6.2806e+005 | 1.8765e+004 |
| 10 | 462.7085 | 0.5373 | 5.7628e+005 | 1.6943e+004 | 22 | 483.4414 | 0.5739 | 5.9594e+005 | 1.7632e+004 |
| 11 | 428.5580 | 0.4783 | 5.4320e+005 | 1.5796e+004 | 23 | 418.2421 | 0.4611 | 5.3336e+005 | 1.5453e+004 |

[b] $f_1$ represents the fuel consumption.
[c] $f_2$ represents the Emissions.
[d] $f_3$ represents the Power Purchase Costs.
[e] s represents the $f_1^{w_1} \times f_2^{w_2} \times f_3^{w_3}$.

**Table 5** The objective function values corresponding to optimal dispatch for $w = (0.7, 0.2, 0.1)$

| t | $f_1$ | $f_2$ | $f_3$ | s | t | $f_1$ | $f_2$ | $f_3$ | s |
|---|-------|-------|-------|---|---|-------|-------|-------|---|
| 0 | 336.5777 | 0.3497 | 4.6416e+005 | 175.5358 | 12 | 409.7109 | 0.4759 | 5.4244e+005 | 217.6069 |
| 1 | 365.8914 | 0.4010 | 4.9722e+005 | 192.5885 | 13 | 428.6827 | 0.5092 | 5.6138e+005 | 228.4539 |
| 2 | 330.3431 | 0.3391 | 4.5699e+005 | 171.9222 | 14 | 436.1619 | 0.5224 | 5.6874e+005 | 232.7263 |
| 3 | 316.4379 | 0.3159 | 4.4070e+005 | 163.8801 | 15 | 436.1620 | 0.5224 | 5.6874e+005 | 232.7263 |
| 4 | 321.5162 | 0.3243 | 4.4670e+005 | 166.8142 | 16 | 425.6928 | 0.5039 | 5.5841e+005 | 226.7418 |
| 5 | 334.8772 | 0.3468 | 4.6221e+005 | 174.5499 | 17 | 435.1604 | 0.5206 | 5.6776e+005 | 232.1517 |
| 6 | 329.8635 | 0.3383 | 4.5643e+005 | 171.6453 | 18 | 539.9436 | 0.7071 | 6.6479e+005 | 291.6141 |
| 7 | 340.5656 | 0.3565 | 4.6871e+005 | 177.8455 | 19 | 548.5038 | 0.7216 | 6.7206e+005 | 296.3643 |
| 8 | 365.8855 | 0.4009 | 4.9721e+005 | 192.5763 | 20 | 541.5830 | 0.7099 | 6.6619e+005 | 292.5262 |
| 9 | 376.7662 | 0.4167 | 5.0728e+005 | 198.4905 | 21 | 507.0145 | 0.6516 | 6.3609e+005 | 273.3143 |
| 10 | 452.2984 | 0.5513 | 5.8447e+005 | 241.9643 | 22 | 472.8450 | 0.5885 | 6.0415e+005 | 253.7271 |
| 11 | 418.5331 | 0.4913 | 5.5129e+005 | 222.6477 | 23 | 408.5325 | 0.4738 | 5.4125e+005 | 216.9289 |

# 4 Conclusions

Geometric weighting method can obtain the non-inferior solutions of multi-objective optimization problem. This method can be used to solve the actual problem.

In electric market environment, load dispatch problem is no longer a pure economic dispatch problem which minimizes only fuel consumption or operating costs, it is necessary to build the multi objective load dispatch mode.

# References

1. Abido, M.A.: Environmental/economic power dispatch using multiobjective evolutionary algorithms. IEEE Transactions on Power Systems 18, 1529–1537 (2003)
2. Attaviriyanupap, P., Kita, H., Tanaka, E., Hasegawa, J.: A hybrid EP and SQP for dynamic economic dispatch with nonsmooth fuel cost function. IEEE Transactions on Power Systems 17, 411–416 (2002)
3. Balamurugan, R., Subramanian, S.: Hybrid integer coded differential evolution - dynamic programming approach for economic load dispatch with multiple fuel options. Energy Conversion and Management 49, 608–614 (2008)
4. Hota, P.K., Chakrabarti, R., Chattopadhyay, P.K.: Economic emission load dispatch through an interactive fuzzy satisfying method. Electric Power Systems Research 54, 151–157 (2000)
5. Hu, Y.D.: Practical Multi-Objective Optimization. Shang Hai Science and Technology, Shang Hai (1990)
6. Irisarri, G., Kimball, L.M., Clements, K.A., Bagchi, A., Davis, P.W.: Economic dispatch with network and ramping constraints via interior point methods. IEEE Transactions on Power Systems 13, 236–242 (1998)

7. Li, D.P., Pahwa, A., Das, S., Rodrigo, D.: A new optimal dispatch method for the day-ahead electricity market using a multi-objective evolutionary approach. In: 2007 39th North American Power Symposium (NAPS 2007), pp. 433–439 (2007)
8. Ma, R., Mu, D.Q., Li, X.R.: Study on fuzzy decision of multi-objective dispatch strategy for daily active power in electricity market. Power System Technology 25, 25–29 (2001)
9. Morgan, L.F., Williams, R.D.: Towards more cost saving under stricter ramping rate constraints of dynamic economic dispatch problems-a genetic based approach. Genetic Algorithms in Engineering Systems: Innovations and Applications 446, 221–225 (1997)
10. Rabbani, M., Aramoon Bajestani, M., Baharian Khoshkhou, G.: A multi-objective particle swarm optimization for project selection problem. Expert Systems with Applications 37, 315–321 (2010)
11. Summanwar, V.S., Jayaraman, V.K., Kulkarni, B.D., Kusumakar, H.S., Gupta, K., Rajesh, J.: Solution of constrained optimization problems by multi-objective genetic algorithm. Computers and Chemical Engineering 26, 1481–1492 (2002)
12. Yang, H.T., Yang, P.C., Huang, C.L.: Evolutionary programming based economic dispatch for units with non-smooth fuel cost functions. IEEE Transactions on Power Systems 11, 112–118 (1996)

# Application of Cross-Correlation in Velocity Measurement of Rainwater in Pipe

Zhanpeng Li and Bin Fang

**Abstract.** According to the diversity of fluid composition and the complexity of flow pattern of rainwater in the pipeline, a flow velocity measurement method based on cross-correlation analysis is proposed, reference flow filed measurement technique in Particle Image Velocimetry. The fast algorithm of cross-correlation analysis is studied. Established a pipeline rainwater calibration device using weir slot. The results show that extracting an average speed from the velocity field can smooth the flow variability and accurately measure the average velocity of rainwater.

**Keywords:** Cross-correlation, Flow measurement, Rainwater.

## 1 Introduction

As the rapid development of urbanization, there is a significant increase in impervious area, and cities face increasing pressure on storm and flood. By monitoring the rainwater flow in the pipe network, we can predict the region where seeper may occur. Due to the rotary mechanical structure, traditional propeller flow meter is easily winded or even damaged by the impurities in the pipe. On the other hand, current ultrasonic flow meters are too expensive to install widely.

Unlike existing methods, we use the solid particles as tracer particles of the rainwater, combining with image processing technology, firstly measure the distribution of two-dimensional flow field by cross-correlation technique, then obtain the average flow rate by data analysis method. Experimental results

Zhanpeng Li and Bin Fang
College of Electronic Information and Control Engineering, Beijing University of Technology, 100 Pingleyuan, Chaoyang District, Beijing, 100124, China
e-mail: `caiffei@emails.bjut.edu.cn,fangbin@bjut.edu.cn`

S. Li (Eds.): Nonlinear Maths for Uncertainty and its Appli., AISC 100, pp. 587–594.
springerlink.com      © Springer-Verlag Berlin Heidelberg 2011

show that this method is suitable for the complex rainwater flow composition, which can measure rainwater flow accurately.

## 2  Principle of Cross-Correlation and Rapid Realization

### 2.1  Cross-Correlation Analysis

Cross-correlation analysis method is a new detection method based on information theory and random process theory. Recent years, with the rapid development of computer technology and signal (image) processing technology, cross-correlation has been successfully applied to many engineering fields, such as image matching, flow measurement and modal analysis and so on.

The flow field image at time t1 is represented as $p(x,y) = I(x,y) + n_1(x,y)$. The relative displacement of the internal flow field is not intense when the time interval $\Delta t$ is small enough. So, flow field at $t_2$ ($t_2 = t_1 + \Delta t$) time can be represented as $q(x,y) = I(x + \Delta x, y + \Delta y) + n_2(x,y)$, where $n_1(x,y), t_2(x,y)$ is random noise in the imaging system. $p(x,y)$ and $q(x,y)$ calculated the cross-correlation function $r_{pq}(\tau_x, \tau_y)$, assuming the noise $n_1(x,y)$ and $n_2(x,y)$ is not relevant to image signal $I(x,y)$, that their cross-correlation function is 0. We can get the following formula [5]:

$$r_{pq}(\tau_x, \tau_y) = \iint p(x,y)q(x + \tau_x, y + \tau_y)dxdy$$
$$= \iint I(x,y)I(x + \Delta x + \tau_x, y + \Delta y + \tau_y)dxdy \tag{1}$$

According to the definition of auto-correlation function, the auto-correlation function of $I(x,y)$ is:

$$r_{II} = \iint I(x,y)I(x + \tau_x, y + \tau_y)dxdy \tag{2}$$

Thus, equation (1) can be transformed into:

$$r_{pq} = r_{II}(\tau_x + \Delta x, \tau_y + \Delta y) \tag{3}$$

Auto-correlation function is an even function and obtain the maximum value at the origin, that satisfies the inequality $r_{II}(\tau_x, \tau_y) \leqq r_{II}(0,0)$. With formula (3), the following inequality holds:

$$r_{pq} \leqslant r_{pq}(-\Delta x, -\Delta y) \tag{4}$$

The location of maximum of cross-correlation function corresponds to the relative displacement between the flow field (cross-correlation peaks is showed in Figure 1), that is the average displacement of tracer particles in the time between $t_1$ and $t_2$. The interval time of the image collection can be determined, that $\Delta t$ is known. Average speed of tracer particles in $\Delta t$ can be calculated using basic definition of speed. The speed of tracer particles can approximated represent the speed of water.

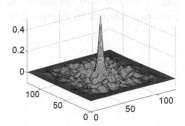

**Fig. 1** Cross-correlation peak

In actual analysis process, the pair of two images need to be meshed, and calculate the cross-correlation function of each grid pair. Use the maximum location of cross-correlation to determine average relative displacement of the grid. Then further calculate the average speed. Scan the whole image and obtain two-dimensional velocity vector distribution.

## 2.2 Fast Implementation of Cross-Correlation Analysis

In order to improve the efficiency of cross-correlation analysis and reduce the computing time, two-dimensional fast Fourier transform can be used to achieve cross-correlation. Suppose two-dimensional Fourier transform of $p(x,y)$, $q(x,y)$ and $r_{pq}(\tau x, \tau y)$ respectively is $P(u,v)$, $Q(u,v)$ and $R(u,v)$. Following equation holds:

$$\begin{cases} P(u,v) = \iint p(x,y)e^{-i(ux+vy)}dxdy \\ Q(u,v) = \iint p(x,y)e^{-i(ux+vy)}dxdy \\ R(u,v) = \iint r_{pq}(x,y)e^{-i(ux+vy)}dxdy \end{cases} \qquad (5)$$

The corresponding frequency domain expression is [2]:

$$R(u,v) = P^*(u,v) \cdot Q(u,v), \qquad (6)$$

where $P^*(u,v)$ is the conjugate function of $P(u,v)$.

Using cross-correlation properties of Fourier transform and fast Fourier transform algorithm, we can only calculate twice Fourier transform and one inverse Fourier transform to achieve the calculation of cross-correlation function $r_{pq}(\tau_x, \tau_y)$. It avoids complex integral operation and greatly improves the computing speed.

# 3   Image Preprocessing and Data Post-Processing

## 3.1   Weighting Function

Since particles may run out of or into the cross-correlation analysis regional, a mutation will occur around the edges of analysis region. Before conducting cross-correlation analysis, flow field images need to add cosine weighting function shown in Figure 2 to reduce the weight of region edge.

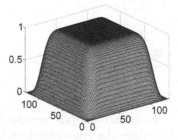

**Fig. 2** Cosine weighting function

## 3.2   Sub-Pixel Fit

In order to get sub-pixel precision, cross-correlation results must make sub-pixel interpolation [3]. Commonly used sub-pixel precision method are centroid method, parabolic fit, Gaussian fit and Whittaker reconstruction. They have different accuracy in different situations. For the characteristics of particle image is similar to a Gaussian distribution, Gaussian curve fitting for sub-pixel interpolation can achieve good results.

## 3.3   Vector Correction

Related to camera image quality, flow instability, the tracer particle (i.e. impurities in the water) too much or too little and other factors, false velocity vector will inevitably be generated in the analysis process [4]. For the mutation between error vector and other vectors around is greatly, using median filter or low pass filter can effectively remove the error vector. Calculate the average of velocity vector field can obtain the average speed of water flow.

## 3.4   Feasibility Analysis of Algorithms

Figure 3 is an image (pixel size is 640×480) of flow filed captured by generic USB camera, bright part of which is the image of solid particles in water. Shift it to the upper right corner of 4 pixels by image processing software. Make cross-correlation analysis between the two images. Figure 4 is the distribution of displacement vector. The average of horizontal displacement is 4.02 and standard deviation is 0.04. The average of vertical displacement is 4.01 and standard deviation is 0.04. The result shows that this method can correctly reflect the displacement of flow field.

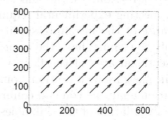

**Fig. 3** Flow image

**Fig. 4** Distribution of displacement vector

## 4 Cross-Correlation Analysis Software Design

After verification of the cross-correlation algorithm by using Matlab, we develop cross-correlation analysis software using MFC in Visual Studio 2008.

Use DirectShow technology to collect the images of flow field which captured by USB camera. DirectShow can be easily used to capture image data form video capture device which support WDM driver model (such as general USB camera), and do corresponding post-processing. DirectShow is based on COM component object mode, composed of many modular components. Developers just need to write applications follow the COM interfaces without regard to the difference of terminal hardware and configuration of hardware, greatly simplifying the program development process [6].

## 5 Experimental Device

In order to simulate the components of urban rain pipeline, you need to add appropriate amount of impurities into the water (such as solid particulate matter, organic matter, etc.). To be able to recycle the rain mixed with impurities - we formed a circulating water channel to simulate the flow of rainwater in pipe by using the plexiglass plate, shown in Figure 5.

Pump pumps the 'rain' to interval 2 from interval 1 continuously, and the 'rain' clockwise flow in the tank. Blocked by the triangular weir, the water level difference formed on both sides of triangular weir. The relationship between water level of right-angle triangle weir and flow rate is [1]:

$$Q = 1.4h^{5/2} \tag{7}$$

Flow can be calculated by measuring the height of water head on the left of the triangular weir. Although many measurements can measure flow, such as Venturi flow meter, electromagnetic flow meter, etc, but usually still use the thin-walled rectangular or triangular weir, because of its high measurement accuracy. Normally, technician use a triangular weir with high accuracy to measure small flows. When $\theta = 90°$, the measurement range of triangular weir is $0.001\sim1.8m^3/s$, the uncertainty of calculated flow rate ranges from 1 to 3%, meeting experimental requirements.

Because of the incompressibility of water, the 'rain' flows through both sides of the channel have the same flow in recycling process. Combined with cross-sectional area of the flow, we can easily calculate the velocity of the 'rain' in measurement section; compare with the velocity obtained by cross-correlation analysis, you can confirm whether the method is valid.

The installation of the sink holes in the rectifier board can reduce the level fluctuation and the measurement error.

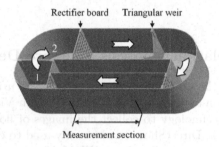

**Fig. 5** Distribution of absolute error

## 6  Experimental Results Analysis

Green light laser slices perpendicular light the area of the flow field, and parallel to the direction of the flow; USB camera collected the image from the side of the flow field which be illuminated; finally, the velocity vector distribution can be got by using the cross-correlation analysis (Figure 6). The left is the real-time video information the camera shot, the right is an image collected, and the white arrow means the velocity vector distributions.

Before carrying on the actual survey, needs to identify the actual size of each single pixel in the image. Put the grid coordinates with actual size the sign (the unit: mm) in water, and photographs an image. Then, use the mouse to click on the corresponding coordinates of the points in order in the program interface, and the camera pats between image picture element and real displacement's proportional relationship has been identified. Figure 7(a) is the actual photographic analyzing result of the flow field mutual correlation. After Figure 7(b) is the result after the vector revising. Calculation of mean can obtain average value.

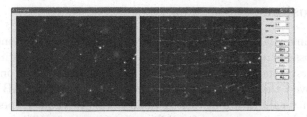

**Fig. 6** Distribution of absolute error

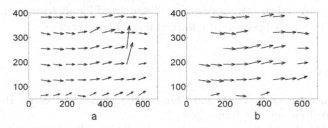

**Fig. 7** Distribution of absolute error

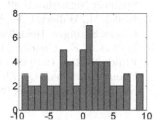

**Fig. 8** Distribution of
relative error

**Table 1** Measurement error under different flow velocity

|                          | 1     | 2     | 3     | 4     | 5     |
|--------------------------|-------|-------|-------|-------|-------|
| Measurements(mm/s)       | 13.36 | 20.89 | 36.62 | 39.37 | 44.51 |
| Cross-correlation (mm/s) | 13.83 | 20.08 | 38.11 | 38.21 | 42.15 |
| Relative error (%)       | 3.52  | 3.88  | 4.07  | 2.95  | 5.30  |

Make the flow velocity calculated from the weir the flow tank as a reference, when the flow velocity in the measurement section is 13.36mm/s, the absolute error distribution between the measured value and the reference values shown in Figure 8, the average velocity measured is 13.83mm/s, the relative error is 3.5%, which can meet the requirements of general measurement.

# 7  Conclusions

This paper investigates a pipeline rainwater velocity measurement method based on cross-correlation analysis technology, expounds the principle of the method and system design scheme. The fast algorithm based on FFT effectively improved the calculation speed. Based on the MATLAB simulation, used a velocity calibration device to validate this method. Experimental results showed the accuracy and feasibility of the method. With no mechanical movement structure and low-costs, this measurement system can be used in piping conditions. The method of extracting average velocity from velocity field can be applied to other occasions where are difficult to measuring using traditional methods because of complex flow characteristics.

# References

1. Flow Measurement by Weirs and Flumes. The Ministry of Water Resources of the People Republic of China (1991)
2. Jensen, K.D.: Flow measurements. Journal of the Brazilian Society of Mechanical Sciences and Engineering 26, 400–419 (2004)
3. Keane, R.D., Adrian, R.J.: Theory of cross-correlation analysis of PIV images. Springer Netherlands 49, 191–215 (1992)
4. Raffel, M., Willert, C.E., Wereley, S.T.: Particle Image Velocimetry - A Practical Guide. Springer, Heidelberg (2007)
5. Willert, C.E., Gharib, M.: Digital particle image velocimitry. Experiments in Fluids 10, 181–193 (1991)
6. Zheng, H., Chen, Q., Zhang, W.: On DirectShow and Its Application. Microcomputth Applications 08 (2001)

# A Decision-Making Model of Adopting New Technology by Enterprises in Industrial Clusters under the Condition of Interaction

Xianli Meng, Xin-an Fan, Baomin Hu, and Xuan Li

**Abstract.** Adopting new technology by enterprises, i.e. technology diffusion, in industrial clusters is an important issue for the development of industrial clusters. Technology diffusion in country industrial clusters is a complicated process and has specific characteristics. As most country industrial clusters are composed of small and medium sized enterprises which situate close to each other, technology diffusion within these industrial clusters is usually carried out through the informal relationships between them. In this paper, the technology diffusion process within country industrial clusters is analyzed based on the herd behavior in the behavior finance theory. A decision-making model of adapting new technologies by enterprises in country industrial clusters under the condition of interaction is proposed.

**Keywords:** Decision-making model, Herd behavior theory, Industrial Clusters, Technology Diffusion.

## 1 Introduction

Industrial clusters have great strengths in providing competitiveness and creativity [1, 2, 3, 4]. One important issue of the emergence and development of industrial clusters is that, with the help of the technology-spreading network established within the clusters, enterprises within a cluster can adopt essential production technologies to develop themselves more conveniently and collectively. The evolution of industrial clusters can be viewed as a process of

Xianli Meng, Baomin Hu, and Xuan Li
School of Management, Hebei University of Technology, Tianjin, P.R. China
e-mail: meng5079510@126.com

Xianli Meng and Xin-an Fan
Hebei University, Baoding, P.R. China

S. Li (Eds.): Nonlinear Maths for Uncertainty and its Appli., AISC 100, pp. 595–601.
springerlink.com                    © Springer-Verlag Berlin Heidelberg 2011

enterprises adopting new technologies, that is, a process of technology diffusion [5,6]. Therefore, the technology diffusion in industrial clusters is vital for the development of industrial clusters. However, the technology diffusion process is a complicated process. For enterprises within the industrial clusters, a decision-making model of adopting new technologies is needed.

Country industrial clusters are a kind of specific clusters. They compete and cooperate with each other. Their geographical proximity and complex informal relationship doomed that their emergence, development and the technology diffusion within them have some specific characteristics. Technology diffusion in country industrial clusters is a complicated process. Because most country industrial clusters are composed of small and medium sized enterprises which situate close to each other, technology diffusion within these industrial clusters is usually carried out through the informal relationships between them [7], which makes the decision-making problem during the technology diffusion process complex. The technology diffusion process within country industrial clusters meets the assumption of the herd behavior theory in behavioral finance, and the problem is similar to the herd behavior theory of investors in capital market. In the light of the information stacking discipline in the herd behavior theory, this paper is trying to build a decision-making model for country industrial clusters to adopt new technologies.

The rest of this paper is organized as follows. Section 2 presents basic assumptions of the country technology diffusion model based on the herd behavior theory. The decision-making model of adapting new technology by enterprises in country industrial clusters under the condition of interaction is proposed in Section 3. The model is discussed in Section 4.

## 2 The Basic Assumptions

In view of the overstrict assumption in traditional financial theory, the herd behavior is generated. It relaxes these assumptions, rationally takes people's psychological factors into consideration and makes the research more in line with the actual situation of the capital market. However, the construction of these theories is based on a series of assumptions. Therefore, the country technology diffusion model based on herd behavior theory is also built on the basis of these assumptions. Specifically speaking, it includes:

(1) Information is visible. This assumption is easily embodied in real life. In the county industrial clusters, all enterprises are concentrated in the same county and are familiar with each other. The complex relationship between staff members in different enterprises and the petticoat influence result that the information transmission between enterprises carries out more through informal relationships. This also increases the visibility of information.

(2) Decision-making is carried out in turns. Generally speaking, technology diffusion starts in one single enterprise but followed by other enterprises in the same country. This is in line with the order assumption of decision-making.

(3) All investors are seeking the maximum of the VNM utility function (anticipated utility function, put forward by Von Neumann and Morgenstern in 1944, takes the weighted average of anticipated utility by investors on uncertain conditions as the judgment standard of investment, instead of the anticipated value of currency result). It is the basic assumption of behavioral finance theory.

## 3 The Model Building

This model assumes the existence of the N investors. We assume an equal amount of investment from each investor. Suppose there are a series of investment, each is represented by numerical figure between $[0, 1]$ and $a(i)$ represents the $ith$ project. The return of project $a(i)$ is $z(i) \in R$. Suppose the existence of $i^*$ is unique, any $i \neq i^*$ ensures $z(i) = 0$ and $z(i^*) = z, z \succ 0$. This indicates that the return of investing in certain project is strictly much more than that of investing on other projects. Investors don't know the specific $i^*$, but they have their own estimation. Each investor gets the information that the value of $i^*$ is $i'$ by the possibility of $\alpha$. Of course, this information is not necessarily true. The possibility of being false, that is the value of $i^*$ is not $i'$, is $1 - a$.

Decision-making is carried out in turns. We assume a certain investor who is chosen randomly makes decision first. The second investor who is also chosen randomly makes decision accordingly. Different from the first investor, investors followed can see the decisions made by the investors before them. However, investors followed can not make sure whether the first investor really obtains certain information. The following game goes on in the same way. Each new decision-maker makes decisions based on the past records. After all decisions are made, these decisions will be tested. If any of them is valid, the investor who makes it will be rewarded. If no one makes a valid decision, none of them will be rewarded.

Suppose that the game structure and Bayesian reason are common knowledge, and they obey the following rules:

(1) When an investor has no information and other investors have chosen $i = 0$ (investors do not make any investment), then the investor also choose $i = 0$.

(2) When there is no difference in making choices between the two ways, following other investors and according to his own information, he always gives priority to the latter one.

(3) When there is no difference in making choices by following different preceding investors, he always gives priority to follow the choice of the investor whose $i$ value is the highest.

If the first investor acquires a certain amount of information, he must make an appropriate decision according to the information; if he doesn't have any information, he will choose $i = 0$ according to (1). This option has the smallest possibility of making mistakes. If the second investor has no information, he will imitate the first investor and choose $i = 0$ too. But if he has information and the first investor doesn't choose $i = 0$, then he will know the first investor must have acquired certain amount of information and that the possibility of their information being right is the same. Thus, there would be no difference for the second investor to make decisions whether by referring to his own information or by following the first investor. According to (2), at that time, investors give priority to their own information.

In this way, the choices of the first two investors will have four possible results: 1) both of them have chosen $i = 0$; 2) the first investor has chosen $i = 0$, while the second has chosen $i \neq 0$; 3) both of them have chosen the same $i \neq 0$; 4) they chose different non zero values. The third investor will observe one of the results. When both of the first two investors choose $i = 0$, the third investor will follow the decision made by the preceding two investors if he has no information, otherwise, he will follow his own information. Under the third condition, if the third investor has no information, he will follow the person who has chosen $i \neq 0$. If both of the first two investors choose different $i$ which is not zero, then according to (3), the third investor will follow the decision made by the investor whose $i$ value is higher. On the other hand, if the third investor has got the information $i'$ and the first two investors haven't chosen the same $i^0(i^0 \neq 0 \; and \; i^0 \neq i')$, he will make his decision according to his own information. If both of the first two investors have chosen the same $i^0$, it proves that the information of the first investor is at least as good as that of the third investor. His decision is probably right, thus the third investor should also choose $i^0$. When the first two investors both have chosen $i = 0$, obviously, the third investor will make the decision according to his own information. When there is only one who makes a choice different from $i = 0$ and $i = i'$, the other will choose $i = 0$. This is the result of (2). When his information coincides with the choice of the preceding investors, it is sure that he will make the decision according to his own information.

Next investor will observe the following three cases:

A number of investors choose a certain option other than $i = 0$, and it is the option with the biggest $i$ inside.

A number of investors choose a certain option other than $i = 0$, but it is not the option with the biggest $i$ inside.

A number of investors choose two certain options other than $i = 0$ and one of them is the option with the biggest $i$ inside.

The first case is similar to the front view. The next investor should follow the options that most investors have chosen before him. In the second and third cases, obviously, the decision-maker who has chosen the option without the biggest $i$ inside is sure to have the same information. It's sure to be the right option and the following investors should choose it.

This view is suitable for all the following investors. In a word, according to (1), (2), (3), the balanced decision-making rules that investors follow are as follows:

1. If the first investor has information, he will make the decision according to his own information; If not, he will choose $i = 0$.

2. If the $k(k \succ 1)$ th investor has information, he will make the decision according to his own information under such conditions: if and only if Condition (1) is satisfied or Condition (1) isn't satisfied but Condition(2) is. Condition (1): his information coincides with the choice of a certain preceding investor; Condition (2): apart from $i = 0$, no choices of the preceding investors are the same.

3. Suppose the $k(k \succ 1)$th investor has information, if the preceding investors have chosen a certain option except the biggest $i$ for many times, then he also chooses it. If the choices of the preceding investors coincide with his information, then he will make the decision according to his own information.

4. Suppose the $k(k \succ 1)$th investor has information, if the preceding investors have chosen a certain option except the biggest $i$ for many times and besides only $i = 0$ has been chosen repeatedly, then he will also choose the biggest $i$ option. But if the choices of the preceding investors coincide with his information, he will choose this option.

5. Suppose the $k(k \succ 1)$th investor doesn't have information, if and only if all the preceding investors have chosen $i = 0$, he will choose $i = 0$. Otherwise, if there is no other repeatedly chosen option except $i = 0$, then he will choose the biggest $i$ option; but if there is a repeatedly chosen option except $i = 0$, he will choose this option too. To be more visual and clearer, see the specific rules in Table 1.

**Table 1** The Balanced Decision-making Rules for Investors

| Information | Choice records | Option |
|---|---|---|
| Information $i$ | No other people choose $i = i_k$ two options except $i = 0$ are chosen by many people | The smaller one of the two |
| | No other people choose $i = i_k$ one option except $i = 0$ is chosen by many people | This one |
| | No other people choose $i = i_k$, no option except $i = 0$ is chosen by many people | $i = i_k$ |
| | Another person chooses $i = i_k$ | $i = i_k$ |
| No Information | Except $i = 0$, two options are chosen by many people | The smaller one of the two |
| | Except $i = 0$, only one option is chosen by many people | This one |
| | One person chooses all the other options except $i = 0$ | The biggest from all of the chosen ones |
| | The others all choose $i = 0$ | $i = 0$ |

The characteristics of the balanced decision-making rule in the model are the externalities of the herd behavior. Even if investors cannot exactly make sure whether the choices of other people are right or not, they give up their own information and follow others. If the first one has information, he will follow his own, so will the second one. But we cannot guarantee that the third will also follow his own. If the first one chooses $i = 0$ and the second one follows him, then the third one will always follow them, so will all the following investors.

When the first person, the second, even the third and the fourth make different options, herd behavior can also occur. After making $k$ different options, for any $k \succ 0$, if next investor has no information, he will choose the option with the biggest $i$ value inside. If certain information of all the following investors coincides with the chosen options, they will choose the same option. Only if the right option is chosen, this will happen. Therefore, if the first information investor or a certain investor after him but before the first none-information investor doesn't make the right option, then the herd behavior of wrong option occurs.

The key of this model is that the choices made by investors are not always the result of the complete collections of their information. If the choice is always the result of the complete collections, potential investors will always know the information that the preceding investors are based on. In that case there will be no herd externality and validity. Potential herd externality exists when the choice of some investors influence the information obtainment of the following investors. Moreover, in this model, the herd behavior has the characteristics of positive feedback, so its balanced pattern is variable and unstable. This shows the variability and unsteadiness of the technology diffusion in industrial clusters.

# 4 Discussion and Conclusions

Based on the herd behavior in the behavioral finance theory, the model builds the one for technology diffusion in industrial clusters. On one hand, this model is built on the basis of a series of assumptions and rules. However, these assumptions and rules are not exactly the same to practical life. On the other hand, this model only takes into account the information stacking up in industrial clusters. It doesn't consider other factors which affect the technology diffusion in industrial clusters, especially the relationship between enterprises, enterprises and government, enterprises and external environment. So the analysis has certain limitation. Nevertheless, this model grasps the essence of technology diffusion in industrial clusters and represents the core of technology diffusion. It also provides some help to improve technology diffusion within industrial clusters.

# References

1. Cooke, S.: Structural competitiveness and learning region. Enterprise and Innovation Management Studies 1, 265–280 (2000)
2. Asheim, B.: Industrial districts as learning region: a condition for prosperity. European Planning Studies 4, 379–400 (1996)
3. Padmore, T., Gibson, H.: Modeling systems of innovation: a framework of industrial cluster analysis in regions. Research Policy 26, 625–641 (1998)
4. Radosevic: Regional innovation system in central and easten Europe: determinants, organization and alignments. Journal of Technology Transfer 27, 87–96 (2002)
5. Wang, J.: Innovative Spaces: Enterprise Clusters and Regional Development. Beijing university press, Beijing (2001)
6. Shang, Y., Zhu, C.: The Theory and Practice of Regional Innovation Systems. China Economic Publishing House, Beijing (1999)
7. Qiu, B.: Small and Medium Enterprises Cluster. Fudan University Press, Shanghai (1999)
8. Zhou, Z.: Behavioral Finance Theory and Application. Tsinghua University Press, Beijing (2004)
9. Ma, C.: Enterprise Development Strategy – Reflections on Community. PhD Thesis, Nankai University, Tianjin (1993)
10. Hu, Z.: National Innovation System – Theoretical Analysis and International Comparison. Social Sciences Academic Press, Beijing (2000)

# Study of Newsboy Problem with Fuzzy Probability Distribution

Xinshun Ma and Ting Yuan

**Abstract.** Newsboy problem, as the famous and basic model in stochastic programming, has numerous applications. Completely known probability distribution for random demand is a basic hypothesis in classical Newsboy model, but this is almost impossible to achieve in practice due to that the probability distribution is generally determined by some approximative methods such as statistical inferences and expertise. Newsboy problem with fuzzy probability distribution is studied in this paper, and two-stage stochastic linear programming model with recourse is set up. Improved L-shaped algorithm is designed to solve the problem. Numerical example demonstrates the feasibility and efficiency of the algorithm.

**Keywords:** Newsboy problem, Two-stage stochastic programming, Fuzzy probability distribution, L-shaped algorithm.

## 1 Introduction

Newsboy problem (also known as Newsvendor problem) is a classical stochastic programming problem [10], and has been widely applied in many fields including supply chain management. With the increased decentralization of manufacturing activities, there is a renewed interest in inventory theory, and Newsboy problem has caused wide public concern in the research works [6]. Newsboy problem and its extensions have received much attention from researchers in inventory managements, the studies range over that multiple

Xinshun Ma and Ting Yuan
Department of Mathematics and Physics,
North China Electric Power University, Baoding 071003,
Hebei Province, P.R. China
e-mail: xsma@ncepubd.edu.cn, luckyyuanting@163.com

S. Li (Eds.): Nonlinear Maths for Uncertainty and its Appli., AISC 100, pp. 603–610.
springerlink.com      © Springer-Verlag Berlin Heidelberg 2011

periods, multiple products, random demand forecasting, lead-time demand etc. [3,11].

More often than not, the aforementioned inventory models developed from Newsboy problem framework are based on a basic hypothesis that demands are stochastic and can be described by random variables with completely known probability distribution. However, in many situations, the probability distribution is determined through some approximate methods such as statistical inferences and expertise, so it is nearly impossible to obtain the probability distribution accurately.

Fuzzy random demand basic methods arise in Newsboy modeling in the literatures [13,5]. Fuzzy random variables are random variables whose values are fuzzy numbers. Since H. Kwakernaak [8] first proposed the concept of fuzzy random variables based on fuzzy set theory [14], fuzzy random theories have received much attention from researchers. In [9], a new concept of fuzzy random variables was introduced based on the credibility measure, the expected value model was presented, and the hybrid intelligent algorithms integrated genetic algorithm and artificial neural network were designed to solve the problem. Based on the above theory, the Newsboy models were presented in [13] under the fuzzy random demand, and simulating based method was used to achieve the approximate solution. The other one type of fuzzy random concept emerged from work in [5], where the random variable obeyed normal distribution whose uncertainty mean was a triangle fuzzy number. Based on this approach, inventory model related to fuzzy random lead-time demand and fuzzy total demand was proposed in this literature.

Another fuzziness with respect to stochastic probability distribution known as linear partial information (LPI) is introduced by E. Kofler [7]. Essentially, LPI-fuzziness is described by establishment of linear equality or inequality constraint. Compared with fuzzy set theory, LPI-fuzziness not only has simpler algorithm, but also is easier to implement in practice. Stochastic linear programming problem in which the stochastic probability distribution has LPI-fuzziness was studied in literature [1], and two approaches including the chance constraint and the recourse method are used to solve the problem. Based on above methodological framework, multi-objective stochastic linear programming with LPI-fuzziness was researched in [2], a compromise programming approach was used to solve the problem.

In this paper, Newsboy problem is studied based on LPI-fuzziness, and two-stage stochastic linear programming with recourse model is set up. Improved L-shaped algorithm is employed to solve the model [12,4], numerical example demonstrates the feasibility and efficiency of the method.

## 2  Problem Statement

The popular Newsboy problem is described as follows. A newsboy goes to the publisher every morning and buys $x$ newspapers at a wholesale price of

$c$, then he sells as many newspapers as possible at the retail price $q_1$. Any unsold newspapers can be returned to the publisher at a return price $q_2$. Demand for newspapers is unknown to the newsboy at the time of purchase, and it is described by a random variable $\xi$ defined on the probability space $(\Omega, \mathscr{A}, P)$, where $\Omega$, $\mathscr{A}$, and $P$ are nonempty set, $\sigma$- algebra of subsets of $\Omega$, and probability measure, respectively.

To formulate the newsboy's profit, suppose that $y_1$ is the number of the effective sales, and $y_2$ is the number of newspapers returned to the publisher at the end of the day. Obviously, both $y_1$ and $y_2$ are associated with the random demand $\xi$. Because of the limitation of purchasing power, the number of newspapers purchased by the newsboy is bounded above by some limit $u$, i.e. $0 \le x \le u$. Newsboy problem is to decide how many newspapers are purchased from publisher to maximize his own profit, where the demand for newspapers varies over days and is described by the random variable $\xi$. As a result, the problem can be formulated as the following two-stage stochastic programming problem:

$$min \quad cx + E[Q(x,\xi)] \tag{1a}$$
$$s.t. \quad 0 \le x \le u, \tag{1b}$$

and

$$Q(x,\xi) = min \quad \{-q_1 \cdot y_1(\xi) - q_2 \cdot y_2(\xi)\} \tag{2a}$$
$$s.t. \quad y_1(\xi) \le \xi, \tag{2b}$$
$$y_1(\xi) + y_2(\xi) \le x, \tag{2c}$$
$$y_1(\xi), y_2(\xi) \ge 0, \tag{2d}$$

where $E[\cdot]$ denotes the mathematical expectation with respect to $\xi$. By introducing the following notations,

$$y(\xi) = \left( y_1(\xi), y_2(\xi), s_1(\xi), s_2(\xi) \right)^T, q = \left( -q_1, -q_2, 0, 0 \right)^T,$$

$$W = \begin{pmatrix} 1 & 0 & 1 & 0 \\ 1 & 1 & 0 & 1 \end{pmatrix}, h(\xi) = \left( \xi, 0 \right)^T, T = \left( 0, -1 \right)^T,$$

where $s_1(\xi)$ and $s_2(\xi)$ are slack variables associated to the inequality constraints (2b) and (2c), respectively, then the second stage problem (2) has the following form:

$$Q(x,\xi) = min \quad q^T \cdot y(\xi) \tag{3a}$$
$$s.t. \quad Wy(\xi) = h(\xi) - Tx, \tag{3b}$$
$$y(\xi) \ge 0. \tag{3c}$$

Assume that the finite discrete probability distribution is taken into consideration, i.e. $\Omega$ and $\mathscr{A}$ are the finite nonempty set with $\Omega = \{\omega_1, \omega_2, \cdots, \omega_n\}$

and the power set of $\Omega$ with $\mathscr{A} = 2^{\Omega}$, respectively. Moreover, assume that $p_i = P(\{\omega = \omega_i\})$ and $\xi_i = \xi(\omega_i)$, for $i = 1, 2, \cdots, n$. In the classical Newsboy problem, all probabilities mentioned above are considered as the completely known values. However, in many situations, these probabilities cannot be explicitly determined due to some practical methods such as statistical inferences and expertise.

Based on linear partial information [7], we assume that $p_i$ almost belongs to some interval $[a_i, b_i]$, i.e. $a_i \lesssim p_i \lesssim b_i, i = 1, 2, \cdots, n$, where $a_i$ and $b_i$ are crisp values. $a_i \lesssim p_i \lesssim b_i$ are fuzzy inequalities, especially, $a_i \lesssim p_i \lesssim a_i$ means that $p_i$ is almost equal to $a_i$. Suppose that the membership function $\mu_i$ associated with the fuzzy inequality $a_i \lesssim p_i \lesssim b_i$ is the following trapezoidal piecewise linear function:

$$
\mu_i(p_i) = \begin{cases} \dfrac{p_i - a_i + d_i^{-}}{d_i^{-}}, & if \quad a_i - d_i^{-} \le p_i \le a_i, \\ 1, & if \quad a_i \le p_i \le b_i, \\ \dfrac{b_i + d_i^{+} - p_i}{d_i^{+}}, & if \quad a_i \le p_i \le b_i + d_i^{+}, \\ 0, & otherwise, \end{cases} \tag{4}
$$

for $i = 1, 2, \cdots, n$, where $d_i^{-}$ and $d_i^{+}$ are vagueness level. By making use of $\alpha$-cut technique, we can transform the fuzzy inequalities into the following set:

$$
\pi = \left\{ \begin{aligned} & P = (p_1, p_2, \cdots, p_n)^T \in R^n | \sum_{i=1}^{n} p_i = 1, p_i \ge 0, \\ & a_i - (1 - \alpha_i^{-})d_i^{-} \le p_i \le b_i + (1 - \alpha_i^{+})d_i^{+}, i = 1, 2, \cdots, n \end{aligned} \right\},
$$

where, $\alpha_i^{-}$ and $\alpha_i^{+}$ are two levels of $\alpha$-cut technique associated with the constraint $p_i \lesssim b_i$ and $a_i \lesssim p_i$, respectively, and which express the decision maker (DM) credibility degree for partial information on probability distribution. The different level values express the different attitudes of the DM in dealing with the uncertain information, for instance, $\alpha_i^{+} = \alpha_i^{-} = 0$ means all possible values of $p_i$ in $[a_i - d_i^{-}, a_i + d_i^{+}]$ will be considered by the DM, and $\alpha_i^{+} = \alpha_i^{-} = 1$ means deterministic information is dealt with in decision making process, and other situations mean some intermediate attitudes of the DM.

Clearly, in the situation mentioned above, the mathematical expectation $E[Q(x, \xi)]$ has uncertain value due to the existence of fuzziness in probability distribution of $\xi$, actually, the two-stage problem cannot be solved. In order to overcome the difficulty, we employ minmax criterion to evaluate the second stage target value, and substitute $\max_{P \in \pi} E[Q(x, \omega)]$ for $E[Q(x, \omega)]$. Consequently, the two-stage stochastic problem with fuzzy probability distribution can be solved through the following deterministic equivalent problem:

$$\min_{x \in X} (c^T x + \max_{P \in \pi} E[Q(x, \xi)]). \tag{5}$$

Here, $X = \{x \in R | 0 \le x \le u\} \cap K$, and $K = \{x \in R|$ for all $i = 1, 2, \cdots, n$, there exists $y \ge 0$, such that $Wy = h(\xi_i) - Tx\}$, and $\max_{P \in \pi} E[Q(x, \xi)] = \max_{P \in \pi} \sum_{i=1}^{n} p_i Q(x, \xi_i)$.

From the polyhedral form for $X$ (see [4]), we can obtain an optimal vector $\bar{P} = (\bar{p}_1, \bar{p}_2, \cdots, \bar{p}_n)^T \in \pi$ such that $\sum_{i=1}^{n} \bar{p}_i Q(x, \xi_i) = \max_{P \in \pi} E[Q(x, \xi)]$, for the given $x$. As a result, we can design the following L-shaped algorithm to solve the problem.

## 3  Algorithm

*Step* 0: Set $r = s = k = 0$.
*Step* 1: Set $k = k + 1$. Solve the following linear program (master problem)

$$min \quad c^T x + \theta \tag{6a}$$
$$s.t. \quad D_l x \ge d_l, l = 1, 2, \cdots, r, \tag{6b}$$
$$E_l x + \theta \ge e_l, l = 1, 2, \cdots, s, \tag{6c}$$
$$0 \le w \le u, \theta \subset R. \tag{6d}$$

Let $(x^k, \theta^k)$ be an optimal solution. If no constraint (6c) is present, $\theta^k$ is set equal to $-\infty$, and $x^k$ is chosen arbitrary from the polyhedral set defined by the constraints (6b) and (6d). Go to step 2.
*Step* 2: For $i = 1, \cdots, n$, solve the following linear program

$$min \quad Z_i = e^T u^+ + e^T u^- \tag{7a}$$
$$s.t. \quad Wy + Iu^+ - Iu^- = h(\xi_i) - Tx^k, \tag{7b}$$
$$y \ge 0, u^+ \ge 0, u^- \ge 0, \tag{7c}$$

where $e = (1, 1, \cdots, 1)^T$, until, for some $i$, the optimal value $Z_i > 0$. Let $\sigma^k$ be the optimal simplex multiplier associated to equality in (7), and define

$$\begin{cases} D_{s+1} = (\sigma^k)^T T, \\ d_{s+1} = (\sigma^k)^T h(\xi_i) \end{cases} \tag{8}$$

to generate a new constraint (called feasibility cut) of type (6b). Set $r = r + 1$, add the constraint set (6b) and return to step 1. If for all $i$, $Z_i = 0$, then go to step 3.
*Step* 3: For $i = 1, \cdots, n$, solve the following linear program

$$min \quad q^T y(\xi_i) \tag{9a}$$
$$s.t. \quad Wy = h(\xi_i) - Tx^k, \tag{9b}$$
$$y(\xi_i) \geq 0 \tag{9c}$$

to obtain $Q(x^k, \xi_i)$ for the given $x^k$. Let $z_i{}^k$ be the optimal simplex multiplier associated with (9b) for $i = 1, \cdots, n$. Solve

$$R(x^k) = \max_{P \in \pi} \sum_{i=1}^{n} p_i Q(x^k, \xi_i) \tag{10}$$

to get optimal solution $\overline{P} = (\overline{p_1}, \cdots, \overline{p_n})^T$, define

$$\begin{cases} E_{s+1} = \sum_{i=1}^{n} \bar{p}_i (z_i{}^k)^T T, \\ e_{s+1} = \sum_{i=1}^{n} \bar{p}_i (z_i{}^k)^T h(\xi_i). \end{cases} \tag{11}$$

If $\theta^k \geq R(x^k) = e_{s+1} - E_{s+1}x^k$, stop, then $x^k$ is an optimal solution. Otherwise, set $s = s + 1$, add $E_{s+1}x^k + \theta^k \geq e_{s+1}$ to the constraint set (6c), and return to step 1.

## 4 Numerical Example

Assume that $c$, $q_1$, and $q_2$ are 0.61, 0.8 and 0.6, respectively, and the upper limit of purchasing power $u$ is specified to 1000. Suppose that the random demand $\xi$ take three possible values $\xi_1 = 90$, $\xi_2 = 100$, and $\xi_3 = 120$, and the corresponding probabilities $p_1$, $p_2$ and $p_3$ are almost equal to 1/3, 1/2, and 1/6, respectively, i.e. $p_1 \cong 1/3$, $p_2 \cong 1/2$ and $p_3 \cong 1/6$. Set $d_i{}^+ = d_i{}^- = 1/6$, $\alpha_i{}^+ = \alpha_i{}^- = 1/2$, for $i = 1, 2, 3$. It follows from $\alpha$-cut technique that

$$\pi = \left\{ P = (p_1, p_2, p_3)^T \in R^3 \Big| \sum_{j=1}^{3} p_j = 1, \atop 1/4 \leq p_1 \leq 5/12, 5/12 \leq p_2 \leq 7/12, 1/12 \leq p_3 \leq 1/4 \right\}. \tag{12}$$

If the initial solution $x$ is specified to 0, the Newsboy problem with fuzzy probability distribution can be solved from L-shaped algorithm stated above. The calculating results are listed in Table 1.

The optimal solution is $x_k = 120.00$ with an optimal value equal to $c^T x_k + \theta^k = -18.300$. This optimal solution is achieved with the probability distribution $\bar{P} = (0.417, 0.500, 0.083)^T$.

If fuzziness is not taken into consideration in the Newsboy problem, three probabilities with respect to three random demands are $p_1 = 1/3$, $p_2 = 1/2$, and $p_3 = 1/6$, respectively. Under the circumstances, the Newsboy problem

**Table 1** Results iterated from L-shaped algorithm for the Newsboy problem

| $k$ | $c^T x_k + \theta^k$ | $x_k$ | $e_{s+1}$ | $E_{s+1}$ |
|---|---|---|---|---|
| 1 | -2426.709 | 1000.000 | -0.000 | 3.037 |
| 2 | -19.420 | 8.00 | -19.500 | 0.600 |
| 3 | -18.525 | 97.50 | -0.000 | 0.800 |
| 4 | -18.471 | 102.86 | -7.500 | 0.717 |
| 5 | -18.300 | 120.00 | -17.500 | 0.617 |

with fuzzy probability distribution degenerates into the classical stochastic problem. We still use L-shaped algorithm [4] to solve the problem, and the optimal objective value is -18.800. Compared with the former result, the difference between -18.300 and -18.800 is 0.500, which is the DM's loss brought from fuzziness.

## 5 Conclusions

Two-stage stochastic newsboy problem with fuzzy probability distribution is modeled in this paper. $\alpha$-cut technique is employed to defuzzify. Improved L-shaped algorithm is designed to solve the problem based on minimax rule. Numerical examples demonstrate the essential character of algorithm. The newsboy model presented in this paper is very simple from point of view in practical application. Based on fuzzy probability distribution, the newsboy models, under more general hypothesizes such as multiple products and multiple periods, will be considered by future research activity.

## References

1. Abdelaziz, F.B., Masri, H.: Stochastic programming with fuzzy linear partial information on probability distribution. European Journal of Operational Research 162, 619–629 (2005)
2. Abdelaziz, F.B., Masri, H.: A compromise solution for the multiobjective stochastic linear programming under partial uncertainty. European Journal of Operational Research 202, 55–59 (2010)
3. Abdel-Maleka, L., Montanarib, R., Morales, L.C.: Exact, approximate, and generic iterative models for the multi-product Newsboy problem with budget constraint. International Journal of Production Economics 91, 189–198 (2004)
4. Birge, J.R., Louveaux, F.: Introduction to Stochastic Programming. Springer, New York (1997)
5. Chang, H.-C., Yao, J.-S., Ouyang, L.-Y.: Fuzzy mixture inventory model involving fuzzy random variable lead time demand and fuzzy total demand. European Journal of Operational Research 169, 65–80 (2006)
6. Khouja, B.: The single period (news-vendor) problem: literature review and suggestions for future research. Omega 27, 537–553 (1999)

7. Kofler, E.: Linear partial information with applications. Fuzzy Sets and Systems 118, 167–177 (2001)
8. Kwakernaak, H.: Fuzzy random variables - I: definition and theorems. Information Sciences 15, 1–29 (1978)
9. Liu, Y., Liu, B.: A class fuzzy random optimization: expected value models. Information Science 155, 89–102 (2003)
10. Porteus, E.L.: The newsvendor problem. International series in operations research and management science (2008), doi:10.1007/978-0-387-73699-0
11. Qin, Y., Wang, R., Vakharia, A.J., et al.: The newsvendor problem: Review and directions for future research. European Journal of Operational Research (2011), doi:10.1016/j.ejor.2010.11.024
12. Van Slyke, R.M., Wets, R.: L-shaped linear programs with applications to optimal control and stochastic programming. SIAM Journal of Applied Mathematics 17(4), 638–663 (1969)
13. Yu, C., Zhao, X., Peng, Y., et al.: Extended newsboy problem based on fuzzy random demand. Systems Engineering 153, 103–107 (2006) (in Chinese)
14. Zadeh, L.: Fuzzy sets. Information and Control 8, 338–353 (1965)

# Blogger's Interest Mining Based on Chinese Text Classification

Suhua Yang, Jianzhuo Yan, Chen Gao, and Guohua Tan

**Abstract.** In this paper, a new blogger's interest mining module is proposed, which is based on Chinese text classification. In fact, the problem of the interest mining is transformed into the problem of Chinese text categorization. Before the Chinese text categorization, the text is pre-processed for the text representation. The Chinese text is represented in vector space model and classified by support vector machine classification, while filter algorithm which filters the unrelated interest text is proposed. After the filtering, the text can get it's interest category. Finally the new module has been made use of to carry out an interest mining experiment, and the other experiment which has not filter algorithm is also carried in order to compare with the new module. The two experimental results show that the support vector machine is a effective algorithm, and the comparing data of the two experiments shows that new module make the interest mining more effective.

**Keywords:** Interest mining, Text classification, Support vector machine, Filter algorithm.

## 1  Introduction

In recent years, weblog has became one of the main information resources, with the rapid development of weblog, the domains of scientific research and industry have been interested in weblog. If you can make full use of the abundant weblog resources and mine the valuable information, it is of great

Suhua Yang, Jianzhuo Yan, Chen Gao, and Guohua Tan
College of Electronic Information and Control Engineering,
Beijing University of Technology, Beijing, China
e-mail: yangsuhua86@126.com, yanjianzhuo@bjut.edu.cn,
gaochen@emails.bjut.edu.cn, tgh@emails.bjut.edu.cn

S. Li (Eds.): Nonlinear Maths for Uncertainty and its Appli., AISC 100, pp. 611–618.
springerlink.com                                     © Springer-Verlag Berlin Heidelberg 2011

practical and research significance to learn the development of internet, improve various internet service and enrich user's internet lives [3].

Mining blogger's interest is the core and basis in personalized services, only blogger's personalized information is well understood, the ideal of personalized services may be achieved. At present, most methods of mining blogger's interests are by classifying weblog's articles [2]. The technique of text classification is mainly based on statistical theory and machine learning, such as Naive Bayes and KNN. The model of Naive Bayes is a probability classification model based upon two assumptions. It requires that the probabilities of all words are independent and the class of the document has got nothing to do with it's length, but the effect is unstable in practical application. KNN(k-Nearest Neighbor algorithm) is a method based on lazy and required learning method the effect of classification is better, but the time of classification is nonlinear, and when the number of training text increasesthe time of classification will sharp increase [5]. SVM(Support Vector Machine) is a new machine learning method advanced by Vapnik according to statistical learning theory, it is similar to structure risk minimization principle, it has splendidly learning ability, it only needs few samples for training a high-performance text classifier. In this paper, Support vector machine is used for text classification, and achieves a satisfying effect.

Although user's interest category could be reasoned by the algorithm based on the text classification, the personalized characteristic of different bloggers make weblog's content disorganized. Each user has their own interest, so the content which they browse is different. We find that not all blog texts could reflect user's interest by observing. This paper calls the article which can not express blogger's interest as unrelated interest article, and calls the article which can express blogger's interest as related interest article. In the method of mining weblogger's interest based on text classification, the wrong classification of unrelated interest article will directly make the mined interesting collection in chaos. Thus the accuracy of interests mining will be reduced. Therefore, Filtering out the unrelated interest articles is greatly important .In this paper, the discriminative value of each interest category is counted. And filtering the unrelated article by setting the threshold in the experiment. Thus it improves the accuracy of interest mining.

## 2 Blogger's Interest Mining

The general framework of interest mining consists of two modules: the preprocessing module and the interest classification module [1]. However, for the classification of blog texts, there often have some unrelated interest texts that greatly affected the accuracy of classification. In view of this problem, we add a interest determination module, It filters out the unrelated interest articles by calculating to improve the accuracy of mining interests. The framework of interest mining proposed in this paper is shown in Fig. 1.

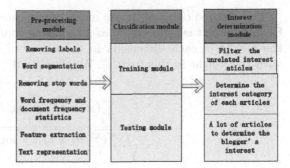

**Fig. 1** The framework of interest mining based on SVM Chinese text classification

The frame mainly has three modules, namely pre-processing module, classification module and interest determination module, which are detailed below.

## 2.1 Pre-processing Module

Blog pages are written by HTML(Hypertext Markup Language), they are semi-structured text files. In addition to plain text, the page itself also contains some of the labels. HTML contains a wealth of information. Before feature extraction ,we need web filtering for getting the body of the page. And then the body of the page is as regular text, we classify the regular texts.

Specific steps include removing labels, word segmentation, removing stop words, word frequency and document frequency statistics, the feature extraction, text representation. Specific steps are as follows:

1. Removing labels: Using regular expressions to remove the label of the source of the document.
2. Word segmentation: Using ICTCLAS(Institute of Computing Technology, Chinese Lexical Analysis System) to segment words.
3. Removing stop words: We created list of terms which were filtered before the word frequency process started. The list includes mainly conjunctions, prepositions or pronouns [4].
4. Word frequency and document frequency statistics: Word frequency statistic for each word which appears in the text, if the first time when the word frequency $F$ is set to 1, while adding a statistical document frequency of each interest category.
5. The feature extraction: Delete the words from the text which can not contribute to or very little contribution to the entry category information, Taking into account the large amount of information carried by nouns, verbs followed by adjectives and adverbs, this article frame realizes uses only retains the noun used for feature selection methods.
6. Text representation: Using vector space model to represent the text.

The method of vector space model representation as follows:
Each blog text is represented as a n-dimensional vector $(w_1, w_2, w_3, \ldots, w_n)$, the weight of each dimension in the vector of this text should correspond with the weight in this text [6]. Weight Set: $W = \{w_i | i \in n\}$

$$w_i = \frac{\sum\limits_{i \in s}(w_i \times tf_i) \times \log(N/n_i)}{\sqrt{\sum\limits_{j}((\sum\limits_{i \in s} w_i \times tf_i) \times \log(N/n_i))^2}} \tag{1}$$

$w_i$ is the corresponding weight of the i-key words, and $tf_i$ is the frequency of the $i$- key words in the page, $N$ is the total number of text contained in the training set, $n_i$ is the number of the text which contains the characteristics.

## 2.2 Classification Module

The next step is Chinese text classification based on SVM. Support Vector Machine (SVM) is a general learning machine, the idea is: the input vector $X$ is mapped to a high dimensional feature space $Z$ by nonlinear mapping which is pre-selected, in which we structure the optimal separating hyperplane. SVM classification function is similar to neural network in form, the output is a linear combination of intermediate nodes, each intermediate node corresponds to a support vector, the dot product is operated between vectors [7]. The expression of the function of SVM for classification of non-linear optimal separating surface as follow:

$$f(z) = \sum_{supvector} a_i{}^* y_i \varphi(z_i)\varphi(z) + b^* = \sum_{supvector} a_i{}^* y_i k(z_i, z) + b^* \tag{2}$$

Therefore, if we adopt the kernel function to avoid the high dimensional feature space for complex operations. The process can be expressed as follows:
First, mapping the input vector $X$: $\psi : R_n \to H$

Mapping into a high dimensional Hilbert space $H$. The kernel function has different forms, Different kernel functions will form a different algorithm. In general, the commonly used kernel functions are the following:

1. Polynomial kernel function:

$$K(z, z_i) = (z \times z_i + 1)^d \tag{3}$$

2. Radial basis function:

$$K(z, z_i) = exp(-\|z - z_i\|^2/\delta^2) \tag{4}$$

3. Neural network kernel function:

$$K(z, z_i) = S(v(z \times z_i) + c) \tag{5}$$

The choice of kernel function has little effect on classification accuracy. But polynomial classifier both for the low-dimensional, high-dimensional, large sample, small sample and so on are applicable, and has a wider domain of convergence, parameter easy to control, etc, this paper choose polynomial classifier as a kernel function [8].

## 2.3  Interest Determination Module

For an article $d$, SVM classification results will produce a discriminative vector, whose elements are the discriminative values belonging to each category. This type is descending ordered based on their discriminative values, and the sorted vector is expressed as $\vec{p} = \{p(d, c_1), p(d, c_2), \ldots, p(d, c_m)\}$, $m$ is the number of interest category.

Use $Pr(d)$ to express the cumulative value of the discriminative vector element, the formula as follows:

$$Pr(d) = \sum_{i=1}^{m} p(d, c_i) \tag{6}$$

Filter text formula:

$$Pr(d) \geq T \tag{7}$$

Training through multiple experiments, the threshold $(T)$ is set as 0.1. If the article does not fits the formula, it will be classified as unrelated interest article.

After the filtering, select $p(d, c_1)$ as the last result of related interest article, if it is a unrelated interest article, it will be filtered. Then we get the blogger's interest by the blogger's more articles.

## 3  Experiments and Result Analysis

### 3.1  Experimental Data

Experimental data includes eight class documentations, and they are news, sports, finance, entertainment, shopping, reading, travel, and military.

The training data is different from testing data, the training data is from the language material database of FuDan University, the testing data is from $http : //blog.sina.com.cn/$.

## 3.2 Experimental Results

The first experiment does not use filter algorithm, the second experiment uses filter algorithm. Performance evaluation of interest classification mainly includes accuracy rate($P$), recall rate($R$) and $F1$ ($F1 = \frac{2 \times P \times R}{P+R}$). The following two sets of experimental data are the different classification results with the same training data sets and testing data sets under different classification algorithms.

1. The result of SVM classification which does not have filter algorithm is shown in Table 1

**Table 1** The result of SVM classification which does not have filter algorithm

| Interest category | Training corpus/Testing corpus | $P(\%)$ | $R(\%)$ | $F1(\%)$ |
|---|---|---|---|---|
| news | 1800/200 | 85.0 | 83.4 | 84.2 |
| sports | 1800/200 | 87.2 | 87.6 | 87.4 |
| finance | 1800/200 | 82.1 | 85.2 | 83.6 |
| entertainment | 1800/200 | 89.2 | 88.8 | 89.0 |
| shopping | 1800/200 | 84.6 | 85.8 | 85.2 |
| reading | 1800/200 | 89.2 | 86.7 | 87.9 |
| travel | 1800/200 | 89.1 | 87.2 | 88.1 |
| military | 1600/200 | 88.7 | 85.7 | 87.7 |

2. The result of SVM classification which has filter algorithm is shown in Table 2

**Table 2** The result of SVM classification which has filter algorithm

| Interest category | Training corpus/Testing corpus | $P(\%)$ | $R(\%)$ | $F1(\%)$ |
|---|---|---|---|---|
| news | 1800/200 | 87.7 | 89.3 | 88.5 |
| sports | 1800/200 | 89.1 | 88.2 | 88.6 |
| finance | 1800/200 | 85.5 | 89.1 | 87.3 |
| entertainment | 1800/200 | 90.2 | 90.9 | 90.5 |
| shopping | 1800/200 | 87.6 | 88.1 | 87.8 |
| reading | 1800/200 | 90.2 | 89.7 | 89.9 |
| travel | 1800/200 | 93.1 | 90.6 | 91.8 |
| military | 1600/200 | 92.2 | 89.7 | 90.9 |

Accuracy rate and recall rate reflect two different aspects of classification quality, while a comprehensive evaluation index of the two aspects is the $F1$ value. As shown in Fig. 2, the figure reflects classification results of the two modules under the composite index $F1$ value.

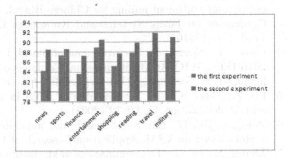

**Fig. 2** The comparison of comprehensive index $F1$ value

## 3.3 Result Analysis

We can clearly see that SVM classification which has filter algorithm is more effective than SVM classification which does not have filter algorithm. It has higher accuracy rate and recall rate. As a comprehensive evaluation index for text classification, $F1$ test value is better to reflect the effects of a good or bad classifier. So as a whole, SVM classification which has filter algorithm is superior for interest mining.

## 4 Conclusions

In this paper, we present a new blogger's interest mining method based on Chinese text classification technology. It improves the interest classification accuracy by adding the unrelated interest text filter algorithm. The effectiveness of our method has been tested on real blog data.

On the other hand, blogger's interest is related to the browsing behavior. If combining the Chinese text classification with the browsing behavior to mining the blogger's interest, it will get a better result. So how to promote blogger's interest mining better, will remain our ongoing efforts to study direction in the future.

**Acknowledgements.** The experiments used Institute of Computer Technology segmentation system interface, Fudan University Corpus and the data of $http$ : $//blog.sina.com.cn/$, a special thanks here.

# References

1. Chau, M., Lam, P., Shiu, B., Xu, J., Jinwei, C.: A blog mining framework. IT Professional 1, 36–41 (2009)
2. Chen, N.: Weblogger's interest mining based on classification technology. Friend of Science Amateurs 3, 155–156 (2010)
3. Hui, Y., Bin, Y., Xu, Z., Chunguang, Z., Zhe, W., Zhou, C.: Community discovery and sentiment mining for Chinese BLOG. In: 2010 Seventh International Conference on Fuzzy Systems and Knowledge Discovery (FSKD), vol. 4, pp. 1740–1745 (2010)
4. Kuzar, T., Navrat, P.: Preprocessing of Slovak Blog Articles for Clustering. In: 2010 IEEE/WIC/ACM International Conference on Web Intelligence and Intelligent Agent Technology (WI-IAT), vol. 3, pp. 314–317 (2010)
5. Lu, Z., Zhao, S., Lin, Y.: Research of KNN in text categorization. Computer and Modernization 11, 69–72 (2008)
6. Pang, J., Bu, D., Bai, S.: Research and implementation of text categorization system based on VSM. Application Research of Computers 9, 23–26 (2001)
7. Zhang, X., Li, Q.: The summary of text classification based on support vector machines. Science and Technology Information 28, 344–345 (2008)
8. Zuo, S., Guo, X., Wan, J., Zhou, Z.: Fast classification algorithm for polynomial kernel support vector machines. Computer Engineering 6, 27–29 (2007)

# Characterization of Generalized Necessity Functions in Łukasiewicz Logic

Tommaso Flaminio and Tomáš Kroupa

**Abstract.** We study a generalization of necessity functions to MV-algebras. In particular, we are going to study belief functions whose associated mass assignments have nested focal elements. Since this class of belief functions coincides with necessity functions on Boolean algebras, we will call them *generalized necessity functions*. Using geometrical and combinatorial techniques we provide several characterizations of these functions in terms of Choquet integral, Lebesgue integral, and min-plus polynomials.

**Keywords:** Necessity function, Belief function, MV-algebra

## 1 Introduction

There are at least two different, yet equivalent, ways to define necessities on Boolean algebras [4]. If the Boolean algebra is the set $2^X$ of all subsets of a given universe $X$, then the first approach consists in axiomatizing a necessity $N : 2^X \to [0,1]$ as a map satisfying $N(X) = 1$, $N(\emptyset) = 0$, and $N(A \cap B) = \min\{N(A), N(B)\}$. According to the second way, a necessity is viewed as a *belief function* [15] defined by a mass assignment $\mu : 2^X \to [0,1]$ such that the class of its *focal elements* $\{ A \subseteq X \mid \mu(A) > 0 \}$ is a chain with respect to the set inclusion. Since the former axiomatic approach can be traced back to Halpern's *belief measures* [6], we will henceforth distinguish

Tommaso Flaminio
Artificial Intelligence Research Institute (IIIA-CSIC),
Campus Universitat Autonoma de Barcelona 08913 Bellaterra, Spain
e-mail: tommaso@iiia.csic.es

Tomáš Kroupa
Institute of Information Theory and Automation of the ASCR,
Pod Vodárenskou věží 4, 182 08 Prague, Czech Republic
e-mail: kroupa@utia.cas.cz

S. Li (Eds.): Nonlinear Maths for Uncertainty and its Appli., AISC 100, pp. 619–626.
springerlink.com                    © Springer-Verlag Berlin Heidelberg 2011

between *necessity measures*, if the former is the case, and *necessity functions* in the latter case. These two ways of introducing necessities on Boolean algebras are equivalent. Specifically, a map $N : 2^X \to [0,1]$ is a necessity measure if and only if $N$ is a necessity function.

Since MV-algebras [1] are among important many-valued generalizations of Boolean algebras, which provide a useful algebraic framework to deal with a certain and a relevant class of fuzzy sets, it is natural to ask what happens when we generalize necessity measures and necessity functions to these algebraic structures. Moreover, it is worth noticing that, as it was already remarked in [3], the generalizations of necessity measures and necessity functions to MV-algebras do not lead to one single concept as in the Boolean case. Hence it makes sense to study those notions separately.

In [5] the authors provide an axiomatic approach to necessity measures on MV-algebras and they show that they are representable by generalized Sugeno integrals. In this paper we characterize generalized necessity functions in the framework of the generalization of belief functions to MV-algebras proposed in [12]. In particular, we are going to use geometrical and combinatorial tools to provide several characterizations for these measures in terms of Choquet integral, Lebesgue integral, and min-plus polynomial.

The paper is organized as follows. In Section 2 we introduce the preliminaries about MV-algebras and states. We recall the theory of belief functions on Boolean algebras together with the equivalence between the two approaches to necessities in Section 3. Section 4 introduces generalized necessity functions with the main characterization (Proposition 3). Due to lack of space we are unable to include proofs; however, we provide examples to clarify main features of the discussed concepts.

## 2 Basic Notions

MV-algebras [1] play the same role for Łukasiewicz logic as Boolean algebras for the classical two-valued logic. An *MV-algebra* is an algebra $(M, \oplus, \neg, 0)$, where $M$ is a non-empty set, the algebra $(M, \oplus, 0)$ is an abelian monoid, and these equations are satisfied for every $x, y \in M$: $\neg\neg x = x$, $x \oplus \neg 0 = \neg 0$, $\neg(\neg x \oplus y) = \neg(\neg y \oplus x)$.

In every MV-algebra $M$, we define the constant $1 = \neg 0$ and the following binary operations: for all $x, y \in M$, put $x \odot y = \neg(\neg x \oplus \neg y)$, $x \vee y = \neg(\neg x \oplus y) \oplus y$, $x \wedge y = \neg(\neg x \vee \neg y)$. For every $x, y \in M$, we write $x \leq y$ iff $\neg x \oplus y = 1$ holds in $M$. As a matter of fact, $\leq$ is a partial order on $M$, and $M$ is said to be *linearly ordered* whenever $\leq$ is a linear order.

*Example 1.* Every Boolean algebra $A$ is an MV-algebra in which the operations $\oplus$ and $\vee$ coincide (similarly, the operations $\odot$ and $\wedge$ coincide). Moreover, in every MV-algebra $M$, the set $B(M) = \{ x \mid x \oplus x = x \}$ of its idempotent elements is the domain of the largest Boolean subalgebra of $M$ (the so-called *Boolean skeleton* of $M$).

*Example 2.* Endow the real unit interval $[0,1]$ with the operations $x \oplus y = \min\{1, x+y\}$ and $\neg x = 1 - x$. Then $([0,1], \oplus, \neg, 0)$ becomes an MV-algebra called the *standard MV-algebra*. In this algebra, $x \odot y = \max\{0, x+y-1\}$, $x \wedge y = \min\{x, y\}$ and $x \vee y = \max\{x, y\}$. The two operations $\odot, \oplus$ are the so-called *Łukasiewicz t-norm* and the *Łukasiewicz t-conorm*, respectively.

*Example 3.* Let $X$ be a nonempty set. The set $[0,1]^X$ of all functions $X \to [0,1]$ with the pointwise operations of the MV-algebra $[0,1]$ is an MV-algebra. In particular, if $X$ is a finite set, say $X = \{1, \ldots, n\}$, then we can identify the MV-algebra $[0,1]^X$ with the $n$-cube $[0,1]^n$ and each $a \in [0,1]^X$ with the $n$-dimensional vector $a = (a_1, \ldots, a_n) \in [0,1]^n$. The set of vertices of $[0,1]^n$ coincides with the Boolean skeleton of $[0,1]^n$.

Throughout the paper, we will assume that $X$ is always finite whenever we write $[0,1]^X$. The MV-algebra $[0,1]^X$ is the natural algebraic framework for studying belief functions in Łukasiewicz logic (cf. [10]). The extensions towards infinite $X$ are possible and mathematically nontrivial (see [11, 12]). Herein we confine to the case of finite $X$ for the sake of clarity.

Normalized and additive maps on MV-algebras (so-called states) were introduced in [7, 13]. States are many-valued analogues of probabilities on Boolean algebras. A *state* on an MV-algebra $M$ is a function $s : M \to [0,1]$ satisfying the following properties:

(i) $s(0) = 0$, $s(1) = 1$,
(ii) $s(x \oplus y) = s(x) + s(y)$, whenever $x \odot y = 0$.

Observe that the restriction of every state $s$ on $M$ to its Boolean skeleton $B(M)$ is a finitely additive probability measure on $B(M)$. Much more is known: every MV-algebra $M$ is (isomorphic to) an MV-algebra of continuous functions over some compact Hausdorff space $X$ (see [1]) and each state on $M$ is the Lebesgue integral with respect to a unique regular Borel probability measure on $X$ (see [8] or [14]). In case of the MV-algebra $[0,1]^X$ with finite $X$, the previous fact can be formulated as follows. Observe that every probability measure on $2^X$ with $X = \{1, \ldots, n\}$ can be represented by a unique vector $\mu$ from the *standard $n$-simplex* $\Delta_n = \{\, \mu \in \mathbb{R}^n \mid \mu_i \geq 0, \sum_{i=1}^n \mu_i = 1 \,\}$.

**Proposition 1 ( [8,14]).** *Let $X = \{1, \ldots, n\}$. If $s$ is a state on $M = [0,1]^X$, then there exists a unique $\mu \in \Delta_n$ such that*

$$s(a) = \sum_{i=1}^n a_i \mu_i, \quad \text{for each } a \in M.$$

*Moreover, the coordinates of $\mu$ are $\mu_i = s(\{i\})$, provided $\{i\}$ is identified with its characteristic function, for each $i \in X$.*

## 3   Necessity Functions

See [15] for an in-depth treatment of Dempster-Shafer theory of belief functions. Let $X$ be a finite set and $M = 2^X$. A *mass assignment* $\boldsymbol{\mu}$ is a function

$2^X \to [0,1]$ satisfying $\mu(\emptyset) = 0$ and $\sum_{A \in 2^X} \mu(A) = 1$. A *belief function* (with the mass assignment $\mu$) is a function Bel : $2^X \to [0,1]$ given by $\mathrm{Bel}(A) = \sum_{B \subseteq A} \mu(B)$, for each $A \in 2^X$. Each $A \in 2^X$ with $\mu(A) > 0$ is said to be a *focal element*. A mass assignment $\mu$ is called *nested* provided the set of its focal elements $\{ A \in 2^X \mid \mu(A) > 0 \}$ is a chain in $2^X$ with respect to the set inclusion. By definition every belief function Bel is uniquely determined by the restriction of its mass assignment to the set of all focal elements. A *necessity function* on $2^X$ is a belief function whose mass assignment is nested. This can be rephrased as follows: a belief function is a necessity function iff its mass assignment determines a finitely additive probability on $2^{2^X}$ that is supported by a chain and vanishing at the singleton $\{\emptyset\}$.

If Bel is a belief function on $2^X$, then the *credal set* of Bel is the following set $\mathcal{C}(\mathrm{Bel})$ of finitely additive probability measures $P$ on $2^X$:

$$\mathcal{C}(\mathrm{Bel}) = \{ P \mid P(A) \geq \mathrm{Bel}(A), A \in 2^X \}.$$

It is well-known that Bel arises as the lower envelope of $\mathcal{C}(\mathrm{Bel})$:

$$\mathrm{Bel}(A) = \bigwedge_{P \in \mathcal{C}(\mathrm{Bel})} P(A), \quad \text{for each } A \in 2^X.$$

*Example 4.* Let $A \in 2^X$ be nonempty and put $\mathrm{Bel}_A(B) = 1$, if $A \subseteq B$, and $\mathrm{Bel}_A(B) = 0$, otherwise. Then $\mathrm{Bel}_A$ is a necessity function whose mass assignment is

$$\mu_A(B) = \begin{cases} 1, & A = B, \\ 0, & \text{otherwise.} \end{cases} \tag{1}$$

The credal set $\mathcal{C}(\mathrm{Bel}_A)$ is just the set of all probabilities whose support is the set $A$. Specifically, this means that $\mathcal{C}(\mathrm{Bel}_A)$ is (affinely isomorphic to) the simplex $\Delta_{|A|}$, where $|A|$ is the cardinality of $A$. Observe that $A \subseteq B$ iff $\mathcal{C}(\mathrm{Bel}_A) \subseteq \mathcal{C}(\mathrm{Bel}_B)$ iff $\Delta_{|A|} \subseteq \Delta_{|B|}$.

In the next proposition we summarize some of the characterizations of necessity functions that appeared in the literature. Our goal is to compare these descriptions with the properties of extensions of necessity functions to MV-algebras in Section 4.

**Proposition 2.** *Let* Bel *be a belief function on* $2^X$ *with the mass assignment* $\mu$. *Then the following are equivalent:*

(i) Bel *is a necessity function,*
(ii) $\mathrm{Bel}(A \cap B) = \mathrm{Bel}(A) \wedge \mathrm{Bel}(B)$, *for each* $A, B \in 2^X$,
(iii) *the set* $\{ \mathcal{C}(\mathrm{Bel}_A) \mid A \in 2^X, \mu(A) > 0 \}$ *is a chain and*

$$\mathcal{C}(\mathrm{Bel}) = \sum_{\substack{A \in 2^X \\ \mu(A) > 0}} \mu(A)\mathcal{C}(\mathrm{Bel}_A), \tag{2}$$

*where the sum and the multiplication in* (2) *are the Minkowski sum of sets and the pointwise multiplication of sets of vectors, respectively,*

(iv) *there exist* $n \in \{1, \ldots, |X|\}$, *a vector* $\alpha = (\alpha_1, \ldots, \alpha_n) \in \Delta_n$ *with all coordinates positive, and a chain of standard simplices* $\Delta_{i_1} \subset \cdots \subset \Delta_{i_n}$, *where* $i_n \le n$, *such that* $\mathcal{C}(\mathrm{Bel})$ *is (affinely isomorphic to) the Minkowski sum* $\sum_{j=1}^{n} \alpha_j \Delta_{i_j}$.

The equivalence of (i) with (ii) was proven in [15]. The properties (iii)-(iv) are a purely geometrical way to describe necessities by the composition of the associated credal sets. This approach has appeared first in [9], where the equivalence of (i) with (iii) was proven in a slightly more general setting. The property (iv) is just a direct reformulation of (iii). Geometrical treatment of belief functions appeared also in [2], where the properties of the set of all belief functions are discussed.

# 4    Generalized Necessity Functions

We will introduce the generalized necessity functions as particular cases of generalized belief functions in Łukasiewicz logic (cf. [10]). The starting point for this research was the generalization of Möbius transform established in a fairly general framework [11]. The interested reader is referred to those papers for further motivation and details.

If $X = \{1, \ldots, n\}$, then by $\mathcal{P}$ we denote the set $2^X \setminus \{\emptyset\}$. Let $M_{\mathcal{P}}$ be the MV-algebra of all functions $\mathcal{P} \to [0, 1]$. We will consider the following embedding $\rho$ of the MV-algebra $M = [0, 1]^n$ into $M_{\mathcal{P}}$:

$$\rho : M \times \mathcal{P} \to [0, 1], \quad \rho_a(A) = \bigwedge_{i \in A} a_i, \quad \text{for each } a \in M, A \in \mathcal{P}.$$

If $a \in M$ is fixed and $\rho_a(\emptyset) := 0$, then observe that function $\rho_a : 2^X \to [0, 1]$ is a necessity measure on $2^X$.

**Definition 1.** Let $M$ be the MV-algebra $[0, 1]^X$. A *state assignment* is a state $\mathbf{s}$ on $M_{\mathcal{P}}$. If $\mathbf{s}$ is a state assignment, then a *(generalized) belief function* $\mathrm{Bel}^*$ on $M$ is given by $\mathrm{Bel}^*(a) = \mathbf{s}(\rho_a)$, $a \in M$. We say that a belief function $\mathrm{Nec}^*$ on $M$ is a *(generalized) necessity function* if the finitely additive probability on $2^{2^{\mathcal{P}}}$ corresponding to its state assignment (via Proposition 1) is supported by a chain.

*Example 5.* Let $A \in \mathcal{P}$ and put $\mathrm{Bel}_A^*(a) = \rho_a(A)$. Clearly, function $\mathrm{Bel}_A^*$ is a necessity function. Its state assignment $\mathbf{s}_A$ is given by $\mathbf{s}_A(f) = f(A)$, for each $f \in M_{\mathcal{P}}$.

*Remark 1.* Following the analogy with Proposition 2(ii), necessity measures on an MV-algebra $M$ have been recently introduced in [5] as mappings $N : M \to [0, 1]$ such that $N(1) = 1$, $N(0) = 0$, and for every $a, b \in M$, $N(a \wedge b) = N(a) \wedge N(b)$. It was observed already in [3] by Dubois and Prade that, in

sharp contrast with the classical case (cf. Proposition 2), necessity functions
are not necessity measures. Indeed, generalized necessity functions do not
satisfy the property $N(a \wedge b) = N(a) \wedge N(b)$, in general: this follows directly
from Definition 1.

Let $\mathrm{Bel}^*$ be a belief function on $M = [0,1]^X$ and let $\mathbf{s}$ be its associated state
assignment. Clearly, for each $A \in \mathcal{P}$, the mass assignment $\mu_A$ from (1) is
an element of $M_{\mathcal{P}}$. As a direct consequence of the definition of state, the
function $\mu_{\mathbf{s}} : 2^X \to [0,1]$ defined by $\mu_{\mathbf{s}}(A) = \mathbf{s}(\mu_A)$ for every $A \in \mathcal{P}$, and
zero otherwise, is a mass assignment. Hence $\mathrm{Bel}^*(a) = \sum_{A \in M_{\mathcal{P}}} \rho_a(A) \mu_{\mathbf{s}}(A)$,
for each $a \in M$, which follows from Proposition 1.

If $\mathrm{Bel}^*$ is a belief function on $M$, then the *credal set* of $\mathrm{Bel}^*$ is the following
set $\mathcal{C}(\mathrm{Bel}^*)$ of states $s$ on $M$:

$$\mathcal{C}(\mathrm{Bel}^*) = \{\, s \mid s(a) \geq \mathrm{Bel}^*(a),\ a \in M \,\}.$$

It can be shown that $\mathrm{Bel}^*$ is the lower envelope of $\mathcal{C}(\mathrm{Bel}^*)$:

$$\mathrm{Bel}^*(a) = \bigwedge_{s \in \mathcal{C}(\mathrm{Bel}^*)} s(a), \quad \text{for each } a \in M. \tag{3}$$

In the following proposition we give several equivalent formulations describing generalized necessity functions within the class of generalized belief
functions. In particular, some of the properties directly correspond to the
respective properties of necessity functions—see Proposition 2.

**Proposition 3 (Characterization of generalized necessity functions).**
*Let $X = \{1, \ldots, n\}$ and $\mathrm{Bel}^*$ be a belief function on the MV-algebra $M = [0,1]^n$ with the state assignment $\mathbf{s}$ and the mass assignment $\mu_{\mathbf{s}}$. Then the following are equivalent:*

*(i) $\mathrm{Bel}^*$ is a necessity function,*
*(ii) there exists a necessity measure $\mathrm{Nec}$ on $2^X$ such that*

$$\mathrm{Bel}^*(a) = \oint a \,\mathrm{d}\,\mathrm{Nec}, \quad a \in M,$$

*where the discrete integral above is the Choquet integral,*
*(iii) the mass assignment $\mu_{\mathbf{s}}$ is nested on a chain $\mathcal{A} \subseteq \mathcal{P}$ such that*

$$\mathrm{Bel}^*(a) = \sum_{A \in \mathcal{A}} \mu_{\mathbf{s}}(A) \rho_a(A), \quad a \in M,$$

*(iv) the mass assignment $\mu_{\mathbf{s}}$ is nested on a chain $A_1 \subset \cdots \subset A_k$ such that*

$$\mathrm{Bel}^*(a) = \bigwedge_{(i_1, \ldots, i_k) \in I} \sum_{j=1}^{k} \mu_{\mathbf{s}}(A_j) a_{i_j}, \quad a \in M,$$

*where $I = A_1 \times \cdots \times A_k$,*

(v) there exists a maximal chain $\mathcal{A} = A_1 \subset A_2 \subset \cdots \subset A_n = X$ in $2^X$ and a mass assignment $\mu$ nested on $\mathcal{A}$ such that

$$\mathrm{Bel}^*(a) = \bigwedge_{s=1}^{n!} \left( \sum_{i=1}^{n} \mu(A_i) \cdot a_{f^{-1}(s)(i)} \right), \quad a \in M,$$

where $f : A_1 \times \ldots \times A_n \to \{1, 2, \ldots, n!\}$ is a bijection,

(vi) the set $\left\{ \mathcal{C}(\mathrm{Bel}_A^*) \mid A \in 2^X, \mu_s(A) > 0 \right\}$ is a chain and

$$\mathcal{C}(\mathrm{Bel}^*) = \sum_{\substack{A \in 2^X \\ \mu_s(A) > 0}} \mu_s(A) \mathcal{C}(\mathrm{Bel}_A^*),$$

(vii) there exist $n \in \{1, \ldots, |X|\}$, a vector $\alpha = (\alpha_1, \ldots, \alpha_n) \in \Delta_n$ with each $\alpha_i \geq 0$, and a chain of standard simplices $\Delta_{i_1} \subset \cdots \subset \Delta_{i_n}$, where $i_n \leq n$, such that $\mathcal{C}(\mathrm{Bel}^*)$ is (affinely isomorphic to) the Minkowski sum $\sum_{j=1}^{n} \alpha_j \Delta_{i_j}$.

Proposition 3, whose proof is omitted due to a lack of space, provides a number of interpretations of necessity functions. In particular, (ii) means that each generalized necessity function is recovered as the Choquet integral extension of a necessity measure. The properties (vi)-(vii) say that the credal set of a generalized necessity function is built from "nested" simplices in a very special way—observe that this is identical with the property of necessity functions on Boolean algebras (Proposition 2(iii)-(iv)). The min-sum formula in (iv) is then a consequence of this geometrization together with (3): when minimizing a linear function given by $a \in [0,1]^n$ over $\mathcal{C}(\mathrm{Bel}^*)$, it suffices to seek the minimum among the elements of any finite set containing the vertices of the convex polytope $\mathcal{C}(\mathrm{Bel}^*)$. Notice that although the equivalence between (iv) and (v) is clear, because in fact (v) is a particular case of (iv), (v) can be easily proved to be equivalent to (iii) by using a combinatorial argument. The results are illustrated with a simple example.

*Example 6.* Let $X = \{1, 2, 3\}$ and $M = [0,1]^X$. Suppose that Nec is the necessity measure on $2^X$ whose mass assignment $\mu$ is defined as $\mu(\{1\}) = \frac{1}{8}$, $\mu(\{1, 2\}) = \frac{4}{8}$, $\mu(X) = \frac{3}{8}$. The necessity function Nec$^*$ associated with Nec via Proposition 3(ii) is then

$$\mathrm{Nec}^*(a) = \tfrac{1}{8} a_1 + \tfrac{4}{8}(a_1 \wedge a_2) + \tfrac{3}{8}(a_1 \wedge a_2 \wedge a_3),$$

for each $a \in [0,1]^3$. Due to Proposition 3(vi), the credal set $\mathcal{C}(\mathrm{Nec}^*)$ can be identified with the Minkowski sum $\frac{1}{8}\Delta_1 + \frac{4}{8}\Delta_2 + \frac{3}{8}\Delta_3$. This is a convex polytope embedded in $\Delta_3$ with the four vertices $(1, 0, 0)$, $(\frac{1}{8}, \frac{7}{8}, 0)$, $(\frac{5}{8}, 0, \frac{3}{8})$, and $(\frac{1}{8}, \frac{4}{8}, \frac{3}{8})$. This means together with Proposition 3(v) that we get the min-sum formula

$$\text{Nec}^*(a) = a_1 \wedge \left(\tfrac{1}{8}a_1 + \tfrac{7}{8}a_2\right) \wedge \left(\tfrac{5}{8}a_1 + \tfrac{3}{8}a_3\right) \wedge \left(\tfrac{1}{8}a_1 + \tfrac{4}{8}a_2 + \tfrac{3}{8}a_3\right).$$

**Acknowledgements.** T. Flaminio acknowledges partial support from the Juan de la Cierva Program of the Spanish MICINN, and partial support from the Spanish project ARINF (TIN2009-14704-C03-03). The work of T. Kroupa was supported by Grants GA ČR 201/09/1891 and Grant No.1M0572 of the Ministry of Education, Youth and Sports of the Czech Republic.

# References

1. Cignoli, R.L.O., D'Ottaviano, I.M.L., Mundici, D.: Algebraic Foundations of many-valued Reasoning. In: Trends in Logic—Studia Logica Library, vol. 7. Kluwer Academic Publishers, Dordrecht (2000)
2. Cuzzolin, F.: A geometric approach to the theory of evidence. IEEE Transactions on Systems, Man, and Cybernetics part C 38, 522–534 (2007)
3. Dubois, D., Prade, H.: Evidence measures based on fuzzy information. Automatica 21, 547–562 (1985)
4. Dubois, D., Prade, H.: Possibility Theory. An approach to computerized processing of uncertainty. Plenum Press, New York (1988)
5. Flaminio, T., Godo, L., Marchioni, E.: On the logical formalization of possibilistic counterpart of states over $n$-valued events. Journal of Logic and Computation, doi:10.1093/logcom/exp012 (in press)
6. Halpern, J.Y.: Reasoning about Uncertainty. The MIT Press, Cambridge (2003)
7. Kôpka, F., Chovanec, F.: D-posets. Math. Slovaca 44, 21–34 (1994)
8. Kroupa, T.: Representation and extension of states on MV-algebras. Archive for Mathematical Logic 45, 381–392 (2006)
9. Kroupa, T.: Affinity and continuity of credal set operator. In: Augustin, T., Coolen, F.P.A., Moral, S., Troffaes, M.C.M. (eds.) Proceedings of the Sixth International Symposium on Imprecise Probability: Theories and Applications, ISIPTA 2009, Durham, UK, pp. 269–276 (2009)
10. Kroupa, T.: From probabilities to belief functions on MV-algebras. In: Borgelt, C., González-Rodríguez, G., Trutschnig, W., Lubiano, M.A., Gil, M.Á., Grzegorzewski, P., Hryniewicz, O. (eds.) Combining Soft Computing and Statistical Methods in Data Analysis. Advances in Intelligent and Soft Computing, vol. 77, pp. 387–394. Springer, Heidelberg (2010)
11. Kroupa, T.: Generalized Möbius transform of games on MV-algebras and its application to Cimmino-type algorithm for the core. In: To appear in Contemporary Mathematics. AMS, Providence (2011)
12. Kroupa, T.: Extension of belief functions to infinite-valued events. Accepted to Soft Computing (2011)
13. Mundici, D.: Averaging the truth-value in Łukasiewicz logic. Studia Logica 55, 113–127 (1995)
14. Panti, G.: Invariant measures in free MV-algebras. Communications in Algebra 36, 2849–2861 (2008)
15. Shafer, G.: A Mathematical Theory of Evidence. Princeton University Press, Princeton (1976)

# On Generating Functions of Two-Dimensional T-Norms

Masaya Nohmi, Aoi Honda, and Yoshiaki Okazaki

**Abstract.** t-norms are significant operations employed in various fields including fuzzy theories. There exist many types of t-norms. t-norms defined on discrete domains and continuous domains have rather different properties. It is known that continuous and strictly monotone t-norms defined on continuous domains have generating functions [1, 4, 5]. Generating functions are important functions that characterize the properties of t-norms. In this article, tow-dimensional t-norms are proposed, and their properties are studied. Furthermore, it is shown that if a tow-dimensional t-norm satisfies particular conditions, it could be decomposed into Cartesian product of two one-dimensional t-norms and each one-dimensional t-norm have a generating function. Applications of two-dimensional t-norms are briefly discussed at the last of this article [2, 3].

**Keywords:** T-norm, Two-dimensional t-norm, Generating function, Strictly monotone operation, Continuous operation.

## 1 Preliminary

Although ordinary t-norms are defined on the unit interval $I = [0, 1]$, in this article, two-dimensional t-norms defined on the closed domain $I \times I$ are studied. For the purpose, order relations are introduced on $I \times I$.

**Definition 1.** (order) For points $\alpha = (x, y)$ and $\alpha' = (x', y')$ in the closed domain $I \times I$, the order relation $\alpha \leq \alpha'$ holds iff the conditions

Masaya Nohmi, Aoi Honda, and Yoshiaki Okazaki
Faculty of Computer Science and Systems Engineering,
Kyushu Institute of Technology
e-mail: nohmi@ai.kyutech.ac.jp, aoi@ces.kyutech.ac.jp,
okazaki@ces.kyutech.ac.jp

S. Li (Eds.): Nonlinear Maths for Uncertainty and its Appli., AISC 100, pp. 627–634.
springerlink.com                                    © Springer-Verlag Berlin Heidelberg 2011

$$x \le x' \text{ and } y \le y' \tag{1}$$

hold.

The order relation defined by Definition 1 is a partial order, but is not a total order. It effects properties of t-norms defined later.

**Definition 2.** (order) For points $\alpha = (x, y)$ and $\alpha' = (x', y')$ in the closed domain $I \times I$, the order relation $\alpha < \alpha'$ holds iff the conditions

$$\alpha \le \alpha' \text{ and } \alpha \ne \alpha' \tag{2}$$

hold.

two-dimensional t-norms on $I \times I$ are defined as follows.

**Definition 3.** (two-dimensional t-norms) An operation

$$T : (I \times I) \times (I \times I) \to (I \times I)$$

is a t-norm, if $T$ satisfies the following four conditions:

(1) commutativity:  $T(\alpha, \alpha') = T(\alpha', \alpha)$,
(2) associativity:  $T(\alpha, T(\alpha', \alpha'')) = T(T(\alpha, \alpha'), \alpha'')$,
(3) monotonicity:  $\alpha \le \alpha' \Rightarrow T(\alpha, \alpha'') \le T(\alpha', \alpha'')$,
(4) boundary conditions:  $T(\alpha, (0, 0)) = (0, 0)$, $T(\alpha, (1, 1)) = \alpha$,

where $\alpha, \alpha', \alpha'' \in I \times I$.

**Definition 4.** (strong monotonicity) If a two-dimensional t-norm $T$ satisfies the condition
$$\alpha < \alpha' \Rightarrow T(\alpha, \alpha'') < T(\alpha', \alpha''), \tag{3}$$
for any $\alpha, \alpha', \alpha'' \in (0, 1] \times (0, 1]$, t-norm $T$ is *strongly monotone* .

Projections $\pi_1 : I \times I \to I$ and $\pi_2 : I \times I \to I$ are defined by the equations

$$\pi_1(\alpha) = x, \quad \pi_2(\alpha) = y, \tag{4}$$

where $\alpha = (x, y) \in I \times I$.

*Example 1.* Let $T_P$ be the operation defined on $I \times I$ by the equation

$$T_P((x, y), (x', y')) = (xx', yy'), \tag{5}$$

where $(x, y), (x', y') \in I \times I$. That is, the operation $T_P$ is the two-dimensional algebraic product. Furthermore, let $\sigma : I \times I \to I \times I$ be the map defined by the equation

$$\sigma((x, y)) = (x, y^{(x+1)/2}). \tag{6}$$

The inverse of $\sigma$ is

$$\sigma^{-1} : I \times I \to I \times I, \quad \sigma^{-1}((x, y)) = (x, y^{2/(x+1)}). \tag{7}$$

Let $T$ be an operation defined by the equation

$$T((x, y), (x', y')) = \sigma^{-1}[T_P(\sigma((x, y)), \sigma((x', y')))]. \tag{8}$$

The the operation $T$ is a two-dimensional t-norm on $I \times I$.

*Proof.* Because the operation $T_P$ is commutative,

$$
\begin{aligned}
T((x, y), (x', y')) &= \sigma^{-1}[T_P(\sigma((x, y)), \sigma((x', y')))] \\
&= \sigma^{-1}[T_P(\sigma((x', y')), \sigma((x, y)))] = T((x', y'), (x, y)),
\end{aligned}
\tag{9}
$$

therefore, the operation $T$ is also commutative. Because the operation $T_P$ is associative,

$$
\begin{aligned}
&T((x, y), T((x', y'), (x'', y''))) \\
&= \sigma^{-1}\Big[T_P\Big(\sigma((x, y)), \sigma[\sigma^{-1}[T_P(\sigma((x', y')), \sigma((x'', y'')))]]\Big)\Big] \\
&= \sigma^{-1}\Big[T_P\Big(\sigma((x, y)), T_P(\sigma((x', y')), \sigma((x'', y'')))\Big)\Big] \\
&= \sigma^{-1}\Big[T_P\Big(T_P(\sigma((x, y)), \sigma((x', y'))), \sigma((x'', y''))\Big)\Big] \\
&= \sigma^{-1}\Big[T_P\Big(\sigma[\sigma^{-1}[T_P(\sigma((x, y)), \sigma((x', y')))]], \sigma((x'', y''))\Big)\Big] \\
&= T(T((x, y), (x', y')), (x'', y'')),
\end{aligned}
\tag{10}
$$

therefore, the operation $T$ is also associative. The operation $T$ could be calculated as

$$T((x, y), (x', y')) = \left(xx', y^{(x+1)/(xx'+1)} y'^{(x'+1)/(xx'+1)}\right). \tag{11}$$

It is shown that the operator $T$ is strongly monotone by calculating the partial differentiation of (11). It is also shown that the operator $T$ satisfies the boundary conditions from the equation (11).

*Example 2.* Let $T_P$ be the algebraic product defined by (5) in *Example 1*. Let $\sigma : I \times I \to I \times I$ be the map defined by the equation

$$\sigma((x, y)) = \left(x, \frac{x+1}{2} y\right). \tag{12}$$

The inverse of $\sigma$ is

$$\sigma^{-1} : I \times I \to I \times [0, 2], \quad \sigma^{-1}((x, y)) = \left(x, \frac{2}{x+1} y\right). \tag{13}$$

Let $T$ be an operation defined by the equation

$$T((x,y),(x',y')) = \sigma^{-1}[T_P(\sigma((x,y)),\sigma((x',y')))]. \tag{14}$$

The operation $T$ is a two dimensional t-norm on $I \times I$.

*Proof.* The commutativity and the associativity of $T$ are shown as same as *Example 1*. The operation $T$ could be calculated as

$$T((x,y),(x',y')) = \left(xx', \frac{(x+1)(x'+1)}{2(xx'+1)}yy'\right). \tag{15}$$

It is shown that the operator $T$ is strongly monotone by calculating the partial differentiation of (15). It is also shown that the operator $T$ satisfies the boundary conditions from the equation (15).

## 2  Main Theorem

**Proposition 1.** *For any $x$, $y$, $y' \in I$,*

$$\pi_1[T((0,y),(x,y'))] = 0. \tag{16}$$

*Proof.* From the monotonicity and the boundary condition of t-norm $T$,

$$T((0,y),(x,y')) \le T((0,1),(1,1)) = (0,1), \tag{17}$$

therefore,

$$\pi_1[T((0,y),(x,y'))] = 0. \tag{18}$$

**Proposition 2.** *For any $x$, $x'$, $y \in I$,*

$$\pi_2[T((x,0),(x',y))] = 0. \tag{19}$$

*Proof.* It is proved as same as Proposition 1.

In this section, it is shown that if a two-dimensional t-norm satisfies the conditions

$$T((1,0),(1,0)) = (1,0), \tag{20}$$
$$T((0,1),(0,1)) = (0,1), \tag{21}$$

it could be decomposed into Cartesian product of two one-dimentional t-norms, and each one-dimentional t-norm have a genarating function. The t-norm mentioned in Example 1 satisfies the condition (20) and (21), but the t-norm mentioned in Example 2 does not satisfy the condition (21).

**Proposition 3.** *Assume that a two-dimensional t-norm $T$ satisfies the condition (21). For any $x$, $x' \in I$, there exists $\hat{x} \in I$ such that the equation*

$$T((x, 1), (x', 1)) = (\hat{x}, 1) \tag{22}$$

holds.

*Proof.* From the condition (21) and the monotonicity of $T$,

$$(0, 1) = T((0, 1), (0, 1)) \leq T((x, 1), (x', 1)), \tag{23}$$

therefore,

$$\pi_2(T((x, 1), (x', 1))) = 1. \tag{24}$$

**Proposition 4.** *Assume that that a two-dimensional t-norm $T$ satisfies the condition (20). For any $y$, $y' \in I$, there exists $\hat{y} \in I$ such that the equation*

$$T((1, y), (1, y')) = (1, \hat{y}) \tag{25}$$

holds.

*Proof.* The proof is same as Proposition 3.

The map

$$T_1 : I \times I \to I, \quad T_1(x, x') = \hat{x} \tag{26}$$

could be defined from the condition (22) in Proposition 3. Similarly, the map

$$T_2 : I \times I \to I, \quad T_2(y, y') = \hat{y} \tag{27}$$

could be defined from the condition (25).

A two dimensional-tnorm $T$ is *continuous* , if $T$ satisfies the condition

$$\lim_{\substack{(x, y) \to (x_0, y_0) \\ (x', y') \to (x'_0, y'_0)}} T((x, y), (x', y')) = T((x_0, y_0), (x'_0, y'_0)) \tag{28}$$

for arbitrary $(x_0, y_0), (x'_0, y'_0) \in I \times I$.

**Theorem 1.** *Assume that a strongly monotone and continuous two-dimensional t-norm $T$ satisfies the conditions (20) and (21). Let $T_1$ and $T_2$ be the maps that are defined by the equations (26) and (27) respectively. The operations $T_1$ and $T_2$ are strongly monotone and continuous one-dimensional t-norms.*

*Proof.* Take $x$, $x' \in I$ arbitrarily. Because t-norm $T$ is commutative, the equation

$$\pi_1[T((x, 1), (x', 1))] = \pi_1[T((x', 1), (x, 1))] \tag{29}$$

holds, therefore, $T_1(x, x') = T_1(x', x)$. That is, the operation $T_1$ is commutative. Take $x$, $x'$, $x'' \in I$ arbitrarily. Because t-norm $T$ is associative, the equation

$$\pi_1[T(T((x, 1), (x', 1)), (x'', 1))] = \pi_1[T((x, 1), T((x', 1), (x'', 1)))] \tag{30}$$

holds, therefore, $T_1\big(T_1(x, x'), x''\big) = T_1\big(x, T_1(x', x'')\big)$. That is, the operation $T_1$ is associative. Assume that $x$, $x'$, $x'' \in I$ and $x < x'$. Because the order relation $(x, 1) < (x', 1)$ holds, the order relation

$$T\big((x, 1), (x'', 1)\big) < T\big((x', 1), (x'', 1)\big) \tag{31}$$

holds from the strong monotonicity of the t-norm $T$. Because the equation

$$\pi_2\big[T\big((x, 1), (x'', 1)\big)\big] = \pi_2\big[T\big((x', 1), (x'', 1)\big)\big] = 1 \tag{32}$$

holds, the inequality

$$\pi_1\big[T\big((x, 1), (x'', 1)\big)\big] < \pi_1\big[T\big((x', 1), (x'', 1)\big)\big] \tag{33}$$

holds. Therefore, $T_1(x, x'') < T_1(x', x'')$. That is, the operation $T_1$ is strongly monotone. Take $x \in I$ arbitrarily. From Proposition 2, the equation

$$\pi_1\big[T\big((x, 1), (0, 1)\big)\big] = 0 \tag{34}$$

holds. Therefore, $T_1(x, 0) = 0$. Furthermore, from the boundary condition of the t-norm $T$, the equation

$$T\big((x, 1), (1, 1)\big) = (x, 1) \tag{35}$$

holds. Therefore, $T_1(x, 1) = x$. Take $x_0$, $x'_0 \in I$ arbitrarily. Because the t-norm $T$ is continuous, the equation

$$\lim_{(x,x')\to(x_0,x'_0)} T\big((x, 1), (x', 1)\big) = T\big((x_0, 1), (x'_0, 1)\big) \tag{36}$$

holds, therefore, the equation

$$\lim_{(x,x')\to(x_0,x'_0)} T_1(x, x') = T_1(x_0, x'_0) \tag{37}$$

holds. That is, the operation $T_1$ is also continuous. The proof for the operation $T_2$ is quite the same as $T_1$.

**Theorem 2.** *Assume that strongly monotone and continuous two-dimensional t-norm $T$ satisfies the condition (20) and (21). There exist two strongly monotone function*

$$h_1 : (0, 1] \to [0, +\infty), \tag{38}$$
$$h_2 : (0, 1] \to [0, +\infty) \tag{39}$$

*and a map*

$$\Lambda : I \times I \to I \times I \tag{40}$$

*and a pseudo-inverse $\tilde{\Lambda}$ of $\Lambda$ such that*

$$T\big((x,y),(x',y')\big) =$$
$$\Lambda\Big[\Big(h_1^{-1}\big(h_1(\tilde{x}) + h_1(\tilde{x}')\big), h_2^{-1}\big(h_2(\tilde{y}) + h_2(\tilde{y}')\big)\Big)\Big], \tag{41}$$
$$\tilde{\Lambda}\big((x,y)\big) = (\tilde{x},\tilde{y}), \quad \tilde{\Lambda}\big((x',y')\big) = (\tilde{x}',\tilde{y}')$$

*holds for any $(x,y)$, $(x',y') \in (0,1] \times (0,1]$.*

*Proof.* Let $T_1$ be the map defined by the equation (26). From Theorem 1, the map $T_1$ is strongly monotone and continuous, therefore, it has a generating function.[1] That is, there exists strongly monotone and continuous function

$$h_1 : (0,1] \to [0,+\infty) \tag{42}$$

such that the equation

$$T_1(x,x') = h_1^{-1}\big(h_1(x) + h_1(x')\big), \quad x,x' \in (0,1] \tag{43}$$

holds. Similarly, let $T_2$ be the map defined by the equation (27), then $T_2$ has a generating function. That is, there exists strongly monotone and continuous function

$$h_2 : (0,1] \to [0,+\infty), \tag{44}$$

such that the equation

$$T_2(x,x') = h_2^{-1}\big(h_2(x) + h_2(x')\big), \quad x,x' \in (0,1] \tag{45}$$

holds. Define the map $\Lambda : I \times I \to I \times I$ by the equation

$$\Lambda\big((x,y)\big) = T\big((x,1),(1,y)\big), \quad (x,y) \in I \times I. \tag{46}$$

Let $\Delta_1 = \{(x,0)\,|\,x \in I\}$. The segment $\Delta_1$ is one of four segments that compose the boundary of the domain $I \times I$. From Proposition 2, $\Lambda\big((x,0)\big) \in \Delta_1$ holds. Furthermore, the equations

$$\Lambda\big((0,0)\big) = T\big((0,1),(1,0)\big) = (0,0), \tag{47}$$
$$\Lambda\big((1,0)\big) = T\big((1,1),(1,0)\big) = (1,0) \tag{48}$$

hold, therefore, $\Lambda(\Delta_1) = \Delta_1$. Similarly, for the segment $\Delta_2 = \{(0,y)\,|\,y \in I\}$, the equation $\Lambda(\Delta_2) = \Delta_2$ holds. For the segment $\Delta'_1 = \{(x,1)\,|\,x \in I\}$, the equation

$$\Lambda\big((x,1)\big) = T\big((x,1),(1,1)\big) = (x,1) \tag{49}$$

holds from the boundary condition of $T$. Furthermore, the equations

$$\Lambda\big((0,1)\big) = T\big((0,1),(1,1)\big) = (0,1), \tag{50}$$
$$\Lambda\big((1,1)\big) = T\big((1,1),(1,1)\big) = (1,1) \tag{51}$$

hold, therefore, $\Lambda(\Delta'_1) = \Delta'_1$. Similarly, for the segment $\Delta'_2 = \{(1, y) \mid y \in I\}$, the equation $\Lambda(\Delta'_2) = \Delta'_2$ holds. Therefore, for the boundary $\Delta = \{\Delta_1, \Delta_2, \Delta'_1, \Delta'_2\}$ of the closed domain $I \times I$, the equation $\Lambda(\Delta) = \Delta$ holds. Therefore, the map $\Lambda : I \times I \to I \times I$ is a surjection, and there exists a pseudo-inverse $\tilde{\Lambda} : I \times I \to I \times I$ such that $\Lambda(\tilde{\Lambda}((x, y))) = (x, y)$. Let

$$\tilde{\Lambda}((x, y)) = (\tilde{x}, \tilde{y}), \qquad \tilde{\Lambda}((x', y')) = (\tilde{x}', \tilde{y}'), \tag{52}$$

then

$$
\begin{aligned}
T\big((x, y), (x', y')\big) &= T\big(\Lambda((\tilde{x}, \tilde{y})), \Lambda((\tilde{x}', \tilde{y}'))\big) \\
&= T\big(T((\tilde{x}, 1), (1, \tilde{y})), T((\tilde{x}', 1), (1, \tilde{y}'))\big) \\
&= T\big(T((\tilde{x}, 1), (\tilde{x}', 1)), T((1, \tilde{y}), (1, \tilde{y}'))\big) \\
&= T\big((T_1(\tilde{x}, \tilde{x}'), 1), (1, T_2(\tilde{y}, \tilde{y}'))\big) \\
&= T\Big(\big(h_1^{-1}(h_1(\tilde{x}) + h_1(\tilde{x}')), 1\big), \big(1, h_2^{-1}(h_2(\tilde{y}) + h_2(\tilde{y}'))\big)\Big) \\
&= \Lambda\Big[\big(h_1^{-1}(h_1(\tilde{x}) + h_1(\tilde{x}')), h_2^{-1}(h_2(\tilde{y}) + h_2(\tilde{y}'))\big)\Big]
\end{aligned}
\tag{53}
$$

holds.

Many applications of two-dimensional t-norms could be considered. In intuitionistic fuzzy set theory [2], a fuzzy set is defined as $A = \{(x, \mu_A(x), \nu_A(x))\}$, where $\mu_A(x) \in I$ is the degree of membership of $x$ in $A$, $\nu_A(x) \in I$ is the degree of non-membership of $x$ in $A$. The degree of membership is defined as a pair of numbers in $I \times I$, therefore, two-dimensional t-norms could be applicable. Furthermore, studies of partially ordered semigroups are important as fundamental studies [3]. The authors are considering them as further themes of our studies.

The authors wish to express their gratitude to the unknown referees for their fruitful comments and advices.

# References

1. Aczel, J.: Sur les opérations définies pour nombres réels. Bulletin de la S. M. F 76, 59–64 (1948)
2. Deschrijver, G., Kerre, E.E.: On the relationship between some extensions of fuzzy set theory. Fuzzy Sets and Systems 133, 227–235 (2003)
3. Fuchs, L., Steinfeld, O.: Principal components and prime factorization in partially ordered semigroups, Annales Universitatis Scientiarum Budapestinensis de Ralando Eötvös nominatae. Sectio mathematica 6, 102–111 (1963)
4. Klement, E.P., Mesiar, R., Pap, E.: Triangular Norms. Kluwer, Dordrecht (2000)
5. Schweizer, B., Sklar, A.: Associative functions and statistical triangle inequations. Publ. Math. Debrecen 8, 169–186 (1961)

# Approximating a Fuzzy Vector Given Its Finite Set of Alpha-Cuts

Xuecheng Liu and Qinghe Sun

**Abstract.** In this paper, with information of a finite set of alpha-cuts of a 2-dimensional fuzzy vector, we propose some approaches in approximating the fuzzy vector by a fuzzy vector or a sequence of fuzzy vectors.

**Keywords:** Fuzzy number, Fuzzy vector, Alpha-cut, Membership function, Convex set, Convex hull.

## 1 Introduction

In any application of fuzzy sets, we first need to elicit their membership functions. It is much more practical to require information of a finite set of alpha-cuts (such as the core and the support) of a fuzzy set than the membership function itself. One question is that how to approximate or construct a fuzzy set with the information of these alpha-cuts. This paper answers the question to a special case of fuzzy sets, 2-dimensional fuzzy vectors, *or fuzzy vectors for short from now on*, defined on the universal set of $\mathbb{R}^2$.

To (1-dimensional) fuzzy numbers, there are two directions in approximation. One is to approximate fuzzy numbers having complicated membership functions with fuzzy numbers having simpler membership functions such as interval numbers (e.g., Grzegorzewski [5], Chanas [3]), fuzzy numbers with triangular and trapezoidal membership functions (e.g., Abbasbandy [1], Ban [2] and Yeh [9]), or more flexible LR fuzzy numbers (Dubois [4]), to name a few.

Xuecheng Liu
Research Unit on Children's Psychosocial Maladjustment,
University of Montreal, Canada
e-mail: xuecheng.liu@umontreal.ca

Qinghe Sun
Department of Mathematics and statistics, University of Montreal, Canada
e-mail: sunqhe@yahoo.com

S. Li (Eds.): Nonlinear Maths for Uncertainty and its Appli., AISC 100, pp. 635–642.
springerlink.com                    © Springer-Verlag Berlin Heidelberg 2011

It is clear that defuzzification is the wildest approximation for fuzzy numbers (e.g., Ma et al [6]).

The other direction in approximating fuzzy numbers is, given a finite set of alpha-cuts of a fuzzy number, how to approximate it or reconstruct it with a new fuzzy number. For such purpose, Wang [8] approximates the fuzzy number with a pairwise approximation of a fuzzy set on $\mathbb{R}^1$ with piecewise linear membership function. Figure 1 from Wang [8] illustrates the notion, and the membership function may be termed as piled trapezoidal membership function.

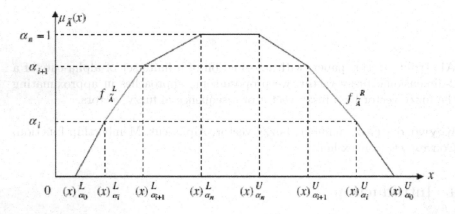

**Fig. 1** Fuzzy number approximation using piecewise linear membership functions (Wang [8])

This paper applies above notion to fuzzy vectors: given a finite set of alpha-cuts of a fuzzy vector, we approximate or reconstruct the fuzzy vector with a fuzzy vector with piled "2-dimensional trapezoidal" membership function. *Without any loss of generality,* in this paper, we consider the case that the given finite set of alpha-cuts consists of only the support and the core.

The rest of this paper is organized as follows. In Section 2, we describe the concepts of fuzzy set, the core and the support of a fuzzy set, fuzzy number, fuzzy vector etc. We also give some notations used in the paper.

In Section 3, we propose some approaches in approximating fuzzy vectors when (1) the core is a singleton; (2) the support and the core are regions enclosed by the projections of two parallel conic sections of a cone; and (3) the core and the support are two rectangles with parallel sides.

In Section 4, first, we use an example to explain that, for more general cores and supports of fuzzy vectors, the approaches used in Section 3 are no longer working. Then we proposal a new approach in approximating fuzzy vectors for any given cores and supports of fuzzy vectors.

In Section 5, we discuss further research questions on approximating 2-dimensional fuzzy vectors and discuss possibilities of the approximating approaches in this paper to even high-dimensional fuzzy vectors.

## 2 Definitions, Assumptions and Notations

**Definition 1.** (**Fuzzy sets, alpha-cuts, cores, supports**, Zadeh [10] and Talaŏvá et al. [7]) A fuzzy set $A$ on $X$ is defined by its membership function $A : X \to [0,1]$. The $\alpha$-cuts $A_\alpha$ ($\alpha \in [0,1]$) are defined by $\{x \in X; \ A(x) \geq \alpha\}$. The $\alpha$-cut $A_1$ is called the core of $A$ and denoted by $\text{Core}(A)$, and the set $\{x \in X; \ A(x) > 0\}$ is called the support of $A$ and denoted by $\text{Supp}(A)$.

**Definition 2.** (**Fuzzy vectors and fuzzy numbers**, Talaŏvá et al [7]) A fuzzy set $A$ on $\mathbb{R}^m$ is called a fuzzy vector, or a fuzzy number when $m = 1$, if (1) $\text{Core}(A) \neq \emptyset$; (2) for all $\alpha \in (0,1]$, $A_\alpha$ are bounded closed convex subsets of $\mathbb{R}^m$; and (3) $\text{Supp}(A)$ is bounded.

**Assumptions and notations.** Throughout this paper:

- We use $A$ to stand for both the fuzzy vector with given support and core and the fuzzy vector used to approximate it;
- We identify $\text{Supp}(A)$ and its closure and denote it by $A_0$;
- We assume that the interior of $A_0$ (to the usual topology in $\mathbb{R}^2$) is not empty. (Otherwise, fuzzy vectors are equivalent to fuzzy numbers or alike);
- We assume that $A_1$ (i.e. $\text{Core}(A)$) is a subset of the interior of $A_0$, i.e., the boundaries of $A_0$ and $A_1$ are disjoint;
- We denote by $\overline{P_1 P_2}$ the segment between $P_1, P_2 \in \mathbb{R}^2$;
- We denote by $d(P_1, P_2)$ the Euclidean distance between $P_1, P_2 \in \mathbb{R}^2$;
- We denote by $P_1 P_2 \cdots P_k$ the polygon with vertices $P_1, P_2, ..., P_k$ in $\mathbb{R}^2$;
- Let $S$ be a bounded solid-region in $\mathbb{R}^2$, we denote its boundary by $\partial(S)$.
- We denote by $L_{CP}$ the ray with vertex $C$ and through the point $P$, and $L_C$ the set consisting of all $L_{CP}$'s;

## 3 Approximating Fuzzy Vectors: Some Special Cases

**The case of singleton cores.** If $A_1$ is a singleton, denoted by $\{C\}$, by our assumption, the point $C$ is an interior point of $A_0$. We approximate the fuzzy vector $A$ by the fuzzy vector whose membership function is the surface of the cone with the vertex $C$ at height 1 and with the boundary of $A_0$ as the perimeter of the cone base. Figure 2 illustrates the case: for $P \in A_1$, $Q$ lies on the surface of the cone and the segment $\overline{QP}$ is orthogonal to the $X_1 X_2$ plane. The membership degree of $A$ at $P$ is the height of $Q$, i.e, $d(Q,P)$. Clearly, the membership function is continuous.

The membership function $A$ can be defined in another equivalent way. In Figure 2, for the point $P$, denote by $P_0$ the point at which the ray $L_{CP}$ and

**Fig. 2** A membership
function of fuzzy vector
with singleton core

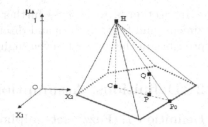

$\partial(A_0)$ intersect. (The point $Q$ must lies on the segment of $\overline{HP_0}$.) The the
membership degree of $A$ at $P$ is $d(P_0, P)/d(P_0, C)$.

**The case of supports and cores paralleling to conic sections.** This
is a case which is an extension of above case. The support and the core are
regions enclosed by the projections of parallel conic sections of a cone. Figure
3 illustrates the case. Similar to the case of singleton cores, the membership
function of the fuzzy vector is defined by the surface of the frustum, that is,
the membership degree is defined as 1 for any point in $A_1$, and as the height
of $Q$, i.e., $d(P, Q)$, for the points $P \in A_0 \setminus A_1$ (the difference of sets $A_0$, $A_1$).

**Fig. 3** An membership
function of a fuzzy vector
with the support and
the core to parallel conic
sections

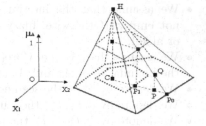

Similar to the case of singleton cores, the membership function of $A$ can be
defined in an equivalent way. For any $P \in A_0 \setminus A_1$, denote by $P_0$ (respectively,
$P_1$) the point at which the ray $L_{CP}$ and $\partial(A_0)$ (respectively, $\partial(A_1)$) intersect.
Then the membership degree of $A$ at $P$ is $d(P_0, P)/d(P_0, P_1)$.

**The case of rectangle supports and cores with parallel sides.** Consider
the case of the support $X_1X_2X_3X_4$ and the core $Y_1Y_2Y_3Y_4$ as rectangles
shown in Figure 4. We do not require that there exists a cone to the support
and the core as the cases discussed above. The most natural way to use
the following approach to approximate the fuzzy vector $A$: for each $P$ in
$A_0 \setminus A_1$, say $P$ in $X_2X_3Y_3Y_2$, draw a line parallel to $\overline{X_2X_3}$, and denote by $P_2$
(respectively $P_3$) the point at which the line and $\overline{X_2Y_2}$ (respectively $\overline{X_3Y_3}$)
intersect. The membership degree of $A$ at $P$ is defined as $d(Y_2, P_2)/d(Y_2, X_2)$
(or, equivalently, $d(Y_3, P_3)/d(Y_3, X_3)$).

**Fig. 4** Rectangle supports and cores with parallel sides

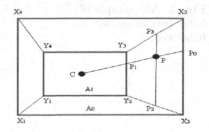

We can define the same membership degree of $A$ at the point $P$ by considering a ray $L_{CP}$ where $C \in A_1$ and $L_{CP}$ and $\overline{Y_2Y_3}$ intersect. Denote by $P_0$ (respectively, $P_1$) the point at which $L_{CP}$ and $\partial(A_0)$ (respectively, $\partial(A_1)$) intersect. The membership degree of $A$ at $P$ is $d(P_0, P)/d(P_0, P_1)$. Note that the choice of $C$ may depend on the point $P$ in this case, unlike the two cases discussed above.

## 4  Approximating Fuzzy Vectors: General Case

In this section, we consider the general case of the supports and the cores of fuzzy vectors. Remind that we have assumed that the boundaries of the support and the core are disjoint (Section 2).

In all of the three special cases discussed in Sections 3, we can always use the "drawing-a-ray" approach in defining the membership functions. It seems natural to apply the same notion to the general case. That is, to select a point $C$ in the interior of $A_1$ and consider all rays with vertex $C$. For any point $P$ in $A_0 \setminus A_1$, denoted by $P_0$ (respectively, $P_1$) the point at which $L_{CP}$ and $\partial(A_0)$ (respectively, $\partial(A_1)$) intersect. Then define the membership degree of $A$ at the point $P$ as $d(P_1, P)/d(P_1, P_0)$. However, the fuzzy set defined in this way usually is *not* a fuzzy vector, as the following example shows.

**An example of fuzzy set on $\mathbb{R}^2$ which is not a fuzzy vector.** In Figure 5, the support $A_0$ is the rectangle with vertices

$$(-6, -8), (6, -8), (6, 8), (-6, 8),$$

and the core $A_1$ is the hexagon with vertices

$$(0, -6), (3/2, -2), (3/2, 2), (0, 6), (-3/2, 2), (-3/2, -2).$$

We select $O(0, 0)$ as the vertex of the rays, and define membership function in the "drawing-a-ray" approach. The $\frac{1}{2}$-cut is not convex, even $A_0$, $A_1$ are symmetric polygons and $O$ is the center of both $A_0$ and $A_1$. $\quad\square$

To the general case, we modify above "drawing-a-ray" approach. The basic idea is to construct a sequence of fuzzy vectors with "2-dimensional

**Fig. 5** An example of fuzzy sets being not a fuzzy vector

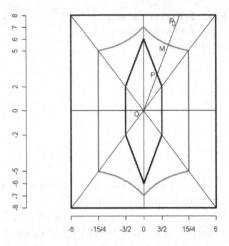

step-functions" whose limit is the fuzzy vector $A$. We explain the idea with above example, where, with "drawing-a-ray" approach, the $\frac{1}{2}$-cut is not convex. We use its convex hull as the $\frac{1}{2}$-cut of $A$ in approximation. In this way, we construct a fuzzy vector with the "2-dimensional step-functions" with 3 alpha-cuts, $A_0$, $A_{\frac{1}{2}}$ and $A_1$, rather than the original 2 alpha-cuts $A_0$ and $A_1$. Then consider the pair of $A_0$, $A_{\frac{1}{2}}$ and the pair $A_{\frac{1}{2}}$ and $A_1$ separately, we can further construct $A_{\frac{1}{4}}$ and $A_{\frac{3}{4}}$. Such process can be repeated as many times as we need, and the approximation will get better and better. Meanwhile, if, after some steps, all newly contracted alpha-cuts are already convex, we come back to the "drawing-a-ray" approach.

We formulate the notion in the following Theorem 1, where we show that the fuzzy vector defined by the limit of the sequence of the fuzzy vectors with 2-dimensional step-functions has an everywhere continuous membership function.

Before we state Theorem 1, we give some notations. Let $S_0 \supseteq S$ be two bounded solid-regions in $\mathbb{R}^2$ and $S$ has nonempty interior, let $C$ be a point in the interior of $S$.

$$H_c(S) \stackrel{\text{def}}{=} \text{the convex hull of } S;$$

$$T(C, S, S_0) \stackrel{\text{def}}{=} \{(P, P_0); \text{ for } L \in L_C, \partial(S), \partial(S_0) \text{ intersect } L \text{ at } P, P_0$$
$$\text{respectively}\};$$

$$M(C, S, S_0) \stackrel{\text{def}}{=} \{M; M \text{ is the mid-point of } P, P_0, (P, P_0) \in T(C, S, S_0)\};$$

$$D(C, S, S_0) \stackrel{\text{def}}{=} \max\{d(P, P_0); (P, P_0) \in T(C, S, S_0)\}.$$

The value $D(C, S, S_0)$ describes the degree of closeness between the boundaries of $S$ and $S_0$: the smaller the value $D(C, S, S_0)$ is, the closer the boundaries of $S$ and $S_0$ are.

We can use Figure 5 to illustrate the notations. The sets $S_0$ and $S$ in defining notations are $A_0$ and $A_1$ in Figure 5 respectively. The set $T(C, S, S_0)$ consists of all "twin" pairs $(P, P_0)$ for all rays with vertex $O$; The set $M(C, S, S_0)$ is the curve of all points $M$'s; The value $D(C, S, S_0)$ is the distance between (say) the points $(3/2, 2)$ and $(6, 8)$, i.e., 7.5.

The sequence of the fuzzy vectors, denoted by $\{A^n\}_{n=0,1,2,\ldots}$, whose limit is used to approximate $A$, is constructed as follows. First, inductively, denote,

$$\text{for } n = 0: \quad A_0^0 \stackrel{\text{def}}{=} A_0, \quad A_1^0 \stackrel{\text{def}}{=} A_1;$$

$$\text{for } n = 1: \quad A_0^1 \stackrel{\text{def}}{=} A_0^0, \quad A_{\frac{1}{2}}^1 \stackrel{\text{def}}{=} H_c(M(C, A_1^0, A_0^0)), \quad A_1^1 = A_1^0;$$

$$\text{for } n = 2: \quad A_1^2 \stackrel{\text{def}}{=} A_1^1, \quad A_{\frac{3}{4}}^2 \stackrel{\text{def}}{=} H_c(M(C, A_1^1, A_{\frac{1}{2}}^1)), \quad A_{\frac{2}{4}}^2 \stackrel{\text{def}}{=} A_{\frac{1}{2}}^1,$$

$$A_{\frac{1}{4}}^2 \stackrel{\text{def}}{=} H_c(M(C, A_{\frac{1}{2}}^1, A_0^1)), \quad A_0^2 = A_0^1,$$

repeat above process, for each $A^n$, we have $2^n + 1$ alpha-cuts, from largest to smallest,

$$A_0^n = A_{0/2^n}^n = A_0, \ A_{1/2^n}^n, \ A_{2/2^n}^n, \ \ldots, \ A_{2^n/2^n}^n = A_1^n = A_1.$$

Then we can define the sequence of fuzzy vectors $\{A^n\}_{n=0,1,2,\ldots}$ as

$$A^n(P) = \begin{cases} 1, & \text{if } P \in A_1^n = A_1, \\ \frac{k}{2^n}, & \text{if } P \in A_{\frac{k}{2^n}}^n \setminus A_{\frac{k+1}{2^n}}^n \quad (k = 2^n - 1, 2^n - 2, \ldots, 1, 0). \end{cases}$$

Above partitioning process can be as fine as we need, which is formalized in Lemma 1.

**Lemma 1.** *Under the assumption that $\partial(A_0)$ and $\partial(A_1)$ are disjoint,*

1. *For $n = 1, 2, \ldots, i = 0, 1, \ldots, 2^n$, $\partial(A_{i/2^n}^n)$ and $\partial(A_{(i+1)/2^n}^n)$ are disjoint.*
2. *For every $\delta > 0$, there exists a positive integer $N(\delta)$ such that*

$$\max\{D(C, A_{i/2^n}^n, A_{(i+1)/2^n}^n); \ i = 0, 1, 2, \ldots, 2^n\} < \delta$$

*when $n \geq N(\delta)$.*

*Proof.* Omitted. □

For every $P \in A_0$, since $A^n(P)$ is increasing and upper bounded by 1, then $\lim_{n \to \infty} A^n(P)$ exists. We obtain a fuzzy set whose membership function is defined by the the limit. By using Lemma 1, we can prove

**Theorem 1.** *The fuzzy set $A$ defined by the limit is a fuzzy vector, and the membership function is continuous over $A_0$.*

*Proof.* Omitted. □

# 5 Summary and Discussion

Given a finite set of alpha-cuts of a fuzzy vector, we discussed how to approximate it with one having "2-dimensional" trapezoidal-like membership functions. Without any loss of generality, assuming that the finite set of alpha-cuts contains the support and the core only,

- we listed 3 special cases that we can use the "drawing-a-ray" approach to approximate the fuzzy vector;
- we used an example to show that the "drawing-a-ray" approach usually does not work in general case;
- and thus, in general case, we constructed a sequence of fuzzy vectors to approximate the fuzzy vector;
- and finally, in general case, we proved that the fuzzy vector defined by the sequence limit has everywhere continuous membership function.

Note that in this paper, we assumed that the boundaries of the support and the core are disjoint. Without it, the continuity of the membership function of the limit fuzzy vector does not hold.

The most natural choice of the vertex of the rays is the centroid the core of the fuzzy vector. Further study is needed in the sensitivity of the choice.

We could extend the discussion from 2-dimensional fuzzy vectors into any multi-dimensional fuzzy vectors. It seems that we have not difficulty for such extension if we "translate" the language in 2-dimensional fuzzy vectors in Sections 3 and 4 into the language in any multi-dimensional fuzzy vectors.

# References

1. Abbasbandy, S., et al.: The nearest trapezoidal form of a generalized left right fuzzy number. Int. J. of Approximate Reasoning 43, 166–178 (2006)
2. Ban, A.: Approximation of fuzzy numbers by trapezoidal fuzzy numbers preserving the expected interval. Fuzzy Sets and Systems 159, 1327–1344 (2008)
3. Chanas, S.: On the interval approximation of a fuzzy number. Fuzzy Sets and Systems 122, 353–356 (2001)
4. Dubois, D., et al.: Fuzzy Sets and Systems: Theory and application. Academic Press, New York (1980)
5. Grzegorzewski, P.: Nearest interval approximation of a fuzzy number. Fuzzy Sets and Systems 130, 321–330 (2002)
6. Ma, M., et al.: A new approach for defuzzification. Fuzzy Sets and Systems 111, 351–356 (2000)
7. Talaǒvá, J., et al.: Fuzzy vectors as a tool for modeling uncertain multidimensional quantities. Fuzzy sets and systems 161, 1585–1603 (2010)
8. Wang, Y.M.: Centroid defuzzification and the maximizing set and minimizing set ranking based on alpha level sets. Computers & Industrial Engineering 57, 228–236 (2009)
9. Yeh, C.T.: On improving trapezoidal and triangular approximations of fuzzy numbers. International Journal of Approximate Reasoning 48, 297–313 (2008)
10. Zadeh, L.A.: Fuzzy Sets. Information and Control 8, 338–353 (1965)

# A New Fuzzy Linear Programming Model and Its Applications

Hsuan-Ku Liu and Berlin Wu

**Abstract.** It is difficult to determine precise values for the parameters in the real world problems. To model the uncertainty, a new fuzzy linear programming model, called $\beta$-tolerance linear programming model, is developed in this paper. When the fuzzy numbers are all triangle fuzzy numbers, the solutions, which propose a selection interval for the decision maker, are obtained.

**Keywords:** Fully fuzzy linear inequations, Fuzzy triangular matrix, Fuzzy programming, Optimization models.

## 1 Introduction

Many decision problems or resources allocation problems are modelled as a linear programming models. However, determining precise values for the parameters in the real world problems is often a difficult task for the decision maker(s). To model the uncertainty, many methods, such as the stochastic programming or the fuzzy programming, are developed. The uncertainty in the stochastic programming is described by a probabilistic random variable.

Many real world problems take place in an imprecise environment in where vagueness in the coefficient may not be of a probabilistic type. In this section, the decision makers could model the imprecise by means of fuzzy parameters [19]. In the last few years, several articles have been devoted to the study of this subject [14], [15]. An early contribution was made by Fang et al. [10]. They focus on the fuzzy linear programming

Hsuan-Ku Liu
Department of Mathematics and Information Education,
National Taipei University of Education, Taiwan
e-mail: hkliu.nccu@gmail.com

Berlin Wu
Department of Mathematical Sciences, National Chengchi University, Taiwan
e-mail: Berlin@nccu.edu.tw

S. Li (Eds.): Nonlinear Maths for Uncertainty and its Appli., AISC 100, pp. 643–650.
springerlink.com                                    © Springer-Verlag Berlin Heidelberg 2011

$$\max c^T x$$
$$\text{s.t.} \quad \tilde{A}x \leq \tilde{b}$$
$$x \geq 0,$$

where $\tilde{A}$ and $\tilde{b}$ are $n \times m$ and $m \times 1$ matrices with fuzzy coefficients, $c$ and $x$ are crisp $n$-dimensional column vectors. Based on the specific ranking of fuzzy numbers, such problems are reduced to a linear semi-infinite programming problems.

Recently, few studies have been made at the full fuzzy linear programming problems of which the coefficients and the variables are all fuzzy numbers. With a special ranking on fuzzy numbers, Hosseinzadeh Lotfi [12] transformed the full fuzzy linear programming to multiple objective linear programming where all variables and parameters are crisp. The fuzzy linear programming with multiple objective linear programming has also been used by Buckley [1] and Xu and Chen [16].

This paper addresses a $\beta$-tolerance fuzzy linear programming model. To find a feasible region of this model, we first consider the the $\beta$-tolerance fuzzy system of inequalities. The advantage is that the programming with triangle fuzzy coefficients can be transformed to a crisp linear programming and its solution is still a triangle fuzzy vector. In the example, the $\beta$-tolerance fuzzy linear programming is applied to model the resources allocation problems in the vagueness environments.

This paper is organized as follows. In Section 2, the basic results of the fuzzy set theory will be discussed. In this section, we provide definitions for the solution of non-negative FFLS. In Section 3, we propose the $\beta$-tolerance relation between two fuzzy numbers and consider the fuzzy linear system of inequalities under the given relation. In Section 4, we develop the $\beta$-tolerance fuzzy linear programming and provide two applications for this programming. In Section 5, we close this paper with a concise conclusion.

## 2  Preliminary and Generalized Definition

This section introduces basic results in fuzzy set theory.

**Definition 1.** A fuzzy number $A = (a, \alpha, \beta)$, $\alpha > 0$, $\beta > 0$ is called a triangle fuzzy number if its membership function has the form:

$$\mu_A(x) = \begin{cases} \frac{x-a+\alpha}{\alpha} & \text{if } x \leq a, \\ 1 & \text{if } x = a, \\ \frac{x-a-\beta}{\beta} & \text{if } x \geq a. \end{cases}$$

where $a$ is the center of $F$ and $\alpha \geq 0$ and $\beta \geq 0$ are the left and right spreads, respectively.

Here, we say that $\tilde{A} = (a, \alpha, \beta)$ is positive if $a - \alpha > 0$ and that two fuzzy numbers $A = (a, \alpha, \beta)$ and $B = (b, \gamma, \delta)$ is equal if $a = b$, $\alpha = \gamma$ and $\beta = \delta$.

For any two fuzzy numbers, we define the following operations [9].

**Definition 2.** Let $\mathcal{F}$ be the set of LR-fuzzy numbers. If $A = (a, \alpha, \beta)$ and $B = (b, \gamma, \delta)$ are two fuzzy numbers in $\mathcal{F}$, then two operations, say addition and multiplication, are defined as:

- Addition
$$(a, \alpha, \beta) \oplus (b, \gamma, \delta) = (a + b, \alpha + \gamma, \beta + \delta).$$

- Approximate multiplication
If $A > 0$ and $B > 0$, then
$$(a, \alpha, \beta) \otimes (b, \gamma, \delta) = (ab, b\alpha + a\gamma, b\beta + a\delta).$$

If $A < 0$ and $B > 0$, then
$$(a, \alpha, \beta) \otimes (b, \gamma, \delta) = (ab, b\alpha - a\delta, b\beta - a\gamma).$$

If $A < 0$ and $B < 0$, then
$$(a, \alpha, \beta) \otimes (b, \gamma, \delta) = (ab, -b\beta - a\delta, -b\alpha - a\gamma).$$

On the other hand, the fuzzy matrix is defined as follows.

**Definition 3.** A matrix $\tilde{A} = (\tilde{a}_{ij})$ is called a fuzzy matrix if each element in $A$ is a fuzzy number.

Following the notation of Dehghan and Hashemi [6], the matrix $\tilde{A}$ can be represented as $\tilde{A} = (A, M, N)$ since $\tilde{A} = (\tilde{a}_{ij})$ and $\tilde{a}_{ij} = (a_{ij}, \alpha_{ij}, \beta_{ij})$. Three crisp matrices $A = (a_{ij})$, $M = (\alpha_{ij})$ and $N = (\beta_{ij})$ are called the center matrix and the right and left spread matrices, respectively.

The symmetric fully fuzzy linear system (SFFLS) is given as form:

$$\begin{aligned}
(\tilde{a}_{11} \otimes \tilde{x}_1) \oplus (\tilde{a}_{12} \otimes \tilde{x}_2) \oplus \cdots \oplus (\tilde{a}_{1n} \otimes \tilde{x}_n) &= \tilde{b}_1, \\
(\tilde{a}_{21} \otimes \tilde{x}_1) \oplus (\tilde{a}_{22} \otimes \tilde{x}_2) \oplus \cdots \oplus (\tilde{a}_{2n} \otimes \tilde{x}_n) &= \tilde{b}_2, \\
&\vdots \\
(\tilde{a}_{n1} \otimes \tilde{x}_1) \oplus (\tilde{a}_{n2} \otimes \tilde{x}_2) \oplus \cdots \oplus (\tilde{a}_{nn} \otimes \tilde{x}_n) &= \tilde{b}_n.
\end{aligned} \quad (1)$$

The matrix form of this linear system is represented as

$$\tilde{A} \otimes \tilde{x} = \tilde{b}$$

where $[\tilde{A}]_\alpha = (\tilde{a}_{ij}) = (A, \underline{A}_\alpha, \overline{A}_\alpha)$, $1 \le i, j \le n$ is an $n \times n$ fuzzy matrix, $\tilde{x} = (\tilde{x}_i) = (x, \underline{x}_\alpha, \overline{x}_\alpha)$ and $\tilde{b} = (\tilde{b}_i) = (b, \underline{b}_\alpha, \overline{b}_\alpha))$, $1 \le i \le n$ are $n \times 1$ fuzzy matrices. Hence the $\alpha$-level set of $\tilde{A} \otimes \tilde{x}$ is

$$[\tilde{A} \otimes \tilde{x}] = (Ax, \underline{A}_\alpha x + A\underline{x}_\alpha, \overline{A}_\alpha x + A\overline{x}_\alpha),$$

as $\tilde{A}$ is positive. This implies that the fuzzy linear system is reformulated as

$$Ax = b,$$
$$\underline{A}_\alpha x + A\underline{x}_\alpha = \underline{b}_\alpha,$$
$$\overline{A}_\alpha x + A\overline{x}_\alpha = \overline{b}_\alpha,$$

for all $0 \leq \alpha \leq 1$. So far, a considerable number of researches, such as [6], [7], [8], has been proposed to find the solution of the linear systems.

In the following sections, we shall consider the linear system of fuzzy inequalities and the fuzzy optimization problems.

## 3   The $\beta$-Tolerance Fuzzy Linear System of Inequalities

To consider the fuzzy inequality system, we first propose a fuzzy relation between two fuzzy numbers.

**Definition 4.** Let $\tilde{a}$ and $\tilde{b}$ be two fuzzy numbers. We say $\tilde{a} \leq_\beta \tilde{b}$ if

$$\sup\{x | x \in [\tilde{a}]_\alpha\} \leq \inf\{x | x \in [\tilde{b}]_\alpha\}$$

for $\alpha \geq \beta$, where $[\tilde{a}]_\alpha$ and $[\tilde{b}]_\alpha$ are $\alpha$-level sets of $\tilde{a}$ and $\tilde{b}$.

Let $\tilde{a}$ and $\tilde{b}$ be two triangle fuzzy numbers, their $\alpha$-level sets are written as

$$[\tilde{a}]_\alpha = (a, (1-\alpha)\underline{a}, (1-\alpha)\overline{a}) \text{ and } [\tilde{b}]_\alpha = (b, (1-\alpha)\underline{b}, (1-\alpha)\overline{b}).$$

This implies

$$\sup\{x | x \in [\tilde{a}]_\alpha\} = a + (1-\alpha)\overline{a} \text{ and } \inf\{x | x \in [\tilde{b}]_\alpha\} = b - (1-\alpha)\underline{b}.$$

For the triangle fuzzy numbers, we say $\tilde{a} \leq_\beta \tilde{b}$ if $(1-\alpha)(\overline{a}+\underline{b}) \leq b - a$, for $\beta \leq \alpha \leq 1$, since $(1-\alpha) \leq (1-\beta)$, for $\beta \leq \alpha \leq 1$. Let $\tilde{A} = (\tilde{a}_{ij})$ be a triangle fuzzy matrix with $\alpha$-level set $[\tilde{A}]_\alpha = (A, (1-\alpha)\underline{A}, (1-\alpha)\overline{A})$ and $\tilde{b} = (\tilde{b}_i)$ be a triangle fuzzy vector with $\alpha$-level set $[\tilde{b}]_\alpha = (b, (1-\alpha)\underline{b}, (1-\alpha)\overline{b})$. Here, $A$ (or $b$) is the center matrix of $\tilde{A}$ (or $\tilde{b}$), and $\underline{A}$ (or $\underline{b}$) and $\overline{A}$ ( or $\overline{b}$) are left and right spread matrices of $[\tilde{A}]_0$ (or $[\tilde{b}]_0$).

Let $[\tilde{x}]_\alpha = (x, (1-\alpha)\underline{x}, (1-\alpha)\overline{x})$ be a triangle fuzzy number. The $\beta$−tolerance fuzzy linear system of inequalities is considered by

$$\tilde{A} \otimes \tilde{x} \leq_\beta \tilde{b}. \tag{2}$$

and is formulated as

$$(Ax, (1-\alpha)\underline{A}x + (1-\alpha)A\underline{x}, (1-\alpha)\overline{A}x + (1-\alpha)A\overline{x}) \leq (b, (1-\alpha)\underline{b}, (1-\alpha)\overline{b}),$$

for all $\beta \leq \alpha \leq 1$. This implies that

$$[(1-\alpha)\overline{A} + A]x \leq [b - (1-\alpha)\underline{b}] - (1-\alpha)A\overline{x}, \ \forall \ \beta \leq \alpha \leq 1. \tag{3}$$

**Definition 5.** The solution of the system (2) is to find the fuzzy vector $[\widetilde{x}]_\alpha = (x, (1-\alpha)\underline{x}, (1-\alpha)\overline{x})$ such that the inequality 3 holds for $\alpha \geq \beta$.

Since $(1-\alpha) \leq (1-\beta)$, it is suffices to solve the system of inequalities

$$[(1-\beta)\overline{A} + A]x \leq [b - (1-\beta)\underline{b}] - (1-\beta)A\overline{x}.$$

## 4  The $\beta$-Tolerance Fuzzy Linear Programming and Their Applications

Many linear programming models are developed under the crisp setting. However, it is difficult to estimate the parameters about the future prices and the market requirements at the current time. To model the uncertainty of future, we shall propose the $\beta$-tolerance fuzzy linear programming models, of which the solutions and the parameters are all fuzzy numbers.

**Definition 6.** Given $0 \leq \beta \leq 1$, the standard statement of the $\beta$-tolerance fuzzy linear programming model has the form

$$\max \widetilde{c} \otimes \widetilde{x},$$
$$s.t. \ \widetilde{A} \otimes \widetilde{x} \leq_\beta \widetilde{b},$$
$$\widetilde{x} \geq 0,$$

where $\widetilde{A}$, $\widetilde{c}$, and $\widetilde{b}$ are fuzzy triangle matrices and $[\widetilde{x}]_\alpha = (x, (1-\alpha)\underline{x}, (1-\alpha)\overline{x})$ is a triangle fuzzy variable.

The fuzzy feasible region is obtained as

$$\{(x, (1-\alpha)\underline{x}, (1-\alpha)\overline{x}) | [(1-\beta)\overline{A} + A]x + (1-\beta)\overline{A}x \leq b - (1-\beta)\underline{b}, \text{ and } x - \underline{x} \geq 0\}$$

and this model is reformulated as the linear programming model with fuzzy objective function:

$$\max \widetilde{c} \otimes \widetilde{x},$$
$$s.t. \ [(1-\beta)\overline{A} + A]x + (1-\beta)A\overline{x} \leq b - (1-\beta)\underline{b},$$
$$x - \underline{x} \geq 0.$$

*Remark 1.* Over the past two decades, a great deal of effort has been made on ranking a set of fuzzy numbers. Most methods, such as [3], [4], [5], [17], [18], rank the fuzzy numbers by integrating the membership function of the fuzzy number. However, for the triangle fuzzy number it suffices to compare the center point, the left spread, and the right spread (c.f. Liu et al. [11]).

*Example 1.* (Resources allocation problems)
Two resources, A and B, are used to produce three main products, C, D, and E. Each unit of resource A yields about 0.3 unit of product C, about 0.4 unit of product D, about 0.2 unit of product E. Each unit of resource B

yields about 0.4 unit of product C, about 0.3 unit of product D, about 0.3 unit of product E. Two resources can purchase up to about 90 units per day of resource A at about \$20 per unit, up to about 60 units per day of resource B at about \$15 per unit. The requirement of products are about 20 units per day of product C, about 15 units per day of product D, and about 05 unit per day of product E. How can these requirements be fulfilled most efficiently?

We define decision variables

$x_1 \triangleq$ units of resource A refined per day.

$x_2 \triangleq$ units of resource B refined per day.

In the crisp setting, this resources allocation problem is formulated as

$$\begin{aligned}
\min \quad & 20x_1 + 15x_2, \\
s.t. \quad & 0.3x_1 + 0.4x_2 \geq 20, \\
& 0.4x_1 + 0.2x_2 \geq 15, \\
& 0.2x_1 + 0.3x_2 \geq 5, \\
& x_1 \leq 90, \ x_2 \leq 60, \ x_1, x_2 \geq 0.
\end{aligned}$$

The solution of above model is obtained as $x_1 = 20$ and $x_2 = 35$.

However, in the future the real requirement or the real price can not be known at the current time. Hence, it may be better to formulate the model by using the $\beta$-tolerance fuzzy linear programming model as follows.

$$\begin{aligned}
\min \quad & \widetilde{20}\widetilde{x}_1 + \widetilde{15}\widetilde{x}_2, \\
s.t. \quad & 0.3\widetilde{x}_1 + 0.4\widetilde{x}_2 \geq_\beta \widetilde{20}, \\
& 0.4\widetilde{x}_1 + 0.2\widetilde{x}_2 \geq_\beta \widetilde{15}, \\
& 0.2\widetilde{x}_1 + 0.3\widetilde{x}_2 \geq_\beta \widetilde{5}, \\
& \widetilde{x}_1 \leq_\beta \widetilde{90}, \ \widetilde{x}_2 \leq_\beta \widetilde{60}, \ \widetilde{x}_1, \widetilde{x}_2 \geq 0.
\end{aligned}$$

Suppose that the spread (uncertain) in this model is 0.5 for each fuzzy number. In practice, the spread should be estimated by using the market data. As a result, this fuzzy model is reformulated as the form.

$$\begin{aligned}
\min \quad & (20x_1 + 15x_2, 2\underline{x}_1 + 15\underline{x}_2 + 0.5x_1 + 0.5x_2, 20\overline{x}_1 + 15\overline{x}_2 + 0.5x_1 + 0.5x_2), \\
s.t. \quad & 0.3x_1 + 0.4x_2 + (1 - \beta)(0.3\underline{x}_1 + 0.4\underline{x}_2) \geq 20 - 0.5(1 - \beta), \\
& 0.4x_1 + 0.2x_2 + (1 - \beta)(0.4\underline{x}_1 + 0.2\underline{x}_2) \geq 15 - 0.5(1 - \beta), \\
& 0.2x_1 + 0.3x_2 + (1 - \beta)(0.2\underline{x}_1 + 0.3\underline{x}_2) \geq 5 - 0.5(1 - \beta), \\
& x_1 + (1 - \beta)\overline{x}_1 \leq 90 - 0.5(1 - \beta), \\
& x_2 + (1 - \beta)\overline{x}_2 \leq 60 - 0.5(1 - \beta), \\
& x_1 - (1 - \beta)\underline{x}_1 \geq 0, \ x_2 - (1 - \beta)\underline{x}_2 \geq 0.
\end{aligned}$$

To find the solution, we should choose a suitable ranking method of the fuzzy numbers for the consideration problems.

When the fuzzy numbers are ranked by their right points, the objective function is written as

$$\min \ (20x_1 + 15x_2) + (20\underline{x}_1 + 15\underline{x}_2 + 0.5x_1 + 0.5x_2).$$

The fuzzy solutions are obtained as $[\tilde{x}_1]_\alpha = (20.5, 0, 0)$ and $[\tilde{x}_2]_\alpha = (35.25, 0, 0)$. Here the left spreads and the right spreads are all less than $10^{-11}$ and are regarded as zero.

When the fuzzy numbers are only ranked by their center points, that is the objective function is written as

$$\max\ 20x_1 + 15x_2,$$

the fuzzy solutions of this $\beta$-tolerance fuzzy linear programming model with $\beta = 0.5$ can be obtained as $[\tilde{x}_1]_\alpha = [(20.5, 0, 63.18(1 - \alpha))]$ and $[\tilde{x}_2]_\alpha = [(35.25, 0, 24.55(1 - \alpha))]$, where $0.5 \le \alpha \le 1$. This implies that the decision maker could select two crisp numbers from $[20.5, 52]$ and $[35.25, 47.5]$ for $x_1$ and $x_2$, respectively.

*Remark 2.* 1. The $\beta$-tolerance model is an extension of the crisp linear programming model. When $\beta = 1$, the $\beta$-tolerance model is equal to the crisp model.
   2. We find that the fuzzy solutions of the $\beta$-tolerance model are depended on their ranking methods. The "goodness" of a selected ranking method depends on the preference of the decision maker used in describing the decision situation.

## 5  Conclusions

For the triangle fuzzy numbers, the main contributions of this paper are (i) to propose a $\beta$-tolerance relation between two fuzzy numbers; (ii) to consider the $\beta$-tolerance fuzzy linear system of inequalities; and (iii) to develop the $\beta$-tolerance fuzzy linear programming for modelling the uncertainty of the real world problems. The coefficients and the variables of the $\beta$-tolerance fuzzy linear programming are all triangle fuzzy numbers. The advantage of this model is that the programming with triangle fuzzy coefficients is transformed to a crisp linear programming. The solution of this model proposes an interval with membership function for the decision maker to make the decision.

## References

1. Buckley, J.J.: Fuzzy programming and Pareto optimal set. Fuzzy Sets and Systems 10, 57–63 (1983)
2. Cai, X., Teo, K.L., Yang, X., Zhou, X.Y.: Portfolio optimization under a Minimax rule. Management Science 16, 957–972 (2000)
3. Chen, L.H., Lu, H.W.: An approximate approach for ranking fuzzy numbers based on left and right dominance. Computers Math. Applic. 41, 1589–1602 (2001)
4. Chen, L.H., Lu, H.W.: The preference order of fuzzy numbers. Computers and Mathematics with Applications 44, 1455–1465 (2002)

 5. Chu, T.C., Tsao, C.T.: Ranking fuzzy numbers with an area between the centroid point and original point. Computers and Mathematics with Applications 43, 111–117 (2002)
 6. Dehghan, M., Hashemi, B.: Solution for the fully fuzzy linear systems using the decomposition procedure. Applied Mathematics and Computation 182, 1568–1580 (2006)
 7. Dehghan, M., Hashemi, B., Ghatee, R.: Computational methods for solving fully fuzzy linear systems. Applied Mathematics and Computation 179, 328–343 (2006)
 8. Dehghan, M., Hashemi, B., Ghatee, R.: Solution of the fully fuzzy linear systems using iterative techniques. Chaos Solutions and Fractals 34, 316–336 (2007)
 9. Dubois, D., Prade, H.: Fuzzy Sets and Systems: theory and applications. Academic Press, London (1980)
10. Fang, S.C., Hu, C.F., Wang, H.F., Wu, S.Y.: Linear programming with fuzzy coefficients in constraints. Computers and mathematics with Applications 37, 63–76 (1993)
11. Liu, H.K., Wu, B., Liu, M.L.: Investors preference order of fuzzy numbers. Computers and Mathematics with Applications 55, 2623–2630 (2008)
12. Hosseinzadeh Lotfi, F., Allahviranloo, T., Alimardani Jondabeh, M., Alizadeh, L.: Solving a full fuzzy linear programming using lexicography method and fuzzy approximate solution. Applied Mathematical Modeling 33, 3151–3156 (2009)
13. Markowitz, H.: Portfolio selection. Journal of Finance 7, 77–97 (1952)
14. Ramik, J., Rimanek, J.: Inequality relation between fuzzy numbers and its use in fuzzy optimization. Fuzzy Sets and Systems 16, 21–29 (1989)
15. Ramik, J., Rommelfanger, H.: A single- and multi-valued order on fuzzy numbers and its use in linear programming with fuzzy coefficients. Fuzzy Sets and Systems 57, 203–208 (1993)
16. Xu, Z., Chen, J.: An interactive method for fuzzy multiple attribute group decision making. Information Sciences 177, 248–263 (2007)
17. Yao, J.S., Wu, K.: Ranking fuzzy numbers based on decomposition principle and signed distance. Fuzzy Sets and Systems 116, 275–288 (2000)
18. Yong, D., Zhu, Z., Liu, Q.: Ranking fuzzy numbers with an area method using radius of gyration. Computers and Mathematics with Applications 51, 1127–1136 (2006)
19. Zimmermann, H.J.: Fuzzy Set Theory and Its Applications. Kluwer Academic, Norwell (1991)

# Interval Relaxation Method for Linear Complementarity Problem

Juan Jiang

**Abstract.** This paper established an interval relaxation method for complementarity problems. We proposed a method for linear complementarity problems,which $M$ is assumed to be an H-matrix with a positive main diagonal. The convergence of these algorithms are proved. Numerical results are presented and show that the algorithms are stable and efficient.

**Keywords:** Complementarity problem, Interval relaxed, Iterative method.

## 1 Introduction

Complementarity problems arise naturally from the mathematical models of various of physical processes, and also applied in engineering and economics. The methods for solving complementarity have achieved an increasing attention in recent years. And there are some methods for solving the problem. For instance, interval method for vertical nonlinear complementarity problem (see [3]), interval method for P0-matrix linear complementarity problem(see [10]), and homophony method for horizontal linear complementarity problem (see [12]), and so on.

Let $f : R^n \longrightarrow R^n$ be a continuously differentiable operator. The complementarity problem is to find a vector $x \in R^n$ such that

$$
\begin{cases}
x \geq 0; \\
f(x) \geq 0; \\
x^T f(x) = 0.
\end{cases}
\tag{1}
$$

Juan Jiang
College of Sciences, China University of Mining and Technology,
Xuzhou 221116, P.R. China
e-mail: jiangjuan217@163.com

S. Li (Eds.): Nonlinear Maths for Uncertainty and its Appli., AISC 100, pp. 651–658.
springerlink.com      © Springer-Verlag Berlin Heidelberg 2011

If $f(x)$ is nonlinear operator,the problem (1) is denoted as nonlinear complementarity problem $NCP(f)$. If $f(x)$ is linear operator, i.e. $f(x) = Mx+q$, then problem (1) is denoted as linear complementarity problem $LCP(M,q)$, where $q = (q_1, q_2, \ldots, q_n) \in R^n, M = (m_{ij} \in R^{n \times n})$.

The following notations on interval mathematics are used throughout our paper. Let $[x] = ([\underline{x_1}, \overline{x_1}], [\underline{x_2}, \overline{x_2}], \ldots, [\underline{x_n}, \overline{x_n}])^T = ([x]_1, [x]_2, \ldots, [x]_n)^T$ be the $n-$ dimensional interval,where $[x]_i \in I(R)$ is a one-dimensional interval. The set of all interval vectors on $[x]$ is denoted as $I(R^n)$. For $\forall [x] \in I(R^n)$, we define

$$m([x]) = (m([x]_1), m([x]_2), \ldots, m([x]_n)),$$
$$W([x]) = (W([x]_1), W([x]_2), \ldots, W([x]_n))^T,$$
$$r([x]) = \frac{W([x])}{2}, i = 1, 2, \ldots, n.$$

More detailed knowledge on interval mathematics can be found in[4-6].

## 2  Main Conclusion and Algorithm

Define

$$g(x) = \max\{0, x - \Omega D f(x)\} \tag{2}$$

where $\Omega$ is relaxation matrix(a nonsingular matrix), $D = diag(d_1, d_2 \ldots, d_n)$ is diagonal matrix,where $d_i > 0, i = 1, 2, \ldots, n$.

**Theorem 2.1.** $x^*$ is the solution of (1), if and only if $x^*$ is a fixed point of $g(x)$.

**Proof.** (1) $x^*$ is a fixed point of $g(x)$, then $x^* = g(x^*) = \max\{0, x^* - \Omega D f(x^*)\}$. If $x^* - \Omega D f(x^*) \geq 0$, then $x^* = x^* - \Omega D f(x^*)$. Therefore, $\Omega D f(x^*) = 0$. Since $\Omega$ is a nonsingular matrix, so $f(x^*) = 0$. It follows that $(x^*)^T f(x^*) = 0$. While if $x^* - \Omega D f(x^*) < 0$, then $x^* = 0$ and $(x^*)^T f(x^*) = 0$, whence $x^*$ is the solution of (1).

(2) If $x^*$ is the solution of (1), then $x^* \geq 0$, $f(x^*) \geq 0$ and $(x^*)^T f(x^*) = 0$. If $x^* = 0$, then $g(x^*) = \max\{0, x^* - \Omega D f(x^*)\} = \max\{0, -\Omega D f(x^*)\} = 0 = x^*$. It can be known that $x^* = 0$ is a fixed point of $g(x)$. If $f(x^*) = 0$, then $g(x^*) = \max\{0, x^* - \Omega D f(x^*)\} = \max\{0, x^*\} = 0 = x^*$. Thus $x^* = 0$ is a fixed point of $g(x)$. The proof is complete.

We introduce an interval operator (see [4])

$$\max\{0, [x]\} = [\max\{0, \underline{x}\}, \max\{0, \overline{x}\}]$$

where $[x]$ is an n-dimensional interval vector,and $\max\{0, [x]\}$ is a componentwise interval operator. It is easy to see that this operator is inclusion-monotonic, that is to say if $[x] \subseteq [y]$, then

$$\max\{0, [x]\} \subseteq \max\{0, [y]\}, r(\max\{0, [x]\}) \le r([x])$$

In [11], Wang Deren introduced the interval relaxation method for nonlinear equations. This paper, we will establish the interval relaxation method for complementarity problems.

Define

$$\Gamma(x, [x], D, \Omega) = \max\{0, x - \Omega Df(x) + (I - \Omega Df'([x]))([x] - x)\}$$

**Theorem 2.2.** Let $[x]$ is n-dimensional interval vector, $f'([x])$ is the interval extension of $f'$ on $[x]$. If

$$\Gamma(x, [x], D, \Omega) = \max\{0, x - \Omega Df(x) + (I - \Omega Df'([x]))([x] - x)\} \subseteq [x] \,(3)$$

where $x \in [x]$ is fixed, $\Omega$, D is defined as above, then $\Gamma(x, [x], D, \Omega)$ include the solution of the problem (1). If any solution $x^*$ of (1) is included in $[x]$, then $x^* \in \Gamma(x, [x], D, \Omega)$.

**Proof.** For $\forall y \in [x]$ then

$$y - \Omega Df(y) \in x - \Omega Df(x) + (I - \Omega Df'([x]))([x] - x)$$

From [7],

$$g(y) = \max\{0, y - \Omega Df(x)\}$$
$$\in \max\{0, x - \Omega Df(x) + (I - \Omega Df'([x]))([x] - x)\}$$

That is to say $\Gamma(x, [x], D, \Omega)$ is interval extension of the mapping $g(\cdot)$ on $[x]$. Then (3) show that $g(\cdot)$ map $[x]$ onto itself,from the continuity of $g(\cdot)$ and the Brouwer fixed point theorem, there is an $x^* \in [x]$ in $g(\cdot)$, where $x^*$ is the solution of (1). For $\forall x^* \in [x]$,

$$x^* = g(x^*) \in \max\{0, x - \Omega Df(x) + (I - \Omega Df'([x]))([x] - x)\}$$

that is

$$x^* \in \Gamma(x, [x], D, \Omega).$$

The proof is complete.

**Corollary 2.1.** Let $\Gamma(x, [x], D, \Omega)$ be defined as (3), if $\Gamma(x, [x], D, \Omega) \cap [x] = \Phi$, then there is no solution of problem (1) in $[x]$.

From theorem 2.2, if we start from an initial $[x]^0$, satisfies(3), then we can conclude a n-dimensional inclusion-monotonic sequence$\{[x]^k\}$, where

$$[x]^{k+1} = \Gamma(x^k, [x]^k, D^k, \Omega^k) \cap [x]^k, k = 0, 1, 2, \ldots,$$

$x^k \in [x]^k$, $D^k$ is a nonsingular matrix, $\Omega^k$ is relaxation matrix. So there is a solution of problem(1)in every $[x]^k$.

**Corollary 2.2.** Let $[x]$ be n-dimensional interval, fix $x \in [x]$, $D$ is a nonsingular matrix, $\Omega^k$ is relaxation matrix. If

$$\Gamma(x, [x], D, \Omega) = \max\{0, x - \Omega D(Mx + q) + (I - \Omega DM)([x] - x)\} \subseteq [x] \quad (4)$$

Then there is a solution $x^*$ of linear complementarity problem in $\Gamma(x, [x], D, \Omega)$. And if $x^* \in [x]$, then $x^* \in \Gamma(x, [x], D, \Omega)$.

Next we will construct the interval iterative algorithm for linear complementarity problem, where $M$ is H-matrix which main diagonal element is positive. Before giving the algorithm, we define the H-matrix and comparison matrix. If $M = (m_{ij}) \in R^{n \times n}$ is H-matrix, then $\exists d = (d_i), d_i > 0$, such that

$$\sum_{j \neq i} |m_{ij}| < |m_{ij}| d (i = 1, 2, \ldots, n).$$

Let $\overline{M} = (\overline{m})_{ij}$, is the comparison matrix of $M$, where

$$(\overline{m})_{ij} = \begin{cases} |m_{ij}|, & i = j, \\ -|m_{ij}|, & i \neq j. \end{cases}$$

Let $D = diag(m_{11}^{-1}, m_{22}^{-1}, \ldots, m_{nn}^{-1})$, $\Omega = diag(\omega_1, \omega_2, \ldots, \omega_n)$ $0 < \omega_i < 1 (i = 1, 2, \ldots, n)$ and the interval $[x]$ include the unique solution $x^*$,

$$\begin{cases} [x]^0 = [x] \\ [x]^{k+1} = [x]^k \cap \max\{0, x^k - \Omega D(Mx^k + q) + (I - \Omega DM)([x]^k - x^k)\} \end{cases} \quad (5)$$

where $[x]^k = m([x]^k)$.

**Theorem 2.3.** Let $M$ be H-matrix which main diagonal element is positive, if the unique solution $x^*$ of the linear complementarity problem is included in $[x]$, then the nested intervals sequence $\{[x^k]\}$, which is generated from (5), converges to $[x^*, x^*]$.

**Proof.** From Corollary 2.2, the nested intervals sequence $\{[x]^k\}, \forall k = 0, 1, 2, \ldots, x^* \in [x]^k$, Denote

$$\mathbf{W} = \begin{bmatrix} 1 - \omega_1 & \cdots & 0 \\ \vdots & \ddots & \vdots \\ 0 & \cdots & 1 - \omega_n \end{bmatrix},$$

Consider

$$[x]^{k+1} \subseteq \max\{0, x^k - \Omega D(Mx^k + q) + (I - \Omega DM)([x]^k - x^k)\}$$

Then

$$r([x]^{k+1}) \leq r(max\{0, x^k - \Omega D(Mx^k + q) + (I - \Omega DM)([x]^k - x^k)\})$$
$$\leq r(x^k - \Omega D(Mx^k + q) + (I - \Omega DM)([x]^k - x^k))$$
$$\leq r((I - \Omega DM)([x]^k - x^k))$$

For $x^k = m([x]^k)$, then

$$(I - \Omega DM)([x]^k - x^k)$$
$$= (I - \Omega DM)(-r(([x]^k) - r([x]^k))$$
$$= [-(W + \Omega(I - D\overline{M})r([x]^k), (W + \Omega(I - D\overline{M})r([x]^k))$$

where $W + \Omega(I - D\overline{M}) \geq 0$. Hence

$$r([x]^{k+1}) \leq r(W + \Omega(I - D\overline{M}))r([x]^k)$$

Denote $A = W + \Omega(I - D\overline{M})$, then

$$\mathbf{A} = \mathbf{W} + \Omega(I - D\overline{M}) = \begin{bmatrix} 1 - \omega_1 & \omega_1 \frac{|m_{12}|}{m_{11}} & \cdots & \omega_1 \frac{|m_{1n}|}{m_{11}} \\ \omega_2 \frac{|m_{21}|}{m_{22}} & 1 - \omega_2 & \cdots & \omega_2 \frac{|m_{2n}|}{m_{22}} \\ \vdots & \cdots & \ddots & \vdots \\ \omega_n \frac{|m_{n1}|}{m_{nn}} & \cdots & \omega_{n-1} \frac{|m_{n(n-1)}|}{m_{nn}} & 1 - \omega_n \end{bmatrix},$$

Because $M$ is H-matrix, so we have

$$\sum_{j=1}^{n} |A_{1j}| = 1 - \omega_1 + \omega_1 \left( \frac{|m_{12}|}{m_{11}} + \cdots + \frac{|m_{1n}|}{m_{11}} \right)$$
$$< 1 - \omega_1 + \omega_1$$
$$= 1$$

And so for $\forall i = 1, 2, \ldots, n$, then $\sum_{j=1}^{n} |A_{ij}| < 1$.
From [5], we know $\rho(A) < 1$, that is

$$\rho(W + \Omega(I - D\overline{M})) < 1$$

So, $r([x])^k \longrightarrow 0$, then we know the nested intervals sequence $r\{[x]\})^k$ converges to $[x^*, x^*]$. The proof is complete.

Based on the above theorems, the detailed steps of the new proposed algorithm for linear complementarity problem is presented as follows, where $\varepsilon$ is precision.

**Algorithm 2.1**
**Step 1** Select $\omega_i$ and initial interval $[x], i = 1, 2, \ldots, n$;
**Step 2** Compute $x = m([x])$, $[y] = x - \Omega D(Mx + q) + (I - \Omega DM)([x] - x)$;

**Step 3** Calculate $\Gamma(x,[x],D,\Omega) = \max\{0,[y]\}$;

**Step 4** Compute $X = \Gamma(x,[x],D,\Omega) \cap [x]$; If $X = \Phi$, print " there is no solution in this interval ," input a new interval and go to **step 2**; or, continue;

**Step 5** If $\|\omega\| < \varepsilon$, then print the approximate interval $X$ and the solution $m(X)$; or,let $[x] = X$, go to **Step 2**;

**Step 6** end.

## 3   Numerical Results

The algorithm has been implemented using Matlab 6.5 on a desktop computer (P IV,2.93GHz,Memory 512M. Numerical results for the following four examples are proposed in the section. The symbols $[x]^0$, $x^*$,$[x]$, $x$, $L$, $\varepsilon = 1e - 15$, $\omega^i = \frac{1}{2}$, $i = 1,2,\ldots$ are denoted as the initial interval, exact solution, optimum solution interval, optimum solution, the number of iteration, precision and relaxation factor, respectively.

**Example 1.** $LCP(M,q)$ *Fathi* [2],

$$M = \begin{pmatrix} 1 & 2 \\ 2 & 5 \end{pmatrix}, q = \begin{pmatrix} -1 \\ -1 \end{pmatrix}, x^* = (1,0)^T, [x]^0 = \begin{pmatrix} [0,2] \\ [0,2] \end{pmatrix}.$$

$$L = 33, X = \begin{pmatrix} [1.000000000000000, 1.000000000000000] \\ [0,0] \end{pmatrix},$$

$$x = \begin{pmatrix} 1.000000000000000 \\ 0 \end{pmatrix}.$$

**Example 2.** $LCP(M,q)$,

$$M = \begin{bmatrix} 3 & 0 & -1 & 0 \\ -1 & 3 & -1 & 0 \\ 0 & -1 & 4 & -2 \\ 1 & -1 & -1 & 5 \end{bmatrix}, q = \begin{bmatrix} -2 \\ 3 \\ -4 \\ 5 \end{bmatrix}, x^* = (1,0,1,0)^T, [x]^0 = \begin{pmatrix} [0,2] \\ [0,2] \\ [0,2] \\ [0,2] \end{pmatrix}.$$

$$L = 34, X = \begin{pmatrix} [1.000000000000000, 1.000000000000000] \\ [0,0] \\ [1.000000000000000, 1.000000000000000] \\ [0,0] \end{pmatrix},$$

$$x = \begin{pmatrix} 1.000000000000000 \\ 1.000000000000000 \\ 1.000000000000000 \\ 0 \end{pmatrix}.$$

**Example 3.** $LCP(M, q)$,

$$M = \begin{bmatrix} 4 & -1 & 0 & 0 \\ -1 & 4 & -1 & 0 \\ 0 & -1 & 4 & -1 \\ 0 & 0 & -1 & 4 \end{bmatrix}, q = \begin{bmatrix} -4 \\ 3 \\ -4 \\ 2 \end{bmatrix}, x^* = (1, 0, 1, 0)^T, [x]^0 = \begin{pmatrix} [0, 2] \\ [0, 2] \\ [0, 2] \\ [0, 2] \end{pmatrix}.$$

$$L = 30, X = \begin{pmatrix} [1.000000000000000, 1.000000000000000] \\ [0, 0] \\ [1.000000000000000, 1.000000000000000] \\ [0, 0] \end{pmatrix},$$

$$x = \begin{pmatrix} 1.000000000000000 \\ 1.000000000000000 \\ 1.000000000000000 \\ 0 \end{pmatrix}.$$

**Example 4.** $LCP(M, q)$ [9],

$$M = \begin{bmatrix} 1 & 2 & 2 & 2 \\ 0 & 1 & 2 & 2 \\ 0 & 0 & 1 & 2 \\ 0 & 0 & 0 & 1 \end{bmatrix}, q = \begin{bmatrix} -1 \\ -1 \\ -1 \\ -1 \end{bmatrix}, x^* = (0, 0, 0, 1)^T, [x]^0 = \begin{pmatrix} [0, 2] \\ [0, 2] \\ [0, 2] \\ [0, 2] \end{pmatrix}.$$

$$L = 29, X = \begin{pmatrix} [0, 0] \\ [0, 0] \\ [0, 0] \\ [1.000000000000000, 1.000000000000000] \end{pmatrix},$$

$$x = \begin{pmatrix} 0 \\ 0 \\ 0 \\ 1.000000000000000 \end{pmatrix}.$$

Trough the results all of above, we can concluded that the proposed algorithm can converge to the solution, and this method can get more accurate result.

## 4 Conclusions

In this paper, a relaxation interval iterative method for complementarity problems has been investigated. Firstly, we transformed this problem into an equivalent problem with the use of fixed point theorem. Then we proposed the corresponding operator and proved the convergence of the algorithm together with relevant properties. Numerical results are presented to show the efficiency of the method.

# References

1. Alefeld, G., Herzberger, J.: Introduction to Interval Computations. Academic press, New York (1983)
2. Fathi, Y.: Computational Complexity of linear complementarity problems associated with positive definite Matices. Math. Programming 17, 335–344 (1979)
3. Han, C., Cao, D., Qin, J.: An interval method for vertical nonlinear complementarity problems. Journal of Nanjing University Mathematical Biquarterly 27(1), 75–81 (2010)
4. Gotz, A., Wang, Z., Shen, Z.: Enclosing solution of linear complementarity problems for H-matrix. Reliable Computing 10(6), 423–435 (2004)
5. Lei, J.: Matric and Application. Mechanical and Industry press, Beijing (2005)
6. Moore, R.E.: Interval Analysis. Prentice-Hall, New Jersey (1966)
7. Moore, R.E.: A test for existence of solutions to nonlinear systems. SIAM J Numer Anal. 14, 611–615 (1977)
8. Moore, R.E.: Methods and Applications of Interval Analysis. SIAM, Philadelphia (1979)
9. Murty, K.G.: Linear Complementarity, Linear and Nonlinear Programming. Hedermann, Berlin (1988)
10. Rui, W., Cao, D., Zhang, Z.: An interval method for the P0-matrix linear complementarity problem. Journal of Nanjing University Mathematical Biquarterly 24(2), 344–350 (2007)
11. Wang, D.: The Method for Nonlinear Equations. Science and Technology press, Shang Hai (1987)
12. Zhao, X.: Homophony method for solving horizontal linear complementarity problem. Journal of Jilin University (Science Edition) 48(5), 766–770 (2010)

# The Representation Theorem on the Category of FTML($\mathcal{L}$)

Jie Zhang, Xiaoliang Kou, and Bo Liu

**Abstract.** The theory of $\mathcal{L}$−Fuzzy topological molecule lattices is the generalization theory of topological molecule lattices. The characteristic of category of $\mathcal{L}$−Fuzzy topological molecule lattices must be given a straightforward description. In this paper, a representation theorem about the category of $\mathcal{L}$−Fuzzy topological molecule lattices is proved: The category of $\mathcal{L}$−Fuzzy topological molecule lattices is equal to that the category of **FSTS**($\mathcal{L}$) which consists of $\mathcal{L}$−fuzzifying scott topological space and the $\mathcal{L}$−fuzzifying continuous mapping of orientation-join preserving and the relation of way-below preserving. Using it as the deduction of this theorem, a representation theorem about the category of topological molecule lattices is obtained.

**Keywords:** Category, Fuzzy topology, Lattice, Mapping.

## 1 Introduction

Based on completely distributive lattices, the theory of topological molecular lattices was constructed in 1992 [5]. And the basic framework of $\mathcal{L}$−fuzzy topological molecular lattices theory ($\mathcal{L}$−smooth topological molecular lattices theory) was constructed from [5] in 2002 [14]. The theory of topological molecule lattices is the generalization theory of $\mathcal{L}$−topological spaces on nonempty set X. The theory of $\mathcal{L}$−Fuzzy topological molecule lattices is the generalization theory of topological molecule lattices. The characteristic of category of $\mathcal{L}$−Fuzzy topological molecule lattices must be given a straightforward description. In this paper, a representation theorem about the category of $\mathcal{L}$− Fuzzy topological molecule lattices is proved: The category of $\mathcal{L}$−Fuzzy topological molecule lattices is equal to that the category of

Jie Zhang, Xiaoliang Kou, and Bo Liu
College of Science, North China University of Technology, Beijing 100144
e-mail: jzhang26@ncut.edu.cn

S. Li (Eds.): Nonlinear Maths for Uncertainty and its Appli., AISC 100, pp. 659–666.
springerlink.com
© Springer-Verlag Berlin Heidelberg 2011

**FSTS($\mathcal{L}$)** which consists of $\mathcal{L}$–fuzzifying scott topological space and the $\mathcal{L}$–fuzzifying continuous mapping of orientation-join preserving and the relation of way-below preserving. From the theorem, a representation theorem about the category of topological molecule lattices is obtained **TML**.

## 2 Preparations

In this paper, $L, \mathcal{L}, L_i$ denote completely distributive lattices ( or molecular lattices). 0 and $1(0 \neq 1)$ denote the smallest element and the greatest element respectively. $M(\mathcal{L})$ denotes the set of all non-zero $\vee$–irreducible (or coprime) elements in $\mathcal{L}$. $P(\mathcal{L})$ denotes the set of all non-unit prime elements in $\mathcal{L}$. Define $\wedge \emptyset = 1$ and $\vee \emptyset = 0$. $\alpha(a), \beta(a)$ are standard the maximal set and the minimal set of $a$ respectively. The others notations and concepts which are not explained come from [5], [14], [11] and [10], respectively.

**Definition 1.** ( [2]) A partially ordered set $P$ is called the continuous partially ordered set, if $P$ is co-perfect(every directed set in $P$ has the least upper bound), and satisfies the conditions as follows:

(i) $\forall x \in P$, $\downarrow x = \{u \in P \mid u \ll x\}$ is an oriented set.

(ii) $\forall x \in P$, $x = \sup \downarrow x$. Where $\forall u, x \in P$, $u \ll x$ if and only if for arbitrary directed set $D \subseteq P$, $x \leq \sup D$ implies $\exists y \in D$ such that $u \leq y$. In this case, the $u$ is called way-below $x$.

(2) Let $P$ be a partially ordered set. $F \subseteq P$ is called a scott closed set, if $F$ is a lower set and it is closed for the operation of directed join. The complementary set of scott closed set is called scott open set. All sets of scott open sets in $P$ consist a topology on $P$. The topology is called the scott topology on $P$, and note it as $\sigma(P)$. All sets of scott closed sets in $P$ consist a co-topology on $P$. Define the co-topology as the scott co-topology on $P$, and note as $\Sigma(P)$.

**Theorem 1.** ([1]) Let $L$ be a molecular lattice. The mapping $\mathcal{T} : L \longrightarrow 2^{M(L)}$ is defined as follows: $\forall A \in L, \mathcal{T}(A) = \{m \in M(L) \mid m \not\leq A\}$. Then the following conclusions are true.

(1) $\mathcal{T}$ is an injection, and $\mathcal{T}(0) = M(L), \mathcal{T}(1) = \emptyset$.

(2) $\forall \{A_i \in L \mid i \in \Delta\} \subseteq L, \mathcal{T}(\bigwedge_{i \in \Delta} A_i) = \bigcup_{i \in \Delta} \mathcal{T}(A_i)$.

(3) $\forall A, B \in L, \mathcal{T}(A \bigvee B) = \mathcal{T}(A) \bigcap \mathcal{T}(B)$.

(4) $\mathcal{T}(L)$ is a molecular lattice and $\mathcal{T}(L) = \sigma(M(L))$. $(M(L), \mathcal{T}(L))$ is the spectral space of $L^{op}$.

(5) If $(L, \eta)$ is a topological molecular lattice, then $\mathcal{T}(\eta) \subseteq \mathcal{T}(L)$ is a topology on $M(L)$. $\forall x \in M(L)$ and $Q \in \eta$, $x \not\leq Q$ if and only if $x \in \mathcal{T}(Q)$, i.e. $Q$ is a closed remote neighborhood of $x$ if and only if $\mathcal{T}(Q)$ is a neighborhood of $x$.

**Theorem 2.** ([6]) (1) Let $P_1, P_2$ be continuous partially ordered sets. Then the mapping $f : P_1 \longrightarrow P_2$ can be expanded to a generalized order homomorphism $f : \Sigma(P_1) \longrightarrow \Sigma(P_2)$ if and only if $f$ preserves orientation-join and *the relation of way-below, i.e.* $f$ *is a scott continuous mapping of preserving way-below relation.*

*(2) Let* $f : L_1 \longrightarrow L_2$ *be a generalized order homomorphism. Then* $f|_{M(L_1)} : M(L_1) \longrightarrow M(L_2)$ *is the mapping of preserving directed join and the relation of way-below.*

*(3) Let* $f : M(L_1) \longrightarrow M(L_2)$ *be a mapping of preserving directed join and the relation of way-below. If* $a \in L_1$, *and define* $L(f)(a) = \bigvee\{f(x) \mid x \in \beta^*(a)\}$, *then* $L(f) : L_1 \longrightarrow L_2$ *is a generalized order homomorphism and* $L(f)|_{M(L_1)} = f$.

*(4) Let* $(L_1, \eta_1)$, $(L_2, \eta_2)$ *be topological molecular lattices,* $f : L_1 \longrightarrow L_2$ *is a generalized order homomorphism. Then* $f$ *is a continuous generalized order homomorphism if and only if* $f|_{M(L_1)} : (M(L_1), \mathcal{T}(\eta_1)) \longrightarrow (M(L_2), \mathcal{T}(\eta_2))$ *is a continuous mapping.*

**Theorem 3.** *(1) Let* $P$ *be a continuous partially ordered set. Define it as* $L(P) = \{\bigcup_{p \in A} \downarrow p : A \subseteq P \text{ and } A =\downarrow A\}$. *Then* $L(P)$ *is a completely distributive lattice (the order is* $\subseteq$*) and* $P \overset{\phi}{\cong} M(L(P))$, *where the mapping* $\phi : P \longrightarrow M(L(P))$ *is defined as* $\forall p \in P$, $\phi(p) = \downarrow p$.

*(2) Let* $L$ *be a completely distributive lattice. Then* $L \overset{\psi}{\cong} L(M(L))$. *Where, the mapping* $\psi : L \longrightarrow L(M(L))$ *is defined as* $\forall a \in L$, $\psi(a) = \beta^*(a) = \bigcup\{\downarrow p \mid p \leq a\}$.

**Definition 2.** ( [7]) Let $X$ be a nonempty set. Then an $\mathcal{L}-$ fuzzy (complement) topology in $2^X$ is called an $\mathcal{L}-$fuzzifying (complement) topology on $X$.

# 3   The Representation Theorem on the Category of FTML($\mathcal{L}$)

**Theorem 4.** *Let* $(L, \mathcal{F})$ *be an* $\mathcal{L}-$*fuzzy topological molecular lattice and* $\mathcal{T} : L \longrightarrow 2^{M(L)}$ *is described in the theorem 1.*

*(1) The mapping* $\mathcal{T}(\mathcal{F}) : 2^{M(L)} \longrightarrow \mathcal{L}$ *is defined as* $\forall A \in 2^{M(L)}$,

$$\mathcal{T}(\mathcal{F})(A) = \begin{cases} \bigvee\{a \in \mathcal{L} \mid A \in \mathcal{T}(\mathcal{F}_{[a]})\}, & A \in \mathcal{T}(L), \\ 0 & , A \notin \mathcal{T}(L). \end{cases}$$

*Then* $\mathcal{T}(\mathcal{F})$ *is an* $\mathcal{L}-$*fuzzifying topology on* $M(L)$ *(i.e. an* $\mathcal{L}-$*fuzzy topology on* $2^{M(L)}$*) which satisfies* $\forall a \in \mathcal{L}, \mathcal{T}(\mathcal{F})_{[a]} = \mathcal{T}(\mathcal{F}_{[a]})$.

*(2) The mapping* $\mathcal{T}'(\mathcal{F}) : 2^{M(L)} \longrightarrow \mathcal{L}$ *is defined as* $\forall A \in 2^{M(L)}, \mathcal{T}'(\mathcal{F})(A) = \mathcal{T}(\mathcal{F})(A')$. *Then* $\mathcal{T}'(\mathcal{F})$ *is an* $\mathcal{L}-$*fuzzifying complement topology on* $M(L)$ *(i.e. an* $\mathcal{L}$*-fuzzy complement topology on* $2^{M(L)}$*) which satisfies* $\forall a \in \mathcal{L}, \mathcal{T}'(\mathcal{F})_{[a]} = \mathcal{T}'(\mathcal{F}_{[a]})$, *where* $\mathcal{T}'(\mathcal{F}_{[a]}) = \{A \in 2^{M(L)} \mid A \in \Sigma(M(L)), A' \in \mathcal{T}(\mathcal{F}_{[a]})\}$.

*Proof.* (1) From theorem 3.5 [14] and theorem 1, for $\forall a \in \mathcal{L}$, $\mathcal{F}_{[a]}$ is a complement topology on $L$ and $\mathcal{T}(\mathcal{F}_{[a]})$ is a topology on $M(L)$. With theorem 3.8 [14], the mapping $T$ is an injection, and satisfies $\mathcal{F}_{[a]} = \bigcap_{b \in \beta^*(a)} \mathcal{F}_{[b]}$. Consequently , $\forall A \in 2^{M(L)}$, $A \in T(\mathcal{F}_{[a]}) \Longleftrightarrow \exists B \in \mathcal{F}_{[a]} = \bigcap_{b \in \beta^*(a)} \mathcal{F}_{[b]}$ such that $A = T(B) \Longleftrightarrow \forall b \in \beta^*(a), \exists B \in \mathcal{F}_{[b]}$ such that $A = T(B) \in T(\mathcal{F}_{[b]}) \Longleftrightarrow A \in \bigcap_{b \in \beta^*(a)} T(\mathcal{F}_{[b]})$. Hence, $T(\mathcal{F}_{[a]}) = \bigcap_{b \in \beta^*(a)} T(\mathcal{F}_{[b]})$. According to theorem 3.9 [14], $T(\mathcal{F})$ is an $\mathcal{L}-$ fuzzy topology on $2^{M(L)}$, which satisfies $\forall a \in \mathcal{L}, T(\mathcal{F})_{[a]} = T(\mathcal{F}_{[a]})$.

(2)  It can be proved by the same way of (1).                                  □

**Definition 3.** Let $P$ be a continuous partially ordered set, and the eigen-function of $\sigma(P)$ is also signified as $\sigma(P)$. Then $\sigma(P)$ is an $\mathcal{L}-$fuzzifying topology on $P$. If $\mathcal{S}$ is an $\mathcal{L}-$fuzzifying topology on $P$, which is satisfied as $\mathcal{S} \preceq \sigma(P)$, then $(P, \mathcal{S})$ is called the $\mathcal{L}-$fuzzifying scott topological space.

**Theorem 5.** Let $(L, \mathcal{F})$ be an $\mathcal{L}-$fuzzy topological molecular lattice. Then $(M(L), T(\mathcal{F}))$ is an $\mathcal{L}-$fuzzifying scott topological space.

*Proof.* Becasue $M(L)$ is a continuous partially ordered set and with the theorem 4, then obtain $T(\mathcal{F}) \preceq \sigma(M(L))$.                          □

**Lemma 1.** Let $P$ be a continuous partially ordered set and the mapping $\mathcal{E} : \Sigma(P) \longrightarrow L(P)$ is defined as $\forall A \in \Sigma(P)$, $\mathcal{E}(A) = \bigcup\{ \downarrow x \mid x \in A\}$. Then

(1) $\mathcal{E}$ is an injection and $\mathcal{E}(\emptyset) = \emptyset$, $\mathcal{E}(P) = \bigcup_{p \in P} \downarrow p$,

(2) $\forall A, B \in \Sigma(P), \{A_i \mid i \in \Delta\} \subseteq \Sigma(P)$, $\mathcal{E}(A \bigcup B) = \mathcal{E}(A) \bigcup \mathcal{E}(B)$, $\mathcal{E}(\bigcap_{i \in \Delta} A_i) = \bigcap_{i \in \Delta} \mathcal{E}(A_i)$.

(3) Let $\tau \subseteq \sigma(P)$ be a topology on $P$. Then $(L(P), \mathcal{E}(\tau'))$ is a topological molecular lattice.

*Proof.* (1)  $\forall A, B \in \Sigma(P)$ and $\mathcal{E}(A) = \mathcal{E}(B)$. If $A \neq B$, there exists $x_0 \in A, x_0 \notin B$. Therefore, obtain $\downarrow x_0 \subseteq \mathcal{E}(A) = \mathcal{E}(B)$. Because $B$ is a scott closed set, it is the lower set of orientation-join preserving . Hence, $\downarrow x_0 \subseteq B$. And $\downarrow x_0$ is the orient set, there is $x_0 = \bigvee \downarrow x_0 \in B$. It is contradictory. Then obtain $A = B$. It is equal to that $\mathcal{E}$ is an injection.

(2)  Let $A, B \in \Sigma(P)$. Then $\mathcal{E}(A \bigcup B) = \bigcup\{ \downarrow x \subseteq P \mid x \in A \bigcup B\} = (\bigcup\{ \downarrow x \subseteq P \mid x \in A\}) \bigcup (\bigcup\{ \downarrow x \subseteq P \mid x \in B\}) = \mathcal{E}(A) \bigcup \mathcal{E}(B)$.

Let $\{A_i \mid i \in \Delta\} \subseteq \Sigma(P)$. Then $\mathcal{E}(\bigcap_{i \in \Delta} A_i) = \bigcup\{ \downarrow x \subseteq P \mid x \in \bigcap_{i \in \Delta} A_i\} \subseteq \bigcup\{ \downarrow x \subseteq P \mid x \in A_i\} = \mathcal{E}(A_i)$. Therefore, $\mathcal{E}(\bigcap_{i \in \Delta} A_i) \subseteq \bigcap_{i \in \Delta} \mathcal{E}(A_i)$. Otherwise, $\forall \downarrow x \in \beta^*(\bigcap_{i \in \Delta} \mathcal{E}(A_i))$, for $i \in \Delta$, $\downarrow x \in \beta^*(\mathcal{E}(A_i))$. There exists $\downarrow y \in \{ \downarrow z \in M(L(P)) \mid z \in A_i\}$ such that $\downarrow x \subseteq \downarrow y$. Then obtain $x \leq y$ and $y \in A_i$. Because $A_i$ is the lower set, get $x \in A_i$ and $x \in \bigcap_{i \in \Delta} A_i$. Furthermore, get $\downarrow x \subseteq \mathcal{E}(\bigcap_{i \in \Delta} A_i)$ and $\mathcal{E}(\bigcap_{i \in \Delta} A_i) \supseteq \bigcap_{i \in \Delta} \mathcal{E}(A_i)$.

(3) From (1), (2) and theorem 3, the conclusion can be proved easily.   □

**Theorem 6.** (1)  Let $(L, \eta)$ be a topological molecular lattice. Then it is homeomorphic with the topological molecular lattice of $(L(M(P)), \mathcal{E}(T'(\eta)))$.

(2)  Let $P$ be a continuous partially ordered set, $\tau \subseteq \sigma(P)$ is a topology on $P$. Then the topological spaces $(P, \tau)$ and $(M(L(P)), T(\mathcal{E}(\tau')))$ is homeo-morphic.

*Proof.* (1) From the theorem 3, the mapping $\psi : L \longrightarrow L(M(L)), \forall a \in L, a \longmapsto \beta^*(a)$ is the isomorphic mapping. So it just needs to prove the continuity and close of $\phi$.

Suppose $A \in \mathcal{E}(T'(\eta))$. Then there exists $B \in \eta$ such that $A = \mathcal{E}(M(L) - T(B))$. Therefore, get $\phi^{-1}(A) = \phi^{-1}(\mathcal{E}(M(L) - T(B))) = \phi^{-1}(\bigcup\{ \downarrow x \mid x \in M(L) - T(B)\}) = \phi^{-1}(\bigcup\{ \downarrow x \mid x \leq B\}) = \phi^{-1}(\beta^*(B)) = B \in \eta$. It shows that $\phi$ is a continuous mapping.

Suppose $A \in \eta$. Then $\phi(A) = \beta^*(A) = \bigcup\{ \downarrow x \mid x \leq A\} = \bigcup\{ \downarrow x \mid x \in M(L) - T(A)\} = \mathcal{E}(M(L) - T(A)) \in \mathcal{E}(T'(\eta))$. So $\phi$ is a closed mapping.

(2) From the theorem 3, the mapping $\psi : P \longrightarrow M(L(P)), \forall a \in P, a \longmapsto \downarrow a$ is an isomorphic mapping. It just needs to prove $\psi$ is continuous and open.

$\forall A \in T(\mathcal{E}(\tau'))$, there exists $B \in \tau$ such that $A = T(\mathcal{E}(P - B))$. Then $\psi^{-1}(A) = \psi^{-1}(T(\mathcal{E}(P - B))) = \psi^{-1}(\{ \downarrow x \mid x \nsubseteq \mathcal{E}(P - B)\}) = \psi^{-1}(\{ \downarrow x \mid x \notin P - B\}) = \{x \mid x \notin P - B\} = B \in \tau$. Hence, $\psi$ is a continuous mapping.

$\forall A \in \tau$, there is $P - A \in \tau'$. Because $\psi(A) = \{ \downarrow x \mid x \in A\} = \{ \downarrow x \mid x \notin P - A\} = \{ \downarrow x \mid \downarrow x \nsubseteq \mathcal{E}(P - A)\} = T(\mathcal{E}(P - A))$, $\psi$ is an open mapping. $\quad\square$

**Theorem 7.** *Let $P$ be a continuous partially ordered set and $(P, \mathcal{S})$ is an $\mathcal{L}-$ fuzzifying scott topological space. Then*

*(1) $(L(P), \mathcal{E}(\mathcal{S}'))$ is an $\mathcal{L}-$fuzzy topological molecular lattice, where the mapping $\mathcal{E}(\mathcal{S}') : L(P) \longrightarrow \mathcal{L}$ is defined as $\forall A \in L(P)$, with $A \in \mathcal{E}(\Sigma(P))$, $\mathcal{E}(\mathcal{S}')(A) = \bigvee\{a \in \mathcal{L} \mid A \in \mathcal{E}(\mathcal{S}'_{[a]})\}$. Otherwise, $\mathcal{E}(\mathcal{S}')(A) = 0$.*

*In an addition, the above theorem satisfies $\forall a \in \mathcal{L}, \quad \mathcal{E}(\mathcal{S}')_{[a]} = \mathcal{E}(\mathcal{S}'_{[a]})$;*

*(2) $(M(L(P)), T(\mathcal{E}(\mathcal{S}'))$ is an $\mathcal{L}-$fuzzifying scott topological space, where the mapping $T(\mathcal{E}(\mathcal{S}')) : 2^{M(L(P))} \longrightarrow \mathcal{L}$ is defined as $\forall A \in 2^{M(L(P))}$, with $A \in \sigma(M(L(P)))$, $T(\mathcal{E}(\mathcal{S}'))(A) = \bigvee\{a \in \mathcal{L} \mid A \in T(\mathcal{E}(\mathcal{S}'_{[a]}))\}$ and $A \notin \sigma(M(L(P)))$, $T(\mathcal{E}(\mathcal{S}'))(A) = 0$; Moreover, it is satisfied as $\forall a \in \mathcal{L}, \ T(\mathcal{E}(\mathcal{S}'))_{[a]} = T(\mathcal{E}(\mathcal{S}'_{[a]}))$;*

*(3) $(P, \mathcal{S})$ with $(M(L(P)), T(\mathcal{E}(\mathcal{S}')))$ is an $\mathcal{L}-$fuzzifying homeomorphic.*

*Proof.* (1) From the definition 2 and the theorem 3.5 [14], $\forall a \in \mathcal{L}, \mathcal{S}_{[a]} \subseteq \sigma(P)$ is a topology on $P$. Though the theorem 6, $\mathcal{E}(\mathcal{S}'_{[a]})$ is a complement topology on the molecular lattice $L(P)$. And with the condition that $\mathcal{E}$ is an injection and $\mathcal{S}'_{[a]} = \bigcap_{b \in \beta^*(a)} \mathcal{S}_{[b]}$, it is proved easily that $\mathcal{E}(\mathcal{S}'_{[a]}) = \bigcap_{b \in \beta^*(a)} \mathcal{E}(\mathcal{S}'_{[b]})$. Therefore, from the theorem 3.5 [14] and the theorem 3.9 [14], get $\mathcal{E}(\mathcal{S}')_{[a]} = \mathcal{E}(\mathcal{S}'_{[a]})$ and $\mathcal{E}(\mathcal{S}')$ is an $\mathcal{L}-$fuzzy complement topology on $L(P)$.

(2) From (1) and the theorem 1, $\forall a \in \mathcal{L}, T(\mathcal{E}(\mathcal{S}'_{[a]}))$ is a topology on $M(L(P))$ which satisfies $T(\mathcal{E}(\mathcal{S}'_{[a]})) \subseteq \sigma(M(L(P)))$. From the theorem 4 and the theorem 2.9 [14], get $T(\mathcal{E}(\mathcal{S}'_{[a]})) = T(\mathcal{E}(\mathcal{S}')_{[a]}) = T(\mathcal{E}(\mathcal{S}'))_{[a]}$ easily. Hence, $T(\mathcal{E}(\mathcal{S}')) \preceq \sigma(M(L(P)))$ is obtained. It is equal to that $T(\mathcal{E}(\mathcal{S}'))$ for $M(L(P))$ is an $\mathcal{L}-$fuzzifying scott topology on $M(L(P))$.

(3) From the theorem 6, $\forall a \in \mathcal{L}$, the mapping $\psi : (P, \mathcal{S}_{[a]}) \longrightarrow (M(L(P)), T(\mathcal{E}(\mathcal{S}'_{[a]}))), \forall x \in P, x \longmapsto \downarrow x$ is a homeomorphic mapping.

From (2) and the theorem 4.9 [14], the mapping $\psi : (P, \mathcal{S}) \longrightarrow (M(L(P)),$
$T(\mathcal{E}(\mathcal{S}')))$ is an $\mathcal{L}-$fuzzifying(fuzzy) homeomorphic mapping.                          □

**Theorem 8.** *Let $(L, \mathcal{F})$ be an $\mathcal{L}-$fuzzy topological molecular lattice. Then*
*(1) $(L(M(L)), \mathcal{E}(T'(\mathcal{F})))$ is an $\mathcal{L}-$fuzzy topological molecular lattice, where*
*the mapping $\mathcal{E}(T'(\mathcal{F})) : L(M(L)) \longrightarrow \mathcal{L}$ is defined as $\forall A \in L(M(L))$:*
*with $A \in \mathcal{E}(\Sigma(M(L)))$, $\mathcal{E}(T'(\mathcal{F}))(A) = \bigvee\{a \in \mathcal{L} \mid A \in \mathcal{E}(T'(\mathcal{F}_{[a]}))\}$ and*
*$A \notin \mathcal{E}(\Sigma(M(L)))$, $\mathcal{E}(T'(\mathcal{F}))(A) = 0$, it satisfies $\forall a \in \mathcal{L}$, $\mathcal{E}(T'(\mathcal{F}))_{[a]} =$*
*$\mathcal{E}(T'(\mathcal{F}_{[a]}))$;*
*(2) $(L, \mathcal{F})$ with $(L(M(L)), \mathcal{E}(T'(\mathcal{F})))$ is an $\mathcal{L}-$fuzzy homeomorphic.*

*Proof.* (1) From the theorem 3, $L(M(L))$ is a completely distributive lattice.
Get $\forall a \in \mathcal{L}$, $\mathcal{E}(T'(\mathcal{F}_{[a]})) = \mathcal{E}(T'(\mathcal{F})_{[a]}) = \mathcal{E}(T'(\mathcal{F}))_{[a]}$ and $\mathcal{E}(T'(\mathcal{F}_{[a]}))$ is a
complement topology on $L(M(L))$ by the way similar to the theorem 7. And
with the theorem 3.5 [14], $(L(M(L)), \mathcal{E}(T'(\mathcal{F})))$ is an $\mathcal{L}-$fuzzy topological
molecular lattice .
(2) From the theorem 6, $\forall a \in \mathcal{L}$, the mapping of $\phi : (L, \mathcal{F}_{[a]}) \longrightarrow$
$(L(M(P)), \mathcal{E}(T'(\mathcal{F}_{[a]})))$, $\forall x \in L, x \longmapsto \beta^*(x)$ is a homeomorphism generalized order homomorphism.
From the theorem 4.9 [14], $\phi : (L, \mathcal{F}) \longrightarrow (L(M(L)), \mathcal{E}(T'(\mathcal{F})))$ is an $\mathcal{L}-$
fuzzy homeomorphism generalized order homomorphism.                          □

**Theorem 9.** *Let $(L_1, \mathcal{F}_1)$, $(L_2, \mathcal{F}_2)$ be $\mathcal{L}-$fuzzy topological molecular lattices*
*and $f : L_1 \longrightarrow L_2$ is a generalized order homomorphism. Then the conclu-*
*sions as follows are equivalent:*
*(1) $f : (L_1, \mathcal{F}_1) \longrightarrow (L_2, \mathcal{F}_2)$ is an $\mathcal{L}-$fuzzy generalized order homomor-*
*phism.*
*(2) $f|_{M(L_1)} : (M(L_1), T(\mathcal{F}_1)) \longrightarrow (M(L_2), T(\mathcal{F}_2))$ is an $\mathcal{L}-$fuzzifying*
*continuous mapping.*
*(3) $\forall a \in \mathcal{L}$, $f|_{M(L_1)} : (M(L_1), T(\mathcal{F}_{1_{[a]}})) \longrightarrow (M(L_2), T(\mathcal{F}_{2_{[a]}}))$ is a con-*
*tinuous mapping.*

**Theorem 10.** *Let $(P_1, \mathcal{S}_1)$, $(P_2, \mathcal{S}_2)$ be $\mathcal{L}-$fuzzifying scott topological spaces*
*and $f : P_1 \longrightarrow P_2$ is a mapping of preserving directed join and the way-below*
*relation. The mapping $L(f) : L(P_1) \longrightarrow L(P_2)$ is defined as $\forall a \in L(P_1)$,*
*$L(f)(a) = \bigcup\{\downarrow f(b) \mid \downarrow b \in \beta^*(a)\}$. Then the conclusions as follows are*
*equivalent:*
*(1) $f : (P_1, \mathcal{S}_1) \longrightarrow (P_2, \mathcal{S}_2)$ is an $\mathcal{L}-$fuzzifying continuous mapping.*
*(2) $f : (P_1, \mathcal{S}_{1_{[a]}}) \longrightarrow (P_2, \mathcal{S}_{2_{[a]}})$ is a continuous mapping.*
*(3) $L(f) : (L(P_1), \mathcal{E}(\mathcal{S}_1')) \longrightarrow (L(P_2), \mathcal{E}(\mathcal{S}_2'))$ is an $\mathcal{L}-$fuzzy continuous*
*generalized order homomorphism.*
*(4) $\forall a \in \mathcal{L}$, $L(f) : (L(P_1), \mathcal{E}(\mathcal{S}_{1_{[a]}}')) \longrightarrow (L(P_2), \mathcal{E}(\mathcal{S}_{2_{[a]}}'))$ is a continuous*
*generalized order homomorphism.*

*Proof.* Prove the definition about $L(f)$ is reasonable firstly.

From the theorem 3, $P_1 \overset{\phi_1}{\cong} M(L(P_1))$ and $P_2 \overset{\phi_2}{\cong} M(L(P_2))$ are true. Define $\widehat{f} = \phi_2 \circ f \circ \phi_1^{-1} : M(L(P_1)) \longrightarrow M(L(P_2))$, $L(f) = L(\widehat{f})$ :

$L(P_1) \longrightarrow L(P_2)$. Then $\forall a \in L(P_1)$, $L(f)(a) = L(\widehat{f})(a) = \bigcup\{\widehat{f}(\downarrow b) \mid \downarrow b \in \beta^*(a)\} = \bigcup\{\phi_2(f(\phi_1^{-1}(\downarrow b))) \mid \downarrow b \in \beta^*(a)\} = \bigcup\{\phi_2(f(b)) \mid \downarrow b \in \beta^*(a)\} = \bigcup\{\downarrow(f(b)) \mid \downarrow b \in \beta^*(a)\}$. Because $\phi_1^{-1}$ and $\phi_2$ are both isomorphic mappings, $\widehat{f}$ is a mapping of preserving directed join and the way-below relation, and $L(f)$ is a generalized order homomorphism which satisfies $L(f)\mid_{M(L(P_1))} = \widehat{f}$.

From the theorem 4.2 [14], (1) $\Leftrightarrow$ (2) and (3) $\Leftrightarrow$ (4) can be proved easily. Therefore, it just needs to prove (2) $\Leftrightarrow$ (4).

$\forall A \in L(P_2)$, $A$ can be proved that it is a lower set and $f^{-1}(A)$ is a lower set; and get $\forall a \in \mathcal{L}$, $A \in \mathcal{S'}_{2_{[a]}} \Longleftrightarrow \mathcal{E}(A) \in \mathcal{E}(\mathcal{S'}_{2_{[a]}})$. Therefore, the equations as follows are true: $L(f)^{-1}(\mathcal{E}(A)) = L(f)^{-1}(\bigcup_{x\in A} \downarrow x) = \bigcup_{x\in A} L(f)^{-1}(\downarrow x)$
$= \bigcup_{x\in A}(\bigcup\{\downarrow b \mid L(f)(\downarrow b) \subseteq \downarrow x\}) = \bigcup_{x\in A}(\bigcup\{\downarrow b \mid \downarrow f(b) \subseteq \downarrow x\}) = \bigcup_{x\in A}(\bigcup\{\downarrow b \mid f(b) \leq x\}) = \bigcup(\bigcup_{x\in A}\{\downarrow b \mid f(b) \leq x\}) = \bigcup\{\downarrow b \mid f(b) \in A\} = \bigcup\{\downarrow b \mid b \in f^{-1}(A)\} = \bigcup_{b\in f^{-1}(A)} \downarrow b = \mathcal{E}(f^{-1}(A))$. Hence, get $\forall a \in \mathcal{L}$, $\forall A \in \mathcal{S'}_{2_{[a]}}$, $f^{-1}(A) \in \mathcal{S'}_{1_{[a]}} \Longleftrightarrow \forall \mathcal{E}(A) \in \mathcal{E}(\mathcal{S'}_{2_{[a]}})$, $L(f)^{-1}(\mathcal{E}(A)) \in \mathcal{E}(\mathcal{S'}_{1_{[a]}})$. It is equal to (2) $\Leftrightarrow$ (4). $\qquad\square$

**Theorem 11.** **FSTS**($\mathcal{L}$) *denotes the category which consists of $\mathcal{L}$−fuzzifying scott topological space and the fuzzifying continuous mapping of orientation-join preserving and the relation of way-below.*

*(1) Defining the mapping $\mathcal{E}$ : **FSTS**($\mathcal{L}$) $\longrightarrow$ **FTML**($\mathcal{L}$) as follows.*

$\forall (P, \mathcal{S}) \in ob(\textbf{FSTS}(\mathcal{L}))$, $\mathcal{E}((P, \mathcal{S})) = (L(P), \mathcal{E}(\mathcal{S}))$;

$\forall f \in Hom((P_1, \mathcal{S}_1), (P_2, \mathcal{S}_2))$, $\mathcal{E}(f) = L(f) : (L(P_1), \mathcal{E}(\mathcal{S}'_1)) \longrightarrow (L(P_2), \mathcal{E}(\mathcal{S}'_2))$. *Then $\mathcal{E}$ is a functor which is from the category **FSTS**($\mathcal{L}$) to the category **FTML**($\mathcal{L}$).*

*(2) Defining the mapping $\mathcal{T}$ : **FTML**($\mathcal{L}$) $\longrightarrow$ **FSTS**($\mathcal{L}$) as follows.*

$\forall (L, \mathcal{F}) \in ob(\textbf{FTML}(\mathcal{L}))$, $\mathcal{T}((L, \mathcal{F})) = (M(L), \mathcal{T}(\mathcal{F}))$,

$\forall f \in Hom((L_1, \mathcal{F}_1), (L_2, \mathcal{F}_2))$, $\mathcal{T}(f) = f\mid_{M(L_1)} : (M(L_1), \mathcal{T}(\mathcal{F}_1)) \longrightarrow (M(L_2), \mathcal{T}(\mathcal{F}_2))$.

*Then $\mathcal{T}$ is a functor which is from the category **FTML**($\mathcal{L}$) to the category **FSTS**($\mathcal{L}$).*

**Theorem 12.** *The functors $\mathcal{T}$ and $\mathcal{E}$ are both complete faithful functors.*

*Proof.* $\forall f, g \in Hom((L_1, \mathcal{F}_1), (L_2, \mathcal{F}_2))$. If $\mathcal{T}(f) = \mathcal{T}(g)$, there is $f\mid_{M(L_1)} = g\mid_{M(L_1)}$. And because $\forall a \in L_1, f(a) = f(\bigvee \beta^*(a)) = \bigvee_{x\in\beta^*(a)} f(x) = \bigvee_{x\in\beta^*(a)} g(x) = g(\bigvee \beta^*(a)) = g(a), f = g$. It is equal to that $\mathcal{T}$ is faithful.

$\forall f \in Hom_{\textbf{FSTP}(\mathcal{L})}(\mathcal{T}(L_1, \mathcal{F}_1), \mathcal{T}(L_2, \mathcal{F}_2))$. From the theorem 9 and 10, there exists the morphism $L(f) \in Hom_{\textbf{FTML}(\mathcal{L})}((L_1, \mathcal{F}_1), (L_2, \mathcal{F}_2))$ such that $\mathcal{T}(L(f)) = L(f)\mid_{M(L_1)} = f$. Hence, $\mathcal{T}$ is the complete functor.

$\mathcal{E}$ is a complete faithful functor which can be proved with the same way. $\qquad\square$

**Theorem 13.** *[10] Let $C$ and $D$ be categories. $G : C \longrightarrow D$ is a functor. Then the conditions as follows are equivalent.*

*(1) The categories of $C$ and $D$ are equivalent.*

*(2)   The functor of G is the complete faithful functor. Fetch $\forall B \in ob(D)$, there exists $A \in ob(C)$ such that $F(A) \cong B$.*

**Theorem 14.** *(The Representation Theorem of the category* **FTML($\mathcal{L}$))*
The category of* **FTML($\mathcal{L}$)** *and category of* **FSTS($\mathcal{L}$)** *are equivalent.*

**Theorem 15.** *If the category of* **STP** *denotes the category which consists of topological space of* $(P, \tau), \tau \subseteq \sigma(P)$ *on the continuous partially ordered set and the continuous mapping of orientation preserving join and the relation of way-below, then the category of* **TML** *and the category of* **STP** *are equivalent.*

**Acknowledgements.** This work was supported by the Beijing Natural Science Foundation Program (1102016), the Beijing Municipal Training Programme for the Talents (2009D005002000006), the Beijing College Scientific Research and entrepreneurship Plan Project, and the College Scientific, technological Activities of North China University of Technology, the Funding Project for Academic Human Resources Development in Institutions of Higher Learning Under the Jurisdiction of Beijing Municipality(PHR(IHLB)A1021).

# References

1. Fan, T.H.: Category of topological molecular lattices. Doctor's paper of SiChuan University (1990)
2. Gierz, G.: A Compendium of Continuous Lattices. Springer, Berlin (1980)
3. Höhle, U., Rodabaugh, E.: Mathematics of Fuzzy Sets: Logic, Topology, and Measure Theory. Kluwer Academic Publishers, Dordrecht (1999)
4. Höhle, U., Šostak, A.P.: General theory of fuzzy topological spaces. FSS 73, 131–149 (1995)
5. Wang, G.J.: Theory of topological molecular lattices. FSS 47, 351–376 (1992)
6. Yang, Z.Q.: Cartesian of topological molecular lattices Category. Doctor's paper of SiChuan University (1990)
7. Ying, M.S.: A new approach for fuzzy topology (I). FSS 39, 303–321 (1991)
8. Ying, M.S.: A new approach for fuzzy topology (II). FSS 47, 221–232 (1992)
9. Ying, M.S.: A new approach for fuzzy topology (III). FSS 55, 193–207 (1993)
10. Zheng, C.Y., Fan, L., Cui, H.B.: Frame and continuous lattices. Press of Capital Normal University, Beijing (2000)
11. Zhang, J.: Lattice Valued Smooth Pointwise Quasi-Uniformity on Completely Distributive Lattice. Fuzzy Systems and Mathematics 2, 30–34 (2003)
12. Zhang, J., Chang, Z., Liu, B., Xiu, Z.Y.: L-fuzzy Co-Alexandrov topology spaces. In: 2008 International Seminar on Business and Information Management, pp. 279–282 (2008)
13. Zhang, J., Shi, F.G., Zheng, C.Y.: On L-fuzzy topological spaces. Fuzzy Sets and Systems 149, 473–484 (2005)
14. Zhang, J., Zheng, C.Y.: Lattice valued smooth topological molecular lattices. Journal of Fuzzy Mathematics 10(2), 411–421 (2002)

# L-Fuzzy Subalgebras and L-Fuzzy Filters of R₀-Algebras

Chunhui Liu and Luoshan Xu

**Abstract.** R₀-algebras are the logic algebras associated to the formal deductive system $L^*$ for fuzzy propositional calculus. In this paper, the concepts of $L$-fuzzy subalgebras and $L$-fuzzy filters of R₀-algebras are introduced. Properties of $L$-fuzzy subalgebras and $L$-fuzzy filters are investigated. characterizations of $L$-fuzzy subalgebras and $L$-fuzzy filters of R₀-algebras are obtained. It is proved that an L-fuzzy set on an R₀-algebra $M$ is an $L$-fuzzy subalgebra of $M$ if and only if for all $t \in L$, every its nonempty t-level section is a subalgebra of $M$. It is also proved that under some reasonable conditions, images and inverse images of $L$-fuzzy subalgebras (resp., $L$-fuzzy filters) of R₀-algebra homomorphisms are still $L$-fuzzy subalgebras (resp. $L$-fuzzy filters).

**Keywords:** R₀-algebras, $L$-fuzzy subalgebras, $L$-fuzzy filters, Fuzzy logic.

## 1 Introduction

With the developments of mathematics and computer science, non-classical mathematical logics have been actively studied [14]. The research interest on the foundation of fuzzy logic has been growing rapidly. Several new logical algebras playing roles of structures of truth values have been introduced and extensively studied [13] [11] [5] [1] [2] [10]. In 1996, Wang [14] and [10] proposed a formal deductive system $L^*$ for fuzzy propositional calculus, and

Chunhui Liu
Department of Elementary Education, Chifeng College, Chifeng 024001,
Inner Mongolia, P.R. China
e-mail: chunhuiliu1982@163.com

Luo-shan XU
Department of Mathematics, Yangzhou University, Yangzhou 225002,
Jiangsu, P.R. China
e-mail: luoshanxu@hotmail.com

S. Li (Eds.): Nonlinear Maths for Uncertainty and its Appli., AISC 100, pp. 667–674.
springerlink.com                      © Springer-Verlag Berlin Heidelberg 2011

introduced a new kind of algebraic structures called $R_0$-algebras. This kind of algebraic structures have been studied to some extends [7] [8].

The concept of fuzzy sets was introduced firstly by Zadeh in [15]. At present, the concept of Zadeh's have been applied to some kinds of algebraic structures. Hoo in [3] [4] applied the concept to MV/BCK-algebras and proposed notions of fuzzy ideals, fuzzy prime ideals and fuzzy Boolean ideals. Liu and Li applied the concept to $R_0$-algebras and proposed notions of fuzzy implicative filters and fuzzy Boolean filters on $R_0$-algebras in [6]. Xu and Qin in [12] applied the concept to lattice implication algebras, proposed the notion of fuzzy lattice implication algebra and discussed some properties of them. In the present paper, to extent the concepts of fuzzy subalgebras and fuzzy filters, we propose the concepts $L$-fuzzy subalgebras and $L$-fuzzy filters of $R_0$-algebras in terms of the concept of $L$-fuzzy sets in [9], where the prefix $L$ a lattice. It should be noticed that when $L = [0,1]$, then $[0,1]$-fuzzy sets are originally meant fuzzy sets. Since $[0,1]$ is a special completely distributive lattice, to investigate properties of $L$-fuzzy subalgebras, sometimes we assume that the prefix $L$ is a completely distributive lattice.

## 2 Preliminaries

**Definition 1.** [14] Let $M$ be a $(\neg, \vee, \wedge, \rightarrow)$ type algebra, where $\neg$ is a unary operation, $\vee$, $\wedge$ and $\rightarrow$ are binary operations. If there is a partial ordering $\leqslant$ on $M$, such that $(M, \leqslant)$ is a bounded distributive lattice, $\vee$ and $\wedge$ are supremum and infimum operations with respect to $\leqslant$, $\neg$ is an order-reversing with respect to $\leqslant$, and the following conditions hold for any $a, b, c \in M$:

(R1) $\neg a \rightarrow \neg b = b \rightarrow a$;

(R2) $1 \rightarrow a = a$, $a \rightarrow a = 1$;

(R3) $b \rightarrow c \leqslant (a \rightarrow b) \rightarrow (a \rightarrow c)$;

(R4) $a \rightarrow (b \rightarrow c) = b \rightarrow (a \rightarrow c)$;

(R5) $a \rightarrow (b \vee c) = (a \rightarrow b) \vee (a \rightarrow c)$, $a \rightarrow (b \wedge c) = (a \rightarrow b) \wedge (a \rightarrow c)$;

(R6) $(a \rightarrow b) \vee ((a \rightarrow b) \rightarrow (\neg a \vee b)) = 1$.

where 1 is the largest element of $M$, then $M$ is called an $R_0$-algebra.

**Lemma 1.** *[10] In an $R_0$-algebra $M$, the following assertions hold for all $x, y, z \in M$:*

*(1) $0 \rightarrow x = 1$ and $x \rightarrow 1 = 1$;*

*(2) $x \leqslant y$ if and only if $x \rightarrow y = 1$;*

*(3) $\neg x = x \rightarrow 0$;*

*(4) $(x \rightarrow y) \vee (y \rightarrow x) = 1$;*

*(5) $x \vee y = ((x \rightarrow y) \rightarrow y) \wedge ((y \rightarrow x) \rightarrow x)$;*

*(6) $x \leqslant y$ implies $y \rightarrow z \leqslant x \rightarrow z$ and $z \rightarrow x \leqslant z \rightarrow y$.*

**Definition 2.** [14] Let $M$ be an $R_0$-algebra and $S$ a non-empty subset of $M$. We say $S$ a subalgebra of $M$ if $S$ is closed under the operations $\neg, \vee$ and $\rightarrow$.

**Definition 3.** [14] Let $M_1$ and $M_2$ are $R_0$-algebras. A map $f : M_1 \to M_2$ is called an $R_0$-algebra homomorphism if $f(\neg x) = \neg f(x)$, $f(x \vee y) = f(x) \vee f(y)$ and $f(x \to y) = f(x) \to f(y)$ for any $x, y \in M_1$.

*Remark 1.* [14] If $f$ is an $R_0$-algebra homomorphism, then $f(0) = 0$, $f(1) = 1$.

**Definition 4.** [9] Let $X$ be a non-empty set and $L$ a lattice. A map $\mathscr{A} : X \to L$ is called an $L$-fuzzy subset of $X$. The set of all $L$-fuzzy subset of $X$ is denoted by $\mathcal{F}_L(X)$. For $\mathscr{A} \in \mathcal{F}_L(X)$ and $t \in L$, the set $\mathscr{A}_t = \{x \in M | \mathscr{A}(x) \geqslant t\}$ is called a $t$−level set of $\mathscr{A}$.

Let $L$ be a lattice, $\mathscr{A}$ and $\mathscr{B}$ two $L$-fuzzy sets of $X$. Define the $L$-fuzzy sets $\mathscr{A} \cup \mathscr{B}$ and $\mathscr{A} \cap \mathscr{B}$ such that for all $x \in X$,

$$(\mathscr{A} \cup \mathscr{B})(x) = \mathscr{A}(x) \vee \mathscr{B}(x), \qquad (\mathscr{A} \cap \mathscr{B})(x) = \mathscr{A}(x) \wedge \mathscr{B}(x).$$

**Definition 5.** Let $M_1$ and $M_2$ be two $R_0$-algebras, $L$ a complete lattice and $f : M_1 \to M_2$ an $R_0$-algebra homomorphism. Then $f$ induces two $L$-fuzzy subsets $f_* : \mathcal{F}_L(M_1) \to \mathcal{F}_L(M_2)$ and $f_*^{-1} : \mathcal{F}_L(M_2) \to \mathcal{F}_L(M_1)$ such that for all $\mathscr{A} \in \mathcal{F}_L(M_1)$ and $\mathscr{B} \in \mathcal{F}_L(M_2)$,

$$\forall y \in M_2, \quad f_*(\mathscr{A})(y) = \vee_{x \in M_1}\{\mathscr{A}(x) | f(x) = y\} = \vee_{x \in f^{-1}(y)}\mathscr{A}(x), \tag{1}$$

$$\forall x \in M_1, \quad f_*^{-1}(\mathscr{B})(x) = (\mathscr{B} \circ f)(x) = \mathscr{B}(f(x)). \tag{2}$$

**Definition 6.** [14] Let $M$ be an $R_0$-algebra. A nonempty subset $F$ of $X$ is called an MP-filter of $X$ if it satisfies (F-1) $1 \in F$ and (F-2) $x \in F$ and $x \to y \in F$ imply $y \in F$ for any $x, y \in X$.

# 3  $L$-Fuzzy Subalgebras of $R_0$-Algebras

**Definition 7.** Let $M$ be an $R_0$-algebra and $L$ a lattice. An $L$-fuzzy subset $\mathscr{A}$ of $M$ is called an $L$-fuzzy subalgebra of $M$ if the following conditions hold.
    (LFS-1) $\mathscr{A}(0) = \mathscr{A}(1)$;
    (LFS-2) $\mathscr{A}(x \vee y) \geqslant \mathscr{A}(x) \wedge \mathscr{A}(y)$ for any $x, y \in M$;
    (LFS-3) $\mathscr{A}(x \to y) \geqslant \mathscr{A}(x) \wedge \mathscr{A}(y)$ for any $x, y \in M$.

**Theorem 1.** *Let $M$ be an $R_0$-algebra and $L$ a lattice. If $\mathscr{A}$ is an $L$-fuzzy subalgebra of $M$, then the following statements hold for any $x, y \in M$.*
    *(LFS-4) $\mathscr{A}(0) = \mathscr{A}(1) \geqslant \mathscr{A}(x)$;*
    *(LFS-5) $\mathscr{A}(x) = \mathscr{A}(\neg x)$;*
    *(LFS-6) $\mathscr{A}(x \wedge y) \geqslant \mathscr{A}(x) \wedge \mathscr{A}(y)$.*

*Proof.* (1) By (R2), (LFS-1) and (LFS-3) we have $\mathscr{A}(0) = \mathscr{A}(1) = \mathscr{A}(x \to x) \geqslant \mathscr{A}(x) \wedge \mathscr{A}(x) = \mathscr{A}(x)$, i.e., (LFS-4) holds.
    (2) On one hand, by (R1), (R2), (LFS-3) and (LFS-4) we have $\mathscr{A}(x) = \mathscr{A}(1 \to x) = \mathscr{A}(\neg x \to \neg 1) \geqslant \mathscr{A}(\neg x) \wedge \mathscr{A}(0) \geqslant \mathscr{A}(\neg x) \wedge \mathscr{A}(\neg x) = \mathscr{A}(\neg x)$. On the other hand, by Lemma 1(3), (LFS-3) and (LFS-4) we have $\mathscr{A}(\neg x) =$

$\mathscr{A}(x \to 0) \geqslant \mathscr{A}(x) \wedge \mathscr{A}(0) \geqslant \mathscr{A}(x) \wedge \mathscr{A}(x) = \mathscr{A}(x)$. So, it follows from the above two aspects that $\mathscr{A}(x) = \mathscr{A}(\neg x)$, i. e., (LFS-5) holds.

(3) By (LFS-2) and (LFS-5) we get $\mathscr{A}(x \wedge y) = \mathscr{A}(\neg(x \wedge y)) = \mathscr{A}(\neg x \vee \neg y) \geqslant \mathscr{A}(\neg x) \wedge \mathscr{A}(\neg y) = \mathscr{A}(x) \wedge \mathscr{A}(y)$, thus (LFS-6) holds. $\qquad\square$

**Theorem 2.** *Let $M$ be an $R_0$-algebra, $L$ a lattice and $\mathscr{A} \in \mathcal{F}_L(M)$. Then $\mathscr{A}$ is an $L$-fuzzy subalgebra of $M$ iff for all $t \in L$, $\mathscr{A}_t$ is a subalgebra of $M$ whenever $\mathscr{A}_t \neq \varnothing$.*

*Proof.* $\Rightarrow$: Suppose that $\mathscr{A} \in \mathcal{F}_L(M)$ is an $L$-fuzzy subalgebra of $M$, $t \in L$ and $\mathscr{A}_t \neq \varnothing$. Then for any $x, y \in \mathscr{A}_t$ we have $\mathscr{A}(x) \geqslant t$ and $\mathscr{A}(y) \geqslant t$. By Definition 7 and Theorem 1 we can get that $\mathscr{A}(\neg x) = \mathscr{A}(x) \geqslant t, \mathscr{A}(x \vee y) \geqslant \mathscr{A}(x) \wedge \mathscr{A}(y) \geqslant t$ and $\mathscr{A}(x \to y) \geqslant \mathscr{A}(x) \wedge \mathscr{A}(y) \geqslant t$, i.e., $\mathscr{A}_t$ is closed under operations $\neg, \vee$ and $\to$. So, $\mathscr{A}_t$ is a subalgebra of $M$ by Definition 2.

$\Leftarrow$: Let $\mathscr{A}_t$ for any $t \in L$ be a subalgebra of $M$ whenever $\mathscr{A}_t \neq \varnothing$. Take $t = \mathscr{A}(1)$. Then $1 \in \mathscr{A}_t \neq \varnothing$ and $\mathscr{A}_t$ is a subalgebra of $M$ by the assumption. Thus $0 = \neg 1 \in \mathscr{A}_t$ and $\mathscr{A}(0) \geqslant t = \mathscr{A}(1)$. Similarly, taking $t = \mathscr{A}(0)$, we have $\mathscr{A}(1) \geqslant \mathscr{A}(0)$ and $\mathscr{A}(0) = \mathscr{A}(1)$. Let $s = \mathscr{A}(x) \wedge \mathscr{A}(y)$ for all $x, y \in M$. Then $\mathscr{A}(x) \geqslant s$ and $\mathscr{A}(y) \geqslant s$ and $x, y \in \mathscr{A}_s \neq \varnothing$. It follows from the assumption that $\mathscr{A}_s$ is a subalgebra of $M$ and $x \vee y, x \to y \in \mathscr{A}_s$. Furthermore, we have $\mathscr{A}(x \vee y) \geqslant s = \mathscr{A}(x) \wedge \mathscr{A}(y)$ and $\mathscr{A}(x \to y) \geqslant s = \mathscr{A}(x) \wedge \mathscr{A}(y)$. So, $\mathscr{A}$ is an $L$-fuzzy subalgebra of $M$ by Definition 7.

**Corollary 1.** *Let $L$ is a lattice. If $\mathscr{A} \in \mathcal{F}_L(M)$ is an $L$-fuzzy subalgebra of $R_0$-algebra $M$, then $S = \{x \in M | \mathscr{A}(x) = \mathscr{A}(0)\}$ is a subalgebra of $M$.*

*Proof.* Obviously, by (LFS-4), $S = \mathscr{A}_{\mathscr{A}(0)}$ and $S \neq \varnothing$ by (LFS-1), thus $S$ is a subalgebra of $M$ By Theorem 2.

**Theorem 3.** *Let $L$ be a complete lattice and $\{\mathscr{A}_i\}_{i \in I}$ a family of $L$-fuzzy subalgebras of $R_0$-algebra $M$. Then $L$-fuzzy subset $\cap_{i \in I}\mathscr{A}$ is also an $L$-fuzzy subalgebra of $M$.*

*Proof.* Since $L$ is a complete lattice and $\{\mathscr{A}_i\}_{i \in I}$ a family of $L$-fuzzy subalgebras of $R_0$-algebra $M$, we have that

$$(\cap_{i \in I}\mathscr{A}_i)(0) = \wedge_{i \in I}\mathscr{A}_i(0) = \wedge_{i \in I}\mathscr{A}_i(1) = (\cap_{i \in I}\mathscr{A}_i)(1),$$

and for any $x, y \in M$ we have

$$(\cap_{i \in I}\mathscr{A}_i)(x \vee y) = \wedge_{i \in I}\mathscr{A}_i(x \vee y) \geqslant \wedge_{i \in I}[\mathscr{A}_i(x) \wedge \mathscr{A}_i(y)]$$
$$= [\wedge_{i \in I}\mathscr{A}_i(x)] \wedge [\wedge_{i \in I}\mathscr{A}_i(y)] = (\cap_{i \in I}\mathscr{A}_i)(x) \wedge (\cap_{i \in I}\mathscr{A}_i)(y),$$

$$(\cap_{i \in I}\mathscr{A}_i)(x \to y) = \wedge_{i \in I}\mathscr{A}_i(x \to y) \geqslant \wedge_{i \in I}[\mathscr{A}_i(x) \wedge \mathscr{A}_i(y)]$$
$$= [\wedge_{i \in I}\mathscr{A}_i(x)] \wedge [\wedge_{i \in I}\mathscr{A}_i(y)] = (\cap_{i \in I}\mathscr{A}_i)(x) \wedge (\cap_{i \in I}\mathscr{A}_i)(y).$$

So, $\cap_{i \in I}\mathscr{A}$ is also an $L$-fuzzy subalgebra of $M$ by Definition 7.

**Corollary 2.** *Let $L$ be a lattice, $\mathscr{A}$ and $\mathscr{B}$ two $L$-fuzzy subalgebras of $R_0$-algebra $M$. Then $L$-fuzzy subset $\mathscr{A} \cap \mathscr{B}$ is also an $L$-fuzzy subalgebra of $M$.*

*Remark 2.* In general, for two $L$-fuzzy subalgebras $\mathscr{A}$ and $\mathscr{B}$, $\mathscr{A} \cup \mathscr{B}$ may not be an $L$-fuzzy subalgebra. For example, let $X = \{a, b, c\}$ and $M = \mathcal{P}(X)$, the power set of $X$, then $M$ is an $R_0$-algebra, where $\neg A = X - A, A \vee B = A \cup B$ and $A \to B = \neg A \vee B$ for any $A, B \in \mathcal{P}(X)$. Take $L = ([0, 1], \max, \min)$ and define two [0,1]-fuzzy subsets $\mathscr{A}$ and $\mathscr{B}$ of $M$ such that

$$\mathscr{A}(\varnothing) = \mathscr{A}(X) = \mathscr{A}(\{a\}) = \mathscr{A}(\{b, c\}) = 1,$$

$$\mathscr{A}(\{b\}) = \mathscr{A}(\{a, c\}) = \mathscr{A}(\{c\}) = \mathscr{A}(\{a, b\}) = t_1,$$

$$\mathscr{B}(\varnothing) = \mathscr{B}(X) = \mathscr{B}(\{b\}) = \mathscr{B}(\{a, c\}) = 1,$$

$$\mathscr{B}(\{a\}) = \mathscr{B}(\{b, c\}) = \mathscr{B}(\{c\}) = \mathscr{B}(\{a, b\}) = t_2,$$

where $0 < t_2 < t_1 < 1$. It is easy to verify that $\mathscr{A}$ and $\mathscr{B}$ are [0,1]-fuzzy subalgebras of $M$. But $\mathscr{A} \cup \mathscr{B}$ is not a [0,1]-fuzzy subalgebra of $M$, for that $(\mathscr{A} \cup \mathscr{B})(\{a\} \vee \{b\}) = t_1 \not\geqslant 1 = (\mathscr{A} \cup \mathscr{B})(\{a\}) \wedge (\mathscr{A} \cup \mathscr{B})(\{b\})$.

This example shows that Theorem 3 is not true for $\vee$-operation in general.

**Theorem 4.** *Let $M_1$ and $M_2$ be two $R_0$-algebras, $L$ a completely distributive lattice and $f : M_1 \to M_2$ an $R_0$-algebra homomorphism.*

*(1) if $\mathscr{A} \in \mathcal{F}_L(M_1)$ is an $L$-fuzzy subalgebra of $M_1$, then $f_*(\mathscr{A})$ is an $L$-fuzzy subalgebra of $M_2$;*

*(2) if $\mathscr{B} \in \mathcal{F}_L(M_2)$ is an $L$-fuzzy subalgebra of $M_2$, then $f_*^{-1}(\mathscr{B})$ is an $L$-fuzzy subalgebra of $M_1$.*

*Proof.* (1) By applying (LFS-4) to $\mathscr{A}$, we have that

$$f_*(\mathscr{A})(0) = \vee_{x \in f^{-1}(0)} \mathscr{A}(x) = \mathscr{A}(0) = \mathscr{A}(1) = \vee_{x \in f^{-1}(1)} \mathscr{A}(x) = f_*(\mathscr{A})(1).$$

So, $f_*(\mathscr{A})$ satisfies (LFS-1) in Definition 7.

Since $f$ is an $R_0$-algebra homomorphism and $L$ is completely distributive, we have by Definition 5 and applying (LFS-2) to $\mathscr{A}$ that for all $y_1, y_2 \in M_2$,

$$
\begin{aligned}
f_*(\mathscr{A})(y_1) \wedge f_*(\mathscr{A})(y_2) &= [\vee_{x_1 \in f^{-1}(y_1)} \mathscr{A}(x_1)] \wedge [\vee_{x_2 \in f^{-1}(y_2)} \mathscr{A}(x_2)] \\
&= \vee_{x_1 \in f^{-1}(y_1), \, x_2 \in f^{-1}(y_2)} \mathscr{A}(x_1) \wedge \mathscr{A}(x_2) \\
&\leqslant \vee_{x_1 \in f^{-1}(y_1), \, x_2 \in f^{-1}(y_2)} \mathscr{A}(x_1 \vee x_2) \\
&\leqslant \vee_{x \in f^{-1}(y_1 \vee y_2)} \mathscr{A}(x) = f_*(\mathscr{A})(y_1 \vee y_2).
\end{aligned}
$$

This shows that $f_*(\mathscr{A})$ satisfies (LFS-2) in Definition 7.

Similar arguments can show that $f_*(\mathscr{A})(y_1 \to y_2) \geqslant f_*(\mathscr{A})(y_1) \wedge f_*(\mathscr{A})(y_2)$ holds for any $y_1, y_2 \in M_2$, i. e., $f_*(\mathscr{A})$ satisfies also (LFS-3). So $f_*(\mathscr{A})$ is an $L$-fuzzy subalgebra of $M_2$ by Definition 7.

(2) Since $f$ is an $R_0$-algebra homomorphism, we have $f(1) = 1$ and $f(0) = 0$. Because that $\mathscr{B} \in \mathcal{F}_L(M_2)$ is an $L$-fuzzy subalgebra of $M_2$, we have that $f_*^{-1}(\mathscr{B})(0) = (\mathscr{B} \circ f)(0) = \mathscr{B}(f(0)) = \mathscr{B}(0) = \mathscr{B}(1) = \mathscr{B}(f(1)) = (\mathscr{B} \circ f)(1) = f_*^{-1}(\mathscr{B})(1)$, and that for any $x, y \in M_1$,

$$f_*^{-1}(\mathscr{B})(x \vee y) = (\mathscr{B} \circ f)(x \vee y) = \mathscr{B}(f(x \vee y)) = \mathscr{B}(f(x) \vee f(y))$$

$$\geqslant \mathscr{B}(f(x)) \wedge \mathscr{B}(f(y)) = f_*^{-1}(\mathscr{B})(x) \wedge f_*^{-1}(\mathscr{B})(y),$$

$$f_*^{-1}(\mathscr{B})(x \to y) = (\mathscr{B} \circ f)(x \to y) = \mathscr{B}(f(x \to y)) = \mathscr{B}(f(x) \vee f(y))$$

$$\geqslant \mathscr{B}(f(x)) \wedge \mathscr{B}(f(y)) = f_*^{-1}(\mathscr{B})(x) \wedge f_*^{-1}(\mathscr{B})(y).$$

So, $f_*^{-1}(\mathscr{B})$ is an $L$-fuzzy subalgebra of $M_1$ by Definition 7.

**Theorem 5.** *Let $M$ be an $R_0$-algebra, $L_1$ and $L_2$ two lattices. Let $f : L_1 \to L_2$ be a lattice homomorphism. If $\mathscr{A}$ is an $L_1$-fuzzy subalgebra of $M$, then $f \circ \mathscr{A}$ is an $L_2$-fuzzy subalgebra of $M$.*

*Proof.* Since $\mathscr{A}$ is an $L_1$-fuzzy subalgebra and $f$ a lattice homomorphism, we have that $(f \circ \mathscr{A})(0) = f(\mathscr{A}(0)) = f(\mathscr{A}(1)) = (f \circ \mathscr{A})(1)$, and

$$(f \circ \mathscr{A})(x \vee y) = f(\mathscr{A}(x \vee y)) \geqslant f(\mathscr{A}(x) \wedge \mathscr{A}(y)) = (f \circ \mathscr{A})(x) \wedge (f \circ \mathscr{A})(y),$$

$$(f \circ \mathscr{A})(x \to y) = f(\mathscr{A}(x \to y)) \geqslant f(\mathscr{A}(x) \wedge \mathscr{A}(y)) = (f \circ \mathscr{A})(x) \wedge (f \circ \mathscr{A})(y).$$

for any $x, y \in M$. So $f \circ \mathscr{A}$ is an $L_2$-fuzzy subalgebra of $M$.

## 4  *L*-Fuzzy Filters of $R_0$-Algebras

**Definition 8.** Let $M$ be an $R_0$-algebra and $L$ a lattice. An $L$-fuzzy subset $\mathscr{A}$ of $M$ is said to be an $L$-fuzzy filter of $M$, if it satisfies
   (LFF-1)  For all $x \in M$, $\mathscr{A}(1) \geqslant \mathscr{A}(x)$ and
   (LFF-2)  For all $x, y \in M$, $\mathscr{A}(y) \geqslant \mathscr{A}(x) \wedge \mathscr{A}(x \to y)$.

**Theorem 6.** *Let $M$ be an $R_0$-algebra, $L$ a lattice and $\mathscr{A}$ an $L$-fuzzy filter of $M$. Then for any $x, y \in M$, $x \leqslant y$ implies $\mathscr{A}(x) \leqslant \mathscr{A}(y)$, that is, $\mathscr{A}$ is order-preserving.*

*Proof.* If $x, y \in M$ and $x \leqslant y$, then $x \to y = 1$ by (R2). By (LFF-2) and (LFF-1), we have that $\mathscr{A}(y) \geqslant \mathscr{A}(x) \wedge \mathscr{A}(x \to y) = \mathscr{A}(x) \wedge \mathscr{A}(1) = \mathscr{A}(x)$.

**Theorem 7.** *Let $M$ be an $R_0$-algebra, $L$ a lattice and $\mathscr{A}$ an $L$-fuzzy subset of $M$. Then $\mathscr{A}$ is an $L$-fuzzy filter of $M$ iff for any $t \in L$ $\mathscr{A}_t$ is an MP-filter whenever $\mathscr{A}_t \neq \varnothing$.*

*Proof.* It proof is similar to that of Theorem2 and we hence omit the details.

**Theorem 8.** *Let $M$ be an $R_0$-algebra and $L$ a lattice, $\mathscr{A} \in \mathcal{F}_L(M)$. Then $\mathscr{A}$ is an $L$-fuzzy filter of $M$ iff $x \to (y \to z) = 1$ implies $\mathscr{A}(z) \geqslant \mathscr{A}(x) \wedge \mathscr{A}(y)$ for all $x, y, z \in M$.*

*Proof.* $\Rightarrow$: Assume that $\mathscr{A}$ is an $L$-fuzzy filter on $M$. Then for all $x, y, z \in M$, we have $\mathscr{A}(z) \geqslant \mathscr{A}(y) \wedge \mathscr{A}(y \to z)$ and by (LFF-2), $\mathscr{A}(y \to z) \geqslant \mathscr{A}(x) \wedge \mathscr{A}(x \to (y \to z))$. If $x \to (y \to z) = 1$, then by (LFF-1), $\mathscr{A}(y \to z) \geqslant \mathscr{A}(x) \wedge \mathscr{A}(1) = \mathscr{A}(x)$. So, $\mathscr{A}(z) \geqslant \mathscr{A}(x) \wedge \mathscr{A}(y)$.

$\Leftarrow$: Assume that $x \to (y \to z) = 1$ implies $\mathscr{A}(z) \geqslant \mathscr{A}(x) \wedge \mathscr{A}(y)$, for all $x, y, z \in M$. Since for any $x \in M$, $x \to (x \to 1) = 1$, we have $\mathscr{A}(1) \geqslant \mathscr{A}(x) \wedge \mathscr{A}(x) = \mathscr{A}(x)$ by the assumption, i.e., (LFF-1) holds. Furthermore, by (R2), for any $x, y \in M$ one has $(x \to y) \to (x \to y) = 1$. It follows by the assumption that $\mathscr{A}(y) \geqslant \mathscr{A}(x) \wedge \mathscr{A}(x \to y)$, i.e., (LFF-2) holds. So, $\mathscr{A}$ is an $L$-fuzzy filter of $M$.

**Theorem 9.** *Let $M$ be an $R_0$-algebra, $L$ a lattice, $\mathscr{A}$ and $\mathscr{B}$ two $L$-fuzzy filters of $M$. Then the $L$-fuzzy subset $\mathscr{A} \cap \mathscr{B}$ is an $L$-fuzzy filter of $M$.*

*Proof.* Suppose $\mathscr{A}$ and $\mathscr{B}$ are two $L$-fuzzy filters and for all $x, y, z \in M$, $x \to (y \to z) = 1$. Then by Theorem 8, we have $\mathscr{A}(z) \geqslant \mathscr{A}(x) \wedge \mathscr{A}(y)$ and $\mathscr{B}(z) \geqslant \mathscr{B}(x) \wedge \mathscr{B}(y)$. So, $(\mathscr{A} \cap \mathscr{B})(z) = \mathscr{A}(z) \wedge \mathscr{B}(z) \geqslant (\mathscr{A}(x) \wedge \mathscr{A}(y)) \wedge (\mathscr{B}(x) \wedge \mathscr{B}(y)) = (\mathscr{A} \cap \mathscr{B})(x) \wedge (\mathscr{A} \cap \mathscr{B})(y)$. It follows from Theorem 8 that $\mathscr{A} \cap \mathscr{B}$ is an $L$-fuzzy filter of $M$.

**Theorem 10.** *Let $M$ be an $R_0$-algebra, $L$ a lattice and $\mathscr{A} \in \mathcal{F}_L(M)$ an $L$-fuzzy filter of $M$, then $F = \{x \in M | \mathscr{A}(x) = \mathscr{A}(1)\}$ is an MP-filter of $M$.*

*Proof.* Clearly, $1 \in F$. If $x \in F$ and $x \to y \in F$ for some $x, y \in M$, then $\mathscr{A}(x) = \mathscr{A}(1)$ and $\mathscr{A}(x \to y) = \mathscr{A}(1)$. Since $\mathscr{A} \in \mathcal{F}_L(M)$ is an $L$-fuzzy filter of $M$, by (LFF-2) we have $\mathscr{A}(y) \geqslant \mathscr{A}(x) \wedge \mathscr{A}(x \to y) = \mathscr{A}(1) \wedge \mathscr{A}(1) = \mathscr{A}(1)$. Thus $y \in F$, showing that $F$ is an MP-filter of $M$.

**Theorem 11.** *Let $M_1$ and $M_2$ be two $R_0$-algebras, $L$ a complete lattice and $f : M_1 \to M_2$ an $R_0$-algebra homomorphism.*
*(1) if $\mathscr{A} \in \mathcal{F}_L(M_1)$ is an $L$-fuzzy filter of $M_1$ and $f$ is an isomorphism, then $f_*(\mathscr{A})$ is an $L$-fuzzy filter of $M_2$;*
*(2) if $\mathscr{B} \in \mathcal{F}_L(M_2)$ is an $L$-fuzzy filter of $M_2$, then $f_*^{-1}(\mathscr{B})$ is an $L$-fuzzy filter of $M_1$.*

*Proof.* (1) Straightforward.
(2) Since $f$ is an $R_0$-algebra homomorphism, we have $f(1) = 1$. It follows from that $\mathscr{B} \in \mathcal{F}_L(M_2)$ is an $L$-fuzzy filter of $M_2$, for any $x \in M_1$ we have that $f_*^{-1}(\mathscr{B})(x) = (\mathscr{B} \circ f)(x) = \mathscr{B}(f(x)) \leqslant \mathscr{B}(1) = \mathscr{B}(f(1)) = (\mathscr{B} \circ f)(1) = f_*^{-1}(\mathscr{B})(1)$. And for any $x_1, x_2 \in M_1$ we have that

$$f_*^{-1}(\mathscr{B})(x_2) = \mathscr{B}(f(x_2)) \geqslant \mathscr{B}(f(x_1)) \wedge \mathscr{B}(f(x_1) \to f(x_2))$$
$$= \mathscr{B}(f(x_1)) \wedge \mathscr{B}(f(x_1 \to x_2)) = f_*^{-1}(\mathscr{B})(x_1) \wedge f_*^{-1}(\mathscr{B})(x_1 \to x_2).$$

Thus $f_*^{-1}(\mathscr{B})$ is an $L$-fuzzy filter of $M_1$.

# 5 Concluding Remarks

In this paper, we introduced the notions of $L$-fuzzy subalgebras and $L$-fuzzy filters of $R_0$-algebras. We also characterized them and discussed their properties. The results obtained in this paper reflect interactions of algebraic method and fuzzifying method in the studies of logic problems. It should be noticed that other types of logic algebras and $L$-fuzzy filters, such as $L$-fuzzy prime filters, $L$-fuzzy boolean filters, $L$-fuzzy implicative filters can also be considered. So, in the future research, it is hoped that more research topics will arise with this work.

**Acknowledgements.** The work is supported by the NSF of China (61074129, BK2010313).

# References

1. Chang, C.C.: Algebras Analysis of Many-valued Logics. Trans. Amer. Math. Soc. 88, 467–490 (1958)
2. Hájek, P.: Metamathematics of Fuzzy Logic. Kluwer Academic Publishers, Dordrecht (1998)
3. Hoo, C.S.: Fuzzy Implicative and Boolean Ideals of MV-algebras. Fuzzy Sets and Systems 41, 315–327 (1994)
4. Hoo, C.S.: Some Fuzzy concepts of BCI, BCK and MV-algebras. Int. J. Approx. Reason 18, 177–189 (1998)
5. Liu, C.H., Xu, L.S.: MP-filters of Fuzzy Implication Algebras. Fuzzy Systems and Math. 2, 1–6 (2009)
6. Liu, L.Z., Li, K.T.: Fuzzy Implicative and Boolean Filters of $R_0$-algebras. Information Sci. 171, 61–71 (2005)
7. Liu, L.Z., Li, K.T.: $R_0$-algebras and Weak Dually Residuated Lattice Ordered Semigroups. Czech Math J. 56, 339–348 (2006)
8. Liu, L.Z., Zhang, X.Y.: States on $R_0$-algebras. Soft Comput. 12, 1094–1104 (2008)
9. Goguen, J.A.: L-fuzzy sets. J. Math. Anal. Appl. 18, 145–174 (1967)
10. Pei, D.W., Wang, G.J.: The Completeness and Application of Formal Systems $L^*$. Science in China (Ser. E) 1, 56–64 (2002)
11. Xu, Y., Qin, K.Y.: On Fuzzy Filters of Lattice Implication algebras. J. Fuzzy Math. 1, 251–260 (1993)
12. Xu, Y., Qin, K.Y.: Fuzzy Lattice Implication Algebras. J. Southweast Jiaotong Univ. 2, 121–127 (1995) (in chinese)
13. Wang, G.J.: On Logic Foundations of Fuzzy Reasoning. Inform. Sci. 117, 47–88 (1999)
14. Wang, G.J.: Non-Classical Mathematical Logic and Approximate Reasoning. Science in China Press, Beijing (2003) (in chinese)
15. Zadeh, L.A.: Fuzzy Sets. Information Control 8, 338–353 (1965)

# Completely Compact Elements and Atoms of Rough Sets

Gaolin Li and Luoshan Xu

**Abstract.** Rough set theory established by Pawlak in 1982 plays an important role in dealing with uncertain information and to some extent overlaps fuzzy set theory. The key notions of rough set theory are approximation spaces of pairs $(U, R)$ with $R$ being an equivalence relation on $U$ and approximation operators $\underline{R}$ and $\overline{R}$. Let $\mathcal{R}$ be the family $\{(\underline{R}X, \overline{R}X) | X \subseteq U\}$ of approximations endowed with the pointwise order of set-inclusion. It is known that $\mathcal{R}$ is a complete Stone lattice with atoms and is isomorphic to the family of rough sets in the approximation space $(U, R)$. This paper is devoted to investigate algebraicity and completely distributivity of $\mathcal{R}$ from the view of domain theory. To this end, completely compact elements, compact elements and atoms of $\mathcal{R}$ are represented. In terms of the representations established in this paper, it is proved that $\mathcal{R}$ is isomorphic to a complete ring of sets, consequently $\mathcal{R}$ is a completely distributive algebraic lattice. An example is given to show that $\mathcal{R}$ is not atomic nor Boolean in general. Further, a sufficient and necessary condition for $\mathcal{R}$ being atomic is thus given.

**Keywords:** Rough set, Atom, Completely compact element, Complete ring of sets.

Gaolin Li
Department of Mathematics, Yangzhou University, Yangzhou 225002, China
Department of Mathematics, Yancheng Teachers College, Yancheng 224002, China
e-mail: ligaolin1981@126.com

Luoshan Xu
Department of Mathematics, Yangzhou University, Yangzhou 225002, China
e-mail: luoshanxu@hotmail.com

S. Li (Eds.): Nonlinear Maths for Uncertainty and its Appli., AISC 100, pp. 675–682.
springerlink.com                                     © Springer-Verlag Berlin Heidelberg 2011

# 1   Introduction

The rough set theory, proposed by Pawlak [7] and developed by numerous mathematicians and computer scientists, is fundamentally important in artificial intelligence and cognitive sciences. It has provide a more general framework to express common sense reasoning and uncertainty reasoning, and received wide attention on research areas in both of the real-life applications and the theory itself. Many important research topics in rough set theory such as various logics related to rough sets, connections between rough sets and fuzzy sets [11, 16, 17], probabilistic approaches to rough sets [8, 14, 15], and algebraic properties of rough sets [1, 3, 4, 5, 6, 9, 10, 12, 13] were presented in the literature.

The key notions of rough set theory are approximation spaces of pairs $(U, R)$ with $R$ being an equivalence relation on $U$ and approximation operators $\underline{R}$ and $\overline{R}$. Let $\mathcal{R}$ be the family $\{(\underline{R}X, \overline{R}X) | X \subseteq U\}$ of approximations endowed with the pointwise order of set-inclusion. It is known that $\mathcal{R}$ is a complete Stone lattice with atoms and is isomorphic to the family of rough sets in the approximation space $(U, R)$. This paper is devoted to investigate algebraicity and completely distributivity of the family of rough sets, or equivalently of the family $\mathcal{R}$ of approximations from the view of domain theory [2, 18]. In terms of some methods and technics of domain theory, completely compact elements and atoms in $\mathcal{R}$ are characterized. With these characterizations, we will prove that $\mathcal{R}$ is a molecular lattice in the sense that every element can be represented as unions of co-primes in $\mathcal{R}$, and that $\mathcal{R}$ is isomorphic to a complete ring of sets. Consequently, $\mathcal{R}$ is a completely distributive algebraic lattice. Besides, We will give an example to show that $\mathcal{R}$ is not atomic nor Boolean in general, and give a sufficient and necessary condition for $\mathcal{R}$ being atomic.

# 2   Preliminaries

In this section, we recall some fundamental notions and results of domain theory and rough set theory. Other used but not stated basic concepts and results please refer to [2, 7].

For a set $U$, a binary relation $R$ on $U$ is called
(1) reflexive if $xRx$ for all $x \in U$;
(2) transitive if $xRy$ and $yRz$ imply $xRz$ for all $x, y, z \in U$;
(3) symmetric if $xRy$ implies $yRx$ for all $x, y \in U$;
(4) antisymmetric if $xRy$ and $yRx$ imply $x = y$ for all $x, y \in U$.

A partially ordered set (poset, for short) is a nonempty set equipped with a partial order of a reflexive, transitive and antisymmetric relation on the set. We will use $\leqslant$ to denote a partial order. A poset $L$ is called a lattice, if for any $x, y \in L$, the least upper bound $x \vee y$ and the greatest lower bound $x \wedge y$

always exist. A poset $L$ is called a complete lattice if every subset $X \subseteq L$ has the least upper bound $\vee X$ and the greatest lower bound $\wedge X$.

**Definition 1.** ( [2,18]) Let $L$ be a poset.

(1) A subset $D$ of $L$ is called directed if it is nonempty and every finite subset of $D$ has an upper bound in $D$.

(2) For two element $x, y \in L$, we say that $x$ is way-below (resp., completely way-below) $y$, written as $x \ll y$ (resp., $x \lhd y$), if for all directed subsets (resp., subsets) $D \subseteq L$, $y \leqslant \vee D$ always implies the existence of $d \in D$ with $x \leqslant d$. The set of all the elements way-below (resp,.completely way-below) $y$ is denoted as $\Downarrow y$ (resp,. $\nabla y$).

(3) An element $k \in L$ is called compact (resp., completely compact) if $k \ll k$ (resp., $k \lhd k$). The set of all compact (resp., completely compact) elements of $L$ is denoted by $K(L)$ (resp., $CK(L)$).

(4) A lattice is called distributive if for all $x, y, z \in L$:

$$x \wedge (y \vee z) = (x \wedge y) \vee (x \wedge z), \quad x \vee (y \wedge z) = (x \vee y) \wedge (x \vee z).$$

(5) A complete lattice $L$ is called completely distributive if for any family $\{x_{i,j} | j \in J, k \in K(j)\}$ in $L$,

$$\wedge_{j \in J} \vee_{k \in K(j)} x_{j,k} = \vee_{f \in M} \wedge_{j \in J} x_{j,f(j)},$$

where $M$ is the set of choice functions defined on $J$ with $f(j) \in K(j)$.

(6) A complete lattice $L$ is called algebraic if for all $x \in L$,

$$x = \vee \{y \in K(L) | y \leqslant x\}.$$

(7) A family $\mathcal{F}$ of subsets of a set is called a complete ring of sets if it is closed under arbitrary intersections and unions.

Let $L$ be a lattice with the small element 0. An element $x^*$ is called a pseudocomplement of $x$ if $x \wedge x^* = 0$ and $x \wedge a = 0$ implies $a \leqslant x^*$. A lattice is called pseudocomplemented if every element has a pseudocomplement. If a lattice $L$ with 0 and 1, the greatest element of $L$, is distributive, pseudo-complemented and satisfies the Stone identity $x^* \vee x^{**} = 1$ for any element $x \in L$, then $L$ is called a Stone lattice. Obviously, every Boolean lattice is a Stone lattice.

For a set $U$, $X \subseteq U$, we write $2^U$ to denote the power set of $U$, and $X^c$ to denoted the complement of $X$ in $U$. If a relation $R$ on $U$ is reflexive, symmetric and transitive, then $R$ is called an equivalence. We use $[x]_R$ or $[x]$ to denote an equivalence class of $R$ containing $x$. For any set $X \subseteq U$, the lower and upper approximations of $X$ are defined respectively as

$$\underline{R}X = \{x \in U | [x]_R \subseteq X\}, \quad \overline{R}X = \{x \in U | [x]_R \cap X \neq \emptyset\}.$$

Lower and upper approximation sets have the following properties.

**Lemma 1.** [7] *Let $(U, R)$ be an approximation space. For all $A, B \subseteq U$ we have*

(1) $\underline{R}A \subseteq A \subseteq \overline{R}A$;

(2) $\underline{R}\emptyset = \overline{R}\emptyset = \emptyset, \underline{R}U = \overline{R}U = U$;

(3) $A \subseteq B \Rightarrow \underline{R}A \subseteq \underline{R}B, \overline{R}A \subseteq \overline{R}B$;

(4) $\overline{R}(A \cup B) = \overline{R}(A) \cup \overline{R}(B), \underline{R}(A \cap B) = \underline{R}(A) \cap \underline{R}(B)$;

(5) $\underline{R}(A^c) = (\overline{R}A)^c, \overline{R}(A^c) = (\underline{R}A)^c$;

(6) $\underline{R}(\underline{R}A) = \overline{R}(\underline{R}A) = \underline{R}A, \overline{R}(\overline{R}A) = \underline{R}(\overline{R}A) = \overline{R}A$.

In [4], the pair $(\underline{R}X, \overline{R}X)$ is called the approximation of $X$, the family of all approximations is denoted by $\mathcal{R}$, i.e., $\mathcal{R} = \{(\underline{R}X, \overline{R}X) | X \subseteq U\}$. The set $\mathcal{R}$ of approximations is ordered by

$$(\underline{R}X, \overline{R}X) \leqslant (\underline{R}Y, \overline{R}Y) \Leftrightarrow \underline{R}X \subseteq \underline{R}Y \, and \, \overline{R}X \subseteq \overline{R}Y.$$

Two subsets of $U$ are said to be roughly equivalent, denoted by $X \equiv Y$, if $\underline{R}X = \underline{R}Y$ and $\overline{R}X = \overline{R}Y$. The equivalence classes of the relation $\equiv$ on $2^U$ are called rough sets. The family of all rough sets of $(U, R)$ is denoted by $\mathcal{R}^*$, i.e., $\mathcal{R}^* = \{[X]_\equiv | X \subseteq U\}$. The set $\mathcal{R}^*$ of rough sets is ordered by

$$[X]_\equiv \leqslant [Y]_\equiv \Leftrightarrow \underline{R}X \subseteq \underline{R}Y \, and \, \overline{R}X \subseteq \overline{R}Y.$$

It is well known that posets $(\mathcal{R}, \leqslant)$ and $(\mathcal{R}^*, \leqslant)$ are isomorphic. So each approximation uniquely determines a rough set.

**Lemma 2.** [19] *Let $\mathcal{R}$ be the family of all approximations of $(U, R)$, $S = \{x \in U | Card[x] = 1\}$. Then for any $(X, Y) \in 2^U \times 2^U$, $(X, Y) \in \mathcal{R}$ iff $X \subseteq Y$ and $(Y - X) \cap S = \emptyset$.*

**Lemma 3.** [4] *Let $\mathcal{R}$ be the family of all approximations of $(U, R)$. Then $(\mathcal{R}, \leqslant)$ is a complete Stone lattice such that*

$$\vee_{i \in I}(\underline{R}X_i, \overline{R}X_i) = (\cup_{i \in I}\underline{R}X_i, \cup_{i \in I}\overline{R}X_i),$$

$$\wedge_{i \in I}(\underline{R}X, \overline{R}X) = (\cap_{i \in I}\underline{R}X, \cap_{i \in I}\overline{R}X),$$

$$(\underline{R}X, \overline{R}X)^* = ((\overline{R}X)^c, (\underline{R}X)^c),$$

*where $(\underline{R}X, \overline{R}X)^*$ is the pseudocomplement of $(\underline{R}X, \overline{R}X)$.*

## 3 Completely Compact Elements in $\mathcal{R}$

The following two propositions show that there are enough completely compact elements in $\mathcal{R}$.

**Proposition 1.** *Let $\mathcal{R}$ be the family of all approximations of $(U, R)$ and $A \subseteq U$. If there exists $x \in U$ such that $A \subseteq [x]_R$, then $(\underline{R}A, \overline{R}A)$ is a completely compact element, i.e., $(\underline{R}A, \overline{R}A) \in CK(\mathcal{R})$.*

*Proof.* Let $\{(\underline{R}X_i, \overline{R}X_i)|i \in I\} \subseteq \mathcal{R}$ with $(\underline{R}A, \overline{R}A) \leqslant \vee_{i \in I}(\underline{R}X_i, \overline{R}X_i)$. We divide the proof into three cases.

(1) If $A = \emptyset$, then $(\underline{R}A, \overline{R}A) = (\emptyset, \emptyset)$ and $(\underline{R}A, \overline{R}A) \leqslant (\underline{R}X_i, \overline{R}X_i)$ for all $i \in I$.

(2) If $A \neq \emptyset$ and $A \subset [x]$, then $(\underline{R}A, \overline{R}A) = (\emptyset, [x]) \leqslant (\cup_{i \in I}\underline{R}X_i, \cup_{i \in I}\overline{R}X_i)$, thus there exists $j \in I$ such that $[x] \subseteq \overline{R}X_j)$, hence $(\underline{R}A, \overline{R}A) \leqslant (\underline{R}X_j, \overline{R}X_j)$.

(3) If $A = [x]$, then $(\underline{R}A, \overline{R}A) = ([x], [x]) \leqslant (\cup_{i \in I}\underline{R}X_i, \cup_{i \in I}\overline{R}X_i)$, thus there is $k \in I$ such that $[x] \subseteq \underline{R}X_k \subseteq \overline{R}X_k$ and $(\underline{R}A, \overline{R}A) \leqslant (\underline{R}X_k, \overline{R}X_k)$.

To sum up, there is some $i \in I$ such that $(\underline{R}A, \overline{R}A) \leqslant (\underline{R}X_i, \overline{R}X_i)$ and by Definition 1 (3), $(\underline{R}A, \overline{R}A)$ is a completely compact element. $\square$

**Proposition 2.** *Every element of $\mathcal{R}$ is a join of a set of completely compact elements.*

*Proof.* Let $(\underline{R}X, \overline{R}X) \in \mathcal{R}$. Then it follows from Lemma 2 that $(\underline{R}X, \underline{R}X)$ and $(\emptyset, \overline{R}X - \underline{R}X)$ are both in $\mathcal{R}$. Furthermore, we have

$$(\underline{R}X, \overline{R}X) = (\underline{R}X, \underline{R}X) \vee (\emptyset, \overline{R}X - \underline{R}X)$$
$$= (\cup_{x \in \underline{R}X}[x], \cup_{x \in \underline{R}X}[x]) \vee (\cup_{x \in \overline{R}X - \underline{R}X}R\{x\}, \cup_{x \in \overline{R}X - \underline{R}X}\overline{R}\{x\})$$
$$= (\vee\{([x], [x])|x \in \underline{R}X\}) \vee (\vee\{(\underline{R}\{x\}, \overline{R}\{x\})|x \in \overline{R}X - \underline{R}X\}).$$

So, we have that $(\underline{R}X, \overline{R}X)$ is a join of some completely compact elements of $\mathcal{R}$ by Proposition 1. $\square$

Next proposition characterizes the completely compact elements of $\mathcal{R}$.

**Proposition 3.** *Let $\mathcal{R}$ be the family of approximations of $(U, R)$ and $A \subseteq U$. Then $(\underline{R}A, \overline{R}A)$ is a completely compact element iff there is $x \in U$ such that $A \subseteq [x]_R$, that is*

$$CK(\mathcal{R}) = \{(\underline{R}A, \overline{R}A)|A \subseteq [x] \text{ for some } x \in U\}.$$

*Proof.* It follows from Proposition 1 that

$$\{(\underline{R}A, \overline{R}A)|A \subseteq [x] \text{ for some } x \in U\} \subseteq CK(\mathcal{R}).$$

Conversely, for any $(\underline{R}X, \overline{R}X) \in CK(\mathcal{R})$, by the proof of Proposition 2, there is $x \in U$ such that $X \subseteq \overline{R}X \subseteq [x]$, thus

$$CK(\mathcal{R}) \subseteq \{(\underline{R}A, \overline{R}A)|A \subseteq [x] \text{ for some } x \in U\}.$$

$\square$

**Theorem 1.** *The family $\mathcal{R}$ is isomorphic to a completely ring of sets.*

*Proof.* Define $\varphi : \mathcal{R} \to 2^{CK(\mathcal{R})}$ such that for all $x \in \mathcal{R}$, $\varphi(x) = \{y \in CK(\mathcal{R})|y \leqslant x\}$. Then by Proposition 2, for any $x, y \in \mathcal{R}$, $x \neq y \Rightarrow \varphi(x) \neq \varphi(y)$. Furthermore, for any $X \subseteq \mathcal{R}$, we have

$$\varphi(\vee X) = \{y \in CK(\mathcal{R})|y \leqslant \vee X\} = \cup_{x \in X}\{y \in CK(\mathcal{R})|y \leqslant x\} = \cup_{x \in X}\varphi(x),$$

$$\varphi(\wedge X) = \{y \in CK(\mathcal{R})|y \leqslant \wedge X\} = \cap_{x \in X}\{y \in CK(\mathcal{R})|y \leqslant x\} = \cap_{x \in X}\varphi(x).$$

This reveals that $\varphi$ is a map into $2^{CK(\mathcal{R})}$ preserving arbitrary sups and infs. So, the image $\varphi(\mathcal{R})$ of $\varphi$ is a complete ring of sets. It is obvious that $\varphi : \mathcal{R} \to \varphi(\mathcal{R})$ is an isomorphism.                                                                                    □

It is well known that every complete ring of sets is a completely distributive lattice. Furthermore, it follows from [2, Corollary I-4.14] that every complete ring of sets is algebraic. So we have the following corollary.

**Corollary 1.** *The family $\mathcal{R}$ is a completely distributive algebraic lattice.*

## 4   Atoms in $\mathcal{R}$

In section 3, it has been shown that there are enough completely compact elements in $\mathcal{R}$, therefor $\mathcal{R}$ is a molecular lattice in the sense of that every element can be represented as unions of co-primes. However, in this section, we shall show that $\mathcal{R}$ is not necessary atomic. The following proposition characterizes the atoms of $\mathcal{R}$.

**Proposition 4.** *Let Atom $(\mathcal{R})$ be the set of all atoms of $\mathcal{R}$. Then*

$$Atom(\mathcal{R}) = \{(\underline{R}\{x\}, \overline{R}\{x\})|x \in U\}.$$

*Proof.* For every $x \in U$, since $\overline{R}\{x\} = [x]$, we have $(\emptyset, \emptyset) < (\underline{R}\{x\}, \overline{R}\{x\})$. Suppose that $(\underline{R}Y, \overline{R}Y) \in \mathcal{R}$ such that $(\emptyset, \emptyset) < (\underline{R}Y, \overline{R}Y) \leqslant (\underline{R}\{x\}, \overline{R}\{x\})$. Then we have $Y \neq \emptyset$ and for every $y \in Y$, $[y] \subseteq \overline{R}Y \subseteq \overline{R}\{x\} = [x]$, thus $\overline{R}Y = \overline{R}\{x\}$. Next, we prove $\underline{R}Y = \underline{R}\{x\}$. If $Card[x] = 1$, then $Y = [x] = \{x\}$, thus $\underline{R}Y = [x] = \underline{R}\{x\}$. If $card[x] > 1$, then $\underline{R}Y \subseteq \underline{R}\{x\} = \emptyset$. Thus $(\underline{R}Y, \overline{R}Y) = (\underline{R}\{x\}, \overline{R}\{x\})$ and we have $\{(\underline{R}\{x\}, \overline{R}\{x\})|x \in U\} \subseteq Atom(\mathcal{R})$.

Conversely, suppose that $(\underline{R}Z, \overline{R}Z) \in Atom(\mathcal{R})$, then $Z \neq \emptyset$ and for every $z \in Z$, $(\emptyset, \emptyset) < (\underline{R}\{z\}, \overline{R}\{z\}) \leqslant (\underline{R}Z, \overline{R}Z)$. Since $(\underline{R}Z, \overline{R}Z) \in Atom(\mathcal{R})$, we have $(\underline{R}Z, \overline{R}Z) = (\underline{R}\{z\}, \overline{R}\{z\})$. So $Atom(\mathcal{R}) \subseteq \{(\underline{R}\{x\}, \overline{R}\{x\})|x \in U\}$.                □

**Corollary 2.** *For the family of approximations $\mathcal{R}$, we have that*

$$Atom(\mathcal{R}) \subseteq CK(\mathcal{R}) \subseteq K(\mathcal{R}).$$

*Proof.* It follows from Proposition 3, Proposition 4 and Definition 1 (3).   □

The following example shows that although $\mathcal{R}$ is a Stone lattice with atoms, $\mathcal{R}$ is not necessary atomic nor Boolean.

*Example 1.* Let $U = \{0, 1, 2\}$, $U/R = \{\{0, 1\}, \{2\}\}$. Then

$$\mathcal{R} = \{(\emptyset, \emptyset), (\emptyset, \{0, 1\}), (\{0, 1\}, \{0, 1\}), (\{2\}, \{2\}), (\{2\}, U), (U, U)\}.$$

$\mathcal{R}$ is presented in Fig.1. It is easy to see that $Atom(\mathcal{R}) = \{(\emptyset, \{0,1\}), (\{2\}, \{2\})\}$ and that $(\{0,1\}, \{0,1\})$ is not the supremum of any subsets of $Atom(\mathcal{R})$. Thus $\mathcal{R}$ is not atomic.

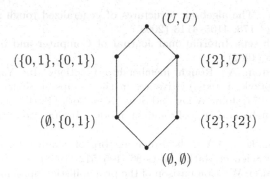

**Fig. 1.**

Now, we give a necessary and sufficient condition under which $\mathcal{R}$ is atomic.

**Proposition 5.** $\mathcal{R}$ *is atomic iff the relation $R$ is an identity.*

*Proof.* $\Leftarrow$: If $R$ is identical, that is for all $x, y \in U$, $xRy \Leftrightarrow x = y$, then it is easy to see that $\mathcal{R} \cong 2^U$ is an atomic Boolean lattice.

$\Rightarrow$: Suppose that $\mathcal{R}$ is atomic. Let $S = \{x \in U | Card[x] = 1\}$. Then $(U, U) = \vee_{x \in U}(\underline{R}\{x\}, \overline{R}\{x\}) = (S, U)$ and $S = U$, showing that for all $x \in U$, $[x] = \{x\}$. So $R$ is the identity.                                         □

By Proposition 5, we see that normally the family $\mathcal{R}$ is a Stone lattice with atoms, but not atomic nor Boolean.

**Acknowledgements.** This work is supported by the NSF of China (61074129), Jiangsu province of China (BK2010313) and Yancheng Teachers College (07YCKL070).

# References

1. Banerjee, M., Chakraborty, M.K.: Algebras from rough sets. In: Pal, S.K., Polkowski, L., Skowron, A. (eds.) Rough-Neural Computing: Techniques for Computing with Words, pp. 157–184. Springer, Berlin (2004)
2. Gierz, G., Hofmann, K., Keimel, K., Lawson, J.D., Mislove, M., Scott, D.: Continuous Lattices and Domains. Cambridge University Press, Cambridge (2003)
3. Järvinen, J.: On the structure of rough approximations. Fundamenta Informaticae 53, 135–153 (2002)
4. Järvinen, J.: The ordered set of rough sets. In: Tsumoto, S., Słowiński, R., Komorowski, J., Grzymała-Busse, J.W. (eds.) RSCTC 2004. LNCS (LNAI), vol. 3066, pp. 49–58. Springer, Heidelberg (2004)

5. Järvinen, J.: Lattice theory for rough sets. In: Peters, J.F., Skowron, A., Düntsch, I., Grzymała-Busse, J.W., Orłowska, E., Polkowski, L. (eds.) Transactions on Rough Sets VI. LNCS, vol. 4374, pp. 400–498. Springer, Heidelberg (2007)
6. Liu, G.L., Zhu, W.: The algebraic structures of generalized rough set theory. Information Science 178, 4105–4113 (2008)
7. Pawlak, Z.: Rough sets. International Journal of Computer and information Science 5, 341–356 (1982)
8. Pawlak, Z., Skowron, A.: Rough membership functions. In: Yager, R.R., Fedrizzi, M., Kacprzyk, J. (eds.) Advances in the Dempster-Shafer Theory of Evidence, pp. 251–271. John Wiley and Sons, New York (1994)
9. Pei, D.W.: On definable concepts of rough set models. Information Sciences 177, 4230–4239 (2007)
10. Pomykala, J., Pomykala, J.A.: The Stone algebra of rough sets. Bulletin of polish Academy of Science: Mathematics 36, 495–512 (1988)
11. Wong, S.K.M., Ziarko, W.: Comparison of the probabilistic approximate calssification and fuzzy set model. Fuzzy Sets and Systems 21, 357–362 (1987)
12. Yang, L.Y., Xu, L.S.: Algebraic aspects of generalized approximation spaces. International Journal of Approximate Reasoning 51, 151–161 (2009)
13. Yao, Y.Y.: Two views of the theory of rough sets in finite universe. International Journal of Approximate Reasoning 15, 291–317 (1996)
14. Yao, Y.Y.: Probabilistic rough set approximations. International Journal of Approximate Reasoning 49, 255–271 (2008)
15. Yao, Y.Y., Wong, S.K.M.: A decision theoretic framework for approximating concepts. International Journal of Man-Machine Studies 37, 793–809 (1992)
16. Zadeh, L.A.: Fuzzy sets. Information and Control 8, 338–353 (1965)
17. Zadeh, L.A.: Toward a generalized theory of uncertainty(GTU)-an outline. Information Science 17, 21–40 (2005)
18. Zeng, C.Y., Fan, L., Cui, H.B.: Introduction to Frames and Continuous Lattice. Cptial Normal University Press, Beijing (2000) (in chinese)
19. Zhang, W.X., Wu, W.Z., Liang, J.Y., Li, D.Y.: Rough Sets Theory and Methods. Science Press, Beijing (2001) (in chinese)

# Weak Approximable Concepts and Completely Algebraic Lattices

Hongping Liu, Qingguo Li, and Lankun Guo

**Abstract.** The notion of a weak approximable concept is introduced in this paper, and the lattice consisting of weak approximable concepts is investigated. It is shown that this concept lattice is a completely algebraic lattice, and each completely algebraic lattice is isomorphic to such a concept lattice.

**Keywords:** Completely compact, Completely algebraic lattice, Formal context, Weak approximable concept.

## 1 Introduction

Formal Concept Analysis (FCA) is a powerful tool for knowledge discovery, information retrieval and data analysis (see [2,4]). In classical FCA, the basis is a formal context, a formal concept and a concept lattice. A formal context is commonly known as two universes with a binary relation between them, and the knowledge is hidden in the formal context. In order to discovery it, the notion of a formal concept is proposed (see [1,3]), the cheerful result is that the concept lattice is a complete lattice and each complete lattice is isomorphic to a concept lattice. Since the definition of a formal concept is too strict for excluding some important information, in [11], Zhang introduced the notion of an approximable concept. It was shown that every approximable concept lattice is an algebraic lattice and each algebraic lattice is isomorphic to such a approximable concept lattice. After that, Hitzler and Krötzsch deeply investigated them on the viewpoint of category ( [6]). Follow Zhang's idea, Lei and Luo developed the representation theory of algebraic lattices by means of the definition of a rough approximable concept in [7].

Hongping Liu, Qingguo Li, and Lankun Guo
College of Mathematics and Econometrics, Hunan University, Changsha, Hunan, 410082, P.R. China
e-mail: hpliu@hnu.edu.cn, liqingguoli@yahoo.com.cn, lankun.guo@gmail.com

S. Li (Eds.): Nonlinear Maths for Uncertainty and its Appli., AISC 100, pp. 683–689.
springerlink.com                                    © Springer-Verlag Berlin Heidelberg 2011

In this paper, we extend the approximable concept with another approximable manner to a formal concept, and pose the notion of a weak approximable concept. Then we study the lattice consisting of weak approximable concepts, and show that it is a special algebraic lattice here called completely algebraic lattice. Moreover, each completely algebraic lattice is isomorphic to such a weak approximable concept lattice. So the representation theorem of completely algebraic lattices is obtained.

The paper is organized as follows. Section 2 lists some necessary definitions and results needed later on. In Section 3, we propose the notion of a weak approximable attribute concept, prove the representation theorem of completely algebraic lattices. In Section 4, we define a weak approximable object concept, and discuss the relationship between the two weak approximable concepts.

## 2 Preliminaries

In this section, we recall some important concepts and well-known results in domains and lattices theory, and also review the basic theory of formal concept analysis.

### 2.1 Domains and Lattices Theory

Let $(P, \leqslant)$ be a poset, a subset $D \subseteq P$ is called directed if any finite set $F \subseteq P$ there is an element $d \in D$ such that $\forall f \in F$, $f \leqslant d$. If $\bigvee D$ exists for any directed set $D \subseteq P$, then $(P, \leqslant)$ is said to be a dcpo, if $\bigvee S$ exists for any subset $S \subseteq P$, then $(P, \leqslant)$ is said to be a complete lattice. $\forall x, y \in P$, we say that $x$ is way below $y$, in symbols $x \ll y$, if for every directed subset $D \subseteq P$, $y \leqslant \bigvee D$ implies $x \leqslant d$ for some $d \in D$. An element satisfying $x \ll x$ is said to be compact. The set of compact elements of $P$ is denoted as $K(P)$. In a poset $(P, \leqslant)$, for any $x \in P$, $\downarrow x$ denotes the set $\{y \in P : y \leqslant x\}$. Unless otherwise stated, $(L, \leqslant)$ always denotes a complete lattice in this paper.

**Definition 1.** [1,3,5,10] Let $(L, \leqslant)$ be a complete lattice and $x, y \in L$. We say that $x$ is completely way below $y$, in symbols $x \lhd y$, if for every subset $S \subseteq L$, $y \leqslant \bigvee S$ implies $x \leqslant s$ for some $s \in S$.

An element satisfying $x \lhd x$ is said to be completely compact. The set of completely compact elements of $L$ is denoted as $CK(L)$.

**Definition 2.** [10] A complete lattice $(L, \leqslant)$ is called a completely algebraic lattice if, for each $a \in L$,

$$a = \bigvee \{k \in CK(L) : k \leqslant a\} = \bigvee (\downarrow a \cap CK(L)).$$

*Example 1.* Let $L = 2^X$ for some non-empty set $X$. Then $(L, \subseteq)$ is a completely algebraic lattice, and $CK(L) = \{\{x\} : x \in X\}$.

**Theorem 1.** *[5, 8, 9] Let $(L, \leqslant)$ be a complete lattice. Then the following conditions are equivalent:*
*(1) $(L, \leqslant)$ is a completely distributive lattice,*
*(2) each $a \in L$ has a minimal family,*
*(3) each $a \in L$ has a maximal family,*
*(4) $\forall a \in L, a = \bigvee \{x \in L : x \lhd a\}.$*

**Lemma 1.** *[10] Let $(L, \leqslant)$ be a complete lattice. Then $(L, \leqslant)$ is a completely algebraic lattice iff it is algebraic and infinitely distributive.*

From definitions and the above lemma , we can obtain the following theorem immediately.

**Theorem 2.** *Let $(L, \leqslant)$ be a complete lattice. Then $(L, \leqslant)$ is a completely algebraic lattice iff it is algebraic and completely distributive.*

*Proof.* Trivial. □

## 2.2 Formal Concepts and Approximable Concepts

In FCA, a formal context $P$ is a triple $(U, V, R)$, where $U$ is a set of objects, $V$ is a set of attributes, and $R$ is a binary relation between $U$ and $V$ with $xRy$ means "object x has attribute y".

**Definition 3.** *[1, 3] Let $P = (U, V, R)$ be a formal context. Define two operators as follows:*

$$\alpha : 2^U \longrightarrow 2^V, A \longrightarrow \{b \in V : \forall a \in A, aRb\},$$

$$\omega : 2^V \longrightarrow 2^U, B \longrightarrow \{a \in U : \forall b \in B, aRb\}.$$

This pair $(\alpha, \omega)$ plays an important role in FCA [3]. A pair of sets $(A, B)$ is called a (formal) concept, if $A \subseteq U$, $B \subseteq V$, $\alpha(A) = B$ and $\omega(B) = A$. $A$ is the extent and $B$ is the intent of $(A, B)$. We use $\mathcal{L}P$ to denote the set of all concepts of $P = (U, V, R)$, and order it by $(A_1, B_1) \leqslant (A_2, B_2) \Leftrightarrow A_1 \subseteq A_2 \Leftrightarrow B_2 \subseteq B_1$. Then it is easy to check that $(\mathcal{L}P, \subseteq)$ is a lattice called concept lattice of $P$.

For any concept $(A, B)$, the extent and the intent are determined by each other, so we can only choose the extent $A$ (the intent $B$) called the extent concept (the intent concept) or just concept for simplicity when considering a concept $(A, B)$.

**Theorem 3.** *[1, 3] Let $P = (U, V, R)$ be a formal context. Then the concept lattice $(\mathcal{L}P, \subseteq)$ is a complete lattice with*

$$\bigwedge_{i \in I}(A_i, B_i) = (\bigcap_{i \in I} A_i, \alpha(\omega(\bigcup_{i \in I} B_i))), \bigvee_{i \in I}(A_i, B_i) = (\omega(\alpha(\bigcup_{i \in I} A_i)), \bigcap_{i \in I} B_i).$$

**Theorem 4.** *[1,3] For every complete lattice $(L, \leqslant)$, there is a formal context $P_L$ such that $L$ is order-isomorphic to $\mathcal{L}P_L$.*

**Definition 4.** [11] Let $P = (U, V, R)$ be a formal context. $Y \subseteq V$ is called an approximable (attribute) concept if for every finite subset $F \subseteq Y$, we have $\alpha(\omega(F)) \subseteq Y$.

**Theorem 5.** *[11] Let $P = (U, V, R)$ be a formal context. Then the set of its approximable concepts with inclusion order forms an algebraic lattice. Conversely, for every algebraic lattice $L$, there is a formal context $P = (U, V, R)$ such that $L$ is isomorphic to the lattice consisting of its approximable concepts under inclusion order.*

## 3   Weak Approximable Attribute Concepts

In this section, we introduce a new approximable concept, and show that this special approximable concept lattice is in one-to-one correspondence with a completely algebraic lattice.

**Definition 5.** Let $P = (U, V, R)$ be a formal context and $Y \subseteq V$ a set of attributes. If $\forall y \in Y$, $\alpha(\omega(y)) \subseteq Y$. Then $Y$ is called a weak approximable attribute concept of $P$.

We use $\mathcal{A}P$ and sometimes $\mathcal{A}$ for simplicity to denote the set of all weak approximable attribute concepts of $P$.

By Definition 5, it is easy to see that a formal concept is an approximable attribute concept, and an approximable attribute concept is a weak approximable attribute concept. But the converse of this may not hold. Let us consider the following example and Example 4.7 in [11].

*Example 2.* Let $P = (U, V, R)$ be a formal context, where $U = \{x, y, z\}$, $V = \{a, b, c\}$, and the information presented in Table 1.

**Table 1** Formal context $P = (U, V, R)$

| elements | a | b | c |
|----------|---|---|---|
| x        | 1 | 0 | 0 |
| y        | 1 | 1 | 1 |
| z        | 0 | 1 | 1 |

Let $Y = \{a, b\} \subseteq V$. Then $\alpha(\omega(a)) = \{a\} \subseteq Y$, and $\alpha(\omega(b)) = \{b\} \subseteq Y$, so $Y$ is a weak approximable attribute concept. But $Y$ is not an approximable attribute concept since $\alpha(\omega(Y)) = \{a, b, c\} \nsubseteq Y$.

**Theorem 6.** *For any formal context $P = (U, V, R)$, the set of its weak approximable attribute concepts $\mathcal{AP}$ forms a completely algebraic lattice under $\subseteq$.*

*Proof.* First, we check that $(\mathcal{AP}, \subseteq)$ is a complete lattice, it suffices to show that $\mathcal{AP}$ is a closure system. Obviously, $\emptyset \in \mathcal{AP}$ and $V \in \mathcal{AP}$ are the least element and the greatest element of $\mathcal{AP}$. Given any subset $\mathscr{F} \subseteq \mathcal{AP}$, and $\forall f \in \bigcap \mathscr{F}$. Then $f \in F$ for each $F \in \mathscr{F}$. Since $F$ is a weak approximable attribute concept, we have $\alpha(\omega(f)) \subseteq F$ for each $F \in \mathscr{F}$. Thus $\alpha(\omega(f)) \subseteq \bigcap \mathscr{F}$. So $\bigcap \mathscr{F} \in \mathcal{AP}$ and it implies $\mathcal{AP}$ is a closure system.

Second, we show that each $Y \in \mathcal{AP}$, $Y = \bigvee \{ \alpha\omega(y) : y \in Y \}$ and $\alpha\omega(y)$ is a completely compact element of $\mathcal{AP}$. Above all, $\forall \mathscr{F} \in \mathcal{AP}$, we have $\bigcup \mathscr{F} \in \mathcal{AP}$ which ensures $\bigvee \mathscr{F} = \bigcup \mathscr{F}$ (Since $\forall x \in \bigcup \mathscr{F}$, there exists some $F \in \mathscr{F}$ such that $x \in F$, so $\alpha\omega(x) \subseteq F \subseteq \bigcup \mathscr{F}$). For $\forall y \in Y \in \mathcal{AP}$, we can easily see that $\alpha\omega(y) \in \mathcal{AP}$, it implies that $\bigcap \{ \alpha\omega(y) : y \in Y \} \in \mathcal{AP}$ and $\bigvee \{ \alpha\omega(y) : y \in Y \} = \bigcup \{ \alpha\omega(y) : y \in Y \}$. It is easy to check that $Y = \bigcup \{ \alpha\omega(y) : y \in Y \}$. Therefore, $Y = \bigvee \{ \alpha\omega(y) : y \in Y \}$. Moreover, $\forall \mathscr{F} \in \mathcal{AP}$ and $\alpha\omega(y) \subseteq \bigvee \mathscr{F}$, we have $y \in \alpha\omega(y) \subseteq \bigcup \mathscr{F}$. Then there exists $F_0 \in \mathscr{F}$ such that $y \in F_0$. Thus $\alpha\omega(y) \subseteq F_0$ since $F_0$ is a weak approximable attribute concept. Hence, $\alpha\omega(y)$ is a completely compact element of $\mathcal{AP}$. Therefore, $(\mathcal{AP}, \subseteq)$ is a completely algebraic lattice by Definition 2.         $\square$

**Theorem 7.** *(Representation Theorem slowromancapi@) For every completely algebraic lattice $(L, \leqslant)$, there is a formal context $P = (U, V, R)$ such that $(L, \leqslant)$ is isomorphic to $(\mathcal{AP}, \subseteq)$.*

*Proof.* Suppose $(L, \leqslant)$ is a completely algebraic lattice. Construct a formal context $P = (U, V, R)$, with $U = L$ and $V = CK(L)$, where $aRb$ iff $b \leqslant a$. We want to show $(\mathcal{AP}, \subseteq)$ is isomorphic to $(L, \leqslant)$.

Firstly note that $\forall y \in V$,

$$
\begin{aligned}
\alpha\omega(y) &= \{ b \in CK(L) : \omega(y) \subseteq Rb \} \\
&= \{ b \in CK(L) :\uparrow y \subseteq \uparrow b \} \\
&= \{ b \in CK(L) : b \leqslant y \} \\
&= \downarrow y \cap CK(L).
\end{aligned}
$$

Hence, for each $Y \in \mathcal{A}$, we have

$$
\begin{aligned}
Y &= \bigvee \{ \alpha\omega(y) : y \in Y \} \\
&= \bigcup_{y \in Y} (\downarrow y \cap CK(L)) \\
&= (\bigcup_{y \in Y} \downarrow y) \cap CK(L) \\
&= \downarrow (\bigvee Y) \cap CK(L).
\end{aligned}
$$

Thus, we can define a mapping $f : (L, \leqslant) \longrightarrow (\mathcal{A}, \subseteq)$ as $f(x) = \downarrow x \cap CK(L)$ for each $x \in L$. From the above analysis, it is easy to check that $f$ is a bijective homomorphism such that $f(a \vee b) = f(a) \vee f(b)$ and $f(a \wedge b) = f(a) \wedge f(b)$. This implies $f$ is a isomorphism and $(L, \leqslant)$ is isomorphic to $(\mathcal{AP}, \subseteq)$.   $\square$

## 4  Weak Approximable Object Concepts

**Definition 6.** Let $P = (U, V, R)$ be a formal context and $X \subseteq U$ a set of objects. if $\forall x \in X$, $\omega(\alpha(x)) \subseteq X$. Then $X$ is called a weak approximable object concept of $P$.

We use $\mathcal{OP}$ and sometimes $\mathcal{O}$ for simplicity to denote the set of all weak approximable object concepts of $P$.

By symmetry and the analysis in Section 3, the following lemma holds immediately.

**Lemma 2.** *For a formal context $P = (U, V, R)$, we have*
*1 $(\mathcal{OP}, \subseteq)$ is a completely algebraic lattice,*
*2 for each completely algebraic lattice $L$, there is a a formal context $P = (U, V, R)$ such that $L$ is isomorphic to $\mathcal{OP}$,*
*3 for any subset $Y \subseteq V$, $\omega(Y)$ is a weak approximable object concept,*
*4 for any subset $X \subseteq U$, $\alpha(X)$ is a weak approximable attribute concept.*

The relationship between $(\mathcal{AP}, \subseteq)$ and $(\mathcal{OP}, \subseteq)$ is stated in the following theorem.

**Theorem 8.** *With respect to a formal context $P = (U, V, R)$, the mappings $\omega : \mathcal{AP} \longrightarrow \mathcal{OP}$ and $\alpha : \mathcal{OP} \longrightarrow \mathcal{AP}$ give rise to a Galois connection.*

*Proof.* It is trivial.   $\square$

## References

1. Davey, B.A., Priestley, H.A.: Introduction to Lattices and Order, 2nd edn. Cambridge University, Cambridge (2002)
2. Düntsch, I., Gegida, G.: Approximable Operators in Qualitative Data Analysis. In: Theory and Application of Relational Structures as Knowledge Instruments. Spinger, Heidelberg (2003)
3. Ganter, B., Wille, R.: Formal Concept Analysis. Springer, Berlin (1999)
4. Gegida, G., Düntsch, I.: Modal-style operators in qualitative data analysis. In: Proceedings of the 2002 IEEE International Conference on Data Mining, pp. 155–162 (2002)
5. Gierz, G., Hofmann, K.H., Keimel, K., Lawson, J.D., Mislove, M., Scott, D.S.: Continuous Lattices and Domains. Cambridge University, Cambridge (2003)
6. Hitzler, P., Krötzsch, M.: A categorical view on algebraic lattices in formal concept analysis. Fundamenta Informaticae 74, 1–29 (2006)
7. Lei, B., Luo, K.: Rough concept lattices and domains. Annals of Pure and Applied Logic 159, 333–340 (2009)

8. Raney, G.N.: Completely distributive complete lattices. Proceedings of the Ameicican Mathematics Society 3, 599–677 (1952)
9. Raney, G.N.: A subdirect-union representation for completely distributive complete lattices. Proceedings of the Ameicican Mathematics Society 4, 518–522 (1953)
10. Winskel, G.: Event structures. Lectures Notes in Computer Science, vol. 255, pp. 325–392 (1987)
11. Zhang, Q., Shen, Q.: Approximable concepts, chu spaces, and information system. Theory and Application of Categories 17, 80–102 (2006)

8. Ranzato, F.: Complete distributive complete lattices. Proceedings of the American Mathematics Society, 3, 590–677 (1952).

9. Raney, G.N.: A subdirect-union representation for completely distributive complete lattices. Proceedings of the American Mathematics Society 4, 518–522 (1953)

10. Winskel, The byte structure. Lecture Notes in Computer Science, vol. 832, pp. 835–879 (1997)

11. Zhang, G., Shen, G.: Approximable concepts, Chu spaces, and information systems. Theory and Applied Categories 17, 80–102 (2006)

# Ideal-Convergence in Quantales

Shaohui Liang

**Abstract.** In this paper, some important properties of points in quantale are discussed. Based on which we constructed the convergence structure on quantale by ideal and point, and some important properties are obtained.

**Keywords:** Quantale, Point, Ideal, Quantale homomorphism.

## 1 Introduction

Quantale was proposed by C.J.Mulvey in 1986 for studying the foundations of quantum logic and for studying non-commutation C*-algebras. The term quantale was coined as a combination of "quantum logic" and "locale" by C.J.Mulvey in [18]. The systematic introduction of quantale theory came from the book [29] , which written by K.I.Rosenthal in 1990. Since quantale theory provides a powerful tool in studying noncommutative structures, it has a wide applications, especially in studying noncommutative C*-algebra theory [20], the ideal theory of commutative ring [21], linear logic [6] and so on. So, the quantale theory has aroused great interests of many scholar and experts, a great deal of new ideas and applications of quantale have been proposed in twenty years [1-5, 7-17, 19, 22-31].

The study of complete Heyting algebra which regards as generalized topology spaces goes back to the work of C.Ehresmann and J.Benabou, and later the theory of locale developed by C.H.Dowker, D.P.Strauss and J.R.Isbell. Some important topological properties on locales were obtained. Locale can be thought of as lattice theoretic generalizations of the lattice of open sets of a topological space. An outstanding introduction of locale theory in

Shaohui Liang
Department of Mathematics,
Xi'an University of Science and Technology, Xi'an 710054, P.R. China
e-mail: Liangshaohui1011@163.com

S. Li (Eds.): Nonlinear Maths for Uncertainty and its Appli., AISC 100, pp. 691–698.
springerlink.com                    © Springer-Verlag Berlin Heidelberg 2011

PT.Johnstone's book Stone Space [8]. In paper, a series of topological properties of locale such as separation Axioms and limit structure in locale by point approach. The study of the paper [11] was introduced convergence and cauchy structure on locales, and given characterization of Hausdorff property in locale by uniqueness of limit.

Quantale can be regard as the non-commutative generalization of frame. The natural question arising in this context is the following: How to introduce convergence structure, separation Axioms, and another properties in quantales? In the paper, we have introduced convergence structures on quantales. We obtained a series of results of properties of quantales, which generalize some results of locales.

## 2  Preliminaries

**Definition 1.** [29] A *quantale* is a complete lattice $Q$ with an associative binary operation "&" satisfying:
$$a\&(\bigvee_{i\in I} b_i) = \bigvee_{i\in I}(a\&b_i) \text{ and } (\bigvee_{i\in I} b_i)\&a = \bigvee_{i\in I}(b_i\&a),$$
for all $a, b_i \in Q$, where $I$ is a set, 0 and 1 denote the smallest element and the greatest element of $Q$, respectively.

A quantale $Q$ is said to be *unital* if there is an element $u \in Q$ such that $u\&a = a\&u = a$ for all $a \in Q$.

**Definition 2.** [29] Let $Q$ be a quantale and $a \in Q$.

(1) $a$ is $right - sided$ if and only if $a\&1 \leq a$.

(2) $a$ is $left - sided$ if and only if $1\&a \leq a$.

(3) $a$ is $two - sided$ if and only if $a$ is both right and left side.

(4) $a$ is *idempotent* if and only if $a\&a = a$.

**Definition 3.** [29] A quantale $Q$ is *commutative* if and only if $a\&b = b\&a$ for all $a, b \in Q$.

**Definition 4.** [29] Let $Q$ and $P$ be quantales. A function $f : Q \longrightarrow P$ is a homomorphism of quantale if $f$ preserves arbitrary sups and the operation "&". If $Q$ and $P$ are unital, then $f$ is unital homomorphism if in addition to being a homomorphism, it satisfies $f(u_Q) = u_P$, where $u_Q$ and $u_P$ are units of $Q$ and $P$, respectively.

**Definition 5.** [29] Let $Q$ be a quantales. A subset $S \subseteq Q$ is a *subquantale* of $Q$ iff the inclusion $S \hookrightarrow Q$ is a quantale homomorphism, i.e., $S$ is closed under sups and "&".

**Definition 6.** [29] Let $Q$ be a quantales. A *quantic nucleus* on $Q$ is a closure operator $j$ such that $j(a)\&j(b) \leq j(a\&b)$ for all $a, b \in Q$.

**Definition 7.** Let $Q$ be a quantales, for any $x \in Q$, defined $x_r^T = \vee\{y \in Q \mid x\&y = 0\}$, $x_r^T$ is said to be *right pseudocomplemented* of $x$.

By the definition of right pseudocomplement, we know that (1) $x_r^T \in R(Q)$, (2) $x \& x_r^T = 0$, (3) $x \leq y \Longrightarrow y_r^T \leq x_r^T$.

**Definition 8.** Let $Q$ be a quantales. A non-empty subset $I$ of $Q$ said to be *ideal* if it satisfies the following conditions:

(i) $1 \notin I$;

(ii) $a \vee b \in I$ for all $a, b \in I$;

(iii) $x \& r \in I$ and $r \& x \in I$ for all $x \in Q, r \in I$;

(iv) $I$ is a down-set.

The set of all ideals of $Q$ is denoted by $Id(Q)$. Let $I$ be a ideal of $Q$, then $I$ is said to be prime if $a, b \in I$ and $a \& b \in I$ imply $a \in I$ or $b \in I$. The set of prime ideal of $Q$ is denoted $PId(Q)$.

**Definition 9.** Let $Q$ be a quantales. A non-empty subset $F$ of $Q$ said to be *filter* if it satisfies the following conditions:

(i) $0 \notin F$;

(ii) $a \in F$, $b \in Q$, $a \leq b$ imply $b \in F$;

(iii) $a, b \in Q$ imply $a \& b \in F$.

The set of all filters of $Q$ is denoted by $Fil(Q)$. The filter $F$ of $Q$ is said to be *prime* if $a \vee b \in F$ imply $a \in F$ or $b \in F$. The set of all prime filters of $Q$ is denoted by $PFil(Q)$.

# 3  Ideal-Convergence in Quantales

Borceux and Vanden Bossche [3] introduced the concept of points of idempotent right-sided quantales, and gave some important results of points. In this section, we generalize the concept of points to quantale, and discussed a series of properties of points of quantales. Base on which we constructed the convergence structure on quantale by ideal and point, and some important properties are obtained.

**Definition 1.** Let $Q$ be a quantale, an element $p \in Q$ is called *prime* iff $p \neq 1$ and $a \& b \leq p$ implies that $a \leq p$ or $b \leq p$. The set of all prime two-sided element of $Q$ is denoted by $TPr(Q)$.

**Definition 2.** Let $Q$ be a quantale, $2 = \{0, 1\}$ is a quantale by taking $x \& y = 0$ with $x = 0$ or $y = 0$ and $1 \& 1 = 1$. A *point* of $Q$ is a onto homomorphism of quantale from $Q$ to 2. We shall denote the all points of $Q$ by $Pt(Q)$.

**Remark 1.** (1) Let $P$ is a point of $Q$, then $\vee p^{-1}(0) \in Pr(Q)$, $p^{-1}(0) = \downarrow (\vee p^{-1}(0))$, $p^{-1}(0)$ and $p^{-1}(1)$ is prime ideal and prime filter of $Q$, respectively.

(2) Let $Q$ be a quantale, there is a one to one correspondence between $TPr(Q)$ and $Pt(Q)$ by function $f : TPr(Q) \longrightarrow Pt(Q)$ such that $r \longmapsto pr(x) = \begin{cases} 0, & x \leq r \\ 1, & x \nleq r \end{cases}$ for all $r \in TPr(Q), x \in Q$.

(3) Let $Q$ be a quantale, $p$ is a point of $Q$, then $x \in P^{-1}(1)$ imply $x_r^T \in P^{-1}(0)$.

**Definition 3.** Let $Q$ be a quantale, $I \in Id(Q)$, $p \in Pt(Q)$.

(1) The point $p$ is called a *cluster point* of $I$ iff $I \subseteq Pt(Q)$.

(2) Ideal $I$ is *converges* to $p$ iff $p$ is a cluster point of $I$ and $x^T \in I$ for all $x \in p^{-1}(1)$.

(3) The point $p$ is a *strongly limit point* of $I$ if $p$ is a cluster point of $I$ and $\forall\, x \in p^{-1}(1)$, there exists $a \in I$ such that $a \vee x = 1$.

**Remark 2.** (i) If $I$ is a prime ideal of qunatale $Q$ and $p$ is a cluster of $I$, then $p$ is a limit point of $I$.

(ii) If $Q$ is a unital quantale with unit 1, then strong limit points $\Longrightarrow$ limit points $\Longrightarrow$ cluster points.

**Example 1.** (1) Let $Q = \{0, a, b, c, 1\}$. The order relation of $Q$ given by figure 1. We define a binary operation "&" on $Q$ satisfying the diagram 1.

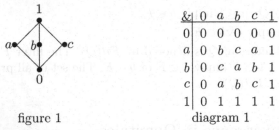

| & | 0 | a | b | c | 1 |
|---|---|---|---|---|---|
| 0 | 0 | 0 | 0 | 0 | 0 |
| a | 0 | b | c | a | 1 |
| b | 0 | c | a | b | 1 |
| c | 0 | a | b | c | 1 |
| 1 | 0 | 1 | 1 | 1 | 1 |

figure 1                                    diagram 1

It is easy to show that $(Q, \&)$ is a quantale. Now, we define $p : Q \longrightarrow 2$ such that $p(x) = \begin{cases} 1, & x \in \{a, b, c, 1\}, \\ 0, & x=0. \end{cases}$ Then $p$ is a onto homomorphism of quantale. Hence $p \in Pt(Q)$. Let $I = \{0\}$, then $I$ be a ideal of $Q$. We can prove that $p$ is a cluster of $I$. Since $p^{-1}(1) = \{a, b, c, 1\}$ and $a^T = b^T = c^T = 1^T = 0 \in I$, therefore $I$ converges to $p$.

(2) Let $Q = \{0, a, b, c, d, 1\}$ be a quantale, the order relation and binary operation "&" on $Q$ given by the following figure 2 and diagram 2.

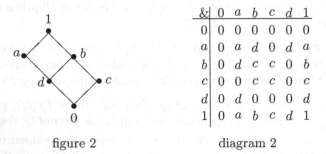

| & | 0 | a | b | c | d | 1 |
|---|---|---|---|---|---|---|
| 0 | 0 | 0 | 0 | 0 | 0 | 0 |
| a | 0 | a | d | 0 | d | a |
| b | 0 | d | c | c | 0 | b |
| c | 0 | 0 | c | c | 0 | c |
| d | 0 | d | 0 | 0 | 0 | d |
| 1 | 0 | a | b | c | d | 1 |

figure 2                                    diagram 2

We know that $b$ is a prime element of $Q$, define mapping $p_b : Q \longrightarrow 2$ such that

$p_b(x) = \begin{cases} 0, & x \le b, \\ 1, & \text{otherwise.} \end{cases}$  By remark 1. we know that $p_b$ is an onto homomorphism of quantale, i.e., $p_b \in Pt(Q)$.

It is easy to verify that $I = \downarrow b = \{0, b, c, d\}$ be a ideal of $Q$, and $I \subseteq p_b^{-1}(0)$, then $p_b$ is a cluster point of $I$. $\forall\, t \in p_b^{-1}(1)$, then $t \in \{a, 1\}$. We know that $a^T = \vee\{0, c\} = c \in I$, $1^T = 0 \in I$, then $I$ converges to $p_b$. Since $I' = \{0, c\}$ is a ideal of $Q$, and $I' \subseteq p_b^{-1}(0)$. Thus $p_b$ is a cluster point of $I'$. By $a^T, 1^T \in I'$, then $I'$ converges to $p_b$. Since $\forall\, s \in p_b^{-1}(1) = \{a, 1\}$ such that $s \vee c = 1$, which implies that $p_b$ is a strongly limit point of $I$ and $I'$.

**Theorem 1.** Let $Q$ be a unital quantale with unit 1, and $I$ be a ideal of $Q$. $I$ is the maximal ideal of $Q$ iff there exist $a \in Q$ such that $a \vee x \ne 1$ for all $x \in I$, then $a \in I$.

**Proof.** Suppose $I$ is not the maximal ideal of $Q$, then there exist $\overline{I} \in Id(Q)$ such that $I \subseteq \overline{I}$, $I \ne \overline{I}$. Thus there exists $m \in \overline{I} \setminus I$. Since $m \vee x \ne 1$ for all $x \in I$, then $m \in I$, which is a contradiction.

Conversely, let $a \in Q$ such that $a \vee x \ne 1$ for all $x \in Q$, but $a \notin I$. Put $I_a = I \cup \downarrow \{a \vee x \mid x \in I\}$. Next, we will prove that $I_a$ is a ideal of $Q$.

(1) It is easy to prove that $I_a$ be a down set and $1 \in Q \setminus I_a$;

(2) $\forall\, y \in Q$, $n \in I_a$. If $n \in I$, then $y\&n$, $n\&y \in I_a$. If $n \in \downarrow \{a \vee x \mid x \in I\}$, then there exists $x_n \in I$ such that $n \le a \vee x_n$. Thus $y\&n \le y\&(a \vee x_n) \le 1\&(a \vee x_n) = a \vee x_n \in I_a$. Therefore $y\&n \in I_a$. Similarly $n\&y \in I_a$. Hence, $I_a$ is a ideal of $Q$ and $I_a \ne Q$, which is a contradiction. We know that $I$ is the maximal ideal of $Q$.  □

**Corollary 1.** Let $Q$ be a unital quantale with unit 1. $I$ be a maximal ideal of $Q$. Then every cluster points of I is the strong limit points of $I$.

Combining the above corollary 1 and Remark 2.(ii), we can obtain the following:

**Theorem 2.** Let $Q$ be a unital quantale with unit 1. Then the following are equivenlent:

(1) Every ideal of $Q$ has cluster points;

(2) Every maximal ideal of $Q$ has limit points;

(3) Every maximal ideal of $Q$ has strongly limit points.

The proof is easy and is omitted.

**Theorem 3.** Let $Q$ be a quantale, $j$ be a nuclei of $Q$. Then the following are true:

(i) $Pr(Q_j) = Pr(Q) \cap Q_j$;

(ii) $Pt(Q_j) = \{p \mid_{Q_j} : p \in Pt(Q),\ \overset{Q}{\vee} p^{-1}(0) \in Q_j\}$, where $p \mid_{Q_j}$ denote the restriction of p to $Q_j$.

**Proof.** (i) $\forall\, r \in Pr(Q_j) \subseteq Q_j$, $x, y \in Q$, if $x\&y \le r$, then $x\&_j y \le r$ by $j(x\&y) \le r$. Since $r \in Pr(Q_j)$, then $j(x) \le r$ or $j(y) \le r$. By $x \le j(x)$ and $y \le j(y)$ we know that $x \le r$ or $y \le r$. Therefore $r \in Pr(Q) \cap Q_j$. So $Pr(Q_j) \subseteq Pr(Q) \cap Q_j$. It is obvious that $Pr(Q) \cap Q_j \subseteq Pr(Q_j)$.

(2) Let $p_0 \in Pt(Q_j)$, then $r_0 = \bigvee_{Q_j} p_0^{-1}(0) = j(\overset{Q}{\bigvee} p_0^{-1}(0)) \in Q_j$ is a prime element, which is corresponds to point $p_0$. By (1) we know that $p_{r_0}$ is the restriction of p to $Q_j$. Thus $Pt(Q_j) \subseteq \{p\mid_{Q_j}: p \in Pt(Q), \overset{Q}{\bigvee} p^{-1}(0) \in Q_j\}$.

Conversely, let $p \in Pt(Q)$ and $r = \overset{Q}{\bigvee} p_0^{-1}(0) \in Q_j$. We only need to show $p\mid_{Q_j}: Q_j \longrightarrow 2$ is a quantale homomorphism.

Firstly, we show that $p\mid_{Q_j}$ preserve operation "&", i.e., $\forall\, x, y \in Q_j$, $p\mid_{Q_j}(x \&_j y) = p\mid_{Q_j}(x)\&p\mid_{Q_j}(y)$.

If $x \&_j y \leq r$, then $x \leq r$ or $y \leq r$. Thus $p\mid_{Q_j}(x \&_j y) = p\mid_{Q_j}(x)\&p\mid_{Q_j}(y)$. If $x \&_j y \not\leq r$, then $x \not\leq r$ and $y \not\leq r$. Otherwise, suppose $x \leq r$ or $y \leq r$, then $x \&_j y \leq x \&_j r \leq r$ or $x \&_j y \leq r \&_j y \leq r$, which is a contradiction. Therefore, $p\mid_{Q_j}(x \&_j y) = p\mid_{Q_j}(x)\&p\mid_{Q_j}(y)$.

$\forall\, A \subseteq Q_j$, if $A = \emptyset$, then $\vee A = j(0) \in Q_j$, and $j(0) \leq r$. We know that $p\mid_{Q_j}(j(0)) \leq p\mid_{Q_j}(r) = 0$, then $p\mid_{Q_j}(j(0)) = 0$, therefore $p\mid_{Q_j}$ preserve empty sups.

Suppose $A \neq \emptyset$, if $\bigvee_{Q_j} A \leq r$, then $p(\bigvee_{Q_j} A) \leq p(r) = 0$. Thus $p\mid_{Q_j}(\bigvee_{Q_j} A) = \bigvee_{a \in A} p\mid_{Q_j}(a)$.

If $\bigvee_{Q_j} A \not\leq r$, then there exists $a \in A$ such that $a \not\leq r$, i.e., $p(a) = 1$. Thus $p\mid_{Q_j}(\bigvee_{Q_j} A) = 1 = \bigvee_{a \in A} p(a)$. Therefore $p\mid_{Q_j}$ preserve arbitrary sups.    $\square$

**Theorem 4.** Let $Q$ be a quantale, $j$ be a nuclei of $Q$, $I \in Id(Q_j)$. Thus $j^{-1}(I) = \{x \in Q \mid j(x) \in I\} \in Id(Q)$.

**Proof.** Firstly, since $j: Q \longrightarrow Q_j$ is a surjective quantale homomorphism. Thus $j(1)$ is the greatest element of $Q_j$. By $I \in Id(Q_j)$ we know that $j(1) \in Q_j \setminus I$, therefore $1 \in Q \setminus j^{-1}(I)$.

Secondly, let $x \in Q$, $y \in j^{-1}(I)$ with $x \leq y$, then $j(x) \leq j(y) \in I$, thus $j(x) \in I$, i.e., $x \in j^{-1}(I)$. Therefore $j^{-1}(I)$ is a down set.

At last, $\forall\, x \in Q$, $y \in j^{-1}(I)$, we have $j(x \& y) = j(x)\&_j j(y) \in I$. Thus $x \& y \in j^{-1}(I)$. Similarly, $y \& x \in j^{-1}(I)$.

Therefore, $j^{-1}(I)$ is a ideal of $Q$ by the proof above.

**Theorem 5.** Let $Q$ be a quantale, $j$ be a nuclei of $Q$, $p \in Pt(Q)$ such that $p\mid_{Q_j} \in Pt(Q_j)$, $I \in Id(Q_j)$. Then the following are true:

(1) $p\mid_{Q_j}$ is a cluster point of $I$ in $Q_j$ iff $p$ is a cluster point of $j^{-1}(I)$ in $Q$;

(2) If $Q_j$ is a dense quotient of $Q$, then $I$ converges to $p\mid_{Q_j}$ in $Q_j$ iff $j^{-1}(I)$ converges to $p$ in $Q$;

(3) If $p^{-1}(1) = p\mid_{Q_j}^{-1}(1)$, then $I$ strong converges to $p\mid_{Q_j}$ in $Q_j$ iff $j^{-1}(1)$ strong converges to $p$ in $Q$.

**Proof.** (1) Let $p \mid_{Q_j}$ is a cluster point $I$ in $Q_j$, then $I \subseteq p \mid_{Q_j}^{-1} (1) = p^{-1}(0) \cap Q_j$. $\forall\, x \in j^{-1}(I)$, then $j(x) \in I$. Since $x \leq j(x)$, thus $x \in p^{-1}(0)$, therefore $p$ is a cluster point of $j^{-1}(I)$ in $Q$.

Conversely, since $j(I) \subseteq I \subseteq Q_j$, thus $I \subseteq j^{-1}(I) \subseteq p \mid_{Q_j}^{-1} (0)$. Hence $I \subseteq p \mid_{Q_j}^{-1} (0)$. Therefore $p \mid_{Q_j}$ is a cluster point of $I$ in $Q_j$.

(2) If $I$ converges to $p \mid_{Q_j}$ in $Q_j$. By (1), we know that $j^{-1}(I) \subseteq p^{-1}(0)$. $\forall\, x \in p^{-1}(1)$, we have $1 = p(x) \leq p(j(x))$, i.e., $j(x) \in p^{-1} \cap Q_j$, thus $j(x)^{T_{Q_j}} \in I$, where $j(x)^{T_{Q_j}}$ denote the right pseudocomplement of $j(x)$ in $Q_j$. Since $j(x)\&_j j(x^T) = j(j(x)\&j(x^T)) \leq j(j(x\&x^T)) = j(j(0)) = j(0) = 0$. Thus $j(x^T) \leq j(x)^{T_j} \in I$. By $I$ be a down set, we know that $j(x^T) \in I$, i.e., $x^T \in j^{-1}(I)$. Thus $j^{-1}$ converges to $p$ in $Q$.

Conversely, if $Q_j$ is a dense quotient of $Q$, then $j(0) = 0$. Let $j^{-1}(I)$ converge to $p$ in $Q$,. By (1) we know that $p \mid_{Q_j}$ is a cluster point of $I$ in $Q_j$. $\forall\, x \in (p \mid_{Q_j})^{-1}(1) = p^{-1}(1) \cap Q_j$, we have $x^{T_{Q_j}} = \bigvee_{Q_j} \{y \in Q_j \mid x\&_j y = 0\} = j(\bigvee_Q \{y \in Q_j \mid x\&_j y = 0\}) \leq j(x^T)$. Since $x^T \in j^{-1}(I)$, and $I$ be a down set, thus $x_{Q_j}^T \in I$. Therefore $I$ converge to $p \mid_{Q_j}$ in $Q_j$.

(3) Let $I$ strongly converges to $p \mid_{Q_j}$. By (1) we know that $p$ is a cluster point of $j^{-1}(I)$ in $Q$. $\forall\, x \in p^{-1}(1)$, since $p^{-1}(1) = p \mid_{Q_j}^{-1} (1)$. Then there exists $a \in I$ such that $a \vee x = 1$. By $I \subseteq j^{-1}(I)$, we know $a \in j^{-1}(I)$. Therefore $j^{-1}(I)$ strongly converges to $p$ in $Q$.

Conversely, let $j^{-1}(I)$ strongly converges to $p$ in $Q$. By (1), we know that $p \mid_{Q_j}$ is a cluster point of $I$ in $Q_j$. $\forall\, y \in p \mid_{Q_j}^{-1} (1)$. Since $p^{-1}(1) = p \mid_{Q_j}^{-1} (1)$, then there exists $b \in j^{-1}(I)$ such that $b \vee y = 1$. By $b \leq j(b)$, we know that $j(b) \vee y = 1$. Therefore $I$ strong converges to $p \mid_{Q_j}$ in $Q_j$.

**Acknowledgements.** This work was supported by the Engagement Award (2010041), Dr. Foundation(2010QDJ024) of Xi'an University of Science and technology.

# References

1. Abramsky, S., Vickers, S.: Quantales,observational logic and process semantics. Math. Struct. Comput. Sci. 3, 161–227 (1993)
2. Berni-Canani, U., Borceux, F., Succi-Cruciani, R.: A theory of quantale sets. Journal of Pure and Applied Algebra 62, 123–136 (1989)
3. Borceux, F., Vanden Bossche, G.: Quantales and their sheaves. Order 3, 61–87 (1986)
4. Brown, C., Gurr, D.: A representation theorem for quantales. Journal of Pure and Applied Algebra 85, 27–42 (1993)
5. Coniglio, M.E., Miraglia, F.: Modules in the category of sheaves over quantales. Annals of Pure and Applied Logic 108, 103–136 (2001)
6. Girard, J.Y.: Linear logic. Theoretical Computer Science 50, 1–102 (1987)
7. Han, S.W., Zhao, B.: The quantic conuclei on quantales. Algebra Universalis 61(1), 97–114 (2009)

8.  Johnstone, P.T.: Stone spaces. Cambridge University Press, London (1982)
9.  Kruml, D.: Spatial quantales. Applied Categorial Structures 10, 49–62 (2002)
10. Li, Y.M., Zhou, M., Li, Z.H.: Projectives and injectives in the category of quantales. Journal of Pure and Applied Algebra 176(2), 249–258 (2002)
11. Liang, J.H.: Convergence and Cauchy Structures on Locales. Acta Mathematica Sinica (Chinses Series) 38(3), 294–300 (1995)
12. Liang, S.H., Zhao, B.: Continuous Quantale and Its Category Properties. Journal of Shanxi Normal University (Natural Science Edition) 36(5), 1–5 (2008)
13. Liang, S.H., Zhao, B.: Resarches on Some Properties of Strongly Fs-poset. Journal of Shandong University (Natural Science Edition) 44(8), 51–55 (2009)
14. Liang, S.H., Zhao, B.: Generalized inverse of quantale matrixs and its positive definiteness. Fuzzy Systems and Mathematics 23(5), 51–55 (2009)
15. Liang, S.H., Zhao, B.: Congruence and nucleus of double quantale module. Journal of Sichuan University (Natural Science Edition) 48(1), 13–18 (2010)
16. Liu, Z.B., Zhao, B.: Algebraic properties of category of quantale. Acta Mathematica Sinica (Chinese Series) 49(6), 1253–1258 (2006)
17. Miraglia, F., Solitro, U.: Sheaves over right sided idempotent quantales. Logic J. IGPL 6(4), 545–600 (1998)
18. Mulvey, C.J.: &,Suppl. Rend. Circ. Mat. Palermo Ser. 12, 99–104 (1986)
19. Mulvey, C.J., Resende, P.: A noncommutative theory of penrose tilings. Internation Journal of Theoretical Physics 44(6), 655–689 (2005)
20. Nawaz, M.: Quantales:quantale sets. Ph.D.Thesis, University of Sussex (1985)
21. Niefield, S., Rosenthal, K.I.: Strong De Morgan's law and the spectrum of a commutative ring. Journal of Algebra 93, 169–181 (1985)
22. Paseka, J.: A note on Girard bimodules. International Journal of Theoretical Physics 39(3), 805–812 (2000)
23. Picado, J.: The quantale of Galois connections. Algebra Universalis 52, 527–540 (2004)
24. Resende, P.: Quantales, finite observations and strong bisimulation. Theoretical Computer Science 254, 95–149 (2001)
25. Resende, P.: Tropological systems are points of quantales. Journal of Pure and Applied Algebra 173, 87–120 (2002)
26. Resende, P.: Topological systems are points of quantales. Journal of Pure and Applied Algebra 173(1), 87–120 (2002)
27. Resende, P.: Sup-lattice 2-forms and quantales. Journal of Algebra 276, 143–167 (2004)
28. Resende, P.: Etale groupoids and their quantales. Advances in Mathematics 208(1), 147–209 (2007)
29. Rosenthal, K.I.: Quantales and Their Applications. Longman Scientific and Technical, London (1990)
30. Rosenthal, K.I.: A general approach to Gabriel filters on quantales. Communications in Algebra 20(11), 3393–3409 (1992)
31. Russo, C.: Quantale modules, with applications to logic and image processing, Ph.D.Thesis.Salerno: university of Salerno (2007)
32. Solovyov, S.: On the category Q-Mod. Algebra universalis 58(1), 35–58 (2008)
33. Sun, S.H.: Remarks on quantic nuclei. Math. Proc. Camb. Phil. Soc. 108, 257–260 (1990)
34. Vermeulen, J.J.C.: Proper maps of locales. Journal of Pure and Applied Algebra 92, 79–107 (1994)
35. Zhao, B., Liang, S.H.: The categorgy of double quantale modules. Acta Mathematica Sinica,Chinese Series 52(4), 821–832 (2009)

# On the Factorization Theorem of a Monotone Morphism in a Topos

Tao Lu and Hong Lu

**Abstract.** In this paper, we investigate the factorization of a monotone morphism between two partially ordered objects in an arbitrary elementary topos by means of diagram proof. And then a new factorization theorem in an arbitrary elementary topos which is similar to the classical one is obtained.

**Keywords:** Partial order object, Monotone morphism, Topos.

## 1 Introduction and Preliminaries

Recall a topos $\mathcal{E}$ is a category which has finite limits and every object of $\mathcal{E}$ has a power object. For a fixed object $A$ of category $\mathcal{E}$, the power object of $A$ is an object $PA$ which represents $\mathrm{Sub}(- \times A)$, so that $\mathrm{Hom}_{\mathcal{E}}(-, PA) \simeq \mathrm{Sub}(- \times A)$ naturally. It says precisely that for any arrow $B' \xrightarrow{f} B$, the following diagram commutes, where $\varphi$ is the natural isomorphism.

$$
\begin{array}{ccc}
\mathrm{Hom}_{\mathcal{E}}(B, PA) & \xrightarrow{\varphi(A,B)} & \mathrm{Sub}(B \times A) \\
{\scriptstyle \mathrm{Hom}_{\mathcal{E}}(f,PA)} \downarrow & & \downarrow {\scriptstyle \mathrm{Sub}(f \times A)} \\
\mathrm{Hom}_{\mathcal{E}}(B', PA) & \xrightarrow{\varphi(A,B')} & \mathrm{Sub}(B' \times A)
\end{array}
$$

**Fig. 1**

Tao Lu
School of Mathematics, HuaiBei Normal University
HuaiBei, AnHui, 235000, P.R. China
e-mail: lutao7@live.com

Hong Lu
School of Information Science and Engineering, Rizhao Polytechnic
Rizhao, ShanDong, 276826, P.R. China
e-mail: luhong0810@sina.com

S. Li (Eds.): Nonlinear Maths for Uncertainty and its Appli., AISC 100, pp. 699–706.
springerlink.com      © Springer-Verlag Berlin Heidelberg 2011

As a matter of fact, the category of sheaves of sets on a topological space is a topos. In particular, the category of sets is a topos. For details of the treatment of toposes and sheaves please see Johnstone [3], Mac and Moerdijk [8], Joyal and Tierney [6], Johnstone and Joyal [5]. For a general background on category theory please refer to [10], [7]

In [8], Lattice and Heyting Algebra objects in a topos are well defined. In this paper we develop our study in the more general and more natural context of partially ordered object and the factorization theorem in categorical sense. More details about lattice and locale please see [1], [2], [9], [4].

## 2 Main Results

**Definition 1.** ( [8]) A subobject $\leq_L \rightarrowtail L \times L$ is called an internal partial order on $L$, provided that the following conditions are satisfied

1) Reflexivity: The diagonal $L \xrightarrow{\delta} L \times L$ factors through $\leq_L \xrightarrow{e_L} L \times L$, as in

**Fig. 2** Reflexivity

2) Antisymmetry: The intersection $\leq_L \cap \geq_L$ is contained in the diagonal, as in the following pullback

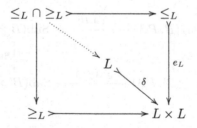

**Fig. 3** Antisymmetry

Where $\geq_L$ is defined as the composite $\leq_L \xrightarrow{e_L} L \times L \xrightarrow{\tau} L \times L$ with $\tau$ as the twist map interchanging the factors of the product.

3) Transitivity: The subobject $C \xrightarrow{\langle \pi_1 ev, \pi_2 eu \rangle} L \times L$ factors through $\leq_L \xrightarrow{e_L} L \times L$, as in

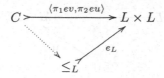

**Fig. 4.** Transitivity

where $C$ is the following pullback

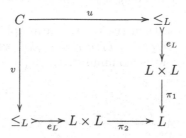

**Fig. 5.** The definition of C

An object $L$ endowed with an internal partial order $\leq_L$ is called a partially ordered object.

Let $L$ and $M$ be two partially ordered objects. We can define the product of partially ordered object $L \times M$ of $L$ and $M$ as the product object $L \times M$ endowed with the "pointwise order" $\leq_L \times \leq_M \rightarrowtail L \times L \times M \times M \simeq L \times M \times L \times M$. Also, a subobject $B$ of a partially ordered object $(L, \leq_L)$ is again a partial order object endowed with the induced partial order $\leq_B$, as in the pullback

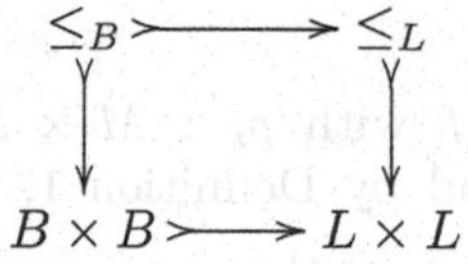

$$
\begin{array}{ccc}
\leq_B & \longrightarrow & \leq_L \\
\downarrow & & \downarrow \\
B \times B & \longrightarrow & L \times L
\end{array}
$$

**Fig. 6.** The induced partial order

Based on the above definitions, we now turn to the discussion of morphisms between partial order objects.

**Definition 2.** ([8]) Let $L, M$ be two partially ordered objects with a pair of morphisms $L \overset{f}{\underset{g}{\rightrightarrows}} M$. Then $f \leq g$ means $L \xrightarrow{\langle f, g \rangle} M \times M$ factors through $\leq_M \overset{e_M}{\rightarrowtail} M \times M$, as in

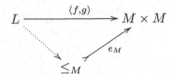

**Fig. 7.** The first definition of $f \leq g$

**Lemma 1.** *Let $L, M$ be two partially ordered objects with a pair of morphisms* $L \underset{g}{\overset{f}{\rightrightarrows}} M$. *Then $f \leq g$ if and only if $fr \leq gr$ for every morphism $A \overset{r}{\longrightarrow} L$.*

*Proof.* $\Rightarrow$. Suppose $f \leq g$, then there exists a morphism $L \overset{k}{\longrightarrow} \leq_M$ such that $\langle f, g \rangle = e_M k$. So $\langle fr, gr \rangle = \langle f, g \rangle r = e_M kr$, which means the outer triangle of Figure 8 below is commutative, i.e., $\langle fr, gr \rangle$ factors through $\leq_M \overset{e_M}{\rightarrowtail} M \times M$.

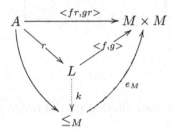

**Fig. 8.** Equivelence of two definitions

$\Leftarrow$. Indeed, in order to verify this, we can take the fixed identity morphism $L \overset{1_L}{\longrightarrow} L$, then $f \leq g$ is obvious. $\qquad\square$

**Corollary 1.** *Let $L, M$ be two partially ordered objects and $L \overset{f}{\longrightarrow} M$ be a morphism. Then $f \leq f$.*

*Proof.* Since $p_i \langle f, f \rangle = p_i \delta f$ with $p_i : M \times M \to M$ ($i = 1, 2$) being projections, $\langle f, f \rangle = \delta f$. And by Definition 1, we know $\delta$ factors through $\leq_M \overset{e_M}{\rightarrowtail} M \times M$. It follows that the outer square is commutative as in the following Figure 9.

So we have that $\langle f, f \rangle$ factors through $\leq_M \overset{e_M}{\rightarrowtail} M \times M$, thus $f \leq f$. $\qquad\square$

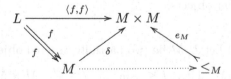

**Fig. 9.** Reflexivity of $f$

**Corollary 2.** *Let $L, M$ be two partially ordered objects and $f, g, h$ morphisms between $L$ and $M$. Then $f \leq g$ and $g \leq h$ imply $f \leq h$.*

**Corollary 3.** *Let $L, M$ be two partially ordered objects and $f : L \to M, g : M \to L$ be morphisms. Then $f \leq g$ and $g \leq f$ imply $f = g$.*

*Proof.* $g \leq f$ implies that $\langle g, f \rangle : L \to M \times M$ can be factored through $\leq_M \rightarrowtail M \times M$, equivalently, $\langle f, g \rangle$ can be factored through $\geq_M \rightarrowtail M \times M$. Thus $\langle f, g \rangle : L \to M \times M$ can be factored through $\delta_M = \leq_M \cap \geq_M \rightarrowtail M \times M$. This shows $f = g$.                                                   □

The above argument shows that for two partially ordered objects $L$ and $M$, the relation $\leq$ defined on the morphism set $\mathrm{Mor}(L, M)$ is a partial order relation.

**Definition 3.** ( [8])Let $L, M$ be two partially ordered objects in $\mathcal{E}$. A morphism $L \xrightarrow{f} M$ is called order-preserving or monotone if the composite $\leq_L \xrightarrow{e_L} L \times L \xrightarrow{f \times f} M \times M$ factors through $\leq_M$, as in

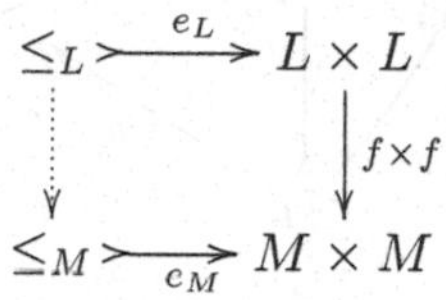

$$\begin{array}{ccc} \leq_L & \xrightarrow{e_L} & L \times L \\ \big\downarrow & & \big\downarrow {f \times f} \\ \leq_M & \xrightarrow[c_M]{} & M \times M \end{array}$$

**Fig. 10.** The definition of a monotone morphism

**Lemma 2.** *A morphism $L \xrightarrow{f} M$ between two partial ordered objects is order-preserving if and only if $r \leq s$ implies $fr \leq fs$ for every pair of parallel morphisms $A \underset{s}{\overset{r}{\rightrightarrows}} L$.*

*Proof.* $\Rightarrow$. We first show $\langle fr, fs \rangle = f \times f \langle r, s \rangle$. This may be pictured as in the following Figure 11, where $p_1, p_2, \pi_1, \pi_2$ are projections.

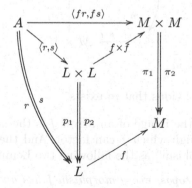

**Fig. 11.** Universal property of product

By the universal property of $M \times M$, it follows that $fp_i = \pi_i f \times f$, $i = 1, 2$. Similarly, $r = p_1\langle r, s \rangle$, $s = p_2\langle r, s \rangle$. Then $fp_i\langle r, s \rangle = \pi_i f \times f\langle r, s \rangle$, so $fr = \pi_1 f \times f\langle r, s \rangle$, $fs = \pi_2 f \times f\langle r, s \rangle$. By the universal property of $M \times M$, we also have $fr = \pi_1\langle fr, fs \rangle$, $fs = \pi_2\langle fr, fs \rangle$. So, $\pi_1 < fr, fs >= \pi_1 f \times f\langle r, s \rangle$, $\pi_2\langle fr, fs \rangle = \pi_2 f \times f\langle r, s \rangle$, thus $\langle fr, fs \rangle = f \times f\langle r, s \rangle$.

Now suppose $r \leq s$, then there exists a morphism $A \xrightarrow{k} \leq_L$ with $\langle r, s \rangle = e_L k$. It follows that the left triangle of in Figure 12 is commutative. Since $f$ is monotone, the right square of the Figure 12 is commutative, i.e., there exists $\leq_L \xrightarrow{m} \leq_M$ such that $f \times f e_L = e_M m$. So $\langle fr, fs \rangle = f \times f\langle r, s \rangle = f \times f e_L k = e_M m k$, which means the outer of the Figure 12 is commutative.

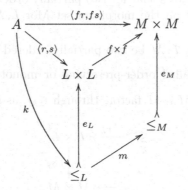

**Fig. 12** The relation between $< fr, fs >$ and $e_M$

Thus, $\langle fr, fs \rangle$ factors through $\leq_M \xrightarrow{e_M} M \times M$.

$\Leftarrow$. It suffices to show there exists $\leq_L \xrightarrow{m} \leq_M$ with $f \times f e_L = e_M m$, as in the Figure 13.

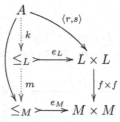

**Fig. 13** The existence of $m$

By Lemma 2, it is obvious that $m$ exists. $\qquad\qquad\square$

It is well known that the image of an arrow $f$ is the smallest subobject (of the codomain $f$) through which $f$ can factor. And the factorization of $f$ is unique "up to isomorphism" as the following two Lemmas show.

**Lemma 3.** ( [8])In a topos, every morphism $f$ has an image $m$ and factors as $f = me$, with $e$ epi.

**Lemma 4.** *If* $f = me$ *and* $f' = m'e'$ *with* $m, m'$ *monic and* $e, e'$ *epi, then each map of the arrow* $f$ *to the arrow* $f'$ *extends to a unique map of* $m, e$ *to* $m', e'$.

*Proof.* A map of the arrow $f$ to the arrow $f'$ is a pair of arrows $r, t$ which make the following square commute.

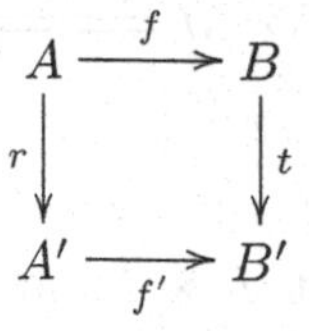

$$
\begin{array}{ccc}
A & \xrightarrow{\ f\ } & B \\
{\scriptstyle r}\downarrow & & \downarrow{\scriptstyle t} \\
A' & \xrightarrow[\ f'\ ]{} & B'
\end{array}
$$

**Fig. 14**

Given such a pair of arrows and the two $e - m$ factorizations, it suffices to construct a unique arrow $s$ from $m$ to $m'$ which makes both squares in the following diagram commute.

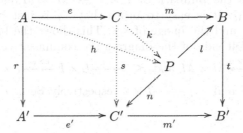

**Fig. 15**

Take the pullback $P$ of $t$ along $m'$, as is shown in the above diagram, then $l$ is monic. By the definition of the pullback and the Figure 14, then, there exists the unique $h$ such that $f$ factors through $l$, i.e., $f = lh$. By the minimal property of the image, then there exists one unique arrow $k$, such that $m = lk$. Because $l$ is monic, then $h = ke$. Let $s = nk$, then $tm$ factors through $C'$ via $s$, as $tm = m's$, and the arrow $s$ is unique because $m'$ is monic. Moreover, we have $se = e'r$ for the same reason, which means the left hand square of the above diagram also commutes.     □

**Theorem 1.** *If a monotone morphism* $L \xrightarrow{\ f\ } M$ *between two partially ordered objects factors as* $f = me$ *with image* $m$. *Then* $m$ *and* $e$ *are monotone morphisms.*

*Proof.* Given $L \xrightarrow{\ f\ } M$, which factors as $L \xrightarrow{\ e\ } I \xrightarrow{\ m\ } M$. The proof is just a matter of observing the corresponding partial order on $I$. Construct the following commutative Figure 16.

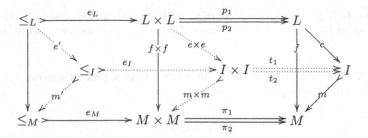

**Fig. 16** The partial order on $I$

By the definition of product $L \times L$, $M \times M$, $I \times I$ with projections $p_i, \pi_i, t_i$ ($i = 1, 2$) respectively, we have $fp_i = \pi_i f \times f$, $ep_i = t_i e \times e$, $mt_i = \pi_i m \times m$, i.e., the front, back, bottom faces of the right side of the diagram are all commutative. Then, $\pi_i f \times f = fp_i = mep_i = \pi_i m \times m \cdot e \times e$, so $f \times f = m \times m \cdot e \times e$, which means the middle triangle is commutative. Since the smallestness of $m \times m$ is obvious, $f \times f = m \times m \cdot e \times e$ is again an epi-momo factorization, i.e., $m \times m$ is the image of $f \times f$.

We take $\leq_I$ as the pullback of $I \times I \to M \times M$ along $e_M$, that is, $\leq_I = (I \times I) \cap \leq_M$. It is easy to prove that $\leq_I$ is just both the induced partial order on $I$ and the image of $\leq_L$. This shows the back and the bottom faces of the left side of the diagram are commutative, in other words, $\leq_I \overset{e_L}{\longrightarrow} I \times I \overset{m \times m}{\longrightarrow} M \times M$ and $\leq_L \overset{e_L}{\longrightarrow} L \times L \overset{e \times e}{\longrightarrow} I \times I$ factor through $\leq_M \overset{e_M}{\longrightarrow} M \times M$ and $\leq_I \overset{e_I}{\longrightarrow} I \times I$ respectively. So $m, e$ are all monotone morphisms.     □

# References

1. Isbell, J.R.: Atomless parts of spaces. Math. Scand. 31, 5–32 (1972)
2. Isbell, J.R.: First steps in descriptive theory of locales. Trans. Amer. Math. Soc. 327(1), 353–371 (1991)
3. Johnstone, P.T.: Sketches of an Elephant, A Topos Theory Compendium. Oxford University Press, Oxford (2002)
4. Johnstone, P.T.: Stone Spaces. Cambridge University Press, Cambridge (1982)
5. Johnstone, P.T., Joyal, A.: Continuous categories and exponentiable toposes. J. Pure Appl. Algebra 25(3), 255–296 (1982)
6. Joyal, A., Tierney, M.: An extension of the Galois theory of Grothendieck. Mem. Amer. Math. Soc. 51(309), 71 (1984)
7. Mac, L.S.: Categories for the Working Mathematician. Springer, New York (1971)
8. Mac, L.S., Moerdijk, I.: Sheaves in Geometry and Logic. Springer, New York (1994)
9. Wei, H., Yingming, L.: Steenrod's theorem for locales. Math. Proc. Cambridge Philos. Soc. 124(2), 305–307 (1998)
10. Wei, H.: Category Theory. Science Press, Beijing (2006) (in Chinese)

# Author Index

Adachi, Tomoko 365
Aknin, Patrice 213
Aoyama, Koji 387

Côme, Etienne 213
Carbonneau, Rene 341
Chen, Guangxia 471
Chen, Wenjuan 411
Chen, Xingxing 549
Chen, Yixiang 357
Chen, Zengjing 19
Cheng, Caozong 411
Cheng, Yue 379
Cherfi, Zohra L. 213
Cote, Sylvana 341
Cui, Li-ai 295

Denœux, Thierry 213
Ding, Hengfei 463
Dong, Jiashuai 395
Du, Jiang 525
Dumrongpokaphan, T. 443

Fan, Xiaodong 379
Fan, Xin-an 595
Fan, Yulian 271
Fang, Bin 587
Fang, Liying 533
Fei, Weiyin 263
Feng, Shuang 189
Flaminio, Tommaso 619
Fukuda, Ryoji 479
Furuichi, Shigeru 317

Gümbel, Martin 205
Gao, Chen 541, 611
Gong, Minqing 517
Gu, Rencai 349
Guan, Li 135, 255
Guo, Lankun 683

Han, Liyan 287
Hao, Ruili 221
Honda, Aoi 627
Hsiao, Kuo Chang 567
Hu, Baomin 595
Hu, Lin 419
Hu, Wei 309
Hu, Yangli 309

Inoue, Hiroshi 255

Jaihonglam, W. 443
Jia, Guangyan 31
Jiang, Juan 651
Jiang, Shaoping 503
Jin, Caiyun 411
Jin, Jianhua 557
Jin, Mingzhong 517
Jiroušek, Radim 179

Kasperski, Adam 197
Kawabe, Jun 35
Kawasaki, Toshiharu 435
Kimura, Yasunori 371
Klement, Erich Peter 325
Kohsaka, Fumiaki 403
Kou, Xiaoliang 659
Kroupa, Tomáš 619

Labuschagne, Coenraad C.A.   51, 231
Li, Chunquan   557
Li, Gaolin   675
Li, Gaorong   509
Li, Hongxing   189
Li, Jun   43, 77
Li, Jungang   161
Li, Qingguo   683
Li, Shoumei   101, 135, 279
Li, Shuangjie   549
Li, Xuan   595
Li, Xuejing   509
Li, Zhanpeng   587
Liang, Shaohui   691
Liang, Yong   263
Liu, Bo   659
Liu, Chunhui   667
Liu, Hongjian   263
Liu, Hongmei   517
Liu, Hongping   683
Liu, Hsuan-Ku   643
Liu, Shaoyue   239
Liu, Xuanhui   295
Liu, Xuecheng   341, 635
Liu, Youming   419
Liu, Zheng   533
Lu, Hong   699
Lu, Tao   699

Ma, Xinshun   603
Malinowski, Marek T.   143
Marraffa, Valeria   51
Meng, Xianli   595
Mesiar, Radko   43, 325
Michta, Mariusz   117, 143
Mitoma, Itaru   125
Miyake, Masatoshi   255
Murofushi, Toshiaki   61

Nakamura, Shinsuke   61
Nguyen, Hung T.   109
Nohmi, Masaya   627

Offwood, Theresa M.   231
Oka, Kaoru   479
Okazaki, Yoshiaki   627
Oukhellou, Latifa   213
Ouncharoen, R.   443

Pan, Min   287
Pap, Endre   43

Qiao, Chen   579
Quaeghebeur, Erik   169

Ren, Fangguo   295
Rui, Shao-ping   427

Sugeno, Michio   1
Sun, Qinghe   635

Takahagi, Eiichiro   93
Takasawa, Toshiyuki   61
Tan, Guohua   541, 611
Tan, Jiyang   239
Tanaka, Tamaki   69, 85
Tao, Dongya   247
Toyoda, Masashi   435
Tremblay, Richard E.   341

Wang, Dabuxilatu   487
Wang, Hongxia   101
Wang, Jingjing   487
Wang, Jinting   161
Wang, Pu   533
Wang, Xia   151
Wang, Yuanheng   395
Watanabe, Toshikazu   69, 85
Wu, Berlin   567, 643
Wu, Dianshuang   333
Wu, Hengyang   357
Wu, Panyu   19

Xia, Dengfeng   263
Xia, Jianming   31
Xie, Tianfa   525
Xie, Yuquan   301
Xu, Dengke   495
Xu, Luoshan   667, 675
Xu, Yong   349

Yan, Jianzhuo   541, 611
Yanagi, Kenjiro   317
Yang, Suhua   541, 611
Yang, Suigen   509
Yang, Wanli   453
Yang, Xiangqun   239
Yao, Kai   567
Yasuda, Masami   77
Ye, Zhongxing   221
Yin, Yue   309
Yu, Mingwei   533
Yuan, Ting   603

Zeng, Wenyi  189
Zhang, Chao  247
Zhang, Guangquan  333
Zhang, Guoli  579
Zhang, Jie  427, 659
Zhang, Jinping  125
Zhang, Jizhou  247
Zhang, Junfei  279

Zhang, Youcai  239
Zhang, Yuxin  463
Zhang, Zhongzhan  495, 525
Zhao, Yanfang  503
Zheng, Suwen  453
Zhou, Ling  77
Zieliński, Paweł,  197

Printed in the United States
By Bookmasters